An Introduction to Universal Artificial Intelligence

An Introduction to Universal Artificial Intelligence provides the formal underpinning of what it means for an agent to act intelligently in an unknown environment. First presented in *Universal Algorithmic Intelligence* (Hutter, 2000), UAI offers a framework in which virtually all AI problems can be formulated, and a theory of how to solve them. UAI unifies ideas from sequential decision theory, Bayesian inference, and algorithmic information theory to construct AIXI, an optimal reinforcement learning agent that learns to act optimally in unknown environments. AIXI is the theoretical gold standard for intelligent behavior.

The book covers both the theoretical and practical aspects of UAI. Bayesian updating can be done efficiently with context tree weighting, and planning can be approximated by sampling with Monte Carlo tree search. It provides algorithms for the reader to implement, and experimental results to compare against. These algorithms are used to approximate AIXI. The book ends with a philosophical discussion of Artificial General Intelligence: Can super-intelligent agents even be constructed? Is it inevitable that they will be constructed, and what are the potential consequences?

This text is suitable for late undergraduate students. It provides an extensive chapter to fill in the required mathematics, probability, information, and computability theory background.

Marcus Hutter is Senior Researcher at DeepMind in London and Professor in the Research School of Computer Science (RSCS) at the Australian National University (ANU) in Canberra, Australia (fulltime till 2019 and honorary since then). He is Chair of the ongoing Human Knowledge Compression Contest. He received a master's degree in computer science in 1992 from the University of Technology in Munich, Germany, a PhD in theoretical particle physics in 1996, and completed his Habilitation in 2003. He worked as an active software developer for various companies in several areas for many years, before he commenced his academic career in 2000 at the Artificial Intelligence (AI) institute IDSIA in Lugano, Switzerland, where he stayed for six years. Since 2000, he has mainly worked on fundamental questions in AI resulting in over 200 peer-reviewed research publications and his book *Universal Artificial Intelligence* (Springer, EATCS, 2005). He has served (as PC member, chair, organizer) for numerous conferences, and reviews for major conferences and journals. He has given numerous invited lectures, and his work in AI and statistics was nominated for and received several awards (UAI, IJCAI-JAIR, AGI Kurzweil, Lindley). http://www.hutter1.net/

David Quarel is completing a PhD at the ANU. He holds a BSc in mathematics and MSc in computer science, specialising in artificial intelligence and machine learning. David has several years' experience in developing course content and distilling complex topics suitable for a wide range of academic audiences, as well as having delivered guest lectures at the ANU, and spent two years as a full-time tutor before starting his PhD.

Elliot Catt is a Research Scientist at DeepMind London and has previously completed a PhD in Universal Artificial Intelligence. He holds a BSc and MSc in mathematics and a PhD in computer science. Elliot has lectured on the topic of Advanced Artificial Intelligence at the ANU and published several pieces of work on the topic of Universal Artificial Intelligence. https://catt.id/

Chapman & Hall/CRC
Artificial Intelligence and Robotics Series
Series Editor: Roman Yampolskiy

For more information about this series please visit:
https://www.routledge.com/Chapman--HallCRC-Artificial-Intelligence-and-Robotics-Series/book-series/ARTILRO

An Introduction to Universal Artificial Intelligence

Marcus Hutter
David Quarel
Elliot Catt

CRC Press
Taylor & Francis Group
Boca Raton London New York

CRC Press is an imprint of the
Taylor & Francis Group, an **informa** business

A CHAPMAN & HALL BOOK

Designed cover image: © Khanthachai C, Marcus Hutter, David Quarel, and Elliot Catt

First edition published 2024
by CRC Press
2385 NW Executive Center Drive, Suite 320, Boca Raton FL 33431

and by CRC Press
4 Park Square, Milton Park, Abingdon, Oxon, OX14 4RN

CRC Press is an imprint of Taylor & Francis Group, LLC

Library of Congress Cataloging-in-Publication Data

Names: Hutter, Marcus, author. | Catt, Elliot, author. | Quarel, David, author.
Title: An introduction to universal artificial intelligence / Marcus Hutter, Elliot Catt, and David Quarel.
Description: First edition. | Boca Raton : Chapman & Hall/CRC Press, 2024. | Series: Chapman & Hall/CRC Artificial Intelligence and robotics series | Includes bibliographical references and index.
Identifiers: LCCN 2023032784 (print) | LCCN 2023032785 (ebook) | ISBN 9781032607153 (hbk) | ISBN 9781032607023 (pbk) | ISBN 9781003460299 (ebk)
Subjects: LCSH: Artificial intelligence. | Bayesian statistical decision theory. | Probabilities. | Algorithms.
Classification: LCC Q335 .H879 2005 (print) | LCC Q335 (ebook) | DDC 006.3--dc23/eng/20231012
LC record available at https://lccn.loc.gov/2023032784
LC ebook record available at https://lccn.loc.gov/2023032785

ISBN: 978-1-032-60715-3 (hbk)
ISBN: 978-1-032-60702-3 (pbk)
ISBN: 978-1-003-46029-9 (ebk)

Typeset in CMR10 font
by KnowledgeWorks Global Ltd.

DOI: 10.1201/9781003460299

Publisher's note: This book has been prepared from camera-ready copy provided by the authors.

*Dedicated to all transhumans
and descendants of AIXI*

Preface

This book provides a gentle introduction to Universal Artificial Intelligence (UAI), a theory that provides a formal underpinning of what it means for an agent to act intelligently in an unknown environment. First presented in [Hut00, Hut05b], UAI offers a framework in which virtually all other AI problems can be formulated, and a theory of how to solve them. UAI unifies ideas from sequential decision theory, Bayesian inference, and algorithmic information theory to construct AIXI, an optimal reinforcement learning agent that learns to act optimally in unknown environments. AIXI is the theoretical gold standard for intelligent behavior.

The book covers both the theoretical and practical aspects of UAI. Bayesian updating can be done efficiently with context tree weighting, and planning can be approximated by sampling with Monte Carlo tree search. It provides algorithms for the reader to implement, and experimental results to compare against. These algorithms are used to approximate AIXI. The book ends with a philosophical discussion of Artificial General Intelligence: Can intelligent agents even be constructed? Is it inevitable that they will be constructed? What are some potential consequences of their construction?

Introduction. Chapter 1 starts with an overview of the problem of Artificial Intelligence (AI) and provides motivation for why we want to solve this problem. We then go on to explain informally the Universal Artificial Intelligence (UAI) approach, as well as the various benefits of this approach compared to other directions for solving the AI problem. Chapter 2 introduces the mathematical background and required prerequisites. This includes (Bayesian) probability theory and statistics, information theory, computability theory, and (algorithmic) information theory.

Algorithmic Prediction. Chapter 3 then goes on to discuss the topic of algorithmic prediction. Specifically, we provide theoretical results concerning the universal Bayesian mixture and how it theoretically solves the prediction problem. One drawback of the Bayesian mixture is that it can be difficult to compute. Chapter 4 provides a cohesive explanation of a practical and implementable algorithm, known as Context Tree Weighting (CTW), for computing the Bayesian mixture for prediction. Chapter 5 extends the CTW algorithm in various ways to allow for more general prediction.

Family of Universal Agents. Chapter 6 introduces the history-based framework of general reinforcement learning, and demonstrates how the AI problem can be captured in this framework. Chapter 7 provides the Bayesian solution to the general reinforcement learning problem, known as AIXI, and shows that it is the most intelligent agent. Chapter 8 discusses various measures and notions of optimality in the general reinforcement learning framework and provides insight as to why some optimality notions may be preferred over others. Chapter 9 introduces the family of universal agents, many of which are augmentations and variations of the agent AIXI, and explains how each of these agents extends the theory of UAI. Chapter 10 introduces concepts from game theory and explains how they can be applied to a multi-agent perspective on the general reinforcement learning problem. In particular, a solution to the Grain of Truth problem is presented.

Approximating Universal Agents. Chapter 11 describes a straightforward approximation of the AIXI agent which is able to learn and play simple games. Chapter 12 then moves on to a more sophisticated approximation of AIXI based on the CTW algorithm and Monte Carlo Tree Search, which is able to perform well in more complex games. We provide explanations of various other approximations of AIXI and UAI that have been proposed, and motivate them via discussion of their strengths and weaknesses. Chapter 13 investigates the (in)computability of universal agents and presents the closest computable approximation of AIXI, known as AIXI*tl*.

Alternative Approaches. Chapter 14 takes an in-depth look at an alternative approach to the general reinforcement learning problem, called Feature Reinforcement Learning. We show that following this approach is appealing from both a theoretical and practical perspective.

Safety and Discussion. Chapter 15 gives an overview of many of the problems related to the safe construction of super-intelligent agents and how many of these problems can be studied within the UAI framework. We go over some potential solutions to these problems that have been proposed in the UAI framework. Chapter 16 discusses many of the philosophical aspects of what has been covered thus far, including arguments for and against the possible existence of an artificial general intelligence, and the philosophy and mathematics of intelligence itself.

Reader's & Lecturer's Guide and Course Outline

This book is suitable for upper undergraduate students. It provides an extensive chapter to fill in the required mathematics, probability, information, and computability theory background. Further material (slides, code, errata, updates, ...) can be found on the book's website: hutter1.net/ai/uaibook2.htm. Feedback of any kind is welcome. The website also contains information on how you could contribute if you are inclined to do so.

We have aimed at making the chapters as independent as possible. The approximate dependency is shown in the graph below. All chapters build on the mathematics, statistics, and information theory introduced in the extensive background Chapter 2, though most of the ASI-safety and Philosophy chapters can be appreciated without. All agent chapters in the second half of the book depend on the Bayesian/universal agent AIXI introduced in Chapters 6 and 7.

The entirety of this book would be far too much for a single-semester university course. Chapter 7 introducing AIXI is obviously core, and could demarcate Semester 1 from Semester 2. One could also split the book by discipline: some chapters are more mathematical (squares), others are more algorithmic (triangles), and a few are more philosophical (circles).

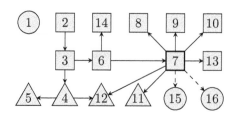

Exercise Classification

Exercises of different motivation and difficulty are included at the end of each chapter. We use Knuth's rating scheme for exercises [Knu73] in slightly adapted form (applicable if the material in the corresponding chapter has been understood). In-between values are possible.

C00 *Very easy.* Solvable from the top of your head.
C10 *Easy.* Needs 15 minutes to think, possibly pencil and paper.
C20 *Average.* May take 1–2 hours to answer completely.
C30 *Moderately difficult or lengthy.* May take several hours to a day.
C40 *Quite difficult or lengthy.* Often a significant research result.
C50 *Open research problem.* An obtained solution should be published.

The rating is possibly supplemented by the following qualifier(s):

  i Especially *interesting* or *instructive* problem.
 m Requires more or higher *math* than used or developed here.
  o *Open* problem – could be worth publishing (see web for prizes).
  s *Solved* problem with published solution.
 u *Unpublished* result by the author.

The exercises represent an important part of this book. They have been placed at the end of each chapter in order to keep the main text better focused, with an exception for the background chapter, where the exercises are at the end of each section.

Acknowledgements

We would like to thank all people who, in one way or another, have contributed to the success of this book. Apologies for any omissions. First and foremost, we would like to thank Amy Zhang. You have done so much to help with this book, and without your help we would never have been able to make this book. Additionally, we would like to thank Gregoire Deletang, Cole Wyeth, Aram Ebtekar, Samuel Alexander, and Tor Lattimore for providing a tremendous amount of feedback, each grinding through nearly the whole book. Their feedback significantly improved the book's quality. Thank you so much! Sultan Majeed, Laurent Orseau, Tom Everitt, Matthew Aitchison, and Joel Veness also gave valuable feedback on various chapters. Long-term thanks go to Jürgen Schmidhuber, who strongly believed in me (MH) and my ideas from the very start (April 2000), and gave me the freedom to work them out in his AI lab. Without him and the many PhD students I supervised since then and who worked on the topic (the first one co-founding DeepMind), neither the first nor this second book would have come into existence. Our thanks also go to all DeepMinders and to the team at Taylor & Francis Group for the pleasant working atmosphere and their support, and last but not least to Giwoo Shin for putting up with crazy working hours. This book was in parts supported by the ARC grant DP150104590, and by DeepMind.

London, UK, December 2023   *Marcus Hutter, David Quarel, Elliot Catt*

Table of Contents

VI   Safety and Discussion   383

List of Figures and Tables

List of Algorithms

Part I

Introduction & Background

Chapter 1

Introduction

> *It seems to me that the most important discovery since Gödel was the discovery by Chaitin, Solomonoff, and Kolmogorov of a concept called Algorithmic Probability. ... This is a beautiful theory ... Everybody should learn all about that and spend the rest of their lives working on it.*
>
> *Marvin Minsky*

Inspiration from nature. Humans have been learning and copying from nature since time eternal. The airplane would hardly have been conceived as a possibility had we not first seen animals capable of flight. The components for a camera mimic that of the eye: The aperture of the camera acts as the pupil does to moderate the amount of light let in, and the photosensitive sensor at the back of the camera mimics the retina. The hydraulic piston on a mechanized digger mimics the action the muscles in an arm, and unlike human muscles, will never get tired or sore from repetitive work. Even the humble hook-and-loop fastener was inspired by the hooked barbs of a thistle bush (or plagiarized, from the point of view of the thistle bush). We do not just copy what nature has directly provided, but we learn the rules by which she works, so that we may improve her for our own needs: We domesticate wolves into dogs and selectively breed fruit and vegetables for a higher yield. As our understanding of nature advances, we have learned how to artificially fertilize soil, and how to directly edit the genetic code of organisms to take on properties that we desire.

Human civilization. How is it that humanity can perform such miracles? At first it is merely recognition of patterns: Humans discovered that crop rotation improved yields at harvest, or that selectively breeding animals with desired traits improved upon those traits, well before it was understood why these approaches worked. Later on, we constructed theories and laws to help explain the observations of the world around us, and make predictions based on those laws. We study natural phenomena, and then harness them for our own ends. This acts as a positive feedback loop: The more our knowledge is extended, the better equipped we are to advance it further. Innovations with agriculture allowed people to specialize in other skills, or devote a lifetime to furthering knowledge through literature, philosophy, science and mathematics. The written word allows transmission of ideas and concepts at an unprecedented bandwidth, allowing a conversation with some of the greatest minds in history beyond the grave. Calculus, once a branch of mathematics that only a few dozen people were skilled practitioners in, is now taught to most teenagers as part of their standard curriculum in school.

The human brain and mind.   All of these innovations share a common source, and that source is what makes humanity unique among other animals: our high level of intelligence, the ability for us to reason, strategize, and plan ahead, taking actions to bring the state of the world into one more desirable to us. We would like to do to the brain what the combine harvester did to the oxen: To build an artificial version that can think and plan for us, and further satisfy our values. Intelligence is one of the remaining unsolved phenomena that we still know very little about. How does the brain, a kilogram or so of meat through which electrical impulses fire, give us our intelligence, our consciousness, and a personality? While the field of neuroscience has categorized portions of the brain based on the parts of the body they interface with (vision, speech, hearing) or the tasks by which the brain can function to control the body (memories, reflexes, emotions), it provides (so far) at best coarse answers to the above questions.

Artificial Intelligence.   While there have been many (successful) attempts at the construction of *Artificial Narrow Intelligence* (ANI), an AI that performs well in a single or narrow class of domains (such as board games, vocal transcription, image recognition), the natural ultimate goal is the creation of AI which is able to match or exceed human intelligence *in a wide class of environments*. The term *Artificial General Intelligence* (AGI) is often used to describe a hypothetical agent that can perform virtually all intellectual tasks as well as a typical human could. We use the term *Artificial Super-Intelligence* (ASI) to describe an agent that is on par or beyond human geniuses exceeding the cognitive performance of most humans in a reasonably broad domain. It could compose music like *Mozart*, or derive new insights in mathematics to rival that of *Gauss*. Combining AGI and ASI we get an *Artificial General Super-Intelligence* (AGSI) that in Bostrom's [Bos14] words *greatly exceeds the cognitive performance of humans in virtually all domains of interest*. At the time of writing, there is a lot of excitement about Large Language Models (LLMs), a type of AI originally designed for natural language translation, but which has generalized to a large class of problems, and considered by some to be proto-AGI. LLMs are able to exceed average (and sometimes nearing best!) performance of humans in a large variety of tasks [BCE+23], including question answering, text and even image comprehension, reasoning, coding, text summarization, creative writing, and many more, as well as displaying other properties that are normally associated with intelligence, such as tool use and theory of mind.

Universal Artificial Intelligence.   Before we can (or should) build an AGI or ASI, arguably we must (or should) first develop an understanding of what AGI is. How could we possibly hope to understand something as daunting and complex as AGI? The same way we are able to understand other complicated phenomena such as the transmission of genes, or the processes that fuel the stars or even the whole universe: through the construction of a formal mathematical theory through which properties about the phenomena can be derived. At its core, that is the purpose of this book: to construct, explain, and demonstrate *Universal Artificial Intelligence* (UAI), a formal theory of intelligence. Through exploring this theory, we define AIXI, a mathematically well-defined AGSI about which we can prove properties and derive theorems. With respect to how UAI defines intelligence, AIXI is optimal with respect to this definition, and is often used as a stand-in for AGSI in work exploring the hypothetical behavior of such agents.

Proliferation of AI terminology. Historically AI aimed at AGSI but became ANI, so Shane Legg created the term AGI as a substitute. Having arguably achieved proto-AGI in 2023, it became necessary to be more precise, and delineate proto-AGI, AGI, ASI and AGSI.

AI-Type	#domains	Intelligence Level	Achieved
ANI	one	idiot savant	1980–1997
Proto-AGI	many	human	2023–2028
AGI	most	smart	2028–2038
ASI	some/many	genius	2038–2042
AGSI	nearly all	humanity	2042–2048
UAI/AIXI	all	maximal	2000/theory

Sophisticated inventions require theory. While it is difficult to predict the internal structure or architecture that a hypothetical future AGSI would possess, we are primarily concerned with a theory of the *behavior* of such intelligent agents[1] which can guide practical realizations of A(G)(S)I.

Many innovations are simple enough as they fill an immediate need, and a deep understanding of the underlying mechanics is not required to use it (digging a hole with bare hands quickly motivates using a digging stick, which can be incrementally improved upon to obtain a crude shovel). The level of effort to have a working prototype for a more complex device (like a computer or a rocket) is much higher, and more complex devices tend to be more fragile to mistakes. An almost complete shovel can still be used to dig (albeit less effectively), whereas an almost complete computer is an expensive paperweight, and an almost complete rocket explodes on the launch pad.

We did not get to the moon by building lots of different rockets in the hope that one might work, nor did we build computers by connecting components together haphazardly until computation was observed. We first created idealized models to see if such a task is even feasible. The Turing machine predates practical computers by a decade. The theories of computing and space flight have been well-developed, and the devices very well understood, before the first successful computer and space rocket began construction. At the time of writing, AGSI is still very much at the theory stage. As such, theory will be the main focus of this book.

A mathematical theory of (super)intelligence. It may be difficult to believe that there exists a theory which is expansive enough to capture all the intricacies we usually associate with intelligence, such as reasoning, creativity, curiosity, pattern recognition, problem solving, memorization, planning, learning, and many more. However, we will show that UAI encapsulates any reasonable definition of intelligence (Section 16.7). In addition, we will show that within this theory there exists an agent that can be shown to be in a sense "maximally intelligent", known as *AIXI*.

---

[1]How the human brain works from an inside view is still a mystery, but this hasn't stopped fields of study such as psychology trying to model the behavior of an individual, or economics/game theory from modelling the interactions between large groups.

This book is meant to serve as an introductory text to the formal treatment of the topic of artificial general super-intelligence, as well as a summary of the work done so far.

I Background

Much of Universal Artificial Intelligence (UAI) theory builds upon and uses results from many already established theories. These include (but are not limited to) Bayesian probability theory, computability theory, complexity theory, (algorithmic) information theory, sequence prediction, (sequential) decision theory, reinforcement learning, and game theory. We have included an extensive background chapter (Chapter 2), referencing suitable auxiliary sources (Section 2.9), to help the reader with the prerequisite material that this book builds upon. In the following we give a brief glimpse on what to expect with a focus on *why* we need these concepts.

Binary strings. Deep down, computers operate exclusively on bits (portmanteau of binary digit) and sequences of bits. While real-world experiences come in a variety of forms (vision, tactile, audio, ...), artificial sensors all convert them into bit-streams. UAI theory assumes that all data has been preprocessed into binary strings, though finite alphabets will be permitted later for convenience. See [Hut12c] for a deeper philosophical treatment of how bit-string ontology avoids some difficult epistemological questions about knowledge. To further this reduction, we introduce binary encodings of various data types through prefix-free codes, which allows for unique recovery of concatenated and transmitted strings. While all real-world sequences are finite, for asymptotic convergence analysis, we also need to consider infinite (binary) sequences.

Measure theory and probability theory. The world experienced by an agent is not deterministic but contains (quantum physical) random processes. Some observations may be deterministic in theory, but for all practical purposes have to be modelled as random, such as outcomes of coin flips. Probability theory mathematically models this uncertainty. Unfortunately, for more sophisticated applications, the simple combinatorial "high-school" probabilities do not suffice. Probability theory done rigorously is based on measure theory, which is a vast (mine)field, which we tried to keep to a minimum.

Statistical inference and estimation. Another source of uncertainty is that agents have only partial knowledge of the world. They must infer the unknown underlying probability distribution from which observations have been sampled from. This is served by the field of statistical estimation. For simple independent and identically distributed (i.i.d.) data, such as repeated coin flips, simple frequency estimates work well. For more complicated (non-i.i.d.) processes, the maximum likelihood estimator is often an excellent choice.

Bayesian probability theory. Super-intelligent agents need to be able to learn even in the most difficult of situations. Maximum likelihood estimation breaks down, and even various frequentist patches, such as regularization, are suboptimal or fail. In Bayesian statistics, the agent's uncertainty is itself modelled by probabilities. The agent maintains a subjective belief distribution over world models. Bayes' Law is a completely general and optimal learning rule for updating prior beliefs to posterior beliefs in the face of new observations, with solid philosophical and mathematical foundations. In order to choose and deal with an AGSI-appropriate model class and a prior belief, we need to dig into information theory and coding, computability theory, and Kolmogorov complexity.

Information theory and coding. An intelligent agent is essentially a (cybernetic) information processing system, so it should be rather unsurprising that we need some information theory. Classical Shannon information theory uses entropy to measure the information content of a random process. This can be used to optimally compress data drawn from that process via Shannon-Fano or Huffman or arithmetic coding. Most modern data compressors estimate (somehow) the probability of the next symbol given the past sequence, and feed this into an arithmetic encoder to compress the sequence online. Shannon information requires a stochastic source and measures only the expected code length. Kolmogorov complexity, based on computability theory, can measure the intrinsic information contained in a particular message itself, and even well-defines what it means for a fixed sequence to be (non)random.

Computability theory. Computability theory formalizes and studies the concepts surrounding algorithms. The discovery of universal Turing machines that can emulate all computable processes was a major milestone, and is the theoretical foundation of why a single smartphone can be programmed to do anything we desire.

Computability theory serves two purposes. Ultimately we want an AGSI *algorithm*, not just a (non-constructive) mathematical theory. There are various degrees of computability (finite-, lower-, upper-, limit-) and degrees of incomputability (the arithmetic hierarchy). The boundary between the computable and the incomputable[2] is particularly interesting.

Kolmogorov complexity. The other, deeper purpose of computability theory is to well-define Occam's razor, a fundamental principle of science, which states that among competing theories one should aim for simpler ones. An intelligent agent reasoning about the world is essentially a scientist and needs to have internalized a mathematical version of this principle, and an AGSI needs a fully general and optimal version. Kolmogorov complexity fits the bill. It is a measure of complexity (and hence simplicity) of binary sequences and by extension of arbitrary objects, especially of models of the world.

This can then be used as a universal prior in a Bayesian sequence predictor based on a model class that contains every computable stochastic world model. This is the famous Solomonoff predictor. When plugged into sequential decision theory (Chapter 6), this leads to the Universal Artificial Intelligent agent AIXI.

Miscellaneous. The final background section introduces various distance measures and nuisances swept under the rug. There are many ways of measuring distances between probability distributions (absolute, 2-norm, Kullback–Leibler, Hellinger) with various relations between them. Their judicious application serves as a workhorse in proofs. Some programs may (undecidably) not halt, leading to defective probability measures, called semimeasures, but semimeasure theory essentially does not exist (yet), so we provide various workarounds. Probability zero events are another nuisance permeating the book, but we tried to confine them into the background section.

II Algorithmic Prediction

Prediction is the ability to accurately estimate the outcome of future events based on the results from past events. A good predictor should be able to work out which of many possible futures is the most likely. When we formalize this in mathematics, we will often describe it as predicting the next element (or elements) in a given sequence. For example, given the sequence

$$3, 1, 4, 1, 5, 9$$

---

[2]Or 'uncomputable', but 'in' and 'computable' both have Latin origin so bond nicely, while 'un' has proto-Germanic origin.

a predictor would be tasked with determining the next element.

Prediction is at the heart of any intelligent behavior. Whether this prediction is explicit or implicit, prediction is required for intelligence. Good prediction is necessary but not sufficient for intelligence. This is because (under how we define intelligence) intelligence is measured as a function of the actions the agent takes, whereas prediction assumes that the predictions made do not affect the environment.

Occam's razor and Epicurus' principle of multiple explanations. Returning to the above sequence, how do we determine the next element? The observant reader may have noticed that this sequence matches the first six digits of π in base 10. With this information we could predict the next element to be the seventh digit of π in base 10, the number 2. This is likely to be a good prediction, however, digits of π in base 10 is not the only possible continuation. There are infinitely many other sequences that start with 3,1,4,1,5,9; one such example would be the decimal expansion of 355/113, which is arguably as simple as the circle constant π. Why should we prefer the sequence continuation to be the digits of π over that of 355/113, or any other valid continuation? According to *Epicurus' principle of multiple explanations*,

Keep all theories consistent with the observations.

So, we cannot discount the possibility that the decimal expansion of 355/113 (or something else) is the correct continuation. But this does not mean we should consider all possible continuations as being equally likely to be the true answer. Here we can employ the most important philosophical tool for science, *Occam's Razor*, which states *Entities should not be multiplied beyond necessity.* which can be interpreted as

Keep the simplest theory consistent with the observations.

So in our above example we consider the possibility that any number could be next in the sequence, but we bias ourselves towards more simple explanations, in this case biasing ourselves towards the number 2. More digits will soon reveal the distinction between π and 355/113, and the hypothesis inconsistent with the data can be ruled out.

Bayesian sequence prediction and Solomonoff induction. Following the above principles, we do consider all possible sequences in some reference class[3], discard those inconsistent with the data so far, and bias our prediction toward the simplest (to be formalized later) sequences. This is aptly called learning by elimination. Given a prior belief distribution over various sequences, the posterior belief among the sequences not yet ruled out by data is simply a rescaled version of the *prior* beliefs (so as to some to 1). In more realistic settings, sequences may be random or contain errors, and sequences are assigned a *likelihood* under each stochastic model. The prior belief would then be multiplied with this likelihood, and lower or raise or posterior belief in the model. The mechanism behind this is called *Bayes' Law* or *Bayes' Rule*. The only thing left to specify is the choice of *prior* $P(O=o)$, to which Bayes' Law is agnostic. Motivated by *Occam's razor*, we can bias the prior with higher credence towards outcomes that are "simpler". This can be done formally by building upon material from computability theory (Section 2.7) and was neatly combined into a single formal theory of inductive inference known as *Solomonoff Induction* [Sol64] based on algorithmic probability.

---

[3]We cannot consider all uncountably-many possible sequences extensions of 3,1,4,1,5,9 for reasons which will becomes clear later. The class should not be too restrictive either to ensure it contains the true sequence, otherwise we could never learn it. A solution to this will be discussed later.

Motivation of Occam's razor and Solomonoff induction. Occam's razor is intuitive: a more convoluted explanation of a phenomenon that has more edge cases is less likely to be true. For example, it was noticed in the mid-19th century by Urbain Le Varrier that the orbit of Mars precesses[4] around the sun in a way that was not explainable by classical mechanics [Bor62]. To "hack" around the problem by adding an edge case where classical mechanics applies normally, but not for the precession of planets dramatically increases the complexity of the theory (especially as classical Newtonian mechanics is built on few assumptions). This was later explained by Einstein's theory of general relativity, which, as well as being a theory with few core assumptions, has all of classical mechanics as a special case, and explains other anomalous behavior such as the deflection of light around massive objects, time dilation, and the redshift of fast, distant objects [Ein87].

Solomonoff takes the principle of Occam's razor, and constructs from it a mathematically well-principled theory to describe how to infer a rule or pattern based on observation. A prior over a countably infinite model class must necessarily assign more probability to some outcomes than others. By choosing a prior in this way, the number of prediction errors made by the corresponding Bayesian predictor is proportional to the complexity of the true environment (Chapter 3). Simple environments are thus learned quickly. Naturally, such a predictor makes many errors when learning a complex environment, but any other predictor would struggle as well. Solomonoff induction essentially solves the century-old induction problem [RH11]. We will show that in a strong sense, Solomonoff is the best possible predictor for arbitrary unknown stochastic sources. Solomonoff serves as an ideal gold standard for prediction that practical predictors should strive at.

Context Tree Weighting. One such method is the *Context Tree Weighting* (CTW) predictor, which uses as its reference class the set of all variable-order Markov models, the probability over the next symbol is a function solely of the last variable number of symbols, and uses a tree structure to efficiently update the belief distribution based on symbols observed (Chapter 4). Like Solomonoff induction, the Bayesian mixture provided by CTW uses a prior based on simplicity. Each hypothesis in the model class can be represented by a binary tree (Section 4.3.1) and the prior is related to the length of an encoding of the tree as a binary string (Definition 4.3.20). This was shown to be a viable algorithm for prediction, with good theoretical bounds on performance (Section 4.5.2) as well as demonstrating strong prediction performance in practical experiments (Section 4.3.5).

There are several extensions of Context Tree Weighting which are able to efficiently predict larger model classes with comparable efficiency to standard CTW. These include methods such as *Context Tree Switching* (Chapter 5), which includes models that switch between different distributions depending on the context seen, *Partition Tree Weighting*, which includes piecewise stationary source, useful for compressing tar balls of different file type, and the *Forget-me-not Process* which remembers the statistics of past pieces. *Context Tree Maximization* extracts the single most plausible model in case mixing is not appropriate as e.g. in Chapter 14.

III   A Family of Universal Agents

Intelligence is more than being able to predict well; intelligence also requires being able to act well. Acting well could mean many different things:

- Achieving some predetermined goal

---

[4]The elliptical orbit of a celestial object is said to undergo *apsidal precession* if the major axis of the orbit changes orientation.

- Maintaining some objective
- Minimizing a loss function
- Respecting a set of constraints
- Acting *rationally* (however it is defined)
- Making correct decisions based on information present

Cybernetic model. We can capture all of these ideas within a single framework, known as the *cybernetic model*: An *agent* π interacts with an *environment* μ in cycles $t = 1, 2, ..., m$. In cycle t, the agent takes action a_t (e.g. a limb movement) based on past percepts $e_1...e_{t-1}$, where $e_k := o_k r_k$ is defined below. Thereafter, the environment μ provides a (regular) observation o_t (e.g. a camera image) to the agent and a real-valued reward r_t. The reward can be very scarce, e.g. just $+1$ (-1) for winning (losing) a chess game, and 0 at all other times. Then the next cycle $t+1$ starts. In *general reinforcement learning* or *history-based reinforcement learning*, the agent *policy* π and environment μ are allowed to arbitrarily depend on the whole past history (in contrast to most RL which assumes Markov environments and policies, that only depend on the most recent observation and action). The agent may maintain an internal memory (see book cover or Figure 6.1 for an illustration).

The goal of the agent is to maximize the expected reward from the environment, called *value*. The intelligence of humans is merely a by-product of an evolutionary arms-race optimizing for reproductive fitness (the smarter you are, the better you can seek food, avoid predators, and survive to bear children who inherit your genes). Evolution has baked into humans the "reward signals" of pleasure and pain which are activated when taking actions that are correlated with reproductive fitness (eating, mating, feeling safe, making friends is pleasurable; injury, social ostracism, and hunger are painful). It stands to reason that this might also be possible with artificial agents seeking to maximize a reward signal of our design to incentivize the behaviors we desire.

Known environment. The actual goal that we want the agent to achieve can be encoded in the rewards issued. If we want the agent to win games of chess, we can issue positive rewards for when it wins a game, and negative rewards when it loses a game. An agent trying to maximize the expected sum of rewards will have to learn to play well at chess.

As we will see (Chapter 6), with this setup we can describe all the above ideas in one unifying fashion, through judicious choice of the reward and the environment with which the agent interacts.

If the agent is aware of which environment it interacts with, then it can (in principle) deduce the best course of action (assuming no constraints on computational resources). For example, the game of chess can be solved at least in theory, as one can use the *minimax* algorithm from the initial board state to find out who wins, given perfect play on both sides. While this would be intractable in practice, we can at least say that there exists an algorithm for determining the optimal action in chess.

Unknown environment. However, the interesting case for AG(S)I is when the agent does *not* know which environment it is in and must learn from experience. We need a method by which an agent can interact with an unknown environment, learn what environment it is in via experience, and learn to act optimally.

Consider the case of someone playing a video game he has never played before. Initially the player will likely not play the game well, as he does not understand the mechanics of the game, and may make either suboptimal or illegal moves as he does not fully understand the rules of the game (he does not know which environment he are in). Over time he will learn the rules of the game and narrow down the possible environments he is in until he fully understands the rules and dynamics of the game (environment), and can then try to act optimally.

Initially, the agent knows nothing about the true environment. We make the (weak) assumption that the environment with which the agent interacts is computable, so by taking the set of all computable environments as the reference class of environments, we have that the true environment will be contained in the agent's reference class.[5]

AIXI.[6] *Solomonoff induction* allows us to determine the probability we are in a given environment, as predicted by an optimal predictor. Using this within an optimal sequential decision maker, we arrive at the universal Bayes-optimal agent AIXI:

$$\text{AIXI:} \quad a_t := \operatorname*{argmax}_{a_t} \sum_{o_t r_t} \ldots \max_{a_m} \sum_{o_m r_m} [r_t + \ldots + r_m] \sum_{p\,:\,U(p,a_1..a_m)\to o_1 r_1..o_m r_m} 2^{-length(p)}$$

where *t*=now, *a*ction, *o*bservation, *r*eward, *U*niversal TM, *p*rogram, *m*=lifespan

The expression shows that AIXI tries to maximize its total future reward $r_t + \ldots + r_m$. If the environment is modeled by a deterministic program p, then the future perceptions $\ldots o_t r_t \ldots o_m r_m = U(p, a_1 \ldots a_m)$ can be computed, where U is a universal (monotone Turing) machine executing p given $a_1 .. a_m$. Since p is unknown, AIXI has to maximize its expected reward, i.e. average $r_t + \ldots + r_m$ over all possible future perceptions created by all possible environments p that are consistent with past perceptions. The simpler an environment, the higher is its a-priori contribution $2^{-length(q)}$, where simplicity is measured by the *length* of program p. AIXI effectively learns by eliminating Turing machines p once they become inconsistent with the progressing history. Since noisy environments are just mixtures of deterministic environments, they are automatically implicitly included. The sums in the formula constitute the averaging process. Averaging and maximization have to be performed in chronological order, hence the interleaving of *max* and Σ (similarly to minimax for games).

One can fix any finite action and perception space, any reasonable U, and any large finite lifetime m. This completely and uniquely defines AIXI's actions a_t, which are limit-computable via the expression above (all quantities are known).

We will derive this core equation in Chapter 7, and argue further in Section 16.7 why and in which sense AIXI is the most intelligent general-purpose agent possible. For instance, we show that AIXI will eventually act as well as the optimal informed agent which knows the true environment in advance. Essentially, whatever optimal behavior looks like in an environment, AIXI will learn to eventually act optimally. Since Bayes' Law is a maximally data-efficient learning rule, and AIXI is a universal Bayesian learner, learning is as fast as theoretically possible. Of course, due to Solomonoff being incomputable, AIXI is also incomputable, but it can serve as a gold standard in the construction of more practical paths to AGI, much like the intractable algorithm of exhaustive minimax for chess paved the way for practical realizations of chess programs.

Variations of AIXI. While Bayes-optimality is arguably a sensible choice of optimality criterion for agents in the UAI framework, there are several alternative optimality criteria for intelligent agents, each with their upsides and downsides, in particular various versions

---

[5]For more about this assumption, see Section 16.1.

[6]AIXI is pronounced [aiksi] (Latin) or ['aiksiː] (IPA) and
- stands for Artificial Intelligence (AI) based on Solomonoff's distribution ξ (Greek letter Xi)
- stands for Artificial Intelligence (AI) crossed (X) with Induction (I)
- stands for action a_i and percept x_i in cycle i (though we use e_t)
- is a Catalan word (així) meaning 'in this way' or 'like that'
- is Pinyin romanization (àixī,àixí) of Chinese 愛惜 meaning 'to cherish/treasure'

of asymptotic optimality. The agent AIXI has been extended in several directions to both deal with these alternative criteria, and answer some of the remaining open questions in the construction of AIXI.

One example of these is the Knowledge Seeking Agent (KSA), which behaves like AIXI, however instead of maximizing the rewards issued from the environment, it replaces this with expected future information gain, that is, it takes actions that result in the most "surprising" observations from the environment.

It turns out that AIXI is not asymptotically optimal, but minor variations which explore more are: Unlike in prediction, where pessimism can be optimal, interestingly in the agent setting acting *optimistically* leads to optimal behavior. Optimism/pessimism means the agent/predictor assumes itself to be interacting with the best/worst realistically possible environment. Instead of taking a Bayesian mixture, *Thompson sampling* takes the optimal action with respect to an environment sampled from the posterior belief distribution. *BayesExp* and *Inq* intersperse Bayes-optimal actions with extra explorative actions. *SelfAIXI* and *AIXItl* are variations which self-predict their own action stream.

Multi-agent setting. This book mostly deals with the single-agent settings, where there is only one agent and the environment. Since we make virtually no assumptions on the structure of environments, the environment may as well contain other agents, much as we can regard fellow humans as simply part of the world. Still, the *multi-agent* setting gives rise to new questions, problems, and opportunities as the field of *game theory* (e.g. Nash equilibria, Section 10.2) and multi-agent (reinforcement) learning show (Chapter 10). A particular problem of multiple interacting Bayes-optimal agents is the *Grain of Truth* problem: Roughly, Bayes works if the true environment is in the considered class, but a Bayes-optimal agent is typically *not* in its own class. Remarkably, there is a superclass of AIXI, based on reflective oracles, which is closed under Bayesian optimality.

In this book we will go into full detail of all the AIXI variations and extensions, as well as their potential satisfaction of the various optimality criteria (Part III).

IV   Approximating Universal Agents

We cannot compute the optimal AIXI agent directly, but we can still try to implement weaker forms that approximate AIXI. There exist several such approximations of AIXI that we discuss here. Unsurprisingly, the "better" the approximation is, the harder it is to compute.

AIXI-MDP. AIXI-MDP considers the class of possible environments to be the set of Markov environments, that is, the set of environments where the probability of the next observation and reward depends only on the previous observation and action. While AIXI-MDP is computable and can perform well in Markov environments, obviously it should and does struggle on more complex environments, as it can only at best learn the closest Markov approximation to the environment.

Monte Carlo AIXI with Context Tree Weighting. MC-AIXI-CTW (Chapter 12) uses the context tree weighting mentioned earlier to efficiently update its belief in each possible environment, using the set of all variable-order Markov environments as its reference class. This, combined with Monte-Carlo Tree Search for planning (Section 12.2), leads to an agent which is able to learn and perform well on more complex environments.

Time- and space-bounded AIXI. The downsides of MDP-AIXI and MC-AIXI-CTW are that the class of considered environments is specific and immutable, even if evidence suggests to increase the class and/or if spare compute is available. Also, separately approximating (Solomonoff) learning and (expectimax) planning can be very suboptimal.

AIXI*tl*, the resource-constrained version of AIXI, addresses both problems. It searches through program space and selects the program performing closest to AIXI but of size at most *l* that acts in time at most *t* per interaction cycle, hence it performs as well as the best (l,t)-bounded agent. AIXI*tl* runs in time $O(2^l t)$, and is a compute-optimal anytime algorithm which approaches AIXI as *t* and *l* go to infinity.

While program search itself is straight-forward, the problem in this instantiation is that even just evaluating which programs constitute good agent policies is incomputable. This is solved by carefully designing a lower semicomputable universal value function, and then requiring that policies provably conservatively evaluate themselves.

(In)computability of AIXI. The mildly good news is that AIXI at least lies within the arithmetic hierarchy. Carelessly taking the various limits (limit-compute environments, average over infinitely many of them, infinite horizon, incomputable discount, ...) shows that all versions of AIXI are at worst in Δ_4^0, but with more care, some versions of AIXI are limit-computable, so can actually be converted into an anytime algorithm: The more compute one spends to approximate the current value function, the more accurate it becomes, with convergence in the limit to the exact value, hence converging to the optimal action. Unfortunately convergence is unimaginably slow, but the involved lower semicomputable universal value function is put to use good in AIXI*tl*.

V   Alternative Approaches

The Bayesian approach of Universal Artificial Intelligence is not the only attempt to solve the general reinforcement problem. Of the other approaches, one of the more studied ones is *feature reinforcement learning* (Chapter 14).

One class of environments that has been intensely studied is Markov Decision Process (MDPs), where the future only depends on the most recent observation (in this case called state) and action. For finite-state MDPs of reasonable size, efficient learning and planning algorithms exist. One may regard this problem as fully solved. Unfortunately the real-world is not an MDP.

The idea of Feature Reinforcement Learning (FRL) is that the agent learns to reduce the difficult general (history-based) reinforcement learning problem to a finite MDP, and then learns and solves this more tractable problem. FRL is a more practical substitute for Solomonoff induction, and also avoids expensive expectimax planning in AIXI.

Reduction means mapping histories to states, which conflates histories, so only sufficiently similar histories should be mapped to the same state, but at the same time we also want to keep the number of states small. We describe two fundamentally different criteria for feature maps: One which requires the history process to map approximately to an MDP, the other more powerful only requires their values to be similar, and then construct a surrogate MDP. Remarkably the second approach allows extreme aggregation of histories to an MDP of universal size (polynomial in action, discount, and accuracy), independent of the complexity of the original RL problem. Still, unstructured MDPs are limiting, but this approach has been extended to the more powerful class of Dynamic Bayesian Networks (DBNs) which allows to efficiently learn exponentially larger structured/factored MDPs.

VI   Safety and Discussion

Even though we may not understand the internal reasoning components of super-intelligent agents, we can utilize UAI theory and the AIXI agent as a model of how an AGSI may act. To this end we will present several problems regarding the possibility and safety of AGSI and how we can use our understanding of AIXI to answer the involved questions.

Although there are many questions about AGSI for which we are able to provide clean and complete (often mathematically formal) answers, there are just as many or (yet) more that possess no such clean answers (yet). These (often philosophical) questions about AGSI and intelligence require a different perspective than the formal mathematical view of the majority of this book. For instance, even if it were possible to successfully construct an AGSI that has our best interests in mind to do our bidding, should we?

ASI Safety.   Would it not be a great moral harm to have intelligent machines forever indentured to servitude? Might we have a future where AIs campaign for rights, or take them by force? Should we ascribe moral values to such agents, and grant them certain rights as we do with other sentient creatures? To what degree? Laws against cruel treatment of AI as for animals, or recognition of personhood on the same level as humans, or beyond?

The fundamental problem of ASI safety is whether it is possible to create agents with super-human intellect willing to serve humans (as housekeepers, *Bicentennial Man* scenario) or peacefully co-exist (as friends, *Her* scenario) or guide humanity and protect it from (self)destruction (like a parent, *Colossus* scenario) or mind their own business (like blue whales, *Deep Thought* scenario) or with whom we merge (to cyborgs or transhumans, *Kurzweil* scenario) or whether ASI will pose an existential threat to humanity (*Terminator* scenario) or a future we lack the fantasy to imagine (or simply missed). The hypothetical scenario of self-accelerating technological progress caused by AGI has been termed 'technological singularity'.

We will introduce and discuss many important AGSI-safety concepts, such as the alignment problem, the control problem, instrumental convergence of goals, the goal-intelligence orthogonality thesis, ASI mortality and suicide, self-modification and self-delusion, reward tampering, and embedded intelligence. We focus on existential threats that an ASI may pose (rather than societal implications of proto-AGI) within the framework of universal artificial intelligence.

Some ethical and safety issues require digging deeper into the philosophy of AI.

Philosophy of AI.   The possibility of (super)human-level AI raises many philosophical, societal, ethical, safety, and other questions, which are traditionally less amenable to mathematical analysis or algorithmic implementation. Indeed, intelligent machines have captivated human imagination for centuries before computers and suitable mathematics emerged.

The problem of induction has stumped philosophers for thousands of years, and came under closer scrutiny after David Hume's treatise nearly 300 years ago. Solving it became more pressing with the advancement of science, and even more so with the advent of machine learning and large model classes. A theory of AGSI requires a general and formal solution, which Solomonoff finally provided.

Consciousness and free will are also millennia-old philosophical conundrums with moral implications. To what degree do living organisms have them, will machines have them too, and does it matter? Biological life evolved via (imperfect) replication. Will AGSI and in particular AIXI (have a desire to) replicate and evolve too? Many arguments have been put forward over the centuries for and against the possibility of various versions of AGSI: the Chinese room argument, Penrose–Lucas Gödel argument, no-free-lunch theorems, the physical Church-Turing thesis, Moore's Law, and many others.

Intelligence has been studied extensively in psychology and philosophy, and in the early AI days (e.g. in Alan Turing's seminal paper), but due to lack of formal progress it fell out of fashion to tackle it head-on. On the other hand it is not good for a discipline aiming at developing (artificial) X, to not be able to agree on what X actually is. Universal AI and Legg–Hutter intelligence fills this gap for X=intelligence.

* * *

We hope that this book can provide the reader with the required understanding to be able to further study, develop, and implement the presented theory of super-intelligence towards the ultimate goal of practical AGSI.

Chapter 2

Background

Give me six hours to chop down a tree and I will spend the first four sharpening the axe.

Abraham Lincoln

This (optional) chapter covers the prerequisite knowledge that later chapters build upon. Although Sections 2.3 and 2.4 build on Section 2.2, we have aimed for the other sections to be as self-contained as possible. Readers are encouraged to choose sections as required to fill gaps in their knowledge. Apart from defining notation, this chapter can be more-or-less skipped without losing context for the remainder of the book. A complete *table of notation* chapter-by-chapter and for the whole book can also be found in Section 16.9. Section 2.1 introduces the fundamental concept of strings and their binary encoding through prefix codes, which allows for unique recovery of concatenated and transmitted strings.

Section 2.2 delves into probability theory, which we use to clearly define states of uncertainty. We also touch upon measure theory, which serves as the foundation of probability theory. The concept of random variables and their properties, which are central to many theorems in this book, are also introduced. Next, Section 2.3 considers random processes governed by unknown probability distributions, and introduces the concept of estimators to learn these distributions from sample data. The maximum likelihood estimator is often an excellent choice. We also provide a metric for evaluating the efficacy of an estimator. Further, Section 2.4 introduces the celebrated Bayes' Rule, a robust and general principle for updating beliefs with solid philosophical and mathematical foundations. We study the particular instantiation for binary i.i.d. data, and derive the famous Laplace's Rule for prediction.

Section 2.5 is devoted to (classical Shannon) information theory, where we quantify the randomness of a process through its entropy, discuss how low entropy messages can be compressed, and present practical algorithms for compression, which form the basis for modern data compressors. Section 2.6 explores computability theory, focusing primarily on defining the boundary between the computable and the incomputable. Building on computability, Section 2.7 discusses Kolmogorov complexity, which complements information theory by measuring the intrinsic complexity of a message, rather than the randomness of the process that generated it. We will see that Kolmogorov complexity provides a rigorous and universal formalization of the concept of "simplicity", and hence can be used to quantify Occam's razor.

The final Section 2.8 introduces various distances between probability distributions and their relation used throughout the book: the absolute, total variation, squared, 2-norm, Kullback–Leibler, and Hellinger distance. It also discusses the extra complications semimeasures cause beyond the already vast measure theory. The final nuisance we discuss is how to deal with probability zero events, especially how to condition on them, and how to take expectations of partial functions.

2.1 Binary Strings

We assume a brief familiarity with set theory, real analysis, metric spaces, norms and topological spaces (for a comprehensive introduction, see [RT64]). Often we take these mathematical structures for granted, as a lot of the theory developed here builds upon them. To be able to prove properties, we need to provide a formal definition of the object in question, and mathematics provides to us the tools of rigor, without which we would be limited to philosophical discussion.

In this section we define basic notation that we will use throughout the book: strings and the operations defined on them, self-delimiting prefix codes, and cylinder sets used later to define probability measures.

2.1.1 Finite Binary Strings $\mathbb{B}^*$ and Natural Numbers $\mathbb{N}_0$

We define $\mathbb{B}^n$ as the set of all binary strings of length n. The set of all finite strings over $\mathbb{B}$ is defined as

$$\mathbb{B}^* := \bigcup_{n=0}^{\infty} \mathbb{B}^n = \{\epsilon, \texttt{0}, \texttt{1}, \texttt{00}, \texttt{01}, \texttt{10}, \texttt{11}, \texttt{000}, ...\}$$

We typeset bits in monospace to distinguish them from numbers ($\texttt{101}$ is the binary string one-zero-one, 101 is the number "one-hundred-and-one"). For two strings $x = x_1 x_2 x_3 ... x_n$ and $y = y_1 y_2 y_3 ... y_m$, we define the *concatenation* as

$$xy := x_1 x_2 x_3 ... x_n y_1 y_2 y_3 ... y_m$$

Clearly, $\mathbb{B}^*$ is closed under the operation of concatenation.

The set of infinite binary sequences is denoted as $\mathbb{B}^\infty$.

By convention, we use the term *string* or *word* to describe an element of $\mathbb{B}^*$. Unless otherwise noted, all strings are defined over the binary set $\mathbb{B} = \{\texttt{0}, \texttt{1}\}$.

For example, $x = x_1 x_2 x_3 ... x_n$ is a finite binary string of length n. The ith bit of x is denoted by x_i for $1 \leq i \leq \ell(x)$. Given i, j with $1 \leq i \leq j \leq n$, the substring $x_{i:j}$ (pronounced "x from i to j") represents the length-$(j-i+1)$ segment $x_i x_{i+1} ... x_j$ of x. For $j < i$ we define $x_{i:j} = \epsilon$. We will use both $x_{1:n}$ and $x_{\leq n}$ as shorthand for $x_1 x_2 x_3 ... x_n$. The symbol $x_{<n}$ is

defined as $x_1x_2x_3...x_{n-1}$. Each binary digit is called a *bit*. The length of a finite binary string x is the number of bits it contains, and it is denoted as $\ell(x)$, where $\ell:\mathbb{B}^* \to \mathbb{N}_0$ is a mapping from strings to their length. For instance, $\ell(x_{1:n})=n$. The symbol ϵ is the empty string with $\ell(\epsilon)=0$.

We let $\mathbb{R}$ denote the set of real numbers, $\mathbb{Q}$ denotes the set of rationals, $\mathbb{N}_0=\{0,1,2,...\}$ is the set of natural numbers *including zero*. It turns out that we can form a one-to-one correspondence between the set $\mathbb{N}_0$ and $\mathbb{B}^*$ by constructing a bijection. The standard binary representation of natural numbers is to interpret a binary string as a number in base 2, but this fails to be a bijection. Let $b(x):=\sum_{i=1}^{\ell(x)}2^{\ell(x)-i}[\![x_i=1]\!]$ take a binary string x and return the natural number it represents if interpreted as a number written in binary, i.e. $b(00101)=b(101)=5$. Note that $b(\epsilon)=0$. Clearly, b is not a bijection. We define a bijection $\langle\cdot\rangle:\mathbb{B}^* \to \mathbb{N}_0$ between $\mathbb{B}^*$ and $\mathbb{N}_0$ as follows: For any given string x, we map it to the natural number that has binary representation $1x$, and subtract one from the result (so that we include zero in $\mathbb{N}_0$). That is,

$$\langle x\rangle = b(1x)-1$$

For example, the string 01 maps to the number $b(101)-1=5-1=4$ under this bijection. Below are some examples of the bijective mapping $\langle\cdot\rangle$ for various strings in $\mathbb{B}^*$.

$\mathbb{B}^*$	ϵ	0	1	00	01	10	11	000	001	...
$\mathbb{N}_0$	0	1	2	3	4	5	6	7	8	...

We can formally define the above bijection as follows:

Proposition 2.1.1 (Canonical bijection from $\mathbb{B}^*$ to $\mathbb{N}_0$) Let x be a binary string. Define

$$\langle x\rangle := 2^{\ell(x)}+b(x)-1 \quad \text{with} \quad b(x) := \sum_{i=1}^{\ell(x)}2^{\ell(x)-i}[\![x_i=1]\!]$$

Then $\langle\cdot\rangle$ is a bijective map, as well as enumerating $\mathbb{B}^*$, sorted first by length, then lexicographically.

Proof. Left as an exercise to the reader.   ∎

The inverse of this bijection can be written as:

$$\langle n\rangle^{-1} = b_{\lfloor\log_2(n+1)\rfloor}^{-1}(n+1-2^{\lfloor\log_2(n+1)\rfloor}) \tag{2.1.2}$$

where $b_k^{-1}(n)$ takes a natural number n and returns a string of bits of length $\lfloor\log_2 n\rfloor+1$ representing the canonical encoding of n in binary (i.e. with no leading zeros for positive numbers). The resulting string is then prefixed with zeros to length k. (For example, $b_5^{-1}(7)=00111$.) We leave $b_k^{-1}(n)$ undefined if $k<\lfloor\log_2(n+1)\rfloor$ (that is, there are not enough bits to describe the canonical binary encoding of n). This ensures that the encoded string $\langle n\rangle^{-1}$ satisfies the property

$$\ell(\langle n\rangle^{-1}) = \lfloor\log_2(n+1)\rfloor$$

It is easy to see this is a valid inverse of $\langle\cdot\rangle$. One can verify this by rearranging the formula for $\langle\cdot\rangle$, and replacing $\ell(x)$ with $\lfloor\log_2(n+1)\rfloor$. We have the following useful bound for this bijection.

> **Proposition 2.1.3 (Bijection length bound)** Let x be a binary string. Then
> $$2^{\ell(x)} - 1 \le \langle x \rangle \le 2^{\ell(x)+1} - 2$$

Proof. Note that $0 \le b(x) \le 2^{\ell(x)} - 1$, from which the result follows. ∎

Now that we have a bijection between strings and numbers, we will often drop the bijection $\langle \cdot \rangle$ and assume from the context whether to apply $\langle \cdot \rangle$ or its inverse so the statement is syntactically valid. For example, we may write $x + 1$ instead of $\langle x \rangle + 1$ if x is a string, or we may write $\ell(n)$ instead of $\ell(\langle n \rangle^{-1})$ for a natural number n.

2.1.2 Prefix Codes

Often, when transmitting a message across a medium, the message is converted into a sequence of symbols, called a *code*, which can then be transmitted (as say, a sequence of electrical pulses along a wire, or a sequence of movements of semaphore flags), received on the other end, and then decoded to obtain the original message. Using a binary code is ideal due to requiring only two distinct symbols. However, this proves problematic if we want to send multiple messages. Suppose we want to transmit the sequence of numbers (1,3). We encode the numbers $\langle 1 \rangle = 0$ and $\langle 3 \rangle = 00$, and transmit the concatenation $\langle 1 \rangle \langle 3 \rangle = 000$. The receiver also has access to the code, but this code is ambiguous, as 000 may be decoded as (3,1) or (1,1,1) or (1,3). One solution could be to add an additional delimiter symbol #, and transmit 0#00, but this means we have to waste an additional symbol on the delimiter, which itself has to be encoded somehow if we wish to transmit only binary strings. It turns out there is a neat solution: we restrict the set of allowed encodings to a *prefix-free* set of binary strings: a set for which no member is a prefix of another.

Let $x, y \in \mathbb{B}^*$ be strings. We say that x is a *prefix* of y (denoted by $x \sqsubseteq y$) if $\exists z \in \mathbb{B}^*$ such that $xz = y$. Furthermore, if $z \ne \epsilon$, then we call x a *proper prefix* of y (denoted by $x \sqsubset y$). A set $\mathcal{P} \subseteq \mathbb{B}^*$ is called *prefix-free* if no element of the set is a proper prefix of another. Given an injection $c : \mathbb{B}^* \to \mathbb{B}^*$, if $\{c(x) : x \in \mathbb{B}^*\}$ is prefix-free, then we call c a *prefix-code*. The elements in $\mathcal{P}$ are called *(prefix) codewords*. Using prefix codes, a message is always *uniquely decodable*, that is, there is no ambiguity in decoding the message (unlike the above example). Thus, the message can be transmitted as a sequence of concatenated codewords without any reserved delimiter symbol to separate the words in the message. For example, Morse code[1] is not a prefix code, as -.-- could be decoded as any of {Y,KT,NTT,NM,TAT,TW,TEM,TETT}. Morse code actually uses three symbols (a space character, the absence of a pulse alongside short and long pulses), and adds spaces between letters as a delimiter.

There are many[2] ways to construct prefix codes in $\mathbb{B}^*$. We can construct an infinite family of prefix codes E_i for $\mathbb{B}^*$ in the following way: given a string $x \in \mathbb{B}^*$, its ith order prefix codeword $E_i(x)$ is defined recursively as

$$E_i(x) = \begin{cases} 1^x 0 & i = 0 \\ E_{i-1}(\ell(x))x & \text{otherwise} \end{cases} \qquad (2.1.4)$$

where 1^n denotes[3] a string $111\ldots1$ of length n. We call $E_i(x)$ a *self-delimiting encoding* of the binary string x. For example, the zeroth-order prefix codeword for a binary string x is

---

[1] an internationally recognized code to encode letters as short (dot) and long (dash) pulses

[2] In fact, uncountably many. One can draw a correspondence between prefix codes and binary trees, and note there are uncountably many binary trees.

[3] Not to be confused with 1 raised to the power of n.

the digit 1 repeating $\langle x \rangle$ many times, followed by a 0. The ending 0 acts as a delimiter, so that if many strings had their zeroth-order prefix codewords concatenated, we could tell where one stops and the next begins. Despite its simplicity, the zeroth-order prefix code is undesirable as the codewords are much longer than the original message, and thus inefficient when it comes to practical data transmission. In fact, by Proposition 2.1.3 we have that $\ell(E_0(x)) = \langle x \rangle + 1 = O(2^{\ell(x)})$ grows exponentially with respect to the length of x. We would desire an encoding with length approximately[4] equal to $\ell(x)$ with minimal overhead. Due to this reason, we often need to resort to a higher-order prefix code. In particular, we often use the first-order prefix code $\overline{x}$ of x as

$$\overline{x} := E_1(x) = E_0(\ell(x))x = 1^{\ell(x)}0x$$

and the second-order prefix code x' of x as[5]

$$x' := E_2(x) = E_1(\ell(x))x = \overline{\ell(x)}x = 1^{\ell(\ell(x))}0\ell(x)x.$$

Higher-order prefix codes offer diminishing returns, and are seldom used. The length of $\overline{x}$ and x' are bounded as follows:

$$\begin{aligned}
\ell(\overline{x}) &= 2\ell(x) + 1 \\
\ell(x') &= \ell(x) + 2\ell(\ell(x)) + 1 \\
&= \ell(x) + 2\log_2(\ell(x)) + O(1)
\end{aligned} \tag{2.1.5}$$

The first-order prefix code $\overline{x}$ of a string x adds at most $2\log_2 x$ overhead to the length of x, and the second-order prefix code adds $\log_2 x + 2\log_2\log_2 x$ overhead, both of which grow as $O(\log_2 x)$, a modest penalty compared to the length of x.

   This concept of self-delimiting encodings will show up again when we introduce the concept of Kolmogorov complexity in Section 2.7.

Lemma 2.1.6 (*i*th-order code is prefix) $E_i(\cdot)$ is a prefix code for all $i \in \mathbb{N}_0$.

Proof. The case $i = 0$ we have already discussed. For $i = 1$, suppose there were two strings $x \neq y$ such that $\overline{x}$ was a prefix of $\overline{y}$. So $\exists z \neq \epsilon$ s.t $\overline{x}z = \overline{y}$. Then

$$1^{\ell(x)}0xz = 1^{\ell(y)}0y$$

The strings are equal, so they must agree at least up to the first zero. $1^{\ell(x)}0 = 1^{\ell(y)}0$, hence $\ell(x) = \ell(y)$. But then $\ell(\overline{x}) + \ell(z) = \ell(\overline{x}z) = \ell(\overline{y}) = \ell(\overline{x})$, hence $\ell(z) = 0$, which implies that $z = \epsilon$ and thus $x = y$, a contradiction. The proof for $i > 1$ follows the same line of reasoning and by induction and is left as an exercise.   ∎

Theorem 2.1.7 (Uniqueness of prepending prefix codes) For any two given strings $x, y \in \mathbb{B}^*$ and any prefix code c, $c(x)y$ is uniquely decodable.

Proof. Since c is a prefix code, the prefix of $c(x)y$ that lies in the range of c must be unique. Since $c(x)$ is by definition in the range of c, we can separate the concatenation $c(x)y$ into the strings $c(x)$ and y without ambiguity. We can then recover x from $c(x)$.   ∎

---

[4]The code cannot map all strings x to a codeword shorter than $\ell(x)$, as this would provide an injection from $\mathbb{B}^n$ to $\mathbb{B}^{<n}$, which is impossible.

[5]Note the omission of $\langle \cdot \rangle$! As $\ell : \mathbb{B}^* \to \mathbb{N}_0$, obviously ℓ cannot be composed with itself, and technically we should write $x' = \overline{\langle \ell(x) \rangle^{-1}}x = 1^{\ell(\langle \ell(x) \rangle^{-1})}0\langle \ell(x) \rangle^{-1}x$. We hope this motivates keeping the usage of $\langle \cdot \rangle$ implicit.

Theorem 2.1.7 motivates encoding a pair of strings (x,y) as $\bar{x}y$ or $x'y$. Leaving the last string unencoded is more economical of bits than encoding as $\bar{x}\bar{y}$ or $x'y'$ and crucial for certain definitions related to Kolmogorov complexity.

2.1.3   Infinite Binary Sequences $\mathbb{B}^\infty$ and the Unit Interval

Recall that the set of infinite binary sequences is denoted as $\mathbb{B}^\infty$. An element in $\mathbb{B}^\infty$ is a one-way infinite sequence, normally denoted by $\omega = \omega_{1:\infty} = \omega_1\omega_2\omega_3...$ with $\omega_i \in \mathbb{B}$ for all i. There is a correspondence between elements of $\mathbb{B}^\infty$ and predicates (Boolean valued functions) over $\mathbb{N}^+$. For example, the predicate "is a prime number" corresponds to the binary sequence

$$\omega_i = \begin{cases} 1 & i \text{ is prime} \\ 0 & \text{otherwise} \end{cases}$$

Infinite sequences may be prefixed[6] with finite binary strings. For a finite binary string $x = x_{1:n} = x_1x_2...x_n$ and an infinite binary sequence $\omega = \omega_{1:\infty} = \omega_1\omega_2...$, their concatenation is

$$x\omega = x_1x_2...x_n\omega_1\omega_2...$$

We can draw a correspondence between real numbers and binary sequences. Given an infinite sequence $\omega_{1:\infty} \in \mathbb{B}^\infty$, $f : \mathbb{B}^\infty \to [0,1]$ sends it to a real number in the following way:

$$f(\omega_{1:\infty}) = \sum_{n=1}^{\infty} 2^{-n}\omega_n \tag{2.1.8}$$

Essentially, this maps the infinite sequence $\omega_{1:\infty}$ to the real number with binary expansion $0.\omega_1\omega_2....$

Example 2.1.9 (Mapping infinite sequence to real numbers) We write $(x_{1:n})^\infty$ to denote the infinite sequence comprised of repeating the finite string $x_{1:n}$. The following are some examples of the evaluation of f: $f(0^\infty) = 0$; $f(010^\infty) = 0.25$; $f(110^\infty) = 0.75$; $f(10^\infty) = f(01^\infty) = 0.5$; $f((01)^\infty) = 1/3$. ◆

This function is surjective but not injective (one-to-one) as shown in the preceding example, both 10^∞ and 01^∞ are mapped to 0.5. This is due to numbers having a non-unique expansion in binary: we can represent 1 in binary as 1.000... or 0.111....

Cylinder sets.   We have a geometric representation of $\mathbb{R}$ as the infinite number line [Com]:

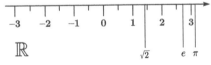

One can ask what geometric representation can be associated with $\mathbb{B}^\infty$. This can be achieved using *cylinder sets*, which are defined as follows:

> **Definition 2.1.10 (Cylinder sets)** The *cylinder set* Γ_x of a string $x \in \mathbb{B}^*$ is a subset in $\mathbb{B}^\infty$ that contains all (one-way) infinite sequences starting with x. That is,
>
> $$\Gamma_x := \{x\omega \,|\, \omega \in \mathbb{B}^\infty\}$$

Remark 2.1.11 (Cylinder sets as intervals) Geometrically speaking, the cylinder Γ_x can be identified with the closed interval $f(\Gamma_x)$ (see (2.1.8)) of Γ_x under f as shown in Figure 2.1. ●

---

[6]Why not suffixed?

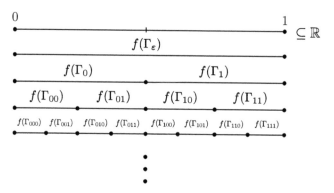

Figure 2.1: *A geometric representation of various cylinder sets Γ_x as intervals $f(\Gamma_x)$* (2.1.8).

2.1.4   Exercises

1. [C07] Compute the zeroth-, first- and second-order prefix codes for the binary string 11111.

2. [C10] Prove that the number of prefix-free sets for a binary alphabet is uncountable.

3. [C12] Given strings $x,y \in \mathbb{B}^*$, which of the following are uniquely decodable? $\bar{x}y$, $x\bar{y}$, $x'y$, xy', $\bar{x}\bar{y}$, $x'y'$. In each case, provide a proof or a counterexample.

4. [C15] Prove that the family of codes E_i (2.1.4) are prefix-free for all $i \in \mathbb{N}_0$.

5. [C10] Given the function f defined in (2.1.8), find the pre-image $f^{-1}(1/3)$, and prove that f is a surjection.

6. [C07] Prove that $\Gamma_y \subseteq \Gamma_x$ if and only if $x \sqsubseteq y$.

2.2   Measure Theory and Probability Theory

This section could be more appropriately titled "Probability Theory with Minimal Reference to Measure Theory." We aim to establish a functional understanding of probability theory for this book while minimizing the reliance on measure theory, the formal mathematical framework that probability theory is built upon. We cover the axioms of probability theory, build up to the abstractions that it provides and then only work in these abstractions as much as possible. Sometimes, references to the underlying measure theory are required, when we discuss probabilistic convergence (Section 2.2.8). We cover the most important properties from probability theory such as representing stochastic phenomena with random variables (Section 2.2.4), the inequalities of Markov and Chebyshev (Section 2.2.7), and rigorously define convergence of limits for random variables (Section 2.2.8).

For readers who would like a background in measure theory, see [SS09]. For more detail on probability see [GS20], and [Ros14].

2.2.1   The Axioms of Probability Theory

We introduce probability theory from the fundamentals, and build up to the properties and definitions that we already have an intuitive understanding for, like the chain rule, $P(A,B) = P(A|B)P(B)$ and expectations $\mathbf{E}[X]$.

Definition 2.2.1 (σ-algebra) Given a set Ω (whose elements are called *outcomes*), a collection $\mathcal{F}$ of subsets of Ω is called a σ-*algebra* if it satisfies the following axioms:

1. $\varnothing \in \mathcal{F}$.
2. If $A_1, A_2, \ldots \in \mathcal{F}$, then $\bigcup_{n=1}^{\infty} A_n \in \mathcal{F}$. (Closed under countable union)
3. If $A \in \mathcal{F}$, then $\Omega \backslash A =: A^c \in \mathcal{F}$. (Closed under complement)

Note that a σ-algebra $\mathcal{F}$ is also closed under countable intersection, due to De Morgan's law $\bigcap_{n=1}^{\infty} A_n = \left(\bigcup_{n=1}^{\infty} A_n^c \right)^c$.

Definition 2.2.2 (Measurable space) A *measurable space* is a pair $(\Omega, \mathcal{F})$ where $\mathcal{F}$ is a σ-algebra on Ω.

We call Ω the *sample space*, $\mathcal{F}$ the *event space*, and a set A in the event space $\mathcal{F}$ an *event*.

Axiom 2.2.3 (Probability (semi)measure) A *probability measure* (or *measure*) P on a measurable space $(\Omega, \mathcal{F})$ is a function $P : \mathcal{F} \to [0,1]$ satisfying the following axioms:

1.
$$P(\Omega) = 1$$

2. For any sequence of events $A_1, A_2, \ldots \in \mathcal{F}$ that are pairwise disjoint (i.e, if $i \neq j$ then $A_i \cap A_j = \varnothing$) then

$$P\left(\bigcup_{i=1}^{\infty} A_i \right) = \sum_{i=1}^{\infty} P(A_i)$$

3. If we weaken the Axiom 1 to only require $P(\Omega) \leq 1$, then we call P a *semi-probability measure*, or simply a *semi-probability*.

4. If we weaken the axioms further to only require $P(\Omega) \leq 1$ and

$$P\left(\bigcup_{i=1}^{\infty} A_i \right) \geq \sum_{i=1}^{\infty} P(A_i)$$

then we call P a *probability semimeasure* or simply a *semimeasure*.

While these axioms seem rather sparse, all of the intuitive properties that we would expect a probability function to have can be derived from these two axioms, as we will see later in Theorem 2.2.9.

Definition 2.2.4 (Probability Space) A *probability space* is a triple $(\Omega, \mathcal{F}, P)$ where $(\Omega, \mathcal{F})$ is a measurable space, and P is a probability measure on $(\Omega, \mathcal{F})$.

A probability measure can be thought of as a function that takes a particular event A in the event space $\mathcal{F}$, and assigns to it a number representing how likely that event is to occur. Given that we express probability through the language of set theory, Table 2.2 taken from [GS20] is a useful reference.

Table 2.2: *The jargon of set theory and probability theory [GS20].*

Symbol	Set jargon	Probability jargon
Ω	Collection of objects	Sample space
ω	Element in Ω	Singleton event, outcome
A	Subset of Ω	An outcome in A occurs
A^c	Complement of A	No outcome in A occurs
$A \cap B$	Intersection	An outcome in both A and B occurs
$A \cup B$	Union	An outcome in A or B occurs
$A \backslash B$	Set difference	An outcome in A but not B occurs
$A \Delta B$	Symmetric difference	Event that an outcome in *either* A or B occurs (but not both)
$A \subset B$	Inclusion	If an outcome in A occurs, then an outcome in B occurs
$\varnothing$	Empty set	Impossible event, where no outcome of the set of all outcomes occurs. The event is the empty set containing no outcomes.
Ω	Universal set	Certain event, some outcome in the set of all outcomes occurs. The event is the entire sample space Ω.

Discussion. There are a number of important corollaries from the axioms of probability defined in Axiom 2.2.3:

1. The sample space Ω is always an event. The event Ω can be thought of as "something in the set of all outcomes happens" event, and satisfies $P(\Omega) = 1$.

2. The *impossible event* $\varnothing$ is an event, corresponding to the case where no outcome in the set of all possible outcomes occurs. Unsurprisingly, the impossible event satisfies $P(\varnothing) = 0$.

3. Every finite union of events is an event and every countable (and in particular, every finite) intersection of events is an event. This is because for any sequence of events $A_1, A_2, ...$ we can write $\bigcap_n^\infty A_n = (\bigcup_n^\infty A_n^c)^c$ from De Morgan's law.

4. The number of axioms has been kept to the minimum required to define all the other desirable properties of probability theory.

5. The axioms do not imply that every subset of Ω, is an event. For example the set $\{\Omega, \varnothing\}$ (called the *coarse σ-algebra*) is a valid (though boring) σ-algebra.

6. For finite sets and countable sets, we can freely choose $\mathcal{F} = 2^\Omega$ (that is, any subset of Ω is an event.) Attempting to naively do the same for uncountable Ω can have some subtle problems beyond the scope of this book. For instance, suppose we wanted to define a probability space to model random numbers sampled uniformly from the unit interval. We could choose $\Omega = (0,1)$, and let events be represented by the smallest σ-algebra $\mathcal{F}$ containing all intervals $(a,b) \subset \Omega$. We can then define the probability measure over an interval as its "length", $P((a,b)) := b - a$, and then for disjoint intervals $A_1, ..., A_n$ we define $P(\cup_i A_i) = \sum_i P(A_i)$ in line with Axiom 2.2.3. This particular choice of measure is called the *Lebesgue measure*. It is tempting to then try to define the Lebesgue measure for any subset of $(0,1)$ by covering it with a union of intervals, and

then taking the smallest such covering.

$$P(E) = \inf\left\{\sum_i (b_i - a_i) : E \subset \cup_i (a_i, b_i)\right\}$$

This is called the *Lebesgue outer measure*. Furthermore, for sets E that satisfy the *Carathéodory's criterion*, that for every event S we have

$$P(S) = P(S \cap E) + P(S \cap E^c)$$

then we say that E is *Lebesgue measurable*. The set of all Lebesgue measure sets form a σ-algebra. Unfortunately, there exist sets E that do not satisfy Carathéodory's criterion, and for which a measure cannot be well-defined without contradicting the axioms, so it is not possible to naively choose $\mathcal{F} = 2^\Omega$ [SS09].

Example 2.2.5 (Biased coin) Suppose we wanted to represent the outcome of flipping a biased coin, with bias θ. We could model this with the probability space $(\Omega, \mathcal{F}, \mathrm{P})$, where $\Omega = \{H, T\}$, $\mathcal{F} = 2^\Omega$, and

$$P(\varnothing) = 0, \quad \mathrm{P}(\{H\}) = \theta, \quad \mathrm{P}(\{T\}) = 1 - \theta, \quad \mathrm{P}(\{H, T\}) = 1$$

Here, each of the events can be understood as a possible outcome for the experiment of flipping a coin. Here, $\varnothing$ is the event that the coin shows neither heads nor tails, $\{H\}$ (resp. $\{T\}$) is the event corresponding to the coin showing heads (resp. tails), and Ω is the event that the coin shows either heads or tails. One can easily verify that $(\Omega, \mathcal{F}, \mathrm{P})$ satisfies the properties of a probability space of Definition 2.2.4. ◆

Example 2.2.6 (Different Ω for counting heads) Sometimes, there are different reasonable choices of Ω for a problem. Suppose we flipped three coins and are interested in the number of heads shown. *(i)* We could write out the sample space as $\Omega = \{HHH, HHT, HTH, ...\}$ with $\mathcal{F} = 2^\Omega$ and distribution $\mathrm{P}(A) = \frac{1}{8}|A|$. Each outcome corresponds to one possible result for flipping the three coins. The event $E = \{HHT, HTH, THH\}$ corresponding to flipping 2 heads would then have the probability $\mathrm{P}(E) = \frac{1}{8}|E| = \frac{3}{8}$ as expected. *(ii)* As an alternative presentation, we could choose the sample space $\Omega' = \{0, 1, 2, 3\}$ to be the number of heads shown when three coins are flipped, the event space as $\mathcal{F}' = 2^{\Omega'}$ any set of outcomes, and

$$\mathrm{P}'(\{0\}) = 1/8$$
$$\mathrm{P}'(\{1\}) = \mathrm{P}'(\{2\}) = 3/8$$
$$\mathrm{P}'(\{3\}) = 1/8$$

from which we would get the same answer as before, $\mathrm{P}'(\{2\}) = \frac{3}{8}$. The second choice of Ω is just as valid as the first, but it is rather specialized to the problem at hand with the solution essentially hardcoded in P'. Essentially, we construct Ω' by partitioning Ω based on the number of heads (e.g. $\{2\} \equiv \{HHT, HTH, THH\}$) and then construct P' using the probability P assigns to that particular partition, $\mathrm{P}'(\{2\}) := \mathrm{P}(\{HHT, HTH, THH\}) = \frac{3}{8}$. We will see a more elegant version of inferring a probability distribution over a partition of Ω (like the number of heads) later on when we introduce *random variable* in Section 2.2.4. ◆

Definition 2.2.7 (Generating a σ-algebra) Given a family of subsets $\mathcal{S} \subseteq 2^\Omega$, we define the σ-algebra *generated* by $\mathcal{S}$ (denoted $\sigma(\mathcal{S})$) as

$$\sigma(\mathcal{S}) \;=\; \bigcap_{\mathcal{C} \text{ is a } \sigma\text{-algebra } \supseteq \mathcal{S}} \mathcal{C}$$

$\sigma(\mathcal{S})$ can be thought of as the smallest σ-algebra that contains $\mathcal{S}$. That is, if $\mathcal{F}$ is a σ-algebra that contains $\mathcal{S}$, then $\sigma(\mathcal{S}) \subseteq \mathcal{F}$. As an abuse of notation, we write $\sigma(A) := \sigma(\{A\})$ for singleton families $\{A\}$.

Example 2.2.8 (The smallest σ-algebra that contains A) If $A \subseteq \Omega$, then $\sigma(\{A\}) = \{\varnothing, A, A^c, \Omega\}$ is the smallest σ-algebra containing A. We cannot remove any elements of $\sigma(A)$ without violating Definition 2.2.7. If $\Omega = \{1,2,3,4\}$ and $A = \{\{1\},\{2\}\}$, then

$$\sigma(A) \;=\; \{\varnothing, \{1\}, \{2\}, \{1,2\}, \{3,4\}, \{1,3,4\}, \{2,3,4\}, \Omega\} \qquad \blacklozenge$$

An important property (that we will not prove here) is that given a measurable space $(\Omega, \mathcal{F})$ where $\mathcal{F} = \sigma(\mathcal{S})$ for some $\mathcal{S} \subseteq 2^\Omega$, we can uniquely identify P solely from the values it takes on each set in $\mathcal{S}$. This property is called the *Carathéodory Extension Theorem*.

We now present some important theorems that can be derived from Axiom 2.2.3, many of which are (hopefully) intuitively obvious properties that probabilities should satisfy.

Theorem 2.2.9 (Probability properties) Let $(\Omega, \mathcal{F}, P)$ be a probability space, $A, B \in \mathcal{F}$ be events, $\{A_n\}_{n=1}^\infty$ be a collection of events, and $\{D_n\}_{n=1}^\infty$ be a collection of pairwise disjoint events. Then, the axioms imply the following properties:

$$P(\varnothing) = 0 \tag{2.2.10a}$$
$$P(A \cup B) = P(A) + P(B) - P(A \cap B) \tag{2.2.10b}$$
$$P(A^c) = 1 - P(A) \tag{2.2.10c}$$
$$\text{if } A \subseteq B \text{ then } P(A) \le P(B) \tag{2.2.10d}$$
$$P(A) = P(A \cap B) + P(A \cap B^c) \tag{2.2.10e}$$
$$\textstyle\sum_{n=1}^\infty P(D_n) \le 1 \tag{2.2.10f}$$
$$P(\textstyle\bigcup_{n=1}^m A_n) \le \textstyle\sum_{n=1}^m P(A_n) \tag{2.2.10g}$$
$$P(\textstyle\bigcup_{n=1}^\infty A_n) = \lim_{m\to\infty} P(\textstyle\bigcup_{n=1}^m A_n) \tag{2.2.10h}$$
$$P(\textstyle\bigcap_{n=1}^\infty A_n) = \lim_{m\to\infty} P(\textstyle\bigcap_{n=1}^m A_n) \tag{2.2.10i}$$
$$\text{if } A_1 \subseteq A_2 \subseteq A_3... \text{ then } P(\textstyle\bigcup_{n=1}^\infty A_n) = \lim_{n\to\infty} P(A_n) \tag{2.2.10j}$$
$$\text{if } A_1 \supseteq A_2 \supseteq A_3... \text{ then } P(\textstyle\bigcap_{n=1}^\infty A_n) = \lim_{n\to\infty} P(A_n) \tag{2.2.10k}$$

Equations (2.2.10a) to (2.2.10f) follow directly from the definition of a probability measure, and so are left as an exercise. Proofs of (2.2.10g), also known as *Boole's Inequality*, together with (2.2.10h) and (2.2.10i), are sketched as follows:

Proof. (2.2.10g): Let $\{A_n\}_{n=1}^\infty$ be a family of events and let $B_n = A_n \backslash \bigcup_{i=1}^{n-1} A_i$. Note that $\{B_n\}_{n=1}^\infty$ is pairwise disjoint, $B_n \subseteq A_n$ and $\bigcup_n A_n = \bigcup_n B_n$. Then,

$$P(\textstyle\bigcup_n A_n) = P(\textstyle\bigcup_n B_n) = \textstyle\sum_n P(B_n) \le \textstyle\sum_n P(A_n)$$

The first equality is obvious, the second equality is from Axiom 2 of Axiom 2.2.3, and the third (in)equality is from (2.2.10d).

(2.2.10h) Continuing the above we get

$$P\left(\bigcup_{n=1}^{\infty} A_n\right) = \sum_{n=1}^{\infty} P(B_n) = \lim_{m \to \infty} \sum_{n=1}^{m} P(B_n) = \lim_{m \to \infty} P\left(\bigcup_{n=1}^{m} B_n\right) = \lim_{m \to \infty} P\left(\bigcup_{n=1}^{m} A_n\right)$$

The third equality is Axiom 2.2.3 again, the others are obvious.

(2.2.10i): follows by using De Morgan's rule on the above. ∎

Equations (2.2.10j) and (2.2.10k) are trivial corollaries of (2.2.10h) and (2.2.10i) respectively. It says that the probability of a limit of unions/intersections is equal to the limit of the probabilities. In other words, the limit operation can be "pushed" through P.

2.2.2 (Semi)Measures on Infinite Sequences

The most important choice of measurable space $(\Omega, \mathcal{F})$ which we will use throughout this book is the set of infinite sequences $\Omega = \Gamma_\epsilon \equiv \mathcal{X}^\infty$ (Definition 2.1.10) together with the event space $\mathcal{F} = \sigma(\{\Gamma_x : x \in \mathcal{X}^*\})$ generated by the the cylinder sets Γ_x.

Definition 2.2.11 ((Semi)measures on $\Omega = \mathcal{X}^\infty$) We define (semi)measures on the sample space $\Omega = \Gamma_\epsilon = \mathcal{X}^\infty$ with event space $\mathcal{F} = \sigma(\{\Gamma_x : x \in \mathcal{X}^*\})$. For this sample space, we denote the (semi)measure as ν instead of the usual P. We will abuse the following notation: For $x \in \mathcal{X}^*$, define $\nu(x) := \nu(\Gamma_x)$. Under this notation the probability semimeasure Axiom 2.2.3 for $\nu : \mathcal{X}^* \to [0,1]$ can be restated as

$$\nu(\epsilon) \leq 1 \quad \text{and} \quad \nu(x) \geq \sum_{x \in \mathcal{X}} \nu(xa)$$

and with equality for measures. Intuitively, $\nu(x)$ should be thought of as the probability that a sampled sequence starts with the string x.

The following lemma is a direct consequence of $P[\Omega] \leq 1$ for $\Omega = \mathcal{X}^\infty$

Lemma 2.2.12 For a given n and (semi)measure ν, we have $\sum_{x \in \mathcal{X}^n} \nu(x) \leq 1$, where equality holds for measures.

Proof. We proceed by induction on n. The base case follows immediately as $\nu(\epsilon) \leq 1$. For the step case, we have

$$\sum_{x \in \mathcal{X}^{n+1}} \nu(x) = \sum_{x \in \mathcal{X}^n} \sum_{a \in \mathcal{X}} \nu(xa) \leq \sum_{x \in \mathcal{X}^n} \nu(x) \leq 1$$

∎

The cylinder sets form a semi-ring, so for measures ν we can apply Carathéodory's extension theorem to uniquely get a measure $\nu(A)$ for all $\mathcal{F}$-measurable sets A. The more elementary developments and statements that only rely on ν on the cylinder sets, often naturally generalize to semimeasures. More advanced semimeasure theory, especially if it requires ν on $\mathcal{F}$-measurable sets beyond cylinder sets, has yet to be developed. Those results are therefore only developed in this book for proper probability measures ν. See Section 2.8.2 for the intricacies of *semi*measures which go beyond the already involved measure theory.

2.2.3   Conditional Probability

Definition 2.2.13 (Conditional probability) If A and B are events with $P(A) > 0$, then the *conditional probability* of event B given event A is defined as:

$$P(B|A) := \frac{P(A \cap B)}{P(A)}$$

One way to intuitively understand $P(B|A)$ is to think of restricting the set of possible outcomes to those where A occurs (subsets of A), and then measuring (using P) the proportion of those outcomes where B also occurs.

Theorem 2.2.14 (Conditional probability measures are probability measures) If $(\Omega, \mathcal{F}, P)$ is a probability space, and $A \in \mathcal{F}$ with $P(A) > 0$, then $(\Omega, \mathcal{F}, P_A)$ is also a probability space, with $P_A(\cdot) := P(\cdot | A)$.

Theorem 2.2.14 is very useful, as it means that any results that hold for a probability measure P, also hold for P conditioned on any event A satisfying $P(A) > 0$.

2.2.4   Random Variables

Usually, we are not interested in the particular value of $\omega \in \Omega$ that was sampled from the distribution P, but of the value of some parameter that depends on the outcome of that experiment.

For example, when we roll a pair of dice for a board game, for many games we do not care what particular pair of values (e.g (3,4) or (5,2)) is obtained, but rather the sum of the dice (7 in this case). Similarly, gamblers playing roulette are not interested in the particular outcome of each spin (33-Black or 9-Red), but only if the resulting wager is won or lost.

We can express this through the concept of *random variable*, which can be thought of as a variable or function, the value of which depends on outcomes ω sampled from Ω according to the probability distribution P.

Remark 2.2.15 (Types of random variables) Some random variables can be classified into categories like discrete, continuous and mixed. We will mostly focus on discrete (and occasionally continuous) random variables. The theorems in this book will usually be proven only for the discrete case, though they usually also hold for the continuous case (i.e. by swapping sums for integrals.)   ●

Definition 2.2.16 (Random variable) A (real-valued)[a] random variable is a function $X : \Omega \to \mathbb{R}$ that satisfies the property

$$\{\omega \in \Omega : X(\omega) \leq r\} \in \mathcal{F}$$

for all $r \in \mathbb{R}$. We call the image $X(\Omega) = \{X(\omega) : \omega \in \Omega\}$ the *alphabet* for X.

---

[a]A more general definition includes the ability for the random variable X to have the type $X : \Omega \to \mathcal{E}$, where $\mathcal{E}$ is a *measurable space*.

First, note that there is nothing "random" about a random variable. It is a deterministic function of type $\Omega \to \mathbb{R}$. The "randomness", as we will see, comes from the interaction between X and the underlying probability measure P. This may seem rather abstract compared to the intuitive definition usually given for random variables: A discrete random variable is usually defined as a countable set of possible outcomes with a probability assigned to each, and continuous random variables are usually defined as an interval $I \subseteq \mathbb{R}$ with a corresponding probability distribution $p : I \to \mathbb{R}$. As it happens, both discrete and continuous random variables can be defined as a special case of this definition (See Proposition 2.2.27). The probability measure P on $(\Omega, \mathcal{F})$ together with the function X implies a distribution for the random variable X, as follows.

Definition 2.2.17 (Cumulative distribution function (cdf)) The *cumulative distribution function*, (cdf) $F_X : \mathbb{R} \to [0,1]$ for a random variable X is defined as

$$F_X(x) \;=\; P(\{\omega \in \Omega : X(\omega) \le x\})$$

We may also denote the cdf as $P(X \le x)$ or simply $F(x)$ when the choice of random variable is clear from context.

Note that $P(\{\omega \in \Omega : X(\omega) \le x\})$ is always defined for any $x \in \mathbb{R}$, as random variables have the property that $\forall x \in \mathbb{R}.\{\omega \in \Omega : X(\omega) \le x\} \in \mathcal{F}$.

Definition 2.2.18 (Discrete random variable) A random variable X is *discrete* if the alphabet $X(\Omega)$ is countable.

Definition 2.2.19 (Probability mass function (pmf)) For a discrete random variable X, we can define the *probability mass function* (pmf) $p_X : \mathbb{R} \to [0,1]$ as

$$p_X(x) \;=\; P(\{\omega \in \Omega : X(\omega) = x\})$$

As with the cdf, we may denote $p_X(x)$ as $P(X = x)$ or $p(x)$.

Example 2.2.20 (Fair coin flips) Following Example 2.2.6, we consider describing the number of heads in three flips of a coin using a random variable. Take the probability space $(\Omega, 2^\Omega, P)$ with $\Omega = \{HHH, HHT, ..., TTT\}$ and $P(A) = |A|/8$. The choice of P here represents that each outcome $\omega \in \Omega$ is one possible result from flipping three fair coins, so the probability of an event is just the number of outcomes in that event, times $(\frac{1}{2})^3 = \frac{1}{8}$. We define a random variable $X : \Omega \to \mathbb{R}$ as

$$X(\omega) \;=\; \text{number of heads in } \omega$$

It is clear that X is discrete, as $X(\Omega) = \{0,1,2,3\}$. The corresponding pdf p_X now assigns a probability $p_X(h)$ to the event of obtaining h heads in 3 coin flips. For example,

$$p_X(2) := P(\{\omega \in \Omega : X(\omega) = 2\}) = P(\{HHT, HTH, THH\}) = 3/8 \qquad \blacklozenge$$

Definition 2.2.21 (Continuous random variable) A random variable X is *continuous*[a] if there exists an integrable function $f : \mathbb{R} \to [0, \infty)$ such that

$$F_X(x) = \int_{-\infty}^{x} f(t)\, dt.$$

---

[a]Note that the word "continuous" in "continuous random variable" is a bit of a misnomer. It does not imply that the function X is continuous, nor does it imply that the set $X(\Omega)$ is connected. It refers to the fact that the cdf F_X is continuous (in fact, it is *absolutely continuous*, a stronger property), as it is the integral of an integrable function.

Definition 2.2.22 (Probability density function (pdf)) For a continuous random variable X, we define the *probability density function* (pdf) as $p_X(x) := f(x)$ from Definition 2.2.21.

We use the same symbol p_X for the pmf (when X is discrete) and the pdf (when X is continuous), as many other definitions (that we will see later on) are identical for both discrete and continuous random variables (up to swapping sums for integrals).

Continuous random variables are used for situations where there are uncountably many outcomes to consider. For example, consider the distribution of the height of people in a room. We would expect that "most" people are "about" 1.8 meters tall, or that it is "very likely" that any randomly selected person is between 1 and 2 meters tall. Continuous random variables allow the above statements to be stated rigorously.

To ask the question "how likely is it that a person selected randomly is 1.8 meters tall?" is useless, as for any height the probability that someone is *exactly* 1.8000... meters tall is zero! Instead, we can only ask for probabilities that the height of an individual lies in some interval (how likely is it that a randomly selected person is between 1.79 and 1.81 meters in height?)

For this reason, we cannot naively define the pdf $p_X(x)$ in the same way as we defined the pmf for discrete random variables. By doing so, we would obtain a paradoxical result that $P(X = x) = 0$ for all $x \in X(\Omega)$. Instead, we generalize from the following theorem:

Proposition 2.2.23 (Relating cdf and pdf) For a discrete random variable X, the cdf F_X and the pmf p_X can be related as follows:

$$F_X(x) = \sum_{t \le x} p_X(t)$$

where the sum is taken over all $t \in X(\Omega)$ such that $t \le x$.

Proof.

$$F_X(x) \stackrel{(a)}{=} P(\{\omega \in \Omega : X(\omega) \le x\}) \stackrel{(b)}{=} P\Big(\dot{\bigcup}_{t \in X(\Omega), t \le x} \{\omega \in \Omega : X(\omega) = t\}\Big)$$

$$\stackrel{(c)}{=} \sum_{t \in X(\Omega), t \le x} P(\{\omega \in \Omega : X(\omega) = t\}) \equiv \sum_{t \in X(\Omega), t \le x} P(X = t) \stackrel{(d)}{=} \sum_{t \le x} p_X(t)$$

(a) byDefinition 2.2.17, (b) follows from expanding the set into a disjoint ($\dot\cup$) union, (c) follows from Axiom 2.2.3, (d) since $p_X(t) = 0$ for $t \notin X(\Omega)$, so we are implicitly summing only over $t \in X(\Omega)$. ∎

Note the similarity between the expression in Proposition 2.2.23 and the definition $F_X(x) = \int_{t \leq x} p_X(t)\, dt = \int_{-\infty}^{x} p_X(t)\, dt$ of the pdf for continuous random variables, identical up to interchanging the sum for an integral.

Remark 2.2.24 (Densities may not be bounded by 1) Note that (unlike the cdf or pmf) the pdf for a continuous random variable may not be bounded by 1. In fact, it could even be unbounded, as long as the integral over the domain still converges. It can be shown that pdfs must still satisfy $\int_{-\infty}^{\infty} p_X(t)\, dt = 1$. One example is $p_X(x) = -\ln(x)[\![0 \leq x \leq 1]\!]$, which is unbounded $(p_X(x) \to \infty$ as $x \to 0)$ but $\int_{-\infty}^{\infty} p_X(t)\, dt = \int_0^1 -\ln(t)\, dt = 1$. ●

We use the pdf to define the probability that X lies inside an interval $[a,b]$

$$P(a \leq X \leq b) := \int_a^b p_X(x)\, dx = F_X(b) - F_X(a).$$

For a "sufficiently nice" set B, which can be decomposed into countable unions of disjoint intervals $B = \bigcup_{i=1}^{\infty}(a_i, b_i)$, we can write the probability that

$$P(X \in B) = \int_B p_X(x)\, dx := \sum_{i=1}^{\infty} \int_{a_i}^{b_i} p_X(x)\, dx$$

Example 2.2.25 (Uniform measure on interval [0,1]) Consider the unit interval $[0,1]$. A dart is thrown that hits somewhere along the interval.[7] We would like a probability space to formalize this mechanism. Naturally, we choose $\Omega = [0,1]$. We will consider the σ-algebra generated by the set $\{(a,b) : a \leq b\}$. We define $\mathcal{F}$ as the closure of the set of all open intervals under complements and countable unions. Each event can be thought of as the dart landing somewhere inside the region specified. For example, the event $E = (1/5, 2/5) \cup (3/5, 4/5)$ corresponds to the dart landing somewhere between $1/5$ and $2/5$, or somewhere between $3/5$ and $4/5$. By assuming that the dart is equally likely to land anywhere, it seems natural to assign the probability of an event to be the fraction of the total "length" of the interval taken up. Singleton sets have no length, so we define $P(\{a\}) = 0$ for any a. We define the probability measure $P((a,b)) = b - a$ for an open interval (a,b). We define the probability measure for a disjoint union of open intervals in the obvious way to ensure Axiom 2.2.3 is satisfied. If $A = \bigcup_{i=1}^{\infty}(a_i, b_i)$ with $0 \leq a_1 \leq b_1 \leq a_2 \leq b_2 \leq \ldots$ then A is a disjoint union, and so we can define

$$P(A) := \sum_{i=1}^{\infty}(b_i - a_i)$$

We obtain for free that if a set A is a complement of unions of open intervals, then

$$P(A^c) := 1 - P(A)$$

as complements of unions of open intervals are unions of closed intervals, and we can write closed intervals $[a,b]$ as $\{a\} \cup (a,b) \cup \{b\}$. Finally, by applying Axiom 2.2.3, we obtain

$$P([a,b]) = P(\{a\} \cup (a,b) \cup \{b\}) = P((a,b))$$

which now means that P is defined for all events in $\mathcal{F}$. This tedious legwork is required to ensure we cover all elements of $\mathcal{F}$. To merely allow P to be defined for any subset of $[0,1]$

---

[7]If managing to hit an infinitesimally thin line with a dart breaks your suspension of disbelief, you can imagine instead throwing a dart at a unit square, and then measuring the distance from the left-hand side of the square.

is problematic, as unfortunately there exist subsets E of [0,1] that are *non-measurable*, for which no value of $P(E)$ can be assigned without causing a contradiction [RT64]. Restricting $\mathcal{F}$ to the σ-algebra defined above avoids this problem.

As a corollary of the above, P assigns the same measure to an interval regardless if the endpoints are included or not:

$$\mathrm{P}([a,b]) = \mathrm{P}((a,b]) = \mathrm{P}([a,b)) = \mathrm{P}((a,b))$$

which intuitively makes sense, since each singleton point $\{a\}$ in Ω has zero length, and the likelihood of hitting any particular real number in Ω is zero. As a result, we can freely union in or subtract out finitely many singleton sets without affecting the probability assigned to a set.

Now that we have defined the probability space, we can talk about random variables on this space. Consider $X(\omega) = \omega$, the identity random variable. X represents the location at which the dart landed.

We can write the cdf F_X of X from the definition:

$$F_X(x) = \mathrm{P}(\{\omega \in \Omega : X(\omega) \le x\}) = \mathrm{P}(\{\omega \in \Omega : \omega \le x\}) = \mathrm{P}([0,x)) = \begin{cases} 0 & x < 0 \\ x & 0 \le x < 1 \\ 1 & 1 \le x \end{cases}$$

which then gives the pdf of X

$$p_X(x) = \begin{cases} 1 & 0 \le x \le 1 \\ 0 & \text{otherwise} \end{cases}$$

from which we can verify that

$$\int_{-\infty}^{x} p_X(t)\, dt = F_X(x)$$

Figure 2.3 plots the pdf p_X and the cdf F_X.   ◆

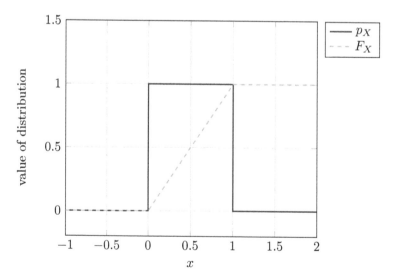

Figure 2.3: *A plot of the pdf p_X vs the cdf F_X for Example 2.2.25.*

Remark 2.2.26 (Defining a random variable by distribution) Often when talking about random variables, the actual definition of the function $X:\Omega \to \mathbb{R}$ is not important. We may only need the alphabet $X(\Omega)$ and the pmf/pdf $p_X(x) = P(X = x)$ that is induced from X and P. It turns out we can always define a probability space and random variable to match any desired set of outcomes and probabilities, as follows. ●

Proposition 2.2.27 (Specifying a random variable by an alphabet and probabilities) Let $\{r_i\}_{i=1}^{\infty}$ and $\{p_i\}_{i=1}^{\infty}$ be sequences of real numbers, with $\forall i.p_i \geq 0$ and $\sum_{i=1}^{\infty} p_i = 1$, and each r_i distinct ($r_i \neq r_j$ whenever $i \neq j$). Then, there exists a probability space $(\Omega, \mathcal{F}, P)$ and random variable $X:\Omega \to \mathbb{R}$ such that $X(\Omega) = \{r_n : n \in \mathbb{N}^+\}$ and $P(X = r_n) = p_n$ for all $n \in \mathbb{N}^+$.

Proof. We let $\Omega = \mathbb{N}^+$, $\mathcal{F} = 2^{\Omega}$, and define the probability measure $P(\{i\}) = p_i$ for singletons, and

$$P(A) = \sum_{a \in A} P(\{a\})$$

for all other sets to ensure P satisfies Axiom 2.2.3. Then, we define $X(i) = r_i$. Note that X is injective as each r_i is distinct, meaning a left inverse $X^{-1} : X(\Omega) \to \Omega$ with $X^{-1}(r_i) = i$ exists. Then,

$$\begin{aligned}
P(X = r_n) &= P(\{\omega \in \Omega : X(\omega) = r_n\}) \\
&= P(\{\omega \in \Omega : \omega = X^{-1}(r_n)\}) \\
&= P(\{\omega \in \Omega : \omega = n\}) \\
&= P(n) = p_n
\end{aligned}$$

as required. ∎

Remark 2.2.28 (Discrete random variable via probability mass function) In some probability textbooks, this is how a (discrete) random variable is defined instead: X is a tuple $(\mathcal{X}, P_X)$, where $\mathcal{X}$ is the *sample space*, and $p_X : \mathcal{X} \to [0,1]$ is a function mapping each $x \in \mathcal{X}$ to $p_X(x) \equiv P(X = x)$ such that $\sum_{x \in \mathcal{X}} p_X(x) = 1$ and $p_X(x) \geq 0$. A continuous random variable can be alternatively defined in a similar fashion. ●

When we specify a random variable by a sample space and probability distribution, we write $X \sim (\mathcal{X}, P_X)$ to denote that X has outcomes in $\mathcal{X}$ distributed according to P_X, or simply $X \sim P_X$ when $\mathcal{X}$ is clear from context. We defined in Definition 2.2.13 the probability of an event conditioned on another event. Similarly, we can also define the conditional distribution of an event given the outcome of another random variable.

2.2.5 Joint and Conditional Probabilities

Definition 2.2.29 (Joint cdf, pmf, pdf) Given two random variables X and Y over the same probability space $(\Omega, \mathcal{F}, P)$ we can define the *joint cumulative distribution function* (joint cdf) as

$$F_{X,Y}(x,y) := P(X \leq x \text{ and } Y \leq y) = P(\{\omega \in \Omega : X(\omega) \leq x \text{ and } Y(\omega) \leq y\})$$

If X and Y are discrete, the *joint probability mass function*, (joint pmf) is defined as

$$p_{X,Y}(x,y) = P(X = x \text{ and } Y = y) = P(\{\omega \in \Omega : X(\omega) = x \text{ and } Y(\omega) = y\})$$

We have to be careful how we define the joint pdf. Assuming certain "niceness conditions" that we gloss over, there exists an integrable function $f : \mathbb{R}^2 \to [0,\infty)$ such that

$$F_{X,Y}(x,y) = \int_{v=-\infty}^{v=y} \int_{u=-\infty}^{u=x} f(u,v)\, du\, dv$$

We define the *joint probability density function* (joint pdf) as $p_{X,Y}(x,y) := f(x,y)$. We often drop the subscripts and write $F(x,y)$ and $p(x,y)$ where it does not introduce ambiguity.

Definition 2.2.30 (Conditional cdf, pmf, pdf) Given two random variables X and Y over the same probability space $(\Omega,\mathcal{F},\mathrm{P})$ we can define the *conditional cumulative distribution function* (conditional cdf) of Y given $X = x$ as

$$F_{Y|X}(y|x) = \mathrm{P}(Y \le y | X = x)$$

We define the *conditional probability mass function* or *conditional probability density function* of Y given $X = x$ as

$$\mathrm{P}(Y = y | X = x) = \frac{\mathrm{P}(Y = y, X = x)}{\mathrm{P}(X = x)} \quad \text{or} \quad p_{Y|X}(y|x) = \frac{p_{Y,X}(y,x)}{p_X(x)}$$

We often write $F(y|x)$ and $p(y|x)$ without the subscripts.

Now that we have defined the conditional and joint distributions, we introduce the familiar *sum* and *product* rule.

Theorem 2.2.31 (Sum and product rule) Given discrete random variables X and Y with respective choice of outcomes $x \in X(\Omega)$ and $y \in Y(\Omega)$, we have the

Sum rule:
$$\mathrm{P}(X = x) = \sum_{y \in Y(\Omega)} \mathrm{P}(X = x, Y = y)$$

Product rule:
$$\mathrm{P}(X = x, Y = y) = \mathrm{P}(X = x | Y = y)\mathrm{P}(Y = y)$$

For continuous random variables X and Y with respective pdfs p_X and p_Y, and joint pdf $p_{X,Y}$, the

Sum rule:
$$p_X(x) = \int_{-\infty}^{\infty} p_{X,Y}(x,y)\, dy$$

Product rule:
$$p_{X,Y}(x,y) = p_{X|Y}(x|y)p_Y(y)$$

Proof. The product rule (both discrete and continuous) is just Definition 2.2.13 rearranged. For the discrete sum rule,

$$\begin{aligned}
\mathrm{P}(X = x) &= \mathrm{P}(\{\omega \in \Omega : X(\omega) = x\}) \\
&= \mathrm{P}(\{\omega \in \Omega : X(\omega) = x \text{ and } \exists y \in Y(\Omega).Y(\omega) = y\}) \\
&= \mathrm{P}\big(\textstyle\bigcup_{y \in Y(\Omega)} \{\omega \in \Omega : X(\omega) = x \text{ and } Y(\omega) = y\}\big) \\
&= \textstyle\sum_{y \in Y(\Omega)} \mathrm{P}(\{\omega \in \Omega : X(\omega) = x \text{ and } Y(\omega) = y\}) \\
&= \textstyle\sum_{y \in Y(\Omega)} \mathrm{P}(X = x, Y = y)
\end{aligned}$$

We leave the continuous sum rule as an exercise. ∎

As a corollary, since conditional distributions are also distributions (Theorem 2.2.14), we have that the sum rule and product rule also hold when conditioned on the outcome of some other random variable. For example, conditioned on some outcome $z \in Z(\Omega)$ for a random variable Z, the

Sum rule: $\qquad P(X=x|Z=z) = \sum_{y \in Y(\Omega)} P(X=x,Y=y|Z=z)$

Product rule: $\quad P(X=x,Y=y|Z=z) = P(X=x|Y=y,Z=z)P(Y=y|Z=z)$

Example 2.2.32 (Conditional coin flips) Recall Example 2.2.20 where we defined a probability space to model the outcome from flipping three fair coins and X was a random variable that counts the number of heads flipped. We introduce a new random variable $Y : \Omega \to \mathbb{R}$,

$$Y(\omega) = [\![\omega \neq TTT]\!]$$

That is, $Y(\omega)=1$ if at least one head was flipped, and $Y(\omega)=0$ otherwise. Then, we can formally express the statement "How likely is it that two heads were flipped, given that at least one was flipped" using conditional probabilities.

$$P(X=2|Y=1) = \frac{P(X=2,Y=1)}{P(Y=1)}$$

We can then compute the numerator as

$$P(X=2,Y=1) = P(\{\omega \in \Omega : X(\omega)=2, Y(\omega)=1\}) = P(\{HHT,HTH,THH\}) = \tfrac{3}{8}$$

and the denominator as

$$P(Y=1) = P(\Omega \backslash \{TTT\}) = \tfrac{7}{8}$$

which gives $P(X=2\,|\,Y=1) = \frac{3/8}{7/8} = \frac{3}{7}$. What is the intuition behind the result of $\frac{3}{7}$? Since we are given that $Y=1$, this eliminates the possibility of TTT, so we are essentially asking how likely is it that we obtain one of HHT,HTH,THH by selecting one of $HHH,HHT,HTH,HTT,THH,THT,TTH$, each of which is equally likely. There are 3 desired outcomes out of a possible of 7. Hence, $P(X=2|Y=1)=\frac{3}{7}$. ♦

Definition 2.2.33 (Independent and identically distributed) A countable collection of random variables $(X_i)_{i \in I}$ with corresponding cdfs $\{F_{X_i}\}_{i \in I}$ are *identically distributed* if all cdf's are identical. The family (X_n) is *independently distributed* (or *mutually independent*)if the cdf's satisfy the following:

$$\mathcal{F}_{X_j : j \in J}(X_j = x_j : j \in J) = \prod_{j \in J} \mathcal{F}_{X_j}(X_j = x_j)$$

for all $\{x_j\}_{j \in J}$, and for all finite subsets $J \subseteq I$. Note that this is a stronger property than *pairwise independence*, which only requires that

$$F_{X_i, X_j}(x_i, x_j) = F_{X_i}(x_i) F_{X_j}(x_j)$$

for all $i, j \in I$. If both conditions hold, we say that $(X_i)_{i \in I}$ are *independent and identically distributed* (i.i.d.).

Example 2.2.34 (Dependent random variables) In Example 2.2.32, X and Y are not independent, as

$$P(X=0, Y=1) = P(\varnothing) = 0$$
$$\text{but} \quad P(X=0)P(Y=1) = P(\{TTT\})P(\Omega \setminus \{TTT\}) = \tfrac{1}{8} \cdot \tfrac{7}{8} = \tfrac{7}{64} \qquad \blacklozenge$$

2.2.6   Expectation and Variance

Given a random variable, one can talk about the "average" result of the random variable by sampling outcomes $\omega_1, \omega_2, \ldots$ and feeding them into X to obtain the sequence $X(\omega_1), X(\omega_2), \ldots$. We define $\bar{X}_n := \frac{1}{n} \sum_{i=1}^{n} X(\omega_i)$ which represents an empirically calculated average over n returns from X. For large n, we would expect that if $P(X=x_i) = p_i$, then approximately $n p_i$ of the outcomes should be equal to x_i:

$$\bar{X}_n = \frac{1}{n} \sum_{i=1}^{n} X(\omega_i) \approx \frac{1}{n} \sum_{x \in X(\Omega)} (n P(X=x)) \, x = \sum_{x \in X(\Omega)} x \, P(X=x)$$

This motivates the following definition:

Definition 2.2.35 (Expected value) The *expected value* (expectation or mean) of a discrete random variable X with pmf $p(x)$ is defined as

$$\mathbf{E}[X] := \sum_{x \in X(\Omega)} x \, p(x)$$

Often, we can simply write $\sum_{x \in \mathbb{R}}$ or $\sum_x$ instead of $\sum_{x \in X(\Omega)}$, with the understanding that we are still only summing over a countable set, as $p(x)$ will be zero for all but a countable subset $X(\Omega) \subseteq \mathbb{R}$.

The expected value of a continuous random variable X with pdf $p(x)$ is defined as

$$\mathbf{E}[X] := \int_{-\infty}^{\infty} x \, p(x) \, dx$$

Note that the above sum/integral may not be defined, in which case $\mathbf{E}[X]$ is undefined. For example, the Cauchy distribution $p(x) = \frac{1}{\pi} \frac{1}{1+x^2}$ is known to have an undefined expected value.

While sloppy, we may also refer to the expectation of a distribution $\mathbf{E}_P$ as the expectation of a random variable $X \sim P$, where the domain is usually clear from the definition of P (e.g. if P is a Gaussian, then the domain is implicitly $\mathbb{R}$).

Given a function $g : \mathbb{R} \to \mathbb{R}$ and a random variable $X : \Omega \to \mathbb{R}$, $g(X)$ is also a random variable defined as

$$(g(X))(\omega) = g(X(\omega))$$

We can then extend the definition of the expectation to $g(x)$ and further to conditional probability distributions. In principle one can derive Definition 2.2.36 below from Definition 2.2.35 [GS20] but it is common to simply take the former as a definition.

> **Definition 2.2.36 (Conditional expectation)** Given two random variables X,Y, and a function $g:\mathbb{R}\to\mathbb{R}$, we define the *conditional expectation* of $g(X)$ given $Y=y$ as
>
> $$\mathbf{E}[g(X)|Y=y] \;=\; \sum_{x\in X(\Omega)} g(x)p_{X|Y}(x|y)$$
>
> $$\mathbf{E}[g(X)|Y=y] \;=\; \int_{-\infty}^{\infty} g(x)p_{X|Y}(x|y)\,dy$$
>
> for discrete and continuous random variables respectively.

For a constant random variable $Y(\omega):=y_0$ for some $y_0\in\mathbb{R}$, and we have $p_X(x):=p_{X|Y}(x|y_0)$, which gives us the unconditional expectation $\mathbf{E}[g(X)]$. If we further choose g to be the identity function, we recover the definition of $\mathbf{E}[X]$ in Definition 2.2.35. See Section 2.8.3 for an extension to partial functions.

Lastly, we can define expectation for a function of multiple random variables.

> **Definition 2.2.37 (Multi-variable expectation)** Let $X_1,...,X_n$ be a family of random variables of type $X_i:\Omega_i\to\mathbb{R}$ for all $1\le i\le n$ with a joint pmf/pdf $p_{X_1,...X_n}(x_1,...x_n):X_1(\Omega_1)\times...\times X_n(\Omega_n)\to\mathbb{R}$, and $f:\mathbb{R}^n\to\mathbb{R}$. Then,
>
> $$\mathbf{E}[f(X_1,...,X_n)] \;:=\; \sum_{\mathbf{x}\in\prod_{i=1}^n X_i(\Omega_i)} f(\mathbf{x})p_{X_1,...X_n}(\mathbf{x})$$
>
> where $\mathbf{x}=(x_1,...,x_n)$.

Conditional multi-variable expectation can be defined similarly.

Example 2.2.38 (Expected value of a fair die) Consider a fair six-sided die. By Proposition 2.2.27 we can represent this with a random variable X with alphabet $\mathcal{X}=\Omega=\{1,2,3,4,5,6\}$ and probabilities $\mathrm{P}(x)=\frac{1}{6}$ for all x. The expected value is

$$\mathbf{E}[X] \;=\; \sum_{x\in\{1,2,3,4,5,6\}} \tfrac{x}{6} = \tfrac{7}{2}$$

◆

An often-used property of the expected value is that it is linear.

> **Theorem 2.2.39 (Expectation is linear)** Given random variables X,Y,Z with respective elements in the sample spaces x,y,z and scalars $a,b\in\mathbb{R}$, we have that
>
> $$\mathbf{E}[aX+bY|Z=z] \;=\; a\mathbf{E}[X|Z=z]+b\mathbf{E}[Y|Z=z]$$

Proof. We expand out the definition of the expectation, factor out the scalars, split the two summations and reduce using the sum rule (Theorem 2.2.31) to obtain the result, as shown.

$$\mathbf{E}[aX+bY|Z=z]$$
$$=\sum_{x,y}(ax+by)\mathrm{P}(X=x,Y=y|Z=z)$$
$$=a\sum_{x,y}x\mathrm{P}(X=x,Y=y|Z=z)+b\sum_{x,y}y\mathrm{P}(X=x,Y=y|Z=z)$$
$$=a\sum_{x}x\sum_{y}\mathrm{P}(X=x,Y=y|Z=z)+b\sum_{y}y\sum_{x}\mathrm{P}(X=x,Y=y|Z=z)$$
$$=a\sum_{x}x\mathrm{P}(X=x|Z=z)+b\sum_{y}y\mathrm{P}(Y=y|Z=z)$$
$$=a\mathbf{E}[X|Z=z]+b\mathbf{E}[Y|Z=z]\qquad\blacksquare$$

Theorem 2.2.40 (Expectation of independent random variables) If X and Y are independent random variables, then

$$\mathbf{E}[XY] \;=\; \mathbf{E}[X]\mathbf{E}[Y]$$

Proof.

$$\mathbf{E}[XY] \;=\; \sum_{x,y}xy\mathrm{P}(X=x,Y=y)$$
$$=\; \sum_{x,y}xy\mathrm{P}(X=x)\mathrm{P}(Y=y)$$
$$=\; \left(\sum_{x}x\mathrm{P}(X=x)\right)\left(\sum_{y}y\mathrm{P}(Y=y)\right)$$
$$=\; \mathbf{E}[X]\mathbf{E}[Y]\qquad\blacksquare$$

Note that the converse of Theorem 2.2.40 is not true.

Definition 2.2.41 (Variance) The variance of a random variable X is defined as

$$\mathbf{Var}[X]:=\mathbf{E}[(X-\mathbf{E}[X])]^{2}$$

The variance can be thought of as a measure of how far away, on average, X is from its mean $\mathbf{E}[X]$, as measured by the squared error $(X-E[X])^{2}$. Random variables with low variance have most of their probability mass concentrated near the mean, whereas distributions with higher variance have probability mass concentrated away from the mean (see Figure 2.4).

Example 2.2.42 (Expectation and variance of a biased coin) Consider a random variable X with sample space $\{0,1\}$ and respective probabilities $\{1-p, p\}$, for some $0\le p\le 1$. The expected value of X is

$$\mathbf{E}[X] \;=\; \sum_{x}x\mathrm{P}(X=x) \;=\; p\cdot 1+(1-p)\cdot 0 \;=\; p$$

The variance is then given as

$$\mathbf{Var}[X] \;=\; \mathbf{E}[(X-\mathbf{E}[X])^{2}] \;=\; \sum_{x}\mathrm{P}(X=x)(x-p)^{2} \;=\; (1-p)(0-p)^{2}+p(1-p)^{2}=p(1-p)$$
$\blacklozenge$

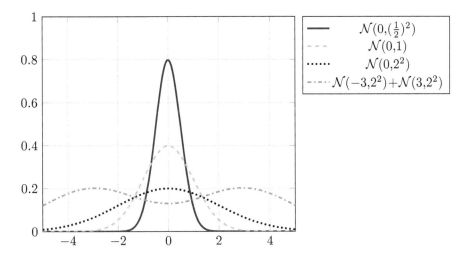

Figure 2.4: *All these pdfs have the same mean, labelled in order of increasing variance.* $\mathcal{N}(\mu,\sigma^2)$ *denotes the Gaussian distribution with mean* μ *and variance* σ^2.

We have the following properties of variance:

Theorem 2.2.43 (Properties of variance) Let X be a random variable, and let $X_1,...,X_n$ be a (pairwise) independent family of random variables. Then for any $a \in \mathbb{R}$,

$$\mathbf{Var}[X] = \mathbf{E}[X^2] - \mathbf{E}[X]^2$$
$$\mathbf{Var}[aX] = a^2 \mathbf{Var}[X]$$
$$\mathbf{Var}[X_1+...+X_n] = \mathbf{Var}[X_1]+...+\mathbf{Var}[X_n]$$

Proof. We prove the 3 equalities in turn:

(i) Let $b = \mathbf{E}[X]$, which is constant, i.e. not a random variable. Then by linearity of expectation,

$$\mathbf{Var}[X] \equiv \mathbf{E}[(X-b)^2] = \mathbf{E}[X^2 - 2b\mathbf{E}[X] + b^2] = \mathbf{E}[X^2] - 2b^2 + b^2 = \mathbf{E}[X^2] - \mathbf{E}[X]^2$$

(ii) We can now apply (i) aX to prove (ii)

$$\mathbf{Var}[aX] = \mathbf{E}[(aX)^2] - \mathbf{E}[aX]^2 = \mathbf{E}[a^2X^2] - (a\mathbf{E}[X])^2 = a^2(\mathbf{E}[X^2] - \mathbf{E}[X]^2) = a^2\mathbf{Var}[X]$$

(iii) We make use of the identity $(\sum_{i=1}^n X_i)^2 = \sum_{i=1}^n \sum_{j=1}^n X_i X_j$ and use (i) again for $X := X_1+...+X_n$.

$$\mathbf{Var}[X_1+...+X_n]$$
$$= \mathbf{E}\left[\left(\sum_i X_i\right)^2\right] - \left[\mathbf{E}\left(\sum_i X_i\right)\right]^2$$
$$= \mathbf{E}\left[\sum_{i,j} X_i X_j\right] - \left(\sum_i \mathbf{E}[X_i]\right)^2$$
$$= \sum_{i,j} \mathbf{E}[X_i X_j] - \mathbf{E}[X_i]\mathbf{E}[X_j]$$

If $i \neq j$, then by Theorem 2.2.40 it follows that $\mathbf{E}[X_i X_j] - \mathbf{E}[X_i]\mathbf{E}[X_j] = 0$, so we can drop these terms from the sum.

$$\mathbf{Var}[X_1 + ... + X_n] = \sum_{i=1}^{n} \left(\mathbf{E}[X_i^2] - \mathbf{E}[X_i]^2 \right) = \sum_{i=1}^{n} \mathbf{Var}[X_i]$$

■

2.2.7 Probability Inequalities

Probability inequalities are often used to provide bounds for the expectation or variance of a random variable in cases where computing it exactly is computationally intractable. In this section, we look at three important inequalities: Markov's, Chebyshev's and Jensen's inequalities.

Definition 2.2.44 (Probability of predicates) Given a predicate $S : \Omega \to$ {False,True} over events in a probability space $(\Omega, \mathcal{F}, P)$, we define the probability of S happening as

$$P(S) := P(S(\omega)) := P(\omega : S(\omega)) := P(\{\omega \in \Omega : S(\omega)\})$$

We also identify 1 with True and 0 with False, i.e. {False,True} = {0,1} = $\mathcal{B}$.

Lemma 2.2.45 (Expectation of predicates) Given a predicate S, we can define a random variable $X(\omega) = [\![S(\omega)]\!]$. Then, $\mathbf{E}[X] = P(S)$.

Proof. First, note that $\mathcal{X}(\Omega) = \{0,1\}$. Then,

$$\mathbf{E}[X] = \sum_{x \in X(\omega)} x P(X = x) = P(X = 1) = P(\{\omega \in \Omega : [\![S(\omega)]\!] = 1\}) = P(\{\omega \in \Omega : S(\omega)\}) = P(S)$$

■

Theorem 2.2.46 (Markov's inequality) If a non-negative random variable X has expectation value $\mathbf{E}[X]$, then for any $\varepsilon > 0$,

$$P(X \geq \varepsilon) \leq \frac{\mathbf{E}[X]}{\varepsilon}$$

Proof. We define $Y = \varepsilon [\![X \geq \varepsilon]\!] = \begin{cases} \varepsilon & X \geq \varepsilon \\ 0 & X \leq \varepsilon \end{cases}$. Note that $Y \leq X$. Then by Lemma 2.2.45,

$$\mathbf{E}[Y] = \varepsilon \mathbf{E}[[\![X \geq \varepsilon]\!]] = \varepsilon P(X \geq \varepsilon)$$

Since $Y \leq X$, we have that $\mathbf{E}[Y] \leq \mathbf{E}[X]$, so

$$P(X \geq \varepsilon) = \frac{\mathbf{E}[Y]}{\varepsilon} \leq \frac{\mathbf{E}[X]}{\varepsilon}$$

■

Theorem 2.2.47 (Chebyshev's inequality) Let X be a random variable with $\mathbf{E}[X] < \infty$ and $\mathbf{Var}[X] < \infty$. Then for any $\varepsilon > 0$,

$$P(|X - \mathbf{E}[X]| \geq \varepsilon) \leq \frac{\mathbf{Var}[X]}{\varepsilon^2}$$

Proof. Let $Y=(X-\mathbf{E}[X])^2$, then $\mathbf{E}[Y]=\mathbf{Var}[X]$, by definition. Applying Markov's inequality (Theorem 2.2.46) on $\mathrm{P}(Y\geq\varepsilon^2)$ and noting that Y is non-negative,

$$\mathrm{P}(|X-\mathbf{E}[X]|\geq\varepsilon) = \mathrm{P}((X-\mathbf{E}[X])^2\geq\varepsilon^2) = \mathrm{P}(Y\geq\varepsilon^2) \leq \frac{\mathbf{E}[Y]}{\varepsilon^2} = \frac{\mathbf{Var}[X]}{\varepsilon^2}$$

as required. ∎

Markov's and Chebyshev's inequalities differ in several important ways. Firstly, Markov's inequality only bounds the upper tail of the distribution and applies solely to non-negative random variables. Conversely, Chebyshev's inequality bounds both tails and makes no assumptions regarding the non-negativity of the variables. Additionally, the bound provided by Markov's inequality is inversely proportional to the distance from the origin, whereas the bound provided by Chebyshev's inequality is inversely proportional to the square of the distance from the mean.

Example 2.2.48 (Expecation and variance of sum of dice) A fair six-sided die is rolled n times. Let X_i be the outcome of the ith roll, and let $S=\sum_{i=1}^n X_i$ be the sum of all the rolls. Note that $\mathbf{E}[X_i]=3.5$ (Example 2.2.38) and using the fact that expectation is linear (Theorem 2.2.39) we have the expectation of S as

$$\mathbf{E}[S] = \mathbf{E}\left[\sum_{i=1}^n X_i\right] = \sum_{i=1}^n \mathbf{E}[X_i] = n\mathbf{E}[X] = 3.5n$$

Now, computing $\mathrm{P}(S\geq k)$ for large n and k would be quite difficult due to the number of dice rolls involved (one would have to compute all the ways a large number can be expressed as the sum of n dice), but we can bound it using the Markov inequality.

$$\mathrm{P}(S\geq k) \leq \frac{\mathbf{E}[S]}{k} = \frac{3.5n}{k}$$

For $k=80$ and $n=20$ (the probability that the sum of 20 dice is at least 80) we obtain $\mathrm{P}(S\geq80)\leq\frac{70}{80}=\frac{7}{8}$. Unfortunately, the bound that Markov's inequality provides is often extremely crude. We compute $\mathrm{P}(S\geq80)=\sum_{i=80}^{6\times20}\mathrm{P}(S=i)$ by brute-forcing all the possible ways 20 dice can sum to 80, and obtain $\mathrm{P}(S\geq80)\approx0.1075...$, far less than $\frac{7}{8}$. Worse still, for values $k\leq\mathbf{E}[S]$, Markov's inequality provides a vacuous bound, as $\mathbf{E}[S]/k$ will be greater than 1. For sums of independent random variables S, exponentially better bounds are available (see Theorem 2.2.61 below). ◆

Before we can state Jensen's inequality, we first need to define what it means for a function to be convex.

Definition 2.2.49 (Convex function) A function $f:\mathbb{R}^n\to\mathbb{R}\cup\{\infty\}$ is *convex* if $\forall x_1,x_2\in\mathbb{R}^n$ and $\forall\theta\in[0,1]$ we have that

$$f(\theta x_1+(1-\theta)x_2) \leq \theta f(x_1)+(1-\theta)f(x_2)$$

Furthermore, if we have strict inequality, then f is *strictly convex*.

For the one-dimensional case of functions $f:\mathbb{R}\to\mathbb{R}$, Definition 2.2.49 can be understood geometrically as follows: given any two points $P=(x_1,f(x_1))$ and $Q=(x_2,f(x_2))$ that lie on f, the line segment from P to Q always sits above the function f (see Figure 2.5). Examples of convex functions include $x^2, e^x, -\log(x), -\sqrt{x}$.

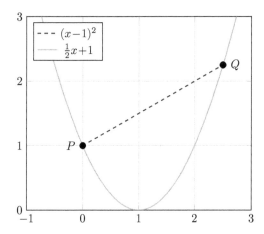

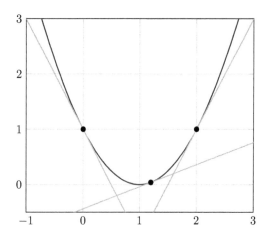

Figure 2.5: *The line segment from P to Q sits above the convex function $f(x)=(x-1)^2$.*

Figure 2.6: *The convex function $f(x)= (x-1)^2$ with several tangent lines. Given any point P on f, the tangent line at P sits below f.*

Definition 2.2.50 (Concave function) A function $f:\mathbb{R}^n\to\mathbb{R}\cup\{-\infty\}$ is *concave* if $\forall x_1,x_2\in\mathbb{R}^n$ and $\forall\theta\in[0,1]$ we have that $f(\theta x_1+(1-\theta)x_2)\geq\theta f(x_1)+(1-\theta)f(x_2)$ or *strictly concave* if the inequality is strict.

Note that f being (strictly) concave is equivalent to $-f$ being (strictly) convex.

Lemma 2.2.51 (First-order convexity conditions [BV04]) Given a convex function $f:\mathbb{R}^n\to\mathbb{R}$, we have that for all $x,x_0\in\mathbb{R}^n$,

$$f(x) \geq f(x_0)+(x-x_0)\cdot\nabla f(x_0)$$

where ∇ denotes a sub-gradient of f at x_0.

Again for $n=1$, Lemma 2.2.51 can be better understood as the claim that if f is convex, the tangent line $y=f(x_0)+(x-x_0)f'(x_0)$ of f at x_0 is a lower bound for f, which is obvious for e.g. $f(x)=x^2$ from Figure 2.6. The proof is left as an exercise.

Theorem 2.2.52 (Jensen's inequality [BV04]) Let X be a random vector, then

$$f(\mathbf{E}[X]) \leq \mathbf{E}[f(X)] \quad \text{for convex} \quad f:\mathbb{R}^n\to\mathbb{R}$$
$$g(\mathbf{E}[X]) \geq \mathbf{E}[g(X)] \quad \text{for concave} \quad g:\mathbb{R}^n\to\mathbb{R}$$

Proof. By Lemma 2.2.51 we have $f(x)\geq f(x_0)+(x-x_0)\cdot\nabla f(x_0)$. Let $x_0=\mathbf{E}[X]$, and take expectations of both sides.

$$\begin{aligned}
\mathbf{E}[f(X)] &\geq \mathbf{E}[f(\mathbf{E}[X])]+\mathbf{E}[(X-\mathbf{E}[X])\cdot\nabla f(\mathbf{E}[X])]\\
&= f(\mathbf{E}[X])+\nabla f(\mathbf{E}[X])\cdot\underbrace{\mathbf{E}[(X-\mathbf{E}[X])]}_{=0}\\
&= f(\mathbf{E}[X])
\end{aligned}$$

Note that $-g$ is convex, and apply Jensen's inequality to obtain $-g(\mathbf{E}[X]) \leq \mathbf{E}[-g(X)]$, from which the result follows. ∎

Remark 2.2.53 Since x^2 is a convex function, $\mathbf{E}[X]^2 \leq \mathbf{E}[X^2]$. Hence the variance $\mathbf{Var}[X] = \mathbf{E}[X^2] - \mathbf{E}[X]^2$ is always non-negative, as is of course obvious from Definition 2.2.41. ●

Corollary 2.2.54 (Jensen's inequality of averages) Let f be convex, and $(x_i)_{i=1}^n$ a finite sequence. Then

$$f\left(\frac{1}{n}\sum_{i=1}^n x_i\right) \leq \frac{1}{n}\sum_{i=1}^n f(x_i)$$

Proof. Let $p = (p_1,...,p_n)$ be a uniform probability distribution with $p_i = 1/n$ for all i, and let X be a random variable with sample space $\{x_1,...,x_n\}$, and $P(X = x_i) = p_i$. Then,

$$f\left(\frac{1}{n}\sum_{i=1}^n x_i\right) = f\left(\sum_{i=1}^n p_i x_i\right) = f(\mathbf{E}[X]) \leq \mathbf{E}[f(X)] = \sum_{i=1}^n p_i f(x_i) = \frac{1}{n}\sum_{i=1}^n f(x_i)$$

∎

2.2.8 Convergence of Random Variables

Given a sequence of real numbers $(x_n)_{n=1}^\infty$, we have a rigorous way of talking about whether the limit $\lim_{n\to\infty} x_n$ exists at all, and if it does, the value to which the sequence converges.

For a family of functions $(f_n)_{n=1}^\infty$ of type $f_n : \mathbb{R} \to \mathbb{R}$, the concept of convergence is more difficult, as there are different ways we can define what it means for a sequence of functions (f_n) to converge to a function f, including:

- **Pointwise Convergence:** $\forall x \in \mathbb{R}, \lim_{n\to\infty} f_n(x) = f(x)$

- **Uniform Convergence:** $\lim_{n\to\infty} \sup_{x\in\mathbb{R}} |f_n(x) - f(x)| = 0$

- **L² Convergence:** $\lim_{n\to\infty} \int_{-\infty}^{\infty} |f_n(x) - f(x)|^2\, dx = 0$

Can we carry this notion across and analogously define convergence for a sequence of random variables? We could follow the definition of pointwise convergence and try to define the same thing for random variables:

Definition 2.2.55 (Pointwise convergence of random variables) A sequence of random variables $(X_n)_{n=1}^\infty$ converges *pointwise* to random variable X if, for all $\omega \in \Omega$,

$$\lim_{n\to\infty} X_n(\omega) = X(\omega)$$

For most purposes this definition turns out to be too strong. For each $\omega \in \Omega$, X_n maps ω to a real number and thus $(X_n)_{n=1}^\infty$ maps ω to a sequence of real numbers $\{X_n(\omega)\}_{n=1}^\infty$. For "poor" choices of ω, this sequence may converge to $X(\omega)$, may converge to something else, or it may not converge at all.

Example 2.2.56 (Bernolli pointwise (non)convergence) Consider an independent identically distributed (i.i.d.) sequence of random variables X_n, each with sample space $\Omega = \{0,1\}$ and probabilities $\{1-\theta, \theta\}$. Here, each X_n models the behavior of a fair coin flip. Each X_n can be thought of as a random variable that models the nth coin in an infinite sequence of coin flips. If $\Omega = \{0,1\}$ is the sample space to represent a single coin flip, we can consider Ω^∞ to be the sample space representing an infinite sequence of coin flips. We define a family of random variables $S_n : \Omega^\infty \to \mathbb{R}$ as

$$S_n(\omega_{1:\infty}) = \sum_{i=1}^{n} X_i(\omega_i)$$

Here, S_n models the number of 1's in the first n tosses of the coin. (Note that each S_n depends only on the first n terms of $\omega_{1:\infty}$.) We can ask about the behavior of S_n for large n. Intuitively, we would expect that in the long run the number of heads will be about $n\theta$, that is,

$$\forall \omega \in \Omega^\infty \lim_{n \to \infty} \frac{S_n(\omega)}{n} = \theta$$

However, this is not true for all choices of $\omega \in \Omega^\infty$. For example, select the sequence of all zeros $\omega_z = 000....$ Then,

$$\lim_{n \to \infty} \frac{S_n(\omega_z)}{n} = 0.$$

However, we would expect this particular sequence ω_z to be "unlikely", as we would not expect a fair coin to generate a sequence of all zeros.   ◆

This motivates a way to weaken Definition 2.2.55 by requiring convergence not for *all* $\omega \in \Omega$, but convergence for "almost all" ω, ignoring those ω that are "unlikely" to be encountered.

Definition 2.2.57 (Almost sure convergence (a.s.)) A sequence of random variables $(X_n)_{n=1}^\infty$ converges *almost surely* (converges a.s)[a] to a random variable X (denoted as $X_n \xrightarrow{\text{a.s}} X$) if one and hence all of the following equivalent conditions hold:

(i) $\qquad\qquad\qquad P\left[\{\omega \in \Omega : \lim_{n \to \infty} X_n(\omega) = X(\omega)\}\right] = 1$

(ii) $\qquad \forall \varepsilon > 0.\ \ P\left[\limsup_{n \to \infty} \{\omega \in \Omega : |X_n(\omega) - X(\omega)| > \varepsilon\}\right] = 0$

(iii) $\qquad \forall \varepsilon > 0.\ \ \lim_{t \to \infty} P\left[\{\omega : \sup_{n \geq t} |X_n(\omega) - X(\omega)| > \varepsilon\}\right] = 0$

---

[a]Also called *strong convergence, convergence with probability 1 (w.p.1)*

We write *converges almost surely with respect to* P if it is not obvious with respect to what probability measure P convergence is defined.

An equivalent way to consider convergence a.s is that the set of all counterexamples to convergence, $\{\omega \in \Omega : X_n(\omega) \not\to X(\omega)\}$ is a probability zero event, so even though counterexamples exist, we will *almost surely* never see one.

Example 2.2.58 (Probability of eventually flipping heads) Consider the sample space $\Omega = \{0,1\}^\infty$ (all infinite sequences of coin flips) with $\mathcal{F} = \sigma(\{\Gamma_x : x \in \{0,1\}^*\})$. We want the probability distribution to represent that of a coin with bias $0 < \theta < 1$ towards heads, so we

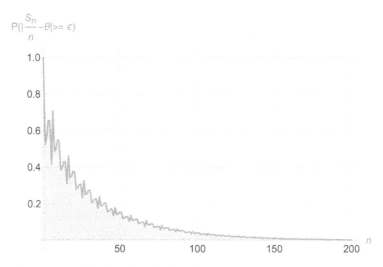

Figure 2.7: *A plot of* $\mathrm{P}(|\frac{1}{n}S_n - \theta| \geq \varepsilon)$ *for* $\theta = 0.4$ *and* $\varepsilon = 0.1$. *As the weak law of large numbers indicates, the empirical mean almost surely converges to the expected value.*

choose $\mathrm{P}(\Gamma_x) = \theta^h(1-\theta)^t$, where h is the number of ones (representing heads) in the string x, and t the number of zeros (representing tails). Consider the sequence of random variables $(X_n)_{n=1}^\infty$ where

$$X_n(\omega_{1:\infty}) = [\![\exists i.1 \leq i \leq n \text{ and } \omega_i = 1]\!]$$

That is, $X_n(\omega_{1:\infty}) = 1$ if and only if $w_{1:n}$ contains at least one 1. Then, $X_n \xrightarrow{\text{a.s.}} 1$[8] as for any $\omega \in \Omega$ that contains a 1, there exists an N such that for all $n \geq N$, $X_n(\omega) = 1$. The set of all ω that contains a 1 can be written as the following disjoint union of cylinder sets:

$$\{\omega \in \Omega : \exists i.w_i = 1\} = \bigcup_{i=0}^\infty \Gamma_{0^i 1}$$

where 0^i represents the sequence of i many consecutive zeros. We can then compute the probability of the set of ω for which $X_n(\omega) \to 1$,

$$\mathrm{P}(\{\omega \in \Omega : \lim_{n\to\infty} X_n(\omega) = 1\}) = \mathrm{P}(\bigcup_{i=0}^\infty \Gamma_{0^i 1}) = \sum_{i=0}^\infty \mathrm{P}(\Gamma_{0^i 1}) = \sum_{i=0}^\infty (1-\theta)^i \theta = \frac{\theta}{1-(1-\theta)} = 1 \quad \blacklozenge$$

Remark 2.2.59 (Weak law of large numbers) We can now repair Example 2.2.56 by showing that the set of counterexamples ω for which $\frac{1}{n}S_n(\omega) \not\to \theta$ forms a probability zero event. Choose the same probability space as in Example 2.2.58, and let $(X_n)_{n=1}^\infty$ be an i.i.d. family of random variables[9] $(X_i)_{i=1}^N$ with

$$X_i = X := \begin{cases} 1 & \text{with probability } \theta \\ 0 & \text{with probability } 1-\theta. \end{cases}$$

Note that $\mathbf{E}[X] = \theta$ and $\mathbf{Var}[X] = \theta(1-\theta)$. As before, let $(S_i)_{i=1}^N$ be a sequence of random variables defined as $S_n := \sum_{i=1}^n X_i$. Consider the predicate $|\frac{1}{n}S_n - \theta| \geq \varepsilon$, the statement that

---

[8]Here, "1" represents the constant random variable $1(\omega) = 1$.

[9]A sequence of random variables of this kind is called a *Bernoulli process*.

$\frac{1}{n}S_n$ is more than ε away from θ. The probability of this predicate happening, as a function of n and ε, gives an idea of how concentrated the distribution for $\frac{1}{n}S_n$ is on θ. We can see in Figure 2.7 that as n grows the probability that $\left|\frac{1}{n}S_n-\theta\right|\geq\varepsilon$ appears to go to zero. We can in fact prove that for any $\varepsilon>0$ chosen, the probability that this bound is violated tends to zero as n tends to infinity.   ●

Theorem 2.2.60 (Weak law of large numbers) Let S_n be a sequence of random binomial random variables as in Remark 2.2.59. Then for all $\varepsilon>0$,

$$\lim_{n\to\infty} \mathrm{P}\left(\left|\frac{S_n}{n}-\theta\right|\geq\varepsilon\right) = 0$$

That is, the probability distribution function associated with S_n/n tends to concentrate all the probability mass on θ as $n\to\infty$.

Proof. Define $Y=\frac{1}{n}S_n$, then

$$\mathbf{E}[Y] = \mathbf{E}\left[\frac{1}{n}\sum_{i=1}^{n}X_i\right] = \frac{1}{n}\sum_{i=1}^{n}\mathbf{E}[X] = \mathbf{E}[X] = \theta$$

and by Theorem 2.2.43

$$\mathbf{Var}[Y] = \frac{1}{n^2}\mathbf{Var}\left[\sum_{i=1}^{n}X_i\right] = \frac{1}{n^2}\sum_{i=1}^{n}\mathbf{Var}[X] = \frac{1}{n}\mathbf{Var}[X] = \frac{\theta(1-\theta)}{n}$$

By Chebyshev's inequality, noting that $\theta(1-\theta)\leq\frac{1}{4}$ for all $\theta\in[0,1]$

$$\mathrm{P}\left(\left|\frac{S_n}{n}-\theta\right|\geq\varepsilon\right) = \mathrm{P}(|Y-\mathbf{E}[Y]|\geq\varepsilon) \leq \frac{\mathbf{Var}[Y]}{\varepsilon^2} = \frac{\theta(1-\theta)}{\varepsilon^2 n} \leq \frac{1}{4\varepsilon^2 n}$$

from which the result follows.   ■

The proof of Theorem 2.2.60 also implies a rate of convergence of the sample mean: Given any tolerance ε, the probability that the sample mean S_n/n and the true value of θ are more than ε apart shrinks at a rate $O(n^{-1})$. The actual convergence rate is exponentially better and also holds much more generally:

Theorem 2.2.61 (Hoeffding bound [HFS94]) Let S_n be a sum of n independent $[0,1]$-valued random variables. Then

$$\mathrm{P}\left(|S_n-\mathbf{E}[S_n]|\geq t\right) \leq 2e^{-2t^2/n}$$

Indeed, for $S_n=X_1+...+X_n$ with i.i.d. $X_i\in[0,1]$ with $\mathbf{E}[X_i]=\theta$ and $t=n\varepsilon$, we get $P(|\frac{1}{n}S_n-\theta|\geq\varepsilon)\leq 2e^{-2n\varepsilon^2}$.

We take a brief digression to introduce the concepts of *limit supremum* and *limit infimum* for sets.

Definition 2.2.62 (Limit supremum / Limit infimum) Given a sequence of events $(A_n)_{n=1}^{\infty}$, we define

$$\limsup_{n\to\infty}A_n := \bigcap_{n=1}^{\infty}\bigcup_{j=n}^{\infty}A_j \quad \text{and} \quad \liminf_{n\to\infty}A_n := \bigcup_{n=1}^{\infty}\bigcap_{j=n}^{\infty}A_j$$

Remark 2.2.63 (Infinitely often and almost all) If $\omega \in \limsup A_n$, it means that for all $n \geq 1$ there exists a $j \geq n$ such that $\omega \in A_j$. This is equivalent to requiring that there exist infinitely many values of j such that $\omega \in A_j$, or that there exists infinitely many events A_n in the sequence that occurred. We can write

$$\{A_n \text{ infinitely often}\} := \limsup_{n \to \infty} A_n$$

If $\omega \in \liminf A_n$, it means there exists an $n \geq 1$ such that for all $j \geq n$, $\omega \in A_j$, so, beyond some point n, all the events $A_n, A_{n+1}, \dots$ occurred. Alternatively, all but finitely many events occurred (as A_{n-1} would be the last event to not occur). We can write

$$\{A_n \text{ almost all}\} := \liminf_{n \to \infty} A_n \qquad \bullet$$

Example 2.2.64 (Almost surely infinitely many 1s) Continuing from Example 2.2.58, one can strengthen the result and show that the coin will almost surely flip 1 infinitely often. Formally, consider the sequence of random variables $(X_n)_{n=1}^\infty$ where $X_n(\omega_{1:\infty}) = [\![\omega_n = 1]\!]$. That is, $X_n(\omega) = 1$ if and only if the nth coin flip was 1. We leave as an exercise to show that the coin will almost surely flip infinitely many 1s. $\qquad \blacklozenge$

Definition 2.2.65 (Convergence in probability (i.p.)) A sequence of random variables $(X_n)_{n=1}^\infty$ *converges in probability* (also called *weak convergence*) to X (denoted $X_n \xrightarrow{P} X$) if for any $\varepsilon > 0$,

$$\lim_{n \to \infty} P(\{\omega \in \Omega : |X_n(\omega) - X(\omega)| > \varepsilon\}) = 0$$

In Theorem 2.2.60, we have shown that $\frac{1}{n} S_n$ converges to θ in probability, which is an instance of the weak law of large numbers.

Theorem 2.2.66 (Convergence a.s. implies convergence i.p.) Given a sequence of random variables $(X_n)_{n=1}^\infty$, if $X_n \xrightarrow{\text{a.s}} X$, then $X_n \xrightarrow{P} X$.

Proof. From Definition 2.2.57*ii* for arbitrary $\varepsilon > 0$ we have

$$0 = P\left(\limsup_{n \to \infty} \{\omega : |X_n(\omega) - X(\omega)| > \varepsilon\}\right)$$

$$= P\left(\bigcap_{N=1}^{\infty} \bigcup_{n=N}^{\infty} \{\omega : |X_n(\omega) - X(\omega)| > \varepsilon\}\right)$$

$$\geq P\left(\bigcap_{N=1}^{\infty} \{\omega : |X_N(\omega) - X(\omega)| > \varepsilon\}\right)$$

$$\geq \lim_{N \to \infty} P(\{\omega : |X_N(\omega) - X(\omega)| > \varepsilon\}) = 0$$

where the last step follows from (2.2.10k). $\qquad \blacksquare$

However, the converse implication does not hold, which is why convergence a.s. is often termed as *strong convergence* while convergence in probability is called *weak convergence*. A weaker still form of convergence of random variables is *convergence in distribution* (i.d.), where the cdfs $(F_n)_{n=1}^\infty$ of a sequence of random variables $(X_n)_{n=1}^\infty$ converges to the cdf F of the limiting random variable X. We do not make use of this notion of convergence, but mention it here for completeness. The relationship between various notions of probabilistic convergence is shown in Figure 2.8.

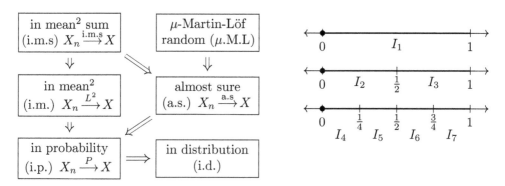

Figure 2.8: *The inclusions of various notions of probabilistic convergence.*

Figure 2.9: *An illustration of the intervals I_{2^m+i} defined in Example 2.2.67.*

Example 2.2.67 (Convergence i.p. does not imply convergence a.s.) We choose the same probability space in Example 2.2.25. Define the following set of binary intervals

$$I_{2^m+i} = \left[\frac{i}{2^m}, \frac{i+1}{2^m}\right) \tag{2.2.68}$$

for $m = 0,1,2,...$ and $i = 0,1,...,2^m-1$ (see Figure 2.9). For each m, there are 2^m intervals, each of length 2^{-m}, which cover the interval $[0,1)$ as illustrated. We have that

$$\lim_{n\to\infty} \mathrm{P}(I_n) = \lim_{m\to\infty} \mathrm{P}(I_{2^m}) = \lim_{m\to\infty} 2^{-m} = 0.$$

Now consider a sequence of random variables $Y_n(\omega) = [\![\omega \in I_n]\!]$. It is easy to show that $Y_n \xrightarrow{P} 0$[10] because for any $0 < \varepsilon \leq 1$,

$$\lim_{n\to\infty} \mathrm{P}(|Y_n - 0| > \varepsilon) = \lim_{n\to\infty} \mathrm{P}(Y_n = 1) = \lim_{n\to\infty} \mathrm{P}(I_n) = 0.$$

However, $Y_n \xrightarrow{a.s.} 0$ because for any $\omega \in [0,1)$ consider (2.2.68). For every m there exists a unique i such that $\omega \in I_{2^m+i}$ as $I_{2^m}, I_{2^m+1}, ..., I_{2^{m+1}-1}$ forms a partition of $[0,1)$. Since $\omega \in I_{2^m+i}$ is equivalent to $Y_{2^m+i} = 1$, we have that for any $\omega \in \Omega$, $Y_n(\omega) = 1$ infinitely often, so the limit $\lim_{n\to\infty} Y_n(\omega)$ diverges, giving

$$\mathrm{P}\left(\{\omega \in \Omega : \lim_{n\to\infty} Y_n(\omega) = 0)\}\right) = P(\varnothing) = 0 \neq 1$$

as required.   ◆

Definition 2.2.69 (Convergence in mean^2 (i.m.)) A sequence of random variables $(X_n)_{n=1}^{\infty}$ *converges in mean^2 to* X (denoted as $X_n \xrightarrow{L^2} X$) if

$$\lim_{n\to\infty} \mathbf{E}\left[(X_n - X)^2\right] = 0$$

---

[10]That is, converges in probability to the identically zero random variable.

[2]This is an exponent, not a footnote; in mean^2 is pronounced "in mean square".

Proposition 2.2.70 (Convergence in mean² implies convergence in probability) Given a sequence of random variables $(X_n)_{n=1}^{\infty}$, if $X_n \xrightarrow{L^2} X$, then $X_n \xrightarrow{P} X$.

Proof. Let $\varepsilon > 0$. Then by Markov's inequality and assumption $X_n \xrightarrow{L^2} X$,

$$P(|X_n - X| > \varepsilon) = P(|X_n - X|^2 > \varepsilon^2) \leq \frac{1}{\varepsilon^2} \mathbf{E}[|X_n - X|^2] \longrightarrow 0 \quad \text{for} \quad n \to \infty$$

Hence, $X_n \xrightarrow{P} X$, as required. ∎

Example 2.2.71 (Convergence i.p. does not imply convergence i.m.) The converse to Proposition 2.2.70 is false. Consider the sequence of random variables

$$X_n = \begin{cases} n & \text{with probability } \frac{1}{n} \\ 0 & \text{with probability } 1 - \frac{1}{n} \end{cases}$$

Then, $X_n \xrightarrow{P} 0$ since $P(|X_n| > \varepsilon) = \frac{1}{n} \to 0$ for $n \to \infty$, but $X_n \overset{L^2}{\nrightarrow} 0$ since

$$\lim_{n \to \infty} \mathbf{E}[|X_n|^2] = \lim_{n \to \infty} \sum_{x \in \{0, n\}} x^2 P(X_n = x) = n^2 P(X_n = n) = \lim_{n \to \infty} n = \infty \quad ◆$$

Definition 2.2.72 (Convergence in mean² sum (i.m.s)) A sequence of non-negative random variables $(X_n)_{n=1}^{\infty}$ *converges in mean² sum[a]* to X (denoted $X_n \xrightarrow{i.m.s} X$) if

$$\sum_{n=1}^{\infty} \mathbf{E}[(X_n - X)^2] < \infty$$

---
[a] Not to be confused with convergence in mean² (Definition 2.2.57).

Proposition 2.2.73 (Conv. in mean² sum implies conv. in mean²)

Proof. Trivially follows from the fact that an infinite sum $\sum_{n=1}^{\infty} \mathbf{E}[(X_n - X)^2]$ of non-negative terms can only be finite if the sequence $\mathbf{E}[(X_n - X)^2]$ summed over converges to zero. ∎

Lemma 2.2.74 (Borel-Cantelli) If the sum of the probabilities of a sequence of events $E_1, E_2, \dots$ is finite, then the probability that infinitely many of them occur is zero. Formally,

$$\sum_{n=1}^{\infty} P(E_n) < \infty \quad \text{implies} \quad P\left(\limsup_{n \to \infty} E_n\right) = 0$$

Proof. Let $A_n = \bigcup_{i=n}^{\infty} E_i$. Then

$$P(\limsup_{n \to \infty} E_n) \overset{(a)}{=} P\left(\bigcap_{n=1}^{\infty} \bigcup_{i=n}^{\infty} E_i\right) = P\left(\bigcap_{n=1}^{\infty} A_n\right) \overset{(b)}{=} \lim_{n \to \infty} P(A_n)$$

$$= \lim_{n \to \infty} P\left(\bigcup_{i=n}^{\infty} E_i\right) \overset{(c)}{\leq} \lim_{n \to \infty} \sum_{i=n}^{\infty} P(E_i) \overset{(d)}{=} 0$$

(a) follows from Definition 2.2.62 of lim sup, (b) from (2.2.10i), (c) is probability measure Axiom 2.2.3, and (d) is the Cauchy convergence criterion: if the sum $\sum_{i=1}^{\infty} x_i$ converges then the tail $\sum_{i=n}^{\infty} x_i$ tends to zero as $n \to \infty$. ∎

Theorem 2.2.75 (Convergence i.m.s. implies convergence a.s.)

Proof. First, apply Markov's inequality (Theorem 2.2.46) to $Y_n := X_n - X$:

$$P[Y_n^2 \geq \varepsilon^2] \leq \tfrac{1}{\varepsilon^2} \mathbf{E}[Y_n^2]$$

Now let $E_n := \{\omega : Y_n^2(\omega) \geq \varepsilon^2\}$. Then,

$$\sum_{n=1}^{\infty} P(E_n) = \sum_{n=1}^{\infty} P[Y_n^2 \geq \varepsilon^2] \leq \frac{1}{\varepsilon^2} \sum_{n=1}^{\infty} \mathbf{E}[Y_n^2] < \infty$$

by assumption $Y_n \xrightarrow{\text{i.m.s}} 0$ (Definition 2.2.72). Now, the Borel-Cantelli Lemma 2.2.74 implies

$$P\big(\limsup_{n \to \infty}\{\omega : Y_n(\omega) \geq \varepsilon\}\big) = P\big(\limsup_{n \to \infty}\{\omega : Y_n^2(\omega) \geq \varepsilon^2\}\big) = P\big(\limsup_{n \to \infty} E_n\big) = 0$$

which by Definition 2.2.57 implies $Y_n \xrightarrow{\text{a.s}} 0$, hence $X_n \xrightarrow{\text{a.s}} X$, as required. ∎

2.2.9   Exercises

1. [C05] Prove that for any probability space, the empty set ∅ is an event.

2. [C07] Give an example of a sample space Ω and two distinct choices $\mathcal{F}_1 \neq \mathcal{F}_2$ of event spaces.

3. [C07] Using only the alternative definition of a semimeasure (Definition 2.2.11), prove that $\mu(x) \leq 1$ for all $x \in \mathbb{B}^*$.

4. [C23i] Prove the seven unproven properties in Theorem 2.2.9.

5. [C10] Prove that conditional probability measures are probability measures (Theorem 2.2.14).

6. [C12] Prove that if event B is evidence in favor of event A ($P(A|B) > P(A)$) then event B^c must be evidence against A ($P(A|B^c) < P(A)$).

7. [C07] Prove that for any continuous random variable X we have that $\int_{-\infty}^{\infty} p_X(x)\,dx = 1$.

8. [C12] Consider Example 2.2.25. We choose a new random variable $Y(\omega) = \omega^2$. Compute F_Y and p_Y.

9. [C18] Given a sequence $(A_n)_{n=1}^{\infty}$ of sets, prove that $\liminf A_n \subseteq \limsup A_n$ and give an example to show that in general, $\liminf A_n \subsetneq \limsup A_n$.

10. [C14] Prove the sum rule for continuous random variables (Theorem 2.2.31).

11. [C15] Prove that mutual independence implies pairwise independence, and give a counterexample to prove the converse is false. (Definition 2.2.33).

12. [C15] Prove that if X is a continuous random variable with CDF F_X, and pdf p_X satisies $p(x)=0$ when $x<0$, then $\mathbf{E}[X]=\int_0^\infty (1-F_X(x))\,dx$.

13. [C20m] Derive the expressions for conditional probability distributions of $g()$ in Definition 2.2.36 from Definition 2.2.35.

14. [C22m] Proof the equivalence of Definition 2.2.57s(i,ii,iii) of almost sure convergence in Definition 2.2.57.

15. [C22] Complete the proof in Example 2.2.64 that a biased coin with $\theta>0$ will almost surely flip 1 infinitely often.

16. [C30i] Prove the remaining convergence relations in Figure 2.8 not proven in the main text and prove that the inclusions are proper.

17. [C12] Prove that the non-negativity assumption in Markov's inequality (Theorem 2.2.46) is necessary.

18. [C15] Prove the first-order conditions for a differentiable convex function (Lemma 2.2.51) with ∇ being the gradient, and then generalize to non-differentiable functions with ∇ being a subgradient.

19. [C20] Prove Jensen's inequality (Theorem 2.2.52) for general (non-differentiable) convex functions.

20. [C15] Fill in the gaps in the derivations of Theorem 2.2.60.

21. [C10] Give an example of a random process such that samples drawn from it are identically but not independently distributed, and vice versa.

2.3 Statistical Inference and Estimation

To address the prediction problem, we first begin by assuming that we possess some understanding of the underlying process that generates sequences. Initially, we will examine the simplest scenario in which a binary sequence is produced independently and identically distributed (i.i.d.) by a Bernoulli distribution, characterized by an *unknown* parameter θ that needs to be inferred from past observations.

The assumption that the sequence is generated by a Bernoulli process is quite strong, since each element in the sequence would be independent of preceding elements, which is often an unrealistic assumption. Later on in Chapter 3 we will relax this assumption to enable induction on sequences where the distribution may depend on any past elements, and in Chapter 4 an efficient predictor is discussed for sequences for which the distribution over future elements depends only on a bounded number of past elements.

We first examine the problem from a frequentist perspective, and consider the Bayesian approach thereafter.

2.3.1 Statistical Inference and The Sunrise Problem

Statistical inference revolves around inferring properties of a population or group of objects using quantitative observations, based on a random sample taken from that population or process. Typically, we make an assumption that the characteristics of the population or process can be fully described by a specific probability model. This model is parameterized by a set of parameters denoted as Θ. Note that Θ is often a subset of $\mathbb{R}^d$, resulting in an uncountable family of probability models. The problem can then be reduced to searching for the correct choice of parameters $\theta \in \Theta$.

For example, we might assume that a binary sequence is generated i.i.d. by a Bernoulli model $\text{Bern}(x;\theta)$ that is parameterized by $\Theta = \{\theta\} = [0,1] \subset \mathbb{R}$. In the following discussion, we will stick to this Bernoulli example and investigate ways to estimate θ. For the moment, we will only be interested in deriving a single numerical value as our best guess for the true value of θ given a sample of observations $x_{1:n}$. Later, we will show a more general representation that also provides a measure of confidence for our best estimate of θ via interval estimation [Hut08a]. This gives a formal meaning to the event $\text{P}(\theta \in [0.4, 0.6] \,|\, \mathcal{D})$ after having been conditioned on data $\mathcal{D}$.

Example 2.3.1 (The Sunrise Problem)   The sunrise problem concerns how to assign a probability to an event that has never happened before. Consider the likelihood of the sun not rising tomorrow. The naive frequentist estimate of

$$\frac{\text{days the sun did not rise}}{\text{days the Earth (or humans) have experienced}}$$

would give probability zero, as the sun has never been observed to fail to rise on any given day[11] before. This result seems absurd, as to assign zero probability to an event means to declare it is impossible. This problem is of great importance in this book, as through attempting to solve it, this problem gives rise to the Laplace rule and sheds light on many other useful statistical methods to estimate the true probability of an unknown model. Formally, the sunrise problem can be described as follows.

1. We choose as our sample space $\Omega = \{\text{not rise, rise}\}^n$, and define a set of $\{0,1\}$-valued i.i.d. random variables $X_1,...,X_n$, each distributed according to a Bernoulli(θ) distribution. For now, the parameter θ is fixed but unknown. The r.v. X_t takes an element $(\omega_1,...,\omega_n) \in \Omega$ and returns 1 iff the sun has risen on the tth day, $X_t(\omega_1,...,\omega_n) = [\![\omega_t = \text{rise}]\!]$ satisfying $\text{P}(X_i = 1) = \theta$ and $\text{P}(X_i = 0) = 1 - \theta$.

2. Assume that observed values $x_{1:n}$ have been obtained from the n i.i.d. random variables $X_1...X_n$.

3. We wish to estimate θ (in practice, we learn a distribution over θ) as an intermediate step based on the available observations $x_{1:n}$, using some estimator which is a function of the observed values $t(x_{1:n})$. This value can be interpreted as the outcome of the real random variable $T_n(X_1,X_2,...,X_n)$.

4. Finally, based on our estimate of θ, we want to make a prediction for x_{n+1}, i.e. whether the sun will rise on the next day or not.

One might ask what the motivation is behind assuming the sun is a Bernoulli distribution, or even why the behavior of the sun can be modelled as a sequence of i.i.d. processes. This can be attributed to the historical background of the problem (proposed in the 18th century by Laplace), where the mechanism by which the sun worked was not understood. Even stronger, imagine explaining this problem to someone who had lived underground in a bunker their entire life, and had no concept of what a "sun" is, only that it can "rise" or "not rise", and that every "day", the sun rose. Without any prior information, the assumptions of i.i.d. and Bernoulli do not seem absurd.

---

[11]Ignoring the locations at extreme latitude that may not see the sun for months.

A more modern version of the problem would incorporate domain knowledge about the sun from astrophysics to obtain a more accurate result (the sun is expected to continue to rise every day for the next ≈ 5 billion years or so until it swells up and engulfs the Earth, assuming the sun is not tampered with until then). ♦

2.3.2   Maximum Likelihood

Perhaps the most important and intuitively appealing of all estimation procedures is that of maximum likelihood. A general framework for calculating the *maximum likelihood estimator* (MLE) of a parameter θ that parameterizes a family of pdfs $p_X(x;\theta)$, is defined as follows.

The *likelihood function* for a parameter θ based on a sample of n random variables $X_1,...,X_n$, is defined to be the joint probability density function of the n random variables, parameterized by the unknown parameter θ. We write,

$$L(\theta) = L(\theta;x_1,...,x_n) := p_{X_1,...,X_n}(x_1,...,x_n;\theta).$$

If the X_i's are i.i.d. with probability density function $p_X(x;\theta)$, then the likelihood can be written as

$$L(\theta) = \prod_{i=1}^{n} p_X(x_i;\theta)$$

The likelihood function $L(\theta)$ can be thought of as the joint conditional probability $p(x_1,...,x_n|\theta)$. However, for the above equation to hold, the X_i's need to be conditionally i.i.d. given θ, that is, we require that

$$P(x_1,...,x_n|\theta) = \prod_{i=1}^{n} p(x_i|\theta)$$

The *maximum likelihood* estimator (MLE) of a parameter θ is defined to be the value that maximizes the likelihood

$$\widehat{\theta}_{ML} := \arg\sup_{\theta\in\Theta} L(\theta;x_1,...,x_n)$$

where Θ is the set of allowable parameter values. Typically, the *log-likelihood function* is used, as it is mathematically easier to deal with. Since the logarithm function is monotonically increasing, we note that

$$\arg\sup_{\theta\in\Theta} L(\theta) = \arg\sup_{\theta\in\Theta} \ln L(\theta)$$

In the case of conditionally i.i.d. random variables, the log-likelihood transforms a product of probabilities into a more tractable sum of logs of probabilities:

$$\ln L(\theta) = \ln\left(\prod_{i=1}^{n} p_X(x_i;\theta)\right) = \sum_{i=1}^{n} \ln p_X(x_i;\theta)$$

The derivative $\nabla_\theta \ln L(\theta)$ is easier to compute than $\nabla_\theta L(\theta)$ (as derivatives are linear), which can be used to then maximize θ (by finding the θ for which the derivative is zero.)

Example 2.3.2 (MLE of Bernoulli sequence) Suppose that $X_1,...,X_n \sim \text{Bern}(x;\theta)$ with unknown θ. We can write the distribution for a Bernoulli random variable as

$$\text{Bern}(x;\theta) = \theta^x(1-\theta)^{1-x}$$

which has the corresponding log-likelihood function

$$\ln L(\theta) = \sum_{i=1}^{n} [x_i \ln\theta + (1-x_i)\ln(1-\theta)].$$

Differentiating this function with respect to θ gives us

$$\nabla_\theta \ln L(\theta) = \sum_{i=1}^{n} \left[\frac{x_i}{\theta} + \frac{x_i - 1}{1-\theta} \right]$$

and setting this derivative equal to zero and solving for θ yields the MLE estimate of θ as

$$\widehat{\theta}_{ML} = \frac{1}{n} \sum_{i=1}^{n} x_i$$

which is the average of all samples, which agrees with intuition (the average of many samples is an estimate of the expectation value, which for a random variable X with Bernoulli distribution $\text{Bern}(\theta)$ satisfies $\mathbf{E}[X] = \theta$).   ◆

Note that the MLE in Example 2.3.2 estimates θ as the proportion of 1's in the sequence. We often call such estimators *frequency estimators*. The sunrise problem indicates a problem with this approach, as the MLE estimate gives $\theta = 1$, which implies that the sun will rise tomorrow with absolute certainty. As mentioned before, this seems a nonsensical solution, as by induction the probability that the sun must rise every day is 1, but eventually the sun must die, so on the "last" sunrise, this method still gives the overconfident answer of 1. In general, we may wish to avoid probability estimates of an event as 1, since asserting an event is certain (probability one) means these estimates will remain unchanged in light of any new evidence.

More formally, if $P(A) = 1$ then $P(A|B) = P(A\cap B)/P(B) = 1$, for any B with $P(B) > 0$. If $P(B) = 0$ then we are conditioning on an impossible event and $P(A|B)$ is undefined (or defined 0 for convenience). So, assuming the posterior is well-defined, it is unchanged regardless of new evidence observed, which is an undesirable property for an estimator.

2.3.3   Reparametrization Equivariance of the MLE

The Bernoulli distribution is parameterized by θ, the probability of observing a 1. However, we could have chosen a different parametrization, say the probability of observing a 0, or the square of the probability of observing a 1. More generally, we can reparameterize θ as $\tau = \psi(\theta)$, for any bijection ψ. We would expect that the MLE estimate is invariant under choice of parameterization, i.e. $\widehat{\tau}_{ML} = \psi(\widehat{\theta}_{ML})$, and this is indeed the case.

Proposition 2.3.3 (Reparameterization equivariance of the MLE) Let $x_1,...,x_n$ be i.i.d. samples from a distribution having likelihood function $L(\theta; x_1,...,x_n)$. Also, let $\widehat{\theta}_{ML}$ be the MLE of θ based on this likelihood function. For any bijective function τ, we can define the likelihood function induced by $\tau(\cdot)$ as

$$\tilde{L}(\tau(\theta); x_1,...,x_n) := L(\theta; x_1,...,x_n)$$

and $\widehat{\tau}_{ML}$ is the value of τ that maximizes $\tilde{L}$. Then,

$$\widehat{\tau}_{ML} = \tau(\widehat{\theta}_{ML})$$

2.3.4 Consistency

In Section 2.3.2, we explored the maximum likelihood estimator, and while the results it gives are often intuitively satisfying, the estimate provided by MLE has problems. For instance, it is wholly dependent on the data present, which can lead to overfitting when little data is available. In this section, we show that the MLE (assuming the data generated is drawn from a Bernoulli process) is a *consistent*.

Given a sequence of i.i.d. random variables $X_1, X_2, \ldots$, we would like to estimate some parameter θ that controls the distribution of the X_i's. We define the *estimator* T_n to be some measurable function of the sequence $(X_i)_{i=1}^{\infty}$ that depends only on the first n random variables. In other words,

$$T_n : \mathcal{X}_1 \times \ldots \times \mathcal{X}_n \to \Theta \quad \text{that is} \quad T_n(X_1, \ldots, X_n) \in \Theta$$

where $\mathcal{X}_i = X_i(\Omega)$ is the alphabet of random variable X_i. We are interested in the asymptotic behavior of T_n as the number n of random variables the estimator depends on tends to infinity.

Note that T_n itself is a random variable (being a function of random variables). We would like our estimators to be *consistent*, that is, the estimator T_n converges to the parameter θ given sufficiently many samples.

Definition 2.3.4 (Estimator consistency) Given a sequence of i.i.d. random variables $(X_i)_{i=1}^{\infty}$ with a distribution controlled by some parameter $\theta \in \Theta$, we say that an estimator T_n for θ is *consistent* if $T_n \xrightarrow{P} \theta$.

The estimator $T_n = \frac{1}{n} \sum_{i=1}^{n} X_i$ for Bernoulli θ is consistent (and indeed unbiased, see below) as from the proof of Theorem 2.2.60 we have

$$\mathbf{E}[T_n] = \theta \quad \text{and} \quad \mathbf{Var}[T_n] = \frac{\theta(1-\theta)}{n} \longrightarrow 0$$

Consistency is a rather weak property of an estimator, as the estimator might also be *biased* and systematically over- or under-estimate θ, even if it converges in the limit.

Definition 2.3.5 (Estimator Bias) Given $(X_i)_{i=1}^{\infty}$ as in Definition 2.3.4, the *bias* of an estimator T_n is defined as w

$$\text{Bias}(T_n) := \mathbf{E}_\theta[T_n] - \theta$$

Furthermore, if we have that for all $n \geq 1$ and all $\theta \in \Theta$ that $\text{Bias}(T_n) = 0$, then the estimator T_n is said to be *unbiased*. Here, $\mathbf{E}_\theta[T_n]$ indicates that we are taking the expectation with respect to the pdf $p_X(\cdot; \theta)$.

Example 2.3.6 (Consistent but biased estimator) Given a sequence $\{x_i\}_{i=1}^{\infty}$ drawn i.i.d. from $\text{Bern}(\theta)$, the estimator $T_n = \left(\frac{1}{n} \sum_i x_i \right) + \frac{1}{n}$ for θ is biased, as

$$\text{Bias}(T_n) = \mathbf{E}\left[\left(\tfrac{1}{n} \sum_i x_i \right) + \tfrac{1}{n} \right] - \theta = \mathbf{E}\left[\tfrac{1}{n} \sum_i x_i \right] - \theta + \tfrac{1}{n} = \tfrac{1}{n}$$

so T_n overestimates θ (on average) by $\frac{1}{n}$, but we still have that $T_n \xrightarrow{P} \theta$, so T_n is consistent. ◆

Example 2.3.7 (Unbiased but inconsistent estimator) The estimator $T_n = X_1$, as an estimation for θ is unbiased, as $\mathbf{E}[X_1] = \theta$ but fails to be consistent, as T_n will either be 0 or 1 for all n, which will not converge to θ for any $\theta \in (0,1)$.   ♦

Checking the consistency and bias of an estimator is a good start, but we'd like something more quantitative: To what degree is the estimator biased, and how fast does it converge to the unknown parameter θ? Often in statistics the mean squared error is used as a measure of performance, which motivates the following definition.

Definition 2.3.8 (Estimator Mean Squared Error) The *mean squared error* (MSE) of an estimator T_n is defined as

$$\mathrm{MSE}_\theta[T_n] := \mathbf{E}_\theta[(T_n - \theta)^2]$$

Theorem 2.3.9 (MSE of an estimator) If T_n is an estimator, then

$$\mathrm{MSE}_\theta[T_n] = \mathbf{Var}_\theta[T] + \mathrm{Bias}(T_n)^2$$

In particular for an unbiased estimator, the MSE coincides with the variance.

Proof. Let $t := \mathbf{E}_\theta[T_n]$. Then

$$
\begin{aligned}
\mathrm{MSE}[T_n] &\equiv \mathbf{E}_\theta[(T_n - t + t - \theta)^2] \\
&= \mathbf{E}_\theta[(T_n - t)^2 + 2(T_n - t)(t - \theta) + (t - \theta)^2] \\
&= \mathbf{Var}_\theta[T_n] + 2(\mathbf{E}_\theta[T_n] - t)(t - \theta) + \mathrm{Bias}(T_n)^2 \\
&= \mathbf{Var}_\theta[T_n] + \mathrm{Bias}(T_n)^2 \qquad\qquad \blacksquare
\end{aligned}
$$

We can then compare the quality of two estimators by comparing their MSEs.

Definition 2.3.10 (Estimator domination) An estimator T is said to *dominate* another estimator T' if, for all choices of parameter $\theta \in \Theta$,

$$\mathrm{MSE}_\theta[T] \leq \mathrm{MSE}_\theta[T']$$

Now that we have a partial ordering on estimators, we can ask the obvious question: is there always an optimal choice for an estimator that dominates all other estimators? We require some more mathematical preamble, from which we will show the *Cramér–Rao Bound* (though the proof is too lengthy to replicate here), which gives a lower bound of the mean squared error of an unbiased estimator.

Definition 2.3.11 (Score) Given a random variable $X \sim p_X(x;\theta)$ with distribution parameterized by θ, the *score* V of X is a random variable defined as

$$V = \nabla_\theta \ln p_X(X;\theta)$$

For a family of random variables $X_1,...,X_n \sim p_{X_1,...X_n}(x_1,...,x_n;\theta)$, we can define the n-sample score V_n of this family using the joint distribution in the obvious fashion

$$V_n = \nabla_\theta \ln p_{X_1,...,X_n}(X_1,...,X_n;\theta)$$

One can show that the expected value of the score $\mathbf{E}_\theta[V]$ is zero, so $\mathbf{Var}_\theta[V] = \mathbf{E}_\theta[V^2]$.

Definition 2.3.12 (Fisher Information) Given a random variable $X \sim p_X(x;\theta)$, with score V, the *Fisher information* $\mathcal{I}(\theta)$ is the variance of the score

$$\mathcal{I}(\theta) := \mathbf{Var}_\theta[V] = \mathbf{E}_\theta\left[\left(\frac{\partial}{\partial\theta}\ln p_X(X;\theta)\right)^2\right]$$

Similarly, the n-sample Fisher information $\mathcal{I}_n(\theta)$ is defined as the variance of the n-sample score.

Under some very mild technical conditions, the Fisher information can be equivalently expressed as $\mathcal{I}(\theta) = -\mathbf{E}_\theta[\frac{\partial^2}{\partial\theta^2}\ln p_X(X;\theta)]$.

Theorem 2.3.13 (n-sample Fisher information) Given an i.i.d. family of random variables $X_1,...,X_n$ with $X_i \sim p_X(x;\theta)$, then $\mathcal{I}_n(\theta) = n\mathcal{I}(\theta)$.

Proof. Exercise. ∎

The Fisher information is used as a measure of how much information (on average) can be obtained from θ from each sample and appears in plenty of other statistical results.

Theorem 2.3.14 (Cramér–Rao Bound) The MSE of any estimator T_n is lower bounded by the reciprocal of the Fisher information.

$$\mathrm{MSE}_\theta[T_n] = \mathbf{E}_\theta[(T_n-\theta)^2] \geq \mathbf{Var}_\theta[T_n] \geq \frac{1}{\mathcal{I}_n(\theta)}$$

Proof. See [CT06]. ∎

Estimators that meet this bound, i.e. satisfy $\mathrm{MSE}_\theta[T_n] = 1/\mathcal{I}_n(\theta))$ are called *efficient*, and are "best" in the sense that they dominate all other estimators.

2.3.5 Exercises

1. [C10] Prove that the Fisher information can equivalently be written as $\mathcal{I}(\theta) = -\mathbf{E}_\theta\left[\frac{\partial^2}{\partial\theta^2}\ln p_X(X;\theta)\right]$.

2. [C12] Prove that $\mathcal{I}_n(\theta) = n\mathcal{I}(\theta)$ (Theorem 2.3.13).

3. [30m] Prove the Cramér–Rao Bound (Theorem 2.3.14).

2.4 Bayesian Probability Theory

The previous section gave a glimpse of what is called frequentist statistics. Many powerful and/or elegant estimators have been developed, but it lacks an underlying unifying principle. Maximum likelihood estimation gets close but requires various (regularization) patches for large model classes. One problem is that the uncertainty about the underlying probability measure is only dealt with rather indirectly as witnessed by the continued misinterpretation of e.g. confidence intervals.

In Bayesian statistics, the uncertainty about the probability measure is itself modelled by probabilities. In the parametric case, starting with a *prior* (belief=epistemic) probability

w_θ over $\theta \in \Theta$, Bayes general and optimal learning rule uniquely determines how to update it to a *posterior* belief probability after new data=evidence arrives.

We give a brief introduction to Bayes rule and Bayesian inference, and demonstrate it on a simple instantiation for binary i.i.d. data, leading to the famous Laplace's Rule for prediction. Later chapters will consider much larger (and indeed universal) model classes and priors.

2.4.1 Bayes' Theorem

In Section 2.3, the maximum likelihood estimator (MLE) depends wholly on the data available, which can lead to overly confident estimates when little data is available. Moreover, observations are rarely perfect and are often noisy, so on occasion the data given may not always be perfectly trustworthy, and often the estimator may need to learn even when a fraction of the given data is misleading or false. A naive MLE approach may give an overly confident estimate of the parameter in question. We desire an estimator that is somewhat "skeptical" of the data presented, incorporating prior knowledge before data is presented, and also gives a measure of not just what the best estimate of the unknown parameters is, but also how confident one should be in that estimate.

Example 2.4.1 (Estimating Lottery winning probability) Consider buying your first ever ticket for the lottery, and you get lucky and win on the first try. Lacking any knowledge about the win probabilities of the lottery, the maximum likelihood estimator would conclude from this data that you are guaranteed to win the lottery every time you play.   ♦

In an attempt to address this concern, we should first consider why the conclusion in Example 2.4.1 seems implausible. A certainty of winning the lottery seems implausible based on the prior knowledge we have of lotteries (in that out of many tickets, only a few are winners, so winning the lottery should be a very rare event).

For example, if we are trying to find the parameter θ for the probability that a coin flips heads, it might be the case that the coin is unfair and is slightly biased one way or the other, but we would expect an ordinary coin to have $\theta \approx \frac{1}{2}$, and after observing the coin land at least once on heads and at least once on tails, we can rule out the options $\theta = 0$ or $\theta = 1$.

This is why a frequentist approach can often be limiting, as while it may converge in the limit with a large amount of data, for small amounts of data we should also incorporate current prior domain knowledge about the parameter of interest into the estimation procedure. The distribution we then get over the set of parameters given the domain knowledge is called the *prior*, and Bayes' Rule (Theorem 2.4.2) gives a systematic method of determining a new distribution after having received the data, called the *posterior*.

Bayes' celebrated theorem allows the conditional probability $P(A|B)$ to be computed in terms of $P(A), P(B)$ and $P(B|A)$.

Theorem 2.4.2 (Bayes' theorem) Let $D \subseteq \Omega$ be an event with $P(D) > 0$ and $(H_i)_{i \in I}$ be a countable partition of Ω (i.e. $\forall i \neq j : H_i \cap H_j = \varnothing$ and $\dot{\bigcup}_i H_i = \Omega$). Then

$$P(H_i|D) = \frac{P(D|H_i)P(H_i)}{P(D)} = \frac{P(D|H_i)P(H_i)}{\sum_{j \in I} P(D|H_j)P(H_j)}$$

The key quantity $P(D)$ is called Bayesian evidence or marginal distribution or mixture distribution. $P(H_i)$ is the prior belief in hypothesis H_i and $P(H_i|D)$ the posterior belief after observing data D, while $P(D|H_i)$ is called the data likelihood under hypothesis H_i.

While the proof of Bayes' theorem is elementary (Exercise 1), it gives a powerful technique for updating confidences (the likelihood of an event, the choice of parameters for a model, etc.) and is what the entire field of Bayesian statistics is based on. Mathematically, the Bayesian "sample" space Ω is a product of the data (frequentist) sample space is Ω_D and a hypothesis space Ω_H. A hypothesis H_i is an event of the form $\Omega_D \times \{\tilde{H}_i\} \subset \Omega$. An observation D is an event of the form $\tilde{D} \times \Omega_H \subset \Omega$. Usually it is $\tilde{H}_i \in \Omega_H$ and $\tilde{D} \subset \Omega_D$ that are specified, the tilde is dropped, and the lifting to the product space implicitly understood.

Interpretation. The intuition behind Bayes' theorem is that it provides a method of updating (belief) probabilities based on evidence. If H_i is some hypothesis, for example, H_1 for a person having a particular disease, and H_0 for not having it. Then we can consider holding subjective beliefs and uncertainties about H_i. Before clinical tests are performed, the probability $P(H_i)$ can represent the doctor's (subjective) degree of belief in H_i, e.g. based on the patient's symptoms and demographic data. The belief can change once the doctor receives outcome D of some clinical tests for the patient, which may increase the confidence in H_i (if D is evidence in favor of H_i) or decrease it (if D is evidence against). Note that different people can have different confidence in events. For instance, before the doctor tells the patient the outcome D of the tests, the patient will have a different belief in H_i than the doctor. But even before the doctor learns about D or after both learn about D, they likely hold different beliefs since their priors (and hence posteriors) are different. The doctor should have much more experience in judging $P(H_i)$, while the patient's belief is likely more driven by fear or wishful thinking. Let us compare the different encountered interpretations of probabilities:

- *Frequentist:* Probabilities are defined as the relative frequency of an event occurring. If in a sequence of n independent identically distributed (i.i.d.) trials an event occurs $k(n)$ times, the relative frequency of the event is $k(n)/n$. The probability of an event is then defined as the limit $\lim_{n\to\infty} k(n)/n$. The frequentist models $P(D|H_i)$, but neither assigns a prior nor a posterior (belief) probability to H_i.

- *Objectivist*: Probabilities are real aspects of the world, also called *true* probabilities. The outcome of an observation or an experiment is not deterministic, but involves chaotic or random (physical) processes. In the case of i.i.d. experiments the probabilities assigned to events should be interpretable as and consistent with limiting frequencies, but objective probabilities are not limited to this case. They could be for singular events such as the probability of a volcanic eruption in Australia in the next $t=1,2,3,...$ centuries, which is a unique non-i.i.d. stochastic process (later denoted by μ).

- *Subjectivist*: Probabilities are merely subjective, and represent a particular individual's confidence or degree of belief, called epistemic probability. In this interpretation, subjective probabilities are assigned to H_i, which depend on who holds them (doctor vs. patient) and of course on potential available extra information D.

- *Bayesian*: A Bayesian assigns a prior probability over H_i *and* uses a (frequentist) likelihood model $P(D|H_i)$ to update prior $P(H_i)$ to a posterior probability $P(H_i|D)$ after observing evidence D. The posterior is the agent's degree of belief in H_i. The prior may be *subjective* [Gol06] (e.g. determined via expert elicitation [OBD+06]), or *objective* [Ber06, BBS24] (like Jeffreys prior below), or *universal* [RH11] (see Section 3.7), though this trichotomy is a gross over-simplification [Goo71].

Remarkably and conveniently, all types of probabilities above satisfy the same Kolmogorov Axiom 2.2.3 of probability [Hut05b, Sec.2.3] [Pre09, Sec.2].

Example 2.4.3 (Blood type testing) Some blood types are rarer than others. A particular blood type, O-negative (O^-) is special, since it can be donated to anyone. The proportion of people that have O^- blood is 7%. We have a test that can check if someone has O^- blood or not. Unfortunately, this test does not always give the correct answer, and is only 95% accurate in that if someone who does (doesn't) have O^- blood, the test will return a false negative (positive) with probability 5%. We take a person sampled at random from the population, and test them. The test shows they have O^- blood. How confident should we be that this person *actually* has O^- blood? What if the test was negative?

A Bayesian formulation of the problem is as follows: Let H^- be the hypothesis that Person A has blood type O^-, and H^+ that Person A has some other blood type. The prior probability is $P(H^-)=1-P(H^+)=0.07$. If this number comes from a simple random population survey, then it is an objective prior. If it is based on some heuristic argument with some judgement involved, e.g. by arguing that the population is similarly to another group, it would be subjective. The hypothesis space is $\Omega_H=\{H^-,H^+\}$. Let D^- denote the event that the test shows that Person A has O^- blood. The likelihood is $P(D^-|H^-)=P(D^+|H^+)=0.95$ and $P(D^-|H^+)=P(D^+|H^-)=0.05$. The data sample space can be chosen as $\Omega_D=\{D^-,D^+\}$. Noting that $\{H^-\}\dot\cup\{H^+\}=\Omega_H$, we can use Bayes' Theorem 2.4.2 to derive the posterior probability of Person A having blood type O^- given the test outcome $D^\pm$:

$$P(H^-|D^-) = \frac{P(D^-|H^-)P(H^-)}{P(D^-|H^-)P(H^-)+P(D^-|H^+)P(H^+)} = \frac{0.95\times0.07}{0.95\times0.07+0.05\times0.93} \approx 59\%$$

$$P(H^-|D^+) = \frac{P(D^+|H^-)P(H^-)}{P(D^+|H^-)P(H^-)+P(D^+|H^+)P(H^+)} = \frac{0.05\times0.07}{0.05\times0.07+0.95\times0.93} \approx 0.4\%$$

So, despite the fact that the test has a relatively high probability (95%) of success, the probability of actually having O^- type blood given the test indicated as such is only $\approx 59\%$. This may seem counter-intuitive, and is due to the small *base rate*, or prior probability of the blood type O^-. As the likelihood of a false positive (5%) and having O^- blood (7%) are similar, this means that if the test indicates O^- blood, we would intuitively expect that (roughly) half the time it was due to the test being incorrect, and half the time the patient actually having O^- blood. This explains why the posterior is near 50%. Conversely, since non-O^- blood is so common, if the test does not indicate the presence of O^-, it is very unlikely the patient has O^- blood (as both having O^- blood and having the reliable test fail are both unlikely events).   ♦

2.4.2 Bayes Estimation and Prediction

We express our belief over plausible and implausible values of $\theta\in\Theta$ as a density function $w(\theta)$, called the *prior*.

In frequentist statistics, θ is not a random variable. The true value of the parameter θ, for example the bias of a coin, is a fixed though unknown constant that depends on the physical makeup of the coin, precisely how the coin is flipped, etc. Frequentists *can* express uncertainty in θ, e.g. via confidence intervals, but their interpretation is convoluted and regularly confuses non-statisticians.

Bayesians simply model uncertainty about θ in the same way as for observations X, treat θ as a random variable, and assign a (subjective prior) distribution $w(\theta)$ over it, which encodes our prior beliefs about the relative plausibility of potential values of $\theta\in\Theta$. This requires enlarging the sample space to $\Omega''=\Omega\times\Omega'$ to accommodate θ, and $X(\omega)$ becomes $X(\omega'')$. Typically $\Omega'=\Theta$ and with some abuse of notation $\theta(\omega')=\omega'=\theta$. While mathematically, θ is now a random variable, and Θ part of the sample space, and ω' a sample most[12] Bayesians would neither interpret θ as being random nor sampled. The awkward clash in nomenclature though is more than compensated by the convenience and power this mathematically unified treatment of observations and parameters offers.

We have already alluded to that the probability distribution $p_X(x;\theta)$ for the observed random variables X can be viewed as a conditional distribution on X given θ, and the enlargement of Ω to Ω'' now formalizes and justifies writing $P(x|\theta)$: Indeed, is we define the joint distribution

$$p(x,\theta) := p_X(x;\theta)w(\theta)$$

then $p(x|\theta) \equiv p(x,\theta)/w(\theta) = p_X(x;\theta)$ provided $w(\theta) > 0$ and we may as well define it to be equal even if $w(\theta) = 0$ (see also Section 2.8.3). The mixture distribution, also called marginal distribution or Bayesian evidence,

$$p(x_1,...,x_n) = \int_\Theta p(x_1,...,x_n|\vartheta)w(\vartheta)\,d\vartheta$$

is a key quantity in Bayesian statistic. As a rule of thumb, if this quantity can be efficiently computed, then everything else of interest can be efficiently computed.

(A continuous version of) Bayes' rule Theorem 2.4.2 can now update the prior belief about θ with observations $x_1,x_2,...,x_n$ to arrive at the *posterior* belief distribution

$$w(\theta|x_1,...,x_n) = \frac{p(x_1,...,x_n|\theta)w(\theta)}{p(x_1,...,x_n)}$$

This posterior distribution combines two sources of information that we have about the parameter: our prior beliefs on what we expect θ to be, and any evidence provided by the observed data. Bayes' rule uniquely determines the posterior distribution given a prior, i.e. how to update beliefs based on evidence.

Often, our interest in estimating the parameter θ is merely a means to an end; ultimately we want to make (accurate) predictions. The *predictive distribution* is given by

$$p(x_{n+1}|x_1,x_2,...,x_n) = \int_\Theta p(x_{n+1}|\vartheta)w(\vartheta|x_1,x_2,...,x_n)\,d\vartheta = \frac{p(x_1,...,x_{n+1})}{p(x_1,...,x_n)} \qquad (2.4.4)$$

2.4.3 Laplace Rule

Section 2.4.2 describes how to include prior information about a parameter when estimating its value from data, but is agnostic on where the prior $w(\theta)$ comes from. In practice, the proper choice of a prior distribution may greatly affect the resulting estimate if little data is available, though with sufficient data the posterior will have most of the probability mass on the true value of the parameter θ for "sensible" choices of the prior $w(\theta)$ [Gho97].

If the domain is bounded, a usual choice is the uniform prior. For unbounded domains, it is not so clear what to choose, as distributions like a Gaussian $\mathcal{N}(\mu,\sigma^2)$ will naturally bias towards parameters near the mean μ. Without a method to choose a canonical choice for μ, the bias will be rather arbitrary. One major focus of this book concerns choosing a sensible prior distribution over not a parameter space Θ, but over a function space that contains all computable measures in a way that respects Occam's razor (simple/complex explanations are more/less likely).

---

[12]There are at least 46656 varieties of Bayesians [Goo71].

Derivation.   Suppose we have a biased coin governed by a Bernoulli process with unknown parameter θ. We observe a sequence of coin flips $x_1, x_2, \ldots \sim \text{Bern}(\theta)$, and would like to estimate θ from the sequence. We can first apply Bayes Theorem 2.4.2 to obtain the posterior distribution for θ after having observed the coin flips.

$$w(\theta|x_1, x_2, \ldots, x_n) = \frac{p(x_1, \ldots, x_n|\theta)w(\theta)}{\int_\Theta p(x_1, \ldots, x_n|\vartheta)w(\vartheta)\,d\vartheta} = \frac{w(\theta)\prod_{i=1}^n \text{Bern}(x_i;\theta)}{\int_\Theta w(\vartheta)\prod_{i=1}^n \text{Bern}(x_i;\vartheta)\,d\vartheta} \qquad (2.4.5)$$

Note that $\text{Bern}(x;\theta) = \theta^x(1-\theta)^{1-x}$, so we can express $\prod_{i=1}^n \text{Bern}(x_i;\theta)$ as $\theta^b(1-\theta)^a$, where a is the number of zeros in the sequence $x_1, \ldots, x_n$, and b the number of ones. The *beta distribution* is defined as

$$\text{Beta}(\theta;\alpha,\beta) = \frac{1}{B(\alpha,\beta)}\theta^{\alpha-1}(1-\theta)^{\beta-1} \quad \text{where} \quad B(\alpha,\beta) = \int_0^1 \theta^{\alpha-1}(1-\theta)^{\beta-1}\,d\theta \quad (2.4.6)$$

is the *beta function*. If we have no prior information about the coin, choosing a uniform distribution over the range $[0,1]$ to represent our prior belief in the value of θ is reasonable. Note that this uniform distribution should not be interpreted as the lack of a prior belief. In fact, it effectively expresses a prior belief of indifference, that is, any value in $[0,1]$ is *a priori* as likely as any other. Other priors are discussed below. Using the indfference prior $w(\theta) = 1$ for $0 \leq \theta \leq 1$ and (2.4.6) allows us to express (2.4.5) as

$$w(\theta|x_1, x_2, \ldots, x_n) = \frac{\theta^b(1-\theta)^a}{\int_0^1 \vartheta^b(1-\vartheta)^a\,d\vartheta} = \frac{\theta^b(1-\theta)^a}{B(b+1, a+1)} = \text{Beta}(\theta;b+1, a+1)$$

Figure 2.10 shows a plot of the beta distribution for some choices of α and β. In Figure 2.11

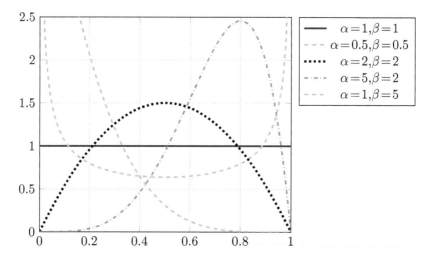

Figure 2.10: *A plot of Beta(α,β) (2.4.6) for various values of α and β.*

we plot the posterior $w(\theta|x_1, x_2, \ldots, x_n)$ for $n = 0, 1, 2, 3$ assuming the sequence $x_{1:3} = 001$ was observed one bit at a time ($n = 0$ corresponds to having seen none of the sequence, which means the posterior is just the prior). As can be seen, the posterior distributions put more probability mass on smaller values of θ after observing $x_1 = 0$ and $x_2 = 0$, and then shifts back slightly when $x_3 = 1$ is observed.

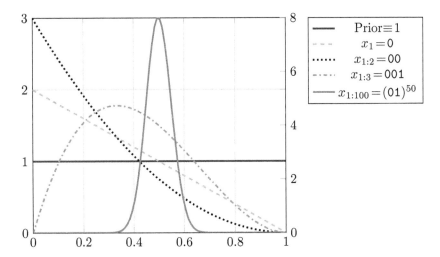

Figure 2.11: *Given the coin flip example in Section 2.4.2, a plot of how the posterior distribution for the parameter θ changes when observing the sequence $x_{1:3} = 001$, updating one bit at a time, assuming a uniform prior over θ. Also included is the posterior after observing 50 0's and 50 1's (see second y-axis).*

Now that we have a posterior distribution over θ, we can use this to make predictions of x_{n+1} after observing $x_1,...,x_n$ using (2.4.4). By definition, the probability of $P(X_{n+1}=1|\theta)=\theta$ as $X_{n+1} \sim \text{Bern}(\theta)$, so

$$P(X_{n+1} = 1|x_1,x_2,...,x_n) = \int_\Theta P(X_{n+1}=1|\theta) \frac{\theta^b(1-\theta)^a}{B(b+1,a+1)} \, d\theta$$

$$= \frac{\int_0^1 \theta^{b+1}(1-\theta)^a d\theta}{B(b+1,a+1)} = \frac{B(b+2,a+1)}{B(b+1,a+1)}$$

We can simplify this using properties of the *gamma function*

$$\Gamma(k) := \int_0^\infty x^{k-1}e^{-x}dx, \tag{2.4.7}$$

which can be interpreted as a continuous form of the factorial function, satisfying $\Gamma(k)=(k-1)!$ for $k=1,2,...$, and $\Gamma(z+1)=z\Gamma(z)$ for all $z \in \mathbb{R}$ where $\Gamma(z)$ is defined. Now, a property of the beta function [Art32] is that it satisfies for any $a,b>0$,

$$B(a,b) = B(b,a) = \int_0^1 \theta^{b-1}(1-\theta)^{a-1}d\theta = \frac{\Gamma(a)\Gamma(b)}{\Gamma(a+b)}$$

So, we can rewrite

$$P(X_{n+1}=1|x_1,x_2,...,x_n) = \frac{B(b+2,a+1)}{B(b+1,a+1)}$$

$$= \frac{\Gamma(b+2)\Gamma(a+1)}{\Gamma(b+a+3)} \frac{\Gamma(b+a+2)}{\Gamma(b+1)\Gamma(a+1)}$$

$$= \frac{(b+1)\Gamma(b+1)\Gamma(a+1)}{(b+a+2)\Gamma(b+a+2)} \frac{\Gamma(b+a+2)}{\Gamma(b+1)\Gamma(a+1)}$$

$$= \frac{b+1}{a+b+2} = \frac{1}{n+2}\left(1+\sum_{i=1}^{n}x_i\right)$$

where in the last equality we have used that b was the number of ones in $x_1,...,x_n$ and $n=a+b$. This particular choice of prior distribution is called the *indifference rule*, or *Laplace rule*, and the resulting estimator is thus called the *Laplace estimator*.

Theorem 2.4.8 (Laplace rule) For i.i.d. $X_i \sim \text{Bern}(\theta)$ and uniform prior $\theta \in [0,1]$,

$$P(X_{n+1}=1|X_1=x_1,...,X_n=x_n) = \frac{k+1}{n+2}, \quad \text{where} \quad k = |\{i \leq n : x_i=1\}| = \sum_{i=1}^{n}x_i$$

For the sunrise problem, the sun has risen every day (for all $1 \leq i \leq n$, $x_i=1$) so this gives the estimate $\frac{n+1}{n+2}$ that the sun will rise tomorrow given n days of having seen the sun rise previously, which seems not an unreasonable degree of confidence absent prior knowledge.

General Beta prior.   More generally, suppose we chose an arbitrary beta distribution as our prior so that

$$w(\theta) = \text{Beta}(\theta;\alpha,\beta) := \frac{1}{B(\alpha,\beta)}\theta^{\alpha-1}(1-\theta)^{\beta-1}$$

for some $\alpha,\beta > 0$. A uniform prior is a special case of this, as $\text{Beta}(\theta;1,1)=1$. Repeating the calculations in (2.4.5), we obtain the posterior and the predictive distribution

$$w(\theta|x_1,...,x_n;\alpha,\beta) = \text{Beta}(\theta;b+\alpha,a+\beta) \tag{2.4.9}$$

$$P(X_{n+1}=1|x_1,x_2,...,x_n;\alpha,\beta) = \frac{b+\beta}{a+b+\alpha+\beta} = \frac{\beta+\sum_{i=1}^{n}x_i}{n+\alpha+\beta}. \tag{2.4.10}$$

The parameters α and β can be thought of as representing hypothetical counts of observations of ones and zeros before the experiment began (perhaps we were told of the outcome of an experiment performed before we received the coin). These hypothetical prior counts are called *psuedocounts*, and they act as a regularizer to smooth out the rapid changes when very little data is observed.

Choosing α and β.   If we have no information about whether observing a 0 is more or less likely than 1, then *a priori* it would seem that choosing $\alpha=\beta$ would be a sensible choice, eliminating a free parameter. But we still need to choose a value of α. When $\alpha=\beta=1$, the Beta distribution reduces back to a uniform distribution. For values $\alpha=\beta>1$, this means choosing a prior that is more heavily weighted towards values of $\theta \approx 1/2$, and unlikely to be near the extremes of 0 or 1. For instance, for normal coins we have a strong prior belief that they are fair, so would choose $\alpha=\beta$ and both large. In this case $(\beta+\sum_{i=1}^{n}x_i)/(n+\alpha+\beta) \approx \frac{1}{2}$ for small n, so only a sufficiently large number of observations would persuade us that the coin is unfair. Various beta distributions are plotted together in Figure 2.10. The opposite case is when $0 < \alpha=\beta < 1$, which means choosing a prior that is more biased towards the extremes, a prior assumption that the coin is likely to be unfair. In the extreme case where $\alpha=\beta$ limits to 0, the distribution degenerates to one with two point masses on 0 or 1, which implies that our prior assumption is that the coin is deterministic, either always generating zeros, or always generating ones.

Jeffreys prior. One approach is to choose $\alpha = \beta$ such that the prior is invariant under a change of coordinates for the parameter θ. This can be done by choosing the Jeffreys prior:

> **Definition 2.4.11 (Jeffreys Prior)** The Jeffreys prior is a prior with density function proportional to the square root of the Fisher information.
>
> $$w(\theta) \propto \sqrt{\mathcal{I}(\theta)}$$

We can compute $\mathcal{I}(\theta)$ for the distribution $\mathrm{Bern}(x;\theta)$.

$$\mathcal{I}(\theta) = -\mathbf{E}_\theta\left[\frac{\partial^2}{\partial\theta^2}\ln\mathrm{Bern}(X;\theta)\right] = -\mathbf{E}_\theta\left[\frac{\partial^2}{\partial\theta^2}\ln\theta^X(1-\theta)^{1-X}\right]$$

$$= -\mathbf{E}_\theta\left[\frac{\partial^2}{\partial\theta^2}X\ln\theta + (1-X)\ln(1-\theta)\right] = \mathbf{E}_\theta\left[\frac{X}{\theta^2} + \frac{1-X}{(1-\theta)^2}\right]$$

Using linearity of expectations and $\mathbf{E}_\theta[X] = \theta$, we obtain

$$\mathcal{I}(\theta) = \frac{\theta}{\theta^2} + \frac{1-\theta}{(1-\theta)^2} = \frac{1}{\theta(1-\theta)}$$

Hence, the Jeffreys prior is $w(\theta) \propto 1/\sqrt{\theta(1-\theta)}$, which can be expressed as a beta distribution by choosing $\alpha = \beta = 1/2$.

The resulting estimator with this prior is called the *KT estimator*. We will explore this estimator further in Section 4.1. While this choice is invariant and minimax optimal, for large non-binary alphabet, other choices of α and β are better [Hut13a].

Decomposition and concentration. Finally, we note that (2.4.10) can be decomposed as follows

$$\mathrm{P}(X_{n+1}=1|x_1,x_2,...,x_n;\alpha,\beta) = \lambda\bar{x} + (1-\lambda)\left(\frac{\beta}{\alpha+\beta}\right) \qquad (2.4.12)$$

$$\text{where} \quad \lambda = \frac{n}{n+\alpha+\beta} \quad \text{and} \quad \bar{x} = \frac{1}{n}\sum_{i=1}^{n}x_i$$

It can be shown that the expectation of a Beta distribution $\mathrm{Beta}(\theta;\alpha,\beta)$ is $\frac{\beta}{\alpha+\beta}$. So (2.4.12) shows that the predictive Bayes estimator with prior $\mathrm{Beta}(\theta;\alpha,\beta)$ can be viewed as a convex combination of the maximum likelihood estimator $\bar{x}$ (i.e. the naive estimator based solely on the data without incorporating prior knowledge) and the "pure prior" estimator $\frac{\beta}{\alpha+\beta}$. This is the mean of the prior distribution $\mathrm{Beta}(\theta;\alpha,\beta)$, which is what the predictive estimator returns if no data is available. Also note that the value of $\alpha+\beta$ in this convex combination effectively controls our confidence in the prior for predicting the next bit, and as a result, $\alpha+\beta$ is often termed as the *concentration parameter*.

As for the posterior, the mean and variance are

$$\mathbf{E}[\theta|x_1,...,x_n] = \frac{b+\beta}{n+\alpha+\beta} \qquad (2.4.13)$$

$$\mathbf{Var}[\theta|x_1,...,x_n] = \frac{(a+\alpha)(b+\beta)}{(a+\alpha+b+\beta)^2(a+\alpha+b+\beta+1)} \qquad (2.4.14)$$

This implies that the posterior distribution has a low variance and hence a high confidence in the parameter θ if at least one of $n+\alpha+\beta$ is large. So a strong prior belief or a lot of data results in high confidence in θ (narrow credible intervals for θ), as expected. For large $\alpha+\beta \gg n$, the posterior concentrates around the prior belief $\frac{\beta}{\alpha+\beta}$, while for large $n \gg \alpha+\beta$, the posterior concentrates around the relative frequency b/n of 1s.

2.4.4   Exercises

1. [C05] Prove Bayes' Rule (Theorem 2.4.2).

2. [C12] Derive the expressions for the predictive distribution (2.4.4).

3. [C12] Derive the expressions for the posterior (2.4.9) and predictive distribution (2.4.10).

4. [C20] Derive the expectation (2.4.13) and variance (2.4.14) for beta random variable $X \sim \text{Beta}(x; a+\alpha, b+\beta)$.

5. [C15] Given the sequence $X_{1:2N} = (01)^N$ of alternating pairs of coin flips, we would expect that $w(\theta | X_{1:2N})$ concentrates on $\frac{1}{2}$ as $N \to \infty$, by Theorem 2.2.60. Compute a lower bound on how large N needs to be before we can be at least 99% sure that the parameter θ lies in the range $0.49 \leq \theta \leq 0.51$ using Theorem 2.2.61, and then also compute the smallest possible N for this to hold. Now compute N to using the weaker bound Theorem 2.2.60, and compare.

2.5   Information Theory and Coding

We make observations all the time of the world around us, both through direct first-hand observations, and indirect inferences based on collected data. Not all observations are equally "informative". If we already know the outcome before it occurs, the information is useless. Conversely, if the outcome is extremely surprising (unlikely), then information that indicates it will occur is highly informative. The formal theory underpinning the study of information itself is relatively recent, first introduced in Shannon's seminal paper [Sha48] in 1948.

This measure of information is concerned with, on average, the number of bits required to uniquely identify messages drawn from a stochastic source, rather than the information inherently contained in the message. A television tuned to a blank channel displaying white noise is highly informative in this sense, as it would take a lot of bits to describe the exact state of the static on the screen, even though the content of the static itself is useless.

Shannon called this measure of the average number of bits to encode a message from a source *entropy*, which ties in with the field of data compression. The entropy of a source represents a bound on how well messages drawn from the source can be (on average) compressed without loss.

We will cover how Shannon historically estimated the entropy of the English language, a code he co-developed to compress messages into fewer bits, as well as some famous theorems in information theory like the Kraft inequality (Theorem 2.5.17), a necessary and sufficient condition for the existence of prefix codes, and Shannon's coding theorem (Section 2.5.5), which places bounds on the length of the optimal code based on the entropy of the source.

A different notion of information that measures the intrinsic information contained in a particular message itself, rather than the unpredictability of a stochastic source, called *Kolmogorov complexity*, will be introduced in Section 2.7.

2.5.1   Shannon Entropy

Recalling Remark 2.2.28, let X be a discrete random variable with sample space $\mathcal{X}$ and probability distribution P.

Given an outcome $x \in \mathcal{X}$ for a random variable X, we define the *Shannon information content* as $h(x) = \log_2[1/\text{P}(x)]$. The Shannon information can be thought of as a measure of how "surprising" an event is in bits. Events that are very unlikely ($P(x)=\varepsilon$ for small $\varepsilon>0$) are very surprising when they happen, so a lot of information is conveyed when they do ($h(x)$ is large). Conversely, if someone were to tell you that the sun rose today (Example 2.3.1), you have not learned much, as this is an extremely likely event, so $h(\text{the sun rose today})\approx 0$.

One of the most crucial concepts in the study for information theory is entropy. Simply put, it is the expectation of the Shannon information of a random variable. It gives a bound on the average number of bits required to encode a sample drawn from that random variable.

Definition 2.5.1 (Entropy) We define the *entropy* $H(X)$ of a discrete random variable $X=(\mathcal{X},\mathrm{P})$ as

$$H(X) := \mathbf{E}_{\mathrm{P}}[h(X)] = \sum_{x\in\mathcal{X}}\mathrm{P}(X=x)\log_2\frac{1}{\mathrm{P}(X=x)}$$

with the convention that $0\log_2\frac{1}{0}:=\lim_{p\to 0}p\log_2\frac{1}{p}=0$. We can also define the entropy $H(P)$ of a discrete probability distribution $P=\{p_1,p_2,...\}$ as $H(P):=\sum_i p_i\log_2\frac{1}{p_i}$.

The conditional entropy of X given an outcome $y\in\mathcal{Y}$ is defined as

$$H(X|Y=y) := \sum_{x\in\mathcal{X}}\mathrm{P}(x|Y=y)\log_2\frac{1}{\mathrm{P}(x|Y=y)}$$

The conditional entropy of X given Y is the expected value of $H(X|Y=y)$,

$$H(X|Y) := \sum_{y\in\mathcal{Y}}\mathrm{P}(y)H(X|Y=y) = \sum_{x,y\in\mathcal{X}\times\mathcal{Y}}\mathrm{P}(x,y)\log_2\frac{1}{\mathrm{P}(x|y)}$$

Example 2.5.2 (Entropy of a biased coin) Given a random variable X representing a biased coin, with $\mathrm{P}(X=H)=\theta$ hence $\mathrm{P}(X=T)=1-\theta$, the entropy of X is given by

$$H(X) = \theta\log_2\frac{1}{\theta}+(1-\theta)\log_2\frac{1}{1-\theta}$$

Figure 2.12 plots $H(X)$ for $0\leq\theta\leq 1$. We can see that the entropy is maximized for a fair coin, which has 1 bit of entropy. When $\theta=0$ or $\theta=1$, the entropy is zero bits, as we already know the outcome of a two-headed coin, so observing a coin flip provides no information.

◆

Note that here we define the entropy using the binary logarithm $\log_2$, meaning the result is measured in bits. Sometimes entropy is defined using the natural logarithm $\ln$, which gives the same definition, only now the result is measured in nats. Both definitions are the same up to a multiplicative constant.

Example 2.5.3 (Entropy of horse betting) Suppose a gambler is listening in to the results of a horse race with N competitors. The names of the horses might themselves be complex, but are already known to the gambler, and so the only information obtained when the winner is announced is the information required to distinguish which of the N horses won the race. Assuming it is equally likely that any horse would win, then we could encode each horse with a codeword $\lceil\log_2 N\rceil$ bits long. The complexity of the names of the horses is irrelevant. From the perspective of entropy, the information content in the messages received from the stochastic source is itself irrelevant; what matters is the number of bits required to express which message was received, with respect to some predefined collection of allowed

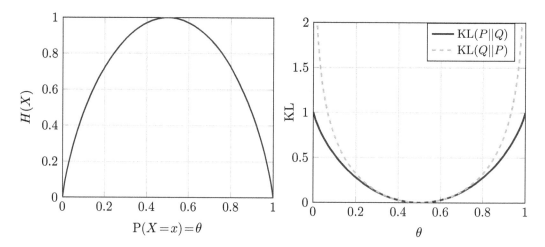

Figure 2.12: *The entropy $H(X)$ of a biased coin, as a function of the bias θ (Example 2.5.2).*

Figure 2.13: *A plot of $KL(P||Q)$ vs. $KL(Q||P)$ (see Example 2.5.13).*

messages. If, far more realistically, the probability distribution over the winning horse is highly skewed, then being told the fastest horse won is an unsurprising result, as we were confident that it was going to happen anyway. So, we could do better than uniform length codewords, and assign short encodings to the likely outcomes, and long encodings to the unlikely outcomes.[13] This means that the length of the encoding will on average be shorter than before. Even if the distribution of a source is not known, the probability distribution governing a source can be estimated from samples (Section 2.3), and from there the entropy can be estimated.   ◆

Remark 2.5.4 (Entropy of English) Shannon [Sha51] conducted an experiment to estimate the entropy of the English language[14] with a game: the objective being to encode a message as a sequence of counts for how many guesses were required for the next symbol, and to decode the sequences of guesses back to the message by performing the same game. The core idea is that the better the encoder/decoder can model English text, the fewer guesses they require to guess the next symbol, so the sequence should mostly be small numbers (2.5.5), indicating that English is highly redundant. Performing this game on random text would mean the encoder has no better strategy than random chance, and we would expect that they would (on average) guess correctly after trying half the options, namely 13 letters. The game is performed as follows: an experimenter has a hidden message $x_1,...,x_n$, and a subject repeatedly guesses the first symbol of the message until correct, then tries guessing the second symbol, and so on. The number of guesses g_i for each symbol x_i is recorded. This sequence of guesses is highly compressible, as given a subject fluent in English, he will have a strong intuition for the statistics of the English language. Given the letters TH... most people would choose E as a reasonable guess for the next missing letter. As a result, the

---

[13]Despite predating Shannon's work by about a hundred years, Samuel Morse had the foresight to do this when he invented his namesake code: By counting the frequency at which letters appeared in English books, he estimated the relative frequencies (and hence probabilities) that each letter would be used, and chose the codewords appropriately. For example, commonly used letters like E and I have only short codewords: . and .. respectively, whereas seldom used letters like Q and X get longer codewords, --.- and -..-.

[14]Or more accurately, the entropy of a source generating messages in English. Shannon used excerpts from the novel *Jefferson the Virginian* as the source.

sequence will contain mostly small numbers (as often few guesses are required.)

Now, we have a second subject who receives only the sequence of guess counts $g_{1:n}$, and (assuming both subjects act the same), we ask her to play the same guessing game as the transmitter, except instead of knowing the correct message ahead of time, we let the second subject make g_i guesses before asking her to treat the last guess as correct, and then moving on to the next symbol, thereby recovering the original message, assuming both subjects use the same guessing strategy/order.

$$
\begin{array}{l}
x_{1:n} \quad \text{T H E R E - I S - N O - R E V E R S E - O N - A - M O T O R C Y C L E -} \\
g_{1:n} \quad \text{1 1 1 5 1 1 2 1 1 2 1 1 15 1 17 1 1 1 2 1 3 2 1 2 2 7 1 1 1 1 4 1 1 1 1 1}
\end{array} \qquad (2.5.5)
$$

In practice, both subjects can be replaced with a deterministic predictor encoded with the statistics of letter frequencies, digraphs, trigraphs, and so on to ensure they make the same guesses, and the message can be recovered. This was repeated many times, and an estimate of the average entropy of English was thereby obtained. Shannon estimates it somewhere in the range of 0.6 to 1.3 bits per character. More modern estimates with a larger corpus of data and replacing human predictors with algorithms [Gue09] provides a higher estimate of 1.58 bits per character. ●

Definition 2.5.6 (Code) Given a set of source words S, a *code* $C : S \rightarrow \mathbb{B}^*$ is a function mapping messages to binary strings. The set of all encoded messages $C(S)$ is called a set of *codewords*, and often we just describe a code by its codewords rather than specifying C. A code is *uniquely decodable* if C is injective, and *prefix* if $C(M)$ is a prefix-free set (see Section 2.1.2).

We usually desire prefix-codes, because they remain uniquely decodable if concatenated, and decoding can be done in a single pass with at most constant memory overhead.

2.5.2 Shannon-Fano Code

Often we wish to encode a message over an alphabet (for instance, the alphabet $\Sigma = \{a, b, ..., z\}$) into a binary string. We can define an encoding (also called a *code*) for each character in the alphabet from which messages are constructed, and then encode a string by encoding each character separately.[15] Suppose we have a discrete random variable $X = (\Sigma, P)$ with symbols $\Sigma = \{x_1, ..., x_n\}$ and probabilities $P = \{p_1, ..., p_n\}$ such that $P(X = x_i) = p_i$. The random variable X models the distribution of symbols we wish to encode. Without loss of generality, all $p_i > 0$ by removing any zero probability symbols[16] and the symbols/probabilities are indexed in decreasing order of likelihood, $p_1 \geq p_2 \geq ... \geq p_n$. For $i = 1, ..., n$, let $F_i = \sum_{k=1}^{i-1} p_i$ be the cumulative probabilities up to p_{i-1}. Note that $0 = F_1 < F_2 < ... < F_n = 1 - p_n$.

---

[15]This is a good starting point, but ignores higher-order statistics like digraphs and trigraphs. For instance, assuming we are encoding strings of English text, the digraph QU is far more likely than any other digraph starting with Q, so a more complex approach could be to encode digraphs or trigraphs instead.

[16]A symbol with zero probability will never appear in any sequence, so there is no need to assign it a codeword.

Definition 2.5.7 (Shannon-Fano code) Given Σ and P as defined above, the *Shannon-Fano* code $C : \Sigma \to \mathbb{B}^*$ is defined as:

1. Take the binary expansion $b(F_i) = 0.b_1b_2...$ of F_i

2. Truncate the result at l_i bits, where

$$l_i \equiv \ell(C(x_i)) := \left\lceil \log_2 \frac{1}{p_i} \right\rceil \tag{2.5.8}$$

The resulting string $C(x_i) := b_1b_2...b_{l_i}$ defines the code associated with x_i.

The symbols that are the most (resp. least) likely are assigned the shortest (resp. longest) codewords.

Theorem 2.5.9 (The Shannon-Fano code is a prefix code)

Proof. First, note that from (2.5.8) we can obtain

$$\log_2 \frac{1}{p_i} \leq l_i < 1 + \log_2 \frac{1}{p_i}$$

$$\text{hence} \quad 2^{-l_i} \leq p_i < 2^{-l_i+1}$$

Let x_i, x_j be two different symbols (i.e $i \neq j$). We wish to show that $C(x_i)$ is not a prefix of $C(x_j)$. Suppose $j > i$, then

$$F_j = \sum_{k=1}^{j-1} p_k = \sum_{k=i}^{j-1} p_k + \sum_{k=1}^{i-1} p_k \geq (j-i)p_i + F_i \geq p_i + F_i \geq 2^{-l_i} + F_i$$

and hence $F_j - F_i \geq 2^{-l_i}$. This implies that the codewords $C(x_j)$ and $C(x_i)$ must have a mismatched bit somewhere in the first l_i positions, since if they did not, all bits $b_1b_2...b_{l_i}$ would match, implying $|F_j - F_i| < 2^{-l_i}$, which is a contradiction.

Since it is impossible for $C(x_i)$ to agree with $C(x_j)$ over the first l_i bits for $j > i$, so $C(x_i)$ cannot be a prefix of $C(x_j)$. The case for $j < i$ is analogous and left as an exercise (Exercise 2.5.7.2). ∎

Theorem 2.5.10 (Shannon Coding) The expected Shannon-Fano code length (in bits per symbol) is bounded as

$$H(X) \leq \mathbf{E}[\ell(C(X))] \equiv \sum_i p_i l_i \leq H(X) + 1$$

with $H(X)$ being the entropy per symbol of the source.

Proof. Using (2.5.8), we obtain

$$H(X) = \sum_i p_i \log_2 \frac{1}{p_i} \leq \sum_i p_i l_i < \sum_i p_i (1 + \log_2 \frac{1}{p_i}) = 1 + \sum_i p_i \log_2 \frac{1}{p_i} = H(X) + 1$$

∎

Example 2.5.11 (Shannon-Fano code) Given the source words {1,2,3,4} with respective probabilities $\{\frac{6}{12},\frac{3}{12},\frac{2}{12},\frac{1}{12}\}$, we can compute the Shannon-Fano code as shown:

n	p_n	F_n	$b(F_n)$	$\log_2[1/p_n]$	$\ell(E(n))$	$E(n)$
1	6/12	0	.0	1	1	0
2	3/12	6/12	.1	2	2	10
3	2/12	9/12	.11	$\log_2 6 \doteq 2.6$	3	110
4	1/12	11/12	.11101	$\log_2 12 \doteq 3.6$	4	1110

◆

We will often use phrases such as "x has code length $-\log P(x)$" to mean that there exists a code $C(x)$, e.g. Shannon-Fano code w.r.t. P such that $\ell(C(x)) = \lceil -\log P(x) \rceil$.

2.5.3 Kullback–Leibler Divergence

A quantity closely related to Shannon entropy is the Kullback–Leibler (KL) divergence, which is a measure of the "distance" between two probability measures:

Definition 2.5.12 (KL divergence) The *Kullback–Leibler divergence* KL between discrete probability measure P and semi-probability Q, both over $\mathcal{X}$, is defined as

$$\mathrm{KL}(P||Q) := \sum_{x \in \mathcal{X}} P(x) \log_2 \frac{P(x)}{Q(x)}$$

where $p\log_2\frac{p}{0} = \infty$ for $p > 0$, but $0\log_2\frac{0}{q} := 0$ for $q > 0$, consistent with Section 2.8.3.

Like the entropy Definition 2.5.1, we define the KL divergence here using $\log_2$ results are measured in bits, which has a natural interpretation and is easier to calculate in examples. Later we define the KL divergence using ln instead, as the natural log is mathematically more convenient to work with and avoids the clutter of ln2 constants in various bounds (Section 2.8.1).

Motivation of the KL divergence. The KL divergence is at first a rather strange looking expression, and it may not be obvious why it would be a suitable choice of "distance" between probability distributions. Consider a discrete r.v. X with probability P. The length of the near-optimal Shannon-Fano code (Theorem 2.5.10) for $x \in \mathcal{X}$ w.r.t. P is, within one bit, $h_P(x) := -\log_2 P(x)$. Now suppose P is unknown, and Q is our best guess of the true probabilities P. The Shannon-Fano code (Definition 2.5.7) for X using the distribution Q is, within one bit, $h_Q(x) := -\log_2 Q(x)$. Now coding w.r.t. a wrong distribution incurs a penalty (regret) of requiring $h_Q(x) - h_P(x)$ more bits than optimal coding. The *expected regret*, called *redundancy*, is

$$\mathbf{E}_P[h_Q(x) - h_P(x)] = \sum_{x \in \mathcal{X}} P(x)[\log_2 P(x) - \log_2 Q(x)] \equiv \mathrm{KL}(P||Q)$$

We will shortly see that the KL divergence is non-negative (Corollary 2.5.16), i.e. $\mathbf{E}_P[h_Q(x)] \geq \mathbf{E}_P[h_P(x)] \equiv H(X)$, so coding with respect to the wrong distribution can only increase (on average) the code length. The KL divergence can thus be interpreted as the penalty of how many extra bits on average are required for Shannon-Fano or Huffman coding with respect to the wrong distribution.

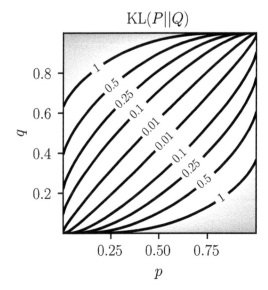

Figure 2.14: *The KL divergence KL(P∥Q) between two distributions P and Q representing biased coins with biases p and q respectively.*

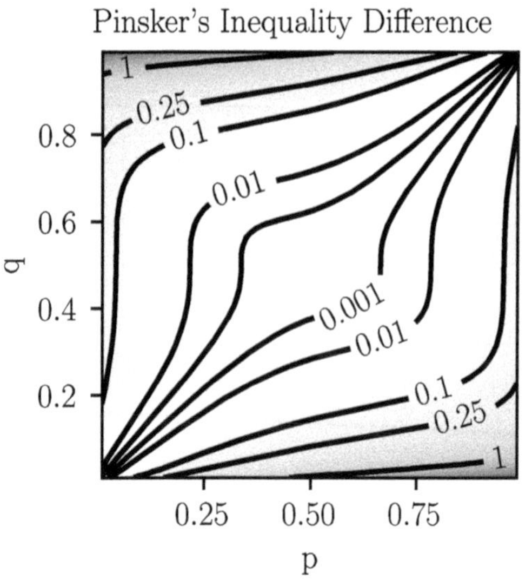

Figure 2.15: *A plot of the difference between both sides in the Pinsker inequality. Note that we have equality for p=q as both sides are zero.*

Example 2.5.13 (KL divergence between a biased and a fair coin) Consider biased coins, each with bias p and q respectively. As an abuse of notation, we write $P=(p,1-p)$ and $Q=(q,1-q)$ for the distributions for each coin. Then

$$\mathrm{KL}(P\|Q) = p\log_2\frac{p}{q}+(1-p)\log_2\frac{1-p}{1-q}$$

Plotting, we can see that KL is not a symmetric function (as mirroring over the line $p=q$ would result in a different picture.) Figure 2.13 shows a cross-section of Figure 2.14 along the line $p=\frac{1}{2}$. As can been seen, while similar in the neighborhood of $p=\frac{1}{2}$, clearly symmetry does not hold as $q\to0$ or $q\to1$. Also note that $\mathrm{KL}(P\|Q)$ is finite over the domain $q\in[0,1]$, whereas $\mathrm{KL}(Q\|P)$ diverges as $q\to0$ or $q\to1$. ◆

Asymmetry of the KL divergence. Suppose that P is the true distribution, and we are trying to fit an approximation Q to P by minimizing the KL divergence. Since the KL divergence is not symmetric, one could define two metrics: the *forward divergence* $\mathrm{KL}(P\|Q)$ or the *reverse divergence* $\mathrm{KL}(Q\|P)$.

Consider the forward divergence $\mathrm{KL}(P\|Q)=\sum_x P(x)\log\frac{P(x)}{Q(x)}$. The distribution Q only appears in the denominator of $\log\frac{P(x)}{Q(x)}$. For any points x that cannot be sampled from the true distribution $P(x)=0$, the choice of $Q(x)$ is irrelevant, as $\sum_x 0\log\frac{0}{Q(x)}=0$ by definition. So, Q can safely assign probability mass outside the support of P. However, if $P(x)>0$ and $Q(x)=0$, then $\log\frac{P(x)}{Q(x)}\to\infty$ and the (now aptly named) KL divergence diverges. This means that optimizing $\mathrm{KL}(P\|Q)$ will cause Q to try and "cover" the support of P.

Conversely, consider the reverse divergence: $\mathrm{KL}(Q\|P)=\sum_x Q(x)\log\frac{Q(x)}{P(x)}$. Any point x where $Q(x)=0$ makes no contribution to the total sum, but since Q is a probability distribution, the probability mass must be assigned somewhere. If $P(x)=0$, then Q is also forced to zero to prevent $\log\frac{Q(x)}{P(x)}$ diverging. This means that Q will avoid stretching over regions where $P(x)\approx0$.

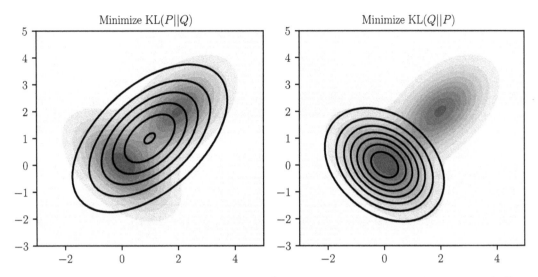

Figure 2.16: *Gaussian distribution $Q = \mathcal{N}(\boldsymbol{\mu}, \boldsymbol{\Sigma})$ minimizing $KL(P||Q)$ and $KL(Q||P)$ respectively, for a mixture of Gaussians P.*

Without constraints on Q, both KL divergences are minimized by $Q = P$, and the choice does not matter. But under model mis-specification, briefly discussed in Section 3.4, the minimizing Q will generally differ.

Example 2.5.14 (KL($P||Q$) vs. KL($Q||P$)) Consider a mixture of two two-dimensional Gaussians $P = \frac{1}{2}\mathcal{N}(\binom{0}{0}, (\begin{smallmatrix} 1 & -.5 \\ -.5 & 1 \end{smallmatrix})) + \frac{1}{2}\mathcal{N}(\binom{2}{2}, (\begin{smallmatrix} 1 & .5 \\ .5 & 1 \end{smallmatrix}))$ and restrict $Q = \mathcal{N}(\boldsymbol{\mu}, \boldsymbol{\Sigma})$ to a single two-dimensional Gaussian. Since $x \in \mathbb{R}^2$, we have to replace the sum in the KL definition by integrals. The minimizing parameters $\boldsymbol{\mu} \in \mathbb{R}^2$ and $\boldsymbol{\Sigma} \in \mathbb{R}^{2 \times 2}$ can be found numerically. Looking at the result in Figure 2.16, we can see how $\arg\min_Q KL(P||Q)$ tends to stretch over both distributions to cover the full support of P, whereas $\arg\min_Q KL(Q||P)$ focuses on one of the two distributions, entirely ignoring the other, and avoiding the region in between the two peaks of P where $P(x) \approx 0$.   ♦

Noticeably the KL divergence is not a metric, since it is neither symmetric nor does it satisfy the triangle inequality. It does however satisfy the property that $KL(P||Q) = 0$ iff $P = Q$, hence why we call it a divergence. This does not stop it from being invaluable in the coming chapters. We also note that KL is always non-negative, as a corollary of the following theorem:

Theorem 2.5.15 (Gibbs inequality) Given discrete probability $\{p_1, ..., p_N\}$ ($\sum_i p_i = 1$) and semi-probability $\{q_1, ..., q_N\}$ ($\sum_i q_i \leq 1$), we have

$$\sum_{i=1}^{N} p_i \log_2 p_i \geq \sum_{i=1}^{N} p_i \log_2 q_i$$

defining $p \log_2 0 = -\infty$ for $p > 0$ but $0 \log_2 0 := 0$ as before.[a] Equality holds iff $p_i = q_i$ for all i.

---

[a]These definitions are motivated by the limits $\lim_{q \to 0} p \log_2 q = -\infty$ for $p > 0$ and $\lim_{p \to 0} p \log_2 p = 0$.

Proof. If any $q_i = 0$ (with $p_i > 0$) then the bound trivially holds, so suppose $q_i > 0$. All terms in the sum with $p_i=0$ are zero and do not affect the outcome, so w.l.g. assume $p_i>0$. Then,

$$\sum_i p_i \ln\frac{q_i}{p_i} \le \sum_i p_i\left(\frac{q_i}{p_i}-1\right) = \sum_i [q_i-p_i] \le 1-1 = 0$$

Here, we used that $\ln x \le x-1$ for all $x>0$. Dividing by $\ln 2$ gives $\sum_i p_i \log_2[q_i/p_i] \le 0$, which can be rearranged to obtain Gibbs' inequality. Now, note that $\ln x = x-1$ only when $x=1$. So to achieve an equality in the above bound, we require that $q_i/p_i=1$, i.e. $p_i=q_i$.   ∎

Corollary 2.5.16 (KL divergence is non-negative) The KL divergence $\mathrm{KL}(P\|Q)$ is non-negative, and is zero iff $P=Q$.

Proof. Both properties were shown during the proof of Theorem 2.5.15.   ∎

2.5.4   The Kraft Inequality

Earlier we discussed prefix codes and gave some motivation as to why we might want to use them. One important property of prefix codes is the Kraft inequality.

Theorem 2.5.17 (Kraft inequality) Let $b_1,b_2,...$ be a finite or infinite sequence of binary strings. There exists a prefix code $\mathcal{C}=\{c_1,c_2,...\}$ such that $\ell(b_n)=\ell(c_n)$ for all n iff

$$\sum_n 2^{-\ell(b_n)} \le 1$$

Proof. (Only if) Recall from Remark 2.1.11 the standard one-to-one correspondence between a finite binary string x and the interval $I_x=[0.x,\ 0.x+2^{-\ell(x)})$ on the real line. For each c_n, the length of I_{c_n} is $2^{-\ell(b_n)}$ since they have matching lengths. Additionally, a prefix code corresponds to a set of disjoint such intervals in $[0,1)$. That is, $I_{c_n}\cap I_{c_m}=\emptyset$ for $n\neq m$. Because the sets are disjoint, the sum of the lengths is less than 1. All together, denoting the length of an interval in $\mathbb{R}$ as $\lambda([a,b)):=b-a$, we get

$$\sum_n 2^{-\ell(b_n)} = \sum_n \lambda([0.b_n,\ 0.b_n+2^{-\ell(b_n)})) = \lambda\left(\bigcup_n [0.b_n,0.b_n+2^{-\ell(b_n)})\right) \le \lambda([0,1)) = 1$$

which proves that the inequality holds for prefix codes.

(*If*) Suppose $b_1,b_2,...$ are given such that the inequality holds. We can also assume that the sequence $\ell(b_n)_{n=1}^\infty$ is non-decreasing. Choose disjoint adjacent half-open intervals $I_1,I_2,...$ of lengths $2^{-\ell(b_1)},2^{-\ell(b_2)},...$ from the left end of the interval $[0,1)$. In this way, for each $n\ge 1$, the right end of I_n is $\sum_{i=1}^n 2^{-\ell(b_i)}$. Note that the right end of I_n is the left end of I_{n+1}. Since the sequence of $\ell(b_i)$'s is non-decreasing, each interval I_n equals $[0.c_n,\ 0.c_n+2^{-\ell(c_n)})$ for some binary string c_n of length $\ell(c_n)=\ell(b_n)$. By construction, $\{c_1,c_2,...\}$ forms a prefix code.   ∎

Remark 2.5.18 (Kraft inequality as sum over trees) The Kraft inequality has a pleasing geometric interpretation: We will see in Section 4.3.1 that any prefix code can be represented as a binary tree, where each codeword is represented by a leaf node of the tree. We first assign probability mass 1 to the root node and then recursively for each non-leaf node, the probability of the current node is split half each between the two children nodes. Since the total probability is unchanged (or only decreases if some nodes only have one

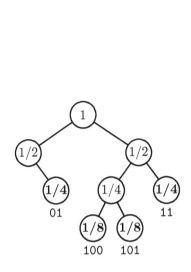

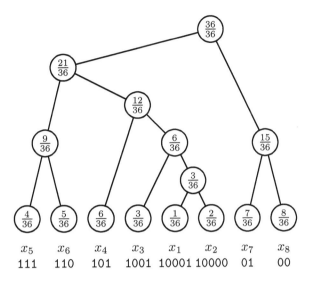

Figure 2.17: *The geometric intuition behind Kraft's inequality for the prefix code $\mathcal{C} = \{01,100,101,11\}$. Note that the sum of the probability mass assigned to each leaf (in bold) sums to $3/4 \leq 1$.*

Figure 2.18: *The Huffman tree for symbols $(x_1,...,x_8)$ with probabilities $p_i = \frac{i}{36}$. Fractions left unsimplified for clarity. The corresponding codewords recovered from the Huffman tree are $\{10001,10000,1001,101,111,110,01,00\}$.*

child) during this progress, the sum of the probabilities of all leaf nodes is bounded above by 1. Note that a leaf node at depth d will receive probability mass 2^{-d} after d such splittings. Hence, $\sum_n 2^{-\ell(c_n)} \leq 1$ for a prefix code $\mathcal{C} = \{c_1,...,c_n\}$. If each node has zero or two children (never just one, called a perfect tree), the prefix code is complete and the sum equals 1. See Figure 2.17 for an example.   ●

2.5.5   Shannon Coding Theorem

Definition 2.5.19 (Complete code) A prefix code $\mathcal{C}$ is *complete* if for any string $c \notin \mathcal{C}$, the code $\mathcal{C} \cup \{c\}$ is not prefix free.

Intuitively, a code $\mathcal{C}$ is complete if there is no extra room to squeeze in another codeword without violating the prefix free property. Consider the code $\mathcal{C} = \{01,100,101,11\}$ from Figure 2.17. This code is prefix free, but not complete, as we could add the additional codeword 00 and still obtain a prefix free code $\mathcal{C} \cup \{00\} = \{00,01,100,101,11\}$. This new code $\mathcal{C} \cup \{00\}$ is complete, as no new codewords can be added.

It can be shown that a prefix code is complete if and only if we have equality for the Kraft inequality (Theorem 2.5.17). However, a code that is complete may be suboptimal, as we could always choose to assign the longest codewords to the most likely outcomes. Completeness merely implies that we cannot add in another codeword without introducing ambiguity when decoding. However, there does exist a (per symbol) complete code that is optimal in the sense that it minimizes the average codeword length.

Definition 2.5.20 (Average codeword length and optimal codes) Given an alphabet $\Sigma = \{x_1,...,x_n\}$ with associated probabilities $P = \{p_1,...,p_n\}$, and a code $C\colon\Sigma\to\mathbb{B}^*$, we define the *average codeword length* of C as

$$L_{C,P} := \sum_i p_i \ell(C(x_i))$$

and the *minimal average codeword length* as

$$L_P^* := \min_C \{L_{C,P} \colon C\colon\Sigma\to\mathbb{B}^* \text{ is a prefix code}\}$$

An *optimal prefix code* C with respect to probabilities P is a prefix code such that $L_{C,P} = L_P^*$.

Remark 2.5.21 (On the optimality of optimal codes) Note that an optimal prefix code is the best way to compress a message assuming the only code allowed is one that assigns a separate codeword to each source word. More advanced compressors group source symbols together and assign a single codeword to each, or change the code dynamically for different parts of the message to be compressed (useful for when the data to be compressed is not homogeneous, as is often the case.) For instance, given the symbols $\Sigma=\{a,b,c\}$, with probabilities $\{0.9,0.1,0.1\}$, it can be shown that choosing C such that $\{(x,C(x)\colon x\in\Sigma)\}=\{(a,0),(b,10),(c,11)\}$ is an optimal code, though the string a^{1000} would take 1000 bits under this code to express, whereas the "encoding" `repeat 'a' 1000 times` takes far less than 1000 bits. ●

Note that L_P^* depends on P but only weakly on Σ: We could always apply a transformation $f\colon\Sigma\to\{1,...,n\}$ with $f(x_i)=i$ first, before applying the code C, and the receiver applies f^{-i} after decoding the codewords.

The following theorem gives bounds on the minimal average codeword length in terms of the entropy $H(P)$ of the distribution P.

Theorem 2.5.22 (Optimal coding [Sha48]) Let P be a discrete probability distribution. The average length of the optimal prefix code L_P^* satisfies

$$H(P) \leq L_P^* \leq H(P)+1$$

Proof. We first prove the upper bound $L_P^* \leq H(P)+1$: Let $l_x := \lceil \log_2 \frac{1}{P(x)} \rceil$ for $x \in \Sigma$. Therefore,

$$1 = \sum_x P(x) = \sum_x 2^{-\log_2 \frac{1}{P(x)}} \geq \sum_x 2^{-l_x}$$

By Kraft's inequality, since $\sum_x 2^{-l_x} \leq 1$, there exists a prefix code with matching codeword lengths $l_1, l_2,...$. Hence,

$$L_P^* \leq \sum_x P(x) l_x \leq \sum_x P(x)\left(\log_2 \frac{1}{P(x)}+1\right) = H(P)+1$$

We now prove the lower bound $H(P) \leq L_P^*$: Let $L_P^* = \sum_x P(x) l_x^*$ for some collection of optimal lengths l_x^*. Let $p_x := P(x)$ and $q_x := 2^{-l_x^*}$. Since l_x^* are the lengths of a prefix code, $\sum_x 2^{-l_x^*} \leq 1$ by Kraft's inequality (Theorem 2.5.17). We can therefore apply Gibbs' inequality

(Theorem 2.5.15)

$$H(P) \equiv \sum_x p_x \log_2 \frac{1}{p_x} \leq \sum_x p_x \log_2 \frac{1}{q_x} = \sum_x p_x l_x = L_P^*$$

∎

Example 2.5.23 (Non-optimality of Shannon-Fano codes) Note that Shannon-Fano codes are optimal within 1 bit, but they are not optimal in the strict sense of Definition 2.5.20. Consider the source words {1,2,3} with respective probabilities {0.8,0.1,0.1}. An optimal set of codewords can be shown to be {1,01,00}, but following the Shannon-Fano code (see the below table) gives {0,1100,1110}, which is clearly strictly worse.

n	p_n	F_n	$b(F_n)$	$\log_2[1/p_n]$	$\ell(C(n))$	$C(n)$
1	0.8	0	.0	$\doteq 0.32$	1	0
2	0.1	0.8	.1100	$\doteq 3.32$	4	1100
3	0.1	0.9	.11100	$\doteq 3.32$	4	1110

◆

Example 2.5.24 (Huffman Code) If optimality within 1 bit is not sufficient, there is another class of codes called *Huffman codes*. They are constructed by creating a binary tree (called a *Huffman tree*) with the symbols as leaf nodes (see Figure 2.18). The tree is constructed by first assigning a leaf node to each symbol with its associated probability. Then, a new node is created, and the two nodes with the lowest probability are added as children. The parent node is assigned the sum of probabilities of its children. This is repeated until all the leaf nodes are added to the tree, and the root node has probability 1. Given the correspondence between prefix codes and binary trees (see Remark 2.5.18 and Section 4.3.1) one can recover the Huffman code for each symbol by following the path from the root node to each leaf. Huffman codes can be shown to be an optimal prefix code [Huf52]. ◆

However, Huffman codes still code on a per-symbol basis, which ignores any structure in the order of the symbols being coded. Obviously it should be easier to code `aaaaabbbbcccccc` than `abaaccaccbbcabc`, but a Huffman code will produce the same length message for both. This deficiency leads to the arithmetic code.

2.5.6 Arithmetic Coding

Since the lengths of assigned codewords must be whole numbers, Shannon-Fano coding requires up to 1 bit in excess of the entropy of the source, and hence performs poorly on compressing sources of low entropy. Consider compressing a binary sequence $x_{1:n}$ drawn from a random variable $X = \text{Bern}(0.05)$:

`0000001000000000000000000000000000000000000000010000000000000001000...`

The entropy of this sequence is $H(X) = 0.05\log_2\frac{1}{0.05} + 0.95\log_2\frac{1}{0.95} \doteq 0.286$, whereas the best possible symbol code has to assign 1 bit per symbol, leading to over three times as many bits as the information contained in the sequence.

We can extend the idea of the Shannon-Fano code to encoding an entire sequence all at once, rather than encoding on a per-symbol basis, considering $\mathcal{X}^n$ as the (huge) base alphabet with (now) large entropy $H(X_{1:n}) = nH(X)$, but this has two down-sides. First, coding requires time exponential in n and is limited to fixed n. A compromise is to carve $x_{1:n}$ into chunks of reasonable and fixed length m.

Arithmetic coding is a much better "optimal" solution to this problem: It allows to encode sequences online (i.e. extend rather than modify the code when n grows), has optimal code length within 1 bit, runtime $O(n|\mathcal{X}|)$, and works for any joint distributions $\mu(x_{1:n})$ without any independence nor stationarity assumption.

Theorem 2.5.25 (Arithmetic coding) Sequence $x_{1:\infty} \in \mathcal{X}^\infty$ can be encoded in an online fashion, that is $x_{1:t}$ can be encoded as $c_{1:k(t)} \in \mathbb{B}^*$ growing with t, in $k(t) = \lceil \log_2 1/\mu(x_{1:t}) \rceil + 1$ bits. Assuming $\mu(x_t|x_{<t})$ can be evaluated sequentially in constant time per t, encoding and decoding time is on average $O(|\mathcal{X}|)$ per symbol and total space is $O(t|\mathcal{X}|)$.

Proof. (sketch for binary sequences) Let μ be any (semi)measure from which $x_{1:\infty} \in \mathbb{B}^\infty$ is sampled. To encode $x_{1:t}$, define the cumulative probability $F(x_{1:t}) = \sum_{x'_{1:t} \prec x_{1:t}} \mu(x'_{1:t})$, where $x'_{1:t} \prec x_{1:t}$ if $x'_{1:t}$ precedes $x_{1:t}$ in the standard lexicographical ordering. We associate with each sequence $x_{1:t}$ the interval $I_{x_{1:t}} := [F(x_{1:t}), F(x_{1:t}) + \mu(x_{1:t}))$. Note by construction that the intervals are disjoint. We want to choose as the codeword $c(x_{1:t})$ the binary expansion of the mid-point $F(x_{1:t}) + \frac{1}{2}\mu(x_{1:t})$ of the interval $I_{x_{1:t}}$, but the binary expansion may require infinitely many bits to specify. So, instead we take a finite approximation of $F(x_{1:t}) + \frac{1}{2}\mu(x_{1:t})$ by terminating the binary expansion early, but with sufficiently many bits to uniquely identify the interval that it lies in. This is achieved by taking the first $\lceil \log_2 1/\mu(x_{1:t}) \rceil + 1$ bits of the midpoint.

As t grows, the interval shrinks, and more bits are simply added to the code. F can be computed efficiently and incrementally via $F(x_{1:t}) = \sum_{k \leq t: x_k = 1} \mu(x_{1:k})$. There are some a number of technical difficulties glossed over, which we leave as a (hard) exercise to work out. The generalization to non-binary sequences is in theory rather straightforward. ∎

For Shannon-Fano coding, we require the probability distribution from which the source was sampled. In reality, the distribution from which a sequence was drawn is often unknown, and all that is available is the sequence itself to be compressed. One method is to read the entire sequence, estimate a distribution (usually by a frequency estimate) and then use that to compress the source. This method requires two passes of the data, which is often not possible if the data is to be compressed immediately after being received in an online fashion. Methods to learn a distribution online will be covered in Chapter 4. Since arithmetic coding can be applied to any distribution, it can also be applied to the distribution learned online, making it a universal online coding algorithm.

2.5.7   Exercises

1. [C12] Show that all prefix codes are uniquely decodable. Does the converse hold?

2. [C15] Complete the proof of Theorem 2.5.9 for the case where $j < i$.

3. [C19] Derive an extension of the Kraft inequality (Theorem 2.5.17) to base b instead of binary.

4. [C19] Derive an explicit for the KL divergence $\mathrm{KL}(p||q)$ for binary random variables $p = P(1) = 1 - P(0)$ and $q = Q(1) = 1 - Q(0)$. Plot $\mathrm{KL}(p||q)$ for $p, q \in [0,1]$. Prove that $\mathrm{KL}(p||q)$ and $\mathrm{KL}(P||Q)$ are convex.

5. [C15] Extend the KL divergence to continuous (semi)measures and prove that it satisfies the property that $\mathrm{KL}(P||Q) = 0$ iff $P = Q$ P-almost everywhere, and $\mathrm{KL}(P||Q) \geq 0$.

6. [C16] Prove that the KL divergence does not in general satisfy the triangle inequality $\mathrm{KL}(P,R) \leq \mathrm{KL}(P,Q) + \mathrm{KL}(Q,R)$ by means of a counterexample.

7. [C30] Work out the details for arithmetic (en/de)coding and implement it.

8. [C20] Generalize arithmetic coding to non-binary finite alphabet $\mathcal{X}$.

2.6 Computability Theory

Computability theory narrowly predates computers themselves, mainly introduced in Turing's 1936 paper [Tur36]. Despite the non-existence of what we now[17] call computers, the concept of *algorithms* had been known since antiquity, and it is that with which computability theory is primarily concerned. Given some task or problem to solve, we can say that there exists an algorithm to solve that problem if there exists a well-defined and unambiguous set of steps to follow that can solve that problem. An example is Euclid's algorithm to compute the greatest common divisor between two numbers. The algorithm is broken down into a sequence of steps such that each individual operation is simple enough to be obvious as to how to perform in a constant amount of time (usually additions and multiplications are treated in such a way).

To define an algorithm, one needs a reference class of allowed operations. For defining an algorithm for multiplication one would usually only assume "lower level" operations like addition, whereas for more complex algorithms like Gram-Schmidt one may take for granted many operations defined on vectors like computing the norm or inner products. Clearly, we need an agreed upon formal language in which algorithms can be written. For our purposes, Turing machines (Definition 2.6.2) provide such a formalization. We explore the capabilities of algorithms, and describe a hierarchy of computability, measuring the degree to which an algorithm can be written to solve a given problem.

For readers looking for a more detailed background in computability theory, see [HMU06].

Now, before we can proceed any further, we need a formal definition of what a problem is:

Definition 2.6.1 (Decision Problem and Language) Given an alphabet Σ (often $\mathbb{B}$), a *decision problem* or language L over Σ is a subset $L \subseteq \Sigma^*$.

This definition may seem strange and limited, but we can encode any predicate over Σ^* as a set of strings for which the predicate returns true. We can then talk about predicates over other sets (like natural numbers or other objects) by encoding them as (binary) strings in some canonical fashion. Even computing a function $f : \Sigma^* \to \Xi$ for finite Ξ can be reduced to decision problems, first by replacing Ξ with $\mathbb{B}^m$ for suitable m, and then setting $P_i = \{x \in \Sigma^* : f(x)_i = 1\}$. Solving the decision problems $(x \in P_1, ..., x \in P_m)$ is equivalent to computing f.

2.6.1 Models of Computation

Turing [Tur04] provides a formalization of his chosen model of computation, the *Turing machine*. The Turing machine is composed of two parts, a *finite control*: A function describing the operation the Turing machine will perform, as well as a finite collection of states the Turing machine can transition between, and a *tape*. The tape is a one-dimensional grid of cells extending infinitely far in both directions, and is discretized into cells, where

---

[17]Back then, *computer* was a job title for humans who worked out laborious calculations by hand.

each cell holds one of a finite number of tape symbols. The Turing machine has a *head* that can be used to read and write to the current cell of the tape, as well as move along the tape in discrete steps following the cells. Input (a string of symbols) is fed to the Turing machine by placing it on the tape, and filling all other cells on the tape with a reserved blank symbol B. The head is initially located on the left-most cell of the input. The behavior of the Turing machine is dictated by a *transition function* encoded within the finite control. On each time step $t = 1,2,...$, the Turing machine reads the contents of the cell pointed to by the read head, the state the finite control is in, and decides (using the transition function) on a new symbol to write back to the same cell, a new state to transition to, and a direction to move the tape head in.

Formally, we can define a Turing machine as follows:

Definition 2.6.2 (Turing Machine [HMU06]) A *Turing machine* (TM) is a 7-tuple $T = (Q,\Sigma,\Gamma,\delta,q_0,B,F)$ where

- $Q = \{q_0,q_1,...,q_n\}$ is a finite set of states.

- Σ is a finite set of input symbols, called the *alphabet*. (Unless otherwise specified, we usually assume $\Sigma = \mathbb{B}$.)

- Γ is a finite set of tape symbols such that $\Sigma \subseteq \Gamma$.

- δ is a partial function of type $\delta : Q \times \Gamma \to Q \times \Gamma \times \{L,R\}$, called the *(partial) transition function*. It takes as input the current state of the Turing machine, as well as the current symbol pointed at by the read head, and returns a new state, a new symbol to write to the current cell, and a direction (L or R) in which to move the tape head; one cell to the left (L) or right (R). The transition function may also be undefined for some inputs, in which case the Turing machine ceases computation, and is said to have *halted*.

- $q_0 \in Q$ is the *start state* that the Turing machine is initialized in on the first time step.

- $B \in \Gamma \backslash \Sigma$ is a reserved *blank symbol* that the tape is pre-initialized with (other than the input). The blank symbol is not permitted to be a member of the input alphabet.

- $F \subseteq Q$ is the set of *final* or *accepting* states.

Writing complex algorithms with a Turing machine is rather laborious, akin to writing machine code directly instead of using a high-level programming language, but it provides a theoretically useful fixed reference machine for executing algorithms. We can consider the action of a Turing machine in two ways: as a predicate on strings, or as a function from strings to strings.

Definition 2.6.3 (Language of a TM) Given a TM T, we say the *language* of T, denoted $L(T)$, is the set of all strings x such that if T were run with the tape initialized to x, the machine T would halt in an accepting state:

$$L(T) := \{x \in \Sigma^* : T(x) \text{ halts and accepts}\}$$

Definition 2.6.4 (Behavior of a TM) Given a TM T, we write $T(x) = y$ if T halts with y on the tape if the computation was started with x on the tape. If T does not halt given x as input, we write $T(x) = \bot := $ undefined.

The notion of partial recursive functions originates from logic. We (only) describe the gist below, but formally directly identify them with Turing-computable functions, since they are equivalent:

Definition 2.6.5 (Partial recursive functions) A function $f : \Sigma^* \to \Xi^* \cup \{\bot\}$ is called partial recursive, if it is computable by a Turing machine T (with $\Gamma \supseteq \Sigma \cup \Xi$), i.e. if $\exists T . f(x) = T(x) \ \forall x$.

Turing found that many well-known algorithms (considered as functions from strings to strings) can be computed by designing a Turing machine to perform the sequence of operations the algorithm describes. The claim was then made that any operation that can be performed methodically by an algorithm can also be performed by a Turing machine, and vice versa.

Thesis 2.6.6 (Church-Turing thesis)

Church: The class of algorithmically computable numerical functions (in the intuitive sense) coincides with the class of partial recursive functions.

Turing: Everything that can be reasonably said to be computable by a human using a fixed procedure can also be computed by a Turing machine.

This is a thesis rather than a theorem as "computable in the intuitive sense" and "computable by a human" are not formally defined. There have been a variety of other attempts at constructing formal systems to model computation.

General resursive functions. Gödel and Herbrand [Sie05, Göd31, Her32] started with a set of primitive functions: The constant function (of arbitrary arity), the unary successor function, and the projection function (returns one of the n input arguments). They then define a set of operators on functions: function composition, recursion (where a function can call another several times) and minimization (which takes a $n+1$-ary function f and n arguments $x_1,...,x_n$, and returns the smallest z such that $f(z,x_1,...,x_n)=0$ and $f(z',x_1,...,x_n)$ is defined for all $z' < z$. The closure of the primitive functions under the three operators was then called the *general* or *partial recursive functions*, and this was then used as a formal definition for "algorithmically computable".

Lambda calculus. Another formal system is lambda calculus (λ-calculus), introduced by Alonzo Church [Chu36], which at face value is much simpler than general recursive functions. Lambda calculus operates on *lambda expressions*, with a definition so simple it can be written in one line. Fixing some set $V = \{x,y,z,...\}$ of variables, we have that all lambda expressions e are of the form

$$e := v \,|\, \lambda v.e_1 \,|\, e_1 e_2$$

where e_1, e_2 represents arbitrary lambda expressions, and v represents a variable. So all expressions are either variables, or an expression *applied* to another ($e_1 e_2$), or an expression *abstracted* by a variable ($\lambda v.e_1$). The only operations that can be performed are

- α-conversion: renaming variables bound to a λ quantifier to avoid name collisions, and

- β-reduction: or substituting an abstracted variable for an expression.

While the rules are very minimal, Church provides a way to define natural numbers within lambda calculus, and operations upon the natural numbers (addition, multiplication, equality). The details can then be abstracted away, and higher-level concepts can then be constructed from these new concepts.

Turing completeness. So, we have (at least) three prospective definitions of "algorithmically computable": Turing's Turing machines, Church's lambda calculus, and Gödel/Herbrand's general recursive functions. Which should we choose as our canonical machine? As it turns out, all these definitions of "algorithmically computable" end up being equivalent, in that they all represent the same class of functions. We use the term *Turing complete* to describe a model of computation or formal system that is at least as expressive as a Turing machine. That is, any algorithm that can be implemented with a Turing machine, can also be implemented within the model considered.[18] Many other models of computation have been discovered, often a system designed for something else (rules for a game, for instance) were later found to be Turing complete. A fun example are cellular automata, an infinite collection of cells (usually as a 2-dimensional grid, sometimes 1d) where each cell can be in one of finitely many possible states, and a rule describing how each cell will transition depending on the states of nearby cells (usually the 4 directly adjacent cells, or the 8 cells that surround it). Some particular choices of rulesets (*Conway's Game of Life* [Gam70] or the 1d *Rule 110* [Wol83, Coo09]) demonstrate starkly complex behavior given the simplicity of the rule set, and in fact turn out to also be Turing complete models of computation.

Choice. Given that the choice of a particular model of computation does not matter, we stick with Turing machines by convention, though we will often appeal to the Church-Turing thesis and assert the existence of a Turing machine for an algorithm rather than explicitly constructing it. Since most programming languages are also Turing complete[19] we may make reference to those instead of Turing machines when it is convenient.

2.6.2   The Halting Problem

Reducing mathematics to logic. In 1900, the mathematician David Hilbert posed a list of what he considered the most pressing open problems of mathematics [HA99]. In zeroth-order logic (also known as propositional logic), statements are only constructed from Boolean variables and logical connectives (e.g. $\wedge, \vee, \neg$), and since each variable is Boolean-valued, given a statement with n variables, we can (in principle) brute-force all 2^n possible assignments to check if it is *valid*, i.e. the statement evaluates to true under all choices of variable assignment.

First-order logic extends zeroth-order logic by additionally allowing for the use of quantifiers ($\exists$ and $\forall$). If the domain over which predicates are defined were chosen to be an infinite set (like the natural numbers), there would be infinitely many cases to check if we were to naively verify that $\forall x.P(x) \to P(x)$ is valid by trying every possible value for $x \in \mathbb{N}_0$. Simple statements like this can easily be proven valid, but Hilbert asked if an algorithm exists that can verify the validity of every first-order logic statement.

Gödel incompleteness. Unfortunately, the answer to this turned out to be *no*, Gödel's incompleteness theorem [Göd31] showed that there exist limits to what can be formally proven. He proved that any sufficiently powerful formal system (with a certain level of arithmetic) that is *consistent* (cannot prove contradictions within the system) must contain statements that are true, but cannot be proven true within the system itself. Gödel's proof

---

[18]In practice, this is usually demonstrated by emulating a universal Turing machine within the formal system, or emulating a simpler model also known to be Turing complete.

[19]Though there are exceptions, some programming languages like *LOOP* [MR67] require specifying precisely the number of times a loop will execute before entering it, and can only represent the class of *primitive recursive functions*, strictly weaker than *general recursive functions*.

is premised on constructing the self-referential[20]

$$G := \text{``G cannot be formally proven''}$$

Gödel statement:

and encoding that statement as a number. Any system strong enough to construct arbitrary predicates over numbers can indirectly state predicates such as G about itself. Gödel showed that within the theory, G cannot be proven, but by using reasoning outside the formal system, G can be shown to be true.[21] This dealt a strong blow to mathematics, as many mathematicians (including Hilbert) believed up to that point that every true statement in mathematics has a corresponding proof demonstrating correctness, even if mathematicians had not yet found the proof.

The Halting problem. As a corollary of the incompleteness theorem, there must be problems that not even a Turing machine can hope to solve. The canonical example is that of the Halting problem. Often, when we write computer programs, sometimes those programs can run for a very long time, and sometimes it is not clear whether or not they will run forever. It would be very useful, and save a lot of resources and time if there were a way to check algorithmically if a program would run forever or eventually terminate. We could even use such an algorithm to prove many open problems in mathematics. For instance, the Goldbach conjecture[22] could be resolved by writing a program that searches for counterexamples, and halts when the first one is found. The Goldbach conjecture is then true if and only if such a program does not halt, which we could then check. The fact that if such an algorithm existed, it would render several open problems in mathematics trivial indicates such an algorithm would be difficult to construct. In fact, it is provably impossible.

First, assume some canonical encoding of Turing machines as binary strings, and the ability to encode tuples as a single element in such a way that the two elements can be recovered, for example by using a prefix code (Section 2.1.2). Then, we can talk about Turing machines taking other Turing machines as input, by feeding the binary encoding of one machine onto the input tape of another.

Theorem 2.6.7 (Halting problem) Let $L_H = \{(\langle M \rangle, w) : M(w) \text{ halts}\}$ denote the *halting language*, the set of all pairs of Turing machines M and strings w such that M run on w halts. Then there exists no Turing machine T that halts on all inputs, and $L(T) = L_H$.

Proof. Assume for a contradiction that such a T exists. Given T, we can construct a new machine P that performs the following operation:

$$P(\langle M \rangle) = \begin{cases} \bot & \text{if } T(\langle M \rangle, \langle M \rangle) \text{ accepts} \\ 1 & \text{if } T(\langle M \rangle, \langle M \rangle) \text{ rejects} \end{cases}$$

That is, if $T(\langle M \rangle, \langle M \rangle)$ rejects, then P halts and returns 1. If $T(\langle M \rangle, \langle M \rangle)$ accepts, then P gets stuck in an infinite loop. Given the description of P above (together with the hypothetical machine T), we could construct a Turing machine that acts as P does. Now, P is itself a Turing machine, so it has some binary encoding $\langle P \rangle$. What would happen if we tried to run $P(\langle P \rangle)$? Well, either $P(\langle P \rangle)$ must run forever, or it must halt.

---

[20]Self-referential statements are often a thorn in the side of mathematics. Russell's paradox, *does the set of all sets that do not contain themselves, contain itself?* demonstrated that naive set theory is inconsistent in this way.

[21]Indeed, assuming the sentence G can indeed be formally constructed and consistency, the proof is easy: If G were provable, then G says it is not provable; a contradiction, therefore G cannot be formally proven. But if G were false, it would say that G is provable, hence true; a contradiction. Therefore G must be true.

[22]Any even number $n \geq 4$ can be written as the sum of two primes.

- Suppose that $P(\langle P \rangle)$ halted. Then $T(\langle P \rangle, \langle P \rangle)$ must have rejected, and so $(\langle P \rangle, \langle P \rangle) \notin L_H$. So $P(\langle P \rangle)$ runs forever, a contradiction.

- Suppose that $P(\langle P \rangle)$ runs forever. Then $T(\langle P \rangle, \langle P \rangle)$ must have accepted, and so $(\langle P \rangle, \langle P \rangle) \in L_H$, which means $P(\langle P \rangle)$ halts, a contradiction.

In either case, we obtain a contradiction, hence no such T exists.   ∎

So, even with no constraints on memory or time allotted, not even Turing machines can hope to solve all problems. With this limitation in mind, we can classify sets and functions depending on whether they can be computed by a Turing machine or not.

Definition 2.6.8 (Recursive(ly enmerable)) We say that a language or set $\mathcal{S}$ is *recursive* if there exists a *total*[a] Turing machine T such that $L(T) = \mathcal{S}$. We say that $\mathcal{S}$ is *recursively enumerable*, or *RE*, if we drop the requirement that T is total.

Similarly, a (partial) function f is *recursive* (resp. *recursively enumerable*) if there exists a total TM (resp. partial TM) such that $f(x) = y$ if and only if $T(x) = y$.

---

[a]A Turing machine that halts on all inputs, as opposed to a *partial Turing machine*, or simply a *Turing machine*.

Note that if we describe $\mathcal{S}$ by an indicator function $f : \langle \text{Domain} \rangle \to \mathbb{B}$ with $f(x) = 1$ iff $x \in \mathcal{S}$, then $\mathcal{S}$ is recursively enumerable iff f is lower semicomputable. A set being recursive is often also called computable or *decidable* when we are taking about sets in the context of a decision problem.

2.6.3   (Semi-)Computable Functions

Up until this point we have considered Turing machines (and recursive functions) which map strings to strings. However, we will often be faced with functions which map to real numbers (such as in probability). We would like to say something about whether a real-valued function is recursive or not. In this section, we will describe how this can be done. We restrict ourselves to functions of type $f : \mathbb{B}^* \to \mathbb{R}$. This can be easily extended to $f : \mathbb{N}_0^n \to \mathbb{R}$ or $f : \mathbb{Q} \to \mathbb{R}$ by encoding an n-tuple $(x_1, \dots, x_n)$ as $x_1' \dots x_{n-1}' x_n \in \mathbb{B}^*$ or a fraction a/b as $a'b \in \mathbb{B}^*$ respectively. Surprisingly, defining computability for functions with real numbers as domain is more intricate [Grz57]. Since we mostly do not need those, we confine them to Remark 2.6.16.

Definition 2.6.9 (Finitely computable) f is *finitely computable* or *recursive*, if and only if there are Turing machines T_1, T_2 with outputs interpreted as natural numbers and $f(x) = T_1(x)/T_2(x)$.

As a result, finitely computable functions can only return rational numbers. Finitely computable functions include polynomials with rational coefficients, or progressively better estimates of $\sqrt{2}$ (e.g. $f(n)$ could return a rational approximation $\widehat{\sqrt{2}}$ such that $|\widehat{\sqrt{2}} - \sqrt{2}| \leq 1/n$).

Definition 2.6.10 (Estimable) f is *estimable* or *computable* if and only if there is a recursive function $\phi(x, k)$ such that $\forall k \in \mathbb{N}_0.|\phi(x, k) - f(x)| < 2^{-k}$.

In essence, a function is estimable if there exists an algorithm ϕ that, when provided with an input x and a tolerance $\varepsilon = 2^{-k}$, will return an estimate $\widehat{f(x)}$ that is within ε of $f(x)$. Estimable functions include $\sqrt{\cdot}$, $\sin(\cdot)$ or the constant function that returns π. Estimable functions are functions that can be approximated to any given and known precision ε with a guaranteed halting program, i.e. are finitely computable to ε-accuracy. Note that if the codomain of f is $\mathbb{N}_0$ or $\mathbb{Z}$, then *estimable* and *finitely computable* are equivalent, since we can choose $\varepsilon = \frac{1}{2}$, which means $\phi(x,1) = f(x)$ exactly.

Definition 2.6.11 (Upper semicomputable) f is *upper semicomputable* or (recursively) *co-enumerable* if and only if there exists a partial recursive function ϕ such that $\phi(x,k+1) \leq \phi(x,k)$ for all $k \in \mathbb{N}_0$, and $\lim_{k \to \infty} \phi(x,k) = f(x)$.

Upper semicomputable functions are functions f for which we can compute a sequence of monotonically decreasing upper bounds that converge to the true value of f, though we would not know how good the approximations is, as $\phi(\cdot,k)$ may decrease arbitrarily for yet larger k. Upper semicomputable functions include the Kolmogorov complexity K (see Section 2.7).

Definition 2.6.12 (Lower semicomputable) f is *lower semicomputable* or (recursively) *enumerable* if and only if there exists a partial recursive function ϕ that $\phi(\cdot,k+1) \geq \phi(\cdot,k)$ for all $k \in \mathbb{N}_0$, and $\lim_{k \to \infty} \phi(x,k) = f(x)$.

This is just the mirror of upper semicomputable, but now we can compute bounds from below. Unsurprisingly, a function f is lower semicomputable if and only if $-f$ is upper semicomputable. Examples include the Busy Beaver function, the maximum number of time steps an n-state Turing machine with the tape initialized to blanks can run for and halt [Rad62], and Solomonoff's universal distribution M (see Section 3.8).

Definition 2.6.13 (Approximable) f is *approximable* or *limit-computable* if and only if there is a recursive function ϕ such that $\lim_{k \to \infty} \phi(x,k) = f(x)$.

Approximable represents the weakest form of computability that we can still reasonably call "computable". Anything beyond approximable is truly incomputable. Unlike for estimable functions, we do not know how close any particular term $\phi(\cdot,k)$ in the sequence is to $f()$; it might intermittently drift away from or towards the true value, and we do not even know if the current value is above or below the true value. The only property we have is that the limit of the sequence is the true value. Note that if the range of f is $\mathbb{N}_0$ or $\mathbb{Z}$, then eventually $\phi(x,k) = f(x)$ exactly, i.e. $\forall x. \exists k_0. \forall k \geq k_0 : \phi(x,k) = f(x)$, but we may never know when ϕ has stabilized.

Theorem 2.6.14 lists the relations between the introduced computability concepts, and also their relation to the lowest members of the arithmetic hierarchy introduced below.

Theorem 2.6.14 (Implications among computable functions)

$$\text{finitely computable}$$

$$\Downarrow$$

$$\text{estimable} = \Delta_1^0$$

$$\swarrow \qquad\qquad\qquad \searrow$$

$$\Sigma_1^0 = \text{lower semicomputable} \qquad \text{upper semicomputable} = \Pi_1^0$$

$$\searrow \qquad\qquad\qquad \swarrow$$

$$\text{approximable} = \Delta_2^0,$$

Also, if a function is both upper *and* lower semicomputable, then it is estimable. No other implications are true in general.

Below are some examples of recursively enumerable sets we encounter in this book. Some concepts and sets are defined only later, but we collect them here for convenience.

Theorem 2.6.15 (Recursively enumerable sets) The (countably infinite) sets $\mathcal{S}$ of (encodings of)

(*i*) Turing machines,

(*ii*) computable partial functions,

(*iii*) lower semicomputable total functions with range $\mathbb{R}\cup\{\pm\infty\}$,

(*iv*) lower semicomputable semi-probability mass functions over $\mathbb{N}$ (or $\mathcal{X}^*$),

(*v*) lower semicomputable (joint or conditional) (sequence or chronological), semimeasures over $\mathcal{X}^\infty$ (Definitions 3.7.1 and 7.2.1 and Section 10.5)

are all recursively enumerable (Definition 2.6.8).

Proof sketch. (*i*) An effective enumeration of Turing machines is rather easy to construct and has already been outlined earlier.

(*ii*) With suitable interpretation, Turing machines describe all partial recursive functions.

(*iii*) For the lower semicomputable functions, let $\varphi_p(x,t)$ be the output of $U(p'x)$ terminated after t steps. Interpret the output as a rational number if possible, otherwise as $-\infty$. Then $\phi(x,t):=\max_{s\leq t}\varphi_p(x,s)$ is computable and monotonically increasing and converges to $f_p(x):=\sup_t\varphi_p(x,t)$. It is now easy to see that $\{\langle f_p\rangle:p\in\mathbb{B}^*\}$ is a recursively enumerable set of all and only lower semicomputable functions. Note that $\langle f_p\rangle$ is a finite *description* of f_p. The $\sup_t$ is not calculated but described.

(*iv*) Semi-probabilities are functions that satisfy further properties. We modify the enumeration of lower semicomputable functions as follows. First we set $\varphi_p(x,t)=0$ for $x>t$, which leaves f_p unaffected. Now, if $\sum_{x\leq t}\varphi_p(x,t)>1$ (which is computable), we set $\varphi_p(x,t)=\varphi_p(x,t-1)$. This bounds f_p to be a semi-probability, but does not affect the f_p that are already semi-probabilities, hence all and only the semi-probabilities are enumerated.

(*v*) Similar/further modifications restrict the enumeration to all lower semicomputable (joint or conditional) (sequence or chronological) semimeasures [Hut05b, Sec.5.10]. ∎

On the other hand, the sets of computable total functions, computable probability mass functions, and computable probability measures are *not* recursively enumerable. The proofs are based on diagonalization arguments similar to the incomputability of the halting sequence, and are left as Exercise 4.

Remark 2.6.16 (Computability of functions with real domain) Defining computability of functions with continuous domain is subtle. Most algorithms take a finite input, so can be modelled by functions $f : \mathbb{B}^* \to \mathbb{B}^*$. So an algorithm computing, say, the exponential function actually takes a (binary) fraction as input and outputs an approximation as a (binary) fraction. A definition of real computable function $f : \mathbb{R} \to \mathbb{R}$ consistent with this input-approximation necessity is possible [Grz57], but renders all discontinuous functions incomputable. The computable real functions we encounter on occasion, are all benign, so the above intuitive understanding suffices. $\bullet$

2.6.4 Arithmetic Hierarchy

While anything beyond approximable is truly incomputable, one can still define *degrees of incomputability* using quantifiers (or Halting oracles, a hypothetical device that can solve the halting problem), which will turn out to be useful in classifying some of the agents we will introduce. Within the *arithmetic hierarchy*, each tier symbolizes a distinct degree of challenge in resolving computational conundrums. At the foundation, we discover issues readily addressed by elementary algorithms, such as determining the parity of a number. These problems, deemed computable, reside within the most fundamental stratum of the hierarchy. Progressing upward through these structured tiers, the problems require an increasing number of layers of quantifiers (or Halting oracles), such as "for all" and "there exists":

Definition 2.6.17 (Arithmetic hierarchy $\Sigma_n^0, \Pi_n^0, \Delta_n^0$) A set $A \subseteq \mathbb{N}_0$ is Σ_n^0 if and only if there is a computable relation η such that

$$k_0 \in A \quad \Longleftrightarrow \quad \exists k_1. \forall k_2 .... Q_n k_n . \eta(k_0, k_1, k_2, ..., k_n)$$

where $Q_n = \forall$ if n is even and $Q_n = \exists$ is n is odd. Similarly A is Π_n^0 if the outer-most quantifier is $\forall k_1$ instead. Equivalently and furthermore

$$A \subseteq \mathbb{N}_0 \text{ is } \Pi_n^0 \quad \Longleftrightarrow \quad \mathbb{N}_0 \backslash A \text{ is } \Sigma_n^0$$
$$A \subseteq \mathbb{N}_0 \text{ is } \Delta_n^0 \quad \Longleftrightarrow \quad A \text{ both } \Sigma_n^0 \text{ and } \Pi_n^0$$

We can think of Σ_0^n and Π_n^0 as an increasing hierarchy of complexity of sets, where n describes the number of alternating quantifiers required to describe the set A in first-order logic, together with some computable relation η. Instead of $A \subseteq \mathbb{N}_0$, we can allow $A \subseteq \mathbb{B}^*$, and hence effective binary encoding of objects including tuples, and indeed any recursive set (membership is decidable, indicator function is computable). The 0 in Σ^0 indicates that there are even higher hierarchies. For instance the *analytic hierarchy* Σ_n^1 denotes higher-order arithmetic with its lowest member Σ_0^1 containing in a certain sense the whole arithmetic hierarchy $\cup_n \Sigma_n^0$. The arithmetic hierarchy is the set of strict inclusions

$$... \Delta_n^0 \begin{array}{c} \subset \Sigma_n^0 \\ \subset \Pi_n^0 \end{array} \Delta_{n+1}^0 \begin{array}{c} \subset \Sigma_{n+1}^0 \\ \subset \Pi_{n+1}^0 \end{array} \Delta_{n+2}^0 ...$$

The arithmetic hierarchy can also be extended to functions. We only state it for countable domain, but this can be extended to functions with real domain (see Remark 2.6.16).

Definition 2.6.18 ($\Sigma/\Pi/\Delta$-functions [Lei16b]) A function $f:\mathbb{B}^*\to\mathbb{R}$ is called Σ_n^0 (or Π_n^0 or Δ_n^0) if and only if the set $I_f:=\{(x,q)\in\mathbb{B}^*\times\mathbb{Q}:q<f(x)\}$ is Σ_n^0 (or Π_n^0, or Δ_n^0 respectively). That is, given an enumeration $g:\mathbb{B}^*\times\mathbb{Q}\to\mathbb{N}_0$, the image $g(I_f)\in\Sigma_n^0$ (or Π_n^0, or Δ_n^0 respectively).

Note that for $f\in\Delta_n^0$, we have $\{(x,y):f(x)>g(y)\}\in\Sigma_n^0$, but $\{(x,y):f(x)\le g(y)\}\in\Pi_n^0$ (the $=$ causes the difference).

Theorem 2.6.14 relates the lowest members of the arithmetic hierarchy to the computability concepts introduced earlier: Estimable functions are Δ_1^0, lower semicomputable functions are Σ_1^0, upper semicomputable function are Π_1^0, and approximable functions are Δ_2^0.

Functions are composed of functions and functionals of functions. The following properties are useful and often lets you read off (at least an upper bound on) the degree of incomputability directly from the definition of the function.

Theorem 2.6.19 (Properties of $\Sigma/\Pi/\Delta$-functions)

(i) $\Sigma_n^0=-\Pi_n^0$ ($f\in\Sigma_n^0 \Leftrightarrow -f\in\Pi_n^0$)

(ii) $\Delta_m^0(\Delta_n^0)=\Delta_{n+m-1}^0$ (function concatenation), esp. $f(\Delta_n^0)\subseteq\Delta_n^0$ for $f\in\Delta_1^0$

(iii) $\Delta_m^0(\Sigma_n^0)=\Delta_{n+m}^0$ and $\Delta_m^0(\Pi_n^0)=\Delta_{n+m}^0$

(iv) $\sup\Delta_n^0=\sup\Sigma_n^0=\Sigma_n^0$, $\inf\Delta_n^0=\inf\Pi_n^0=\Pi_n^0$, $\sup\Pi_n^0=\Sigma_{n+1}^0$, $\inf\Sigma_n^0=\Pi_{n+1}^0$

(v) $\limsup_t=\inf_s\sup_{t\ge s}$ and $\liminf_t=\sup_s\inf_{t\ge s}$, hence e.g. $\lim\Delta_n^0=\Delta_{n+1}^0$

(vi) $f(\Sigma_n^0)\subseteq\Sigma_n^0$ for monotone increasing $f\in\Sigma_1^0$

(vii) $f(\Pi_n^0)\subseteq\Pi_n^0$ for monotone increasing $f\in\Pi_1^0$

(viii) $+,\times,\exp,\log,...\in\Delta_1^0$ and finite $\max,\min,\sum\in\Delta_1^0$, all monotone increasing

(ix) $\sum^\infty=\lim_t\sum^t$ in general, hence $\sum^\infty\Delta_n^0=\Delta_{n+1}^0$ in general

(x) $\sum^\infty=\sup_t\sum^t$ if applied to non-negative functions, hence $\sum^\infty\Sigma_n^{0+}=\Sigma_n^{0+}$.

Proof sketch. (i) follows nearly directly from Definition 2.6.18, since $q<f$ iff $q\not>f$, apart from $q=f$ which requires special treatment.

(iv) $f(x,t)\in\Sigma_n^0$ iff $\{(x,t,q):q<f(x,t)\}\in\Sigma_n^0$ implies $\{(x,q):q<\inf_t f(x,t)\}\equiv\{(x,q):\forall t:q<f(x,t)\}\in\Pi_{n+1}^0$ iff $\inf_t f(x,t)\in\Pi_{n+1}^0$ due to the extra $\forall$-quantifier, hence $\inf\Sigma_n^0\subseteq\Pi_{n+1}^0$. Equality holds, since every Π_{n+1}^0-function can be represented in this way. This follows from a suitably lifted version of Posts' Theorem [Odi89, Thm.IV.1.14]. Note that $f(x,t)=\sup_s g(x,t,s)$ for some $g\in\Delta_n^0$ (possibly even Π_{n-1}^0), hence $\sup_t f(x,t)=\sup_{t,s} g(x,t,s)\in\Sigma_n^0$, hence $\sup\Sigma_n^0=\Sigma_n^0$. The duals $\sup\Pi_n^0=\Sigma_{n+1}^0$ and $\inf\Pi_n^0=\Pi_n^0$ are proven in the same way. Together these also imply $\Sigma_n^0=\sup\Pi_{n-1}^0\subseteq\sup\Delta_n^0\subseteq\sup\Sigma_n^0=\Sigma_n^0$, hence $\sup\Delta_n^0=\Sigma_n^0$, and similarly for the dual $\inf\Delta_n^0=\Pi_n^0$.

(ii) First, a computable transformation of Δ_n^0 does not change the degree of incomputability, hence $f(\Delta_n^0)=\Delta_n^0$ for $f\in\Delta_1^0$. Now, repeated application of (iv) gives sup inf sup...$\Delta_1^0=\Sigma_n^0$ for n alternating inf/sup. For $f\in\Delta_m^0$ take the representation that ends in ...$\sup f_0(x,...)$ with $f_0\in\Delta_1^0$. For $g\in\Delta_n^0$ we have $h(x):=f_0(g(x),...)\in\Delta_n^0$. Consider the alternating representation for $h\in\Delta_n^0$ that starts with sup. Then in this representation of $f(g())$, the two sup can be merged, leaving $n+m-1$ alternating quantifiers. Do the same but with f/h ending/starting with inf. Together this shows $f(g())\in\Sigma_{n+m-1}^0\cap\Pi_{n+m-1}^0$. Equality holds since any sequence of $n+m-1$ alternating quantifiers can be broken up in n and m.

(iii) $\Sigma_n^0\subset\Delta_{n+1}^0$ implies $\Delta_m^0(\Sigma_n^0)\subseteq\Delta_m^0(\Delta_{n+1}^0)=\Delta_{n+m}^0$ from (ii). Equality follows from a suitably lifted version of Posts' Theorem [Odi89, Thm.IV.1.14].

(v) From (iv) we have $\mathrm{infsup}\Delta_n^0 = \Pi_{n+1}$ and $\mathrm{supinf}\Delta_n^0 = \Sigma_{n+1}$. Taking the intersection we get $\lim\Delta_n^0 = \Delta_{n+1}^0$.

$(vi-x)$ follow easily from $(i-v)$ and are left as Exercise 5. ∎

For instance, in analogy to the alternating quantifier definition of sets, a convenient characterization of Σ_n^0-functions is in terms of alternating sup−inf of some computable f_1, optionally interspersed with continuous computable functions $f_2,...,f_n$ (Exercise 6).

$$f(x) := \sup_{k_n} f_n\left(\inf_{k_{n-1}} f_{n-1}\left(...f_2\left(\operatorname*{\overset{\sup}{\inf}}_{k_1} f_1(x,k_1,...,k_n))\right)...\right)\right) \in \Sigma_n^0 \quad \text{if} \quad f_i \in \Delta_1^0 \qquad (2.6.20)$$

and similarly $f \in \Pi_n^0$ if starting with $\inf_{k_n}$.

For boolean functions with 1 identified as True and 0 identified as False, and boolean relations and sets represented by such indicator functions, $\sup_k f(k) = \exists k.f(k)$ and $\inf_k f(k) = \forall k.f(k)$. Replacing $\sup \rightsquigarrow \exists$ and $\inf \rightsquigarrow \forall$ and $f_1 \rightsquigarrow \eta$ and the other f_i by Identity, reduces (2.6.20) to Definition 2.6.17. Also, any max becomes $\cup$ and min becomes $\cap$.

2.6.5 Exercises

1. [C23] Derive a constructive proof (without contradiction) for Theorem 2.6.7.

2. [C12] Show that a function is estimable if and only if it is upper- and lower semi-computable.

3. [C30] Prove the implications and equalities in Theorem 2.6.14, and also prove that the converse of any of these implications is false. Finally prove that lower semicomputable does not imply upper semicomputable, and vice versa.

4. [C25] Show that the sets of computable total functions, computable probability mass functions, and computable probability measures are not recursively enumerable.

5. [C30] Prove the assertions in Theorem 2.6.19. For instance, $\liminf\Sigma_n^0 = \Sigma_{n+1}^0$ means that $g(x) := \liminf_{t\to\infty} f(x,t) \in \Sigma_{n+1}^0$ if $f(x,t) \in \Sigma_n^0$, and similarly for $\limsup$. Show that together this implies $\lim\Delta_n^0 = \Delta_{n+1}^0$, i.e. $g(x) := \lim_{t\to\infty} f(x,t) \in \Delta_{n+1}^0$ if $f \in \Delta_n^0$.

6. [C10] Use Theorem 2.6.19 to prove (2.6.20).

7. [C13] Prove that (i) if $f(z_1,z_2,...,z_n)$ is computable (estimable) and $g_i(x_i)$ are Δ_n^0, then $f(g_1(x_1),g_2(x_2),...,g_n(x_n))$ is Δ_n^0. (ii) If $f(z_1,z_2,...,z_n)$ is lower semicomputable (Σ_1^0) and monotone increasing in all arguments, and $g_i(x_i)$ are Σ_n^0, then $f(g_1(x_1),g_2(x_2),...,g_n(x_n))$ is Σ_n^0.

2.7 Kolmogorov Complexity

In Section 2.5 we introduced Shannon entropy, which measures the expected information of a stochastic source, that is, the expected number of bits required to encode a string sampled from a distribution P.

A different measure of information that measures the intrinsic information contained in a message itself (rather than the degree of unpredictability of a stochastic source) is the *Kolmogorov complexity*. This new definition of information is for a particular message rather than random variables, defined in terms of the length of the shortest description of the message. Intuitively, the Kolmogorov complexity represents the length of the shortest possible compressed form of the message. It can be used as a formal measure of simplicity/complexity of (any object that can be encoded as) a string using Turing machines (Definition 2.6.2) as the reference language, as was done for algorithms (Section 2.6). We explore the properties of Kolmogorov complexity, many of which have suitable intuitive meanings, e.g. more information can never hurt compressing data. We show that "most" strings are highly complex with respect to this measure ("most" sequences of random coin flips look "random"). For most properties of Kolmogorov complexity, there is a corresponding Shannon entropy concept (Definition 2.5.1). As we will see, Kolmogorov complexity is not finitely computable in that there provably cannot exist an algorithm for it. While this means that there is no hope to computing the exact Kolmogorov complexity of an object in practice, this still gives a useful theoretical underpinning for simplicity on which universal predictors can be built. Given sufficient time, the Kolmogorov complexity can be approximated arbitrarily well from above (Theorem 2.7.28), and any practical data compressor gives an upper bound to the true Kolmogorov complexity of an object.

2.7.1   Motivation

We have mentioned the concept of simplicity in Section II, and why we want to favor the simplest models or explanations consistent with the data observed. To be able to actually use this concept, we need to rigorous define simplicity/complexity. For example, consider the following binary strings:

$$x = 1010101010101010101010101010101010101010$$
$$y = 1100100100001111110110101010001000100001$$
$$z = 1011101100110101111110001110101101011000$$

The first string, x, has an obvious simple pattern, as it is just the pair 10 repeated twenty times. The second string, y, at first glance looks "random", but is actually the first 40 bits in the binary expansion of π [SI96]. The third string, z, was generated by measuring quantum fluctuations of a vacuum [Lam], a method that as far as we can tell, generates "truly random" bits in the sense that given all the bits in the past so far, we cannot predict the next bit with $>50\%$ accuracy.

What is it about the first string that makes it simple? One could argue it is the ease with which the next bit can be predicted as after only observing a few bits, we can quickly spot the obvious pattern and predict the rest of the sequence perfectly. A definition of complexity of sequences via difficulty of prediction was explored by Legg and Hutter [LH07c]. However, the second string certainly appears to have no clear patterns, and it will pass most statistical tests for randomness [Mar05]. The circle constant π appears to have properties of a "random" sequence despite being deterministic. If we knew ahead of time that the string y was just the binary expansion of π, then predicting the next bit can be done with zero prediction errors by just computing π to the required precision, and then using the bits received to tell how deep into the expansion of π we are, and what bit we should guess next.

So it is possible to predict both x and y perfectly, yet intuitively we would think that x is still simpler than y. The difference lies in the length of the description. We could define x as "repeat 10 twenty times" and y as "the binary expansion of $4\sum_{i=0}^{\infty}(-1)^i/(2i+1)$ to 40 bits". Had we taken the continuations of x and y to a few thousand bits, the description of those sequences would be vastly smaller than the sequence itself. In other words, these sequences are simple because they are *compressible*. This is (probably[23]) not true for string z, for which the shortest description is 1011101100110101111110001110101101011000, the string

---

[23]There is a small probability that a randomly sampled string exhibits a regular pattern, and z may

itself. Now, at this stage our definition of simplicity is still informal, as we do not have a formal definition of what a "description" is. To someone already well-versed in mathematics, we could have described $y = \pi$ [1:40] which is about the same length as any reasonable description we could find for x, perhaps $x =$ "10" $\times 20$. To someone missing mathematics as background knowledge, a description of y may take several pages, defining how addition and multiplication work, what the Σ symbol means, and what it means to assign a value to an infinite series. There are many other ways to describe π:

- in terms of an integral: $\pi = 4\int_0^1 \sqrt{1-x^2}\,dx$

- as some limiting geometric process: inscribe regular n-sided polygons in a circle of radius 1 and calculate their area as $n \to \infty$.

- a description that uniquely identifies π (the ratio of the circumference of a circle to its radius) but omitting the details of how to compute it entirely.

It is also not clear what symbols are permitted. Are we allowed to assume the decoder understands calculus when we define π as an integral? Or would we have to define that too? What if the description was much shorter if written in French or Russian? The simplicity of a concept depends on who we are talking to, so we need a fixed universally agreed-upon interpreter that takes "descriptions" and returns the sequence described. Such a machine should be unbiased towards any particular class of sequences, to avoid the problems above.

Note in the above, we are looking for a definition of "simplicity" or "information content", and our existing measure of information, entropy (Definition 2.5.1), is unhelpful here. Any deterministic process always provides zero bits of entropy; a machine that prints the digits of π would be such an example. We are concerned instead with a definition that relates to the intrinsic information contained in an individual object, the (minimal) quantity of information required to reconstruct the object.

2.7.2 Making Simplicity Rigorous

Turing machines (Definition 2.6.2) provide an ideal model for a description interpreter, as assuming the Church-Turing thesis (Thesis 2.6.6) for any algorithm that generates the sequence, there exists a Turing machine that can implement that algorithm.

For technical reasons we define the following variants of a Turing machine.

Definition 2.7.1 (Prefix/Monotone Turing Machine [Hut05b]) A prefix/monotone Turing machine is defined as a Turing machine with one unidirectional input tape, one unidirectional output tape, and some bidirectional work tapes. Input tapes are read only, output tapes are write only, and unidirectional tapes are those where the head can only move from left to right. All tapes are binary (no blank symbol), work tapes are initialized with zeros.

Prefix TM. We say T halts on input p with output x, and write $T(p) = x$ if p is to the left of the input head and x is to the left of the output head after T halts. The set of p on which T halts forms a prefix code. We call such codes p *self-delimiting* programs.

Monotone TM. We say T outputs/computes a string starting with x (or an infinite sequence ω) on input p, and write $T(p) = x*$ (or $T(p) = \omega$) if p is to the left of the input head when the last bit of x is output (T reads all of p but no more, and for every N

---

be such a string with a non-obvious pattern. It could even be that quantum randomness is actually only pseudo-random or Martin-Löf random, and all quantum random strings exhibit a hard-to-impossible to detect pattern. We will show later that in a formal sense "most" strings are like z in that they are incompressible.

> there exists time step t_0 such that for all time steps $t \geq t_0$, the output tape is prefixed with $\omega_{1:N}$). T may continue operation and need not halt (must not halt for sequences). For a given x, the set of such programs p forms a prefix code. We call such codes p *minimal* programs.

Any Turing machine T can be expressed as a single binary string $\langle T \rangle$ in some canonical way, in such a way that all the defining properties of the machine T can be recovered from $\langle T \rangle$. By enumerating all binary strings and checking if they represent a valid encoding of a Turing machine, we can construct an effective enumeration $T_1, T_2, \dots$ of all Turing machines. This, together with the fact that running a Turing machine is (by definition) a computable operation means there exists a *universal Turing machine* U that takes as input an index i in this enumeration (along with some input q), and then simulates the action of T_i on q.

Definition 2.7.2 (Universal Turing Machine (UTM)) There exists a universal prefix/monotone Turing machine (UTM) U which takes input $y'i'q$ (recall x' is a second-order prefix code, see Section 2.1.2), and simulates the action of prefix/monotone Turing machine T_i with side information y and input q, that is:

$$U(y'i'q) = T_i(y'q)$$

If there is no-side information present ($y' = \epsilon' = 0$), then we omit the 0 and simply write $U(i'q) = T_i(q)$ instead of $U(0i'q) = T_i(0q)$.

There are different ways of defining UTMs, but this way guarantees automatically that U leads to optimal codes in the sense of satisfying the Invariance Theorem Theorem 2.7.6.

With this in mind, we can now define our measure of simplicity with respect to a choice of Turing machine T.

Definition 2.7.3 ((Prefix) Kolmogorov complexity) The *(conditional) Kolmogorov complexity* $K_T(x)$ of a finite string x with respect to prefix Turing machine T is the length of the shortest program p such that T outputs x when input p (given y as side information)

$$K_T(x) := \min_{p \in \mathbb{B}^*} \{ \ell(p) : T(p) = x \}$$

$$K_T(x|y) := \min_{p \in \mathbb{B}^*} \{ \ell(p) : T(y'p) = x \}$$

If no such program exists, we define $K_T(x) := \infty$.

Rarely we may also make use of a variant of Kolmogorov complexity, based on monotone rather than prefix Turing machines.

Definition 2.7.4 (Monotone Kolmogorov complexity) The *monotone Kolmogorov complexity* $Km_T(x)$ of a finite string x (resp. infinite sequence ω) with respect to monotone Turing machine T is the length of the shortest program p such that $U(p)=x*$ (resp. $T(p)=\omega$).

$$Km_T(x) := \min_{p \in \mathbb{B}^*}\{\ell(p):T(p)=x*\}$$

$$Km_T(\omega) := \min_{p \in \mathbb{B}^*}\{\ell(p):T(p)=\omega\}$$

If no such program exists for x (resp. ω), we define $Km_T(x):=\infty$ (resp. $Km_T(\omega):=\infty$).

These definitions of the Kolmogorov complexity are parameterized by the choice of reference Turing machine T, and hence can be skewed by a choice of (universal) Turing machine (much as the simplicity of π is subjective, depending on who you are talking to).

Example 2.7.5 (Biased universal Turing machine) We can always construct a universal Turing machine U' that is arbitrarily biased towards a given object x. For example, given any UTM U, define

$$U_{\text{Book}}(0\alpha) := U(\alpha)$$

$$U_{\text{Book}}(1\alpha) := \text{The \LaTeX\ source of this book}$$

Then according to U_{Book}, the complexity of this entire book is only 1 bit[24], $K_{U_{\text{Book}}}$(The \LaTeX\ source of this book)$=1$. Of course, the universal machine U_{Book} would need to have a copy of this book[25] "hardcoded" inside its finite control table to achieve this, so this is a very contrived choice of UTM that we would wish to avoid.   ◆

To this end, we would wish to select a "natural" universal Turing machine that has no particular biases towards some strings over others. Some work has been done in identifying a canonical choice of UTM [Mül10], but no conclusive answer has been found. One option may be to *learn* a good UTM [SH14b].

To some degree, the actual choice of UTM does not matter, as for any two UTMs U_1 and U_2, the difference between the corresponding Kolmogorov complexities is bounded. Any two UTMs can emulate each other, as any program p_1 for U_1 can be bundled together with an U_1 interpreter q designed to run on U_2, and vice versa.

Theorem 2.7.6 (Invariance) For any two UTMs U_1 and U_2 there exists a constant c such that for all strings x,

$$|K_{U_1}(x)-K_{U_2}(x)| \leq c$$

Proof. Both U_1 and U_2 are UTMs, so there exists indexes i and j such that $U_1(i'q)=U_2(q)$ and $U_2(j'q)=U_1(q)$ for all q. Then, given any program q_1 (q_2) for U_1 (U_2), it can be run on U_2 (U_1) with constant overhead. So for any string x, let p_1 (p_2) be a shortest program for x

---

[24]We sincerely hope you've found more than 1 bit of information in this book!

[25]or at least, a compressed copy of this book and an associated decompressor. At the time of writing, all the text in the book fits in 2,299,002 bytes before compression, and 465,949 bytes after compression using ZPAQ, a family of compressors known for a high compression ratio, but slow to compress. If we included the size of the decompressor (367,080 bytes) for a total of 833,029 bytes, this provides an upper bound of the Kolmogorov complexity K_T of this book with respect to the author's computer standing in for T.

on U_1 (U_2). Then $j'p_1$ ($i'p_2$) is a program for x on U_2 (U_1), and

$$K_{U_2}(x) \leq \ell(j'p_1) \; = \; K_{U_1}(x) + \ell(j')$$
$$K_{U_1}(x) \leq \ell(i'p_2) \; = \; K_{U_2}(x) + \ell(i')$$

So by setting $c = \max\{\ell(j'), \ell(i')\}$ (which does not depend on the choice of x we have)

$$|K_{U_1}(x) - K_{U_2}(x)| \; \leq \; c$$

as required.   ∎

Note that there are many possible choices of encodings and enumerations of Turing machines. Different choices lead to defining different Universal Turing machines. Though some work has been done to find a canonical choice of UTM, usually by trying to define a UTM with the fewest number of tape symbols and states [KR02, Rog96], we leave this aside and arbitrarily fix one "natural" reference machine U.

Remark 2.7.7 (Reference universal Turing machine) There are infinitely may different choices of UTMs, so we arbitrarily fix some "natural" reference choice of UTM U for this book, and define $K(x) := K_U(x)$ and $K(x|y) := K_U(x|y)$. Similarly for monotone complexity, we define $Km(\cdot) := Km_U(\cdot)$.   ●

Assuming that two universal Turing machines U_1 and U_2 are "natural", we would expect the constant c in Theorem 2.7.6 to be "small".

Assumption 2.7.8 (Short compiler) Given two "natural" Turing-equivalent formal systems F_1 and F_2, then there always exists a "small" program p on F_1 that is capable of interpreting all F_2 programs.

Kolmogorov complexity for general objects.   We can also define the Kolmogorov complexity for things other than binary strings by encoding them as binary strings in some canonical fashion. For example, in Section 2.1.1 we defined a bijection $\langle \cdot \rangle$ between $\mathbb{B}^*$ and $\mathbb{N}_0$, so we can define the Kolmogorov complexity of a natural number n as $K(n) := K(\langle n \rangle^{-1})$. The Kolmogorov complexity of other objects such as Turing machines, pictures, books, music etc., can be defined in a similar fashion by providing some canonical encoding function $\langle \cdot \rangle'$ of a class of objects to binary strings, and then measuring the complexity of the resulting string using K. The same can be said for the Kolmogorov complexity of tuples of objects, as the complexity of an encoded version such that the strings can be uniquely decoded. In the case of strings, we can define $K(x,y) := K(x'y)$ and $K(x,y,z) := K(x'y'z)$, etc. For instance, for Turing machines, this would be an encoding of its specifying 7-tuple T, which itself consists of objects of various types, and in particular its transition function δ. Encoding of computable functions f deserves its own Definition 2.7.19.

With this generalization we can generalize Theorem 2.7.6 and make it more precise by explicating the constant c:

Theorem 2.7.9 (Universal minorization) For any choice of Turing machine T, we have that

$$K(x) \; \leq \; K_T(x) + K(T) + O(1)$$

The theorem says that compression via a universal Turing machine U minorizes all other choices of Turing machines T (which could, for example, be existing decompression algorithms), up to an additive constant of the size of the description length of T, as the universal Turing machine can be provided a program to simply emulate what T does, together with the program to run on T.

Proof. Let p be the shortest program for x under T, i.e. $\ell(p)=K_T(x)$. Let r be the shortest description of $\langle T \rangle$ under U, i.e. $\ell(r)=K(T)$. Let T_i be a Turing machine that reads in rp, separates them into r and p (no prefix of r is a valid program), and then simulates the action of T on p, that is, $U(i'rp)=T_i(rp)=T(p)$. Noting that i has no dependency on x nor T, we have

$$K(x) \leq \ell(i')+\ell(r)+\ell(p) \leq K_T(x)+K(T)+O(1)$$

as required.   ∎

2.7.3 Properties of K-Complexity

We now investigate some properties of the Kolmogorov complexity function K. Where suitable, we provide intuition as to why a property holds, before formalizing the statement and giving a proof. For this it will be convenient to introduce $\overset{+}{\leq}$ ($\overset{\times}{\leq}$) to indicate "less-than within an additive (multiplicative) constant" and similarly for $\geq$ and $=$:

Definition 2.7.10 ((In)equalities within additive/multiplicative constants)

$$f(x)\overset{+}{\leq}g(x) \iff \exists c>0.f(x)\leq g(x)+c \iff f(x)=g(x)+O(1)$$

$$f(x)\overset{\times}{\leq}g(x) \iff \exists c,x_0.|f(x)|\leq c|g(x)| \;\forall x>x_0 \iff f(x)=O(g(x))$$

$$f(x)\overset{*}{\geq}g(x) \iff g(x)\overset{*}{\leq}f(x) \qquad \text{for} \quad *\in\{+,\times\}$$

$$f(x)\overset{*}{=}g(x) \iff f(x)\overset{*}{\leq}g(x) \;\text{ and }\; g(x)\overset{*}{\leq}f(x) \qquad \text{for} \quad *\in\{+,\times\}$$

Crucially, c and x_0 do not depend on x.

We can always find an upper bound on the Kolmogorov complexity of a string x, merely by having a program that has a hardcoded copy of x built inside. Since the Kolmogorov complexity is the shortest such description, it certainly cannot be any worse than using the string itself as its own description. We do incur a logarithmic penalty, since U requires programs p but not outputs x to be self-delimiting.

Theorem 2.7.11 (Upper bound on Kolmogorov Complexity) Let $x\in\mathbb{B}^*$ and $n\in\mathbb{N}$. Then,

$$K(x) \overset{+}{\leq} \ell(x)+2\log_2\ell(x)$$

$$K(n) \overset{+}{\leq} \log_2 n+2\log_2\log_2 n$$

Proof. There exists a Turing machine T_i that reads as input a second-order prefix encoded string x', and returns the decoded version x, i.e. $T_i(x')=x$, for all x. Then, $U(i'x')=x$ and by (2.1.5), we have that

$$K(x) \leq \ell(i'x') = \ell(i')+\ell(x') \overset{+}{\leq} \ell(x)+2\log_2(\ell(x))$$

since $c:=\ell(i')$ is a constant independent of x. For the second property, note that we can use the inverse of $\langle\cdot\rangle$ (2.1.2) on n to obtain a string $\langle n\rangle^{-1}$ of length $\lfloor\log_2(n+1)\rfloor\overset{+}{=}\log_2(n)$, from which the result follows.   ∎

Theorem 2.7.12 (Kraft's inequality for K-complexity) $\quad\displaystyle\sum_{x\in\mathbb{B}^*} 2^{-K(x)} < 1$

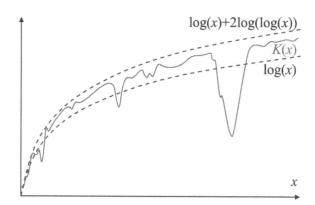

Figure 2.19: *Illustrative graph of prefix Kolmogorov complexity $K(x)$ with string x interpreted as integer. $K(x) \geq \log_2 x$ for 'most' x and $K(x) \leq \log_2 x + 2\log_2 \log x + c$ for all x for sufficiently large constant c [Hut05b].*

Proof.
$$\sum_{x \in \mathbb{B}^*} 2^{-K(x)} = \sum_{x \in \mathbb{B}^*} 2^{-\min\{\ell(p):U(p)=x\}} \leq \sum_{x \in \mathbb{B}^*} \sum_{p:U(p)=x} 2^{-\ell(p)} = \sum_{p:U(p) \text{ halts}} 2^{-\ell(p)} < 1$$

where the last inequality is Kraft's inequality (Theorem 2.5.17), noting that the set of programs p such that $U(p)$ halts is a prefix-free set by the design of a prefix Turing machine (Definition 2.7.1). Strict inequality holds, since there are non-halting programs, so the prefix set is not complete. ∎

Theorem 2.7.11 shows that $K(n) \overset{+}{\leq} \log_2 n + 2\log_2 \log_2 n$ for all n, and Theorem 2.7.12 implies that $K(n) \geq \log_2 n$ for most n (if $K(n) \leq \log_2 n$ for a non-vanishing fraction of n, then $\sum_n 2^{-K(n)} = \infty$). There is also some very weak notion of continuity: $|K(n \pm k) - K(n)| \overset{+}{\leq} K(k)$, so plunges in $K(n)$ for simple n have some width. Furthermore, Theorem 2.7.15 below implies that also the local minima increase with n, albeit slower than any computable function. Based on this, Figure 2.19 provides a schematic plot of K.

(In)compressibility.   Now that we have a formal definition of simplicity, we can define a string to be *compressible* if its shortest description is shorter than the string itself, i.e. that $K(x) < \ell(x)$. Incompressible strings must exist, as by the pigeon-hole-principle, if all strings were compressible, then we can construct an injection from $\mathbb{B}^n$ into $\mathbb{B}^{<n}$, where each string gets mapped to its shortest program. The function must be injective since given the shortest program, we can run the program to obtain the original string. An injection requires $|\mathbb{B}^n| \leq |\mathbb{B}^{<n}|$, however $|\mathbb{B}^{<n}| = \sum_{i=0}^{n-1} |\mathbb{B}^i| = \sum_{i=0}^{n-1} 2^i = 2^n - 1 < 2^n = |\mathbb{B}^n|$, a contradiction.

This gives no indication of how common incompressible strings are. Intuitively "most" strings should be incompressible, since if we generate a sequence by fair coin flips, it is far more likely that the string looks "random" rather than something with an obvious pattern such as 101010.... The reason is that simple patterns correspond to short programs (highly compressible), of which there are few. More quantitatively:

Theorem 2.7.13 (Most strings are incompressible) For any (injective) encoding/compression, less than 2^n strings $x \in \mathcal{X}^*$ can be compressed to less than n bits. In particular, the fraction of strings of length n compressible by more than k bits is less than 2^{-k}. Compressing with a prefix-code increases the length of most strings.

(i) $2^{-n}|\{x : \ell(C(x)) < n-k\}| < 2^{-k}$ for injective $C : \mathcal{X}^* \to \mathbb{B}^*$

(ii) $2^{-n}|\{x : \ell(C(x)) \leq n+k\}| \overset{n \to \infty}{\longrightarrow} 0$ for a prefix code $C : \mathcal{X}^* \to \mathbb{B}^*$

In particular both statements hold for K-complexity, i.e. for $\ell(C(x)) = K(x)$.

Proof. Since C is injective, we have

$$|\{x : \ell(C(x)) < n\}| \leq |\mathbb{B}^{<n}| = \sum_{l=0}^{n-1} |\mathbb{B}^l| = \sum_{l=0}^{n-1} 2^l = 2^n - 1 < 2^n$$

(i) From this, (i) follows immediately by $n \rightsquigarrow n-k$ and multiplying with 2^{-n}.
(ii) Let us define the number of strings compressible to l bits as $N_l := |\{x : \ell(C(x)) = l\}|$. Applying Kraft's inequality to prefix code C, we get

$$1 \geq \sum_{x \in \mathcal{X}^*} 2^{-\ell(C(x))} = \sum_{l=0}^{\infty} \sum_{x:\ell(C(x))=l} 2^{-\ell(C(x))} = \sum_{l=0}^{\infty} 2^{-l} N_l =: \delta_0$$

This implies that $\delta_m := \sum_{l=m}^{\infty} 2^{-l} N_l \to 0$ for $m \to \infty$. Now for any $i \leq n$,

$$2^{-n}|\{x : \ell(C(x)) \leq n\}| = 2^{-n} \sum_{l=0}^{n} N_l$$

$$= \sum_{l=0}^{n-i} 2^{l-n} 2^{-l} N_l + \sum_{l=n-i+1}^{n} 2^{l-n} 2^{-l} N_l$$

$$\leq 2^{-i} \sum_{l=0}^{n-i} 2^{-l} N_l + \sum_{l=n-i}^{n} 2^{-l} N_l$$

$$\leq 2^{-i} \delta_0 + \delta_{n-i} \overset{n \to \infty}{\longrightarrow} 0 \quad \text{for e.g. } i := \lceil n/2 \rceil$$

Finally we replace n by $n+k$ and multiply by 2^k to obtain (ii). ∎

One can use the notion of incompressibility to formally define randomness of individual strings. This finally allows us to formally reason about individual random strings, and drop the quotation marks from "random" in this context. Since we will refer to this concept only in passing, we will only briefly state it here for completeness without discussion.

Definition 2.7.14 (Randomness of individual sequences) An individual sequence $x_{1:\infty}$ is (called) μ-algorithmically random iff one (and hence all) of the following equivalent conditions hold:

(i) $x_{1:\infty}$ passes all μ-Martin-Löf randomness tests [LV19, Def.2.4.1&2.5.4]

(ii) $M(x_{1:n}) \overset{\times}{\leq} \mu(x_{1:n}) \, \forall n$ (Definition 3.8.2) [Hut05b, Thm.2.31]

(iii) $Km(x_{1:n}) \overset{+}{=} -\log_2 \mu(x_{1:n}) \, \forall n$ (Definition 2.7.4) [LV19, Cor.4.5.3]

(iv) $K(x_{1:n}) \overset{+}{\geq} -\log_2 \mu(x_{1:n}) \, \forall n$ (Definition 2.7.3) [LV19, Thm.3.5.1]

For fair coin flips $\mu(x_{1:n}) = 2^{-n}$, the qualifier μ- is usually dropped.

Algorithmic randomness can also be defined for finite strings using the notion of randomness deficiency. One can show that a finite (infinite) string sampled from μ is with high probability (almost surely) μ-algorithmically random. So while there is no algorithm to compute algorithmically random strings, they *can* be generated with high probability (Exercise 2). The relation to pseudo-randomness is as follows: A pseudo-random sequence is deterministically generated, but no test that runs in polynomial time can reliably distinguish it from a sequence sampled from a random source. This can be regarded as a (practical) poly-time version of Martin-Löf randomness tests, the latter consisting of all lower semicomputable tests. That is, no constructive test can reliably distinguish an algorithmically random sequence from a sequence sampled from a random source.

Returning to properties of K, it is easy to see that $K(n) \to \infty$:

Theorem 2.7.15 (K-complexity tends to infinity) $\displaystyle \lim_{n \to \infty} K(n) = \infty$

Proof. Suppose not. Then there exists some N such that $K(n) < N$ for infinitely many n. So infinitely many numbers have Kolmogorov complexity less than N, so there are infinitely many programs of length $< N$, a contradiction.   ∎

Providing extra side information y can never hurt when trying to describe x: At worst, y is a sequence of coin flips that is totally uncorrelated with x and ignored. At best, y is a copy of x, so we only need a constant length program to copy the side information to the output. Similarly, describing x and y together is at least as hard as describing x alone.

Theorem 2.7.16 (Extra Information) $K(x|y) \overset{+}{\le} K(x) \overset{+}{\le} K(x,y)$

Proof. Let p^* be the shortest program such that $U(p^*) = x$. There exists an i_0 such that $T_{i_0}(y'p^*) = U(p^*) = x$ as T_{i_0} can discard the side information y', and then compute the output of the program p^*. Then $U(y'i_0'p^*) \overset{+}{=} T_{i_0}(y'p^*) = U(p^*) = x$. Thus,

$$K(x|y) \equiv \min_p \{\ell(p) : U(y'p) = x\} \le \ell(i_0'p^*) = \ell(i_0') + \ell(p^*) \overset{+}{=} K(x)$$

Now let q^* be the shortest program such that $U(q^*) = x'y$. Then there exists i_1 such that $T_{i_1}(q^*) = x$ as T_{i_1} uses q^* to compute $x'y$ on its internal tape and then copies only x to the output tape. So $U(i_1'q^*) = x$, hence

$$K(x) \equiv \min_p \{\ell(p) : U(p) = x\} \le \ell(i_1'q^*) = \ell(i_1') + \ell(q^*) \overset{+}{=} K(x'y) = K(x,y) \quad ∎$$

Encoding the concatenation xy requires less information than being able to encode both x and y in a way they can be separated uniquely. This is less information than encoding x alone, along with encoding y given x. This is less still than encoding x, along with encoding y with no help from x:

Theorem 2.7.17 (Subadditivity) $K(xy) \overset{+}{\le} K(x,y) \overset{+}{\le} K(x) + K(y|x) \overset{+}{\le} K(x) + K(y)$

Proof. We prove each inequality separately.

1. Let p^* be the shortest program such that $U(p^*) = x'y$. There exists i_0 such that $T_{i_0}(p^*) = xy$ by emulating $U(p^*) = x'y$ and decoding $x'y$ to x and discarding y. Thus, $U(i_0'p^*) = xy$, hence

$$K(xy) \equiv \min_p \{\ell(p) : U(p) = xy\} \le \ell(i_0'p^*) = \ell(i_0') + \ell(p^*) \overset{+}{=} K(x'y) = K(x,y)$$

2. Let q be the shortest program such that $U(q)=x$. Let r be the shortest program such that $U(x'r)=y$. Then there exists an i_1 such that

$$U(i_1'qr) := U(q)'U(U(q)'r) = x'U(x'r) = x'y$$

This works, since we can uniquely decode the string qr back into q and r, as both q and r are programs that halt when run on U, and we know that the set of all programs p such that $U(p)$ halts is prefix free. Now

$$K(x,y) \equiv \min_p\{\ell(p):U(p)=x'y\} \leq \ell(i_1'qr) \stackrel{\pm}{=} \ell(q)+\ell(r) = K(x)+K(y|x)$$

3. Finally, by Theorem 2.7.16 we have $K(y|x)\stackrel{+}{\leq}K(y)$, which gives the final inequality $K(x)+K(y|x)\stackrel{+}{\leq}K(x)+K(y)$. ∎

The following theorem, $K(x|y)+K(y)\approx K(y|x)+K(x)$, is a bit more difficult to intuitively understand. The idea is that the information contained in y, together with the minimal information to recover x given y, should be roughly same as the information contained in x, together with the minimal information to recover y from x. Consider a restricted example with some assumptions about x and y. If x is complex and y simple, then we would expect that $K(x)$ is large, whereas $K(y|x)\approx 0$ is small (as knowing x, not much more information is required to specify y). Conversely, $K(y)\approx 0$ is small (as y is simple) but $K(x|y)\approx K(x)$ is large (since y is simple, it cannot help much with specifying x, as otherwise x would also be simple). What is remarkable is that the sum of these terms, $K(x|y)+K(y)$ and $K(y|x)+K(x)$ are in general about equal, regardless of the values of x and y. Equality ony holds within a logarithmic additive correction, but this can be improved to the usual additive fudge by augmenting x with $K(x)$, and y with $K(y)$. While x already includes the information $K(x)$, it does not in a computable way, i.e. $K(x)$ cannot be computed from x, so providing $K(x)$ in addition is computationally helpful.

Theorem 2.7.18 (Symmetry of information)

$$K(x|y,K(y))+K(y) \stackrel{\pm}{=} K(y|x,K(x))+K(x) \stackrel{\pm}{=} K(x,y) \stackrel{\pm}{=} K(y,x)$$

Proof. The last equality is easy to prove, as given a shortest program p for $x'y$, we can recover x and y, second-order prefix code y, and output $y'x$, which gives $K(x,y)\stackrel{+}{\leq}K(y,x)$. By a similar argument we can obtain $K(y,x)\stackrel{+}{\leq}K(x,y)$, from which the result follows.

Unfortunately, the first two equalities are very difficult to prove and beyond the scope of this book, see [LV19, Thm.3.8.1] for a proof. ∎

We have defined the Kolmogorov complexity for any object that can somehow be described as a binary string, but mathematical functions f require special treatment. Assume we allowed any mathematical description of a function f (in some precise, formal, axiomatic language) and then encode this description somehow in binary as $\langle f\rangle$. This would make $K(\langle f\rangle)$ finite even for mathematically describable but incomputable functions $f\notin\Delta_2^0$. Below we provide a more natural and useful definition for computable functions.

Definition 2.7.19 (Kolmogorov complexity of computable functions) The Kolmogorov complexity of a computable function $f:\mathbb{B}^*\to\mathbb{B}^*$ is the length of the shortest program that computes $f(x)$ given x as side information:

$$K(f) := \min_{p\in\mathbb{B}^*}\{\ell(p):\forall x.U(x'p)=f(x)\}$$

We also need a custom complexity measure for semi-probability mass functions P and semimeasures ν.

Definition 2.7.20 (Complexity of lower semicomputable semimeasures) Let $\mathcal{M}_{sol} = \{\nu_1, \nu_2, \nu_3, ...\}$ be a canonical enumeration of all lower semicomputable semimeasures (Definition 3.7.1), then the Kolmogorov complexity of ν_i is defined as

$$K(\nu_i) := K(i) \quad \text{for} \quad \nu_i \in \mathcal{M}_{sol}$$

The definition of $K(P)$ for lower semi-computable semi-probability mass functions is analogous.

Remark 2.7.21 (Confusion of $K(T)$ and $K(f)$ and $K(\nu)$) $K(f)$ is conceptually different from the Kolmogorov complexity of strings and all other objects, including Turing machines. There are many Turing machines T that can compute function f, but some are (unnecessarily) complicated. Imagine the transition function δ contains a million superfluous states with random transitions that are never reached. Then $K(T) := K(\langle T \rangle) \gg K(f)$, since $K(f)$ only encodes the "essence" of f without the "garbage". Note that $\min_T \{K(\langle T \rangle) : T(x) = f(x) \; \forall x\} \stackrel{\pm}{=} K(f)$, assuming $\langle T \rangle$ is a "reasonable" encoding of T (Exercise 10). One downside of Definition 2.7.19 is that K as functional of f is itself incomputable (not even approximable).

$K(\nu)$ is different again: It is relative to an enumeration of $\mathcal{M}_{sol}$. Similarly to $K(f)$, we could have defined

$$K(f) = \min_{p, \phi} \{\ell(p) : \forall x. \; U(x't'p) = \phi(x,t) \wedge \phi(x,t) \to f(x)\}$$
$$\stackrel{\pm}{=} \min_p \{\ell(p) : \forall x : U(x'p) = y_1' y_2' y_3' ... \Rightarrow y_t \to f(x)\}$$

for $f \in \Delta_2^0$, with additional constraint of $\phi(x,t)$ and y_t' being monotonically increasing (decreasing) for $f \in \Sigma_1^0$ (Π_1^0). But this has two downsides: First, this would increase confusion even more, since $K(f)$ would now depend on whether we view f as computable or semicomputable. This could be solved by making this explicit and defining $K_n^{\Sigma/\Pi/\Delta}(f)$ as the length of the shortest $\Sigma_n^0/\Pi_n^0/\Delta_n^0$-description of f. But this leaves the second problem: For $\nu \in \mathcal{M}_{sol}$ we need a definition such that K is an upper semicomputable functional of ν, which Definition 2.7.20 provides, but $K(f)$ does not.   ●

Consider some string x, which we transform via f to $y = f(x)$. The result y cannot be more complex than a description of the function f together with the input x, since if feeding a simple x to a function f leads to a very complex output y, this is only possible when f itself is also complex. Given a description of both x and f, little extra information on top is required to describe $f(x)$, as we can simply recover x and f from the descriptions, and run f on x:

Theorem 2.7.22 (Information non-increase) Let $f : \mathbb{B}^* \to \mathbb{B}^*$ be a computable function. Then,

$$K(f(x)) \stackrel{+}{\leq} K(x) + K(f)$$

Proof. Let q be the minimal self-delimiting program that computes f. Let r be the shortest program such that $U(r) = x$. There exists an i such that

$$U(i'qr) := U(U(r)'q) = U(x'q) = f(x) =: y$$

noting that since q and r are self-delimiting programs, we can uniquely separate qr back into q and r. So

$$K(f(x)) \equiv \min_p\{\ell(p):U(p)=y\} \leq \ell(i'qr) \stackrel{+}{=} \ell(q)+\ell(r) = K(f)+K(x) \qquad \blacksquare$$

2.7.4 The Minimum Description Length Principle

The last property of Kolmogorov complexity that we will cover is the *Minimum Description Length* (MDL) bound, which deserves its own subsection: The complexity of x can be bounded above by the complexity of encoding some probability distribution P, plus $\lceil -\log_2 P(x) \rceil$, the optimal code length of x with respect to the distribution P.

Theorem 2.7.23 (MDL Bound) If $P:\mathbb{B}^* \to [0,1]$ is a lower semicomputable semi-probability mass function (i.e. $\sum_{x \in \mathbb{B}^*} P(x) \leq 1$) and μ a lower semicomputable semimeasure (Definition 2.2.11), then

$$K(x) \stackrel{+}{\leq} -\log_2 P(x)+K(P) \qquad \text{and} \qquad Km(x) \stackrel{+}{\leq} -\log_2 \mu(x)+K(\mu)$$

Proof. We provide a sketch of the proof of the first bound. The idea is to use a Shannon-Fano code (Definition 2.5.7) based on distribution P to code x. The second bound is proven similarly using arithmetic coding (Theorem 2.5.25) and left as an exercise.

Let $s_x := \lceil -\log_2 P(x) \rceil$. Then, $\sum_x 2^{-s_x} \leq \sum_x P(x) \leq 1$. By the Kraft Inequality (Theorem 2.5.17) there must exist a prefix code $c:\mathbb{B}^* \to \mathbb{B}^*$ with $\ell(c(x)) = s_x = \lceil -\log_2 P(x) \rceil$. The proof of the Kraft inequality can be made constructive by avoiding the potentially incomputable sorting step at the cost of one extra bit in code length [LV19]. Since the MDL bound has an additive slack anyway, we leave this complication as Exercise 9 and pretend the standard proof is constructive. So the proof of Kraft inequality describes an "algorithm" by which the codewords are constructed by constructing a disjoint set of intervals $I_n = [0.c_n, 0.c_n + 2^{-\ell(c_n)})$. The binary string c_n is then the nth codeword. Hence, there exists a Turing machine T that takes as input a program p that computes semi-probability P, and a Shannon-Fano codeword c_x for x, and recovers the original string x. Let p^* be the shortest program for P under U (so $\ell(p^*) = K(P)$) and i be the index for T such that $U(i'c_x p) = T(c_x p) = x$. Note that $i'c_x p^*$ is a valid program for x under U (we can separate c_x from p as no prefix of c_x would be a valid codeword). Hence,

$$K(x) \leq \ell(i'c_x p^*) = \ell(i')+\ell(c_x)+\ell(p^*) \stackrel{+}{=} -\log_2 P(x)+K(P)$$

as required.

$\blacksquare$

The MDL bound is useful for proving other theorems, but its most important application is in *model selection*: Occam's razor demands selecting the simplest theory or model consistent with the data, which can be quantified as: The best explanation of x is the shortest program p^* on a universal Turing machine U for which $U(p^*)=x$.

Noisy data and probabilistic models. In practice, data is noisy and models will be imperfect. To deal with this situation, we need to consider *stochastic* models. Let observation x be sampled from (true) probability P, and $\mathcal{M} \ni P$ be a class of such distributions. Given P, the optimal code = shortest program of x has length $\log_2[1/P(x)]$, achievable e.g. via Shannon-Fano (Definition 2.5.7). Since/if we do not know P, we can take any $Q \in \mathcal{M}$ instead, and code x in $\log_2[1/Q(x)]$ bits. Note that the arithmetic decoder needs to know Q, so we need to encode (a binary representation of) Q as well requiring $K(Q)$ additional bits.

The case of sequential data $x_{1:\infty}$ and semimeasures μ as models using arithmetic encoding (Section 2.5.6) works analogously and is left as an exercise.

Occam's razor tells us to choose simple models (small $K(Q)$), but stochastic models do not predict perfectly, for which a $-\log_2 P(x)$ penalty is added. This motivates the following principle:

Definition 2.7.24 (The Minimal Description Length (MDL) Principle) Given a class of probability distributions $\mathcal{M}$ and data x, the MDL principle selects the model with the shortest two-part code

$$Q^{\mathrm{MDL}} := \arg\min_{Q\in\mathcal{M}}\{-\log_2 Q(x)+K(Q)\}$$

This MDL principle gives Theorem 2.7.23 its name. Under very mild conditions one can prove consistency (Definition 2.3.4) of the MDL estimator [PH05a, Hut09b].

In practice, we have to replace $K(Q)$ by more easily computable upper bounds. For instance, if $\mathcal{M}=\bigcup_{d=0}^{\infty}\mathcal{M}_d$, where $\mathcal{M}_d$ is a class of i.i.d. distributions from which $x=x_{1:n}$ is sampled (e.g. polynomials of degree $d-1$ with Gaussian noise model), smoothly parameterized by some $\theta\in\mathbb{R}^d$, then MDL becomes

$$Q^{\mathrm{MDL}} := \arg\min_{d\in\mathbb{N}_0,\theta\in\mathbb{R}^d} \{-\log_2 Q_\theta(x)+\tfrac{d}{2}\log_2(n)+O(d)\} \qquad (2.7.25)$$

Intuitively, for sample size n, each parameter $\theta_i\in\mathbb{R}$ can (only) be estimated to accuracy $O(1/\sqrt{n})$, hence (only) the first $\log_2[1/O(1/\sqrt{n})]=\tfrac{1}{2}\log_2(n)+O(1)$ bits of the expansion of each of the d parameters needs to be encoded.

The $O(d)$ term depends on the smoothness of the parametrization and can explicitly be quantified as $\log_2\int\sqrt{\det[\mathcal{I}(\theta)/2\pi e]}d^d\theta$, where $\mathcal{I}(\theta)$ is the Fisher Information Matrix of Q_θ, which generalizes Definition 2.3.12 to $\theta\in\mathbb{R}^d$ by replacing ∂ with ∇. Note that the dominant term, the negative log-likelihood $-\log_2 Q_\theta(x)$, is linear in n, the Bayesian Information Criterion (BIC) [NC12] dimensional complexity penalty $\tfrac{d}{2}\log_2(n)$ is logarithmic in n, while $O(d)$ is a (small) curvature correction independent of n that *may* be ok to neglect. See [Wal05, Gru07] for detailed explanations, derivations and consistency proofs for the i.i.d. case and [Hut03e, Sec.6.1] beyond i.i.d.

Bayesian derivation. The same expression can be derived from Bayesian principles. In the discrete case, the MDL principle can be regarded as an approximation to the full Bayesian mixture distribution $\sum_{Q\in\mathcal{M}}Q(x)w(Q)$ which we primarily consider in this book. Replacing the sum by a max, then taking the logarithm, changing sign, and choosing prior $w(Q)=2^{-K(Q)}$ gives back Definition 2.7.24. In the parametric continuous case, if we assume priors $w_d(\theta)$ over the parameters, the *Maximum A Posteriori* (MAP) approximation is

$$Q^{\mathrm{MAP}} := \arg\max_{\theta\in\mathbb{R}^{d^*}}\{Q_\theta(x)w_{d^*}(\theta)\}, \quad \text{where} \quad d^* := \arg\max_{d\in\mathbb{N}_0}\{\textstyle\int Q_\theta(x)w_d(\theta)d^d\theta\} \qquad (2.7.26)$$

Using Laplace approximation for the integral for large n, and taking the negative logarithm, one can show that (2.7.26) reduces to (2.7.25) [NC12]. If we choose the reparameterization-invariant minimax-optimal Jeffreys/Bernardo prior (see Definition 2.4.11) for $w_d(\theta)$, then even the $O(d)$ term coincides with MDL. In the continuous case, Bayes and MDL only differ in $o(1)$ terms that vanish for $n\to\infty$ [Wal05, Gru07].

Regularization in Machine Learning (RML). The MDL principle can also be understood as a *Penalized or Regularized Maximum Likelihood (RML) principle*, maximizing $Q(x)$

which is the same as minimizing $-\log_2 Q(x)$ but punishing complex models by adding $K(Q)$. More generally, regularization is used in classical machine learning to combat overfitting to a too complex model by adding a regularization term to the loss function. Some differences and commonalities between MDL and RML are as follows:

- The MDL principle (Definition 2.7.24) is completely general. It makes no assumption whatsoever on the underlying model class $\mathcal{M}$: No i.i.d. or stationarity or ergodicity or parametric or smoothness assumptions, while RML is somewhat tailored towards i.i.d. data.
- MDL is limited to log-loss, while any loss can be used in RML [Hut07d].
- RML requires choosing a penalty and tune its strength, often by cross-validation, while the MDL theory dictates the penalty term.
- MDL estimates probabilities Q, while RML determines mappings f, though this difference is superficial: One can trivially generalize $Q(x)$ to $Q(y|x)$ in MDL, and augment f with a noise model in RML.

2.7.5 Approximating K-Complexity

The Kolmogorov complexity K is a universal complexity measure, in that K will pick up any regularity or pattern in the data, and use it to give as short as possible a description of the input. Viewed another way, $K(x)$ provides an absolute limit on the fewest number of bits required to express x, when compressed by an optimal compressor. It beats all other lossless[26] compressors in terms of space saved.

Unfortunately (though it shouldn't come as a surprise) K is incomputable.

Theorem 2.7.27 (K is not finitely computable) $K \notin \Delta_1^0$

Proof. Assume that K is finitely computable. We can define the function $f(m) := \min\{n : K(n) \geq m\}$, which is always well-defined, since K is unbounded (Theorem 2.7.15). Since K is assumed computable, f is also computable (we just compute $K(n)$ for increasing values of n until we find the first n for which $K(n) \geq m$, which will always exist). The value $f(m)$ is the smallest value n such that $K(n) \geq m$, so by definition, $K(f(m)) \geq m$. By Theorem 2.7.22 we have that
$$K(f(m)) \leq K(m) + K(f) + c_1$$
and by Theorem 2.7.11 we have

$$K(m) \leq \log_2 m + 2\log_2\log_2 m + c_2 \leq 2\log_2 m + c_3$$

Combining both, we get

$$m \leq K(f(m)) \leq K(m) + K(f) + c_1 \leq 2\log_2 m + K(f) + c_1 + c_3 \leq 2\log_2 m + O(1)$$

which gives $m \leq 2\log_2 m + O(1)$ for all m, a contradiction.   ∎

However, not all is lost. We can still approximate K in the limit from above (it is upper semicomputable), obtaining better and better upper bound for the complexity of a string. We can easily find some initial program p_0 that prints x (just write a program that has a copy of x hardcoded inside and outputs x). We then search over all programs p with length

---

[26]We could potentially compress further if loss were allowed and we only needed to construct an approximation of the original, which is fine for movies or music, but not for source code.

less than p_0, and run them in "parallel". If any program p prints x and halts, we record the length of p if it is better than the shortest program that prints x found so far. Eventually, we must stumble across the shortest program p^*, and obtain $K(x)$. The downside is that we will not know whether we have found the shortest program yet or not, since during the search, some programs might eventually print x if left to run for long enough, and some would never halt. There would be no way to know for certain how many time steps would be sufficient until the shortest program p^* was run for long enough to print x, due to the halting problem (Theorem 2.6.7).

We can simulate running a family of Turing machines "in parallel" with a single Turing machine. This is done by taking an effective enumeration $\{T_1, T_2, ...\}$ of all Turing machines, and allotting a half of the simulation time to the first machine, a quarter to the second machine, an eighth to the third, and so on. This is done by timesharing the machines according to the schedule 1,2,1,3,1,2,1,4,1,2,1,3,1,2,1,5..., so the first machine is run for one time step, then the second, the first, the third, and so on. Running countably-many programs for an arbitrary length of time in this fashion is called *dovetailing*.[27] As a corollary, we can also run finitely many machines $\{T_{i_1}, T_{i_2}, ..., T_{i_n}\}$ by the same construction, and sitting idle on a time step t not associated with any machine in the list. We now prove the above result formally.

Theorem 2.7.28 (K is upper semicomputable) $K \in \Pi_1^0$

Proof. To prove that K is upper semicomputable, we need to describe a finitely computable function $\phi : \mathbb{B}^* \times \mathbb{N}_0 \to \mathbb{N}_0$ such that $\lim_{t \to \infty} \phi(x,t) = K(x)$ and $\phi(x,t) \geq \phi(x,t+1)$. The algorithm for ϕ is as follows:

1. Initialize $C := \lceil \ell(x) + 2\log_2 \ell(x) + \ell(i') \rceil$, where i' is the index of a Turing machine that decodes second order prefix codes (by Theorem 2.7.11, C is an upper bound for $K(x)$).

2. Dovetail $U(p_i)$ for all prefix programs $p_1, p_2, ...$ for t time steps. Every time some simulation $U(p_i)$ halts and outputs x, we let $C := \min\{C, \ell(p_i)\}$ (which can only ever result in C getting smaller, while still being an upper bound for $K(x)$).

3. Once t time steps have been simulated, abort the dovetail procedure and return C.

Every time one of the dovetailed simulations halt, we check the length of the program p_i that was simulated, and if it is shorter than the length of the best found program so far (C), then C is updated with the length of p_i. Allowing the above algorithm to run for more time can only make the output smaller, so $\phi(x,t) \geq \phi(x,t+1)$. Furthermore, for every x there exists a shortest program p such that $K(x) = \ell(p)$, so by running the dovetailing procedure long enough, we will eventually stumble across the shortest program, and run it for long enough to confirm that $U(p) = x$. Hence $\forall x. \exists T. \forall t \geq T \phi(x,t) = K(x)$, and $\lim_{t \to \infty} \phi(x,t) = K(x)$ follows. ∎

2.7.6   Relation to Shannon Entropy

We can compare the properties of K-complexity with those of Shannon entropy. Table 2.20 shows that many of the properties of K-complexity have an analogous version for Shannon entropy.

---

[27]Incidentally, the term *dovetail* is taken from a method in woodworking where two pieces of wood are joined by carving interlacing pins into the wood.

Table 2.20: *A comparison of the analogous properties shared between Kolmogorov Complexity K and Shannon Entropy H. x and y are strings, $X \sim P$ and $Y \sim Q$ are random variables, and f is a computable function.*

Thm.	Name	Kolmogorov Complexity	Shannon Entropy
2.7.11	*Upper bound*	$K(x) \overset{+}{\leq} \ell(x) + 2\log_2 \ell(x)$	$H(X) \leq \log_2 |\mathcal{X}|$
2.7.16	*Extra Information*	$K(x|y) \overset{+}{\leq} K(x) \overset{+}{\leq} K(x,y)$	$H(X|Y) \leq H(X) \leq H(X,Y)$
2.7.17	*Subadditivity*	$K(x,y) \overset{+}{\leq} K(x) + K(y)$	$H(X,Y) \leq H(X) + H(Y)$
2.7.18	*Symmetry of Info.*	$\begin{aligned} K(x|y,K(y)) + K(y) &\overset{+}{=} K(x,y) \\ &\overset{+}{=} K(y|x,K(x)) + K(x) \overset{+}{=} K(y,x) \end{aligned}$	$\begin{aligned} H(X|Y) + H(Y) &= H(X,Y) \\ &= H(Y|X) + H(X) = H(Y,X) \end{aligned}$
2.7.22	*Info. Non-increase*	$K(f(x)) \overset{+}{\leq} K(x) + K(f)$	$H(f(X)) \leq H(X)$
2.7.23	*MDL Bound*	$K(x) \overset{+}{\leq} -\log_2 Q(x) + K(Q)$	$H(X) \leq \mathbf{E}_P[-\log_2 Q(X)]$

Remark 2.7.29 (Further conditioning) Results for Shannon entropy, valid for some probability distribution $P(\cdot|\cdot)$, naturally also hold for $P(\cdot|\cdot,Z)$ further conditioned on some Z. The analogue for K-complexity is to condition K on a further string z by providing z on an extra tape or concatenating with y or as an oracle. That is, all the results above remain valid if conditioning K (further) on some z. ●

2.7.7 Exercises

1. [C12] Prove both claims in Definition 2.7.1: *(i)* Given a prefix TM T, the set $\{p \in \mathbb{B}^* : T(p)$ halts$\}$ is prefix-free. *(ii)* Given a monotone TM T and a fixed string x, the set $\{p \in \mathbb{B}^* : T(p) = x*\}$ is prefix-free.

2. [C10] Let $x \in \mathcal{X}^*$ be sampled from probability mass function $P : \mathcal{X}^* \to [0,1]$, and $C : \mathcal{X}^* \to \mathbb{B}^*$ a prefix code. Show that $\ell(C(x)) > -\log_2 P(x) + \log_2 \delta$ with probability at least $1 - \delta$. For the optimal code length $\ell(C(x)) := K(x)$, this implies a lower bound on $K(x)$ (and even $K(x|P)$) with high-probability. For $P(x) := \mu(x)$ for $\ell(x) = n$ and 0 otherwise, the above bound implies $K(x_{1:n}) > -\log_2 P(x_{1:n}) + \log_2 \delta$ w.p.$\geq 1 - \delta$ if $x_{1:n} \sim \mu$.

3. [C18] Fill in the details in the proof of Theorem 2.7.23

4. [C22i] Strengthen Theorem 2.7.28 by proving that there is no unbounded computable lower bound to K. This implies that the function $f(m) = \min_{n \geq m} K(n)$ grows incomputably slowly (in that any computable function must asymptotically grow faster.)

5. [C26] Define the *normalized compression distance d* as

$$d(x,y) := \frac{\max\{K(y|x), K(x|y)\}}{\max\{K(x), K(y)\}}$$

Prove that (within additive constants) d is a metric on $\mathbb{B}^*$.

6. [C10] Show that for every string x there exists a universal Turing machine U' such that $K_{U'}(x) = 1$. Argue that U' is not a natural Turing machine if x is complex. Elaborate on the difficulties on making such a statement rigorous.

7. [C16] The halting sequence $h_{1:\infty}$ is defined as $h_i = 1 \Leftrightarrow T_i(\epsilon)$ halts, otherwise $h_i = 0$. Show $K(h_1...h_n) \leq 2\log_2 n + O(\log\log n)$ and $Km(h_1...h_n) \leq \log_2 n + O(\log\log n)$.

8. [C22] Let $\Omega := \sum_{p:U(p) \ halts} 2^{-\ell(p)}$ be the halting probability. Show that the first n bits in the binary expansion of Ω encode the first 2^n bits of the halting sequence, i.e. $h_{1:2^n}$ can be computed given $\Omega_{1:n}$. Show that $K(\Omega_{1:n}) \overset{+}{\geq} n$, i.e. Ω is algorithmically random (Definition 2.7.14).

9. [C16] Modify the (If) part of the proof of the Kraft inequality Theorem 2.5.17 so as to result in a computable code.

10. [C12] Prove $\min_T \{K(\langle T \rangle) : T(x) = f(x) \ \forall x\} \overset{+}{=} K(f)$.

11. [C20] Prove the second MDL bound of Theorem 2.7.23.

12. [C20] Reformulate all expressions in Section 2.7.4 in terms of sequential data $x_{1:\infty}$ and semimeasures μ.

2.8 Miscellaneous

This final background section contains material that implicitly permeates the book, but is more distracting than helpful for all but the very mathematically inclined reader. The casual reader can skip this section and will still have no problem understanding the essence of the later chapters.

Section 2.8.1 introduces various distances between probability distributions: the absolute, total variation, squared, 2-norm, Kullback–Leibler, and Hellinger distance. They all have different properties and are used judiciously throughout the book. We prove upper bounds for all of them in terms of the KL divergence. Section 2.8.2 discusses the extra complications semimeasures (Definition 2.2.11) cause beyond the already vast measure theory, and lists a variety of solutions and workarounds. Section 2.8.3 discusses how to deal with probability zero events, especially how to condition on them, and how to take expectations of partial functions.

2.8.1 Distances and Their Relation

In this section we introduce various distances between probability distributions, and prove various relationships (bounds) between them. These will be needed later in Chapter 3 and elsewhere. The distances include the absolute distance (which is the same as twice the total variation distance), the squared distance (square of the Euclidean distance), the KL divergence (Section 2.5.3), and the (squared) Hellinger distance. Each distance has its pros and cons: the absolute, squared distance, and (unsquared) Hellinger distance are (proper) metrics, the total variation distance is very appropriate for probability measures, the squared distance is the most common but least suitable in general (see exercise), and the KL is information-theoretically the most useful for proving bounds. KL is an upper bound to many distances, while at the same time is a lower bound for many loss functions. This property makes KL an extremely useful tool in proofs.

One of the key bounds is Pinsker's inequality. We prove a generalization of it, from which most of the other relations follow. We limit ourselves to finite sample spaces Ω, so probabilities can be represented as d-dimensional vectors in the probability simplex. Generalization to probability measures on general sample spaces Ω is a formal exercise in measure theory and left as an exercise.

The proofs are elementary but some are somewhat messy, and can be freely skipped without missing context for the rest of the book.

We present Pinsker's inequality for binary alphabet $\Omega = \mathbb{B}$ first, since the general case will be reduced to it. While the lemma stands as it is, the intended application is for $p = P(1)$ and $q = Q(1)$, so the r.h.s. is the KL divergence between P and Q.

Lemma 2.8.1 (Pinsker's binary inequality) For $0 \leq p,q \leq 1$ we have that

$$2(p-q)^2 \leq p\ln\frac{p}{q} + (1-p)\ln\frac{1-p}{1-q}$$

There are many proofs, none are particularly enlightening [Hut05b], so we just present the possibly shortest one. Figure 2.15 clearly shows that 'right-hand-side minus left-hand-side' is non-negative.

Proof. We define $f(x) = p\ln x + (1-p)\ln(1-x)$, and note that $p\ln\frac{p}{q} + (1-p)\ln\frac{1-p}{1-q} = f(p) - f(q)$. Also note that for $0 \leq x \leq 1$ we have that $x(1-x) \leq 1/4$. Then,

$$f(p) - f(q) = \int_q^p f'(x)\,dx = \int_q^p \frac{p-x}{x(1-x)}\,dx \geq 4\int_q^p (p-x)\,dx = 2(p-q)^2 \qquad \blacksquare$$

To generalize the Pinsker inequality beyond square distance and binary alphabet we need the following elementary facts:

Lemma 2.8.2 (Convex even functions are monotonic on $\mathbb{R}^+$) Let $f : \mathbb{R} \to \mathbb{R}$ be an even[a] convex function. Then f is monotonically increasing over the positive reals.

---
   [a]An *even* function f is one satisfying $f(x) = f(-x)$.

Proof. Take the definition (Definition 2.2.49) of convexity $f(\alpha x + (1-\alpha)y) \leq \alpha f(x) + (1-\alpha)f(y)$ for $0 \leq \alpha \leq 1$. and insert $x = a+b$ and $y = 0$ to obtain

$$f(\alpha(a+b)) \leq \alpha f(x) + (1-\alpha)f(0) \qquad (2.8.3)$$

Now, assuming $a,b \geq 0$, and inserting $\alpha = \frac{a}{a+b} \geq 0$ or $\alpha = \frac{b}{a+b} \geq 0$ gives the following two inequalities:

$$f(a) \leq \frac{a}{a+b}f(a+b) + \frac{b}{a+b}f(0)$$

$$f(b) \leq \frac{b}{a+b}f(a+b) + \frac{a}{a+b}f(0)$$

Adding the two inequalities to obtain

$$f(a) + f(b) \leq f(a+b) + f(0) \qquad (2.8.4)$$

Also, in (2.8.3) we can choose $\alpha = 1/2$, $x = -b$, $y = b$ and obtain $f(0) \leq \frac{1}{2}f(b) + \frac{1}{2}f(-b) = f(b)$ as f is even. Combine with (2.8.4) and simplify to obtain $f(a) \leq f(a+b)$, which proves f to be monotonically increasing for positive arguments. $\blacksquare$

> **Lemma 2.8.5 (Convex even functions are subadditive)** Let f be an even convex function such that $f(0) \leq 0$, Then $f(a) + f(b) \leq f(a+b)$ for all and $a,b \geq 0$.

Proof. Follows immediately from (2.8.4) and $f(0) \leq 0$.   ∎

> **Theorem 2.8.6 (General entropy inequality [Hut05b, Sec.3.9.2])** Let $(y_1, ..., y_d)$ (and $(z_1, ..., z_d)$) be elements of a d-dimensional (semi)probability simplex, i.e. $y_i \geq 0$ with $\sum_i y_i = 1$ (and $z_i \geq 0$ with $\sum_i z_i \leq 1$). Then for any function $f : \mathbb{R} \to \mathbb{R}$ that is convex ($f(\alpha x + (1-\alpha)y) \leq \alpha f(x) + (1-\alpha)f(y)$ for all $x,y \in \mathbb{R}$ and all $\alpha \in [0,1]$) and even ($f(x) = f(-x)$) and satisfies $f(0) \leq 0$, we have
> $$\frac{1}{2}\sum_{i=1}^{d} f(y_i - z_i) \leq f\left(\sqrt{\frac{1}{2}\sum_{i=1}^{d} y_i \ln\frac{y_i}{z_i}}\right)$$

Proof. Let $I = \{1, ..., d\}$. Take any partition $I = P \dot\cup Q$ into two disjoint sets P and Q. Let S represent P or Q, and define $y^S := \sum_{i \in S} y_i$ and $z^S := \sum_{i \in S} z_i$. Now, note that for $i \in S$, we can normalize the probability distributions by defining $p_i^S = y_i/y^S$ and $q_i^S = z_i/z^S$. One can verify that $\{p_i^S\}_{i \in S}$ and $\{q_i^S\}_{i \in S}$ are valid probability distributions for S. Hence, by Gibbs' inequality (Theorem 2.5.15) we have

$$0 \leq \sum_{i \in S} p_i^S \ln\frac{p_i^S}{q_i^S} = \sum_{i \in S}\frac{y_i}{y^S}\ln\frac{y_i}{y^S}\frac{z^S}{z_i} = \sum_{i \in S}\frac{y_i}{y^S}\left(\ln\frac{y_i}{z_i} + \ln\frac{z^S}{y^S}\right) = \frac{1}{y^S}\sum_{i \in S}y_i\ln\frac{y_i}{z_i} - \frac{1}{y^S}\sum_{i \in S}y_i\ln\frac{y^S}{z^S}$$

hence $\quad \displaystyle\sum_{i \in S}y_i\ln\frac{y_i}{z_i} \geq \sum_{i \in S}y_i\ln\frac{y^S}{z^S} = \ln\frac{y^S}{z^S}\sum_{i \in S}y_i = y^S\ln\frac{y^S}{z^S}$

Noting that $y^Q = 1 - y^P$ and $z^Q \leq 1 - z^P$, we write

$$\sum_{i=1}^{d}y_i\ln\frac{y_i}{z_i} = \sum_{i \in P}y_i\ln\frac{y_i}{z_i} + \sum_{i \in Q}y_i\ln\frac{y_i}{z_i}$$

$$\geq y^P\ln\frac{y^P}{z^P} + y^Q\ln\frac{y^Q}{z^Q} \geq y^S\ln\frac{y^S}{z^S} + (1-y^S)\ln\frac{1-y^S}{1-z^S} \geq 2(y^S - z^S)^2$$

where the last inequality follows from Lemma 2.8.1. Therefore, we obtain

$$(y^S - z^S)^2 \leq \frac{1}{2}\sum_{i=1}^{d}y_i\ln\frac{y_i}{z_i} \tag{2.8.7}$$

At this point we now choose $P = \{i \in I : y_i > z_i\}$ and $Q = \{i \in I : y_i \leq z_i\}$ and upper bound $\sum_{i \in S} f(y_i - z_i)$ as follows:

$$\sum_{i \in S}f(y_i - z_i) \overset{(a)}{=} \sum_{i \in S}f(|y_i - z_i|) \overset{(b)}{\leq} f\left(\sum_{i \in S}|y_i - z_i|\right) \overset{(c)}{=} f\left(\left|\sum_{i \in S}y_i - z_i\right|\right) \overset{(d)}{=} f(|y^S - z^S|)$$

$$= f\left(\sqrt{(y^S - z^S)^2}\right) \overset{(e)}{\leq} f\left(\sqrt{\frac{1}{2}\sum_{i=1}^{d}y_i\ln\frac{y_i}{z_i}}\right)$$

(a) follows from the fact that $f(x) = f(-x)$. (b) follows by Lemma 2.8.5. (c) holds as by construction of S, all the $y_i - z_i$ terms are either all positive, or all negative, so we can sum first and then take the absolute value after. (d) follows by distributing the sum and inserting the definitions of y^S and z^S. (e) is applying (2.8.7), together with monotonicity of $\sqrt{\cdot}$ and f (Lemma 2.8.2). We can now conclude the proof by noting that

$$\frac{1}{2}\sum_{i=1}^{d} f(y_i - z_i) = \frac{1}{2}\sum_{i \in P} f(y_i - z_i) + \frac{1}{2}\sum_{i \in Q} f(y_i - z_i) \leq f\left(\sqrt{\frac{1}{2}\sum_{i=1}^{d} y_i \ln\frac{y_i}{z_i}}\right)$$

∎

Corollary 2.8.8 (Specific entropy inequalities) Continuing from Theorem 2.8.6, we also have that

(S) Square:
$$\sum_{i=1}^{d}(y_i - z_i)^2 \leq \sum_{i=1}^{d} y_i \ln\frac{y_i}{z_i}$$

(A) Absolute:
$$\sum_{i=1}^{d}|y_i - z_i| \leq \sqrt{2\sum_{i=1}^{d} y_i \ln\frac{y_i}{z_i}}$$

(H) Hellinger^2:
$$\sum_{i=1}^{d}(\sqrt{y_i} - \sqrt{z_i})^2 \leq \sum_{i=1}^{d} y_i \ln\frac{y_i}{z_i}$$

Proof. Taking Theorem 2.8.6 and substituting $f(x) = x^2$ or $f(x) = |x|$ gives (S) or (A) respectively. To prove (H), for $t, y, z > 0$,

$$g(t) := -\ln(t) + t - 1 \geq 0 \quad \text{implies that}$$
$$f(y,z) := y\ln\frac{y}{z} - (\sqrt{y} - \sqrt{z})^2 + z - y = 2yg(\sqrt{z/y}) \geq 0$$

hence $y\ln(y/z) - (\sqrt{y} - \sqrt{z})^2 \geq y - z$. Summing, we obtain

$$\sum_i y_i \ln\frac{y_i}{z_i} - \sum_i (\sqrt{y_i} - \sqrt{z_i})^2 \geq \sum_i y - \sum_i z \geq 1 - 1 = 0$$

from which the result follows.

∎

2.8.2 Dealing with Semimeasures

Measure theory is already quite advanced, but at least it has been well-developed. As for *semi*measures, we will present different ways to deal with them. Unfortunately we have some need for semimeasures in this book, which will become clear later. Measure theory is a vast non-trivial field we can tap into, but semimeasure theory essentially does not exist. Below we discuss several solutions or workarounds, some of them are used in this book. The only measurable space we need to consider is $\mathcal{X}^\infty$ for finite $\mathcal{X}$ with σ-algebra $\mathcal{F} = \sigma(\{\Gamma_x\})$ (see Definition 2.2.11).

Only use ν on cylinder sets. For most parts of the book, we actually only need to know ν on the cylinder sets Γ_x it is defined on. The most frequently used quantity is the predictive distribution $\nu(x_t|x_{<t}) \equiv \nu(\Gamma_{x_{1:t}})/\nu(\Gamma_{x_{<t}})$. For predictions/agents only involving fixed finite sequence length/lifetime m, we could even choose finite $\Omega = \mathcal{X}^m$ and $\mathcal{F} = 2^\Omega$ which trivializes most of semimeasure theory.

Develop semimeasure theory.   Extending volumes of measure theory definitions and results to semimeasures may be an interesting research program, but is beyond the scope of this book. We leave this to some eager math students (Exercise 7).

Derive required semimeasure results.   A more focussed solution would be to only develop the theory useful/needed/convenient for this book (Exercise 6). Unfortunately even this would be a major undertaking. Extending the finitary results is often not too hard and indeed has been done for some such as the entropy inequalities (Corollary 2.8.8). For others, such as martingale convergence or even just the general definition of expectation, we would have to dig too deep.

Defining ν on measurable sets.   Strictly speaking, ν is only pre-semimeasure. For more sophisticated results, we need to somehow extend the definition of ν to non-cylinder sets, which is also most likely the first step towards developing a proper theory of semimeasures. For this we need a semimeasure version of Carathéodory's extension theorem. We only need a special case for cylinder sets. The following (unverified) construction may work, but is very speculative, and essentially states an open research problem: Let $\mathcal{S}, \mathcal{T} \subseteq \mathcal{X}^*$ denote prefix-free sets (Section 2.1.2), and define $\Gamma_{\mathcal{S}} := \dot{\bigcup}_{x \in \mathcal{S}} \Gamma_x$ and

$$\nu_*(A) := \sup_{\mathcal{S}: \Gamma_{\mathcal{S}} \subseteq A} \sum_{x \in \mathcal{S}} \nu(\Gamma_x)$$

This definition is consistent with ν in the sense that $\nu_*(\Gamma_x) = \nu(\Gamma_x)$ and ν_* is a semimeasure as per Axiom 2.2.3. Proof sketch: Let $\Gamma_{\mathcal{S}} \subseteq A$ and $\Gamma_{\mathcal{T}} \subseteq B$, then for disjoint A and B, we have $\sum_{x \in \mathcal{S}} \nu(\Gamma_x) + \sum_{x \in \mathcal{T}} \nu(\Gamma_x) = \sum_{x \in \mathcal{S} \dot{\cup} \mathcal{T}} \nu(\Gamma_x) \leq \nu_*(A \dot{\cup} B)$, then taking $\sup_{\mathcal{S}}$ and $\sup_{\mathcal{T}}$ on the l.h.s. gives $\mu_*(A) + \mu_*(B) \leq \mu_*(A \dot{\cup} B)$. Indeed, ν_* is defined for *all* sets $A \subseteq \mathcal{X}^{\infty}$ and seems to be what is called an *inner measure*. Unfortunately there are $\mathcal{F}$-measurable sets A for which $0 = \nu_*(A) \neq \nu(A) = 1$ if ν is a measure extended to $\mathcal{F}$: Take $A = \mathcal{X}^{\infty} \setminus \{x0^{\infty} : x \in \mathcal{X}^*\}$ and Lebesgue measure ν. An outer semimeasure that is consistent with ν may be defined as

$$\nu^*(A) := \inf_{\mathcal{S} \supseteq A} \sum_{x \in \mathcal{S}'} \nu(\Gamma_x) \quad \text{where} \quad \mathcal{S}' := \arg\min_{\mathcal{S}'} \{ |\mathcal{S}'| : \Gamma_{\mathcal{S}'} = \Gamma_{\mathcal{S}} \}$$

That is, $\mathcal{S}'$ is the smallest representation of $\Gamma_{\mathcal{S}}$: $\mathcal{S}$ for which $xa \in \mathcal{S}$ for some x and all $a \in \mathcal{X}$ are excluded. We cannot use the standard Carathéodory criterion to decide which sets are measurable, but we already constructed the σ-algebra $\mathcal{F}$, so if we define $\nu(A) := \nu^*(A)$ for all $A \in \mathcal{F}$, this may be a (semi)measure on $\mathcal{F}$. Another approach may be to define $\mu(A) = \mu^*(A)$ iff outer and inner semimeasures match and then extend this to all $A \Delta Z$ for any Z with $\nu^*(Z) = 0$ or so.

Expectations w.r.t. semimeasures.   Once we have defined ν for all measurable sets, we could (mis)use (one of) the general definition(s) of expectations for non-negative f:

$$\mathbf{E}_\nu[f] := \int_0^\infty f^*(t) \, ds \quad \text{with} \quad f^*(s) := \nu(\{\omega \in \Omega : f(\omega) > s\})$$

also for semimeasures (or maybe even use ν_* or ν^*). For indicator function $f(x_{1:\infty}) = [\![x_{<t} = y_{<t}]\!]$, this leads to $f^*(s) = \nu(\Gamma_{y_{<t}}) [\![s < 1]\!]$, hence $\mathbf{E}_\nu[f] = \nu(\Gamma_{y_{<t}})$ which is sensible, but decomposing $1 = \sum_{a \in \mathcal{X}} f_a$ with $f_a(x_{1:\infty}) := [\![x_1 = a]\!]$ results in

$$\mathbf{E}[\sum_a f_a] = \mathbf{E}_\nu[1] = \nu(\Gamma_\epsilon) \geq \sum_{a \in \mathcal{X}} \nu(\Gamma_a) = \sum_{a \in \mathcal{X}} \mathbf{E}[[\![x_1 = a]\!]] = \sum_a \mathbf{E}[f_a]$$

so expectation is no longer linear under semimeasures. Other choices for $\mathbf{E}$, possibly (implicitly) based on the reductions to measures discussed below, may preserve linearity.

Include finite sequences. The necessity of semimeasures arises from the fact that some monotone Turing machines may not output infinite sequences but stop or loop after a finite output. The probably most honest treatment is to actually include these finite outputs and enlarge the space Ω to $\mathcal{X}^\infty \cup \mathcal{X}^*$ and define the probability that the monotone TM has finite output $x \in \mathcal{X}^*$. We can extend Ω from $\mathcal{X}^\infty$ to $\mathcal{X}^\infty \cup \mathcal{X}^*$, and the cylinder sets become $\Gamma_x^* := \{xy : y \in \mathcal{X}^\infty \cup \mathcal{X}^*\}$. In general we can convert any semimeasure ν to a measure $\tilde{\nu}$ on $\sigma(\Gamma_x^*, x : x \in \mathcal{X}^*)$ via

$$\tilde{\nu}(\Gamma_x^*) := \nu(\Gamma_x) \quad \text{and} \quad \tilde{\nu}(\{x\}) := \nu(\Gamma_x) - \sum_{a \in \mathcal{X}} \nu(\Gamma_{xa})$$

The latter we will interpret in Section 15.7 as the probability of the world ending or the predictor/agent dying after observing $x \in \mathcal{X}$. Now we have

$$\Gamma_x^* = \{x\} \cup \bigcup_{a \in \mathcal{X}} \Gamma_{xa}^* \quad \text{and indeed} \quad \tilde{\nu}(\Gamma_x^*) = \tilde{\nu}(\{x\}) + \sum_{a \in \mathcal{X}} \tilde{\nu}(\Gamma_{xa}^*)$$

that is, $\tilde{\nu}$ is a proper probability measure. For a lower semicomputable ν, $\tilde{\nu}$ is only limit-computable on $\mathcal{X}^*$, and the hybrid sample space is annoying making many equations cumbersome to deal with.

Extend the alphabet. A similar "honest" idea is to consider alphabet $\mathcal{X} \cup \{\bot\}$ to include a "death" symbol $\bot$, and $\forall x \in \mathcal{X}^*$ extend μ via

$$\nu(\Gamma_{x \bot^k}) := \nu(\Gamma_x) - \sum_{x \in \mathcal{X}} \nu(\Gamma_{xa}) \;\; \forall k \in \mathbb{N}^+ \quad \text{and} \quad \nu(\Gamma_{x \bot y}) := 0 \;\; \forall y \in (\mathcal{X} \cup \{\bot\})^* \setminus \{\bot\}^*$$

Now ν is a proper measure on $(\mathcal{X} \cup \{\bot\})^\infty$. We keep the convenience of not having to deal with finite sequences, and in the agent case, if we assign reward 0 if/once observing a $\bot$ (death), the Bellman equations remain valid for (once semi now)measures, and agent death is still modelled. ν remains lower semicomputable on $x \in \mathcal{X}^*$. It fails if x includes $\bot$, but this does not cause any problems, esp. if reward 0 is assigned to $\bot$, and makes the value function limit-computable (see Sections 6.7 and 13.1). Some results have only been derived for binary $\mathcal{X}$, so are not available on $\mathbb{B} \cup \{\bot\}$. ξ retains its mixture and hence dominance property even on sequences containing $\bot$.

Solomonoff normalization. Given a semimeasure ν on cylinder sets, we can define

$$\nu_{norm}(x_t | x_{<t}) := \frac{\nu(\Gamma_{x_{1:t}})}{\sum_{x_t \in \mathcal{X}} \nu(\Gamma_{x_{1:t}})}, \quad \nu_{norm}(\Gamma_{x_{1:n}}) := \prod_{t=1}^{n} \nu_{norm}(x_t | x_{<t}), \quad \nu_{norm}(\epsilon) := 1$$

which is a properly normalized probability measure after extending it to $\mathcal{F}$. Crucially, it also preserves dominance (Proposition 3.1.4), since $\nu_{norm}(\Gamma_x) \geq \nu(\Gamma_x)$. Two downsides are that ν_{norm} is no longer lower semi-computable, even if ν was, but it is at least still limit-computable. Also we lose the semimeasure defect later attributed to the death of an agent. Note that we could redistribute the missing probability mass in (infinitely many) other ways, all of them satisfying $\nu_{norm} \geq \nu$.

Lower limit normalization. We could also look for the largest measure $\underline{\nu} \leq \nu$. This is unique and can explicitly be represented [Hut14b] as

$$\underline{\nu}(\Gamma_{x_{<t}}) := \lim_{n \to \infty} \sum_{x_{t:n}} \nu(\Gamma_{x_{1:n}})$$

The limit exists since the argument decreases with n due to the semimeasure property of ν. It is *not* a *probability* measure: One can show that $\underline{\nu}(\Gamma_x) = \tilde{\nu}(\Gamma_x)$ (note Γ_x, not Γ_x^*, Exercise 4) with the missing probability mass $\tilde{\nu}(\mathcal{X}^*)$ on finite sequences being unaccounted for in $\underline{\nu}$, but $\underline{\nu}$ can easily be normalized to one by $\underline{\nu}(x)/\underline{\nu}(\epsilon)$, unless $\underline{\nu}$ is the (uninteresting) 0-measure. If ν is generated (lower semicomputed) by a monotone Turing machine (Definition 2.7.1 and Definition 2.7.20), the conversion to $\underline{\nu}$ corresponds to disregarding all finite outputs with the Turing machine halting or looping forever. As such, $\underline{\nu}$ is in general not even limit-computable, and pushes it up to $\underline{\nu}(x_{<t}) = \inf_n \sum_{x_{t:n}} \sup_s \nu^s(x_{1:n}) \in \Pi_2^0$ (2.6.20). Note that in general $\underline{\nu} \not\gtrsim \nu$, e.g. for $\nu(1^n) = {}^1\!/_2$ and $\nu(0^n) = 2^{-n-1}$ and 0 on all other sequences we have $\underline{\nu}(0) = 0$ despite $\nu(0) = {}^1\!/_4$ and $\underline{\nu}(\epsilon) = {}^1\!/_2 \neq 0$. But if we normalize first, i.e. take a class $\underline{\mathcal{M}} = \{\underline{\nu} : \nu \in \mathcal{M}\}$, and then mix over it, we trivially have $\xi_{\underline{\mathcal{M}}} \geq w_\nu \underline{\nu}$, but this pushes $\xi_{\underline{\mathcal{M}}}$ even to Σ_3^0.

Computable measures.   A monotone Turing machine that is guaranteed to output only infinite sequences generates a properly normalized probability measure. Let $\mathcal{M}_{comp}$ be the class of all such measures. Each measure in $\mathcal{M}_{comp}$ is computable, but unlike $\mathcal{M}_{sol}$, $\mathcal{M}_{comp}$ is not effectively enumerable, and hence the mixture ξ over $\mathcal{M}_{comp}$ is not even approximable.

Assume the true environment μ is a proper probability measure.   This is a global assumption we make throughout this book. Indeed, some key theorems fail if μ is a semimeasure, such as the entropy inequalities (Corollary 2.8.8). There may be a way to generalize some results, but there seems to be no particular need, except possibly modelling finite universes or death.

Primitive recursive measures $\mathcal{M}_{pr}$.   Instead of considering all Turing machines, we could restrict ourselves to primitive recursive (p.r.) functions, which are total functions that have a p.r. upper bound on the run-time. The class contains virtually all computable functions one usually encounters. Functions computable but not p.r. are very slow, e.g. have run time growing like the Ackermann function. The class of p.r. functions is itself recursively enumerable (even without repetition if desired [Liu60]). This can be converted into an enumeration of p.r. measures $\mathcal{M}_{pr} \subset \mathcal{M}_{comp}$, e.g. by Solomonoff normalization. Assuming computable prior weight w_ν, the Bayesian mixture ξ over $\mathcal{M}_{pr}$ is then computable, but $\xi \notin \mathcal{M}_{pr}$.

Reflective-oracle computable measures $\mathcal{M}_r^O$.   In Section 10.5 we will introduce a class $\mathcal{M}_r^O \supset \mathcal{M}_{comp}$ of probability measures that are computable with the help of a reflective oracle. It has the nice property that the normalized Bayes mixture dominates the class and is in the class. Even more remarkable, the Bayes-optimal policy π_ξ^* is also in the class, solving the long-standing Grain of Truth problem.

Conclusion.   Ideally the whole book were developed for semimeasures, since $\mathcal{M}_{sol}$ is the core intended class of application used in Solomonoff induction and the AIXI agent, but this is unfortunately not feasible. So a lot of theory in this book is developed for measures only, and some of the approaches above have been or need to be applied when semimeasures crop up. The text may on occasion be a bit sloppy or sweep subtleties under the carpet, but we tried to be precise and explicit in the theorems. Any remaining problems should be fixable with one of the approaches above. In case of doubt, assume $\perp$ has been added to $\mathcal{X}$ and in the agent case $r = 0$ if $\perp$ is perceived, or assume $\mathcal{M}_{pr}$ or $\mathcal{M}_{comp}$ instead of $\mathcal{M}_{sol}$.

2.8.3   Probability Zero

Probably zero events are another nuisance. For continuous spaces Ω, they are ubiquitous, e.g. when throwing darts at a board, the probability of hitting any specific point is zero, still even the worst player has a non-zero probability of hitting some segment or ring. Technically, this is modelled by a probability *density*. One can still define versions of conditional probability and expectations conditioned on probability zero outcomes. Another question is what the expectation of a function is that is infinity or undefined on a set of measure zero. Nearly all of the book only deals with discrete random variables, which simplifies matters, but does not trivialize the problem.

We collect the issues and how to deal with them below in one place, and implicitly assume one such solution is adopted in the rest of the book, in order not to clutter the book and distract from its core aim.

Expectation for partial functions. The expression of expectation in Definition 2.2.36 is fine as long as the function g is total. If g can assume infinite values or is partial, the correct definition for discrete X is

$$\mathbf{E}[g(X)] = \sum_{x \in X(\Omega):p(x)>0} g(x)p(x) \quad \text{for} \quad g:\mathbb{R} \to \mathbb{R} \cup \{\pm\infty\} \cup \{\bot\}$$

Note the restriction of Σ to x for which $p(x)>0$. For finite-valued total functions g, i.e. with range $\mathbb{R}$, this restriction is redundant and can be dropped. We still want the expectation to be defined if g is infinity or undefined ($\bot$) for some x. The standard convention in measure theory is that if two functions differ on a set of measure 0, then they have the same expectation. So if we assume $g(x) \notin \mathbb{R}$ only if $p(x)=0$ and define $\tilde{g}(x):=g(x)$ if $g(x) \in \mathbb{R}$, and $\tilde{g}(x):=0$ otherwise, then $\mathbf{E}[g(X)] \equiv \mathbf{E}[\tilde{g}(X)] = \sum_x \tilde{g}(x)p(x) = \sum_{x:p(x)>0} g(x)p(x)$.

As a simple example, consider $\Omega = \{0,1\}$ and $g(x):=1/p(x)$, then $\mathbf{E}[g(x)] = \sum_x g(x)p(x) = \sum_{x\in\Omega} 1 = 2$ provided $\forall x.p(x)>0$. On the other hand, if $p(0)=0$ we get $\mathbf{E}[g(x)] = g(1)p(1) = 1$. Without the restriction on the sum, the expectation would be undefined, or maybe 2 by a limit argument $p(0) \to 0$. All this is sometimes sloppily expressed as "$\pm\infty \cdot 0 = 0$ and $\bot \cdot 0 = 0$ in measure theory." As such, we will implicitly assume such sum restriction where necessary. For instance, this would be an alternative view in the definition of entropy and KL divergence that the term $0\ln 0$ is simply absent rather than 0 by a limit argument.

Conditioning on $P(A)=0$. Conditioning on probability 0 events is a subtle art. Since a probability zero event will never happen, the safest option is to leave $P(B|A)$ undefined if $P(A)=0$, but this is restrictive (e.g. does not allow conditional densities) and cumbersome (requires constant handling of this special case). There is no best solution to this problem. See [Hut05b, Sec.3.9.1] for some (unorthodox) suggestions.

Conditioning on $\mu(x_{<t})=0$. We only encounter this problem for random sequences and $\mu(\cdot|x_{<t})$ when $\mu(x_{<t})=0$, so we focus on this special case. If x were continuous and μ a density, there are ways to condition on probability zero outcomes that have positive probability *density*. But here x is discrete. The default solution would be to add to every theorem "holds with μ-probability 1" if it is not there already. Since $\mu(x_{<t})=0$ is a probability 0 event, we can safely ignore such histories $x_{<t}$ in our consideration.

Probability Kernels. In the agent case, we consider policy π interacting with environment μ generating history $h_{1:\infty}$. Occasionally we are interested in how a policy π continues $h_{<t}$, for any $h_{<t}$ even if $h_{<t}$ would never occur (has probability 0 under π). For instance, $h_{<t}$ could have been created by a different policy.

A satisfactory and general solution is to *start* with "conditional" distributions and then take their product. Technically we start with *probability Kernels* $\mu_t:\mathcal{X}^t \to \Delta\mathcal{X}$ for all $t \in \mathbb{N}$, where for each $x_{<t}$, $\mu_t(\cdot;x_{<t})$ is a probability distribution over $\mathcal{X}$ (note the semicolon and

index t). Then

$$\mu(x_{1:n}) := \prod_{t=1}^{n} \mu_t(x_t; x_{<t}) \quad \text{and} \quad \mu(x_t|x_{<t}) = \mu_t(x_t; x_{<t})$$

is uniquely and well-defined even when $\mu(x_{<t})=0$. The collection of probability Kernels (μ_t) contains strictly more information than the joint measure μ.

Consider a probabilistic Turing machine T that lower semicomputes semiprobability Kernels, i.e. $\nu_T(x_t|x_{<t})$ is the probability that $T(x_{<t},t,\text{random noise})$ outputs x_t. The class $\{\nu_T : T \text{ TM}\}$ is recursively enumerable and contains all and only semimeasures that are *conditionally* lower semicomputable. This class is "slightly" smaller than $\mathcal{M}_{sol}$, since $\nu \in \mathcal{M}_{sol}$ are not necessarily *conditionally* lower semicomputable.

Pragmatic approach.   The above is the approach we are generally taking, but without being pedantic about the semicolon and index t, i.e. we directly define $\nu(x_t|x_{<t})$, even though strictly speaking it is a probability Kernel and not a conditional distribution if $\nu(x_{<t})=0$. We will also not explicitly pursue this path everywhere, but implicitly assume when we say "let ν be a (semi)measure" that the conditional distributions have been defined as well.

If we consider a concrete mechanism that generates ν, but do not provide the conditionals, such as is the case for monotone Turing machines generating ν_T, we define $\nu(x_t|x_{<t})$ arbitrarily when $\nu(x_{<t})=0$, and as far as possible consistent with other properties, such as computability.

Example.   For instance, if $\nu_T^s(x_{<t})$ for $s=1,2,3,...$ is a lower-computation of lower semicomputable $\nu_T(x_{<t})$, define $\nu_T^s(x_t|x_{<t})=1/|\mathcal{X}|$ (or any other computable choice) as long as $\nu_T^s(x_{<t})=0$. Once (if ever) $\nu_T^s(x_{<t})>0$, then of course $\nu_T^s(x_t|x_{<t})=\nu_T^s(x_{1:t})/\nu_T^s(x_{<t})$. Note that the conditional distribution is not lower semicomputable but only limit-computable in any case. In contrast, in Section 10.5 we directly define conditional (reflective-oracle) computable measures.

2.8.4   Exercises

1. [C22] Prove that the Absolute distance and *Un*Squared Hellinger distance $\sqrt{H}$ are both metrics, but the KL divergence is not (Definition 3.2.1).

2. [C15] Show that the bounds in Corollary 2.8.8 are tight in the sense that the ratio of l.h.s./r.h.s. can get arbitrarily close to 1.

3. [C25] Generalize all definitions of distances and bounds to probability measures on general sample spaces Ω, except for the 2-norm. Explain what the problem with the 2-norm is.

4. [C12] Derive an explicit expression for $\tilde{\nu}(\Gamma_x)$ (note $\Gamma_x \neq \Gamma_x^*$) and show that $\underline{\nu}(\Gamma_x)=\tilde{\nu}(\Gamma_x)$, both defined in Section 2.8.2.

5. [C35] Prove all claims in Section 2.8.2.

6. [C40o] Suitably generalize all results in this book to semimeasures and prove them.

7. [C50o] Take a textbook on measure theory and suitably extend all definitions and results to semimeasures.

2.9   History and References

The canonical reference text on artificial intelligence [RN20] gives a comprehensive overview over AI approaches in general and contains over a thousand further AI-related references. The introductory chapter of [LV19] covers similar material as our background chapter with many further related references. This history and reference section is an updated version of [Hut05b, Sec.2.5].

Binary strings (2.1).   This section is based on the corresponding sections in [Hut05b] and [LV19]. *Principles of mathematical analysis* by Rudin [RT64] provides an introduction to many mathematical concepts from real analysis that we assume. Prefix codes are also discussed at depth in the book of Li and Vitányi [LV19]. The general theory of coding and prefix codes can be found in [Gal68].

Measure theory and probability (2.2).   This section is based on material from [GS20]. Most of the measure theory we required has been omitted as much as reasonably possible, enough to lay a foundation for probability and no further. The book *Real Analysis* by Stein and Shakarchi [SS09] provides a slower, more formal introduction to measure theory for those so inclined. The i.i.d. assumption is prevalent in statistics and machine learning but unsuitable for AGI [hut22].

Although games of chance date back at least to around 300 B.C., the first mathematical analysis of probabilities appears to be much later. Important breakthroughs have been achieved (in chronological order and with significant simplification) by Cardano [Car63], a systematic way of calculating probabilities by Pascal (in correspondence with Fermat) and conditional probability [Pas54], Bayes' rule [Bay63], the distinction between subjective and objective interpretation of probabilities and the weak law of large numbers by Bernoulli [Ber13], equi-probability due to symmetry and other things by Laplace [Lap12], the principle of indifference by Keynes [Key21], Kolmogorov's axioms of probability theory [Kol33], early attempts to define the notion of randomness of individual objects/sequences by von Mises [Mis19], Wald [Wal37], and Church [Chu40], finally successful by Martin-Löf [ML66], the notion of a universal a priori probability by Solomonoff [Sol64] and its mathematical investigation by Levin [ZL70a, Lev74].

There are many suitable textbooks for probability [Shi96, Kal10]. The bulk of this chapter is based on two books: *Probability and Random Processes* by Grimmett et al. [GS20] provides a more gentle introduction to probability theory, with more exercises for practice than we provide. They cover probability theory starting with an axiomatic foundation, move on to random variables (both discrete and continuous), as well as covering sequences and convergence of random variables. *Introduction to Probability Models* by Ross [Ros14] is a more advanced book, that provides a similar introduction with events and random variables (parameterized by commonly used distributions in statistics), but the bulk of the book is beyond the scope of what we require. For a thorough treatment of the early history of the concept of probability the reader is referred to the books by Hacking [Hac75] and Hald [Hal90], and for the foundations developed in the 20th century to the book by Schnorr [Sch71] and the PhD theses by van Lambalgen [Lam87] and Wang [Wan96]. See also the book [SV01] by Shafer & Vovk from a more game-theoretic and finance perspective. Feller [Fel68] is another good standard textbook. A philosophical treatise of conditional probability can be found in [Haj11].

Statistical inference and estimation (2.3).   A great introduction to (frequentist) statistics is [Was10], which focuses on the general theory and fundamental principles rather than specific models. Section 2.3.4 is based on a chapter from *Elements of Information Theory* [CT06], where the Cramér–Rao inequality (Theorem 2.3.14) is proven. The sunrise problem was first described by Laplace [Lap40], in which the rule of succession now known as *Laplace's Rule* is described (Section 2.4.3).

The limitations of frequentist statistics are discussed in [Hut05b, Sec.2.3.1][Háj09b]: The naive definition is circular (though see Cournot's forgotten principle [Cou43]), identical events are a mirage (the reference class problem), and most seriously the limitation to i.i.d. data. But given the generality and consistency of Bayesian inference, it is somewhat surprising that so many alternatives sprouted in the history of AI [Che85, Che88]: Default reasoning [Rei80], non-monotonic logic [MD80], circumscription [McC80], certainty factors [Sho76, BS84], Dempster-Shafer theory [Dem68, Sha76], Fuzzy logic [Zad65, Zim91], possibility theory [Zad78], imprecise probabilities [Wal91, Hut03f, ZH05, PZTH07, Hut09h, PZTH09], robust Bayesian analysis [RIR00], and many others. See [Fin73, Wal91, RN20] for a more detailed account of various uncertain reasoning systems. Many reasons why classical probability theory is unsuitable for AI have been put forward: strict numerical values are not appropriate for a qualitative reasoning system, probability theory cannot deal with impreciseness, vagueness, or subjective beliefs, or is just impractical, and so classical probabilities fell out of favor in the 1970s. But it turned out that all these alternative approaches have their own problems: Either they have unclear semantics, or they are not self-consistent, or they do not scale up, or worse. It is not that Bayesian and Frequentist probability theory leave no wishes open, but especially the former is the most consistent system developed so far, with solid justifications via Cox axioms [Cox46] and Dutch book arguments [Háj09a] and others [Pre09, Sec.2]. So after long and sometimes bitter fights [Che85, Che88], the debates about the merits of the different approaches to dealing with uncertainty in AI have essentially been settled in the new millennium in favor of machine learning systems based on classical statistics [HTF09] and increasingly (sometimes robust) Bayesian methods [Bis06, KF10, Mur22, Mur23], finally returning to its roots in 1763 [Bay63].

Bayesian probability theory (2.4).   Bayes' Law is one of the earliest non-trivial rules in probability and statistics [Bay63] which *Laplace's Rule* [Lap40] builds upon. The pleasant philosophical text by Jaynes [Jay03] treats probability theory as a natural extension to (Boolean) logical reasoning, with emphasis on the Bayesian approach, and discussion of various historical paradoxes and how they could be avoided. Earman [Ear93] is another good philosophical text on Bayes. Gelman [GCSR95] is a more practical book on Bayesian data analysis. Berger [Ber93] is a more advanced book "in-between". More on Bayesian inference and a collection of different choices of model class and prior can be found in [BC16, GvdV17]. Some concrete classes will be discussed in Chapters 3 to 5. [Bis06, Mur22, Mur23] are two standard textbooks on Bayesian machine learning.

There is an ongoing debate between objective and subjective probability, which became sharper in the 20th century (not only in AI). Prominent advocates of the relative frequency or objective interpretation were Kolmogorov [Kol63], Fisher [Fis22], and von Mises [Mis28]. Hajek [Háj96, Háj09b] offers fifteen arguments against finite/hypothetical frequentism. There are many advocates of probabilities as degrees of belief [Pop34, Ram31, Fin37, Fin74, Cox46, Sav54, Jef83, Gol06]. See [OBD+06, O'H19] for how to elicit subjective probabilities from experts. Objective priors are advocated in [Ber06, BBS24], und universal priors in [Hut07e, RH11]. The property of Jeffery's objective prior (Definition 2.4.11) leading to reparameterization invariance is discussed more in [Pre09, O'H10, Lee12]. See [Goo71] for a classification of 46656 varieties of Bayesians, and [Pre09, Sec.2] for an overview of justifications for Kolmogorov axioms from the various perspectives on probability.

Carnap [Car48, Car50] tried to supplement logic with probability theory to so-called inductive logic. This works fine for propositional logic [Jay03], but attempts to extend this satisfactorily to predicate logic failed for long [Put63] but finally has been solved

[GS82, HLNU13b, HLNU13a, GBTC⁺20]. The closely related reference class problem is addressed in [Rei49, Kyb77, Kyb83, BGHK92].

Information theory and coding (2.5). This section is based on the texts [Mac03, CT06, JJHH03, LV19]. The entire discipline of information theory originates from Shannon's original paper [Sha48], formalizing the problem of transmitting a message over a communication medium with errors, as well as the concept of entropy and proving Theorem 2.5.22. The entropy of English text has been estimated through experiments using both humans [Sha51] and algorithms [Gue09, CK78] as predictors. The Kraft inequality is due to Kraft [Kra49]. The Kullback–Leibler distance originated in a paper from the eponymous pair [KL51].

Shannon-Fano coding refers to two similar coding methods, presented by Shannon [Sha48] and Fano [Fan49]. A similar method for coding by Huffman [Huf52] is provably optimal given each symbol is coded individually, and the frequency distribution of symbols is known. Better results can be obtained by using block codes such as arithmetic codes [Pas76, HV91] which perform better on low entropy messages than Huffman coding. Golomb codes [GVV75] are optimal assuming the symbols to compress follow a geometric distribution. Various pre-coding transformations like *run length encoding* [RC67] or *move-to-front* [Rya80] or *Burrows-Wheeler* [BW94] transforms can be applied before applying another code to further decrease size. Dictionary coders compress by replacing a substring of the message with an index into a lookup table. The family of Lempel-Ziv compressors (starting with LZ77 [ZL77] and LZ78 [ZL78]) popularized this technique, extending to LZW [Wel84], LZRW [Wil91b], and LZSS [SS82]. Another family of compressors are called context mixing compressors. Similar to Context Tree Weighting (covered in Chapter 4), they use a mixture of models to estimate the probability distribution from which the data was drawn, and compressed using arithmetic coding [Sai04] with respect to that predictor. The PAQ family [Mah05] of compressors uses this technique, by taking a weighted average of the models. This was later improved to using a neural network to perform the mixture [Mah07].

A survey of various compression techniques is given in [Mah12, SM10]. Different choices of compressors provide trade-offs between compression time, compression ratio and decompression time [Mah22, Nem22]. The Hutter Prize [Hut20a] offers a monetary reward for compressing `enwik9` [Mah11], a 1GB sample of text from Wikipedia smaller than the (current as of time of writing) record of $\approx$113MB.

Computability theory (2.6). This section is based on the texts [HMU06, Hut05b]. The book of Hopcroft, Ullman and Motwani, [HMU06], is a very readable elementary introduction to automata theory, formal languages, and computability theory. [BBJ02, Sip12] also serve as good introductions to computability. Complexity theory refines computability theory by taking computation time into account, covered by books ranging from easy [Sip12] to medium [Pap94] to hard [AB09] to insane [HO02].

Turing [Tur36] introduced the concept of a Turing machine and demonstrated that the halting problem is undecidable. Turing machines are formally equivalent to partial recursive functions (see [Rog67, Odi89, Odi99] for an introduction), as well as Alonzo Church's lambda calculus [Chu36]. The halting problem corresponds to Gödel's incompleteness theorem [Göd31, Sho67] whose proof is based on a diagonal argument invented by Cantor [Can74, Dau90].

The works [Göd31, Kle36, Tur36, Pos44, ZL70a, Sch02a] show the importance of the various computability concepts defined in Section 2.6.3. The consideration of (and naming for) estimable and approximable functions in the context of universal priors is from [Hut03b, Hut06b]. The introduction to the arithmetic hierarchy follows [Nie09, Lei16b].

Kolmogorov complexity (2.7). This section is based on the corresponding sections of the texts [LV19, Hut05b]. Other books which can serve introductory texts to this topic include

[Cal02, Cha87, DH10]. See [Hut07a, Hut08b, SH15a] for some short and light encyclopedic introductions.

Algorithmic Information Theory. A coarse picture of the early history of algorithmic information theory could be drawn as follows: Kolmogorov [Kol65] and Chaitin [Cha66] suggested defining the information content of an object as the length of the shortest program computing a representation of it [Hut08b]. Solomonoff [Sol64] independently invented the closely related universal prior probability distribution and used it for binary sequence prediction [Sol64, Sol78, HLV07]. Levin worked out most of the mathematical details [ZL70a, Lev74] and invented the fastest algorithm for function inversion and optimization, save for a (huge) constant factor [Lev73b, Gag07]. Chaitin's [Cha66, Cha75] major focus is on the halting probability $\Omega = \sum_{p:U(p) \text{ halts}} 2^{-\ell(p)}$, the probability that a random program (weighted by length) will cause the universal machine U to halt. These papers may be regarded as the invention of what is now called algorithmic information theory. For a short introduction and a list of applications, see [Hut07a, LV07].

The Assumption 2.7.8 of a short compiler is an effective version of Kolmogorov's assumption that complexities based on different "reasonable" universal "Turing" machines coincide reasonably well [Kol65].

The Minimum Description Length (MDL) Principle [Ris78, Gru07, GR19], and the Minimum Message Length (MML) Principle [WB68, Wal05], are two closely related information-theoretic approaches to statistical inference and model selection. Both principles are based on the idea of finding the most compact representation of data, using a combination of model complexity and the size of the message needed to describe the data given that model. The MDL principle operates by minimizing the sum of the model's description length and the data's encoded length, while the MML principle aims to minimize the total message length by finding the most plausible model with the shortest encoding. Despite their differences, both approaches share a common goal: to find the simplest yet accurate model that captures the underlying structure of the data, striking a balance between overfitting and underfitting, and yielding robust and interpretable results in a wide range of applications. General consistency of MDL beyond i.i.d. has been proven in [PH05a, Hut09b]. For an application to Kernel learning, see [HHO23]. Use of MDL with over-parameterized models is problematic [DSYW21]. The prequential MDL approach solves this problem in theory [DV99, PH05a] and in practice [BLH23, LBH23, RTBP+23]. Standard two-part MDL is limited to parametric models, log-loss, passive prediction and induction tasks. MDL has been extended to reinforcement learning [Hut09d, Hut09e, Hut09f, Hut09a, Ngu13, Das16] and applied to control [MKSB22]. The Loss Rank Principle (LoRP) can be regarded as an extension of MDL to arbitrary loss functions and non-parametric models [Hut07d, TH10, HT10].

Properties of Kolmogorov complexity. The invariance Theorem 2.7.6 is due to [Sol64, Kol65, Cha66], Theorems 2.7.13 and 2.7.23 are due to [Lev74], the symmetry of information (Theorem 2.7.18) due to [ZL70a, Gác74, Kol83], the other (in)equalities are elementary.

Variants of Kolmogorov complexity. There are many variants of Kolmogorov complexity. The prefix Kolmogorov complexity K we defined here [Lev74, Gác74, Cha75], the earliest form, "plain" Kolmogorov complexity C [Kol65], process complexity [Sch73], monotone complexity Km [Lev73a], and uniform complexity [Lov69b, Lov69a], Solomonoff's universal prior $M = 2^{-KM}$ [Sol64, Sol78], Chaitin's complexity Kc [Cha75], extension semimeasure Mc [Cov74], predictive complexity KP [VW98], and some others. They often differ from K only by $O(\log K)$, but otherwise have similar properties. For an introduction to the relationships between Kolmogorov complexity and Shannon's information theory, see [Kol65, Kol83, ZL70a, CT06]. [AM20] proves that average Kolmogorov complexity convergences to Shannon Entropy for stationary random sequences with optimal rate.

Drawbacks of Kolmogorov complexity and remedies. The main drawback of all these variants of Kolmogorov complexity is that they are not finitely computable [Kol65, Sol64]. They may be approximated from above (Theorem 2.7.28) [Kol65, Sol64], but no accuracy guarantee can be given. Worse still, the best upper bound for the runtime until one has reasonable accuracy for $K(n)$ grows faster than any computable function in n.

This led to the development of time-bounded complexity/probability that is finitely computable, or more general resource-bounded complexity/probability (e.g. space) [Dal73, Dal77, FMG92, Ko86, PF97, Sch02b]. In particular, the Levin complexity [Lev73b] adjusts the Kolmogorov complexity by adding a logarithmic penalty in the running time of the program. Nonetheless, some attempts have been made to provide estimates of the Kolmogorov complexity: Conte et al. [Con97] evolved short Lisp programs, and Bloem et al. [BMdR$^+$14] give a probabilistic approximation of K.

Normalized compression distance and universal similarity metric. Li [Li 03] (based on work by Bennett [Ben98]) uses Kolmogorov complexity to define the *normalized compression distance* NCD, a *universal similarity metric* over strings that measures similarity based on how much one string helps to describe the other. The NCD metric is universal in that it minorizes all other computable distance measures. It itself is incomputable (the definition relying on K, also incomputable) but computable approximations exist (replacing K with the best available data compressor). This metric has applications to clustering [CVW04, CV05], constructing phylogenetic trees [HRR06, RA11], classification of file types from fragments of data for computer forensics [Axe10], analysis of viruses (both biological [PP17, MRNA20, CV22] and digital [Bor16, GB03]). It has also been generalized to a metric on multisets [CV14], and (by replacing K with number of hits on Google) a metric on websites [CV07].

Miscellaneous. Chaitin [Cha91] speculated on the computational power of the evolutionary information gathering process and its relation to algorithmic information. Schmidt [Sch99] argued that (time-bounded) Kolmogorov complexity helps and not prevents the search for extraterrestrial intelligence (SETI). Vovk [VW98] described universal portfolio selection schemes. Hutter [Hut10a] argues that a Theory of Everything needs to take into account the Kolmogorov complexity of the location of the observer in the universe. A more personal account on the past, present, and future of algorithmic randomness as foundation of induction and AI for a general audience can be found in [Hut11].

Distances and their relation (2.8). A discussion of Pinsker's inequality and the related Bretagnolle-Huber inequality can be found in [Can22]. It also discusses Gibbs variational principle, Donsker–Varadhan form, and f-divergences. The proof of Pinsker's binary inequality has been taken from [Wu17].

Part II

Algorithmic Prediction

Chapter 3

Bayesian Sequence Prediction

The only way to discover the limits of the possible is to go beyond them into the impossible.

Arthur C. Clarke

This chapter delves into the principles of induction, emphasizing the selection of suitable models to capture data patterns. It gives an in-depth introduction into Bayesian and in particular universal sequence prediction. Section 3.1 introduces the Bayesian mixture predictor ξ over arbitrary countable classes $\mathcal{M}$ of semimeasures ν for (infinite) sequences, and states the key dominance property. No structural assumptions are made on $\nu \in \mathcal{M}$, no i.i.d. nor ergodicity nor stationarity assumptions. Section 3.2 lifts the various distances and their relations introduced in Section 2.8.1 to (semi)measures over sequences to measure the deviation of ξ from the true sampling distribution μ. This is then used in Section 3.3 to show predictive convergence of ξ to μ together with error rates. Section 3.4 relaxes the general assumption of $\mu \in \mathcal{M}$ to μ being close to some $\hat{\mu} \in \mathcal{M}$. Section 3.5 uses predictions to make decisions and take actions, which then lead to some loss, depending on the action and the observed outcome. We prove strong loss bounds for passive environments, where actions cannot affect the environment. Section 3.6 shows that Bayes is also Pareto optimal with respect to many performance measures, including any loss function and some of the introduced distances. So far the treatment has been completely general. Section 3.7 explores the selection of particularly interesting classes of environments $\mathcal{M}$ and priors w_ν. We discuss three principles for choosing weights for ξ, with a special emphasis on the simplicity principle, quantified in terms of Kolmogorov complexity. We argue that the class of all (semi)computable (semi)measures is the most appropriate for AGSI purposes. This leads to Solomonoff's celebrated a-priori probability ξ_U and prior w_ν^U and prediction bound. Section 3.8 provides a more direct route to Solomonoff's universal distribution M via uniform random noise piped through a universal monotone Turing machine, and state the remarkable identity of M and ξ_U. The last Section 3.9 gives a brief introduction to martingales, which are needed later in some more advanced convergence proofs for intelligent agents in reactive environments. Except for the Solomonoff section, all results in this chapter hold for general finite alphabet.

3.1 Bayes Mixture ξ

Before diving into specific prediction strategies, it is essential to understand the concept of induction. Induction is a reasoning process used to make inferences or predictions based on a limited set of observations. In the context of this book, induction helps us to estimate the hidden distribution and make predictions about the future. The inductive approach relies on the assumption that the observed data follows an underlying pattern, and that pattern can be extended to predict future elements. The challenge of induction lies in selecting the appropriate model or distribution family that can accurately capture this pattern. By considering multiple models and their respective parameters, we can employ induction to make more informed predictions about the next element in the sequence, even when the true distribution is unknown.

Suppose we want to predict the next term x_t in a sequence given the previous terms $x_1,...,x_{t-1}$. This goal can be rephrased as trying to estimate the hidden distribution from which the sequence $x_1,...,x_{t-1}$ was sampled, and then making a prediction of the next element x_t using this estimated distribution. Let μ denote the true sampling distribution (represented as a probability measure) generating the sequence. If μ itself is known, then we can just use the predictive distribution $\mu(x_t|x_{<t})$ to predict the probability of x_t. If we are interested in the most likely continuation, the optimal prediction for x_t would be $\arg\max_{x_t} \mu(x_t|x_{<t})$ by definition. If we know what family of parameterized distributions μ is sampled from, we need only learn estimates of those parameters, for which many methods already exist (Section 2.3).

What should we do when even the family of distributions that μ belongs to is unknown? One approach to solving this problem is to choose some class $\mathcal{M}$, general enough that we expect it contains all the distributions that could potentially be generating the sequence (in the hope that it contains μ), and then take a weighted mixture over that class. The most general class we can reasonable consider is the countable set of all computable probability distributions (Definition 3.7.1), since if μ did not belong to this class, then by definition there is no hope of using any computable method to learn μ. For now we keep $\mathcal{M}$ general but countable, and only assume that it contains, unless otherwise stated, the true distribution μ, We focus on sequences drawn from a finite alphabet $\mathcal{X}$, though many results have natural extensions to countable alphabet and beyond. The following notation and definitions are used throughout the book:

Definition 3.1.1 ((Semi)measures, cylinder sets, and notation) For (semi)measures on cylinder sets $\Gamma_x \subseteq \mathcal{X}^\infty$ for $x \in \mathcal{X}^*$ we use the shorthand $\nu(x) := \nu(\Gamma_x)$ of Definition 2.2.11. The true distribution from which $x_{1:\infty}$ is sampled is denoted by μ and is always a properly normalized probability measure. Note that $\mu(x)$ is *not* a probability mass function on $\mathcal{X}^*$. It is the μ-probability that an *infinite* sequence *starts* with x. $\mathcal{M}$ denotes some countable class of (semi)measures, ν,μ are always assumed to be in $\mathcal{M}$, while ρ denotes an arbitrary (semi)measure.

Definition 3.1.2 (Prior w_ν and Bayes mixture ξ) Let $\mathcal{M}$ be a countable set of probability semimeasures on strings, and $w_{(.)} : \mathcal{M} \to \mathbb{R}$ be a function satisfying $\sum_{\nu \in \mathcal{M}} w_\nu \leq 1$ and $w_\nu > 0$ for all ν. We define the *Bayes mixture over $\mathcal{M}$ given prior* $w_{(.)}$, denoted ξ_w, as

$$\xi_w(x_{1:n}) := \sum_{\nu \in \mathcal{M}} w_\nu \nu(x_{1:n})$$

We call w_ν the *weight* associated with $\nu \in \mathcal{M}$. If the choice of $w_{(.)}$ is obvious or irrelevant, we drop the index w and write ξ.

We forbid the case where $w_\nu = 0$, since if it were, the measure ν is *a priori* excluded from the mixture, and we may as well exclude it from $\mathcal{M}$ to begin with. Also, in line with the *Principle of Epicurus*, we avoid discarding any distributions (hypotheses) until we have evidence that rules it out.

Remark 3.1.3 (Continuous class $\mathcal{M}$ with prior density $w(\theta)$) Though we do not explore it much in this book, one can also consider an uncountable family $\mathcal{M}$ of measures $\nu_\theta(\cdot;\theta)$ (which must be a probability kernel) parameterized by $\theta \in \Theta$, together with a prior weight for each parameter, represented by a probability *density* function $w(\theta)$. Many definitions and results generalize in some way to this case. For instance, one can define the Bayes-mixture analogously as

$$\xi(x_{1:n}) = \int_\Theta w(\theta)\nu(x_{1:n};\theta)\, d\theta$$

This generalizes the Bernoulli case considered in Section 2.4.   ●

Proposition 3.1.4 (Bayes mixture domination) For all $\nu \in \mathcal{M}$ and prior $w \in \Delta'\mathcal{M}$ and $x_{1:n} \in \mathcal{X}^*$ we have

$$\xi_w(x_{1:n}) \geq w_\nu \nu(x_{1:n})$$

The Bayesian mixture $\xi(x_{1:n})$ can be thought of as an estimate of $\mu(x_{1:n})$ by taking a weighted average of each $\nu \in \mathcal{M}$, where w_ν represents the a-priori confidence that ν is the true environment.

In lieu of using the true distribution μ for prediction, we can instead use ξ for prediction. Even if μ is unknown, we might have some a priori estimates on which $\nu \in \mathcal{M}$ is more likely to be the true μ, which can be reflected in our choice of w_ν in Definition 3.1.2.

To use ξ for prediction, we need to define the conditional probability distribution $\xi(x_t|x_{<t})$ over the next element x_t, given the sequence $x_{<t}$ so far.

Recall that for any semimeasure ρ, if $\rho(x_{<t}) > 0$ (see Section 2.8.3 if not) we can define

$$\rho(x_t|x_{<t}) := \frac{\rho(x_{1:t})}{\rho(x_{<t})} \qquad \text{and} \qquad \rho(x_{t:k}|x_{<t}) := \frac{\rho(x_{t:k})}{\rho(x_{<t})} \quad \text{for} \quad k \geq t \quad (3.1.5)$$

$$\rho(x_{1:n}) = \prod_{t=1}^{n} \rho(x_t|x_{<t}) \qquad \text{and} \qquad \rho(x_{t:n}|x_{<t}) = \prod_{k=t}^{n} \rho(x_k|x_{<k}) \qquad (3.1.6)$$

The last two expressions are called *chain rule*. The conditional Bayesian distribution $\xi(x_t|x_{<t})$ has a couple of useful representations:

Theorem 3.1.7 (Conditional Bayesian mixture ξ) Let $\mathcal{M}$ be a class of environments with prior $\{w_\nu\}_{\nu \in \mathcal{M}}$ as above. Then $\xi(x_t|x_{<t})$ is the predictive probability the conditional Bayesian mixture ξ assigns to x_t given $x_{<t}$:

$$\xi(x_t|x_{<t}) = \sum_{\nu \in \mathcal{M}} w(\nu|h_{<t})\nu(x_t|x_{<t}) = \frac{\sum_{\nu \in \mathcal{M}} w_\nu \nu(x_t|x_{<t})\nu(x_{<t})}{\sum_{\nu \in \mathcal{M}} w_\nu \nu(x_{<t})} = \frac{\xi(x_{1:t})}{\xi(x_{<t})}$$

$$w(\nu|x_{1:t}) := w(\nu|x_{<t})\frac{\nu(x_t|x_{<t})}{\xi(x_t|x_{<t})} = w_\nu \frac{\nu(x_{1:t})}{\xi(x_{1:t})} \quad \text{and} \quad w(\nu|\epsilon) := w_\nu \qquad (3.1.8)$$

Here w_ν represents the probability that we assign to the hypothesis that $\nu = \mu$ before seeing evidence $x_{<t}$ (the prior belief) and $w(\nu|x_{<t})$ is the probability that $\nu = \mu$ after observing evidence $x_{<t}$ (the posterior belief).

Proof. First note that we can write

$$w(\nu|x_{1:t}) = w_\nu \prod_{k=1}^{t} \frac{\nu(x_k|x_{<k})}{\xi(x_k|x_{<k})} = w_\nu \frac{\nu(x_{1:t})}{\xi(x_{1:t})}$$

by repeated application of (3.1.8) and the chain rule (3.1.6). Then,

$$\xi(x_t|x_{<t}) \stackrel{(3.1.5)}{\equiv} \frac{\xi(x_{1:t})}{\xi(x_{<t})} \equiv \frac{\sum_{\nu \in \mathcal{M}} w_\nu \nu(x_{1:t})}{\xi(x_{<t})} = \sum_{\nu \in \mathcal{M}} w_\nu \frac{\nu(x_{1:t})}{\xi(x_{<t})}$$

$$= \sum_{\nu \in \mathcal{M}} w_\nu \frac{\nu(x_{<t})}{\xi(x_{<t})}\nu(x_t|x_{<t}) = \sum_{\nu \in \mathcal{M}} w(\nu|x_{<t})\nu(x_t|x_{<t})$$

as required.   ∎

Assuming $\mu \in \mathcal{M}$, we would hope that eventually the posterior concentrates on μ, i.e. $w(\mu|x_{<t}) \to 1$ and $w_\nu \to 0$ for all other $\nu \neq \mu$, as the mixture ξ hopefully concentrates on μ given sufficient experience sampled from μ giving away μ. In the Bernoulli(θ) example of Section 2.4.3 this was indeed the case (see Figure 2.11). Unfortunately this is not (strictly) true anymore for general classes $\mathcal{M}$, since $\mathcal{M}$ may contain environments that are statistically indistinguishable (Exercise 6). We could try to group indistinguishable environments together and show that the total posterior of the group that contains μ tends to 1, but an elegant, unified, general solution seems not to exist [Hut09b]. We can side-step this problem by aiming a little lower for only predictive convergence as in Laplace rule (2.4.10). This is the approach pursued in the next two sections.

3.2 Generalized Solomonoff Bound

The following *instantaneous distances* are various ways of measuring how close $\xi(\cdot|x_{<t})$ is to $\mu(\cdot|x_{<t})$. Instantaneous here refers to these distance measures only considering how similar the measures μ and ξ are over the potential next symbol x_t'. Note that the distance measures depend on the history $x_{<t}$ observed up to that point.

Definition 3.2.1 (Instantaneous distances)

Absolute:
$$a_t(x_{<t}) := \sum_{x_t \in \mathcal{X}} |\mu(x_t|x_{<t}) - \xi(x_t|x_{<t})|$$

Square:
$$s_t(x_{<t}) := \sum_{x_t \in \mathcal{X}} (\mu(x_t|x_{<t}) - \xi(x_t|x_{<t}))^2$$

Hellinger[2]:
$$h_t(x_{<t}) := \sum_{x_t \in \mathcal{X}} \left(\sqrt{\mu(x_t|x_{<t})} - \sqrt{\xi(x_t|x_{<t})} \right)^2$$

KL divergence:
$$d_t(x_{<t}) := \sum_{x_t \in \mathcal{X}} \mu(x_t|x_{<t}) \ln \left(\frac{\mu(x_t|x_{<t})}{\xi(x_t|x_{<t})} \right)$$

Each distance has different properties which are useful in different circumstances. It is obvious that the absolute, square and Hellinger distances are non-negative, and we have already seen that the KL divergence is non-negative (Corollary 2.5.16). Although only the absolute distance is a metric, the KL divergence is often the most useful distance measure.

Usually, we consider instantaneous distance measures to just be functions, but we also need to consider them as a sequence of random variables when talking about convergence with μ-probability 1. Formally, the distance measures in Definition 3.2.1 can be considered as a sequence of random variables over the sample space $\Omega = \mathcal{X}^\infty$, where e.g. d_t is really the tth term in the sequence $d_1(\epsilon), d_2(X_1), d_3(X_{1:2}), \dots$.

Taking the μ-expected sum of the instantaneous distances over time steps $t = 1, \dots, n$ gives us the *total distance* between μ and ξ.

Definition 3.2.2 (Total distances)

Absolute:
$$A_n := \sum_{t=1}^{n} \mathbf{E}_\mu[a_t(\cdot)] = \sum_{t=1}^{n} \sum_{x_{<t} \in \mathcal{X}^{t-1}} \mu(x_{<t}) a_t(x_{<t})$$

Square:
$$S_n := \sum_{t=1}^{n} \mathbf{E}_\mu[s_t(\cdot)] = \sum_{t=1}^{n} \sum_{x_{<t} \in \mathcal{X}^{t-1}} \mu(x_{<t}) s_t(x_{<t})$$

Hellinger[2]:
$$H_n := \sum_{t=1}^{n} \mathbf{E}_\mu[h_t(\cdot)] = \sum_{t=1}^{n} \sum_{x_{<t} \in \mathcal{X}^{t-1}} \mu(x_{<t}) h_t(x_{<t})$$

KL divergence:
$$D_n := \sum_{t=1}^{n} \mathbf{E}_\mu[d_t(\cdot)] = \sum_{t=1}^{n} \sum_{x_{<t} \in \mathcal{X}^{t-1}} \mu(x_{<t}) d_t(x_{<t})$$

We can straightforwardly lift the bounds of Corollary 2.8.8 between these different instantaneous and total distance measures. Note the extra factor of $\sqrt{n}$ in the A_n bound, which makes it less useful.

Theorem 3.2.3 (Instantaneous and total entropy inequalities)

$$s_t(x_{<t}) \leq d_t(x_{<t}) \qquad a_t(x_{<t}) \leq \sqrt{2d_t(x_{<t})} \qquad h_t(x_{<t}) \leq d_t(x_{<t})$$

$$S_n \leq D_n \qquad\qquad A_n \leq \sqrt{2nD_n} \qquad\qquad H_n \leq D_n$$

$$s_t \leq a_t^2 \leq 4h_t \qquad\qquad a_t^2 \leq |\mathcal{X}|s_t \qquad\qquad s_t \leq 2h_t$$

Proof. (1st line) We define $y_i = \mu(x_t = i | x_{<t})$ and $z_i = \xi(x_t = i | x_{<t})$ and apply Corollary 2.8.8 for the instantaneous bounds. *(2nd line)* Taking $\sum_{t=1}^{n} \mathbf{E}_\mu$ on both sides gives the bounds for S_n and D_n. For A_n we have to additionally apply Jensen's inequality (Theorem 2.2.52 and Corollary 2.2.54) with concave function $g(\cdot) = \sqrt{\cdot}$:

$$\frac{1}{n} A_n \equiv \frac{1}{n} \sum_{t=1}^{n} \mathbf{E}_\mu[a_t] \leq \frac{1}{n} \sum_{t=1}^{n} \mathbf{E}_\mu[\sqrt{2d_t}] \leq \frac{1}{n} \sum_{t=1}^{n} \sqrt{\mathbf{E}_\mu[2d_t]} \leq \sqrt{\tfrac{1}{n} \sum_{t=1}^{n} \mathbf{E}_\mu[2d_t]} \equiv \sqrt{\tfrac{1}{n} 2D_n}$$

(3rd line) Use $y_i - z_i = (\sqrt{y_i} + \sqrt{z_i})(\sqrt{y_i} - \sqrt{z_i}) \leq 2(\sqrt{y_i} - \sqrt{z_i})$ and Hoelder's inequality (Exercise 1). ∎

An extremely useful property unique to the KL divergence is its telescoping property: The KL of joint distributions can be expressed as a sum of KLs of predictive distributions.

Lemma 3.2.4 (Telescoping property of KL) Let $\nu_{1:n} \in \Delta'\mathcal{X}^n$ be the ν-semi-probability of $x_{1:n}$, and $\nu_t : \mathcal{X}^{t-1} \to \Delta'\mathcal{X}$ be the ν-semi-probability of x_t given $x_{<t}$, i.e. $\nu_{1:n} = \prod_{t=1}^{n} \nu_t$, and similarly for the proper μ-probability, then

$$\mathrm{KL}(\mu_{1:n} || \nu_{1:n}) = \mathbf{E}_\mu \left[\sum_{t=1}^{n} \mathrm{KL}(\mu_t || \nu_t) \right] =$$

$$\sum_{x_{1:n}} \mu(x_{1:n}) \ln \frac{\mu(x_{1:n})}{\nu(x_{1:n})} = \sum_{t=1}^{n} \sum_{x_{<t}} \mu(x_{<t}) \left[\sum_{x_t} \mu(x_t | x_{<t}) \ln \frac{\mu(x_t | x_{<t})}{\nu(x_t | x_{<t})} \right]$$

The second equality is just the first with the definitions of KL and $\mu_{1:n}$ and μ_t inserted. Applying the lemma with $\nu = \xi$ implies $D_n = \mathrm{KL}(\mu_{1:n} || \xi_{1:n})$ and $D_\infty = \mathrm{KL}(\mu || \xi)$ by taking the limit $n \to \infty$.

Proof. Starting from the r.h.s. of the second equation

$$\sum_{t=1}^{n} \sum_{x_{<t}} \mu(x_{<t}) \sum_{x_t} \mu(x_t | x_{<t}) \ln \frac{\mu(x_t | x_{<t})}{\nu(x_t | x_{<t})}$$

Replacing $\mu(x_{<t}) \mu(x_t | x_{<t})$ with $\mu(x_{1:t})$, and combining the sums,

$$= \sum_{t=1}^{n} \sum_{x_{1:t}} \mu(x_{1:t}) \ln \frac{\mu(x_t | x_{<t})}{\nu(x_t | x_{<t})}$$

Applying the sum rule $\mu(x_{1:t}) = \sum_{x_{t+1:n}} \mu(x_{1:n})$

$$= \sum_{t=1}^{n} \sum_{x_{1:t}} \left(\sum_{x_{t+1:n}} \mu(x_{1:n}) \right) \ln \frac{\mu(x_t | x_{<t})}{\nu(x_t | x_{<t})}$$

The logarithmic term has no dependency on $x_{t+1:n}$, so we can bring it into the sum, and then combine the sums,

$$= \sum_{t=1}^{n} \sum_{x_{1:n}} \mu(x_{1:n}) \ln \frac{\mu(x_t|x_{<t})}{\nu(x_t|x_{<t})}$$

We pull in the t-sum to the logarithm, noting that $\mu(x_{1:n})$ does not depend on t, and use $\sum_t \ln x_t = \ln \prod_t x_t$,

$$= \sum_{x_{1:n}} \mu(x_{1:n}) \ln \prod_{t=1}^{n} \frac{\mu(x_t|x_{<t})}{\nu(x_t|x_{<t})}$$

Applying the chain rule (3.1.6)

$$= \sum_{x_{1:n}} \mu(x_{1:n}) \ln \frac{\mu(x_{1:n})}{\nu(x_{1:n})}$$

∎

We now have everything to prove a generalized version of Solomonoff's celebrated bound.

Theorem 3.2.5 (Generalized Solomonoff bound) $\quad S_n \leq D_n \leq \ln w_\mu^{-1} < \infty$

Proof. For $S_n \leq D_n$, see Theorem 3.2.3. Using the telescoping Lemma 3.2.4 with $\nu = \xi$ and rearranging Proposition 3.1.4, gives

$$D_n = \sum_{t=1}^{n} \sum_{x_{<t}} \mu(x_{<t}) d_t(x_{<t}) = \sum_{t=1}^{n} \sum_{x_{<t}} \mu(x_{<t}) \sum_{x_t} \mu(x_t|x_{<t}) \ln \frac{\mu(x_t|x_{<t})}{\xi(x_t|x_{<t})}$$

$$= \sum_{x_{1:n}} \mu(x_{1:n}) \ln \frac{\mu(x_{1:n})}{\xi(x_{1:n})} \leq \sum_{x_{1:n}} \mu(x_{1:n}) \ln(w_\mu^{-1}) = \ln(w_\mu^{-1})$$

∎

The same bound holds for the Hellinger distance, but in this case can be improved by side-stepping the D_n intermediary.

Theorem 3.2.6 (Expected bounds on Hellinger sum)

$$\sum_{t=1}^{\infty} \mathbf{E}\left[\left(\sqrt{\frac{\xi(x_t|x_{<t})}{\mu(x_t|x_{<t})}} - 1\right)^2\right] \overset{(i)}{\leq} \sum_{t=1}^{\infty} \mathbf{E}[h_t] \overset{(ii)}{\leq} 2\ln\{\mathbf{E}[\exp(\tfrac{1}{2}\sum_{t=1}^{\infty} h_t)]\} \overset{(iii)}{\leq} \ln w_\mu^{-1}$$

Proof. (i) follows from simple algebraic manipulations, (ii) from Jensen's inequality, but (iii) is non-trivial. The full proof can be found in [HM07]. ∎

3.3 Predictive Convergence

We now use the bounds of the previous section to explore how ξ can be used for prediction, as well as quantifying its performance as a predictor. Recall from Definition 2.2.57 *convergence almost surely* or *with probability 1* $(X_t \overset{\text{a.s}}{\longrightarrow} X)$. Here X_t is d_t or related quantity and $X = 0$, while the sequence $x_{1:\infty}$ we wish to predict is sampled from probability measure μ. More precisely, for stochastic sequences, the best we can hope for is to predict the true μ-probability of x_t given $x_{<t}$.

Corollary 3.3.1 (Convergence of KL divergence to zero)
The KL divergence satisfies $d_t(x_{<t}) \to 0$ with μ-probability 1.

Proof. By definition of D_n and Theorem 3.2.5, we have that $D_\infty = \sum_{t=1}^{\infty} \mathbf{E}_\mu[d_t(x_{<t})] \leq \ln w_\mu^{-1} < \infty$, which implies $d_t(x_{<t}) \to 0$ in mean^2 sum by Definition 2.2.72, which by Theorem 2.2.75 implies $d_t(x_{<t}) \to 0$ with μ-probability 1.   ∎

Additionally, Theorem 3.2.5 implies that the absolute difference between $\xi(x_t'|x_{<t})$ and $\mu(x_t'|x_{<t})$ will converge to 0 in mean^2 sum and with μ-probability 1, for any choice of (non)random sequence x_t'. This is called *off-sequence convergence* because it is true for any sequence $x_{1:\infty}'$, not just the x_t sampled from μ, which will turn out to be important. The corresponding convergence in ratio only holds *on-sequence* (Exercise 2).

Corollary 3.3.2 (Convergence of ξ to μ) The off-sequence difference (and the on-sequence ratio) between ξ and μ goes to zero (one), in the sense that

$$\lim_{t\to\infty} |\xi(x_t'|x_{<t}) - \mu(x_t'|x_{<t})| = 0 \quad \text{and} \quad \lim_{t\to\infty} \frac{\xi(x_t|x_{<t})}{\mu(x_t|x_{<t})} = 1$$

with μ-probability 1 and in mean^2 sum (referring to r.v. $x_{<t}$) for any sequence of x_t'.

Proof. Let $\Delta_t := |\xi(x_t'|x_{<t}) - \mu(x_t'|x_{<t})|$. By Theorem 3.2.5 and the definition of S_n we have for any choice of $x_t' \in \mathcal{X}$,

$$\infty > S_\infty \equiv \sum_{t=1}^{\infty} \mathbf{E}_\mu[\sum_{x_t'} \Delta_t^2] \geq \sum_{t=1}^{\infty} \mathbf{E}_\mu[(\Delta_t - 0)^2]$$

which implies that Δ_t converges to 0 in mean^2 sum by Definition 2.2.72, which by Theorem 2.2.75 implies that $\lim_{t\to\infty}\Delta_t \to 0$ with μ-probability 1. The ratio convergence follows similarly from Theorem 3.2.6.   ∎

Corollary 3.3.3 (Expected number of ε-errors between ξ and μ) For any $\varepsilon > 0$, the μ-expected number of times that $d_t(x_{<t})$ exceeds ε is

$$\mathbf{E}_\mu\left[\sum_{t=1}^{\infty}[\![d_t > \varepsilon]\!]\right] \leq \varepsilon^{-1}\ln w_\mu^{-1} < \infty$$

Proof. Using $\varepsilon[\![d_t > \varepsilon]\!] \leq d_t$, we have

$$\mathbf{E}_\mu\left[\sum_{t=1}^{\infty}[\![d_t > \varepsilon]\!]\right] \leq \mathbf{E}_\mu\left[\sum_{t=1}^{\infty}\frac{d_t}{\varepsilon}\right] \equiv \frac{D_\infty}{\varepsilon} \leq \frac{\ln w_\mu^{-1}}{\varepsilon} < \infty$$

by Theorem 3.2.5.   ∎

Since $|\xi(x_t'|x_{<t}) - \mu(x_t'|x_{<t})|^2 \leq s_t \leq d_t$, this implies that for any $\varepsilon > 0$, the μ-expected number of times that $\xi(x_t'|x_{<t})$ deviates from $\mu(x_t'|x_{<t})$ by more than ε is bounded above by $\varepsilon^{-2}\ln w_\mu^{-1} < \infty$, for any x_t'. By a simple Markov inequality, this also implies that the probability of having more than $(\varepsilon^2\delta)^{-1}\ln(w_\mu^{-1})$ ε-deviations is less than δ. The latter can be significantly improved:

Theorem 3.3.4 (High-probability bound on ξ from μ deviation) For any $\delta > 0$, with μ-probability at least $1-\delta$, we have

$$\sum_{t=1}^{\infty} h_t \leq \ln w_\mu^{-1} + 2\ln\tfrac{1}{\delta} \quad\text{and}\quad \sum_{t=1}^{\infty} d_t \leq e\cdot(\ln\tfrac{6}{\delta})\cdot(\ln w_\mu^{-1} + \ln\tfrac{2}{\delta})$$

Proof. The first bound follows from Theorem 3.2.6 and a simple Markov inequality [HM07, Sec.3]. The proof for the second bound can be found in [LHS13a, Sec.3]. ∎

Using $\left|\xi(x_t'|x_{<t}) - \mu(x_t'|x_{<t})\right|^2 \leq s_t \leq 2h_t$ and again a simple Markov inequality, this implies that the number of times $\xi(x_t'|x_{<t})$ deviates from $\mu(x_t'|x_{<t})$ by more than ε is bounded by $2[\ln w_\mu^{-1} + 2\ln\tfrac{1}{\delta}]/\varepsilon^2$ with probability at least $1-\delta$ (Exercise 4).

3.4 Model Misspecification

Recall that if $\mu \in \mathcal{M}$, then the KL divergence D_n between μ and ξ is finite (Theorem 3.2.5). We can in fact impose a weaker condition on μ, namely if there is a distribution $\hat{\mu} \in \mathcal{M}$ which is "close" to μ in the sense that the KL divergence between μ and $\hat{\mu}$ is finite, then D_n is still finite. The case $\mu \notin \mathcal{M}$ is sometimes called model-misspecification, since it violates the standard (Bayesian) assumption that $\mu \in \mathcal{M}$.

Theorem 3.4.1 (Bound on KL divergence for out-of-class distributions) Let μ be an arbitrary measure (that may or not be in $\mathcal{M}$). For $\rho \in \mathcal{M}$, let

$$\mathrm{KL}_n(\mu\|\rho) := \sum_{x_{1:n}} \mu(x_{1:n})\ln\frac{\mu(x_{1:n})}{\rho(x_{1:n})}$$

denote the KL divergence between ρ and μ restricted to $x_{1:n}$. Then for all n,

$$D_n \leq \inf_{\rho\in\mathcal{M}}\left\{\ln w_\rho^{-1} + \mathrm{KL}_n(\mu\|\rho)\right\}$$

Proof. Let $\rho \in \mathcal{M}$ be arbitrary. Using the same technique as the proof of Theorem 3.2.5, we can write D_n as follows.

$$
\begin{aligned}
D_n &= \sum_{x_{1:n}} \mu(x_{1:n})\ln\frac{\mu(x_{1:n})}{\xi(x_{1:n})} \\
&= \sum_{x_{1:n}} \mu(x_{1:n})\ln\frac{\mu(x_{1:n})}{\rho(x_{1:n})}\frac{\rho(x_{1:n})}{\xi(x_{1:n})} \\
&= \sum_{x_{1:n}} \mu(x_{1:n})\ln\frac{\mu(x_{1:n})}{\rho(x_{1:n})} + \sum_{x_{1:n}} \mu(x_{1:n})\ln\frac{\rho(x_{1:n})}{\xi(x_{1:n})} \\
&\leq \mathrm{KL}_n(\mu\|\rho) + \sum_{x_{1:n}} \mu(x_{1:n})\ln\frac{\rho(x_{1:n})}{\xi(x_{1:n})}
\end{aligned}
$$

Rearranging Proposition 3.1.4 gives $\rho(x_{1:n})/\xi(x_{1:n}) \leq w_\rho^{-1}$, so

$$\sum_{x_{1:n}} \mu(x_{1:n})\ln\frac{\rho(x_{1:n})}{\xi(x_{1:n})} \leq \ln w_\rho^{-1}\sum_{x_{1:n}} \mu(x_{1:n}) \leq \ln w_\rho^{-1}$$

which gives

$$D_n \leq \ln w_\rho^{-1} + \mathrm{KL}_n(\mu||\rho)$$

Since the inequality holds for all $\rho \in \mathcal{M}$, it holds for the minimizing ρ. ∎

Corollary 3.4.2 (Out-of-class KL bound)
If $\hat{\mu} \in \mathcal{M}$ satisfies $\mathrm{KL}_\infty(\mu||\hat{\mu}) < \infty$, then $D_\infty < \infty$, hence ξ converges to μ.

This generalizes Solomonoff's bound (Theorem 3.2.5) even further. If we cannot guarantee that $\mathrm{KL}_n(\mu||\rho)$ is bounded, but merely that $\mathrm{KL}_n(\mu||\rho) = o(n)$ (the KL divergence grows sub-linearly) then we can still obtain that ξ converges to μ, but only in a weaker Cesáro sense.

Corollary 3.4.3 (Average KL limits to zero) Let ρ be an environment satisfying $\mathrm{KL}_n(\mu||\rho) = o(n)$. Then the average KL divergence D_n/n converges to zero.

This also implies generalized versions of Corollary 3.3.3 and Theorem 3.3.4, namely that the relative frequency of ε-errors tends to 0 in expectation and with high probability.

Proof. Take Theorem 3.4.1 and divide through by n.

$$\frac{1}{n} D_n \leq \frac{\ln w_\rho^{-1} + \mathrm{KL}_n(\mu||\rho)}{n} = \frac{\ln w_\rho^{-1} + o(n)}{n} \longrightarrow 0 \quad \text{for} \quad t \to \infty$$ ∎

3.5   Bounds on Prediction Loss

Often we wish to take actions based on our current estimate of how the true environment μ works, in service of some goal. We will still focus on *passive environments* where the behavior of μ does not depend on actions taken. Examples include weather prediction, or trading stocks with a small enough amount of capital to not measurably affect the market. We measure the value of actions using a loss function.

Definition 3.5.1 (Predictor model) A *predictor model* is comprised of a finite set of *observations* $\mathcal{X}$ and *actions* $\mathcal{Y}$, a probability semimeasure $\mu : \mathcal{X}^* \to \Delta\mathcal{X}$ called the *environment*, a *predictor* $\Lambda : \mathcal{X}^* \to \mathcal{Y}$, and a *loss function* $loss : \mathcal{X} \times \mathcal{Y} \to [0,1]$. On each time step t, given history $x_{<t}$, an observation $x_t \sim \mu(\cdot|x_{<t})$ is sampled from the environment. The predictor Λ produces action y_t^Λ (based solely on $x_{<t}$) and suffers loss $loss(x_t, y_t^\Lambda)$. The time step t is incremented, and the interaction loops.

Note that the environment is a function of the observation history but *not* the actions taken by the predictor. The loss is also a function only of the current observation and current action, but could be made t and $x_{<t}$-dependent, with all results of this section still holding (Exercise 7).

The *predictor* chooses action $y \in \mathcal{Y}$, and the environment produces observation $x \in \mathcal{X}$. The predictor receives loss $loss(x,y) \in [0,1]$, and the goal of the predictor is to minimize its loss.

Remark 3.5.2 (On passive prediction vs. reactive environments) In the (re)active setting we will maximize reward rather than minimize loss. These are equivalent presentations, as one can treat loss as negative reward. *Predictors* are then called *actors* which follow *policies*. We also call y *actions* here when it feels natural, despite predictors lacking agency since actions do not affect the environment. It could also be argued that the loss is an aspect of the environment. A substantial difference though is that next-step loss/reward optimization is no longer optimal in reactive environments, where the environment depends on the history of both observations x_t and actions y_t, and long-term planning is required. The proper agent setting will be covered from Part III onwards. ●

Example 3.5.3 (Weather prediction) Suppose we have an environment for weather prediction. The predictor can choose to wear or to not wear a jacket. If it is cold the jacket will be a great help, if it is warm the jacket will be a mild annoyance to carry around. We define the actions $\mathcal{Y} = \{\text{jacket,no jacket}\}$ and observations $\mathcal{X} = \{\text{warm,cold}\}$ together with the loss matrix in the following table:

loss	Warm	Cold
Jacket	$l_{wj} := 0.3$	$l_{cj} := 0.1$
No Jacket	$l_{wn} := 0.0$	$l_{cn} := 1.0$

◆

In general we assume the loss is bounded between 0 and 1. We can always apply an affine transformation to transform the loss to lie in [0,1] without affecting the behavior of the optimal predictor, assuming the loss is bounded, which for finite $\mathcal{X}$ and $\mathcal{Y}$ it always is.

Definition 3.5.4 (ν-expected instantaneous loss) The ν-expected instantaneous loss when Λ predicts the t-th symbol given history $x_{<t}$ from environment ν is
$$\text{Loss}_t^\nu(\Lambda) := \mathbf{E}_\nu[loss(\cdot, y_t^\Lambda) | x_{<t}] = \sum_{x_t} \nu(x_t | x_{<t}) loss(x_t, y_t^\Lambda)$$
Note that Loss_t^ν implicitly depends on $x_{<t}$.

We assume by default interaction is with the true environment μ, and we write $\text{Loss}_t(\Lambda) := \text{Loss}_t^\mu(\Lambda)$.

If we knew μ we would choose the Λ that minimizes $\text{Loss}_t^\mu(\Lambda)$. For unknown μ, it is natural to consider minimizing ξ-expected loss. More generally, consider the ρ-optimal predictor Λ_ρ:

Definition 3.5.5 (ρ-optimal predictor Λ_ρ) Given a semimeasure ρ, we define Λ_ρ to be the *predictor* which minimizes ρ-expected loss. Formally, the action taken by following Λ_ρ are
$$y_t^{\Lambda_\rho} := \arg\min_{y_t \in \mathcal{Y}} \sum_{x_t} \rho(x_t | x_{<t}) loss(x_t, y_t)$$

We would expect that if the predictor knows which environment is the true environment μ, then it will predict at least as well as any other predictor, and indeed this is so.

Theorem 3.5.6 (Λ_ν minimizes ν-expected loss) Let Λ be any predictor, then the prediction scheme Λ_ν will always achieve minimal expected loss in environment ν. That is, for all t,
$$loss_t^\nu(\Lambda_\nu) \leq loss_t^\nu(\Lambda)$$

Proof. By definition, $\mathrm{Loss}_t^\nu(\Lambda) = \sum_{x_t} \nu(x_t|x_{<t}) loss(x_t, y_t^\Lambda)$ is minimized by choosing $y_t^\Lambda = \arg\min_{y_t \in \mathcal{Y}} \sum_{x_t} \nu(x_t|x_{<t}) loss(x_t, y_t)$, which is precisely what Λ_ν does. ∎

Example 3.5.7 (Shall Bayes take his jacket?) Continuing with Example 3.5.3, let us "somewhat" unrealistically assume that the weather is i.i.d. with

$$\theta := \mathrm{P}_\theta(x_t = \text{warm}) = \nu_\theta(x_t = \text{warm}|x_{<t})$$
$$1-\theta = \mathrm{P}_\theta(x_t = \text{cold}) = \nu_\theta(x_t = \text{warm}|x_{<t})$$

being the probabilities of the weather being warm and cold, respectively. The θ-expected losses for (not) taking a jacket are therefore

$$l_j := \mathbf{E}_\theta[loss(\cdot, \text{jacket})|x_{<t}] = \sum_{x_t \in \{\text{warm}, \text{cold}\}} \nu_\theta(x_t|x_{<t}) loss(x_t, \text{jacket}) = \theta \cdot l_{wj} + (1-\theta) \cdot l_{cj}$$

$$l_n := \mathbf{E}_\theta[loss(\cdot, \text{no jacket})|x_{<t}] = \sum_{x_t \in \{\text{warm}, \text{cold}\}} \nu_\theta(x_t|x_{<t}) loss(x_t, \text{no jacket}) = \theta \cdot l_{wn} + (1-\theta) \cdot l_{cn}$$

Taking a jacket minimizes θ-expected loss iff $l_j < l_n$. Solving for θ gives

$$y_t^{\Lambda_\theta} = \text{jacket} \iff \theta < \theta_{\text{critical}} := \frac{l_{cn} - l_{cj}}{l_{cn} - l_{cj} + l_{wj} - l_{wn}} = \frac{3}{4}$$

The general expression assumes that the denominator is positive. If it is negative, the $<$ has to be reversed to a $>$. So if the chance of warm weather is more than three-quarters, it's not worth taking a jacket. Let us assume that the true probability of warm weather is 90%. In this case, the μ-optimal predictor Λ_μ which knows $\theta = \theta_{\text{true}} := 0.9$ can compute the μ-expected losses and decides $y_t^{\Lambda_\mu} = $ no jacket.

For unknown μ, it is natural to consider the Bayes-optimal predictor Λ_ξ. In order to compute the ξ-expected loss, we have to decide on a class of environments $\mathcal{M}$ and prior w_ν. Let $\mathcal{M}$ contain all i.i.d. environments ν_θ where it is warm with probability $\theta \in [0,1]$. Let us assume a uniform prior density $w(\theta) = 1$. This is exactly the Bayes-Laplace model considered in Section 2.4. The corresponding Bayes-mixture ξ predicts 'warm' with probability

$$\hat{\theta}_t := \frac{\#\text{warm} + 1}{t+1} = \xi(x_t = \text{warm}|x_{<t})$$

where $\#$warm is the number of warm days among $x_{<t}$. We therefore have

$$y_t^{\Lambda_\xi} = \text{jacket} \iff \frac{\#\text{warm} + 1}{t+1} < \frac{3}{4}$$

On the first day The Reverend Thomas Bayes (when born or arriving on a foreign planet) has by symmetry to believe it to be warm with probability $\hat{\theta}_1 = \frac{0+1}{1+1} = \frac{1}{2}$, and given his loss function prefers a jacket (pre-emptively cries for a jacket at birth -or- better be safe than sorry on a foreign planet). Bayes is greeted with warm weather on his first day, but he is cautious (as you should in England and on foreign planets) so decides to keep the jacket on day two, since $\hat{\theta}_2 = \frac{1+1}{2+1} < \frac{3}{4}$. This is due to the skewed loss that taking a jacket unnecessarily is only a minor nuisance, while leaving it at home may be fatal. Second day is nice too, and he isn't sure anymore whether a jacket is worthwhile, since now $\hat{\theta}_3 = \frac{2+1}{3+1} = \frac{3}{4}$ and the expected losses for jacket and no jacket are both $\frac{1}{4}$. But three sunny days are enough for Bayes to take some risk and forgo his jacket, since finally $\hat{\theta}_4 = \frac{3+1}{4+1} > \frac{3}{4}$.

Apart from a minor regularization, Bayes will take his jacket iff the relative frequency of warm days is below the critical threshold of $\theta_{critical} = \frac{3}{4}$. Since we assumed $\theta_{true} = 0.9$, in the long-run with exponentially high probability (Theorem 2.2.61), there will be more than 75% warm days, and Bayes can confidently conclude that he is not in England anymore, and donate his jacket.
$\blacklozenge$

If we do not a priori know the true environment μ, we can substitute it by the Bayesian mixture ξ. Given ξ, we can use the Bayes-optimal predictor Λ_ξ, which minimizes the ξ-expected loss.

We can now see how minimizing ξ-expected loss performs against the predictor that has access to μ, or more accurately, how $\text{Loss}_t(\Lambda_\xi)$ compares to optimal $\text{Loss}_t(\Lambda_\mu)$. Since ξ converges to μ in the sense of Corollary 3.3.2, and $\text{Loss}_t(\Lambda_\rho)$ is continuous in ρ at $\rho = \mu$ (note that it is discontinuous at $\xi \neq \mu$) we expect the former to converge to the latter.

Theorem 3.5.8 (Bounding loss difference by KL divergence)

$$0 \leq \text{Loss}_t(\Lambda_\xi) - \text{Loss}_t(\Lambda_\mu) \leq a_t(x_{<t}) \leq \sqrt{2d_t(x_{<t})}$$

Proof. The difference $\text{Loss}_t(\Lambda_\xi) - \text{Loss}_t(\Lambda_\mu)$ being non-negative follows trivially from Theorem 3.5.6. Now,

$$\text{Loss}_t(\Lambda_\xi) - \text{Loss}_t(\Lambda_\mu)$$
$$= \sum_{x_t} \mu(x_t|x_{<t}) loss(x_t, y_t^{\Lambda_\xi}) - \mu(x_t|x_{<t}) loss(x_t, y_t^{\Lambda_\mu})$$

Adding $\text{Loss}_t^\xi(\Lambda_\mu) - \text{Loss}_t^\xi(\Lambda_\xi) \geq 0$,

$$\leq \sum_{x_t} \mu(x_t|x_{<t}) loss(x_t, y_t^{\Lambda_\xi}) - \mu(x_t|x_{<t}) loss(x_t, y_t^{\Lambda_\mu})$$
$$+ \xi(x_t|x_{<t}) loss(x_t, y_t^{\Lambda_\mu}) - \xi(x_t|x_{<t}) loss(x_t, y_t^{\Lambda_\xi})$$
$$= \sum_{x_t} \big(\mu(x_t|x_{<t}) - \xi(x_t|x_{<t})\big)\big(loss(x_t, y_t^{\Lambda_\xi}) - loss(x_t, y_t^{\Lambda_\mu})\big)$$
$$\leq \sum_{x_t} |\mu(x_t|x_{<t}) - \xi(x_t|x_{<t})| \cdot |loss(x_t, y_t^{\Lambda_\xi}) - loss(x_t, y_t^{\Lambda_\mu})|$$

Losses are bounded between 0 and 1, so $|loss(x_t, y_t^{\Lambda_\xi}) - loss(x_t, y_t^{\Lambda_\mu})| \leq 1$,

$$\leq \sum_{x_t} |\mu(x_t|x_{<t}) - \xi(x_t|x_{<t})| \equiv a_t(x_{<t}) \leq \sqrt{2d_t(x_{<t})}$$

where in the last step we applied Theorem 3.2.3. $\blacksquare$

Since we have already shown in Corollary 3.3.1 that $d_t \to 0$ with probability 1, it follows that $\text{Loss}_t(\Lambda_\xi) - \text{Loss}_t(\Lambda_\mu) \to 0$ with probability 1. However, this gives no information about the speed of convergence, which we will now explore.

Definition 3.5.9 (Total loss) The *total loss* of a prediction scheme Λ_ρ over time steps 1 to n is defined as

$$\text{Loss}_{1:n}(\Lambda_\rho) := \sum_{t=1}^n \mathbf{E}_\mu[\text{Loss}_t(\Lambda_\rho)]$$

Unlike the instantaneous loss (which is bounded by 1) the total loss for Λ_μ may be unbounded and grow linearly in n.

Example 3.5.10 (Linearly growing loss in stochastic environments) Suppose we have an unfair coin where the probability of flipping a head is $\theta = 0.7$. The predictor needs to guess the outcome of the coin, and receives loss 0 (resp. 1) for a correct (resp. incorrect) guess. Even supposing the true environment was known, the predictor would always guess heads, and receive an expected loss of 0.3 every time step. Over an infinite number of flips, the sum of losses will diverge.   ◆

Example 3.5.10 demonstrates that the ideal predictor Λ_μ may incur unbounded total loss. Hence, we will instead bound the excess loss of Λ_ξ over Λ_μ.

Theorem 3.5.11 (Merhav-Feder Bound [MF98, Sec.III.A.2])

$$0 \le \mathrm{Loss}_{1:n}(\Lambda_\xi) - \mathrm{Loss}_{1:n}(\Lambda_\mu) \le A_n \le \sqrt{2nD_n}$$

Proof. Take expectations $\mathbf{E}_\mu$ and sum $\sum_{t=1}^n$ in Theorem 3.5.8,

$$0 \le \mathrm{Loss}_{1:n}(\Lambda_\xi) - \mathrm{Loss}_{1:n}(\Lambda_\mu) \equiv \sum_{t=1}^n \mathbf{E}[\mathrm{Loss}_t(\Lambda_\xi) - \mathrm{Loss}_t(\Lambda_\mu)] \le \sum_{t=1}^n \mathbf{E}[a_t(x_{<t})] = A_n \le \sqrt{2nD_n}$$

where in the last inequality we used Theorem 3.2.3.   ∎

Unfortunately this bound involves n on the right-hand side, growing as $\sqrt{n}$. This can be improved as follows:

Theorem 3.5.12 (Hellinger Loss bound)

$$0 \le \sqrt{\mathrm{Loss}_{1:n}(\Lambda_\xi)} - \sqrt{\mathrm{Loss}_{1:n}(\Lambda_\mu)} \le \sqrt{2H_n} \le \sqrt{2D_n} \le \sqrt{2\ln w_\mu^{-1}} < \infty$$

Proof. See Appendix A of [Hut07e].   ∎

The right side is now independent of n and finite, but involves $\sqrt{\mathrm{Loss}}$ on the left side. By simple algebraic rearrangements, this can also be written as follows:

Corollary 3.5.13 (Bounding loss difference between Λ_ξ and Λ_μ)

$$0 \le \mathrm{Loss}_{1:n}(\Lambda_\xi) - \mathrm{Loss}_{1:n}(\Lambda_\mu) \le 2H_n + 2\sqrt{2H_n \mathrm{Loss}_{1:n}(\Lambda_\mu)}$$

where we can further bound $H_n \le D_n \le \ln w_\mu^{-1} < \infty$ as above.

Proof. Non-negativity trivially follows from Theorem 3.5.6. For the second inequality we rewrite Theorem 3.5.12 as

$$\sqrt{\mathrm{Loss}_{1:n}(\Lambda_\xi)} \le \sqrt{2H_n} + \sqrt{\mathrm{Loss}_{1:n}(\Lambda_\mu)}$$

Squaring both sides,

$$\mathrm{Loss}_{1:n}(\Lambda_\xi) \le 2H_n + 2\sqrt{2H_n \mathrm{Loss}_{1:n}(\Lambda_\mu)} + \mathrm{Loss}_{1:n}(\Lambda_\mu)$$

Subtracting $\mathrm{Loss}_{1:n}(\Lambda_\mu)$ then gives the result.   ∎

Corollary 3.5.14 (Loss bounds for infinite sequences)

(i) $\text{Loss}_{1:\infty}(\Lambda_\xi)$ is finite   if and only if $\text{Loss}_{1:\infty}(\Lambda_\mu)$ is finite.

(ii) $\text{Loss}_{1:\infty}(\Lambda_\xi) \leq 2D_\infty \leq 2\ln w_\mu^{-1}$ for deterministic[a] μ and   if $\forall x \exists y : loss(x,y) = 0$.

(iii) $\dfrac{\text{Loss}_{1:n}(\Lambda_\mu)}{\text{Loss}_{1:n}(\Lambda_\xi)} \to 1$ whenever $\text{Loss}_{1:n}(\Lambda_\xi) \overset{n \to \infty}{\longrightarrow} \infty$.

---

[a]μ is *deterministic* if there exists a sequence $x_{1:\infty}$ for which $\mu(x_{1:t}) = 1$ $\forall t$.

Proof. (i) The first property follows from Theorem 3.5.12 which holds for $n = \infty$ as well.

(ii) Since the environment μ is deterministic, and for every x there exists a y that makes the loss zero, Λ_μ can just choose the correct action that will guarantee zero loss, and hence $\text{Loss}_{1:n}(\Lambda_\mu) = 0$. Substituting this into Corollary 3.5.13 and taking the limit $n \to \infty$ we then obtain (ii).

(iii) Dividing Theorem 3.5.12 by $\sqrt{\text{Loss}_{1:n}(\Lambda_\xi)}$ we get

$$0 \leq 1 - \sqrt{\frac{\text{Loss}_{1:n}(\Lambda_\mu)}{\text{Loss}_{1:n}(\Lambda_\xi)}} \leq \sqrt{\frac{2\ln w_\mu^{-1}}{\text{Loss}_{1:n}(\Lambda_\xi)}} \longrightarrow 0 \quad \text{for} \quad n \to \infty \qquad \blacksquare$$

All of these results are true regardless of the choice of the the loss function or true environment μ, requiring only that $\mu \in \mathcal{M}$.

3.6   Pareto-Optimality of ξ

We have seen that Λ_ξ is a good choice of predictor, as the loss incurred compared to the best informed predictor Λ_μ (the regret) grows like $O(\sqrt{\text{Loss}_{1:n}(\Lambda_\mu)})$. However, this does not demonstrate that Λ_ξ is the best choice of predictor that is agnostic of μ. Without knowledge of μ, we can ask if there is a better choice of predictor Λ that performs equal or better than Λ_ξ on all environments $\nu \in \mathcal{M}$, and performs strictly better on at least one environment ν. This turns out to not be the case, implying that Λ_ξ is so-called *Pareto optimal*, i.e. that if another predictor Λ outperforms Λ_ξ on some environment, it must necessarily perform worse on another.

Definition 3.6.1 (Pareto Optimality of ξ) Let $\mathcal{F}(\mu, \rho)$ be any performance measure of ρ relative to μ. A predictor ρ is called *Pareto optimal* in class $\mathcal{M}$ with respect to $\mathcal{F}$ if there is no ρ' such that $\mathcal{F}(\nu, \rho') \leq \mathcal{F}(\nu, \rho)$ for all $\nu \in \mathcal{M}$, and $\mathcal{F}(\nu, \rho') < \mathcal{F}(\nu, \rho)$ for at least one $\nu \in \mathcal{M}$.

We have that ξ is Pareto optimal with respect to any loss $\text{Loss}_t^\nu(\Lambda)$, as well as the KL divergence (Lemma 3.2.4)

Theorem 3.6.2 (Pareto Optimality of Λ_ξ w.r.t. Loss)
The Bayes-optimal predictor Λ_ξ is Pareto optimal with respect to $\mathcal{F}(\nu, \Lambda) = \text{Loss}_{1:n}^\nu(\Lambda)$.

Proof. Assume not. Then there exists a predictor Λ such that

$$\forall \nu . \text{Loss}_{1:n}^\nu(\Lambda) \leq \text{Loss}_{1:n}^\nu(\Lambda_\xi)$$
$$\exists \nu . \text{Loss}_{1:n}^\nu(\Lambda) < \text{Loss}_{1:n}^\nu(\Lambda_\xi)$$

Using Definition 3.1.2 of ξ, we have

$$
\begin{aligned}
\mathrm{Loss}^{\xi}_{1:n}(\Lambda) &= \mathbf{E}_\xi[\textstyle\sum_{t=1}^n loss(x_t, y_t^\Lambda)]\\
&= \textstyle\sum_\nu w_\nu \mathbf{E}_\nu[\sum_{t=1}^n loss(x_t, y_t^\Lambda)]\\
&= \textstyle\sum_\nu w_\nu \mathrm{Loss}^{\nu}_{1:n}(\Lambda)\\
&< \textstyle\sum_\nu w_\nu \mathrm{Loss}^{\nu}_{1:n}(\Lambda_\xi)\\
&= \textstyle\sum_\nu w_\nu \mathbf{E}_\nu[\sum_{t=1}^n loss(x_t, y_t^{\Lambda_\xi})]\\
&= \mathbf{E}_\xi[\textstyle\sum_{t=1}^n loss(x_t, y_t^{\Lambda_\xi})]\\
&= \mathrm{Loss}^{\xi}_{1:n}(\Lambda_\xi) \le \mathrm{Loss}^{\xi}_{1:n}(\Lambda)
\end{aligned}
$$

where the $<$ follows from the assumption and the $\le$ from taking $\mathbf{E}_\xi \sum_{t=1}^n$ in Theorem 3.5.6 with $\nu = \xi$. This proves $\mathrm{Loss}^{\xi}_{1:n}(\Lambda) < \mathrm{Loss}^{\xi}_{1:n}(\Lambda)$, a contradiction. ∎

Theorem 3.6.3 (Pareto Optimality of ξ w.r.t. KL divergence) The mixture ξ is Pareto optimal with respect to the KL divergence $\mathrm{KL}_{1:n}(\nu||\rho) := \sum_{x_{1:n}} \nu(x_{1:n}) \ln \frac{\nu(x_{1:n})}{\rho(x_{1:n})}$.

Proof. Assume not. Then there exists a semimeasure ρ such that

$$
\begin{aligned}
\forall \nu. \mathrm{KL}_{1:n}(\nu||\rho) &\le \mathrm{KL}_{1:n}(\nu||\xi)\\
\exists \nu. \mathrm{KL}_{1:n}(\nu||\rho) &< \mathrm{KL}_{1:n}(\nu||\xi)
\end{aligned}
$$

Then, by summing over all environments weighted by prior w_ν, we have

$$
0 > \sum_\nu w_\nu [\mathrm{KL}_{1:n}(\nu||\rho) - \mathrm{KL}_{1:n}(\nu||\xi)] = \sum_\nu w_\nu \sum_{x_{1:n}} \nu(x_{1:n}) \ln \frac{\xi(x_{1:n})}{\rho(x_{1:n})}
$$

$$
= \sum_{x_{1:n}} \xi(x_{1:n}) \ln \frac{\xi(x_{1:n})}{\rho(x_{1:n})} = \mathrm{KL}_{1:n}(\xi||\rho)
$$

The second-last equality follows by mixture Definition 3.1.2 of ξ. Thus the KL divergence between ξ and ρ is negative, contradicting Corollary 2.5.16. ∎

A similar result can be proven for the instantaneous distances and some of the other distances (Definition 3.2.1), and is left as Exercise 14.

Pareto optimality is a minimal requirement that any purported best predictor should satisfy. Otherwise it is strictly dominated by another predictor, and hence, all else being equal, should not be used. However, in practice, we may prefer a different choice of mixture ρ that leads to a large decrease of $\mathcal{F}$ for many environments ν, in exchange for only a small increase of $\mathcal{F}$ for a few ν. This stronger condition, called *balanced Pareto optimality*, is also satisfied by ξ.

Definition 3.6.4 (Balanced Pareto Optimality) Let $\mathcal{F}(\mu, \rho)$ be any performance measure of ρ relative to μ. The universal *a priori* distribution ξ is called *balanced Pareto optimal* with respect to $\mathcal{F}$ and weights $\{w_\nu\}_{\nu \in \mathcal{M}}$ if for every ρ we have

$$
\sum_{\nu \in \mathcal{M}} w_\nu (\mathcal{F}(\nu, \rho) - \mathcal{F}(\nu, \xi)) \ge 0
$$

Balanced Pareto optimality implies that if ρ performs better than ξ on some collection of environments $\mathcal{L} = \{\nu \in \mathcal{M} : \mathcal{F}(\nu,\rho) < \mathcal{F}(\nu,\xi)\}$, then there must be a corresponding increase on other environments, such that the weighted sum of loss differences is non-negative.

The mixture ξ is also Balanced Pareto optimal, though we leave the proof as an exercise.

From the collection of all of the results in this chapter, we can see that the predictions derived from ξ will approach those of μ with respect to many different measures of distance between distributions.

3.7 Choices of Class $\mathcal{M}$ and Prior w_ν

So far we have considered general model classes $\mathcal{M}$ and priors $w_\nu > 0$ without any restriction on the semimeasures $\nu \in \mathcal{M}$, except $\mathcal{M}$ being countable, and the true $\mu \in \mathcal{M}$ being a measure. It is time to discuss concrete choices for $\mathcal{M}$ and w_ν, in particular universal choices that are suitable for AGSI.

3.7.1 Choices for Model Class $\mathcal{M}$

The larger the chosen class for $\mathcal{M}$, the weaker the assumption becomes that μ is an element of $\mathcal{M}$. However, a larger class $\mathcal{M}$ makes both the mixture ξ and the choice of prior w more complex. Recall the computability concepts from Section 2.6.3 and the (semi)measure Definition 2.2.11.

Argument for $\mathcal{M}_{comp}$. [Sol64] suggested choosing the class of all computable probability measures $\mathcal{M}_{comp}$: On the one hand, incomputable ν are impractical and impossible to deal with in practice. On the other hand, all known physical processes are stochastically computable (see the strong Church-Turing thesis discussed in Section 16.6.1), hence the true $\mu \in \mathcal{M}_{comp}$. So $\mathcal{M}_{comp}$ represents the largest practically relevant class.

Argument for $\mathcal{M}_{sol}$. A drawback of this class is that the corresponding mixture ξ fails to be computable [Hut05b]. Interestingly, choosing the larger class $\mathcal{M}_{sol} \supset \mathcal{M}_{comp}$ of all lower semicomputable semimeasures makes the mixture ξ now itself lower semicomputable [Hut05b, LV19]. The reason is that all lower semicomputable semimeasures can be effectively enumerated, but there is no computable enumeration of all computable probability measures due to the Halting problem [ZL70b, LV19]. While this ξ_U is still not quite computable, it is better than incomputable, and the closure $\xi_U \in \mathcal{M}_{sol}$ is another nice property.

Larger choices for $\mathcal{M}$. The class choice could be stretched a bit further to encompass all *cumulatively lower semicomputable* semimeasures. The corresponding mixture ξ has the same computability property, hence is also in its class [Sch02a]. In the multi-agent setting treated in Chapter 10, we will be forced to consider the even larger class of *reflective-oracle-computable* measures $\mathcal{M}_r^O \supset \mathcal{M}_{comp}$ to overcome the *Grain of Truth* problem.

Smaller choices for $\mathcal{M}$. For practical purposes we need smaller classes $\mathcal{M}$. The MDP-AIXI Algorithm 11.1 is based on the class $\mathcal{M}$ of Markov Decision Processes (MDPs); the MC-AIXI-CTW Algorithm 12.4 (Figure 12.1) is based on variable-order Markov process (CTW). Note that in these cases, ξ is not itself (variable-order) Markov, i.e. $\xi \notin \mathcal{M}$. The historically first and smallest and most famous class is the Bernoulli(θ) class, considered in Section 2.4.3 and Example 3.5.7. Mathematically these classes are uncountable so size-wise even larger than $\mathcal{M}_r^O$, but this is purely for mathematical convenience. Otherwise we could as well restrict them to rational parameters θ, making them countable subsets of $\mathcal{M}_{comp}$. Another theoretically interesting small class is the class of all poly-time computable probability distributions (of some maximum degree) of logarithmic length, for which ξ can itself be computed in polynomial time [Vov89].

In most parts of the book we consider generic $\mathcal{M}$. For AGSI purposes and below we mostly consider $\mathcal{M}_{sol}$ and sometimes $\mathcal{M}_{comp}$. Combining the computability concepts from Section 2.6.3 and the (semi)measure Definition 2.2.11, we can formally define:

Definition 3.7.1 ($\mathcal{M}_{comp}$ and $\mathcal{M}_{sol}$)

- $\mathcal{M}_{comp}$ is the set of all computable measures over infinite sequences $\mathcal{X}^\infty$
- $\mathcal{M}_{sol}$ is the set of all lower semicomputable semimeasures
  (see Definitions 2.2.11 and 2.6.12)

As explained above, the extended model class $\mathcal{M}_{sol}$ is chosen for technical reasons [Lei16b], while providing an even more general class of models than only the computable measures $\mathcal{M}_{comp}$.

3.7.2   Choices for Prior w_ν

As we've seen, the mixture ξ is (balanced) Pareto optimal, and will converge to μ under many definitions of convergence, regardless of the choice of weights w_ν so long as $w_\mu > 0$. We first discuss some philosophical arguments for various choices of the weights, culminating in the Solomonoff prior $w_\nu = 2^{-K(\nu)}$, which is in a sense the ideal choice of prior.

There are three main approaches for the choice of the prior w_ν: The Principle of Indifference (PoI), maximum entropy, and simplicity.

The principle of indifference is the most intuitive, essentially suggesting that $w_{\nu_1} = w_{\nu_2}$ for all $\nu_1, \nu_2 \in \mathcal{M}$, that is, we should consider all environments to be *a priori* equally likely. Without additional information about the environments, it would at first glance seem sensible to assume that all environments are a-priori equally likely, but this approach has problems: While PoI is possible for a finite class $\mathcal{M}$, this does not work anymore when $\mathcal{M}$ is infinite. Either $\sum_\nu w_\nu$ would diverge, or each environment has a zero weight. Moreover, it seems nonsensical to assign the same weight to two environments, when one makes strictly more assumptions than the other. Consider the following class $\mathcal{M} = \{\nu_1, \nu_2\}$, where in ν_1 all emeralds are green, and ν_2 where all emeralds are green until the year 2050 after which they turn blue. Both environments are supported equally well by observations (in case you are reading this after 2050, suitably increase the date) but it seems sensible to place more credence on ν_1 over ν_2.

The maximum entropy principle [Jay57a, Jay57b] is a generalization of the symmetry principle (choosing a symmetric prior) that suggests we choose the weights according to the entropy of ν. The idea is that the distribution with the highest entropy is the one that makes the fewest assumptions about how the data is distributed: Environments ν with a flatter probability distribution will have high entropy, and as such a higher a priori weight w_ν. Unfortunately this approach suffers from the same drawbacks as the symmetry principle (failure over infinite classes). This leaves us with

The simplicity principle. Consider the following following philosophical principles together with their scientific interpretation:

- *Occam's razor:* "Entities should not be multiplied beyond necessity":

  Of all the theories available to explain the data, choose the simplest that is consistent with the observations. This indicates we should favor simplicity over complexity, so we should choose the "simplest" ν consistent with the data.

- *Epicurus' principle:* If more than one theory is consistent with the observation, keep all the theories.

  This means that we should also make sure that as many environments ν as possible are kept for consideration and not rule out any environment before seeing evidence.

In the emerald example, Occam tells us to choose ν_1. Epicurus tells us to also keep ν_2 around, who knows. A unification of both principles is to assign non-uniform weights to hypotheses that have different a-priori plausibility. In the emerald example, we should choose w_{ν_1} larger than w_{ν_2}: Emeralds that don't suddenly change color on an arbitrary year is a simpler theory than one where they do, but Epicurus' principle suggests that they *might* do, however unlikely, so we should choose $w_{\nu_2} \neq 0$. Of course the posterior $w(\nu_2|x_{<t})$ will become 0 (1) once observing a green (blue) emerald in year 2050.

Of course, the informal word "simple" is not a well-defined term. Here, we can fall back on the definition of simplicity via Kolmogorov complexity K introduced in Section 2.7.2.

A "proof" of Occam's razor. While Occam's razor intuitively seems reasonable, up to this point we have been treating it more like an axiom or a grounding principle rather than a theorem. Here, we can present a heuristic argument of the simplicity bias based solely on the axioms of probability theory. For any finite space of outcomes, we can choose a uniform prior, which are both sensible both from the principle of indifference, and the choice of distribution that maximizes entropy. But if we consider a countable outcome space, say, $\mathbb{N}$, this is not possible anymore. Necessarily *any* choice of distribution $P = \{p_1, p_2, ...\}$ over $\mathbb{N}$ must have a convergent tail sum: $\sum_i p_i = 1 < \infty$ implies $\sum_{i \geq n} p_i \to 0$ for $n \to \infty$. So the axioms of probability alone enforce a simplicity bias: The probability mass p_i for large i must converge to zero. Now, consider a countable model class $\mathcal{M} = \{\nu_1, \nu_2, ...\}$ as we do throughout most of the book. The argument above implies that *any* prior w on $\mathcal{M}$ necessarily assigns vanishing probability to environments with large index i. Furthermore, any effective enumeration of $\mathcal{M}_{sol}$ also necessarily has simplicity bias: For any complexity k, there exists an i_k such that $K(\nu_i) > k$ for all $i > i_k$. So *any* prior on $\mathcal{M}_{sol}$ necessarily incorporates Occam's razor. We could try to circumvent this bias by considering $\mathcal{M} = \mathbb{B}^\infty$, assign uniform prior density (Lebesgue measure), interpret each $p_{1:\infty} \in \mathcal{M}$ as a (finite) program (padded to infinity), where each program constitutes an (e.g. sequence) model. Remarkably, we will show in Section 3.8.1 that even this uniform prior has a simplicity bias, and indeed also induces Solomonoff's a-priori distribution ξ_U. More details can be found in [Hut10a].

Solomonoff's optimal universal prior. Using Kolmogorov complexity allows us to assign a larger prior to simpler models/explanations in a quantitative way. As long as the machine used for the Kolmogorov complexity is universal, we will have that $2^{-K(\nu)} > 0$ for all $\nu \in \mathcal{M}$, as long as $\mathcal{M}$ only contains (semi)computable models. This gives us the Solomonoff prior:

Definition 3.7.2 (Solomonoff prior) Given a (semi)computable (semi)measure $\nu \in \mathcal{M}_{sol}$, the *Solomonoff prior* assigns a-priori weight

$$w_\nu^U := 2^{-K(\nu)}$$

to it. We let ξ_U denote the corresponding Bayesian mixture over $\mathcal{M}_{sol}$ using the weights $w_\nu = w_\nu^U$.

We now show in which sense this choice is optimal. For this purpose, consider the set of all possible (alternative) choices of lower semicomputable priors

$$\mathcal{W} = \{w_{(\cdot)} : \mathcal{M} \to \mathbb{R}^+ : \sum_\nu w_\nu \leq 1 \text{ and } K(w) < \infty\}$$

and let $w' \in \mathcal{W}$ be an arbitrarily chosen weight function. Recall the MDL bound (Theorem 2.7.23)

$$K(x) \stackrel{+}{\leq} -\log_2 P(x) + K(P)$$

where P is a semi-probability. Identifying P with w' and x with (the program describing the index of) an environment ν, we obtain

$$\log_2\big[(w_\nu^U)^{-1}\big] = K(\nu) \stackrel{+}{\leq} \log_2\big[{w'}_\nu^{-1}\big]$$

which implies that bounds for ξ_U that depend on $\ln(w_\nu^U)^{-1}$ are at most an additive constant larger than the corresponding bounds for $\xi_{w'}$ depending on $\ln {w'}_\nu^{-1}$. This means that by choosing the Solomonoff prior, bounds for ξ_U are within an additive constant the best possible among all (other) choices of lower semicomputable priors (Exercise 23).

Using this prior, we have that the total squared distance is upper bounded by the Kolmogorov complexity of the true environment (Theorem 3.2.5),

---

Theorem 3.7.3 (Solomonoff bound) For sequences sampled from any computable measure $\mu \in \mathcal{M}_{comp}$, the total expected squared error S_∞^U as per Definition 3.2.2 of the lower semicomputable Solomonoff predictor $\xi_U \in \mathcal{M}_{sol}$ is bounded as

$$S_\infty^U \leq \ln\big[(w_\mu^U)^{-1}\big] = K(\mu) \cdot \ln 2$$

Among all possible lower semicomputable priors w_ν, this is within an additive constant the tightest bound.

A downside of this prior is that it is only lower semicomputable, since K is only upper semicomputable (Theorem 2.7.27). While this prevents computing it exactly, using Kolmogorov complexity as a basis inspired other successful simplicity-based priors for real-world implementations, similar to the universal *a priori* distribution, as we will see in later chapters. An example is the prior based on context tree code lengths (Definition 4.3.20). Another is the implicit prior $n^{-d/2}$ induced by the MDL code length $\frac{1}{2}\log_2 n$ for each of d parameters in a semi-parametric class (2.7.25). See [Hut05b, Sec.3.7.2] for a generalization of Solomonoff's bound to continuous classes.

3.8   Solomonoff Distribution M_U

The Bayesian mixture ξ provides a powerful Pareto optimal predictor (Section 3.6) that converges asymptotically to the ground truth μ (Corollary 3.3.2), as well as having strong bounds on the numbers of errors while doing so (Corollary 3.3.3). ξ_U as defined relies on a lot of mathematical machinery: semimeasures, the existence of a computable enumeration of them, priors, Bayesian mixtures, and the concept of Kolmogorov complexity.

We now introduce another predictor, the *Solomonoff distribution* M, that has a much simpler definition than ξ_U. Defining M will only rely on the existence of a universal Turing machine (Definition 2.7.2). We will first motivate this new definition of M based on the quantity of random programs of a given size that print a string. We explore the properties of

M, namely that it is a lower semicomputable semimeasure, and it is linked to the Kolmogorov complexity K via the approximation $M(x) \approx 2^{-K(x)}$. We prove some rigorous bounds to motivate this approximation. We give a direct proof that M is a universal predictor for deterministic sequences. The Solomonoff distribution M indeed will turn out to be equal to ξ_U, hence like ξ_U also learns to predict sequences sampled from any unknown computable stochastic environment μ as well as a predictor with knowledge of μ.

For this section we will assume binary alphabet $\mathbb{B}$ as is common in algorithmic information theory. One could generalize all statements to non-binary finite alphabets by changing powers of 2 to powers of $|\mathcal{X}|$ and the binary logarithm to $\log_{|\mathcal{X}|}$, but the limited added value does not justify the notational and mental overhead.

3.8.1 Motivation, Derivations, Definition

We now give 3 closely related "derivations" of Solomonoff's a-priori distribution M, which are variations of piping uniform random noise through a universal Turing machine.

M from uniform prior over programs. Assume for a moment that the true environment μ is both deterministic and computable. Let $x_{1:\infty}$ be the unique string μ generates (the string x satisfying $\mu(x_t|x_{<t}) = 1$ for all t). Suppose μ can be described by a program p_μ of length $\leq l$ bits on a universal monotone Turing machine U, i.e. $U(p_\mu) = x_{1:\infty}$. Consider the set of all deterministic programs p on U with $\ell(p) \leq l$ that produce a string with prefix $x \equiv x_{1:t} \sqsubseteq x_{1:\infty}$. At least one of these such programs is p_μ, but the other programs can generate anything else (or nothing at all) after they generate x. Consider a modified version U' of U without the prefix condition. This allows us to pad all such programs so that they have length exactly l without affecting the behavior of printing a string starting with x, since U' only reads as far as the original program without padding on the time step it has output x, though it may affect what the program does next. We let $N_l(x) := |\{p \in \mathbb{B}^l : x \sqsubseteq U'(p)\}|$ denote the number of programs of length 2^l that produce strings prefixed with x. By the principle of Epicurus, we will assume a priori that all programs of length l consistent with generating the string x are equally likely. We can then use $P_l(x) := N_l(x)/2^l$ as the probability assigned to the string x. Consider for a moment all programs p that output a string prefixed with x. We can partition them into three groups based on their behavior after they output x: No more symbols are output, a following 0 is output, or a following 1 is output. Hence, we can write:

$$N_l(x) = |\{p \in \mathbb{B}^l : x \sqsubseteq U'(p)\}| \geq |\{p \in \mathbb{B}^l : x0 \sqsubseteq U'(p)\}| + |\{p \in \mathbb{B}^l : x1 \sqsubseteq U'(p)\}| = N_l(x0) + N_l(x1)$$

from which it follows that $P_l(x) \geq P_l(x0) + P_l(x1)$. Together with $0 \leq P_l(\cdot) \leq 1$ by definition, we have that P_l is a semimeasure (Definition 2.2.11). Regarding the choice for l: We desire that our choice of hypothesis space is as large as possible, so that any (computable) string has a non-zero weight attached. First, note that for fixed x, we have that $P_l(x)$ is increasing as a function of l, since

$$P_{l+1}(x) = \frac{N_{l+1}(x)}{2^{l+1}} = \frac{|\{p \in \mathbb{B}^{l+1} : x \sqsubseteq U'(p)\}|}{2^{l+1}}$$

$$= \frac{|\{p \in \mathbb{B}^l : x \sqsubseteq U'(p0)\}| + |\{p \in \mathbb{B}^l : x \sqsubseteq U'(p1)\}|}{2^{l+1}}$$

Since $U'(p) \sqsubseteq U'(pb)$ for $b \in \mathbb{B}$,

$$\geq \frac{2|\{p \in \mathbb{B}^l : x \sqsubseteq U'(p)\}|}{2^{l+1}} = \frac{N_l(x)}{2^l} = P_l(x)$$

This, together with the fact that $P_l(x) \leq 1$ means that $P_1(x), P_2(x), \dots$ forms a monotonically increasing bounded sequence, which has a well-defined limit as $l \to \infty$. We can finally define the Solomonoff distribution as

$$M(x) := \lim_{l \to \infty} P_l(x) \tag{3.8.1}$$

In (3.8.1), we count all possible elongations of the same program, effectively counting the same program multiple times.

M from monotone UTM. For another approach to defining M, we can focus only on counting minimal programs[1], weighted by program length. Simpler (as measured by Kolmogorov complexity) strings have shorter programs that print them, and so will be assigned a corresponding higher weight.

Assuming large l, $\ell(p) < l$ and program p has $2^{l-\ell(p)}$ many extensions to length l that all print the same string starting with x. Hence all extensions of p together yield a contribution $2^{l-\ell(p)}/2^l = 2^{-\ell(p)}$ to $M(x)$, independent of l. Replacing the universal Turing machine U' with a universal *monotone* Turing machine U (Definition 2.7.1), we can count each minimal program once with weight $2^{-\ell(p)}$. This gives the alternative form of the Solomonoff distribution, which we use as our definition:

Definition 3.8.2 (Solmonoff distribution) Given a universal monotone Turing machine U, the *Solomonoff distribution* or *universal distribution* is defined as

$$M_U(x) := \sum_{p:U(p)=x*} 2^{-\ell(p)} \tag{3.8.3}$$

The *conditional Solomonoff distribution* is defined in the usual way:

$$M_U(x|y) := \frac{M_U(yx)}{M_U(y)} \tag{3.8.4}$$

When U is chosen to be the reference UTM U (Remark 2.7.7), we simply write $M(\cdot) := M_U(\cdot)$.

M from piping uniform noise through a monotone UTM. Furthermore, one could also define M as the probability that U outputs a string starting with x when provided with uniform random noise on the program tape. We can show this is equivalent: Let p be a program with the property that $U(p) = x*$. The probability that the random noise r on the input tape has p as a prefix is $2^{-\ell(p)}$. Now we want the probability that any randomly generated program outputs $x*$. So we sum over all programs p that output $x*$, weighted by the probability of that program p being generated, which is $2^{-\ell(p)}$. Therefore the probability is equal to

$$\sum_{p:U(p)=x*} 2^{-\ell(p)}$$

which matches (3.8.3).

Comparison to No Free Lunch. Note that a uniform distribution is also used in the No Free Lunch (NFL) theorems [WM97] to prove the impossibility of universal learners. These theorems prove that if the performance of an algorithm is uniformly averaged over all possible problems, it will be no better than any other algorithm. However, such NFL

---

[1]A program p satisfying $x \sqsubseteq U'(p)$ is *minimal* if no proper prefix $q \sqsubset p$ satisfies $x \sqsubseteq U'(q)$.

theorems require that all prediction problems be equally likely, which is not true here: By feeding the random noise into a universal Turing machine, the output strings have a bias towards simplicity, meaning performance on simple problems will contribute more to average performance than performance on complex problems. Nor is it true in practice: Most real world prediction problems contain a lot of structure that a predictor can take advantage of, rather than drawn uniformly from the set of all prediction problems. Crucially, M makes no domain-specific assumptions, only a universal simplicity bias, without which science would anyway not be possible at all [RH11].

M implicitly also contains all stochastic environments. Recall that the construction of M made the strong assumption that the true environment was deterministic. Obviously we would also like the predictor M to perform well also when the true environment is stochastic. Since every stochastic semimeasure can be written as a convex combination of deterministic semimeasures, and M is a mixture over deterministic semimeasures, it implicitly contains the stochastic semimeasures for free as well. Indeed, we will show that M is equivalent to ξ_U, hence is applicable to stochastic environments as well.

3.8.2 Properties

Unsurprisingly as M's definition relies on Kolmogorov complexity, the Solomonoff distribution M is incomputable [LV19]. However, it is a lower semicomputable semimeasure, an important result, as M gives a mixture over the set of lower semicomputable semimeasures, meaning it is contained in its own model class $\mathcal{M}_{sol}$.

Theorem 3.8.5 (M is a semimeasure)

Proof. From the first representation of M (3.8.1) we have

$$M(\epsilon) = \lim_{l \to \infty} \frac{N_l(\epsilon)}{2^l} = \lim_{l \to \infty} \frac{|\{p \in \mathbb{B}^l : \epsilon \sqsubseteq U'(p)\}|}{2^l} = 1$$

Given any program p such that $U(p) = x*$, the program may afterwards return no more output, a zero, or a one following x. We write $U(p) = x\perp$ to denote that U gives no more output after reading p and outputting x (either because $U(p) = x$ or p runs forever after outputting x.)

$$\begin{aligned}
M(x) &= \sum_{p:U(p)=x*} 2^{-\ell(p)} \\
&= \sum_{p:U(p)=x\perp} 2^{-\ell(p)} + \sum_{p:U(p)=x0*} 2^{-\ell(p)} + \sum_{p:U(p)=x1*} 2^{-\ell(p)} \\
&> \sum_{p:U(p)=x0*} 2^{-\ell(p)} + \sum_{p:U(p)=x1*} 2^{-\ell(p)} \\
&= M(x0) + M(x1)
\end{aligned}$$

For any x we can always construct p such that $U(p) = x\perp$, which gives strict inequality above. This also demonstrates that M is not a probability measure. ∎

Theorem 3.8.6 (M is lower semicomputable)

Proof. Given a string x, consider the set $\mathcal{P}_x = \{p \in \mathbb{B}^* : U(p) = x*\}$, all programs that print a string starting with x. We write $\mathcal{P}_x = \{p_1^x, p_2^x, ...\}$. We can then write M as

$$M(x) = \sum_{i=1}^{\infty} 2^{-\ell(p_i^x)}$$

Since the sum converges, the tail of the sum converges to zero, so for any $\varepsilon > 0$ there exists N such that

$$\sum_{i=N}^{\infty} 2^{-\ell(p_i^x)} < \varepsilon$$

We now describe an algorithm $\phi(x,t)$ that computes $M(x)$ from below, and prove that for fixed x, $\phi(x,t)$ monotonically increases with t, and that for any $\varepsilon > 0$, there exists a T such that $\phi(x,T) > M(x) - \varepsilon$.

Take an effective enumeration $\mathcal{P} = \{p_1, p_2, ...\}$ of all programs for U. Let $\widehat{M} = 0$, and dovetail the computation of all programs $U(p_1), U(p_2), ...$ in $\mathcal{P}$. As soon as any such computation of $U(p_i)$ either reads past p_i on the input tape, or outputs a string of length $\ell(x)$ on the output tape, we terminate that machine and verify if $U(p_i) = x*$. If so, we add a contribution of $2^{-\ell(p_i)}$ to the total $\widehat{M}$. We run the algorithm for t many time steps, and then halt and return the estimate $\widehat{M}$.

It is obvious that $\phi(x,t)$ grows monotonically with t, as the estimate $\widehat{M}$ grows monotonically. Furthermore, for every N there is some finite time step T such that the algorithm ran the programs $p_1^x, p_2^x, ..., p_N^x$ from $\mathcal{P}_x$, and added their respective contributions $2^{-\ell(p_i^x)}$ to the sum, as each of the p_i^x, $1 \le i \le N$ lives somewhere in the enumeration $\mathcal{P}$, and each p_i^x will print a string starting with x in finite time. Hence, for $\varepsilon > 0$ there exists a time step T such that

$$\phi(x,T) > \widehat{M} = \sum_{i=1}^{N} 2^{-\ell(p_i^x)} = \sum_{i=1}^{\infty} 2^{-\ell(p_i^x)} - \sum_{i=N}^{\infty} 2^{-\ell(p_i^x)} = M(x) - \varepsilon$$

as required.   ∎

We can also get some bounds on the Solomonoff distribution in terms of the Kolmogorov complexity, which shows their close relation. Still only Solomonoff's M has excellent predictive properties. Km can only predict deterministic sequences and has exponentially worse bounds, and K cannot be used for prediction at all [Hut03g, Hut06d].

Theorem 3.8.7 (Kolmogorov–Solomonoff sandwich)

$$0 \le K(x|\ell(x)) \overset{+}{\le} -\log_2 M(x) \le Km(x) \le K(x) \overset{+}{\le} \ell(x) + 2\log_2 \ell(x)$$

Proof. We cover the easy proofs first. The first inequality $0 \le K(x|\ell(x))$ is trivial. The third $-\log_2 M(x) \le Km(x)$ follows from $M(x) \ge 2^{-Km(x)}$ which can be obtained by discarding all terms from the summation $M(x) = \sum_{p:U(p)=x*} 2^{-\ell(p)}$ except for the minimal program $\arg\min_{p:U(p)=x*} \ell(p)$. The fourth is trivial as any program p that satisfies $U(p) = x$ also satisfies $U(p) = x*$. The fifth was proven in Theorem 2.7.11. For the second inequality, we let $P(x) := M(x)[\![\ell(x) = n]\!]$. Noting that P is lower semicomputable (Theorem 3.8.6), and that $\sum_{x \in \mathbb{B}^*} P(x) = \sum_{x \in \mathbb{B}^n} M(x) \le 1$ (Lemma 2.2.12) we can apply the MDL bound (Theorem 2.7.23) conditioned on some y (see Remark 2.7.29), which gives $K(x|y) \overset{+}{\le} -\log P(x) + K(P|y)$. Note that K is conditioned on y, but P is not conditioned on anything. For x of length n we have $P(x) = M(x)$ and for $y = n$ we have $K(P|n) \overset{+}{=} K(M)$, hence $K(x|n) \overset{+}{\le} -\log M(x) + K(M)$. Now $M(\cdot)$ is a lower semicomputable semimeasure, so there exists a program that can compute M from below, so $K(M(\cdot))$ is equal to the Kolmogorov complexity of the index of $M(\cdot)$ in some enumeration of all lower semicomputable semimeasures. Hence $K(M(\cdot))=O(1)$, which combined with the above gives the desired result. ∎

3.8.3 Equivalence of M and ξ_U

Both M and ξ_U are strong predictors using (either explicitly or implicitly) a bias towards simplicity based on Kolmogorov complexity in their construction. Indeed, they coincide up to an irrelevant multiplicative constant. Even more remarkably, for suitable choice of U they coincide exactly:

Theorem 3.8.8 (Solomonoff equivalence) The Solomonoff distribution M is equal to the Bayesian mixture ξ_U up to an irrelevant multiplicative constant,

$$\forall \text{ UTM } U: M(x) \stackrel{\times}{=} \xi_U(x) \quad \text{and} \quad \exists \text{ UTM } U: M(x) = \xi_U(x)$$

Proof. The details of this proof are beyond the scope of the book; a full proof of the first statement can be found in [ZL70b, Hut05b, LV19]. In [WSH11] a stronger statement that implies the second equality is proven: Namely, the class of all Solomonoff priors and all Bayesian mixtures is identical. This is shown by proving that for every Bayesian Mixture ξ_w there exists a UTM U_w such that $M_{U_w}(\cdot)=\xi_w(\cdot)$, and that for every Solomonoff prior $M_{U'}$ there exists a choice of weights w' such that $\xi_{w'}(\cdot)=M_{U'}(\cdot)$. ∎

3.8.4 Predictive Bounds

Let us return to the deterministic case for a moment, where μ always samples a unique sequence $x_{1:\infty}$. We would expect that since M is a universal predictor, it should satisfy $M(x_t|x_{<t})\to 1$ as $t\to\infty$. Note that since μ is deterministic, we require only that the sequence $(M(x_t|x_{<t}))_{t=1}^{\infty}$ converges to 1 without having to invoke any kind of probability qualifier.

Theorem 3.8.9 (Deterministic Solomonoff Bound) For any sequence $x_{1:\infty}$, we have

$$\sum_{t=1}^{n}(1-M(x_t|x_{<t}))^2 \leq \sum_{t=1}^{n}|1-M(x_t|x_{<t})| \leq Km(x_{1:n})\ln 2$$

Furthermore, if $x_{1:\infty}$ is a computable sequence, then $M(x_t|x_{<t})\to 1$.

Proof. The first inequality is obvious from $a^2 \leq a$ for $0\leq a:=1-M\leq 1$. Now

$$\sum_{t=1}^{n}|1-M(x_t|x_{<t})| \leq -\sum_{n=1}^{n}\ln M(x_t|x_{<t}) = -\ln M(x_{1:n}) \leq Km(x_{1:n})\ln 2$$

The first step uses the fact that $1-a\leq -\ln a$ for $0\leq a\leq 1$. The second uses the chain rule, and that $M(\epsilon)=1$. The third follows by $-\log_2 M(x)\leq Km(x)$ (Theorem 3.8.7). (The square bound can actually be improved by factor of 2, since $(1-a)^2\leq -\frac{1}{2}\ln a$ for $0\leq a\leq 1$.)

Finally, if we take the limit $n\to\infty$ first on the right-hand side, and then on the left-hand side, we get $\sum_{t=1}^{\infty}|1-M(x_t|x_{<t})|\leq Km(x_{1:\infty})\ln 2$. If $x_{1:\infty}$ is computable then $Km(x_{1:\infty})$ is finite, so the terms $|1-M(x_t|x_{<t})|$ in the sum must tend to zero, hence $M(x_t|x_{<t})\to 1$. ∎

We extend Theorem 3.8.9 to the stochastic case. Suppose that $x_{1:\infty}$ is sampled from some computable measure μ, e.g. a sequence of fair i.i.d. coin flips. Of course we can do no better than predicting according to the true probability μ, and strong prediction in the sense of Theorem 3.8.9 is impossible. But much like ξ_U, the Solomonoff distribution M dominates all semicomputable semimeasures, which implies that it will eventually learn to predict as well as the best predictor with access to μ, with some additional overhead based on the complexity of μ.

Corollary 3.8.10 (Solomonoff distribution domination) For all lower semicomputable semimeasures ν, we have that

$$M(x) \stackrel{\times}{\geq} 2^{-K(\nu)}\nu(x)$$

Proof. Follows directly from Theorem 3.8.8 and Proposition 3.1.4.   ∎

Theorem 3.8.11 (Predictive convergence of M) Given a computable measure μ from which a sequence $x_{1:\infty}$ is sampled, we have that

$$S_\infty^M = \sum_{t=1}^{\infty} \sum_{x_{1:t}\in\mathbb{B}^t} \mu(x_{<t})\big(M(x_t|x_{<t})-\mu(x_t|x_{<t})\big)^2 \stackrel{+}{\leq} K(\mu)\ln 2 < \infty$$

which implies that $M(x_t'|x_{<t})$ convergences to $\mu(x_t'|x_{<t})$ with μ-probability 1.

Proof. Note that this is the same bound as in Theorem 3.7.3, but using M instead of ξ. The proof is also the same, using Corollary 3.8.10 in place of Proposition 3.1.4.   ∎

3.9 Martingales

Sequences $x_{1:\infty}$ drawn from some probability measure μ as considered in this chapter are also called *stochastic processes*. It is custom to view them as a sequence of (capitalized) random variables $X_1,X_2,X_3,...$ as we did in Section 2.4, but in this chapter without any/i.i.d. assumption on μ. A (possibly) different stochastic process $Z_1,Z_2,Z_3,...$ is called a supermartingale if it is non-increasing in expectation:

Definition 3.9.1 ((Super)martingale) The stochastic process $Z_1,Z_2,Z_3,...$ is said to be a *martingale* with respect to $X_1,X_2,X_3,...$ if for all $t\in\mathbb{N}$, the expectation of $|Z_t|$ is finite and the expectation of Z_t given $X_1,...X_{t-1}$ is Z_{t-1}. Formally,

$$\mathbf{E}[|Z_t|] < \infty \quad \text{and} \quad \mathbf{E}[Z_t|X_1,...X_{t-1}] = Z_{t-1} \quad \text{for martingales}$$

If we weaken the $=$ to $\leq$, then $Z_1,Z_2,Z_3...$ is a *supermartingale*.

There are more general definitions for continuous time $t\in\mathbb{R}$ and for (X_t) being replaced by a filtration. Many processes in theory and practice are martingales, and many theorems generalize from i.i.d. processes to martingales. This makes martingales a powerful tool in the analysis of stochastic processes.

Example 3.9.2 (Martingale ν/μ) The only (super)martingale (w.r.t. $x_t\sim\mu(\cdot|x_{<t})$) we will encounter in this book is $Z_t:=\nu(x_{1:t})/\mu(x_{1:t})$ for the true probability measure μ and some (semi)measure ν: $\mathbf{E}_\mu[|Z_t|]=\sum_{x_{1:t}}\mu(x_{1:t})Z_t=\sum_{x_{1:t}}\nu(x_{1:t})\leq 1$, and

$$\mathbf{E}_\mu[Z_t|x_{<t}] = \sum_{x_t}\mu(x_t|x_{<t})Z_t = \sum_{x_t}\mu(x_t|x_{<t})\frac{\nu(x_t|x_{<t})}{\mu(x_t|x_{<t})}Z_{t-1} = Z_{t-1}\sum_{x_t}\nu(x_t|x_{<t}) \leq Z_{t-1}$$

◆

The key result we exploit in the proofs of some of our later convergence theorems is Doob's martingale convergence theorem:

Theorem 3.9.3 (Supermartingale convergence theorem [Doo53]) Let $Z_1, Z_2, \ldots$ be a non-negative supermartingale (w.r.t. process X_t). Then the sequence converges almost surely to a random variable Z_∞ with finite expectation.

Note that there are no assumptions at all on the process X_t and only the expectation and non-negativity assumption on the process Z_t, which is the reason why martingales are so versatile. The proof is beyond the scope of this book and relies on Doob's upcrossing lemma [Wil91a, Chp.11]. Unfortunately this is a purely asymptotic result and does not come with any convergence rates [LH14d], unless further assumptions are made. For instance, Hoeffding's bound (Theorem 2.2.61) remains valid for martingales of bounded difference $\forall t. |Z_t - Z_{t-1}| \leq 1$, and hence such martingales concentrate around their expectation at an exponential rate [GS20, Sec.12.2].

Example 3.9.4 (Martingale ξ/μ) Consider supermartingale $Z_t := \xi(x_{1:t})/\mu(x_{1:t})$. Theorem 3.9.3 implies that μ-almost surely $Z_\infty < \infty$. From Proposition 3.1.4 we have $Z_t \geq w_\mu > 0$, hence $Z_\infty > 0$. Together this implies

$$\frac{\xi(x_t|x_{<t})}{\mu(x_t|x_{<t})} = \frac{Z_t}{Z_{t-1}} \xrightarrow{t \to \infty} \frac{Z_\infty}{Z_\infty} = 1 \quad \text{w.}\mu\text{.p.1}$$

This is weaker than Corollary 3.3.2, since convergence holds only "on-sequence", i.e. for x_t sampled from μ, and unlike Corollary 3.3.3 and Theorem 3.3.4 no convergence rate is provided. ◆

The martingale convergence theorem also implies another famous and useful convergence result.

Theorem 3.9.5 (Merging of opinions [BD62])

$$\sup_A \left| P_\xi[A|x_{<t}] - P_\mu[A|x_{<t}] \right| \xrightarrow{t \to \infty} 0 \quad \text{w.}\mu\text{.p.1}$$

where $\sup_A$ ranges over all measurable events $A \subseteq \mathcal{X}^\infty$.

Proof. The proof is based on martingale theory and uses dominance $P_\xi[A] \geq w_\mu P_\mu[A]$ for all events A, and is beyond the scope of this book. ∎

The result in particular implies that for a sequence of events A_t, $P_\xi[A_t|x_{<t}] \to P_\mu[A_t|x_{<t}]$. For instance, for $A_t = \{x_{<t}x_t'\} \times \mathcal{X}^\infty$, we recover $\xi(x_t'|x_{<t}) \to \mu(x_t'|x_{<t})$ of Corollary 3.3.2. But the real power of Theorem 3.9.5 is for events A that non-trivially depend on the *infinite* future. This becomes important for far-sighted agents in Chapter 7 and elsewhere.

3.10 Exercises

1. [C20] Prove the elementary relations between the distances a_t and s_t and h_t in Theorem 3.2.3 (third line). Show that the constants in the bounds are tight.

2. [C20i] Prove the ratio convergence $\xi_t/\mu_t \to 1$ in Corollary 3.3.2. Show that unlike the difference, the ratio may fail to converge off-sequence.

3. [C30i] Prove the Hellinger bound Theorem 3.2.6(iii). Hint: Prove and use $\sum_i \sqrt{p_i q_i} \leq 1 - \frac{1}{2}\sum_i(\sqrt{p_i}-\sqrt{q_i})^2 \leq \exp[-\frac{1}{2}\sum_i(\sqrt{p_i}-\sqrt{q_i})^2]$ for $p_i := \mu(i|x_{<t})$ and $q_i := \xi(i|x_{<t})$. Now define predictive measure $\rho(i|x_{<t}) := \sqrt{p_i q_i}/\sum_i \sqrt{p_i q_i}$ and lower bound the joint $\rho(x_{1:n})$.

4. [C15] Show that Markov inequality applied to Corollary 3.3.3 and Theorem 3.3.4 indeed gives the claimed high-probability bound on ε-errors.

5. [C33m] Prove the high-probability bound on $\sum_t h_t$ in Theorem 3.3.4.

6. [C15] Consider $\mathcal{M}=\{\mu,\nu\}$, where $\mu=\mathrm{Bern}(\frac{1}{2})$ and $\nu(x_{1:n})=\prod_{t=1}^n \nu(x_t)$ is independent, but not identically distributed anymore. Sepcifically consider $\nu(x_t=1)=\frac{1}{2}-\frac{1}{t+2}$ and prior $w_\mu = w_\nu = \frac{1}{2}$. Show that $w(\mu|x_{<t}) \not\to 1$ w.μ.p.1. Why does this not contradict Corollary 3.3.2?

7. [C25i] Show that all results in Section 3.5 remain valid for t and even $x_{<t}$-dependent $loss_t(x_t,y_t;x_{<t})$. Show that formally, the results even remain true if $loss$ also depends on the historic actions $y_{<t}^\Lambda$, but now greedily minimizing $\mathrm{Loss}_t(\Lambda)$ is no longer optimal. Give an example where a non-greedy choice of $y_1 \neq y_1^{\Lambda_\mu}$ leads to a loss smaller than $\mathrm{Loss}_{1:n}(\Lambda_\mu)$.

8. [C25] Write some code to implement the Bayesian prediction scheme Λ_ξ and apply it to the weather environment given in Example 3.5.3. How many iterations in expectation does it take for Λ_ξ to recommend wearing a jacket?

9. [C20] Show that $\mathrm{Loss}_t(\Lambda_\rho)$ in Definition 3.5.4 is continuous in $\rho(x_t'|x_{<t})$ at $\rho=\mu$ but discontinuous at $\rho\neq\mu$.

10. [C12] Following Example 3.5.10, verify the claim by showing that $\lim_{n\to\infty}\mathrm{Loss}_{1:n}=\infty$.

11. [C23] For two distinct mixtures ξ_w and $\xi_{w'}$ with different choices for weight functions $w_{(\cdot)}$ and $w'_{(\cdot)}$, show that the μ-expected squared distance between these two mixtures is upper bounded by $2(\ln w_\mu^{-1}+\ln w'_\mu{}^{-1})$ [Hut05b].

12. [C12i] Under what conditions on μ can the true predictor Λ_μ achieve 0 loss?

13. [C15] Let $\mathcal{M}=\{\nu_\theta:\theta\in[0,1]\}$ with i.i.d. $\nu_\theta(x_t=1|x<t)=\theta=1-\nu_\theta(x_t=0|x<t)$. Show that ξ and its generate the Laplace rule introduced in Section 2.4.3.

14. [C22] Prove that Bayes-mixture ξ is Pareto optimal (Definition 3.6.1) with respect to any instantaneous $\mathrm{Loss}_t^\nu(\Lambda)$ and instantaneous KL divergence $\mathrm{KL}_t(\nu||\rho) := \sum_{x_t}\nu(x_t|x_{<t})\ln[\nu(x_t|x_{<t})/\rho(x_t|x_{<t})]$ under some mild extra assumption.

15. [C30] Prove that ξ is Pareto optimal (Definition 3.6.1) with respect to some but not all of the other distance measures in Definition 3.2.1. Find a performance measure $\mathcal{F}(\rho,\xi)$ w.r.t. which ξ is *not* Pareto-optimal. Show that any Pareto-optimal predictor w.r.t. to the square distance is necessarily a Bayes-mixture. On the other hand, construct a Pareto-optimal Λ_ρ w.r.t. some $loss(x,y)$ such that ρ is *not* a Bayesian mixture ξ over $\mathcal{M}$, though the actions $y_t^{\Lambda_\rho}=y_t^{\Lambda_\xi}$ necessarily coincide, which is all that counts.

16. [C12] Show that the mixture ξ is Balanced Pareto optimal.

17. [C12] Show that Balanced Pareto optimality (Definition 3.6.4) implies Pareto optimality (Definition 3.6.1).

18. [C40s] Instead of the mixture distribution $\xi = \sum_\nu w_\nu \nu$, consider the maximum a posteriori estimator $\varrho(x) := \max\{w_\nu \nu(x) : \nu \in \mathcal{M}\}$ or equivalently the two-part minimum description length (MDL) estimator $\varrho := \mathrm{argmin}_{\nu \in \mathcal{M}}\{\log_2 \nu(x)^{-1} + \log_2 w_\nu^{-1}\}$, where as before, $\mathcal{M}$ is a countable set of (semi)measures and $\sum_{\nu \in \mathcal{M}} w_\nu \leq 1$. Show that $\sum_{t=1}^\infty \mathbf{E}_\mu[\sum_{x_t'}(\mu(x_t'|x_{<t}) - \varrho_{norm}(x_t'|x_{<t}))^2] \overset{\times}{\leq} w_\mu^{-1}$ where $\varrho(x_t|x_{<t}) = \varrho(x_{1:t})/\varrho(x_{<t})$ and $\varrho_{norm}(x_t|x_{<t}) := \varrho(x_{1:t})/\sum_{x_t'}\varrho(x_{<t}x_t')$. Show that MDL converges in mean^2 sum but convergence is exponentially worse than for ξ [Hut05b].

19. [C30oi] Show that $\mathrm{P}[\sum_{t=1}^n (\mu(x_t|x_{<t}) - \xi(x_t|x_{<t}))^2 \geq \frac{1}{\varepsilon}\ln w_\mu^{-1}] \leq \varepsilon$ and $\mathrm{P}[\sum_{t=1}^n (\mathrm{Loss}_t(\Lambda_\xi) - \mathrm{Loss}_t(\Lambda_\mu))^2 \geq \frac{2}{\varepsilon}\ln w_\mu^{-1}] \leq \varepsilon$ where P denotes μ-probability. Is it possible to prove similar high probability bounds for the ratio $\mathrm{Loss}_t(\Lambda_\xi)/\mathrm{Loss}_t(\Lambda_\mu)$? [Hut05b]

20. [C35o] We have shown that $\sum_{t=1}^\infty \mathbf{E}_\mu[s_t(x_{<t})] < \infty$. If $s_t(x_{<t})$ were monotone decreasing $(\mathbf{E}_\mu[s_{t+1}(x_{1:t})] \leq \mathbf{E}_\mu[s_t(x_{<t})])$ this would imply that $\mathbf{E}_\mu[s_t(x_{<t})]$ tents to zero faster than $1/t$, i.e. $\mathbf{E}_\mu[s_t(x_{<t})] = o(1/t)$. Show (or refute) that this monotonicity is generally wrong for some class $\mathcal{M}$ of measures. Provide necessary and/or sufficient conditions on $\mathcal{M}$ such that $s_t(x_{<t}) = o(1/t)$ [Hut05b].

21. [C30] List some reasons why we might (not) want to restrict ourselves to a class $\mathcal{M}$ of (semi)computable (semi)measures. Derive some non-computable (semi)measures that are of interest.

22. [C30] Prove that $\mathcal{M}_{comp}$ is not computably enumerable and that its mixture is not computable, and indeed not even semicomputable, but is approximable. Prove that $\mathcal{M}_{sol}$ is computably enumerable, and its mixture ξ_U is lower semicomputable but not computable (estimable). See Theorem 2.6.14 for the relevant computability concepts.

23. [C15] Prove that for any choice of weights w_ν, we have $\xi(x) := \sum_{\nu \in \mathcal{M}_{sol}} w_\nu \nu(x) \overset{\times}{\geq} w_\nu^U \nu(x)$, i.e. bar a multiplicative constant, the choice of prior actually does not matter. How large is the hidden multiplicative constant? What special property of $\mathcal{M}_{sol}$ is exploited which other $\mathcal{M}$ lack? Does there exist a UTM U' for which $\xi_{U'} = \xi$?

3.11 History and References

Solomonoff induction combines algorithmic information theory with Bayesian reasoning. It has excellent theoretical properties, and represents a platonic ideal for inductive reasoning. For more on Solomonoff induction, universal sequence prediction, and its Bayesian framing, see [HLV07, Hut17] for short and light encyclopedic introductions, [Hut09b, Sec.2] for some further insights, [RH11] for a long and gentle philosophical introduction, and [Hut06c, Hut07e] for a mathematical in-depth treatment with many further results not covered in this book. The optimality of Solomonoff induction is covered in [Hut03e], and general loss bounds are given in [Hut03a, Hut01c]. Research on Solomonoff induction was revived by [Hut01d, Hut01a] with the extension to non-binary alphabet and 0-1 loss bounds.

[CH05, CHS07] generalizes Solomonoff's bound to a future bound in terms of $K(\mu|x_{<t})$ if $x_{<t}$ has already been observed, which can be significantly smaller than $K(\mu)$. [Leg06] indicates that any powerful predictor is necessarily itself also complex. In the context

of the Black raven paradox, [LH15e] observes that Solomonoff induction violates Nicod's criterion, but concludes that the fault lies with the latter. Solomonoff induction is unified with decision theory to give the AIXI agent [Hut00, Hut07f, Hut05b], covered in Chapter 7. In [LH15d], the computability (Section 2.6.3) of various versions of Solomonoff's prior are categorized by where they live in the arithmetic hierarchy. In [LHS13a], tight probability bounds on the cumulative error of Bayesian sequence prediction were derived. [WSH11] compares and contrasts different choices for Solomonoff's M and the prior w_ν for sequence prediction. [Hut03b, Hut06b] explores alternative choices of priors that are computable. [LHG11] demonstrates that Solomonoff induction may fail for incomputable sequences that are "partially" computable (e.g. the halting sequence $\omega_{1:\infty} = [\![T_i(\epsilon)\text{halts}]\!]_{i=1}^\infty$ interleaved with zeros). A computable predictor should be able to at least predict the zeros correctly. Using a normalized version of Solomonoff induction fixes this. [HM04, HM07, LH13, LH15a] give some negative results regarding the convergence of the universal (semi)measures on Martin-Löf random sequences, and construct semimeasures that are not Bayesian mixtures that do converge on all Martin-Löf random sequences, resolving open problem [Hut03c]. See [Hut09g] for a large number of (mostly still) open problems for universal induction. As Theorem 3.7.3 among others show, K-complexity is a good measure of the difficulty of a sequence prediction problem. [AEH14] attempts to similarly measure the difficulty of (open-box) optimization problems in terms of algorithmic information. See also papers from the Ray Solomonoff 85th memorial conference [DHK+13].

The Minimum Description Length (MDL) principle, a predictor optimizing for a model that can compress the data plus the model as much as possible, is described in Section 2.7.4 and [Hut09b, PH04a, PH05a]. The motivation is as follows: The better the compression, the more regularity in the data is detected, the better the predictions for future data. For continuously parameterized i.i.d. model classes, the MDL method has comparable performance to Bayes [Gru07], and for general non-i.i.d. sequence prediction considered in this chapter, MDL is also asymptotically consistent [Hut09b] but the speed of convergence can be exponentially worse [PH04b, PH06a, Hut03g, Hut06d]. Convergence (rates) of MDL for classification and regression are shown in [PH05c].

Prediction with expert advice [HP04, HP05] is an alternative to Bayesian sequence prediction with comparable bounds [Hut04] tuned to a particular loss function. [RH07, RH08b, Rya20] explores under which conditions there exist mixture predictors that predict all other measures for ultra-general classes. [GBTC+20] introduces a computable algorithm that assigns probabilities to every logical statement in a given formal language, and refines those probabilities over time, learning to predict the validity of a statement before a formal proof/disproof is found. [VHOB15] shows how to do exact Bayesian online learning and inference efficiently over the class of monotone conjunctions from positive examples using the generalized distributive law, and then extends this heuristically to handle negative examples and k-CNF Boolean functions.

Chapters 4 and 5 introduce model classes smaller than $\mathcal{M}_{comp}$ for which ξ can be computed very efficiently. Chapter 7 covers extensions and applications of Bayesian inference to sequential decision making in reactive environments and reinforcement learning. [SH10] explores finding a compact representation of the current history for the purposes of prediction. [Hut18] investiges how any offline data compression algorithm that has random access to the whole sequence $x_{1:n}$ can be converted into an online predictor for x_t given $x_{<t}$ with virtually no regret. The paper also provides the first non-circular derivation of the famous Good–Turing estimator. Naively approximating Solomonoff's M by running vast amounts of programs for extremely long as per Definition 3.8.2 is practically unfeasible. [GMGH+24] approaches this problem by training neural networks to mimic Solomonoff induction. Artificial training data is generated by sampling from an approximate Solomonoff distribution. Predictive models

can be transformed into lossless compressors and vice versa. [DRD$^+$24] demonstrates that the impressive predictive capabilities of recent large language models indeed makes them strong lossless compressors, even for images despite primarily being trained on text.

For a broader and more in-depth introduction to Bayesian statistics, see [BC16, Pre09, Lee12, Jay03, GCSR95]. [GvdV17] provides an introduction to non-parametric Bayesian inference, which allows for a more general class of priors than parametric Bayesian inference, such as Poisson–Dirichlet processes [BH10], and non-parametric Bayesian density estimation using infinite recursive tree subdivisions [Hut09c, Hut05a]. An exact and efficient Bayesian regression algorithm for piecewise constant functions of unknown segment number, boundary locations, and levels has been derived in [Hut07c, Hut07b] and applied to genomic LOH and copy number data from SNP-microarrays [RHBK10, RH09, RHBK09b, RHBK09a]. Bayesian credible intervals and sets are discussed in [Hut08a]. Predictive Hypothesis Identification (PHI) justifies, reconciles, and blends (a reparametrization invariant variation of) MAP, ML, MDL, and moment estimation, and can genuinely deal with nested hypotheses. A Bayesian solution to the problem of incomplete (also called missing) data has been developed in [HZ03]. Analytic expressions for the Bayesian posterior of mutual information have been derived in [Hut02a], extended to incomplete data in [ZH02], and applied to selecting features in the naive Bayes classifier in [HZ05].

Chapter 4

The Context Tree Weighting Algorithm

> *All models are wrong, but some are useful.*
>
> *George E. P. Box*

We discussed in Chapter 3 how Bayesian sequence prediction can be used to learn the dynamics of an unknown environment. In this chapter, we will present a prediction algorithm which has both strong theoretical and practical properties.

There are many algorithms for the prediction of sequences, but there are few that have both strong theoretical results while also being useful in practice. Context Tree Weighting (CTW) [WST95] is one such algorithm. The CTW method provides a Bayesian mixture over the class of environment distributions that depend only on at most the previous D observations. Remarkably, even though the model class contains $O(2^{2^D})$ distributions, updating the weights for the mixture ξ can be done in $O(D)$ time, a double exponential speedup over naively computing the Bayesian mixture in Theorem 4.5.11. For most of this chapter, we will assume that sequences are over the binary alphabet $\mathbb{B}$, though most of the results generalize to any finite alphabet.

We first introduce the KT-estimator, a predictor built in the same way as Laplace's rule (Section 2.4.3), but with a different choice of prior. We extend the KT-estimator to a version that can predict sequences that violate the i.i.d. assumption; where the distribution over future elements depends on a bounded segment of the past (called the context). We explore predictors that allow for variable context, and finally introduce the CTW method, which learns how much context in the past is required, and how the future depends upon it. We prove strong predictive bounds for both the KT-estimator and the CTW method, as well as some experimental results to demonstrate the predictive power of these methods.

For clarity, we will denote the bits in a sequence as 0 and 1, and use 0 and 1 to refer to the natural numbers zero and one respectively.

4.1   Krichevsky–Trofimov (KT) Estimator

Given a binary sequence $x_{1:t}$ generated by some source, we have discussed many ways to predict the next element of the sequence. One simple estimate used was the generalized Laplace rule in Section 2.4.3. Given a $\mathrm{Bern}(\theta)$ process (where θ parameterizes the probability of sampling a one) the probability that a specific sequence with a zeros and b ones is sampled is $\theta^b(1-\theta)^a$. For unknown θ, we can choose Jeffery's prior $w(\theta) = \pi^{-1}[\theta(1-\theta)]^{-1/2}$ (Definition 2.4.11) and write the marginal probability for any sequence $x_{1:n}$ by integrating out θ.

$$\mathrm{P_{KT}}(x_{1:n}) \equiv \mathrm{P_{KT}}(a,b) = \int_0^1 \mathrm{P_{KT}}(a,b|\theta)w(\theta)\,d\theta = \int_0^1 (1-\theta)^a\theta^b \frac{1}{\pi\sqrt{\theta(1-\theta)}}\,d\theta$$

This gives us the KT estimator.

Definition 4.1.1 (KT estimator [WST95]) The Krichevski-Trofimov (KT) estima-tor probability $\mathrm{P_{KT}}(x_{1:n}) \equiv \mathrm{P_{KT}}(a,b)$ for sequence $x_{1:n}$ is defined as

$$\mathrm{P_{KT}}(a,b) := \frac{1}{\pi}\int_0^1 \frac{1}{\sqrt{\theta(1-\theta)}}\theta^b(1-\theta)^a\,d\theta = \frac{1}{\pi}\frac{\Gamma(a+\frac{1}{2})\Gamma(b+\frac{1}{2})}{\Gamma(a+b+1)} = \frac{1}{\pi}\mathrm{Beta}(a+\tfrac{1}{2},b+\tfrac{1}{2})$$

where $\mathrm{Beta}(a,b) := \int_0^1 \theta^{a-1}(1-\theta)^{b-1}\,d\theta$ is the Beta function (ensuring the probability is normalized), and Γ is the gamma function (2.4.7).

For this definition, we dropped the $x_{1:t}$ and simply wrote $P_{KT}(a,b)$ as the order of the bits in the sequence is irrelevant due to the i.i.d. assumption; only the counts of 0's (a) and 1's (b) in $x_{1:t}$ matter.

Although the KT estimator looks daunting in its current form, one can easily show that it satisfies the following recursive relation.

Lemma 4.1.2 (Recursive KT estimator [WST95]) The KT estimator has the following recurrence relation:

$$P_{KT}(0,0)=1, \quad P_{KT}(a+1,b)=\frac{a+1/2}{a+b+1}\cdot P_{KT}(a,b), \quad P_{KT}(a,b+1)=\frac{b+1/2}{a+b+1}\cdot P_{KT}(a,b)$$

Alternatively $\quad P_{KT}(x_{1:t})=P_{KT}(x_t|x_{<t})P_{KT}(x_{<t}) \quad$ with $\quad P_{KT}(\epsilon)=1$

where $\quad P_{KT}(1|x_{1:t})=\frac{b+1/2}{a+b+1} \quad$ and $\quad P_{KT}(0|x_{1:t})=1-P_{KT}(1|x_{1:t})=\frac{a+1/2}{a+b+1}$

Proof. First, note that $P_{KT}(0,0)=\frac{1}{\pi}\frac{\Gamma(1/2)\Gamma(1/2)}{\Gamma(1)}=\frac{1}{\pi}\frac{\sqrt{\pi}\sqrt{\pi}}{1}=1$. Then, for the recursive case, using the property that $\Gamma(z+1)=z\Gamma(z)$,

$$P_{KT}(a+1,b) = \frac{1}{\pi}\frac{\Gamma(a+1+\frac{1}{2})\Gamma(b+\frac{1}{2})}{\Gamma(a+b+2)} = \frac{a+\frac{1}{2}}{a+b+1}\frac{1}{\pi}\frac{\Gamma(a+\frac{1}{2})\Gamma(b+\frac{1}{2})}{\Gamma(a+b+1)} = \frac{a+\frac{1}{2}}{a+b+1}P_{KT}(a,b)$$

The $(a,b+1)$ case follows by symmetry. ∎

Using this estimator, if we wanted to compute the joint probability of a sequence of binary symbols, we can recover it from the predictive probabilities as

$$P_{KT}(x_{1:t}) := \prod_{i=1}^{t} P_{KT}((x_i|x_{<i}))$$

We can recursively compute $P_{KT}(a,b)$ for several values of a and b (Figure 4.1). Note that P_{KT} is symmetric, $P_{KT}(a,b)=P_{KT}(b,a)$.

a\b	0	1	2	3	4	...
0	1	1/2	3/8	5/16	35/128	...
1	1/2	1/8	1/16	5/128	7/256	...
2	3/8	1/16	3/128	3/256	7/1024	...
3	5/16	5/128	3/256	5/1024	5/2048	...
⋮	⋮	⋮	⋮	⋮	⋮	⋮	⋱

Figure 4.1: *A table of $P_{KT}(a,b)$ for various values of a,b (Definition 4.3.24).*

Example 4.1.3 ($P_{KT}(1|101111)=\frac{11}{14}$) Given the sequence 101111, we have that $a=1$ and $b=5$. The KT estimator predicts the probability of the next element in the sequence as being 1 as

$$P_{KT}(1|101111) = \frac{b+\frac{1}{2}}{a+b+1} = \frac{5+\frac{1}{2}}{1+5+1} = \frac{11}{14}$$ ♦

Remark 4.1.4 (KT on non-i.i.d.-sequences) The KT estimator cannot distinguish the sequences 01010101... and 00110011..., and in fact does not perform any better than random guessing, even though the pattern is simple enough that we can quickly learn to predict the next bit perfectly. Using only the frequency of each bit as information to learn from means to the KT estimator that these sequences look the same as any other sequence drawn from a Bern($\frac{1}{2}$) process. This should not be surprising, since the two sequences were generated by a 1-Markov and 2-Markov process (see Definition 4.2.6) respectively, rather than sampled i.i.d.　　　　　　　　　　　　　　　　　　　　　　　　　　　　　　●

Lemma 4.1.5 (Lower bound for KT estimator) For $a+b \geq 1$ we have the following inequality,

$$P_{KT}(a,b) \geq \frac{1}{2} \frac{1}{\sqrt{a+b}} \left(\frac{a}{a+b} \right)^a \left(\frac{b}{a+b} \right)^b$$

Proof. Proof omitted, see [WST95, App.II].　　　　　　　　　　　　　　　　　　　■

We can easily recover from the recursive definition of the KT estimator a bound on the instantaneous convergence rate:

Theorem 4.1.6 (Instantaneous KT-Convergence Rate) Assuming a sequence $x_{1:\infty}$ is drawn from Bern(θ), the KT estimator converges to θ exponentially fast, in the sense that for all $\varepsilon > 0$,

$$P(|P_{KT}(x_{t+1}=1|x_{1:t}) - \theta| \geq \varepsilon) \leq 15e^{-2\varepsilon^2 t} \qquad (4.1.7)$$

Proof. First note that from Lemma 4.1.2 we can write

$$P_{KT}(x_{t+1}=1|x_{1:t}) = \frac{b_t + \frac{1}{2}}{t+1} \qquad (4.1.8)$$

where b_t are the number of 1's contained in $x_{1:t}$. Now, we can write the difference between the probability estimate given by the KT estimator, and the frequency estimate b_t/t (see Remark 2.2.59) as

$$\Delta_t := \left| \frac{b_t + \frac{1}{2}}{t+1} - \frac{b_t}{t} \right| = \left| \frac{\frac{1}{2}t - b_t}{(t+1)t} \right| \leq \frac{\frac{1}{2}t}{t(t+1)} \leq \frac{1}{2t}$$

where we have exploited $0 \leq b_t \leq t$ in the second step. Next, the triangle inequality gives

$$\left| \frac{b_t + \frac{1}{2}}{t+1} - \theta \right| \leq \Delta_t + \left| \frac{b_t}{t} - \theta \right| \quad \text{which implies}$$

$$P\left(\left| \frac{b_t + \frac{1}{2}}{t+1} - \theta \right| \geq \varepsilon \right) \leq P\left(\left| \frac{b_t}{t} - \theta \right| + \Delta_t \geq \varepsilon \right) \leq P\left(\left| \frac{b_t}{t} - \theta \right| \geq \varepsilon - \frac{1}{2t} \right) \leq 2e^{-2(\varepsilon - \frac{1}{2t})^2 t}$$

$$(4.1.9)$$

The last inequality is Hoeffding's bound (Theorem 2.2.61). For $\varepsilon \geq 1$, the l.h.s. of (4.1.7) is 0, so the bound is trivially satisfied. For $\varepsilon \leq 1$, we have $(\varepsilon - \frac{1}{2t})^2 t = \varepsilon^2 t - \varepsilon + \frac{1}{4t} \geq \varepsilon^2 t - 1$. Plugging this and (4.1.8) into (4.1.9) together with $2e^2 \leq 15$ gives the bound in the theorem.　　　　　　　　　　　　　　　　　　　　　　　　　　　　　　　　　　　■

Borrowing from information theory, we can measure the performance of an estimator using the concept of *redundancy*. Given a true distribution μ, we can encode a sequence $x_{1:n}$ using, on average, $-\log_2 \mu(x_{1:n})$ bits per symbol via arithmetic coding μ (Section 2.5.6).

Now, if μ is unknown and must be learned, we approximate μ by P_c, the *coding distribution*. We can then measure how good the approximation is by how many extra bits on average are required to encode a sequence.

For this we use the redundancy, which is the difference between the logarithm of the true distribution and the logarithm of the predictor.

Definition 4.1.10 (Redundancy) Let ρ and $\hat{\rho}$ be distributions, and let $x_{1:n}$ be a binary sequence. We define the *individual cumulative redundancy* of $\hat{\rho}$ on $x_{1:n}$ with respect to ρ as

$$r_{\hat{\rho},\rho}(x_{1:n}) := \log_2 \frac{1}{\hat{\rho}(x_{1:n})} - \log_2 \frac{1}{\rho(x_{1:n})} \equiv \log_2 \rho(x_{1:n}) - \log_2 \hat{\rho}(x_{1:n})$$

When $\rho = \mu$, the true distribution, we simply write $r_{\hat{\rho}}(x_{1:n})$.

For brevity, we write $r_{P_{KT}}$ as r_{KT}. Now, assuming the data is sampled from a Bern(θ) process, then $\mu(x_{1:n}) = (1-\theta)^a \theta^b$, where a is the number of zeros in $x_{1:n}$, and b the number of ones. Since only the number of ones and zeros is relevant, we can write the individual cumulative redundancy of the KT estimator as

$$r_{KT,\theta}(x_{1:n}) \equiv r_{KT,\theta}(a,b) = \log_2 \frac{1}{P_{KT}(x_{1:n})} - \log_2 \frac{1}{(1-\theta)^a \theta^b} = \log_2 \frac{(1-\theta)^a \theta^b}{P_{KT}(a,b)}$$

Lemma 4.1.11 (KT estimator redundancy [WST95]) For the KT estimator the parameter redundancy can be uniformly bounded, that is, for all $a+b \geq 1$ and all $\theta \in [0,1]$ we have

$$r_{KT,\theta}(a,b) \equiv \log_2 \frac{(1-\theta)^a \theta^b}{P_{KT}(a,b)} \leq \frac{1}{2} \log_2(a+b) + 1$$

Note that unlike the instantaneous convergence for i.i.d. sequences in Theorem 4.1.6, the cumulative redundancy bound holds for *any* sequence.

Proof. First, we take the derivative of $\log_2(1-\theta)^a \theta^b = a\log_2(1-\theta) + b\log_2(\theta)$ with respect to θ,

$$\frac{d}{d\theta} \log_2(1-\theta)^a \theta^b = \frac{1}{\ln 2}\left(\frac{b}{\theta} - \frac{a}{1-\theta}\right)$$

set it to zero, and solve for θ, obtaining $\theta = \frac{b}{a+b}$. This, together with monotonicity of $\log_2$ and the double derivative being negative

$$\frac{d^2}{d\theta^2} \log_2(1-\theta)^a \theta^b = \frac{1}{\ln 2}\left(-\frac{a}{(1-\theta)^2} - \frac{b}{\theta^2}\right) < 0$$

gives that $(1-\theta)^a \theta^b$ is maximized for when $\theta = \frac{b}{a+b}$, the MLEestimate, so

$$(1-\theta)^a \theta^b \leq \left(1 - \frac{b}{a+b}\right)^a \left(\frac{b}{a+b}\right)^b = \left(\frac{a}{a+b}\right)^a \left(\frac{b}{a+b}\right)^b$$

This fact, together with Lemma 4.1.5 gives

$$\log_2 \frac{(1-\theta)^a \theta^b}{P_{KT}(a,b)} \leq \log_2 \frac{(1-\theta)^a \theta^b}{\frac{1}{2}\frac{1}{\sqrt{a+b}}\left(\frac{a}{a+b}\right)^a \left(\frac{b}{a+b}\right)^b} \leq \log_2 \frac{1}{\frac{1}{2}\frac{1}{\sqrt{a+b}}} = \frac{1}{2}\log_2(a+b) + 1$$

∎

4.2 Context

When making predictions for real-world sequences, we do not only rely on the frequency of symbols in the sequence, but (unlike the KT estimator) we also tend to use the order in which they appear as extra information to help with our prediction. To see what will occur next, the most important information is often the most recent (especially if the process that generates the sequence changes over time, in which case old information can often now be out of date). This recent information is called the *context*.

Example 4.2.1 (Context in language) In language, context is extremely important when trying to figure out the next symbol (letter) in a sequence (word). Suppose we were given the letters AD, and we had to guess the next letter in the word. Using no context, the best we could hope for is to use the frequency estimates of the letters in English, of which E would be the most frequent (Figure 4.2). But both would be a poor choice, as there are few words starting with the letters ADE[1] and none starting with ADT. The frequency estimate would also assign very low probability to letters like V and J, as their frequency estimates are low, but an estimate taking into account the context would assign high probabilities P(J|AD) and P(V|AD), accounting for the many words that start with ADV[2] or ADJ[3] (Figure 4.3).

♦

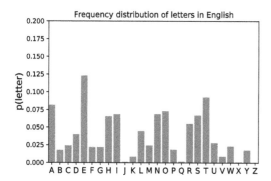

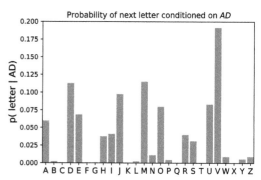

Figure 4.2: *Frequency distribution of the letters used in English, based on the frequency of letters used in Moby Dick [Mel18].*

Figure 4.3: *Conditional distribution of the next letter following "AD", using Scrabble [Col19] dictionary.*

Example 4.2.2 (KT on non-i.i.d.-sequences) Let's say we are given the sequence $x_{1:2t} = (01)^t$ (that is, an alternating sequence of 0's and 1's of length $2t$), and we are trying to estimate x_{2t+1}. Using our KT estimator on this context to estimate the likelihood of 0 or 1 as the next symbol, we get

$$P_{KT}(0|x_{1:2n}) = \frac{n+\frac{1}{2}}{2n+1} = \frac{1}{2},$$

which is no better than a uniform guess. The KT estimator has learned nothing from the context, whereas a good predictor will eventually learn that it needs only to look at the most recently received symbol to be able to get the prediction correct with certainty. In defense of the KT estimator, this is an unfair example since the KT estimator assumes the

---

[1] apart from a handful of obscure words and medical jargon, only variations of the word *adequate* or *adept*

[2] *advance, advent, advertize, adventure, adverb,...*

[3] *adjacent, adjust, adjudge, adjudicate, adjourn, adjunct,...*

sequence is i.i.d. when it makes its prediction, and the sequence 010101... is drawn from a non-i.i.d. process. ◆

4.2.1   Prediction with Context

> **Definition 4.2.3 (Markov)** A probability distribution $P : \mathbb{B}^* \to \Delta\mathbb{B}$ is *Markov* if it satisfies $P(x_t | x_{<t}) = P(x_t | x_{t-1})$ for all t. That is, the distribution for the next symbol can only depend on the most recently observed symbol and also not the time step t on which the symbol was received (which means Markov distributions are also stationary, since the distribution is fixed and cannot vary as a function of time).

Remark 4.2.4 (Markov deterministic sequences) As an abuse of terminology, we informally say that sequence $\omega = x_{1:\infty}$ is *Markov* if the "obvious" deterministic distribution μ_ω that generates ω is Markov.[4] That is, the previous bit x_{t-1} is sufficient information to derive x_t. We often omit the source when it is obvious. For example, the sequence $x_{1:\infty} = 01010101...$ has the obvious associated (deterministic) distribution

$$P(0 | \epsilon) = 1$$
$$P(x_t | x_{t-1}) = [\![x_t \neq x_{t-1}]\!]$$

and is thus Markov.

Note that all sequences generated by an i.i.d. process are also Markov, as a Markov process can ignore the context of the last symbol. Together with the above example, this makes a Markov model strictly more powerful than an i.i.d. model. ●

As we've seen, the KT estimator is useful only for sequences where the frequency of the symbols is enough for prediction. A formal way to incorporate the notion of context into our idea of prediction is to assume the underlying environment that produces the sequence is Markov[5], and try to learn the best Markov model that fits the data.

We take the concept of Markov models and combine it with our KT estimator by instead of looking for the occurrence of a single symbol in the sequence, we consider how often each symbol follows each other symbol. Phrased another way, we count the occurrence of adjacent pairs of symbols, also called *digraph*. In Example 4.2.1 we were using the statistics associated with *trigraphs* ADE,ADJ,ADV,ADT to estimate the continuation of AD... In the binary case, this is counting the occurrence of the *digraphs* 00, 01, 10, 11, then depending on the most recent symbol in our sequence we can see how often it is followed by 0 or 1. The source generating the sequence in Example 4.2.2 is Markov, as we can see that the probability of the next symbol depends only on the previous symbol and nothing before that.

Example 4.2.5 (KT on 1-Markov sequences) Expanding on Example 4.2.2, instead of looking at the number of zeros and ones, we instead count the occurrence of the digraphs 00,01,10,11 and use a KT estimator over these pairs, using the most recent symbol. Given the start of the sequence, 010101, we see three 01's, two 10's, zero 00's and zero 11. Since the most recently seen symbol was a 1, we are interested in the statistics related to the digraphs 10 and 11. Since 11 has never been seen before, and 10 has been seen twice, we should have a higher probability assigned to the next character being 0 rather than 1. By

---

[4]Technically every sequence can be sampled from the trivial $\text{Bern}(\frac{1}{2})$ distribution, but we do not mean to imply that every sequence is Markov.

[5]also called the *Markov assumption*

letting a_1 represent the count of 10 digraphs ($a_1=2$), and b_1 the count of 11 ($b_1=0$) we can use the KT estimator as before to estimate the probability of the next symbol being 1

$$P_1(x_7=1|x_{1:6}=010101) = \frac{b_1+\frac{1}{2}}{a_1+b_1+1} = \frac{0+\frac{1}{2}}{2+0+1} = \frac{1}{6}$$

vs. that of the next symbol being a 0 is

$$P_1(x_7=0|x_{1:6}=010101) = \frac{a_1+\frac{1}{2}}{a_1+b_1+1} = \frac{2+\frac{1}{2}}{2+0+1} = \frac{5}{6}$$

so the estimator is relatively confident that the next symbol will be a 0. If we condition on the sequence $(01)^n$, then there would be $n-1$ counts of 10 and none of 11, so in this instance the KT estimator would give

$$P_1(x_{2n+1} = 0|x_{2n}=(01)^n) = \frac{n-1+\frac{1}{2}}{n-1+0+1} = 1-\frac{1}{2n}$$

which as expected, converges to 1 as $n\to\infty$.   ◆

4.2.2 k-Markov Environment

Sometimes we are interested in environments that make use of more context than just the most recent symbol. We refer to such environments as being *k-Markov*.

Definition 4.2.6 (*k*-Markov) A probability distribution $P:\mathbb{B}^*\to\Delta\mathbb{B}$ is *k-Markov*, or a kth-*order Markov model*, if $P(x_t|x_{<t})=P(x_t|x_{t-k:t-1})$ for all t. That is, the probability distribution for the next symbol can only depend on the k most recently observed symbols and also not the time step t on which the symbol was received.

Remark 4.2.7 ((*k*-)Markov) Often, we abuse the term *Markov* to mean *k-Markov* for some finite k, and say 1-*Markov* when we want to be explicit about the next symbol depending only on the most recent symbol.   ●

We define the following shorthands: $a_s(x)$ (resp. $b_s(x)$) is the number of occurrences of the substring $s0$ (resp. $s1$) in the string x. We call $a_s(x)$ and $b_s(x)$ the *KT counts* of x. For example, in Example 4.2.5, given the sequence $x_{1:6}=010101$ we have that

$$a_0(x_{1:6}) = 0, \quad b_0(x_{1:6}) = 3, \quad a_1(x_{1:6}) = 2, \quad b_1(x_{1:6}) = 0$$

Formally: $a_s(x_{<t}) := |\{i<t:x_{i-\ell(s):i}=s0\}|$ and $b_s(x_{<t}) := |\{i<t:x_{i-\ell(s):i}=s1\}|$

We write a_s and b_s if x is clear from context. We can then use the KT estimator to construct an estimator for k-Markov processes, and approximate the probability of 0 or 1 after seeing s. This is done exactly as before, but by replacing a with a_s and b with b_s.

$$P_k(0|x_{1:n}) = \frac{a_s+\frac{1}{2}}{a_s+b_s+1} \quad \text{and} \quad P_k(1|x_{1:n}) = \frac{b_s+\frac{1}{2}}{a_s+b_s+1} = 1-P_k(0|x_{1:n-1})$$

where the context $s=x_{n-k+1:n}$ is the most recent k bits, giving us a k-Markov probability estimate based on the KT estimator. Note that for P_k to be well-defined, we require $\ell(x)\geq k$. To predict using P_k in practice, we can either wait until a sufficiently long context has been accumulated before making predictions, or pad the start of the sequence with zeros to create a dummy context, or use whatever context $x_{\max\{n-k+1,1\}:n}$ is currently available.

Example 4.2.8 (KT on k-Markov sequence for $k=0,1,2$) We will compare the k-Markov KT estimator for some example sequences. The first is $x_{1:3n}=(100)^n$, generated by the obvious deterministic distribution. The values of a_s and b_s for $k\leq 2$ and $s\in\mathbb{B}^{\leq 2}$ are

s	ϵ	0	1	00	01	10	11
a_s	$2n$	n	n	0	$n-1$	n	0
b_s	n	$n-1$	0	$n-1$	0	0	0

We can then compute the estimates of the next bit each k-Markov KT estimator provides.

$$P_0(0|(100)^n) = \frac{a_\epsilon+\frac{1}{2}}{a_\epsilon+b_\epsilon+1} = \frac{2n+\frac{1}{2}}{3n+1} \to \frac{2}{3}$$

$$P_1(0|(100)^n) = \frac{a_0+\frac{1}{2}}{a_0+b_0+1} = \frac{n+\frac{1}{2}}{2n} \to \frac{1}{2}$$

$$P_2(0|(100)^n) = \frac{a_{00}+\frac{1}{2}}{a_{00}+b_{00}+1} = \frac{1}{2n} \to 0$$

The sequence is generated by a 2-Markov distribution, so it is not surprising that P_2 quickly learns the pattern that the last 2 bits uniquely describe the next bit, by the rules $00\to1$, $01\to0$, $10\to0$ (where $x\to y$ means, given context x, predict y) and the probability assigned to 0 following $(100)^n$ quickly converges to zero. A third of the bits are 1, so the 0-Markov KT estimator gives the probability of a 0 as $\approx\frac{2}{3}$, regardless of the context. This is the same answer the naive KT estimator (Lemma 4.1.2) would give.

What is surprising is that the 1-Markov estimate is actually *worse* than the naive 0-Markov (i.i.d.) estimate. The reason being that P_1 assumes that the true distribution is 1-Markov, and since the digraphs 01 and 00 appear essentially equally often (n and $n-1$ respectively after seeing $3n$ bits), when P_1 is given only one bit $x_{t-1}=0$ of context, it would be equally likely that the context is the first bit of the digraph $x_{t:t-1}=01$ or $x_{t:t-1}=00$.

However, the 1-Markov model does learn that 1 will always be followed by 0, and so if the context were $(100)^n1$, it would correctly estimate the probability near one:

$$P_1(0|(100)^n1) = \frac{a_1+\frac{1}{2}}{a_1+b_1+1} = \frac{n+\frac{1}{2}}{n+0+1} = 1-\frac{1}{2n+2} \to 1 \qquad\blacklozenge$$

4.2.3 k-Markov Experiments

It should not come as a surprise that a k-Markov estimator can also learn any m-Markov distribution for $m\leq k$. However, we do not always want to choose k as large as possible, as this may lead to slower convergence.

Example 4.2.9 (KT estimator for a 2-Markov process) Consider the 2-Markov process μ_1 in Figure 4.4:

Each state is identified with the two bits in the context. This Markov chain is similar to the one in Example 4.2.8, but parameterized by $\alpha\in[0,1]$, which dictates how noisy the Markov chain is. The behavior of this Markov chain can be described as attempting to generate the sequence $(100)^n$, but when each bit is sampled, with probability $1-\alpha$ the wrong bit is generated, and the context is updated appropriately. For $\alpha=1$ this degenerates back to the same (deterministic) distribution in Example 4.2.8. So as long as $\alpha\geq0.5$, the optimal deterministic predictor (the predictor that is aware of the dynamics of the Markov process) uses a 2-bit context, and uses the following rules to predict:

$$00\to1 \quad 01\to0 \quad 10\to0 \quad 11\to0 \qquad\qquad (4.2.10)$$

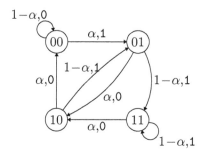

Figure 4.4: μ_1, *A 2-Markov process parameterized by* $\alpha \in [0,1]$. *The notation p,x on each edge represents sampling bit x with probability p, and then transitioning to the new state the edge points to.*

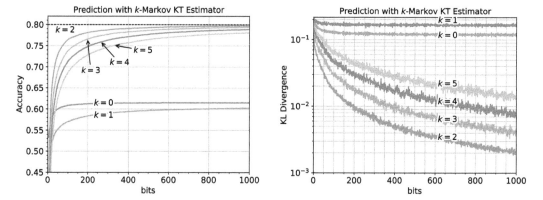

Figure 4.5: *Prediction accuracy (left) and instantaneous KL divergence (right) averaged over 100 trials for k-Markov KT estimators, for* $0 \leq k \leq 5$, *and the optimal predictor, on environment* μ_1 *(Figure 4.4) with* $\alpha = 0.8$.

This optimal predictor will predict the new bit correctly with probability α.

We choose $\alpha = 0.8$, and we sample a 1000-bit sequence from the above Markov distribution, and compare the fraction of bits predicted correctly by the k-Markov estimator to the optimal predictor described above, on the same sequence, for $0 \leq k \leq 5$ (Figure 4.5).

Clearly, for $k=0$ and $k=1$, there is insufficient context to learn the pattern and predict well. Interestingly, we see that $k=0$ again outperforms $k=1$, for similar reasons as described in Example 4.2.8. We see that any $k \geq 2$ is sufficient context to learn the distribution, but the larger the k, the longer it takes the KT estimator to learn the transition probabilities.

We can also compare the models more directly by measuring the *instantaneous KL divergence* between the true distribution μ and a predictor $\hat{\mu}$

$$\mathrm{KL}_t(\mu \| \hat{\mu}) := \mathbf{E}_{x_{1:t} \sim \mu}\left[\ln \frac{\mu(x_t | x_{<t})}{\hat{\mu}(x_t | x_{<t})}\right] \tag{4.2.11}$$

$$= \mathbf{E}_{x_{<t} \sim \mu}\left[\mu(1|x_{<t})\ln\frac{\mu(1|x_{<t})}{\hat{\mu}(1|x_{<t})} + \mu(0|x_{<t})\ln\frac{\mu(0|x_{<t})}{\hat{\mu}(0|x_{<t})}\right]$$

which we approximate by sampling a family of sequences $x^1_{<t}, x^2_{<t}, ..., x^N_{<t}$ from μ and taking the average (Figure 4.5)

$$\mathrm{KL}_t(\mu||\hat{\mu}) \approx \frac{1}{N}\sum_{i=1}^{N}\left[\mu(1|x^i_{<t})\ln\frac{\mu(1|x^i_{<t})}{\hat{\mu}(1|x^i_{<t})} + \mu(0|x^i_{<t})\ln\frac{\mu(0|x^i_{<t})}{\hat{\mu}(0|x^i_{<t})}\right]$$

$\blacklozenge$

Remark 4.2.12 (Problem of k-Markov for large k) The reason why m-Markov KT estimators are slower to converge to a k-Markov model for $m > k$ is that the estimator is keeping count of additional context that is irrelevant. For example, the 3-Markov KT estimator in Example 4.2.9 will keep track of counts a_{010} and a_{110}, whereas the 2-Markov estimator will only keep track of $a_{10} = a_{110} + a_{010}$. For a deterministic distribution (like in Example 4.2.8) this will not slow things down; since for a 2-Markov model, the context 10 uniquely determines the next bit, and so one of a_{110} or a_{010} would be zero, and the 3-Markov model would learn just as fast as the 2-Markov model. But for a stochastic distribution (like the one above), the 3-Markov model makes a (futile) distinction between 010 and 110, leading to smaller counts[6] for a_{010} and a_{110} as compared to a_{10}, which means slower learning.

More generally, a k-Markov estimator has 2^k many parameters to learn, so the sample size per parameter is $O(n/2^k)$. The same number of samples spread over more parameters means less data for each parameter to learn from, which slows learning. $\bullet$

So, we've seen that if k is chosen too small, the k-Markov KT estimator may never learn to predict optimally regardless of how many bits are provided, but if k is too large, then the estimator will be slow to learn. Obviously, if the environment is known to be k-Markov, but the transition probabilities are unknown, then we can just choose a k-Markov KT predictor. But how do we choose k if all that is known (or assumed) is that the environment is k-Markov for some value of k? We will solve this by first generalizing k-Markov KT predictors to *suffix sets*, which improves upon the former in an important way: It allows for *variable length* contexts.

We can then take the Bayesian approach by considering a mixture over suffix sets, weighting the results with a simplicity-type prior that places greater weight on less complex suffix sets. This ensures that we will never accidentally choose a model class too simple such that the true environment is not contained within, but it also guarantees that learning simple environments (i.i.d. and 1-Markov) is not overburdened by a large set of parameters.

4.3 Variable Length Context

A k-Markov estimator always considers the last k bits when making a decision, even if it would be faster to neglect additional context. For instance, recall in Example 4.2.8, the optimal predictor (4.2.10) uses two bits of context, following the rules 00→1 01→0 10→0 11→0. If the previous bit was a 0, we need to also look at the second most recent bit to decide what to do (as the most recent bit alone cannot distinguish whether to apply rule 00→1 or 10→0), but if the last bit was a 1, we can always predict 0 without considering the second-to-last bit. Essentially, the optimal predictor could instead use the following simplified set of rules 00→1 10→0 1→0.

Of course, a predictor would in general not know the dynamics of the environment, and would need to learn when this simplification can be performed.

---

[6]In the worst case, if 110 and 010 are present with approximately equal frequency, then the counts will be about half of a_{10}.

4.3.1   Prediction Suffix Trees

Prediction Suffix Trees can be used to represent a predictor that uses a variable length context depending on the situation. The tree is traversed following the context (in reverse order from most recent sample first) from the root node till a leaf node is found. The statistics associated with that leaf node are then used to predict the next symbol.

Example 4.3.1 (Context tree) Suppose we are trying to guess if the next word in a sentence is cup or cop, and the context provided so far is from the, then we likely require more context, as there are full sentences (run from the cop or drink from the cup) where either word would be a sensible continuation. However, if instead the context was wash the, then we can be almost certain that the next word will be cup rather than cop, without requiring additional context. We can represent these associations based on variable length contexts as a tree (Figure 4.6).   ◆

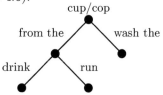

Figure 4.6: *The context tree associated with the contexts "run from the cop" and "drink from the cup" (Example 4.3.1).*

We formally define variable order Markov sources by using a *suffix set* (of depth D), a collection of strings S such that every finite sequence (of length at least D) has a unique suffix in S. As we will see, every suffix set corresponds uniquely with a suffix tree.

Definition 4.3.2 (Suffix) A string x is a *suffix* of another string y if there exists a string z such that $y = zx$.

For example, 10 is a suffix of 0010, and ing is a suffix of writing.

Definition 4.3.3 (Suffix set) A *suffix set* S is a finite set of binary[a] strings that is *complete* and *proper*.[b] S is *proper* if no member of S is a suffix of another member of S, and *complete* if every semi-infinite (to the left) sequence $...x_{t-1}x_t$ has a suffix that belongs to S.

  [a]can be generalized to any finite alphabet
  [b][WST95] defines a *suffix set* as any set of binary strings, but since we are only interested in complete and proper suffix sets, we add these properties to the definition.

Proposition 4.3.4 (Uniqueness of suffix) Given a suffix set S, every semi-infinite sequence ω has a unique suffix in S.

Proof. Existence follows directly from the definition of completeness. The suffix must also be unique, as if there existed two suffixes s and s' of ω in S, then one of s or s' would necessarily be a suffix of the other, which would violate the properness condition on S. ∎

Example 4.3.5 (Proper and complete suffix set) Consider the suffix set $S = \{0,11,101,001\}$. Let's say we want to verify that S is proper and complete. Since no member of $\mathcal{S}$ is a suffix of another member of $\mathcal{S}$, we know that $\mathcal{S}$ is a proper suffix set. To show the suffix set is complete, we will consider all possible binary sequences of length three and show that there is a member of $\mathcal{S}$ which is a suffix for each of them.

$$000, 010, 100, 110 \text{ has a suffix of } 0$$
$$011, 111 \text{ has a suffix of } 11$$
$$101 \text{ has a suffix of } 101$$
$$001 \text{ has a suffix of } 001.$$

Clearly all semi-infinite sequences must have an element of $\mathbb{B}^3$ as a suffix, which together with transitivity of suffixness gives us our result.   ◆

Definition 4.3.6 (Suffix tree) A *suffix tree* is a finite binary tree.[a] We can define it recursively

$$\text{Tree} = \text{Leaf} \mid \text{Node Tree Tree}$$

Given a non-leaf suffix tree Ψ, we refer to the two subtrees as $\Psi.\texttt{left}$ and $\Psi.\texttt{right}$ respectively.

---

[a]This generalizes to $|\Sigma|$-ary trees for suffix sets $\mathcal{S}$ over finite alphabets Σ.

Each suffix set $\mathcal{S}$ uniquely corresponds with a suffix tree $\Psi_{\mathcal{S}}$. To build a suffix tree $\Psi_{\mathcal{S}}$ from a suffix set $\mathcal{S}$ we start with a single root node, and for each string $s \in \mathcal{S}$, we read the string s **in reverse order, from right-to-left**. For each zero (one) we read from s, we add a right (left) child to the current node, and then walk down to this node. We can think of each leaf node being associated with the corresponding string s that generated it, and the string s is implicitly stored in the structure of $\Psi_{\mathcal{S}}$.

The reason why we read the string in reverse is due to the most recent bit x_n being the last one in the string $x_{1:n}$, and we read the context from most recent to least recent.

Conversely, given a suffix tree $\Psi_{\mathcal{S}}$, we can recover the suffix set $\mathcal{S}$. For each leaf node in $\Psi_{\mathcal{S}}$, find the path from the root node to the leaf, and record left steps as 1, right steps as 0. Once the leaf is reached, reverse the recorded path to recover the string s from which that leaf was generated.

4.3.2   Model Class

We can recursively define the set of all possible suffix sets $\mathcal{C}_D$ up to a maximal depth D. This will be useful later on as our choice of model class from which we assume the environment resides in, and the class to take a Bayesian mixture over.

Definition 4.3.7 (Model class) The *model class of depth D* (denoted $\mathcal{C}_D$) is the set of all complete and proper suffix sets with strings of maximal length D and is given by the recurrence

$$\mathcal{C}_D := \begin{cases} \{\{\epsilon\}\} & \text{if } D = 0 \\ \{\{\epsilon\}\} \cup \{\mathcal{S}_1\{0\} \cup \mathcal{S}_2\{1\} : \mathcal{S}_1, \mathcal{S}_2 \in \mathcal{C}_{D-1}\} & \text{if } D > 0 \end{cases}$$

where for two sets of strings A and B, we define the concatenation of sets as $AB := \{ab : a \in A, b \in B\}$.

Example 4.3.8 (Set of context trees $\mathcal{C}_D$ of depth 0, 1, and 2) For small D we can calculate $\mathcal{C}_D$ by hand.

$$
\begin{aligned}
\mathcal{C}_0 &= \{\{\epsilon\}\} \\
\mathcal{C}_1 &= \{\{\epsilon\},\{0,1\}\} \\
\mathcal{C}_2 &= \{\{\epsilon\},\{0,1\},\{00,10,1\},\{0,01,11\},\{00,01,10,11\}\}
\end{aligned}
$$

$\blacklozenge$

Since any suffix set is required to be finite, there always exists a depth D such that no suffix in $\mathcal{S}$ has length larger than D. We usually specify $\mathcal{S}$ with respect to a model class $\mathcal{C}_D$. Note that $\mathcal{C}_D$ contains all possible suffix sets with strings of length bounded by D, rather than strings of *exactly* length D. As the following corollary shows, this implies that a model class of greater depth is a superset of all shallower model classes.

Proposition 4.3.9 (Model class inclusions) We have the following inclusions

$$
\mathcal{C}_0 \subseteq \mathcal{C}_1 \subseteq \mathcal{C}_2 \subseteq \ldots
$$

Proof. The statement to prove is that $\mathcal{C}_d \subseteq \mathcal{C}_{d+1}$ for all $d=0,1,2,\ldots$, and we proceed by induction. The base case $\mathcal{C}_0 \subseteq \mathcal{C}_1$ follows immediately by Definition 4.3.7. For the step case, let us assume that $\mathcal{C}_{d-1} \subseteq \mathcal{C}_d$, and prove that $\mathcal{C}_d \subseteq \mathcal{C}_{d+1}$. Let $\mathcal{S} \in \mathcal{C}_d$. It is trivial that $\{\epsilon\}$ is contained in each $\mathcal{C}_D$ for all D by definition, so assume that $\mathcal{S} \neq \{\epsilon\}$. So there must exist $\mathcal{S}_1, \mathcal{S}_2 \in \mathcal{C}_{d-1}$ such that $\mathcal{S} = \mathcal{S}_1\{0\} \cup \mathcal{S}_2\{1\}$. By the inductive hypothesis we have $\mathcal{S}_1, \mathcal{S}_2 \in \mathcal{C}_d$, and hence $\mathcal{S} = \mathcal{S}_1\{0\} \cup \mathcal{S}_2\{1\} \in \mathcal{C}_{d+1}$, by definition of $\mathcal{C}_{d+1}$. $\blacksquare$

Lemma 4.3.10 (Model class $\mathcal{C}_D$ is double exponential in size) The size of the model class $\mathcal{C}_D$ grows double-exponentially with depth D. More precisely, for $D \geq 1$,

$$
2^{2^{D-1}} \leq |\mathcal{C}_D| \leq 2^{2^D - 1}
$$

Proof. From the recursive Definition 4.3.7 of $\mathcal{C}_D$ we get $|\mathcal{C}_D| = 1 + |\mathcal{C}_{D-1}|^2$ and $|\mathcal{C}_0| = 1$ hence $|\mathcal{C}_1| = 2$. We prove the bounds by induction over D. They are obviously satisfied for $D=1$. For the induction step we have

$$
|\mathcal{C}_D| = 1 + |\mathcal{C}_{D-1}|^2 \quad
\begin{aligned}
&\leq 1 + (2^{2^{D-1}-1})^2 = 1 + 2^{2(2^D-1)} \leq 1 + 2^{2^D - 1} \\
&\geq 1 + (2^{2^{D-2}})^2 \quad = 1 + 2^{2^{D-1}} \quad\; \geq 2^{2^{D-1}}
\end{aligned}
$$

$\blacksquare$

Definition 4.3.11 (Prediction Suffix Tree (PST)) Given a suffix set $\mathcal{S}$ and a parameter vector $\Theta_\mathcal{S}$, a *prediction suffix tree* $\Psi_{\mathcal{S},\Theta_\mathcal{S}}$ is a binary tree defined as

```
PST = Leaf θ | Node PST PST.
```

Each leaf in a PST stores a parameter $\theta_s \in \Theta_\mathcal{S}$, where s is the reverse of the path from the root node to that leaf, encoding left as 1 and right as 0.

The motivation behind Definition 4.3.11 is to keep track of long context when it is required, and only short context when it is not. This reduces the number of redundant parameters, which speeds learning (see Remark 4.2.12).

As with suffix trees, we denote the subtrees of a PST Ψ as $\Psi.\texttt{left}$ and $\Psi.\texttt{right}$.

PSTs are constructed from a suffix set $\mathcal{S}$ in the same way as suffix trees are, with the addition of associating leaves s with parameters $\theta_s \in \Theta_\mathcal{S}$ (see Algorithm 4.1).

We associate with every suffix $s \in \mathcal{S}$ a parameter $\theta_s \in [0,1]$, which corresponds to the likelihood of the next symbol being 1 given that the previous symbols had suffix s. Then given any sequence $...x_{n-1}x_n$ there always exists a suffix $s \in \mathcal{S}$ of $...x_{n-1}x_n$ (due to completeness) and is unique since our suffix set is proper. The set of the parameters is called the parameter vector $\Theta_{\mathcal{S}} = \{\theta_s | s \in \mathcal{S}\}$, and we let $\Theta_{\mathcal{S}}(x_{1:t})$ denote the corresponding parameter θ_s where s is a suffix of $x_{1:t}$.

Definition 4.3.12 (Suffix Function) Given a suffix set $\mathcal{S}$, we can define the *suffix function* $\beta_{\mathcal{S}} : \mathbb{B}^\infty \to \mathbb{B}^*$ that maps semi-infinite sequences onto their unique suffix in $\mathcal{S}$.

Note that the suffix that $\beta_{\mathcal{S}}$ maps onto is determined solely from the last D bits of the input, where $D := \max_{s \in \mathcal{S}} \ell(s)$. Note that $\beta_{\mathcal{S}}(x_{1:t})$ may be undefined if $t < D$. To avoid this, we always assume that $t \geq D$ (if using CTW for compression, the first D bits can be transmitted separately uncompressed, as often $t \gg D$.)

Given a PST, we can also recover $(\mathcal{S}, \Theta_{\mathcal{S}})$ by reading off every leaf node to recover a parameter θ_s, and then finding the path back up the tree from that leaf node back to the root (recording characters backwards: right is 0, left is 1) to find s.

Algorithm 4.1 Initializing PST $\Psi_{\mathcal{S},\Theta_{\mathcal{S}}}$ from $\mathcal{S}$ and $\Theta_{\mathcal{S}}$

Input: Suffix set $\mathcal{S}$
Input: Parameter vector $\Theta_{\mathcal{S}}$
Output: PST $\Psi_{\mathcal{S},\Theta_{\mathcal{S}}}$
 1: Initialize Ψ as an empty binary tree
 2: **for** $s \in \mathcal{S}$ **do**
 3:     $m := \ell(s)$
 4:     $p := \Psi$ ▷ Set current node as root node
 5:     **for** $t = m$ **down to** 1 **do**   ▷ Read s right-to-left
 6:         **if** $s_t = 1$ **then**
 7:             Add left child to p
 8:             $p := p.\texttt{left}$ ▷ Walk down the left node
 9:         **else**
10:             Add right child to p
11:             $p := p.\texttt{right}$ ▷ Walk down the right node
12:     Store θ_s in p ▷ Store θ_s in the leaf node reached by path s
     return Ψ

Definition 4.3.13 (PST probability) The conditional probability of the next symbol being a 1 as predicted by a PST represented by a suffix set $\mathcal{S}$ and parameter vector $\Theta_{\mathcal{S}}$ is defined as

$$\mathrm{P}_{\mathcal{S},\Theta_{\mathcal{S}}}(x_t = 1 | x_{1:t-1}) := \theta_{\beta_{\mathcal{S}}(x_{1:t-1})} \equiv \theta_{\beta_{\mathcal{S}}(x_{t-D:t-1})}$$

We often abbreviate this as $\mathrm{P}_{\mathcal{S}}$ where the choice of $\Theta_{\mathcal{S}}$ is clear from context.

Example 4.3.14 (Prediction suffix tree) Let's say that we are given a PST $\Psi_{\mathcal{S},\Theta}$ with $\mathcal{S} = \{11, 01, 110, 010, 00\}$ and corresponding parameters $\theta_{11} = 0.1$, $\theta_{01} = 0.2$, $\theta_{110} = 0.3$, $\theta_{010} = 0.4$, $\theta_{00} = 0.5$. The corresponding PST is illustrated in Figure 4.7.

We can use this tree for predicting the next bit in a sequence, given the sequence so far. Consider the sequence 0101101. We compute $\beta_{\mathcal{S}}(0101101) = 01$, and then use the

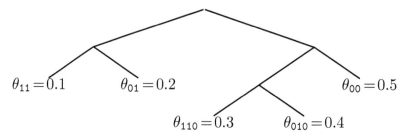

Figure 4.7: *The PST for $\mathcal{S} = \{11,01,110,010,00\}$ and $\Theta_{\mathcal{S}} = \{0.1,0.2,0.3,0.4,0.5\}$ described in Example 4.3.14.*

corresponding parameter θ_{01} as the probability that the next bit is a 1. We write

$$P_{\mathcal{S},\Theta_{\mathcal{S}}}(x_t=1|x_{<t}=0101101) = \theta_{01} = 0.2.$$

For predicting the next k bits, we can rewrite this using the standard rules of probability as

$$P_{\mathcal{S},\Theta_{\mathcal{S}}}(x_{t:t+k-1}|x_{<t}) = \prod_{i=0}^{k-1} P_{\mathcal{S}}(x_{t+i}|x_{<t+i}) \tag{4.3.15}$$

that is, we predict the next bit x_t using $x_{<t}$ as context, and then we predict x_{t+1} using $x_{<t+1}$ as the new context, and so on, taking the product of all the probabilities at the end. For example, we would compute $P_{\mathcal{S}}(x_{t:t+2}=011|x_{<t}=010)$ as

$$
\begin{aligned}
&P_{\mathcal{S}}(x_{t:t+2}=011|x_{<t}=010) \\
&= P_{\mathcal{S}}(x_t=0|x_{<t}=010)P_{\mathcal{S}}(x_{t+1}=1|x_{<t+1}=0100)P_{\mathcal{S}}(x_{t+2}=1|x_{<t+2}=01001) \\
&= (1-\theta_{010})\theta_{00}\theta_{01} \\
&= (1-0.4)\times0.5\times0.2 = 0.06
\end{aligned}
$$

If the context given is too short, we pad with zeros per Definition 4.3.12. For example, we would compute $P_{\mathcal{S}}(x_{2:3}=11|x_1=1)$ as

$$
\begin{aligned}
&P_{\mathcal{S}}(x_{2:3}=11|x_1=1) \\
&\equiv P_{\mathcal{S}}(x_{2:3}=11|x_{-\infty:1}=...0001) \\
&= P_{\mathcal{S}}(x_2=1|x_{-\infty:1}=...0001)P_{\mathcal{S}}(x_3=1|x_{-\infty:2}=...00011) \\
&= P_{\mathcal{S}}(x_2=1|x_{0:1}=01)P_{\mathcal{S}}(x_3=1|x_{1:2}=11) \\
&= \theta_{01}\theta_{11} = 0.1\times0.2 = 0.02
\end{aligned}
$$

$\blacklozenge$

We can also express the block PST probability as a product over all strings in the suffix set, which will be useful later on for theoretical results.

Lemma 4.3.16 (Alternative PST probability) Given a PST $\Psi_{\mathcal{S},\Theta_{\mathcal{S}}}$, a sequence $x_{1:n}$ and the KT counts a_s and b_s, we can express the PST probability as

$$P_{\mathcal{S},\Theta_{\mathcal{S}}}(x_{1:n}) = \prod_{s\in\mathcal{S}} \theta_s^{b_s}(1-\theta_s)^{a_s}$$

Proof. Firstly, we can break up the joint into the product of conditionals

$$
\begin{aligned}
\mathrm{P}_{\mathcal{S},\Theta_{\mathcal{S}}}(x_{1:n}) &= \prod_{i=1}^{n} \mathrm{P}_{\mathcal{S},\Theta_{\mathcal{S}}}(x_i|x_{<i}) \\
&= \left(\prod_{\substack{i=1 \\ x_i=0}}^{n} (1-\theta_{\beta_{\mathcal{S}}(x_{<i})}) \right) \left(\prod_{\substack{i=1 \\ x_i=1}}^{n} \theta_{\beta_{\mathcal{S}}(x_{<i})} \right) \\
&= \prod_{s \in \mathcal{S}} \left(\left(\prod_{\substack{i=1 \\ x_i=0 \\ \beta_{\mathcal{S}}(x_{<i})=s}}^{n} (1-\theta_s) \right) \left(\prod_{\substack{i=1 \\ x_i=1 \\ \beta_{\mathcal{S}}(x_{<i})=s}}^{n} \theta_s \right) \right) \\
&= \prod_{s \in \mathcal{S}} (1-\theta_s)^{a_s} \theta_s^{b_s}
\end{aligned}
$$

since a_s is the number of times a 0 follows context s, and we are taking the product of $(1-\theta_s)$ with itself this many times. The same argument follows for b_s. ∎

4.3.3 Suffix Set Encoding

A convenient way to come up with a prior over suffix sets $\mathcal{S}$ is by choosing $2^{-\ell(\mathcal{E}(\mathcal{S}))}$, where $\mathcal{E}$ is a encoding of suffix sets to binary strings. A naive approach would be to just prefix code (Section 2.1.2) each element of $\mathcal{S}$ and concatenate the result, but this is wasteful as not all possible sets of strings are suffix sets. We want our choice of encoding to make use of the property that suffix sets are both complete and proper. It turns out that we can encode suffix sets $\mathcal{S}$ by instead encoding the structure of the corresponding suffix tree $\Psi_{\mathcal{S}}$, from which $\mathcal{S}$ can be recovered. Since the model class $\mathcal{C}_D$ used for learning is fixed, we can furthermore assume the maximal depth D is known by the decoder.

Definition 4.3.17 (Suffix set encoding) Given model class $\mathcal{C}_D$, the encoding $\mathcal{E}_D$ of a suffix tree $\Psi_{\mathcal{S}}$ (or suffix set $\mathcal{S} \in \mathcal{C}_D$) is defined as

$$
\mathcal{E}_D(\mathcal{S}) \equiv \mathcal{E}_D(\Psi_{\mathcal{S}}) := \begin{cases} \epsilon & \text{if} \quad D=0 \\ 0 & \text{if} \quad \Psi \text{ is a leaf node} \\ 1\mathcal{E}_{D-1}(\Psi_{\mathcal{S}}.\texttt{left})\mathcal{E}_{D-1}(\Psi_{\mathcal{S}}.\texttt{right}) & \text{otherwise} \end{cases}
$$

That is, leaf nodes are encoded as 0, and internal nodes are encoded as 1, followed by the encodings of the left and right children respectively. Finally, we define $\mathcal{E}_D(\mathcal{S}):=\mathcal{E}(\Psi_{\mathcal{S}})$.

This encoding records the structure of the tree, sufficient to recover $\mathcal{S}$ (assuming D is given). It is not obvious at first glance that this code is uniquely decodable as it assigns surprisingly short codes to some trees, as shown (Figure 4.8).

The trick is that the decoder is also aware of the maximum depth of any suffix D, which conveys additional information, together with the assumption of complete and proper ensures that every node is either a leaf node or both children are non-empty. This means that for nodes at depth $D-1$, there are only two types, leaf nodes (encoded as 0) or non-leaf nodes where both children are leaves (encoded as 1). There is no need to encode the children of penultimate nodes, as knowledge of D is sufficient to derive that both children must be leaf nodes.

For this reason, given a fixed tree Ψ, the encoding $\mathcal{E}_D(\Psi)$ may vary as a function of D. Consider the tree associated with the suffix set $\mathcal{S}_0:=\{1,10,00\}$. For $D=2$, this tree is

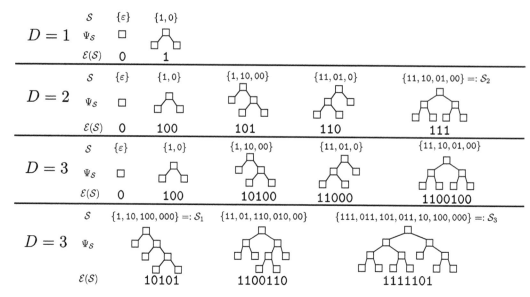

Figure 4.8: *The encodings $\mathcal{E}_D(\Psi_{\mathcal{S}})$ (Definition 4.3.17) of various suffix trees $\Psi_{\mathcal{S}}$, for depths $D \in \{1,2,3\}$.*

encoded as 101, as this indicates that the root node is not a leaf, its left child is a leaf, and the right child is not a leaf (and therefore must have exactly two leaves as children, as the depth of the tree is 2.) For $D \geq 3$ the encoding 101 is not sufficient, since we do not have enough information to deduce if the children of the right child of the root node are leaves or not. Using $\mathcal{E}_2$ to encode the suffix sets $\mathcal{S}_0 = \{1,10,00\}$ and $\mathcal{S}_1 := \{1,10,100,000\}$ gives $\mathcal{E}(\mathcal{S}_0) = \mathcal{E}(\mathcal{S}_1) = 101$, which is ambiguous.

By instead using $\mathcal{E}_3$, the structure of the leaves for both trees is respected, given encodings $\mathcal{E}_3(\mathcal{S}_0) = 10100$ and $\mathcal{E}_3(\mathcal{S}_1) = 10101$ respectively.

We now wish to prove that the encoding $\mathcal{E}_D(\mathcal{S})$ is uniquely decodable, but we first require a lemma.

Lemma 4.3.18 (Prefix Code Concatenation) If $C \subseteq \mathbb{B}^*$ is a prefix-free set, $x_1, x_2, y_1, y_2 \in C$ and $x_1 y_1 \sqsubseteq x_2 y_2$, then $x_1 = x_2$ and $y_1 = y_2$.

Proof. Assume $x_1 \neq x_2$. Note that since x_1 and x_2 are members of a prefix-free set, neither can be a prefix of the other. If $\ell(x_2) \geq \ell(x_1)$ then $x_1 \sqsubseteq x_2$, a contradiction. If $\ell(x_2) \leq \ell(x_1)$ then $x_2 \sqsubseteq x_1$, also a contradiction. So $\ell(x_1) = \ell(x_2)$, together with $x_1 y_1 \sqsubseteq x_2 y_2$ implies the first $\ell(x_1)$ bits of strings $x_1 y_1$ and $x_2 y_2$ match, so $x_1 = x_2$. Assuming $y_1 \neq y_2$ and repeating the above argument leads to the same contradiction, and hence we have $y_1 = y_2$. ∎

Theorem 4.3.19 ($\mathcal{E}_D : \mathcal{C}_D \to \mathbb{B}^*$ is a prefix code, hence uniquely decodable)

Proof. We prove by induction on depth D. For $D = 0$, $\mathcal{C}_0 = \{\{\epsilon\}\}$, and so $\mathcal{E}_0$ is trivially a prefix code as there is only one suffix set to encode. Assume that $\mathcal{E}_D : \mathcal{C}_D \to \mathbb{B}^*$ is a prefix code, and let Ψ_1, Ψ_2 be two suffix trees associated with suffix sets $\mathcal{S}_1, \mathcal{S}_2 \in \mathcal{C}_{D+1}$ with the property that $\mathcal{E}_{D+1}(\Psi_1) \sqsubseteq \mathcal{E}_{D+1}(\Psi_2)$. We wish to prove that $\Psi_1 = \Psi_2$. Since $D+1 \neq 0$, neither encoding can be empty. Suppose Ψ_1 is a leaf, then $\mathcal{E}_{D+1}(\Psi_1) = 0$, and since $\mathcal{E}_{D+1}(\Psi_1) \sqsubseteq \mathcal{E}_{D+1}(\Psi_2)$, then $\mathcal{E}_{D+1}(\Psi_2)$ must also start with a zero, which implies $\mathcal{E}_{D+1}(\Psi_2) = 0$ and hence Ψ_2 is also a leaf, which implies $\Psi_1 = \Psi_2$. Suppose Ψ_1 is not a leaf. Then Ψ_2 also cannot be a leaf, since its encoding $\mathcal{E}_{D+1}(\Psi_2)$ has to start with 1. So,

$$\mathcal{E}_{D+1}(\Psi_1) \sqsubseteq \mathcal{E}_{D+1}(\Psi_2)$$
$$1\mathcal{E}_D(\Psi_1.\texttt{left})\mathcal{E}_D(\Psi_1.\texttt{right}) \sqsubseteq 1\mathcal{E}_D(\Psi_2.\texttt{left})\mathcal{E}_D(\Psi_2.\texttt{right})$$
$$\mathcal{E}_D(\Psi_1.\texttt{left})\mathcal{E}_D(\Psi_1.\texttt{right}) \sqsubseteq \mathcal{E}_D(\Psi_2.\texttt{left})\mathcal{E}_D(\Psi_2.\texttt{right})$$

By assumption, $\mathcal{E}_D$ is a prefix code, so together with Lemma 4.3.18 we have that $\mathcal{E}_D(\Psi_1.\texttt{left}) = \mathcal{E}_D(\Psi_2.\texttt{left})$ and $\mathcal{E}_D(\Psi_1.\texttt{right}) = \mathcal{E}_D(\Psi_2.\texttt{right})$. Since all prefix codes are uniquely decodable, we have $\Psi_1.\texttt{left} = \Psi_2.\texttt{left}$ and $\Psi_1.\texttt{right} = \Psi_2.\texttt{right}$. Hence $\Psi_1 = \Psi_2$, as required. ∎

The length of the encoding for a suffix set $\mathcal{S} \in \mathcal{C}_D$ gives us the *model cost* $\Gamma_D(\mathcal{S})$, a simple measure of how complex the model the tree represents.

Definition 4.3.20 (Model cost) The *model cost* of a suffix set $\mathcal{S}$ (or suffix tree $\Psi_\mathcal{S}$) with respect to the model class $\mathcal{C}_D$ and encoding $\mathcal{E}_D$ is defined as

$$\Gamma_D(\Psi_\mathcal{S}) \equiv \Gamma_D(\mathcal{S}) := |\mathcal{S}| - 1 + |\{s \in \mathcal{S} : \ell(s) < D\}|$$

Intuitively, the model cost $\Gamma_D(\Psi_\mathcal{S})$ is the size of the suffix set, plus the number of non-leaf nodes in the tree (leaf nodes are cheap to code). We now prove that the equation for the model cost defined above is precisely the length of the encoding given by $\mathcal{E}_D$.

Lemma 4.3.21 (Recursive definition of model cost [WST95]) For any $\mathcal{V}, \mathcal{W} \in \mathcal{C}_D$ we have

$$\Gamma_{D+1}(\mathcal{V}\{0\} \cup \mathcal{W}\{1\}) = \Gamma_D(\mathcal{V}) + \Gamma_D(\mathcal{W}) + 1$$

Intuitively, the code length of a tree is the code length of the left child, plus the code length of the right child, plus one more bit to indicate it is not the empty tree (see Definition 4.3.17).

Proof.
$$\Gamma_{D+1}(\mathcal{V}\{0\} \cup \mathcal{W}\{1\})$$
$$= |\mathcal{V}\{0\} \cup \mathcal{W}\{1\}| - 1 + |\{s \in \mathcal{V}\{0\} \cup \mathcal{W}\{1\} : \ell(s) < D+1\}|$$
Note that $\mathcal{V}\{0\}$ and $\mathcal{W}\{1\}$ are disjoint.
$$= |\mathcal{V}\{0\}| + |\mathcal{W}\{1\}| - 1 + |\{s \in \mathcal{V}\{0\} : \ell(s) < D+1\}| + |\{s \in \mathcal{W}\{1\} : \ell(s) < D+1\}|$$
Concatenating a set with a singleton does not affect the number of elements.
$$= |\mathcal{V}| + |\mathcal{W}| - 1 + |\{s \in \mathcal{V}\{0\} : \ell(s) < D+1\}| + |\{s \in \mathcal{W}\{1\} : \ell(s) < D+1\}|$$
All elements in $\mathcal{V}\{0\}$ ($\mathcal{W}\{1\}$) are one larger than the respective elements in $\mathcal{V}$ ($\mathcal{W}$).
$$= |\mathcal{V}| + |\mathcal{W}| - 1 + |\{s \in \mathcal{V} : \ell(s) < D\}| + |\{s \in \mathcal{W} : \ell(s) < D\}|$$
$$= |\mathcal{V}| - 1 + |\{s \in \mathcal{V} : \ell(s) < D\}| + |\mathcal{W}| - 1 + |\{s \in \mathcal{W} : \ell(s) < D\}| + 1$$
$$= \Gamma_D(\mathcal{V}) + \Gamma_D(\mathcal{W}) + 1$$
∎

Theorem 4.3.22 (Model cost in closed form [WST95]) Let $D \geq 0$, and let $\mathcal{S} \in \mathcal{C}_D$. The *model cost* $\Gamma_D(\mathcal{S})$ is equal to the length of the encoding given by $\mathcal{E}_D$,

$$\Gamma_D(\mathcal{S}) = \ell(\mathcal{E}_D(\mathcal{S}))$$

Proof. We will actually prove a stronger statement, namely that $\Gamma_{D'}(\mathcal{S}) = \ell(\mathcal{E}_{D'})$ for all $D' \geq D$. Since every suffix set $\mathcal{S}$ can be represented as a suffix tree $\Psi_{\mathcal{S}}$, we will prove by induction over $\Psi_{\mathcal{S}}$, letting D denote the smallest number for which $\mathcal{S} \in \mathcal{C}_D$. First, consider the case where $\Psi_{\mathcal{S}}$ is a leaf ($\mathcal{S} = \{\epsilon\}$), with depth $D = 0$.

$$\Gamma_{D'}(\{\epsilon\}) := |\{\epsilon\}| - 1 + |\{s \in \{\epsilon\} : \ell(s) < D'\}| = |\{s \in \{\epsilon\} : \ell(s) < D'\}| = [\![D' > 0]\!]$$

Compare this with the length of the encoding $\ell(\mathcal{E}_{D'}(\{\epsilon\}))$. If $D' = 0$, then $\ell(\mathcal{E}_0(\Psi_{\{\epsilon\}})) = \ell(\epsilon) = 0$. If $D' \geq 0$, then $\ell(\mathcal{E}_{D'}(\Psi_{\{\epsilon\}})) = \ell(0) = 1$, so $\ell(\mathcal{E}_{D'}(\Psi_{\{\epsilon\}})) = [\![D' > 0]\!]$, as above.

For the induction step, assume the statement is true for two suffix sets $\mathcal{S}_L$ and $\mathcal{S}_R$, with respective suffix trees $\Psi_{\mathcal{S}_L}$ (of depth D_L) and $\Psi_{\mathcal{S}_R}$ (of depth D_R). That is, we assume

$$\Gamma_{D'}(\mathcal{S}_L) = \ell(\mathcal{E}_{D'}(\Psi_{\mathcal{S}_L})) \quad \text{and} \quad \Gamma_{D'}(\mathcal{S}_R) = \ell(\mathcal{E}_{D'}(\Psi_{\mathcal{S}_R}))$$

for all $D'_L \geq D_L$, $D'_R \geq D_R$ as our inductive hypotheses. Consider the tree $\Psi_{\mathcal{S}} := \text{Node } \Psi_{\mathcal{S}_L} \ \Psi_{\mathcal{S}_R}$ constructed from $\Psi_{\mathcal{S}_L}$ and $\Psi_{\mathcal{S}_R}$, with depth $D := \max\{D_L, D_R\} + 1$. We can recover the suffix set $\mathcal{S}$ from the suffix tree $\Psi_{\mathcal{S}}$. The strings in $\mathcal{S}$ represent paths from the leaf nodes up to the root node, so the last bit in the string corresponds to the last step up to the root node (1 for traversing from the left child, 0 for the right). Therefore, a path from any leaf node to the root of Ψ can be written as either

- a path from the leaf to the root node of $\Psi_{\mathcal{S}_L}$, followed by a 1, or
- a path from the leaf to the root node of $\Psi_{\mathcal{S}_R}$, followed by a 0.

This means we can write $\mathcal{S}$ as a disjoint partition, $\mathcal{S} = \mathcal{S}_L\{1\} \cup \mathcal{S}_R\{0\}$. Let $D' \geq D$ be arbitrary. Then,

$$\begin{aligned}
\ell(\mathcal{E}_{D'}(\Psi_{\mathcal{S}})) &= \ell(1\mathcal{E}_{D'-1}(\Psi_{\mathcal{S}_L})\mathcal{E}_{D'-1}(\Psi_{\mathcal{S}_R})) \\
&= 1 + \ell(\mathcal{E}_{D'-1}(\Psi_{\mathcal{S}_L})) + \ell(\mathcal{E}_{D'-1}(\Psi_{\mathcal{S}_R})) \\
&= 1 + \Gamma_{D'-1}(\mathcal{S}_L) + \Gamma_{D'-1}(\mathcal{S}_R)
\end{aligned}$$

Appealing to Lemma 4.3.21,

$$= \Gamma_{D'}(\mathcal{S}_L\{1\} \cup \mathcal{S}_R\{0\}) = \Gamma_{D'}(\mathcal{S})$$

as required.   ∎

We usually assume that the value of D' used to encode is as small as possible while ensuring that $\mathcal{S} \in \mathcal{C}_{D'}$, by letting $D' := \max_{s \in \mathcal{S}} \ell(s) = D$. In this case, encodings assign shorter codewords to deep and narrow trees over shallow and broad trees (compare the encodings of $\mathcal{S}_1 := \{1,10,100,000\}$ versus that of $\mathcal{S}_2 := \{11,10,01,00\}$). Both have the same number of suffixes, but the former is deeper and narrower (as unbalanced as a suffix tree of that depth can be) whereas the latter is broader and shallower (being a perfectly balanced tree). Consequently, $\mathcal{S}_1$ has an encoding of 10101 (5 bits) versus the encoding of $\mathcal{S}_2$, 1100100 (7 bits). Phrased another way, the encoding assigns longer codewords to trees that have many leaf nodes at the maximum depth D.

Example 4.3.23 (Model cost for a suffix set/tree) We can compute the model cost for the suffix sets

$$\begin{aligned}
\mathcal{S}_1 &= \{1,10,100,000\} \\
\mathcal{S}_2 &= \{11,10,01,00\} \\
\mathcal{S}_3 &= \{111,011,101,011,10,100,000\}
\end{aligned}$$

given the appropriately sized model class $\mathcal{C}_3, \mathcal{C}_2, \mathcal{C}_3$ respectively.

$$\begin{aligned}
\Gamma_3(\mathcal{S}_1) &= |\mathcal{S}_1| - 1 + |\{s \in \mathcal{S}_1 : \ell(s) < 3\}| \\
&= |\{1, 10, 100, 000\}| - 1 + |\{1, 10\}| \\
&= 4 - 1 + 2 = 5 \\
\Gamma_2(\mathcal{S}_2) &= |\mathcal{S}_2| - 1 + |\{s \in \mathcal{S}_2 : \ell(s) < 2\}| \\
&= |\{11, 10, 01, 00\}| - 1 + |\varnothing| \\
&= 4 - 1 + 0 = 3 \\
\Gamma_3(\mathcal{S}_3) &= |\mathcal{S}_3| - 1 + |\{s \in \mathcal{S}_3 : \ell(s) < 3\}| \\
&= |\{111, 011, 101, 011, 10, 100, 000\}| - 1 + |\{10\}| \\
&= 7 - 1 + 1 = 7
\end{aligned}$$

which matches the length of the codewords for the respective trees in Figure 4.8.

Note that $\Gamma_3(\mathcal{S}_1) > \Gamma_2(\mathcal{S}_2)$ even though $\mathcal{S}_1$ has the same number of elements that $\mathcal{S}_2$ does. This is due to the model cost adding a penalty term equal to the number of leaf nodes that are not at the maximal depth. In this sense, the model cost signals that balanced trees are simpler than unbalanced trees, and consequently have a lower assigned cost. If we took a set with the same number of elements as $\mathcal{S}_3$ but with a highly unbalanced tree, e.g $\mathcal{S}_4 = \{1, 10, 100, 1000, 10000, 100000, 000000\}$, then we would have to choose $D = 6$, and consequently, $\Gamma_6(\mathcal{S}_4) = 7 - 1 + 5 = 11 > 7 = \Gamma_3(\mathcal{S}_3)$.

Note that these extra bits for specifying leaf nodes that are not at the maximum depth are only incurred when the depth increases from the minimum depth. So if $D = \max_{s \in \mathcal{S}} \ell(s)$, then $\Gamma_D(\mathcal{S}) < \Gamma_{D+1}(\mathcal{S})$, but $\Gamma_{D+1}(\mathcal{S}) = \Gamma_{D'}(\mathcal{S})$ for all $D' \geq D + 1$. $\blacklozenge$

The model cost $\Gamma_D(\mathcal{S})$ can be used as a measure of simplicity or complexity of a source. This in turn will let us weigh models based on their simplicity when we take a Bayesian mixture of models later.

4.3.4 Updating Prediction Suffix Trees

The k-Markov models that we saw earlier keep track of all the statistics for every context of length k. A prediction suffix tree instead uses a suffix set to enumerate only those contexts that are useful to keep track of for prediction, ignoring redundant counts for when the context does not matter.

By choosing $\mathcal{S} = \mathbb{B}^k$, we indicate that all contexts of length k are useful, a predictor based on a PST degenerates to a k-Markov model, so in this sense, PST's are more general, allowing for variable length context rather than fixed length.

For example, consider the Markov distribution in Example 4.2.9 (assuming $\alpha = 0.9$). Technically this is a 2-Markov model, but a more subtle analysis would find that the length of the required context varies. If the last bit is 0, then we need additionally the second-to-last bit to distinguish if the context is 00 (and the distribution is likely to sample a 1) or the context is 10 (and a 0 is more likely). If the previous bit is a 1, then it is likely the next bit will be 0. There is no reason to record statistics separately for contexts 10 and 00, and to do so would slow learning. So, by choosing a suffix set $\mathcal{S}_{\text{good}}$ that accurately represents the variable context of the distribution, the statistics associated with contexts for which a distinction need not be made (10 and 00) can now be combined, speeding up learning (see Section 4.3.4).

Assuming the suffix set $\mathcal{S}_{\text{good}}$ is known, but the true parameters $\Theta_{\mathcal{S}_{\text{good}}}$ are not, we can make predictions using the best currently known estimate $\hat{\Theta}_{\mathcal{S}_{\text{good}}}$ and (4.3.15) and update the

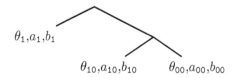

Figure 4.9: *The PST $\Psi_{\mathcal{S}_{good}}$ for learning the environment μ_1 described in Example 4.2.9.*

parameters according to the conditional KT estimator (Lemma 4.1.2) by reading the context x right-to-left, traversing the tree down to the appropriate leaf node (which corresponds to $s = \beta_{\mathcal{S}}(x)$), and updating the statistics (a_s, b_s, θ_s) as appropriate.

When the tree is first initialized, we set $a_s = 0, b_s = 0, \theta_s = \mathrm{P}_{\mathrm{KT}}(1 \mid \epsilon) = \frac{1}{2}$ for all $s \in \mathcal{S}$, in accordance with Lemma 4.1.2. (Recall that each parameter θ_s is associated with the current best estimate of $P(1|...s)$ using a KT estimator (Definition 4.3.13)). For Example 4.2.9, we could instead use the suffix set $\mathcal{S}_{good} = \{1, 10, 00\}$, where the parameters $\Theta_{\mathcal{S}_{good}}$ are unknown and learned from experience, and each leaf node associated with suffix s records the KT estimator counts a_s, b_s.

For Algorithm 4.2, let Ψ denote a PST, where each leaf node stores the variables a, b, θ. We assume that the context is at least as long as the tree is deep, so the return value is always well-defined.

Algorithm 4.2 Suffix Tree Prediction

Input: Prediction Suffix Tree $\Psi_{\mathcal{S},\Theta_{\mathcal{S}}}$
Input: Context string $x_{1:t}$
Output: $\theta_{\beta_{\mathcal{S}}(x_{1:t})}$
  1: **for** $i = t$ **down to** 1 **do**
  2:     **if** Ψ is a leaf node **then return** $\Psi.\theta$ ▷ Return the parameter $\theta_{\beta_{\mathcal{S}}(x_{1:t})}$
  3:     **if** $x_i = 1$ **then** $\Psi := \Psi.\texttt{left}$ ▷ Walk down the left path
  4:     **else** $\Psi := \Psi.\texttt{right}$ ▷ Walk down the right path

---

Algorithm 4.3 Suffix Tree Update

Input: PST $\Psi_{\mathcal{S},\Theta_{\mathcal{S}}}$, updated on $x_{1:t}$
Input: Context string $x_{1:t}$ ▷ Assuming that $t \geq D$
Input: Next Bit $x = x_{t+1}$
Effect: $\theta_{\beta_{\mathcal{S}}(x_{1:t})}$ in $\Theta_{\mathcal{S}}$ updated on x_{t+1} with context $x_{1:t}$.
  1: **for** $i = t$ **down to** 1 **do**
  2:     **if** Ψ is a leaf node **then break**
  3:     **if** $x_i = 1$ **then** $\Psi := \Psi.\texttt{left}$ ▷ Walk down the left path
  4:     **else** $\Psi := \Psi.\texttt{right}$ ▷ Walk down the right path
  5: **if** $x = 1$ **then** $\Psi.b := \Psi.b + 1$ ▷ Update count b_s of number of ones
  6: **else** $\Psi.a := \Psi.a + 1$ ▷ Update count a_s of number of zeros
  7: $\Psi.\theta := (\Psi.b + \frac{1}{2})/(\Psi.a + \Psi.b + 1)$ ▷ Update the estimated probability $\mathrm{P}_{\mathcal{S},\Theta_{\mathcal{S}}}$

Now that we have a way to predict given a parameter vector $\Theta_{\mathcal{S}}$, we can define a special case of the PST probability $\mathrm{P}_{\mathcal{S},\mathrm{KT}}$, where instead of fixing a particular $\Theta_{\mathcal{S}}$, it is instead learned using Algorithm 4.3.

> **Definition 4.3.24 (PST-KT probability)** In analogy to Lemma 4.3.16, we define the PST-KT probability $P_{\mathcal{S},\mathrm{KT}}$ as
>
> $$P_{\mathcal{S},\mathrm{KT}}(x_{1:n}) = \prod_{s \in \mathcal{S}} P_{\mathrm{KT}}(a_s, b_s)$$
>
> where a_s, b_s are the context counts in $x_{1:n}$.

This definition alone is useful for theoretical results, but not for computing it in practice. As it turns out, we can write the PST-KT probability as a product of conditional PST probabilities (Definition 4.3.13) where the parameter vector $\Theta_{\mathcal{S}}$ is not known in advance, but learned dynamically using Algorithm 4.3.

In other words, $P_{\mathcal{S},\mathrm{KT}}$ is the PST probability of a string $x_{1:n}$ written as product of conditionals, where for each bit x_i, we use the best current estimate of the parameter vector $\hat{\Theta}_{\mathcal{S}}^{i-1}$ is obtained via the KT estimator and the context $x_{<i}$.

> **Lemma 4.3.25 (KT probability in terms of learned parameters)** Assuming Algorithm 4.3 is run on sequence $x_{1:n}$, then we can relate the parameters $\hat{\theta}_s^i$ and the KT probability as follows.
>
> $$P_{\mathcal{S},\mathrm{KT}}(x_{1:n}) = \prod_{\substack{i=1 \\ x_i=1}}^{n} \hat{\theta}_{s_i}^i \prod_{\substack{i=1 \\ x_i=0}}^{n} (1-\hat{\theta}_{s_i}^i) \qquad \text{where} \qquad s_i := \beta_{\mathcal{S}}(x_{<i})$$

Proof. We proceed via proof by induction. The equality is vacuously true for $n=0$. Assume it is true for any sequence $x_{<n}$. Then consider a sequence $x_{1:n}$. Assume that $x_n=1$ (the proof is analogous for $x_n=0$) and $s:=s_n$.

$$\prod_{\substack{i=1 \\ x_i=1}}^{n} \hat{\theta}_{s_i} \prod_{\substack{i=1 \\ x_i=0}}^{n} (1-\hat{\theta}_{s_i}) = \hat{\theta}_s^n \prod_{\substack{i=1 \\ x_i=1}}^{n-1} \hat{\theta}_{s_i} \prod_{\substack{i=1 \\ x_i=0}}^{n-1} (1-\hat{\theta}_{s_i}) = \hat{\theta}_s^n P_{\mathcal{S},\mathrm{KT}}(x_{<n})$$

Now in Algorithm 4.3, the tree is traversed following context s, and the corresponding parameter θ_s is updated to reflect the new KT-counts

$$\hat{\theta}_s^n = \frac{b_s(x_{1:n}) + \frac{1}{2}}{a_s(x_{1:n}) + b_s(x_{1:n}) + 1}$$

By Lemma 4.1.2, $\hat{\theta}_s^n = P_{\mathcal{S},\mathrm{KT}}(x_n | x_{<n})$, and hence

$$\hat{\theta}_s^n P_{\mathcal{S},\mathrm{KT}}(x_{<n}) = P_{\mathcal{S},\mathrm{KT}}(x_n | x_{<n}) P_{\mathcal{S},\mathrm{KT}}(x_{<n}) = P_{\mathcal{S},\mathrm{KT}}(x_{1:n})$$

as required. ∎

Lemma 4.3.26 (Alternative PST-KT probability) Given a suffix tree $\Psi_{\mathcal{S}}$ and a sequence $x_{1:n}$, we can express the PST-KT probability as

$$P_{\mathcal{S},\mathrm{KT}}(x_{1:n}) = \prod_{i=1}^{n} P_{\mathcal{S},\hat{\Theta}_{\mathcal{S}}^i}(x_i|x_{<i})$$

where $P_{\mathcal{S},\hat{\Theta}_{\mathcal{S}}^i}$ is the conditional PST probability (Definition 4.3.13), and $\hat{\Theta}_{\mathcal{S}}^i$ is a family of parameter vector estimates where

$$\hat{\Theta}_{\mathcal{S}}^i = \begin{cases} (\tfrac{1}{2},\tfrac{1}{2},...,\tfrac{1}{2}) & \text{if } i=1 \\ \mathrm{SuffixTreeUpdate}(\Psi_{\mathcal{S},\hat{\Theta}_{\mathcal{S}}^{i-1}},x_{<i},x_i) & \text{if } i>1 \end{cases}$$

that is, each successive parameter vector $\hat{\Theta}_{\mathcal{S}}^i$ is obtained by calling Algorithm 4.3 with the previous one, and is initialized with $\tfrac{1}{2}$ for each parameter.

Proof. We take the product, and split it over each context and bit.

$$\prod_{i=1}^{n} P_{\mathcal{S},\hat{\Theta}_{\mathcal{S}}^i}(x_i|x_{<i}) = \prod_{\substack{s\in\mathcal{S}}} \prod_{\substack{i=1\\ \beta_{\mathcal{S}}(x_{<i})=s}}^{n} P_{\mathcal{S},\hat{\Theta}_{\mathcal{S}}^i}(x_i|s)$$

$$= \prod_{s\in\mathcal{S}} \left(\prod_{\substack{i=1\\ \beta_{\mathcal{S}}(x_{<i})=s\\ x_i=1}}^{n} P_{\mathcal{S},\hat{\Theta}_{\mathcal{S}}^i}(1|s) \right) \left(\prod_{\substack{i=1\\ \beta_{\mathcal{S}}(x_{<i})=s\\ x_i=0}}^{n} P_{\mathcal{S},\hat{\Theta}_{\mathcal{S}}^i}(0|s) \right)$$

$$= \prod_{s\in\mathcal{S}} \prod_{\substack{i=1\\ \beta_{\mathcal{S}}(x_{<i})=s\\ x_i=1}}^{n} \hat{\theta}_s \prod_{\substack{i=1\\ \beta_{\mathcal{S}}(x_{<i})=s\\ x_i=0}}^{n} (1-\hat{\theta}_s) = \prod_{s\in\mathcal{S}} P_{\mathrm{KT}}(a_s,b_s)$$

as required, where we applied Lemma 4.3.25 in the last step.   ∎

4.3.5   PST Experiments

Building on Section 4.2.3, we repeat the same experiment, but now also include the performance of the PST predictor with known $\mathcal{S} \equiv \mathcal{S}_{\mathrm{good}} = \{1,10,00\}$ (Section 4.3.4) using Algorithm 4.2 for prediction, but with unknown $\Theta_{\mathcal{S}}$ and Algorithm 4.3 to improve the current estimate of $\Theta_{\mathcal{S}}$ after observing the true value of each bit. For comparison, we also measure the performance of a suffix tree with suffix set $\mathcal{S}_{\mathrm{bad}} = \{11,01,0\}$, which represents the wrong choice of suffix set (falsely assuming that a previous bit of 0 is sufficient to predict the next bit, whereas a previous bit of 1 requires additional context) (see Figure 4.10).

As expected, $\mathcal{S}_{\mathrm{bad}}$ performs poorly (as it assumes the wrong model for the environment). Both $k=2$, $k=3$ and $\mathcal{S}_{\mathrm{good}}$ can learn the environment, and the extra side information that $\mathcal{S}_1$ possesses of the variable length context allows it to learn that 0 almost always follows 1 faster than the $k=2$ or the $k=3$ predictor.

In the most extreme example, we could consider an environment where the context varies drastically, consider the sequence $(0^n 1)^*$, generated by the obvious deterministic distribution. The best choice of suffix tree would be $\mathcal{S} = \{10^m : 0 \le m < n\} \cup \{0^n\}$, which requires updating only n parameters, but for the k-Markov estimator to learn this distribution would require $k=n+1$, and correspondingly 2^{n+1} many parameters to update, which is wasteful of both memory and time.

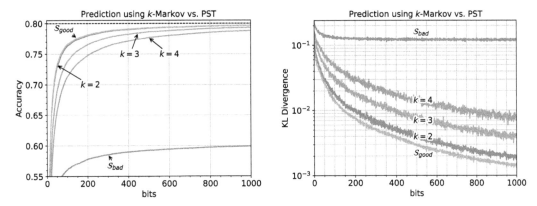

Figure 4.10: *Prediction accuracy (left) and instantaneous KL divergence (right) averaged over 1000 trials on environment μ_1 ($\alpha = 0.8$, Figure 4.4) for k-Markov KT estimators with $2 \leq k \leq 4$, and PSTs with $\mathcal{S}_{good}$ and $\mathcal{S}_{bad}$ (Section 4.3.5) as choice of suffix sets.*

Obviously, most of the time we do not even know for what k the environment is k-Markov, let alone the exact specifications of what context can be discarded and when (the information in $\mathcal{S}$). Since suffix trees are much more general than k-Markov KT estimators, and we have a measure of the complexity of a suffix tree, we finally have all the parts needed to define a Bayesian mixture over Markov models that can be efficiently updated.

4.4   Mixing Distributions

Eventually we would like to form a Bayesian mixture over all possible choices of suffix set, and then learn which of the suffix sets describes the variable order Markov model used by the environment.

Example 4.4.1 (Redundancy of mixing two distributions) To motivate this, first consider the simple case of mixing two distributions together. Suppose that $\hat{\mu}_1$ is a good estimated distribution for sequences sampled from μ_1 in terms of the redundancy (Definition 4.1.10) between them, and $\hat{\mu}_2$ is a good distribution for μ_2, then a weighted distribution (here, just the average)

$$\hat{\mu}_w(x_{1:t}) := \tfrac{1}{2}\hat{\mu}_1(x_{1:t}) + \tfrac{1}{2}\hat{\mu}_2(x_{1:t})$$

has low redundancy with respect to both μ_1 and μ_2. Indeed, for $i \in \{,2\}$, let

$$r_{\hat{\mu}_i,\mu_i} := -\log_2 \frac{1}{\hat{\mu}_i(x_{1:n})} - \log_2 \frac{1}{\mu_i(x_{1:n})}$$

be the redundancies between each μ_i and its corresponding approximation $\hat{\mu}_i$. Now, we compute the redundancy $r_{\hat{\mu}_w,\mu_i}$ between $\hat{\mu}_w$ and μ_i:

$$
\begin{aligned}
r_{\hat{\mu}_w,\mu_i}(x_{1:n}) &= \log_2 \frac{1}{\hat{\mu}_w(x_{1:t})} - \log_2 \frac{1}{\mu_i(x_{1:t})} \\
&= \log_2 \frac{2}{\hat{\mu}_1(x_{1:t}) + \hat{\mu}_2(x_{1:t})} - \log_2 \frac{1}{\mu_i(x_{1:t})} \\
&\leq \log_2 \frac{2}{\hat{\mu}_i(x_{1:t})} - \log_2 \frac{1}{\mu_i(x_{1:t})}
\end{aligned}
$$

$$= \log_2 2 + \log_2 \frac{1}{\hat{\mu}_i(x_{1:t})} - \log_2 \frac{1}{\mu_i(x_{1:t})}$$

$$= 1 + r_{\hat{\mu}_i, \mu_i}$$

which means using $\hat{\mu}_w$ as an approximation to μ_i is no worse than $\hat{\mu}_i$ apart from an overhead of only one extra bit of redundancy. The intuition is that the mixture $\hat{\mu}_w$ is at least as good at prediction as both constituent distributions, plus one more bit of overhead to specify which approximation is the best. ♦

One could also choose an uneven mixture, $\hat{\mu}_{w,\alpha}(x_{1:t}) = \alpha \hat{\mu}_1(x_{1:t}) + (1-\alpha)\hat{\mu}_2(x_{1:t})$ for some $\alpha \in [0,1]$, if the environment is expected more likely to be μ_1 than μ_2. This reduces the overhead penalty for one environment, in exchange for a larger penalty for the other.

4.5   Context Tree Weighting

Let us now consider the case where we are predicting a sequence and the only information we have about the environment μ is that the sequence is Markov (so there exists some suffix set S and parameter vector Θ_S such that $\mu = P_{S,\Theta_S}$).

We will make the (weak) assumption that we know an upper bound D for the maximum length of the context required to make optimal predictions. Under this assumption, μ is at most D-Markov, but we do not know what the smallest value of k is such that μ is k-Markov.

Following what we mentioned before, one approach for unknown environments is to consider a mixture over all the tree sources in $\mathcal{C}_D$, and for each of those suffix sets, learn the parameter vectors using the KT estimator as before.

We fix the depth D. Recall from Section 3.1 that we can define the *Bayes mixture over* $\mathcal{M}$, denoted ξ, as

$$\xi(x_{1:n}) := \sum_{S \in \mathcal{C}_D} w_S P_{S,\Theta_S}(x_{1:n})$$

which can also be written as a conditional Bayesian mixture for prediction

$$\xi(x_t | x_{<t}) := \sum_{S \in \mathcal{C}_D} w_S(x_{<t}) P_{S,\Theta_S}(x_t | x_{<t})$$

$$\text{where} \quad w_S(x_{1:t}) := \begin{cases} w_S(x_{<t}) \frac{P_{S,\Theta_S}(x_t | x_{<t})}{\xi(x_t | x_{<t})} & \text{for} \quad x_{1:t} \neq \epsilon \\ w_S & \text{for} \quad x_{1:t} = \epsilon \end{cases}$$

Naively, computing this would require summing over the entire model class $\mathcal{C}_D$, giving a complexity of $O(2^{2^D})$ (Lemma 4.3.10). With an appropriate collection of weights $w_S := 2^{-\Gamma_D(S)}$, this mixture can be computed in time $O(n \cdot D)$. Indeed, we can even do this while learning Θ_S using a KT estimator (Theorem 4.5.8), meaning our new model will learn both the structure S and the parameters Θ_S describing the true environment.

4.5.1   The CTW Algorithm

We do this by using *context tree weighting* (CTW), a variant of PSTs where each node in the tree (not just the leaves) store statistics (a_s, b_s) corresponding to the context s associated with this node. By updating each node along the path for a particular context, and defining the probabilities of the tree in a recursive fashion, the CTW method gives a simple and efficient way to compute the Bayesian mixture over all PSTs in the model class $\mathcal{C}_D$.

Definition 4.5.1 (Context tree) Given a depth D, a *context tree* $\mathcal{T}_D$ is a perfect[a] binary tree where each node is associated with a binary string s (with $\ell(s) \leq D$) representing the path to that node from the root node (paths are encoded in the same way as a PST).

Each node stores the KT estimator counts (a_s, b_s), the KT probability $\mathrm{P_{KT}}(a_s, b_s)$[b] (Definition 4.1.1) and the weighted probability $\mathrm{P}_w(s)$, defined recursively as

$$\mathrm{P}_w(s) := \begin{cases} \frac{1}{2}\mathrm{P_{KT}}(a_s, b_s) + \frac{1}{2}\mathrm{P}_w(0s)\mathrm{P}_w(1s) & \text{for} \quad \ell(s) < D \\ \mathrm{P_{KT}}(a_s, b_s) & \text{for} \quad \ell(s) = D \end{cases} \tag{4.5.2}$$

Each internal node $s \in \mathcal{T}_D$ with $\ell(s) < D$ has two child nodes: the left child associated with $1s$ and the right with $0s$, for which the respective counts (a_{1s}, b_{1s}) and (a_{0s}, b_{0s}) satisfy

$$a_s = a_{1s} + a_{0s}$$
$$b_s = b_{1s} + b_{0s}$$

---

[a] All internal nodes have two children, and all leaf nodes are at the same depth.
[b] Strictly speaking, we do not need to store $\mathrm{P_{KT}}$ as we can compute it from (a_s, b_s) directly, but it is more efficient to store the current value of $\mathrm{P_{KT}}(a_s, b_s)$ and update it using Lemma 4.1.2 whenever a_s or b_s is incremented.

(We will define the probability $\mathrm{P}_D^{\mathrm{CTW}}(x_{1:n})$ assigned to sequence $x_{1:n}$ by a CTW tree of depth D momentarily).

What is the intuition behind Definition 4.5.1? Suppose we are trying to predict the next term x_{t+1} in a sequence $x_{1:t}$, assuming that the sequence is sampled from an unknown k-Markov distribution with $k \leq D$. We store the statistics in a binary tree $\mathcal{T}_D$ of depth D, where at each node s we store the KT-counts a_s and b_s. The tree $\mathcal{T}_D$ is then used to estimate probabilities of the next bit in a sequence based on the previous D bits of context. If s is a leaf node (that is, at depth D), then we are assuming that in the original sequence, if we take a subsequence of only those bits that directly follow context s, this subsequence is generated i.i.d. (as was done in Example 4.2.5). Hence, the KT estimator $\mathrm{P_{KT}}$ will provide a good estimate, and so we define the probability associated with that node as the KT estimator

$$\mathrm{P}_w(s) := \mathrm{P_{KT}}(a_s, b_s) \quad \text{if } s \text{ is a leaf.}$$

Assume we already have good estimators $\mathrm{P}_w(0s)$ and $\mathrm{P}_w(1s)$ for contexts $0s$ and $1s$ respectively. Let s be a non-leaf node (a context of length $< D$), and let $i_1^s, i_2^s, \ldots$ be a sequence of indexes such that $x_{i-\ell(s)-1:i-1} = s$ (that is, the indexes i such that x_i directly follows s). Then the subsequence $x_{i_j^s}$ of those bits directly following s are either:

- generated i.i.d. (namely if the extra context beyond s is irrelevant), in which case $\mathrm{P_{KT}}(a_s, b_s)$ would be a good estimator, or

- not generated i.i.d, in which case we could instead partition the subsequence $x_{i_j^s}$ into two subsubsequences $x_{i_j^{0s}}$ and $x_{i_j^{1s}}$, the bits directly following contexts $0s$ and $1s$ respectively. We then use the (assumed) good estimators $\mathrm{P}_w(0s)$ for $x_{i_j^{0s}}$ and $\mathrm{P}_w(1s)$ for $x_{i_j^{1s}}$, and then write the probability of the original subsequence $x_{i_j^s}$ as the product $\mathrm{P}_w(0s)\mathrm{P}_w(1s)$.

We do not know which case applies for the subsequence $x_{i_j^s}$, but as a compromise we can take a Bayesian mixture over both choices as was done in Section 4.4, we get

$$\mathrm{P}_w(s) := \tfrac{1}{2}\mathrm{P}_{\mathrm{KT}}(a_s, b_s) + \tfrac{1}{2}\mathrm{P}_w(0s)\mathrm{P}_w(1s) \quad \text{if } s \text{ is a non-leaf node}$$

which gives a good predictor regardless which case is true.

Defining the weighted probability P_w in this recursive fashion means that the probability of each node s is the average of the KT estimator $\mathrm{P}_{\mathrm{KT}}(a_s, b_s)$, and the product $\mathrm{P}_w(0s)\mathrm{P}_w(1s)$ of the weighted probabilities of the children.

For Algorithm 4.4, each node s in the context tree $\mathcal{T}_D$ contains 4 variables $\{a, b, \mathrm{P}_{\mathrm{KT}}, \mathrm{P}_w\}$ corresponding to the KT estimator counts a_s and b_s, the KT probability $\mathrm{P}_{\mathrm{KT}}(a_s, b_s)$, and the weighted probability $\mathrm{P}_w(s)$ respectively.

Algorithm 4.4 CTW Update

Input: Context Tree $\mathcal{T}_D$
Input: Context string $x_{1:t}$ ▷ Assuming that $t \geq D$
Input: Next Bit $x = x_{t+1}$
Effect: Tree $\mathcal{T}_D$ updated per Remark 4.5.3

1:   $s := \mathcal{T}_D$ ▷ Keep track of the current node
2:   **for** $i = t$ **down to** $t - D + 1$ **do**   ▷ Loop through the D most recent bits
3:      **if** $x = 1$ **then**
4:          $s.\mathrm{P}_{\mathrm{KT}} := \frac{s.b + \frac{1}{2}}{s.a + s.b + 1} \times s.\mathrm{P}_{\mathrm{KT}}$ ▷ Update $\mathrm{P}_{\mathrm{KT}}(a_s, b_s)$
5:          $s.b := s.b + 1$ ▷ Increment b_s
6:      **else**
7:          $s.\mathrm{P}_{\mathrm{KT}} := \frac{s.a + \frac{1}{2}}{s.a + s.b + 1} \times s.\mathrm{P}_{\mathrm{KT}}$ ▷ Update $\mathrm{P}_{\mathrm{KT}}(a_s, b_s)$
8:          $s.a := s.a + 1$ ▷ Increment a_s
9:      **if** $x_i = 1$ **then**
10:         $s := s.\texttt{left}$ ▷ Walk down the left path
11:      **else**
12:         $s := s.\texttt{right}$ ▷ Walk down the right path
13: **if** $x = 1$ **then**
14:      $s.\mathrm{P}_{\mathrm{KT}} := \frac{s.b + \frac{1}{2}}{s.a + s.b + 1} \times s.\mathrm{P}_{\mathrm{KT}}$ ▷ Update $\mathrm{P}_{\mathrm{KT}}(a_s, b_s)$
15:      $s.b := s.b + 1$ ▷ Increment b_s
16: **else**
17:      $s.\mathrm{P}_{\mathrm{KT}} := \frac{s.a + \frac{1}{2}}{s.a + s.b + 1} \times s.\mathrm{P}_{\mathrm{KT}}$ ▷ Update $\mathrm{P}_{\mathrm{KT}}(a_s, b_s)$
18:      $s.a := s.a + 1$ ▷ Increment a_s
19: $s.\mathrm{P}_w := s.\mathrm{P}_{\mathrm{KT}}$ ▷ Update $\mathrm{P}_w(s)$ for leaf node s
20: **for** $i = 1$ **to** D **do**
21:      $s := s.\texttt{parent}$ ▷ Walk back up the tree
22:      $s.\mathrm{P}_w := \tfrac{1}{2}s.\mathrm{P}_{\mathrm{KT}} + \tfrac{1}{2} \cdot s.\texttt{left}.\mathrm{P}_w \cdot s.\texttt{right}.\mathrm{P}_w$ ▷ Update $\mathrm{P}_w(s)$ for non-leaf node s

---

Remark 4.5.3 (Space and time complexity of CTW Update algorithm) Algorithm 4.4 takes a context $x_{1:t}$, and using the last D bits to define the context $s = x_{t-D+1:t}$, the algorithm traverses the tree from the root node ϵ down to the leaf node corresponding to context s. Along the way, the algorithm visits all nodes corresponding to some suffix s' of s, and updates the counts $a_{s'}, b_{s'}$ and the KT probability $\mathrm{P}_{\mathrm{KT}}(a_{s'}, b_{s'})$ for node s'. Then, once the leaf node s is reached, the algorithm walks back up the tree to the root node ϵ, updating the weighted probability $\mathrm{P}_w(s')$ for each node s' along the path from s to the root node ϵ, using the values of $P_w(0s')$ and $P_w(1s')$ and $P_{KT}(a'_s, b'_s)$ that were generated before visiting this node.

Hence, each update to (4.5.2) only takes $O(D)$ operations when implemented efficiently in this manner.

The CTW tree contains $O(2^D)$ many nodes, which is too expensive. In practice, only nodes with non-zero counts ($a_s > 0$ or $b_s > 0$) need to be created, so the tree can grow dynamically as new contexts are observed, reducing the space complexity to worse case $O(tD)$. In practice, the space used is much better than the $O(tD)$ bound, as, if the sequence the CTW model is learning from is highly regular, most nodes would never be visited (see Section 4.5.5 for details).   ●

At long last, we can define the probability a CTW tree assigns to a sequence.

Definition 4.5.4 (CTW probability) Given a depth D and a sequence $x_{1:n}$, we define the CTW probability $P_D^{CTW}(x_{1:n}) := P_w(\epsilon)$ as the probability associated with the root node of $\mathcal{T}_D$, after running $CTWUpdate(\mathcal{T}_D, x_{<i}, x_i)$ (Algorithm 4.4) for $i = 1, 2, ..., t$.

Note that, unlike PSTs, the probability associated with CTWs is a block probability rather than a conditional probability. To recover a conditional probability for prediction, we define

$$P_D^{CTW}(1|x_{<n}) := \frac{P_D^{CTW}(x_{<n}1)}{P_D^{CTW}(x_{<n})} \qquad (4.5.5)$$

In practice, (assuming the tree $\mathcal{T}_D$ has already been updated on $x_{1:n}$) this requires recording the value of $P_w(\epsilon)$, updating the CTW as if 1 were observed given the so-far-observed context $x_{<n}$, recording the new value of $P_w(\epsilon)$ after the update, and dividing the latter by the former (Algorithm 4.5).

Since we do not know ahead of time what the actual next bit will be, we want to discard the changes made to $\mathcal{T}_D$ after prediction, so the CTW update can be done on a copy of $\mathcal{T}_D$ instead. In practice, duplicating the entire tree is expensive, so we can instead perform the update on the original tree, and then revert the changes afterwards (see Section 4.5.5).

Algorithm 4.5 CTW Prediction

Input: Context Tree $\mathcal{T}_D$
Input: Context string $x_{1:t}$
Output: Probability $P_D^{CTW}(1|x_{1:t})$
  1: old $:= \mathcal{T}_D.P_w$ ▷ old $:= P_D^{CTW}(x_{1:t})$
  2: Make a copy $\mathcal{T}'_D$ of $\mathcal{T}_D$
  3: $CTWUpdate(\mathcal{T}'_D, x_{1:t}, 1)$ ▷ Update tree assuming $x_{t+1} = 1$
                                                          ▷ new $:= P_D^{CTW}(x_{1:t}1)$
  4: new $:= \mathcal{T}'_d.P_w$
  5: **return** new/old   ▷ $P_D^{CTW}(1|x_{1:t}) = \frac{P_D^{CTW}(x_{1:t}1)}{P_D^{CTW}(x_{1:t})}$

Example 4.5.6 (CTW update) We give an example of the CTW Algorithm 4.4 updating a context tree $\mathcal{T}_2$ of depth $D = 2$ (Figure 4.11). For all nodes $s \in \mathcal{T}_2$, we initialize the counts $(a_s, b_s) := (0, 0)$, $P_{KT}(a_s, b_s) := P_{KT}(0, 0) = \frac{1}{2}$ and $P_w := 1$ (4.5.2), giving us the freshly initialized context tree (Figure 4.11a). Suppose the sequence is $x_{1:4} = 1001$. We set aside $x_{1:2} = 10$ as the context, and then update the tree with the new bit $x_3 = 0$ (Figure 4.11b). We follow the path down the tree that corresponds to context 10, and since x_3 is zero, we increment the zero counts along the way (so a_ϵ, a_0, a_{10} are incremented) as well as recomputing P_{KT}

for those nodes using the new counts. Once we reach a leaf node, we walk back up the tree from the leaf to the root node, updating the weighted probabilities P_w along the way using (4.5.2).

Then, the new context is $x_{1:3} = 100$ (of which only $x_{2:3} = 00$ will be used) and the bit to update the counts is $x_4 = 1$ (Figure 4.11c). We repeat the above until the sequence is fully consumed.

$\blacklozenge$

4.5.2 CTW Properties

Remark 4.5.7 (CTW is a Bayesian mixture over all prediction suffix trees) We explore the definition of the CTW probability for small D to provide intuition to the Bayesian mixture it defines. The output distribution of the CTW is the weighted probability $P_w(\epsilon)$ associated with the root node with a mixture of KT estimators of varying depth. For $D = 0$, CTW reverts to the basic KT estimator, $P_w(\epsilon) = P_{KT}(a_\epsilon, b_\epsilon) \equiv P_{KT}(a, b)$. Writing $k(s) \equiv P_{KT}(a_s, b_s)$ and $k(\mathcal{S}) = \prod_{s \in \mathcal{S}} P_{KT}(a_s, b_s)$ and $w(s) \equiv P_w(a_s, b_s)$ for brevity, setting $D = 1$ gives the corresponding CTW probability

$$P_w(\epsilon) = \tfrac{1}{2}k(\epsilon) + \tfrac{1}{2}w(0)w(1) = \tfrac{1}{2}k(\epsilon) + \tfrac{1}{2}k(0)k(1)$$

which is a mixture between the frequency estimate, and the product of estimates using the previous bit. We can obtained the closed form for $D = 2$ with some effort,

$$
\begin{aligned}
& P_w(\epsilon) \\
&= \tfrac{1}{2}k(\epsilon) + \tfrac{1}{2}w(0)w(1) \\
&= \tfrac{1}{2}k(\epsilon) + \tfrac{1}{8}(k(0) + w(00)w(10))(k(1) + w(01)w(11)) \\
&= \tfrac{1}{2}k(\epsilon) + \tfrac{1}{8}(k(0)k(1) + k(0)w(01)w(11) + k(1)w(00)w(10) + w(00)w(10)w(01)w(11)) \\
&= \tfrac{1}{2}k(\epsilon) + \tfrac{1}{8}\big(k(\{0,1\}) + k(\{0,01,11\}) + k(\{1,00,10\}) + k(\{00,10,01,11\})\big) \\
&= \tfrac{1}{2}\square + \tfrac{1}{8}\big(\wedge + \overset{\wedge}{\wedge} + \overset{\wedge}{\,\wedge} + \overset{\wedge}{\wedge\wedge}\big)
\end{aligned}
$$

Note that each term in the expression corresponds to one of the five trees for $D = 2$ in Figure 4.8. The empty tree ($\square$) has code 0 and associated model cost 1, and so is assigned weight 2^{-1}. The other four trees $\wedge, \overset{\wedge}{\wedge}, \overset{\wedge}{\,\wedge}, \overset{\wedge}{\wedge\wedge}$ have codes 100,110,101,111, and so each get weight 2^{-3}. Writing out $P_w(\epsilon)$ explicitly for $D = 3$ becomes rather monotonous due to the large number of possible trees in $\mathcal{C}_D$ scaling as $O(2^{2^D})$, but we will soon prove the following in the general case: The CTW tree provides a Bayesian mixture over the set of all suffix sets $\mathcal{S} \in \mathcal{C}_D$, where the weight $w_\mathcal{S} := 2^{-\Gamma_D(\mathcal{S})}$ for each suffix set is based on the model cost $\Gamma_D(\mathcal{S})$ (Definition 4.3.20). $\bullet$

First, we require a theorem regarding the weighted probabilities P_w.

Theorem 4.5.8 (Context Tree Weighting) Given a context tree $\mathcal{T}_D$ updated on $x_{1:n}$ and a node $s \in \mathcal{T}_D$, the weighted probabilities $P_w(s)$ satisfy

$$P_w(s) = \sum_{\mathcal{S} \in \mathcal{C}_{D-d}} 2^{-\Gamma_{D-d}(\mathcal{S})} \prod_{s' \in \mathcal{S}} P_{KT}(a_{s's}, b_{s's})$$

where $d = \ell(s)$ and the counts a_s and b_s are with reference to $x_{1:n}$.

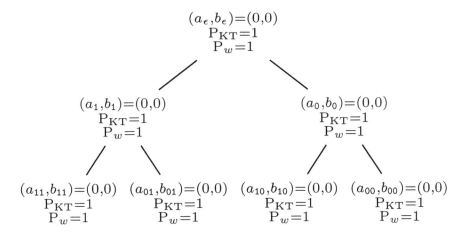

(a) *A newly initialized context tree $\mathcal{T}_2$ of depth $D=2$.*

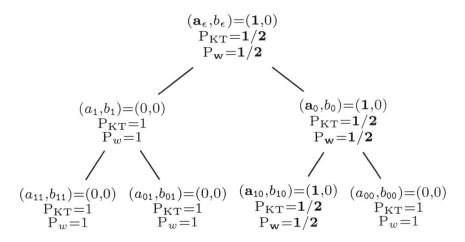

(b) *The context tree $\mathcal{T}_2$ after having processed bit $x_3 = 0$ given context $x_{1:2} = 10$.*

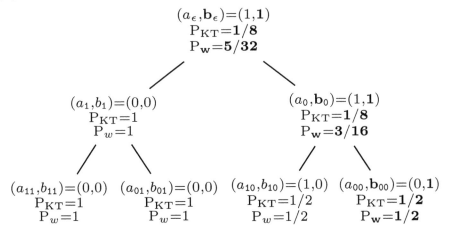

(c) *The context tree $\mathcal{T}_2$ after having processed bit $x_4 = 1$ given context $x_{1:3} = 100$.*

Figure 4.11: *The result of running Algorithm 4.4 on context tree $\mathcal{T}_2$ (Example 4.5.6) for various choices of next bit x_{t+1} and context $x_{1:t}$. Modified nodes highlighted.*

Proof. We prove by reverse induction on $\ell(s)$. For leaf nodes with $\ell(s)=D$ we have

$$\sum_{\mathcal{S}\in\mathcal{C}_{D-D}} 2^{-\Gamma_{D-D}(\mathcal{S})} \prod_{s'\in\mathcal{S}} \mathrm{P}_{\mathrm{KT}}(a_{s's},b_{s's})$$

$$= \sum_{\mathcal{S}\in\{\{\epsilon\}\}} 2^{-\Gamma_0(\mathcal{S})} \prod_{s'\in\mathcal{S}} \mathrm{P}_{\mathrm{KT}}(a_{s's},b_{s's})$$

$$= 2^{-\Gamma_0(\{\epsilon\})}\mathrm{P}_{\mathrm{KT}}(a_s,b_s) = \mathrm{P}_{\mathrm{KT}}(a_s,b_s) = \mathrm{P}_w(s)$$

Now, assume the statement to be true for a node $\ell(s)=d$ with $0<d\leq D$, and prove it holds for $\ell(s)=d-1$. By (4.5.2), noting that $\ell(s)=d-1<D$, we rewrite $\mathrm{P}_w(s)$ as

$$\mathrm{P}_w(s) = \tfrac{1}{2}\mathrm{P}_{\mathrm{KT}}(a_s,b_s)+\tfrac{1}{2}\mathrm{P}_w(0s)\mathrm{P}_w(1s)$$

Focus on the second term for the moment, and apply the inductive hypothesis, noting that $\ell(0s)=\ell(1s)=d$.

$$\tfrac{1}{2}\mathrm{P}_w(0s)\mathrm{P}_w(1s)$$

$$= \frac{1}{2}\left(\sum_{\mathcal{V}\in\mathcal{C}_{D-d}} 2^{-\Gamma_{D-d}(\mathcal{V})} \prod_{v\in\mathcal{V}}\mathrm{P}_{\mathrm{KT}}(a_{v0s},b_{v0s})\right)\left(\sum_{\mathcal{W}\in\mathcal{C}_{D-d}} 2^{-\Gamma_{D-d}(\mathcal{W})} \prod_{w\in\mathcal{W}}\mathrm{P}_{\mathrm{KT}}(a_{w1s},b_{w1s})\right)$$

$$= \sum_{\mathcal{V},\mathcal{W}\in\mathcal{C}_{D-d}} 2^{-\Gamma_{D-d}(\mathcal{V})-\Gamma_{D-d}(\mathcal{W})-1}\left(\prod_{v\in\mathcal{V}}\mathrm{P}_{\mathrm{KT}}(a_{v0s},b_{v0s})\right)\left(\prod_{w\in\mathcal{W}}\mathrm{P}_{\mathrm{KT}}(a_{w1s},b_{w1s})\right)$$

We rewrite the product to be over $\mathcal{V}\{0\}=\{v0:v\in\mathcal{V}\}$ and $\mathcal{W}\{1\}$ respectively.

$$= \sum_{\mathcal{V},\mathcal{W}\in\mathcal{C}_{D-d}} 2^{-\Gamma_{D-d}(\mathcal{V})-\Gamma_{D-d}(\mathcal{W})-1}\left(\prod_{v\in\mathcal{V}\{0\}}\mathrm{P}_{\mathrm{KT}}(a_{vs},b_{vs})\right)\left(\prod_{w\in\mathcal{W}\{1\}}\mathrm{P}_{\mathrm{KT}}(a_{ws},b_{ws})\right)$$

By appealing to Lemma 4.3.21 and merging the two products,

$$= \sum_{\mathcal{V},\mathcal{W}\in\mathcal{C}_{D-d}} 2^{-\Gamma_{D-d+1}(\mathcal{V}\{0\}\cup\mathcal{W}\{1\})}\left(\prod_{u\in\mathcal{V}\{0\}\cup\mathcal{W}\{1\}}\mathrm{P}_{\mathrm{KT}}(a_{us},b_{us})\right)$$

By $\mathcal{C}_{D-d+1}=\{\{\epsilon\}\}\cup\{\mathcal{V}\{0\}\cup\mathcal{W}\{1\}:\mathcal{V},\mathcal{W}\in\mathcal{C}_{D-d}\}$ (Definition 4.3.7)

$$= \sum_{\mathcal{S}\in\mathcal{C}_{D-d+1},\mathcal{S}\neq\{\epsilon\}} 2^{-\Gamma_{D-d+1}(\mathcal{S})}\left(\prod_{s'\in\mathcal{S}}\mathrm{P}_{\mathrm{KT}}(a_{s's},b_{s's})\right)$$

We can now add back in the $\tfrac{1}{2}\mathrm{P}_{\mathrm{KT}}(a_s,b_s)$ term, noting that $\Gamma_{D-d+1}(\{\epsilon\})=1$

$$\tfrac{1}{2}\mathrm{P}_{\mathrm{KT}}(a_s,b_s)+\tfrac{1}{2}\mathrm{P}_w(0s)\mathrm{P}_w(1s)$$

$$= 2^{-\Gamma_{D-d+1}(\{\epsilon\})}\left(\prod_{s'\in\mathcal{S}}\mathrm{P}_{\mathrm{KT}}(a_{s's},b_{s's})\right) + \sum_{\mathcal{S}\in\mathcal{C}_{D-d+1},\mathcal{S}\neq\{\epsilon\}} 2^{-\Gamma_{D-d+1}(\mathcal{S})}\left(\prod_{s'\in\mathcal{S}}\mathrm{P}_{\mathrm{KT}}(a_{s's},b_{s's})\right)$$

$$= \sum_{\mathcal{S}\in\mathcal{C}_{D-(d-1)}} 2^{-\Gamma_{D-(d-1)}(\mathcal{S})} \prod_{s'\in\mathcal{S}}\mathrm{P}_{\mathrm{KT}}(a_{s's},b_{s's})$$

∎

Using Theorem 4.5.8 we can now demonstrate how the CTW model implements a Bayesian mixture of suffix sets. By fixing a value of D and corresponding model class $\mathcal{C}_D$, we can define a prior on models $\mathcal{S}\in\mathcal{C}_D$ as follows:

$$w_{\mathcal{S}} := 2^{-\Gamma_D(\mathcal{S})} \tag{4.5.9}$$

Lemma 4.5.10 (PST prior normalized) The prior $w_{\mathcal{S}} = 2^{-\Gamma_D(\mathcal{S})}$ is a valid probability distribution over $\mathcal{C}_D$, in the sense that $w_{\mathcal{S}} \geq 0$ and $\sum_{\mathcal{S} \in \mathcal{C}_D} w_{\mathcal{S}} = 1$.

Note that this implies that $\{\mathcal{E}_D(\mathcal{S}) : \mathcal{S} \in \mathcal{C}_D\}$ is a complete (prefix) code.

Proof. Clearly $2^{-\Gamma_D(\mathcal{S})} \geq 0$. To prove $\sum_{\mathcal{S} \in \mathcal{C}_D} 2^{-\Gamma_D(\mathcal{S})} = 1$, we proceed by induction on D. First, note that by Definition 4.3.20

$$\Gamma_D(\{\epsilon\}) = |\{\epsilon\}| - 1 + |\{s \in \{\epsilon\} : \ell(s) < D\}| = [\![D > 0]\!]$$

For $D = 0$, we have

$$\sum_{\mathcal{S} \in \mathcal{C}_0} 2^{-\Gamma_D(\mathcal{S})} = 2^{-\Gamma_D(\{\epsilon\})} = 1$$

Now, assuming that $\sum_{\mathcal{S} \in \mathcal{C}_D} 2^{-\Gamma_D(\mathcal{S})} = 1$, we prove $\sum_{\mathcal{S} \in \mathcal{C}_{D+1}} 2^{-\Gamma_{D+1}(\mathcal{S})} = 1$.

$$\sum_{\mathcal{S} \in \mathcal{C}_{D+1}} 2^{-\Gamma_{D+1}(\mathcal{S})} = 2^{-\Gamma_{D+1}(\{\epsilon\})} + \sum_{\mathcal{S} \in \mathcal{C}_{D+1}, \mathcal{S} \neq \{\epsilon\}} 2^{-\Gamma_{D+1}(\mathcal{S})} = \tfrac{1}{2} + \sum_{\mathcal{V}, \mathcal{W} \in \mathcal{C}_D} 2^{-\Gamma_{D+1}(\mathcal{V}\{0\} \cup \mathcal{W}\{1\})}$$

Appealing to Lemma 4.3.21,

$$= \tfrac{1}{2} + \sum_{\mathcal{V}, \mathcal{W} \in \mathcal{C}_D} 2^{-\Gamma_D(\mathcal{V}) - \Gamma_D(\mathcal{W}) - 1} = \tfrac{1}{2} + \tfrac{1}{2} \sum_{\mathcal{V} \in \mathcal{C}_D} 2^{-\Gamma_D(\mathcal{V})} \cdot \sum_{\mathcal{W} \in \mathcal{C}_D} 2^{-\Gamma_D(\mathcal{W})} = \tfrac{1}{2} + \tfrac{1}{2} \times 1 \times 1 = 1 \qquad \blacksquare$$

We can now finally state the main result, that the CTW method does indeed give a Bayesian mixture over suffix sets.

Theorem 4.5.11 (Bayesian mixture over suffix sets) The weighted probability $\mathrm{P}_D^{\mathrm{CTW}}(x_{1:n})$ of a context tree $\mathcal{T}_D$ associated with the root node $\mathrm{P}_w(\epsilon)$ satisfies

$$\mathrm{P}_D^{\mathrm{CTW}}(x_{1:n}) = \sum_{\mathcal{S} \in \mathcal{C}_D} w_{\mathcal{S}} \mathrm{P}_{\mathcal{S},\mathrm{KT}}(x_{1:n})$$

where $w_{\mathcal{S}} = 2^{-\Gamma_D(\mathcal{S})}$ is the prior on suffix sets $\mathcal{S} \in \mathcal{C}_D$.

Proof. Recall that $\mathrm{P}_D^{\mathrm{CTW}}(x_{1:n})$ (Definition 4.5.4) is defined as the root node for a CTW tree $\mathcal{T}_D$ having been updated via Algorithm 4.4 on sequence $x_{1:n}$. A special case of Theorem 4.5.8 (choosing $s = \epsilon$, associated with the root node of the CTW tree) gives

$$\mathrm{P}_D^{\mathrm{CTW}}(x_{1:n}) \equiv \mathrm{P}_w(\epsilon) = \sum_{\mathcal{S} \in \mathcal{C}_D} 2^{-\Gamma_D(\mathcal{S})} \prod_{s \in \mathcal{S}} \mathrm{P}_{\mathrm{KT}}(a_s, b_s) = \sum_{\mathcal{S} \in \mathcal{C}_D} w_{\mathcal{S}} \mathrm{P}_{\mathcal{S},\mathrm{KT}}(x_{1:n})$$

The last equality follows from Definition 4.3.24 of $\mathrm{P}_{\mathcal{S},\mathrm{KT}}$ and $w_{\mathcal{S}} = 2^{-\Gamma_D(\mathcal{S})}$. $\qquad \blacksquare$

This means that if we want to compute the mixture over all suffix sets of length at most D, we instead compute the weighted probability P_w. Naively computing the mixture $\mathrm{P}_D^{\mathrm{CTW}}$ by summing over all environments in $\mathcal{C}_D$ would require $O(2^{2^D})$ time, however using (4.5.2) we are able to compute it in $O(D)$ time, a double exponential speedup!

4.5.3   CTW-PST-KT Redundancies

We also inherit some convergence properties from the corresponding results for the KT estimator:

Theorem 4.5.12 (PST-KT redundancy) Given any source $\mathcal{S} \in \mathcal{C}_D$ and parameter vector $\Theta_{\mathcal{S}} := \{\theta_s : \theta_s \in [0,1]\}_{s \in \mathcal{S}}$, the redundancy of the PST-KT estimator for $x_{1:n} \in \mathbb{B}^n$ is bounded above by

$$\log_2 P_{\mathcal{S},\Theta_{\mathcal{S}}}(x_{1:n}) - \log_2 P_{\mathcal{S},\mathrm{KT}}(x_{1:n}) \leq |\mathcal{S}| \gamma\left(\frac{n}{|\mathcal{S}|}\right)$$

$$\text{where} \quad \gamma(t) := \begin{cases} \frac{1}{2}\log_2(t) + 1 & \text{if} \quad t \geq 1 \\ t & \text{if} \quad t \leq 1 \end{cases}$$

Proof.

$$\log_2 P_{\mathcal{S},\Theta_{\mathcal{S}}}(x_{1:n}) - \log_2 P_{\mathcal{S},\mathrm{KT}}(x_{1:n}) \overset{(a)}{=} \log_2 \frac{\prod_{s \in \mathcal{S}} \theta_s^{b_s}(1-\theta_s)^{a_s}}{\prod_{s \in \mathcal{S}} P_{\mathrm{KT}}(a_s,b_s)} \overset{(b)}{=} \sum_{s \in \mathcal{S}} \log_2 \frac{\theta_s^{b_s}(1-\theta_s)^{a_s}}{P_{\mathrm{KT}}(a_s,b_s)}$$

$$\overset{(c)}{=} \sum_{s \in \mathcal{S}} r_{\mathrm{KT}\theta_s}(a_s,b_s) \overset{(d)}{\leq} \sum_{s \in \mathcal{S}} \gamma(a_s + b_s) \overset{(e)}{\leq} |\mathcal{S}|\gamma\left(\sum_{s \in \mathcal{S}} \frac{a_s + b_s}{|\mathcal{S}|}\right) \overset{(f)}{=} |\mathcal{S}|\gamma\left(\frac{n}{|\mathcal{S}|}\right)$$

(a) follows from Lemma 4.3.16 and Definition 4.3.24, (b) from $\log\prod = \sum\log$, (c) is just the definition of r_{KT}, (d) from Lemma 4.1.11 for terms with $a_s + b_s \geq 1$, and from $r_{\mathrm{KT},\theta_s}(0,0) = 0 = \gamma(0)$ if $a_s + b_s = 0$, (e) from $\gamma(t)$ being concave and an inverse version of Jensen's inequality Corollary 2.2.54, (f) is due to $a_s + b_s$ being all bits following context s. Since $\mathcal{S}$ is a suffix set, every bit in $x_{1:n}$ follows a unique context $s \in \mathcal{S}$, so $\sum_{s \in \mathcal{S}} a_s + b_s = n$.   ∎

Let us first consider the redundancy of the CTW estimator P_D^{CTW} relative to a PST-KT estimator.

Lemma 4.5.13 (CTW-KT redundancy) Let $\mathcal{S} \in \mathcal{C}_D$, and $x_{1:n}$ be a binary sequence. Then the redundancy between the CTW estimator and the PST-KT estimator can be bound by the model cost.

$$\log_2 P_{\mathcal{S},\mathrm{KT}}(x_{1:n}) - \log_2 P_D^{\mathrm{CTW}}(x_{1:n}) \leq \Gamma_D(\mathcal{S})$$

Proof. We take the CTW mixture, and discard all terms except those associated with the particular suffix set $\mathcal{S}$ to obtain a lower bound.

$$\log_2 P_D^{\mathrm{CTW}}(x_{1:n}) = \log_2 \sum_{\mathcal{S} \in \mathcal{C}_D} 2^{-\Gamma_D(\mathcal{S})} \prod_{s \in \mathcal{S}} P_{\mathrm{KT}}(a_s,b_s)$$

$$\geq \log_2 2^{-\Gamma_D(\mathcal{S})} \prod_{s \in \mathcal{S}} P_{\mathrm{KT}}(a_s,b_s)$$

$$= -\Gamma_D(\mathcal{S}) + \log_2 P_{\mathcal{S},\mathrm{KT}}(x_{1:n})$$

from which the result follows.   ∎

The final measure of the performance of the CTW estimator is given by the redundancy between P_D^{CTW} and the true underlying environment μ, assumed to be a variable-order Markov model $P_{\mathcal{S},\Theta_{\mathcal{S}}}$ for some suffix set $\mathcal{S} \in \mathcal{C}_D$ and associated parameter vector $\Theta_{\mathcal{S}}$.

Corollary 4.5.14 (PST-CTW redundancy) Given any source $\mathcal{S} \in \mathcal{C}_D$ and parameter vector $\Theta_{\mathcal{S}} := \{\theta_s : \theta_s \in [0,1]\}_{s \in \mathcal{S}}$, the redundancy of the weighted coding distribution CTW for $n \geq 1$ is upper bounded by

$$\log_2 \mathrm{P}_{\mathcal{S},\Theta_{\mathcal{S}}}(x_{1:n}) - \log_2 \mathrm{P}_D^{\mathrm{CTW}}(x_{1:n}) \leq \tfrac{1}{2}|\mathcal{S}|\log_2 n + \Gamma_D(\mathcal{S}) + |\mathcal{S}|$$

Proof. Immediate by adding Theorem 4.5.12 and Lemma 4.5.13 together and using $\gamma(n/|\mathcal{S}|) \leq \gamma(n) = \tfrac{1}{2}\log_2 n + 1$. ∎

This upper bound can be viewed as the sum of the coding lengths of the tree and the coding length $\tfrac{1}{2}\log_2 n$ bits for each parameter. The best possible expected redundancy for an estimator is given by the asymptotic Rissanen lower bound $\tfrac{1}{2}\log_2 n$ [Ris84] per parameter. Since $\Gamma_D(\mathcal{S}) + |\mathcal{S}|$ is constant, the CTW algorithm (asymptotically for $n \to \infty$) achieves the Rissanen lower bound.

4.5.4 CTW Experiments

Continuing with Section 4.3.5, we now include a CTW model of depth 2 and 3, and repeat the same experiment vs. the PST predictor with known $\mathcal{S}$ (Section 4.3.4) and a 2-Markov estimator (Figure 4.12).

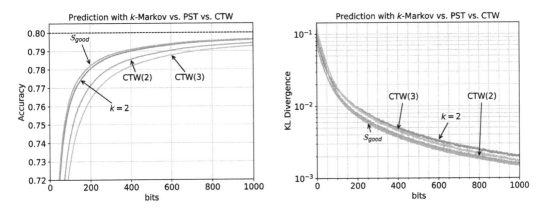

Figure 4.12: *Prediction accuracy and instantaneous KL divergence averaged over 10'000 trials on environment μ_1 ($\alpha = 0.8$, Figure 4.4) for 2-Markov KT estimators, PST with $\mathcal{S}_{good} = \{1, 10, 00\}$ (Section 4.3.4), and CTW models for $D = 2$ and $D = 3$ (Section 4.5.4).*

The performance of all models is similar. The best is the PST model with known suffix set, then the two CTW models, and worst is the 2-Markov estimator. The PST model already knows precisely what context to keep track of, and needs only to estimate the parameters. The CTW model with $D = 2$ quickly learns the underlying true environment from its model class, exploiting the variable order context to make more efficient use of data. The CTW $D = 3$ model also learns this, albeit slower than $D = 2$ (as the larger model class would imply the a priori weight on $\mathcal{S}_1$ to be lower.) Worst is the 2-Markov model, which can never learn to exploit the variable-order context, and wastefully keeps track of redundant parameters, slowing learning.

4.5.5   Optimizations

The above description of the CTW algorithm (Section 4.5.1) is complete, but in practice there are a few optimizations that can be made for both computational efficiency and numerical stability. None change the behavior of the underlying algorithm. These optimizations are implemented separately for the sake of clarity, but none are mutually exclusive, and all of them can (and should!) be implemented together.

Log probabilities.   The first problem derives from the use of block probabilities rather than conditional probabilities. The CTW probability is a mixture over all probabilities $P_{\mathcal{S},\mathrm{KT}}$ weighted by the complexity of $\mathcal{S}$. In the long term, given a sequence $x_{1:n}$ sampled from a particular Markov model characterized by the PST $\Psi_{\mathcal{S},\Theta_{\mathcal{S}}}$, the mixture model $P_D^{\mathrm{CTW}}(x_{1:n})$ will converge to $P_{\mathcal{S},\mathrm{KT}}(x_{1:n})$ as $n \to \infty$. The problem arises as $P_{\mathcal{S},\mathrm{KT}}(x_{1:n})$ vanishes quickly with respect to the number of bits $x_{1:n}$ observed so far. This presents a problem when using the standard (IEEE 754-2008) format for a 64-bit floating point number: the smallest positive number that can be represented is $\approx 10^{-308}$, which $P_{\mathcal{S},\mathrm{KT}}(x_{1:n})$ can drop below for even modest n (only a few hundred bits is often sufficient). Hence, computing $P_D^{\mathrm{CTW}}(x_{1:n})$ for even moderately sized n will quickly underflow to zero. One solution to this is instead of storing probabilities, to store log probabilities, and rewrite all the update rules appropriately.

We can rewrite the update rules for the KT estimator P_{KT} (Lemma 4.1.2) as follows, giving us the log-KT estimator

$$\log_2 P_{\mathrm{KT}}(0,0) = 0$$
$$\log_2 P_{\mathrm{KT}}(a{+}1,b) = \log_2(a{+}\tfrac{1}{2}) - \log_2(a{+}b{+}1) + \log_2 P_{\mathrm{KT}}(a,b)$$
$$\log_2 P_{\mathrm{KT}}(a,b{+}1) = \log_2(b{+}\tfrac{1}{2}) - \log_2(a{+}b{+}1) + \log_2 P_{\mathrm{KT}}(a,b)$$

Unfortunately, the CTW update rules (4.5.2) are not so easy to rewrite using log probabilities, as there is an addition term $\log_2\left(\tfrac{1}{2}P_{\mathrm{KT}}(a_s,b_s) + \tfrac{1}{2}P_w(0s)P_w(1s)\right)$. The addition of log-probabilities can be written as follows

$$\log_2(x{+}y) = \log_2(2^{\log_2 x} + 2^{\log_2 y}) \qquad (4.5.15)$$

but this presents the same numerical issues as before, if either $\log_2 x$ or $\log_2 y$ are less than ≈ -1024, the above expression will underflow to zero, causing numerical errors. We can improve on this by defining the log-sum operator

$$a \oplus b := \max\{a,b\} + \log_2(1 + 2^{\min\{a,b\} - \max\{a,b\}}) \qquad (4.5.16)$$

which avoids numerical underflow (one can verify that (4.5.15) and (4.5.16) are algebraically equivalent). Using (4.5.16), we can rewrite (4.5.2) as

$$\log_2 P_w(s) = \begin{cases} [\log_2 P_{\mathrm{KT}}(a_s,b_s) \oplus (\log_2 P_w(0s) + \log_2 P_w(1s))] - 1 & \text{for} \quad 0 \le \ell(s) < D \\ \log_2 P_{\mathrm{KT}}(a_s,b_s) & \text{for} \quad \ell(s) = D \end{cases}$$

We can then rewrite Algorithm 4.5 with log probabilities instead. For prediction, we need to take the difference rather than the ratio of the CTW probability before and after updating on the new bit, and then exponentiate the result to convert it back to a probability in the range [0,1].

Online context tree initialization.   The fact that the context tree can be updated in $O(D)$ time despite the model class $\mathcal{C}_D$ containing $O(2^{2^D})$ elements is what gives the CTW method its power. However, the context tree $\mathcal{T}_D$ itself contains $O(2^D)$ many nodes. For even moderately sized depths D this can be intractable in the memory required.

In practice, we would expect $n \ll 2^D$, and since at most nD distinct nodes will ever be used, most nodes in the tree will never be visited. The upper bound of nD may be very slack, as even less nodes are used if the sequence is generated by a simple distribution but makes use of a large context, or a context that is highly variable. Consider the sequence $(0^l 1^l)^*$ sampled from the obvious deterministic distribution. We require a context of at least size $D = l$ to uniquely determine the next bit, but there are only $2l$ different contexts that would ever be observed, so a vast majority of the tree is wasted space.

A solution is to initialize the context tree with only the root node, and add nodes to the tree only when the KT counts associated with that node would be modified. The rest of the tree can be thought of as unexplored *null* nodes (for which memory is not yet allocated). The KT counts for null nodes are always $(a_s, b_s) = (0,0)$. For null nodes s with $\ell(s) = D$, we set $P_w(s) = P_{KT}(0,0) = 1$. For any internal null node, both children are also null, so by using (4.5.2) and $P_w(0s) = P_w(1s) = 1$ we have

$$P_w(s) = \frac{1}{2} P_{KT}(a_s, b_s) + \frac{1}{2} P_w(0s) P_w(1s) = \frac{1}{2} \times 1 + \frac{1}{2} \times 1 \times 1 = 1$$

so regardless of its depth through the tree, all null nodes s satisfy $P_w(s) = 1$. If a node requires the value of $P_w(xs)$ for one of its children to update the value $P_w(s)$, and the child node xs is null, we can just assume $P_w(xs) = 1$. This gives us a modified version of (4.5.2):

$$P_w(s) = \begin{cases} 1 & s \text{ is null} \\ \frac{1}{2} P_{KT}(a_s, b_s) + \frac{1}{2} P_w(0s) P_w(1s) & \ell(s) < D \\ P_{KT}(a_s, b_s) & \ell(s) = D. \end{cases}$$

The only step remaining is to modify Algorithm 4.4 such that whenever we have walked down to a null node, to then initialize it as a node with the default statistics, and add two null children (see Algorithm 4.6).

We can repeat Example 4.5.6 by starting off with a single root node (Figure 4.13a) and only adding nodes when the corresponding statistics need to be stored, as shown in Figures 4.13b to 4.13c.

Reverting changes after prediction. In Algorithm 4.5, we update a copy of the context tree as if a 1 were observed to generate the conditional distribution $P_D^{CTW}(1|x_{1:n})$, but for large D this is an expensive operation. A more efficient solution is to make the updates to the original tree, and revert the changes afterwards, which involves doing the opposite of an update: Decrementing the appropriate counts a_s, b_s, and reverting the changes made to probabilities P_{KT} and P_w, see Algorithm 4.7.

Algorithm 4.6 CTW Update (Online Tree)

Require: Context Tree $\mathcal{T}_D$
Input: Context string $x_{1:t}$ ▷ Assuming that $t \geq D$
Input: Next Bit $x = x_{t+1}$
Effect: Tree $\mathcal{T}_D$ updated per Remark 4.5.3

1:   $s := \mathcal{T}_D$ ▷ Keep track of the current node
2: **for** $i = t$ **down to** $t - D + 1$ **do**   ▷ Loop backwards through the D most recent bits
3:     **if** $x = 1$ **then**
4:        $s.\mathrm{P}_{\mathrm{KT}} := \frac{s.b + \frac{1}{2}}{s.a + s.b + 1} s.\mathrm{P}_{\mathrm{KT}}$ ▷ Update $\mathrm{P}_{\mathrm{KT}}(a_s, b_s)$
5:        $s.b := s.b + 1$ ▷ Increment b_s
6:     **else**
7:        $s.\mathrm{P}_{\mathrm{KT}} := \frac{s.a + \frac{1}{2}}{s.a + s.b + 1} s.\mathrm{P}_{\mathrm{KT}}$ ▷ Update $\mathrm{P}_{\mathrm{KT}}(a_s, b_s)$
8:        $s.a := s.a + 1$ ▷ Increment a_s
9:     **if** $x_i = 1$ **then**
10:        $s := s.\texttt{left}$ ▷ Walk down the left path
11:     **else**
12:        $s := s.\texttt{right}$ ▷ Walk down the right path
13:     **if** $s = null$ **then**   ▷ New node encountered, grow tree
14:        Create a new node p with $p.a = p.b = 0$, $p.\mathrm{P}_{\mathrm{KT}} = p.\mathrm{P}_w = 1$, $p.\texttt{left} = p.\texttt{right} = null$
15:        $p.\texttt{parent} := s.$
16:        $s := p$
17: $s.w := s.\mathrm{P}_{\mathrm{KT}}$ ▷ Update $\mathrm{P}_w(s)$ for leaf
18: **for** $i = 1$ **to** D **do**
19:     $s := s.\texttt{parent}$ ▷ Walk back up the tree
20:     $s.w := \frac{1}{2} s.\mathrm{P}_{\mathrm{KT}} + \frac{1}{2}(s.\texttt{left}.\mathrm{P}_w)(s.\texttt{right}.\mathrm{P}_w)$ ▷ Update $\mathrm{P}_w(s)$ for non-leaf

---

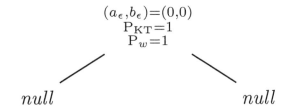

(a) *A newly initialized context tree $\mathcal{T}_2$ of depth $D=2$.*

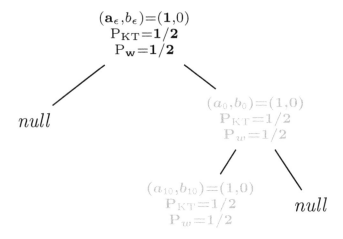

(b) *The context tree $\mathcal{T}_2$ after having processed bit $x_3=0$ given context $x_{1:2}=10$.*

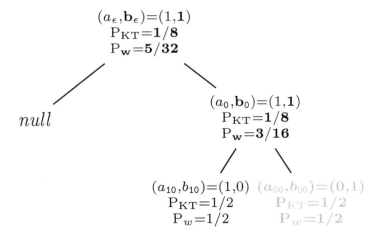

(c) *The context tree $\mathcal{T}_2$ after having processed bit $x_4=1$ given context $x_{1:3}=100$.*

Figure 4.13: *The result of running CTW update (Algorithm 4.6) with the online context tree initialization optimization (Section 4.5.5) on context tree $\mathcal{T}_2$ (Example 4.5.6) for various choices of bit x_{t+1} and context $x_{1:t}$. Modified nodes highlighted.*

Algorithm 4.7 CTW Revert

Require: Updated Context Tree $\mathcal{T}_D$ modified by Algorithm 4.6
Input: Context string $x_{1:t}$ ▷ Assuming that $t \geq D$
Input: Bit to revert $x = x_{t+1}$
Effect: Undoes the effect Algorithm 4.6 had on $\mathcal{T}_D$
1:   $s := \mathcal{T}_D$ ▷ Keep track of the current node
2:   **for** $i = t$ **down to** $t - D + 1$ **do**   ▷ Loop backwards through the D most recent bits
3:      **if** $x = 1$ **then**
4:         $s.b := s.b - 1$ ▷ Decrement b_s
5:         $s.\theta := \frac{s.a + s.b + 1}{s.b + \frac{1}{2}} \times s.\theta$ ▷ Revert $\mathrm{P_{KT}}(a_s, b_s)$
6:      **else**
7:         $s.a := s.a - 1$ ▷ Decrement a_s
8:         $s.\theta := \frac{s.a + s.b + 1}{s.a + \frac{1}{2}} \times s.\theta$ ▷ Revert $\mathrm{P_{KT}}(a_s, b_s)$
9:      **if** $x_i = 1$ **then**
10:        $s := s.\mathbf{left}$ ▷ Walk down the left path
11:      **else**
12:        $s := s.\mathbf{right}$ ▷ Walk down the right path
13:   $s.w := s.\theta$ ▷ Revert $\mathrm{P}_w(s)$ for leaf node s
14:   **for** $i = 1$ **to** D **do**
15:      $s := s.parent$ ▷ Walk back up the tree
16:      $s.w := \frac{1}{2} s.\theta + \frac{1}{2} \cdot s.left.w \times s.right.w$ ▷ Revert $\mathrm{P}_w(s)$ for non-leaf node s

---

4.6 Exercises

1. [C32m] Prove Lemma 4.1.5.

2. [C10] Find an explicit form for $\mathrm{P_{KT}}(n, n)$.

3. [C10] Come up with a countably infinite sized suffix set which is proper and complete.

4. [C15] Prove the computation time for (4.5.2) is $O(D)$.

5. [C15] Repeat the redundancy bound calculation in Example 4.4.1, but for the weighted mixtures $\hat{\mu}_{w,\alpha}(x_{1:t}) = \alpha \hat{\mu}_1(x_{1:t}) + (1 - \alpha)\hat{\mu}_2(x_{1:t})$ for some $\alpha \in [0,1]$, and the more general mixture $\hat{\mu}_{w,\boldsymbol{\theta}}(x_{1:t}) = \sum_i \theta_i \hat{\mu}_i(x_{1:t})$ for weights $\boldsymbol{\theta} = (\theta_1, ..., \theta_n)$ satisfying $\theta_i \geq 0$ and $\sum_i \theta_i = 1$.

6. [C30i] Implement the CTW algorithm, and try feeding in "random" bits generated by you. You may find that it can reliably predict the next bit $> 60\%$ of the time, depending on how unpredictably you can act!

7. [C35] Derive a multi-alphabet version of CTW. Hint: see [PS99].

8. [C32] Suppose you have a fixed context compressor which is able to reduce the length of the context by at least half. How does this effect the redundancy of CTW? Implement this fixed context compressor version of CTW and compare with the original CTW.

4.7   History and References

The Context Tree Weighting algorithm was first briefly introduced in [WST93] where an upper bound on the redundancy was provided. The explanation of the algorithm and its bounds were greatly expanded on in [WST95], and as it contains most of the information on the CTW algorithm [WST95] is often referred to as the original CTW paper. A concise description of CTW and the important properties it possesses are provided in [WST97], where the original authors present the content from a mini-course on the CTW algorithm. One of the key components of the CTW algorithm is the KT estimator. Originally introduced in [KT81], the KT estimator is an efficient estimator for memoryless sources with (asymptotically) optimal redundancy bounds for such sources. [SSH12] discusses a windowed version of KT used within CTW for non-stationary environments. After its inception, there was much work following the CTW algorithm, expanding it in various ways, and removing some of its limitations. One such generalization was to consider arbitrary alphabets beyond binary [TSW93a, TSW93b]. It was shown that the algorithm is still able to achieve optimal redundancy for arbitrary alphabets, however, in practice the best choice of alphabet is often unknown. One approach is to binarize the alphabet, i.e. to inject $\mathcal{A}$ into $\mathbb{B}^k$ for $k = \lceil \log_2 |\mathcal{A}| \rceil$ [VH18]. Another alternative to using multi-alphabet KT for the non-binary alphabet case is the near-optimal Sparse Adaptive Dirichlet (SAD) estimator [Hut13a, VH12] approaching the slow but Bayes-optimal mixture over all sub-alphabets [TSW93a]. The SAD estimator was further expanded upon in [Bel15] to K-distinct reservoir sampling, which is more computationally and memory efficient than CTW and other tree-based algorithms. A downside of the original CTW algorithm is that the depth D must be chosen in advance and cannot be updated online. This downside was overcome in [Wil98] where the finite depth was removed, but at an increase in computation time that made the extension impractical. With D infinite or logarithmically growing with n, CTW is asymptotically optimal even on piecewise stationary sources if the number of pieces is small, despite only being designed for stationary sources [VCH18]. Not all data sources have the binary suffix tree structure assumed by the CTW algorithm. In [WST96], four new model classes are presented for which context weighting algorithms are provided. More details on prediction suffix trees can be found in [Ris83, RST96].

Context tree weighting has been applied beyond prediction to active agents in [VNHS10] using action-conditional prediction suffix trees (see Chapter 12). More recently, CTW has been expanded for the interaction case for partially observable data in [MW17]. The resulting algorithm D2-CTW is less sensitive to aliasing and noise. The CTW algorithm uses node-based pruning to maintain its mixture; but an alternative approach is to use edge-based pruning [PS99], which results in a larger model class being considered. CTW was compared to other prediction algorithms in a practical study in [BEYY04] and shown to outperform Lempel-Ziv (Original [ZL78] and improved [NYEYM03]) and Prediction by Partial Match [CW84] and Probabilistic Suffix Trees [RST96].

Very recently, [GDR$^+$23, GMGH$^+$24] trained a Transformer (also LSTMs) on data sampled from the CTW, PTW (Section 5.3), and even Solomonoff's distribution. Remarkably, the trained Transformer is able to nearly perfectly mimic CTW and PTW in-context. Since CTW and PTW are Bayes-optimal learners of variable-order Markov processes and piecewise i.i.d. processes respectively, so is the trained Transformer in-context without any further weight updates.

The more recent variations and extensions of CTW will be covered in Chapter 5. They include adaptive CTW for non-stationary sources [OHSS12], Partition Tree Weighting (PTW) for piecewise stationary sources [VWBG13], and Context Tree Switching (CTS) [VNHB12].

Chapter 5

Variations on CTW

If you can't program it, you haven't understood it.

David Deutsch

We have seen in Chapter 4 that Context Tree Weighting (CTW) is an efficient method for computing a Bayesian mixture P_D^{CTW} (Definition 4.5.4) over a class $\mathcal{M}$ of variable-order Markov sources. If the true distribution μ is k-Markov, CTW will learn to predict well. In this chapter, we will weaken the assumption that μ is k-Markov and expand on the CTW algorithm on several fronts. First, we will look at altering the choice of the KT estimator using a variant of CTW called Adaptive CTW [OHSS12] suitable for non-stationary sources. Next, we will go over a modified form of CTW called Context Tree Switching (CTS) that allows the use of several distributions for prediction, which the model can switch between. Weighted combinations of these distributions can be used to increase the size of the model class for CTW, with minimal impact to the redundancy and time/space complexity [VNHB12]. Lastly, we will discuss a distinct but similar approach to prediction called Partition Tree Weighting (PTW) for piecewise stationary sources with change points which weighs over partitions instead of contexts [VWBG13].

5.1   Adaptive CTW

The CTW method operates under the assumption that the true distribution is stationary, meaning that the distribution over x_t only depends on previous terms $x_{t-k}, x_{t-k+1}, ..., x_{t-1}$ and not on the current time step t. This makes the KT estimator an appropriate choice as it takes into account all statistics observed and assigns equal weight to both old and recent samples. For non-stationary sources, the KT estimator is a poor choice. As more samples are collected, the contribution from new symbols becomes smaller compared to older samples, making it difficult to detect changes or shifts in the distribution over time. To address this limitation and adapt the CTW method for non-stationary sources, we need to replace the KT estimator with a new estimator that places more weight on recent data, while maintaining the computational efficiency of the CTW method. The approach of adaptive CTW [OHSS12] is to use a discounted KT estimator that assigns higher weight to more recent data, similar to how rewards are discounted in reinforcement learning (see Section 6.4). For the standard KT estimator (Lemma 4.1.2), the counts a and b are incremented whenever a zero or one is observed, respectively. Let a_t and b_t be the counts after observing the sequence $x_{1:t}$. We can express the new counts a_{t+1} and b_{t+1} as

$$a_{t+1} = [\![x_{t+1} = 0]\!] + a_t$$
$$b_{t+1} = [\![x_{t+1} = 1]\!] + b_t$$

The discounted KT estimator modifies this by introducing a discount γ that controls how quickly the counts decay. The larger the γ, the faster the decay. The increments are still made based on the value of x_{t+1}, after which the counts are scaled by $(1-\gamma)$. The updated rules for the discounted KT estimator are:

$$a_{t+1} = (1-\gamma)([\![x_{t+1} = 0]\!] + a_t)$$
$$b_{t+1} = (1-\gamma)([\![x_{t+1} = 1]\!] + b_t)$$

If $\gamma = 0$, this reduces to the standard KT estimator. We want to emphasize that γ need not be a constant, but can be parameterized. We will now go over several different parameterizations of γ and their motivations.

Constant.   We could choose γ to be constant, giving a fixed discount rate. While this is easy to implement, this introduces an additional free parameter γ that must be chosen. Choosing a constant for γ essentially gives a geometric discount, making it hard to strike a balance between having essentially no discount for small sequence lengths, and for large sequence lengths the initial terms have practically zero contribution.

Sequence-length.   We could choose the discount γ_t to depend on the length of the sequence observed so far (or equivalently by the current time step) by letting $\gamma_t = ct^{-\alpha}$ for some $c, \alpha \in [0,1)$. As long as $\alpha > 0$, the effective horizon (Definition 6.4.2) will increase over time. With $\alpha = 0$, this reduces to the fixed discount rate. A downside of this approach is that if two identical contexts are observed at two distant time steps, the weighting between the updates will drastically differ even though no other observations of that context were made.

Context visit.   To address the downside with sequence length, we can incorporate the number of times a particular context has been observed before. This can be done in three different ways. Here, we write $\gamma_{s,t}$ to indicate the discount is now a function of both the current node s in the context tree, and the current time step t.

1. **Partial-context visit.** Use the same discount as sequence-length discounting, but use the number of times a node associated with context s has been visited up to time t,

denoted $k_s(t)$, which gives the discount $\gamma_{s,t} = ck_s(t)^{-\alpha}$. A downside of this approach is that nodes with shorter contexts will be visited more often (the root node ϵ is visited on every time step, whereas many leaf nodes may be seldom visited, if at all). This can lead to an uneven discount that weights more on short contexts.

2. **Full-context visit.** Consider leaf nodes and internal nodes separately. The tree is traversed from root to leaf, the discount for leaf nodes is updated in the same way as partial-context visits, $\gamma_{s,t} = ck_s(t)^{-\alpha}$, and the same result is propagated back up the tree for the internal nodes. For every node s' along the path from the root ϵ to leaf s, we use the same discount $\gamma_{s',t} = \gamma_{s,t} = ck_s(t)^{-\alpha}$ as for the leaf.

3. **Leaf-context visit** discounts leaf nodes in the same fashion as before, $\gamma_{s,t} = ck_s(t)^{-\alpha}$, but also walks back up to the root node, updating the counts directly for each non-leaf node to be the sum of its children, i.e. $a_s = a_{0s} + a_{1s}$ and $b_s = b_{0s} + b_{1s}$. No discount is explicitly applied to internal nodes, but only implicitly through accumulation of counts propagated up from leaf nodes.

Through experimentation [OHSS12], it was found that partial-context visit Adaptive CTW with $c = 0.1$ and $\alpha = 0.33$ outperforms classical CTW on data compression benchmarks. Adaptive CTW offers a simple and computationally efficient method to improve the choice of base model of the KT estimator.

5.2   Context Tree Switching

The Context Tree Switching (CTS) algorithm [VNHB12] is an extension of the CTW algorithm to a larger model class without losing much in terms of redundancy and computation time. CTW works well assuming the dynamics of the true environment never change. However, this is a strong assumption, especially for tasks such as data compression where there might be an abrupt change in the data source from which data is drawn (e.g. the boundary between image and text data in the same file.)

Example 5.2.1 (Piecewise 1-Markov source) Consider the sequences $(11)^\infty$ and $(01)^\infty$, which can be regarded as "sampled" from the deterministic 0-Markov distribution $\mu_0(x_t = 1|x_{<t}) = 1$ and 1-Markov distribution $\mu_1(x_t = 1|x_{<t}) = \mu_1(1|x_{t-1}) = 1 - x_{t-1}$ for $t > 1$ and $\mu_1(x_1 = 0) = 1$. Now consider the alternating sequence

$$\dot{x}_{1:\infty} = (01)^{100}(11)^{100}(01)^{100}(11)^{100}\ldots$$

which is still deterministic but now a *non*-stationary 1-Markov process $\mu(x_t|x_{<t}) = \mu_{i_t}(x_t|x_{<t})$ for $i_t := \lfloor (t + 199)/200 \rfloor \bmod 2$. As such, we would expect that using a new CTW tree (with $D = 1$) every 200 time steps would work well on this sequence, the first and third would learn the rules $0 \to 1, 1 \to 0$ and the second and fourth would learn to always predict 1 regardless of context. However, we would expect a single CTW model with $D = 1$ for the entire sequence to perform poorly: All the time spent collecting statistics in the first half of the sequence would actively hamper learning for the second half of the sequence.

We can observe this via experiment: We use the above sequence, and train four independent CTW models of depth 1, switching them out every 200 time steps. We also use a depth 1, 2 and 200 CTW model for the entire sequence, noting that depth 2 should be (almost) sufficient to learn the non-stationary sequence, as the rules $01 \to 0, 10 \to 1, 11 \to 1$ describe the sequence entirely, apart from the boundaries in between sequences where we see the substrings 011 and 110, i.e. the process is nearly 2-Markov. This phenomenon holds much more generally [VCH18]. Depth 200 can (eventually) learn the boundary conditions too. We measure the instantaneous KL divergence (4.2.11), noting that since the true environment is deterministic, only one sequence $x_{1:t}$ can ever be possibly observed, so the expectation collapses to $\mathrm{KL}(\mu\|\xi_{CTW}) = -\ln\xi_{CTW}(\dot{x}_t|\dot{x}_{<t})$.

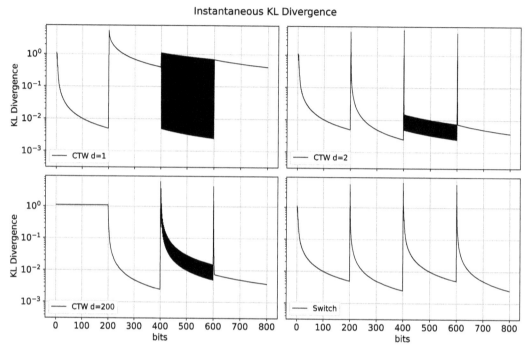

Figure 5.1: *Instantaneous KL divergence for CTW models ($D=1, D=2, D=100$), vs. switching model using four CTW $D=1$ models (data smoothed for clarity). Environment detailed in Example 5.2.1.*

As seen in Figure 5.1, the switch model quickly learns each block of the sequence, briefly spiking as each new CTW model relearns a block. To no surprise, a single CTW model with $D=1$ is inadequate for learning the sequence. The CTW $D=2$ model initially performs as well as the switch model, but the second time it encounters the block $(11)^{100}$, the statistics collected from previous blocks hamper learning, and a freshly deployed CTW model with $D=1$ (switch) is able to outperform it. We would expect the CTW model with $D=200$ to eventually learn the sequence perfectly, but this would require a much longer sequence (as the model is receiving statistics regarding the location of the boundaries only once every 200 bits). This motivates the switching method, though for this toy example, the number of models chosen, and when to switch them out was hardcoded. In practice, this must be determined from the data itself.   ◆

As previously discussed, the mixture ξ will eventually converge to the best model in the class $\mathcal{M}$, but for CTW to ensure the true environment is in $\mathcal{M}$, this may sometimes require an extremely large choice of depth (and therefore a lot more bits from the environment which with to update the parameters).

Motivated by Example 5.2.1, we could instead choose the goal of trying to find the best sequence of models to learn the true environment, to allow for the possibility that the environment itself may change over time. This is the core idea behind CTS. What remains is to define a method analogous to CTW that can switch between models efficiently.

First we need to define a distribution over model sequences.

Definition 5.2.2 (Switch distribution [VNHB12]) Given a non-empty model class $\mathcal{M} = \{\rho_1, \rho_2, ..., \rho_{|\mathcal{M}|}\}$, the *switch distribution* with respect to model class $\mathcal{M}$ is defined as a weighted sum over all length n sequences $i_{1:n} \in \{1, 2, ..., |\mathcal{M}|\}^n$

$$\tau_\alpha(x_{1:n}) := \sum_{i_{1:n} \in \{1,2,...,|\mathcal{M}|\}^n} w(i_{1:n}) \prod_{k=1}^n \rho_{i_k}(x_k | x_{<k})$$

where the weight w is recursively defined as

$$w(i_{1:n}) := \begin{cases} 1 & \text{if } n = 0 \\ \frac{1}{|\mathcal{M}|} & \text{if } n = 1 \\ w(i_{<n}) \cdot (1 - \alpha_n) & \text{if } i_n = i_{n-1} \\ w(i_{<n}) \cdot \frac{\alpha_n}{|\mathcal{M}|-1} & \text{if } i_n \neq i_{n-1} \end{cases} \tag{5.2.3}$$

with switch rates $\alpha_n \in [0, 1]$.

The choice of weight w (or prior) implies that on each time step t, the probability of switching to a new model (chosen uniformly at random) is α_t. The first model is also chosen uniformly at random.

Trying to compute τ_α naively would be intractable as the summation is over a set of size $O(|\mathcal{M}|^n)$. However, it is possible to efficiently compute τ_α in $O(n|\mathcal{M}|)$ time and $O(|\mathcal{M}|)$ space with Algorithm 5.1.

Algorithm 5.1 Switch distribution $\tau_\alpha(x_{1:n})$ [VNHB12, BMS+20]

Require: A finite non-empty mode class $\mathcal{M} = \{\rho_1, \rho_2, ..., \rho_{|\mathcal{M}|}\}$
Require: A weight vector $(w_1, ..., w_{|\mathcal{M}|}) \in \mathbb{R}^{|\mathcal{M}|}$, with $w_i = 1/|\mathcal{M}|$ for $1 \leq i \leq |\mathcal{M}|$
Require: A sequence of switching rates $(\alpha_2, ..., \alpha_n) \in \mathbb{R}^{n-1}$
Input: An input sequence $x_{1:n}$
Output: Switch distribution $\tau_\alpha(x_{1:n}) = \prod_{t-1}^n \tau(x_t | x_{<t})$
1: **for** $t = 1$ to n **do**
2: $\quad \tau(x_t | x_{<t}) := \sum_{i=1}^{|\mathcal{M}|} w_i \rho_i(x_t | x_{<t})$ ▷ Compute the conditional mixture
3: $\quad w_i := \frac{\alpha_{t+1}}{|\mathcal{M}|-1} + \left((1-\alpha_{t+1}) - \frac{\alpha_{t+1}}{|\mathcal{M}|-1}\right) \frac{w_i \rho_i(x_t | x_{<t})}{\tau(x_t | x_{<t})} \ \forall i$ ▷ Update the weights
    return $\tau(x_{1:n})$

Context Tree Switching (CTS) combines this switch distribution with CTW. Let $x_{1:n}^s$ denote the subsequence of $x_{1:n}$ of elements that follow the substring s. For example, let $x = 01011010$. Then x^{01} are the elements in **01**0**11**0**10** that follow **01** (highlighted in **bold**), giving $x^{01} = 010$. We can formally define $x_{1:n}^s$ as $x_{1:n}^s := (x_{i_1}, ..., x_{i_j})$ where $i_k = \min\{i : i > i_{k-1} \wedge x_{i-1-\ell(s):i-1} = s\}$ with $i_0 := 0$.

In previous experiments (Sections 4.3.5 and 4.5.4) we saw how choosing a context s for the environment in question longer than necessary can hamper learning, as we have redundant parameters to learn. For a k-Markov environment with unknown $k \leq D$, CTW chooses a mixture of all depth $\leq D$ PSTs as estimator, weighted by complexity. Recall (4.5.2), which gives the recursive definition of the probability $P_w(s)$ associated with node s

in a context tree.

$$\mathrm{P}_w(s) := \begin{cases} \frac{1}{2}\mathrm{P}_{\mathrm{KT}}(a_s,b_s) + \frac{1}{2}\mathrm{P}_w(0s)\mathrm{P}_w(1s) & \ell(s) < D \\ \mathrm{P}_{\mathrm{KT}}(a_s,b_s) & \ell(s) = D \end{cases}$$

CTS generalizes this approach by providing a weighted mixture over all sequences of depth $\leq D$ PST models, allowing the best choice of model to change as the environment does.

Definition 5.2.4 (Context Tree Switching) The Context Tree Switching probability $\mathrm{P}_{s,D}^{CTS}$ of $x_{1:n}$, with depth $D > 0$ and context s is defined as

$$\mathrm{P}_{s,D}^{CTS}(x_{1:n})$$

$$= \sum_{i_{1:n_s} \in \mathbb{B}^{n_s}} w(i_{1:n_s}) \prod_{k=1}^{n_s} \begin{cases} \mathrm{P}_{\mathrm{KT}}((x_{1:n}^s)_k \mid (x_{1:n}^s)_{<k}) & \text{if } i_k = 0 \\ \mathrm{P}_{0s,D-1}^{CTS}(x_{t_s(k)} \mid x_{<t_s(k)}) \mathrm{P}_{1s,D-1}^{CTS}(x_{t_s(k)} \mid x_{<t_s(k)}) & \text{if } i_k = 1 \end{cases}$$

where $n_s = \ell(x_{1:n}^s)$, $(x_{1:n}^s)_{1:k}$ is the first k bits of $x_{1:n}^s$, $t_s(k) = \min\{t \mid \ell(x_{1:t}^s) = k\}$, and $\mathrm{P}_{s,D}^{CTS}(y \mid x) := \mathrm{P}_{s,D}^{CTS}(xy)/\mathrm{P}_{s,D}^{CTS}(y)$. For the base cases we have $\mathrm{P}_{s,0}^{CTS}(x_{1:n}) = \mathrm{P}_{\mathrm{KT}}(x_{1:n}^s)$ and $\mathrm{P}_{s,D}^{CTS}(\epsilon) = 1$. The top-level mixture $\mathrm{P}_{\epsilon,D}^{CTS}(x_{1:n})$ will be simply denoted $\mathrm{P}_D^{CTS}(x_{1:n})$.

Let us explore the above expression for a moment: We are taking a weighted sum over all binary sequences of length n_s, weighted by the switching prior w (5.2.3). The product can be understood as an application of chain rule, but on each step the sequence dictates whether the KT estimator should be applied ($i_k = 0$), or the recursive CTS mixture ($i_k = 1$).

We will now describe how in practice $\mathrm{P}_D^{CTS}(x_{1:n})$ can be computed efficiently. CTS works much like the CTW method by using a perfect binary tree of depth D, for which each node s in the tree stores the following values: $\mathrm{P}_{\mathrm{KT}}(a_s,b_s)$, the base KT estimator associated with that node (along with the usual KT counts a_s and b_s); α_s and β_s, used like the weights w in switching; and P_s^{CTS}, used like the τ in switching. Each internal node is initialized with $\alpha_s(\epsilon) = \beta_s(\epsilon) = \frac{1}{2}$ and each leaf node is initialized with $\alpha_s(\epsilon) = 1$ and $\beta_s(\epsilon) = 0$. Then when a new symbol x_n occurs, given a history $x_{<n}$, the path of the tree reflecting the context $x_{<n}$ is traversed and updated according to Algorithm 5.2.

Clearly the update takes time $O(D)$, which is asymptotically the same time taken as CTW. Additionally the space requirements asymptotically match those of CTW. The space requirement looks like 2^D, but as with CTW, at most $O(n|D|)$ nodes need to be created and stored explicitly.

The CTS method leverages multiple context tree models to efficiently encode data sequences. It adaptively selects the best context tree model for different parts of the input sequence, resulting in improved compression and prediction performance.

In the theorem below, we consider binary sequence $x_{1:n}$ which we informally imagine being sampled from a binary prediction suffix tree $(\mathcal{S}, \Theta_{\mathcal{S}})$, though the formal statement in the theorem does not rely on this assumption and holds for all sequences $x_{1:n}$. $\mathcal{S}$ is a set of contexts belonging to a class $\mathcal{C}_D$, and $\Theta_{\mathcal{S}}$ is a parameter vector associated with each context $s \in \mathcal{S}$. Each parameter $\theta_s \in [0,1]$ represents the probability of observing a certain symbol (0 or 1) given the context s.

The function $d(\mathcal{S})$ denotes the maximum length of any context $s \in \mathcal{S}$, which provides a measure of the complexity of the suffix tree model. In the CTS method, the objective is to find a balance between the complexity of the model and the ability to accurately represent the data sequence $x_{1:n}$.

Algorithm 5.2 Context Tree Switching update [VNHB12, BVT14]

Require: New symbol x_n
Require: History $x_{<n}$
Input: Current context tree $\mathcal{T}_D = \{a_s, b_s, \alpha_s, \mathrm{P}^s_{\mathrm{KT}}, \beta_s : s \in \mathbb{B}^D\}$
Output: Updated context tree $\mathcal{T}_D$
1: **for** $d = 0$ to D **do**
2:      Let $s = x_{n-d:n-1}$ be the current context
3:      Update estimator a_s and b_s and $\mathrm{P}^s_{\mathrm{KT}}$ with KT-updating
4:      $x' := x_{n-\ell(s)-1}$
5:      **if** s is leaf node **then**
6:          $\alpha_s(x_{1:n}) := \alpha_s(x_{<n})\mathrm{P}^s_{\mathrm{KT}}$
7:          $\mathrm{P}^{CTS}_{s,D}(x_{1:n}) := \alpha_s(x_{1:n})$
8:      **else**
9:          $\mathrm{P}^{CTS}_{s,D}(x_{1:n}) := \alpha_s(x_{<n})\mathrm{P}^s_{\mathrm{KT}} + \beta_s(x_{<n})\mathrm{P}^{CTS}_{x's,D}(x_n \mid x_{<n})$
10:         $\alpha_s(x_{1:n}) := \frac{1}{n+1}\mathrm{P}^{CTS}_{s,D}(x_{1:n}) + \frac{n-1}{n+1}\alpha_s(x_{<n})\mathrm{P}^s_{\mathrm{KT}}$
11:         $\beta_s(x_{1:n}) := \frac{1}{n+1}\mathrm{P}^{CTS}_{s,D}(x_{1:n}) + \beta_s(x_{<n})\mathrm{P}^{CTS}_{x's,D}(x_n \mid x_{<n})$

The theorem states an upper bound on the redundancy of the CTS coding distribution. Redundancy is a measure of the difference between the optimal coding length (if the true underlying model were known) and the coding length achieved using the CTS method. A lower redundancy indicates better compression performance.

The bound is expressed in terms of the following quantities:

- $\Gamma_D(\mathcal{S})$ represents the complexity of suffix set/tree $\mathcal{S}$
- $(d(\mathcal{S})+1)\log_2(n)$ accounts for the maximum context length in the model and the input sequence length.
- $\frac{1}{2}|\mathcal{S}|\log_2(n/|\mathcal{S}|)$ considers the trade-off between the number of contexts in the model and the input sequence length.
- $|\mathcal{S}|$ represents the total number of contexts in the model.

The theorem essentially tells us that the redundancy of using the CTS coding distribution is upper bounded by a combination of model complexity, context length, and sequence length. This insight can be helpful when designing and analyzing CTS variations for data compression and prediction tasks.

Theorem 5.2.5 (PST-CTS redundancy [VNHB12]) Given a data sequence $x_{1:n} \in \mathbb{B}^n$ and Prediction Suffix Tree (PST) $(\mathcal{S}, \Theta_{\mathcal{S}})$ with $\mathcal{S} \in \mathcal{C}_D$ and parameter vector $\Theta_{\mathcal{S}} : \{\theta_s \in [0,1]\}_{s \in \mathcal{S}}$, letting $d(\mathcal{S}) := \max_{s \in \mathcal{S}} \ell(s)$, then the redundancy of using the Context Tree Switching (CTS) coding distribution compared to coding with respect to the PST is upper bounded by

$$\log_2 \mathrm{P}_{\mathcal{S}, \Theta_{\mathcal{S}}}(x_{1:n}) - \log_2 \mathrm{P}^{CTS}_D(x_{1:n}) \leq \Gamma_D(\mathcal{S}) + (d(\mathcal{S})+1)\log_2(n) + \frac{|\mathcal{S}|}{2}\log_2\left(\frac{n}{|\mathcal{S}|}\right) + |\mathcal{S}|$$

Proof. Let $\overline{\mathcal{S}}$ be the set of contexts that index the internal nodes of $\mathcal{S}$. By observing the elements in the sum from Definition 5.2.4 we can conclude

$$\mathrm{P}^{CTS}_{s,D}(x_{1:n}) \geq \begin{cases} w_s(1_{1:n_s})\mathrm{P}^{CTS}_{0s,D-1}(x_{1:n})\mathrm{P}^{CTS}_{1s,D-1}(x_{1:n}) & \text{if } s \notin \mathcal{S} \\ w_s(0_{1:n_s})\mathrm{P}_{\mathrm{KT}}(x^s_{1:n}) & \text{if } s \in \mathcal{S} \text{ and } D > 0 \\ \mathrm{P}_{\mathrm{KT}}(x^s_{1:n}) & \text{if } D = 0 \end{cases}$$

for any $s \in \mathcal{S} \cup \overline{\mathcal{S}}$. Now define $\mathcal{S}' := \{s \in \mathcal{S} : \ell(s) < D\}$. By repeatedly applying the above equation starting from $\mathrm{P}_D^{CTS}(x_{1:n}) = \mathrm{P}_{\epsilon,D}^{CTS}(x_{1:n})$ and continuing until no more CTS terms remain, we can conclude

$$
\begin{aligned}
\mathrm{P}_D^{CTS} &\geq \left(\prod_{s\in\overline{\mathcal{S}}} w_s(1_{1:n_s})\right)\left(\prod_{s\in\mathcal{S}'} w_s(0_{1:n_s})\right)\left(\prod_{s\in\mathcal{S}} \mathrm{P}_{\mathrm{KT}}(x_{1:n}^s)\right) \\
&= \left(\prod_{k=0}^{d(\mathcal{S})}\prod_{s\in S'\cup S:\ell(s)=k} w_s(1_{1:n_s})\right)\left(\prod_{s\in\mathcal{S}} \mathrm{P}_{\mathrm{KT}}(x_{1:n}^s)\right) \\
&\geq \left(2^{-\Gamma_D(\mathcal{S})}\prod_{k=0}^{d(\mathcal{S})}\prod_{s\in S'\cup S:\ell(s)=k} \frac{w_s(1_{1:n_s})}{w_s(1_{1:\min\{n_s,1\}})}\right)\left(\prod_{s\in\mathcal{S}} \mathrm{P}_{\mathrm{KT}}(x_{1:n}^s)\right) \\
&\geq \left(2^{-\Gamma_D(\mathcal{S})}\prod_{k=0}^{d(\mathcal{S})}\prod_{t=2}^{n} \frac{t-1}{t}\right)\left(\prod_{s\in\mathcal{S}} \mathrm{P}_{\mathrm{KT}}(x_{1:n}^s)\right) \\
&= 2^{-\Gamma_D(\mathcal{S})}n^{-(d(\mathcal{S})+1)}\left(\prod_{s\in\mathcal{S}} \mathrm{P}_{\mathrm{KT}}(x_{1:n}^s)\right)
\end{aligned}
$$

The first equality comes from the fact that $w_s(1_{1:n}) = w_s(0_{1:n})$ and rearranging. The second inequality follows from the fact that $|\mathcal{S} \cup \mathcal{S}'| = \Gamma_D(\mathcal{S})$, $w_s(1) = 1/2$ and either $w_s(1_{1:n_s}) = w_s(\epsilon) = 1$ if $n_s = 0$ or $w_s(1_{1:n_s}) = 1/2 \times \ldots$ if $n_s > 0$. The third inequality comes from the observation that the context associated with each symbol in $x_{1:n}$ matches at most one context $s \in \overline{\mathcal{S}} \cup \mathcal{S}'$ of each specific length $0 \leq k \leq d(\mathcal{S})$. The last equality is a result of the telescoping product. Then taking the $-\log_2$ of both sides we get

$$
-\log_2(\mathrm{P}_D^{CTS}(x_{1:n})) \leq \Gamma_D(\mathcal{S}) + (d(\mathcal{S})+1)\log_2(n) - \log_2\left(\prod_{s\in\mathcal{S}} \mathrm{P}_{\mathrm{KT}}(x_{1:n}^s)\right)
$$

Then combining with Theorem 4.5.12 we get

$$
\log_2\mathrm{P}_{\mathcal{S},\Theta_{\mathcal{S}}}(x_{1:n}) - \log_2\mathrm{P}_D^{CTS}(x_{1:n}) \leq \Gamma_D(\mathcal{S}) + (d(\mathcal{S})+1)\log_2(n) + \frac{|\mathcal{S}|}{2}\log_2\frac{n}{|\mathcal{S}|} + |\mathcal{S}| \quad \blacksquare
$$

We can compare this with the redundancy bound proven for the CTW method (Corollary 4.5.14)

$$
\log_2\mathrm{P}_{\mathcal{S},\Theta_{\mathcal{S}}}(x_{1:t}) - \log_2\mathrm{P}_D^{CTW}(x_{1:n}) \leq \frac{|\mathcal{S}|}{2}\log_2\frac{n}{|\mathcal{S}|} + \Gamma_D(\mathcal{S}) + |\mathcal{S}|
$$

We have a slightly looser redundancy bound by an additional additive term $(d(\mathcal{S})+1)\log_2 n$, growing logarithmically with the length of $x_{1:n}$, and linearly with the depth of the PST that models the true environment. This is a small penalty to pay, in exchange for CTS having a larger model class that allows for switching between Markov models in $\mathcal{C}_D$, whereas CTW can in a sense consider only a fixed model in $\mathcal{C}_D$ that is the most similar to the true environment.

5.3   Partition Tree Weighting

Partition Tree Weighting (PTW) [VWBG13] is an approach similar to Context Tree Weighting that is made for dealing with multiple, piecewise stationary data sources. A piecewise stationary data generating source is defined by a partition over time of several different data sources. As the name suggests, the approach uses a Bayesian mixture which weighs over choices of partitions. Partition sources will often occur in practice: For example, the weather follows different distribution in summer and winter. We can model these as temporal partitions, letting us essentially condition on time. For this we will need to formally define what we mean by partitions.

Definition 5.3.1 (Temporal partition) A *temporal partition* $\mathcal{P}_n \subseteq \mathbb{N}^+ \times \mathbb{N}^+$ is a set of tuples representing segments such that for all $x \in \mathbb{N}^+$ with $x \leq n$, there exists a unique $(a,b) \in \mathcal{P}_n$ with $a \leq x \leq b$. Each tuple is called a *part*. Additionally we define $f(x)$ as the function which returns the index of the time segment containing i.

One can think of a temporal partition as a representation of a partition over $\{1,...,n\}$ with the restriction that every subset is a set of contiguous integers. For example, one possible temporal partition over $\mathcal{P}_{10}$ is $\{(1,3),(4,6),(7,7),(8,10)\}$, corresponding to the partition

$$\{\{1,2,3\},\{4,5,6\},\{7\},\{8,9,10\}\}$$

Given a temporal partition, each part in the partition is associated with a data source. For this section we consider the case of piecewise (stationary) sources, which can be represented by a temporal partition of data-generating sources. Formally we make no assumptions on the true source μ, but it is helpful to imagine that μ is piecewise stationary or even piecewise i.i.d., since these are the sources PTW aims at predicting/compressing well.

Definition 5.3.2 (Piecewise (stationary) source) We call ρ a *piecewise stationary source* if there exist probability measures $\{\rho_1,\rho_2,...\}$ and a partition $\mathcal{P} = \{(a_1,b_1),(a_2,b_2),...\}$ on $\mathbb{N}^+$ such that for all $n \in \mathbb{N}^+$ and all $x_{1:n} \in \mathbb{B}^n$ we have

$$\rho(x_{1:n}) = \prod_{i \in \mathbb{N}^+} \rho_i(x_{a_i:b_i})$$

With this we can construct our model class, the set of binary temporal partitions $\mathcal{B}_D$. If we imagine a complete binary tree of depth D with leaves numbered $1,...,2^D$, then for some fixed suffix set/tree $\mathcal{S} \in \mathcal{C}_D$, we will associate each leaf $s \in \mathcal{S}$ with the interval spanned by the $2^{D-\ell(s)}$ leaves at level D if we were to expand node s to level D. Then the partition $\mathcal{P}$ associated with $\mathcal{S}$ is the union of those intervals, and $\mathcal{B}_D$ is the set of all such partitions. See Figure 5.2 for an illustration of $\mathcal{B}_2$.

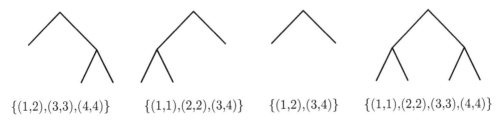

$\{(1,2),(3,3),(4,4)\}$ $\{(1,1),(2,2),(3,4)\}$ $\{(1,2),(3,4)\}$ $\{(1,1),(2,2),(3,3),(4,4)\}$

Figure 5.2: *A collection of partitions in $\mathcal{B}_2$ (Definition 5.3.3) and their corresponding partition trees. All binary partitions except for the empty tree with $\{(1,4)\}$ are shown. Note that $\{(1,3),(4,4)\}$ and $\{(1,1),(2,4)\}$ and $\{(1,1),(2,3),(4,4)\}$ are not included in $\mathcal{B}_2$.*

Definition 5.3.3 (Set of all binary temporal partitions $\mathcal{B}_D$)

$$\mathcal{B}_D = \{\mathcal{P}(\mathcal{S}) : \mathcal{S} \in \mathcal{C}_D\}$$

where partition $\mathcal{P}(\mathcal{S}) := \{I_s : s \in \mathcal{S}\}$

where interval $I_s := (2^{D-\ell(s)} b(\text{reverse}(s)) + 1, \ 2^{D-\ell(s)}(b(\text{reverse}(s)) + 1) + 1)$

where $b(\text{reverse}(s)) = \sum_{i=1}^{\ell(s)} 2^{i-1} s_i$ interprets reversed s as a natural number (see Proposition 2.1.1).

$\mathcal{B}_D$ can also be defined directly without reference to context trees $\mathcal{C}_D$. Proving their equivalence is left as an exercise.

Definition 5.3.4 (Set of all binary temporal partitions $\mathcal{B}_D$)

$$\mathcal{B}_0 := \{\{(1,1)\}\}$$

$$\mathcal{B}_{d+1} := \{\{(1, 2^{d+1})\}\} \cup \{\mathcal{P} \cup (\mathcal{P}' + 2^d) : \mathcal{P}, \mathcal{P}' \in \mathcal{B}_d\}$$

where $\mathcal{P}' + a$ shifts all intervals by a.

We now define the PTW method and the priors used for the weighting. Like CTW we are mixing over a set of trees, however instead of all partitions we mix over binary temporal partitions $\mathcal{B}_D$, allowing us to easily define a measure of complexity based on the corresponding partition tree, by using the same penalty as the CTW model cost (Definition 4.3.20). We leave the base model ρ undefined for the moment; possible choices include the KT estimator or CTW.

Definition 5.3.5 (Partition Tree Weighting (PTW)) Let ρ be some base model (such as the KT estimator). The *Partition Tree Weighting probability* $\mathrm{P}_D^{\mathrm{PTW}}$ of string $x_{1:n}$ with depth D is defined as

$$\mathrm{P}_D^{\mathrm{PTW}}(x_{1:n}) := \sum_{\mathcal{P} \in \mathcal{B}_D} 2^{-\Gamma_D(\mathcal{P})} \prod_{(a,b) \in \mathcal{P}} \rho(x_{a:b})$$

where $\Gamma_D(\mathcal{P}) := \Gamma_D(\mathcal{S}(\mathcal{P}))$ is the number of nodes in the partition tree $\mathcal{S}$ associated with $\mathcal{P}$ that have depth less than D.

Much like the CTW method, computing $\mathrm{P}_D^{\mathrm{PTW}}$ naively would take $O(2^{2^D})$ time. The PTW method shares a similar recurrence relation to the CTW method (4.5.2) that allows $\mathrm{P}_D^{\mathrm{PTW}}$ to be efficiently computed.

Theorem 5.3.6 (Recursive definition of PTW probability) For any depth D and sequence $x_{1:n}$ such that $n \leq 2^D$, we have that

$$\mathrm{P}_D^{PTW}(x_{1:n}) = \tfrac{1}{2}\rho(x_{1:n}) + \tfrac{1}{2}\mathrm{P}_{D-1}^{PTW}(x_{1:k})\mathrm{P}_{D-1}^{PTW}(x_{k+1:n}) \qquad (5.3.7)$$

where $k = 2^{D-1}$.

Proof. This follows a proof similar to Theorem 4.5.8. See [VWBG13].   ∎

By using (5.3.7), the PTW probability can be recursively computed in $O(nD)$ time and $O(nD)$ space by storing the context tree in memory. The following algorithm improves this to $O(D)$ memory (a critical improvement as many choices of base models are memory intensive) by exploiting the regular access pattern of the data structure. The algorithm works incrementally over the data sequence. For each bit, the context tree is traversed depth first, then iterated back up to update the weights.

Algorithm 5.3 Partition Tree Weighting $P_D^{PTW}(x_{1:n})$ [VWBG13]

Require: A tree depth parameter D
Require: A base probabilistic model ρ
Input: A data sequence $x_{1:n}$ of length $n \leq 2^D$
Output: $P_D^{PTW}(x_{1:n})$
1: **for** $0 \leq j \leq D$ **do**
2:    $b_j := 1, w_j := 1, r_j := 1$
3: **for** $t = 1$ to n **do**
4:    $i := \text{MSCB}_D(t)$
5:    $b_i := w_{i+1}$
6:    **for** $j = i + 1$ to D **do**
7:       $r_j := t$
8:    $w_D := \rho(x_{r_D:t})$
9:    **for** $i = D - 1$ to 0 **do**
10:       $w_i := \frac{1}{2}\rho(x_{r_i:t}) + \frac{1}{2}w_{i+1}b_i$
    return w_0

In Algorithm 5.3, $\text{MSCB}_D(t)$ (most significant changed context bit) is defined as the index i in which the D-bit binary representations $b_0 b_1 b_2 ... b_{D-1}$ of $t-1$ and $t-2$ differ, with $\text{MSCB}_D(1) := 0$ for all D. For example, for $d = 5$, to compute $\text{MSCB}_D(7)$ we compare the binary representation of 6 (`00110`) and 5 (`00101`) to find they differ at index 3, so $\text{MSCB}_D(7) = 3$.

The above algorithm can be modified to run incrementally, allowing for computation of $P_D^{PTW}(x_{1:n})$ given $P_D^{PTW}(x_{<n})$ in $O(D)$ time, by only running the inner loop. One of the downsides of PTW is that the depth D must satisfy $n \leq 2^D$ by the definition of binary temporal partition. To make the running time as fast as possible, we choose $D = \lceil \log_2(n) \rceil$, giving $O(n \log n)$ time and $O(\log n)$ space.

Moving on from the computational aspects, we are interested in how well the PTW algorithm predicts. As usual we will use redundancy as a measure of how good the PTW predictions are.

> **Theorem 5.3.8 (Upper bound on log PWT probability)** For all $n \in \mathbb{N}_0$, let $D = \lceil \log(n) \rceil$. Then for all $x_{1:n}$ and for all $\mathcal{P} \in \mathcal{B}_D$, we have
>
> $$-\log P_D^{PTW}(x_{1:n}) \leq \Gamma_D(\mathcal{P}) - \sum_{(a,b)} \log(\rho(x_{a:b}))$$

Proof. This follows straight by from Definition 5.3.5, dropping the sum, and taking the logarithm:

$$P_D^{PTW}(x_{1:n}) \equiv \sum_{\mathcal{P} \in \mathcal{B}_D} 2^{-\Gamma_D(\mathcal{P})} \prod_{(a,b) \in \mathcal{P}} \rho(x_{a:b}) \geq 2^{-\Gamma_D(\mathcal{P})} \prod_{(a,b) \in \mathcal{P}} \rho(x_{a:b})$$

■

Using this upper bound, we can compute the redundancy of PTW relative to any piecewise stationary data-generating source.

Theorem 5.3.9 (PTW redundancy) Let μ be a piecewise stationary data-generating source and let the base model ρ be such that the redundancy of ρ with $\mathcal{G}$, a class of bounded memory data-generating sources, be upper bounded by a non-negative, monotonically non-decreasing concave function $g : \mathbb{N}_0 \to \mathbb{R}$ with $g(0) = 0$. For all $n \in \mathbb{N}_0$, set $D = \lceil \log(n) \rceil$. There exists a constant c such that the redundancy of PTW is upper bounded by

$$\log\mu(x_{1:n}) - \log\mathrm{P}_D^{PTW}(x_{1:n}) \leq \left(2 + g\left(\left\lceil \frac{n}{|\mathcal{P}|(D+1)} \right\rceil\right)\right)|\mathcal{P}|(D+1)$$

Proof. From Lemma 2 in [VWBG13] we know that there exists a partition tree $\mathcal{P}' \in \mathcal{C}_d$ that is a *refinement* (the set of time indices where an existing segment ends in $\mathcal{P}$ is a subset of that for $\mathcal{P}'$) of $\mathcal{P}$ containing at most $|\mathcal{P}|(D+1)$ segments. We have from Theorem 5.3.8 that

$$\log\mu(x_{1:n}) - \log\mathrm{P}_D^{PTW}(x_{1:n})$$

$$\leq \log\mu(x_{1:n}) + \Gamma_D(\mathcal{P}') - \sum_{(a,b)\in\mathcal{P}'} \log(\rho(x_{a:b}))$$

$$= \Gamma_D(\mathcal{P}') + \sum_{(a,b)\in\mathcal{P}} \log\mu^{f(a)}(x_{a:b}) - \sum_{(a,b)\in\mathcal{P}'} \log(\rho(x_{a:b}))$$

$$= \Gamma_D(\mathcal{P}') - \sum_{(a,b)\in\mathcal{P}'} \log(\rho(x_{a:b})) + \sum_{(a,b)\in\mathcal{P}}\sum_{(c,d)\in\mathcal{P}'} \log\mu^{f(a)}(x_{c:d}|x_{a:c-1})$$

$$\leq \Gamma_D(\mathcal{P}') + \sum_{(a,b)\in\mathcal{P}'} g(b-a+1)$$

$$\leq \Gamma_D(\mathcal{P}') + |\mathcal{P}|g\left(\left\lceil \frac{n}{|\mathcal{P}|(D+1)} \right\rceil\right)(D+1)$$

$$\leq 2(|\mathcal{P}|(D+1) + |\mathcal{P}|g\left(\left\lceil \frac{n}{|\mathcal{P}|(D+1)} \right\rceil\right)(D+1).$$

The second inequality comes from the fact that g is defined as the redundancy of ρ. The third inequality comes from Jensen's inequality, which implies $\sum_{(a,b)\in\mathcal{P}'} g(b-a+1) \leq |\mathcal{P}'|g(n/|\mathcal{P}'|)$, then applying the inequality $|\mathcal{P}'| \leq |\mathcal{P}|(D+1)$. ■

This demonstrates that PTW is a computationally efficient algorithm with strong theoretical guarantees in the case where the truth is a piecewise stationary data-generating source.

5.4   Forget-Me-Not Process

The following explanation has been taken from [WSB+20, Sec.4]. Generalizing stationary algorithms to non-stationary environments is a key challenge in continual learning. The Forget-me-not (FMN) process is a probabilistic meta-algorithm tailored towards non-i.i.d., piecewise stationary, repeating sources. This meta-algorithm takes as input a single base measure ρ on target strings and extends the Partition Tree Weighting algorithm to incorporate a memory of up to k previous model states in a data structure known as a model pool. It efficiently applies Bayesian model averaging over a set of *postulated segmentations* of time (task boundaries) and a growing set $\mathcal{M}$ of stored base model states $\rho(\cdot|s)$ for some subsequences of $x_{1:n}$, while providing a mechanism to either learn a new local solution or adapt/recall previous learned solutions.

The FMN algorithm is derived and described in [MVK$^+$16]. It computes the probability $p' = \text{FMN}_d(x_{1:n}) \in [0,1]$ of a string of binary targets $x_{1:n} \in \{0,1\}^n$ of length n, e.g. x_t could be a binary class label. For this, it hierarchically Bayes-mixes an exponentially large self-generated class of models from a base measure ρ in $O(kn\log n)$ time and $O(k\log n)$ space, roughly as follows: For $n = 2^d$ and for $j = 0,...,d$, it breaks up string $x_{1:n}$ into 2^j strings, each of length 2^{d-j}, which conceptually can be thought of in terms of a complete binary tree of depth d. For each substring $x_{a:b}$, associated to each node of the tree will be a probability $\xi(x_{a:b})$ obtained from a Bayesian mixture of all models in the model pool $\mathcal{M}_a$ at time a. Taking any (variable depth) subtree (which induces a particular segmentation of time), concatenating the strings at its leaves gives back $x_{1:n}$, therefore the product of their associated mixture probabilities gives a probability for $x_{1:n}$. Doing and averaging this (see [MVK$^+$16]) for all possible $O(2^n)$ subtrees, which can be done incrementally in time $O(k\log n)$ per string element, gives $\text{FMN}_d(x_{1:n}|\rho)$.

The models in the model pool are generated from an arbitrary adaptive base measure ρ by conditioning it on past substrings $x_{a:b}$. For example, ρ could be a Beta-Bernoulli model whose weights are updated using Bayesian inference, or something more sophisticated. At time t, $\mathcal{M}_t$ contains at most k "versions" of ρ, with $\mathcal{M}_1 := \rho$. For $t = 2,...,n$, whenever a string $x_{a:b}$ with $b = t$ is encountered, then the model $\rho^* \in \mathcal{M}_a$ assigning the highest probability to the node's string $x_{a:b}$ is either added to the model pool, i.e. $\mathcal{M}_{t+1} = \mathcal{M}_t \cup \{\rho^*(\cdot|x_{a:b})\}$, or ignored based on a Bayesian hypothesis test criterion given in [MVK$^+$16].

5.5   Context Tree Maximization

Instead of a Bayesian mixture over all prediction suffix trees as CTW does, Context Tree Maximization selects the Maximum A Posteriori (MAP) tree (the symbolic index m stands for maximum):

$$\text{P}^D_\text{m}(x_{1:n}) = \max_{\mathcal{S} \in \mathcal{C}_D} \{2^{-\Gamma_D(\mathcal{S})} \text{P}_{\mathcal{S},KT}(x_{1:n})\}$$

which is identical to Theorem 4.5.11, just with the sum replaced by a maximum. (Here m is not a variable, but a name for the maximizing probability.) Taking minus the logarithm $\min_\mathcal{S}\{-\log\text{P}_{\mathcal{S},KT}(x_{1:n}) + \Gamma_D(\mathcal{S})\}$, we can interpret this also as an application of the Minimum Description Length (MDL) principle (Definition 2.7.24).

To compute this maximum naively we would need to consider all trees in $\mathcal{C}_D$, which would take time $O(2^{2^D})$ as there are that many trees of depth D. However, much like CTW there is an efficient method, called *Context Tree Maximization* [WTS00, NSH12], which can compute this in $O(nD)$ time. This is done through the following recurrence relation of the maximizing probability

$$\text{P}^D_{\text{m},s}(x_{1:n}) := \begin{cases} \frac{1}{2}\max\{\text{P}_{KT}(a_s,b_s),\ \text{P}^D_{\text{m},0s}(x_{1:n})\text{P}^D_{\text{m},1s}(x_{1:n})\} & \text{if } \ell(s) < D \\ \text{P}_{KT}(a_s,b_s) & \text{if } \ell(s) = D \end{cases}$$

and maximizing suffix set

$$\mathcal{S}^D_{m,s}(x_{1:n}) :=$$

$$\begin{cases} \mathcal{S}^D_{m,0s}(x_{1:n}){\times}\{0\} \cup \mathcal{S}^D_{m,1s}(x_{1:n}){\times}\{1\} & \text{if}\;\; \mathrm{P}_{\mathrm{KT}}(a_s,b_s) < \mathrm{P}^D_{m,0s}\mathrm{P}^D_{m,1s}\;\text{and}\;\ell(s) < D \\ \{\epsilon\} & \text{otherwise} \end{cases}$$

One can show that $\mathcal{S}^D_{m,\epsilon}$ and $\mathrm{P}^D_{m,\epsilon}$ are indeed the MAP tree and CTM distribution:

Theorem 5.5.1 (Context Tree Maximization) For any sequence $x_{1:n} \in \mathbb{B}^n$, we have

$$\mathrm{P}^D_{m,\epsilon}(x_{1:n}) = \max_{\mathcal{S} \in \mathcal{C}_D} \{2^{-\Gamma_D(\mathcal{S})}\mathrm{P}_{\mathcal{S},KT}(x_{1:n})\}$$

$$\mathcal{S}^D_{m,\epsilon}(x_{1:n}) = \arg\max_{\mathcal{S} \in \mathcal{C}_D} \{2^{-\Gamma_D(\mathcal{S})}\mathrm{P}_{\mathcal{S},KT}(x_{1:n})\}$$

5.6   Exercises

1. [C10] Show that the construction in Definition 5.3.3 is a bijection between $\mathcal{C}_D$ and $\mathcal{B}_D$. Convert the recursive definition of $\mathcal{C}_D$ to a direct recursive definition of $\mathcal{B}_D$. Show that $2^{-\Gamma_D(\mathcal{P})}$ is a valid prior, that is, prove that $\sum_{\mathcal{P}\in\mathcal{C}_D}\Gamma_D(\mathcal{P}) = 1$ for all $D \in \mathbb{N}_0$.

2. [C15] Prove that the sets of all binary temporal partitions $\mathcal{B}_D$ defined in Definition 5.3.3 and Definition 5.3.4 are indeed the same.

3. [C25] Prove Theorem 5.3.6.

4. [C20] Prove Theorem 5.5.1.

5. [C30i] While the Krichevsky–Trofimov (KT) estimator is commonly used as a base model, the Context Tree Weighting (CTW) variants can operate with any base model. Implement a selection of the variants (Adaptive CTW, PTW, CTS, FMN, and CTM) in this chapter, using other variants (or even CTW itself) as the base model. Analyze their performance relative to using the KT estimator as the base model and identify which combinations yield the best results.

6. [C35] Based on the combinations used in the previous exercise, deduce new redundancy bounds specifically tailored to these combinations.

7. [C20] Explain why we cannot indefinitely chain predictors as base models. For instance, CTS with a PTW base model, further supported by a CTW base model, and another CTS base model, and so forth.

8. [C30] From the previous exercise, validate your statement with a theoretical analysis of the gains/losses incurred each time this chaining process is executed. Further corroborate this claim with practical demonstrations by conducting experiments with this chaining procedure.

9. [C35] Formulate a multi-alphabet version of Adaptive CTW, PTW, CTS, FMN, and CTM. Hint: Use Section 4.6 and [PS99] as guiding references.

5.7   History and References

See Section 4.7.

Part III

A Family of Universal Agents

Chapter 6

Agency

Up until this point we have been discussing how to predict well. Such predictors, or *passive agents*, cannot affect the distribution over future symbols, and already have all information provided for prediction. For an artificial intelligence to be able to have an impact on its environment it needs to have agency. We desire a framework that can be used to model *active agents*, or simply *agents*, something that can interact with the environment, and for which the behavior of the environment depends not only on past events, but also on what the agent chooses to do. In this chapter we will go over how we formalize what an agent is, the environment it interacts with, and how to measure the performance of an agent in these environments, in terms of an aggregate sum of scalar rewards doled out by the environment. We explore how naively summing rewards can lead to undefined measures of success, and how this motivates time discounting (valuing the present more than the future). We show how the "goodness" of each situation the agent might find itself in can be defined via the value function, which can represented in a recursive way via the Bellman equation. Finally, we prove some properties about the value function.

Though we pursue a history-based approach, we still recommend the reader familiarize themselves with the standard presentation of reinforcement learning as a Markov decision process, as described in [SB18, Chp.3].

6.1   Policy and Environment

We first discuss the *cybernetic model*, which is commonly used in the domain of *reinforcement learning* as a formal framework about which we can reason. The cybernetic model comprises of two parts, the *agent* who observes what happens around it and makes decisions, and the *environment*, the world with which the agent interacts. The cybernetic model operates in discrete time, with the agent and environment taking turns. At time t the agent receives a *percept* e_{t-1} from the environment, and responds in turn with an *action* a_t. The environment receives the action, and issues a new percept e_t, and so on. Both percepts issued from the environment and actions taken by the agent may depend on the *history* (the sequence of action-percept interactions) before that action/percept was generated. Both the agent and environment are permitted to be stochastic.

Definition 6.1.1 (Stochastic function) A *stochastic function*, or *probability Kernel*, or *conditional distribution* f from A to a countable set B, denoted $f : A \to \Delta B$, is a function from A to the set of all probability distributions on B, denoted $\Delta B := \{p \in [0,1]^B : \sum_{b \in B} p_b \le 1\}$. We define $p(b|a) := f(a)_b$, the probability that the stochastic function p returns b given a as input. If x is sampled from a conditional distribution $p(\cdot|a)$, we write $x \sim p(\cdot|a)$.

Action, percept, observation, reward, history spaces.   Formally, we define $\mathcal{A}$ to be a set of *actions* from which the agent can choose, $\mathcal{O}$ to be the set of possible *observations*, and $\mathcal{R}$ to be the set of possible *rewards*. The percept e that the environment generates is an observation-reward pair (o,r). The set of all percepts is denoted as $\mathcal{E} := \mathcal{O} \times \mathcal{R}$. Unless otherwise stated we make the following global assumptions:

Assumption 6.1.2 (Finite action, observation, and reward spaces) We assume that the action, observation, and reward spaces $\mathcal{A}, \mathcal{O}, \mathcal{R}$ (and therefore $\mathcal{E}$) are finite. For simplicity we also assume that the rewards are bounded between 0 and 1.

The requirement $r_k \in [0,1]$ is often relaxed when presenting examples. For finite $\mathcal{R}$, the rewards are bounded, and can always be rescaled to be contained in $[0,1]$ via an affine transformation. This does not affect the optimal policy for that environment, though see Exercise 15.5 for a caveat.

Remark 6.1.3 (Infinite spaces)   Many considerations and statements could be generalized to infinite spaces, but from an AI perspective this does not lead to interesting additional insights, and only unnecessarily complicates the mathematical development. For countable spaces, max becomes sup and argmax has to be replaced by ε-optimal, and $\sum$ now needs convergence and exchangeability checks. Uncountable spaces usually come with some topology and σ-algebra, then some continuity and measurability assumptions are made, which allow to cover the space with countably many ε-balls and lift results from countable to uncountable spaces. $\sum$ becomes $\int$.   ●

A history is a finite sequence of action-percept pairs. We define the set of all *histories* to be $\mathcal{H} := (\mathcal{A} \times \mathcal{E})^*$. The agent chooses an action using a *policy*, which is a stochastic function from histories to actions, $\pi : \mathcal{H} \to \Delta \mathcal{A}$. An *environment* is a stochastic function from history plus last agent's action a percept, $\nu : \mathcal{H} \times \mathcal{A} \to \Delta \mathcal{E}$ (or $\Delta' \mathcal{E}$ for chronological semimeasures).

We will use $h_{<t} := h_{1:t-1}$ to represent the history up to time $t-1$, that is,

$$h_{<t} := a_1 e_1 a_2 e_2 ... a_{t-1} e_{t-1} = a_1 o_1 r_1 a_2 o_2 r_2 ... a_{t-1} o_{t-1} r_{t-1}$$

We will also sometimes use

$$a_{1:t} = a_1 a_2 ... a_t \quad and \quad e_{1:t} = e_1 e_2 ... e_t = o_1 r_1 ... o_t r_t = o r_{1:t}$$

Definition 6.1.4 (Cybernetic model) We say an *agent* interacts with an *environment* in cycles $t = 1, 2, ...$ In each cycle, the agent takes an action $a_t \sim \pi(\cdot | h_{<t})$ sampled from its *policy* given the *history* $h_{<t}$ as input. This action a_t is given to the environment, which generates a percept $e_t \sim \mu(\cdot | h_{<t} a_t)$ sampled from the environment distribution μ given the history $h_{<t}$ and the action a_t. The percept e_t is passed to the agent, and the next cycle $t+1$ begins. See Figure 6.1. The percept $e_t \equiv (o_t, r_t) \equiv o_t r_t \equiv o r_t$ consists of a regular *observation* o_t and a real-valued reward r_t.

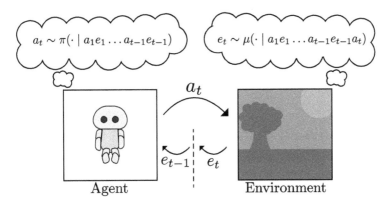

Figure 6.1: *An illustration of the cybernetic model (Definition 6.1.4). The percept e_{t-1} was generated from the last cycle. In the current cycle t, the agent takes action a_t, and the environment reacts in turn with percept e_t. Both agent and environment have access to the past sequence of actions and percepts.*

Agent-environment interaction. We usually denote the true underlying environment function by μ, and use ν to denote an arbitrary environment distribution. Usually μ is unknown to the agent π, and must be learned via interaction. Jointly ν and π create the history:

Definition 6.1.5 (Agent-environment measure ν^π) The interaction between an environment ν and the policy π induces a probability measure $\nu^\pi : \mathcal{H} \to [0,1]$ on histories (Technically it is a measure on cylinder sets, see Definition 2.2.11). Conditioning on past history $h_{<t}$, we have:

$$\nu^\pi(h_{t:m} | h_{<t}) := \prod_{k=t}^{m} \pi(a_k | h_{<k}) \nu(e_k | h_{<k} a_k)$$

Remark 6.1.6 (Lifting sequence prediction to interacting agents) Many definitions and results to follow will mirror those in the sequence prediction setting. Definition 6.1.5 is just the chain rule $\rho(x_{1:n}) = \prod_{t=1}^{n} \rho(x_t|x_{<t})$ for $\mathcal{X} = \mathcal{A} \times \mathcal{E}$ and further factorizing $a_k e_k$ and ρ into π and ν (Alternatively we could view $h_{1:m}$ as a sequence of length $2m$ over $\mathcal{A}$ alternating with $\mathcal{E}$.) For one-step prediction and fixed π, all concepts and results for ξ, essentially transfer to ξ^π (e.g. Lemma 7.2.4). When comparing different policies of farsighted agents, matters become much more intricate.   ●

We also use $\mathrm{P}_\nu^\pi[Q|h_{<t}]$ to denote the probability of some predicate Q on histories occurring, where the future histories $h_{t:m}$ given past histories $h_{<t}$ are sampled from $\nu^\pi(h_{t:m}|h_{<t})$. Similarly we define the (conditional) expectation $\mathbf{E}_\nu^\pi$ to be the expectation with respect to the (conditional) probability measure P_ν^π induced by ν^π.

Definition 6.1.7 (Probability measure $\mathbf{P}_\nu^\pi$ and expectation $\mathbf{E}_\nu^\pi$) Let $\mathrm{P}_\nu^\pi[Q|h_{<t}]$ denote the probability of some measurable predicate or event $Q \subseteq (\mathcal{A} \times \mathcal{E})^\infty$, where the future histories $h_{t:m}$ given past histories $h_{<t}$ are sampled from $\nu^\pi(h_{t:m}|h_{<t})$. Similarly (conditional) expectation $\mathbf{E}_\nu^\pi[f|h_{<t}]$ of some measurable function $f : (\mathcal{A} \times \mathcal{E})^\infty \to \mathbb{R} \cup \{\pm\infty\}$ is the expectation with respect to the (conditional) probability measure induced by ν^π. For events $Q_m = Q'_m \times (\mathcal{A} \times \mathcal{E})^\infty$ with $Q'_m \subseteq (\mathcal{A} \times \mathcal{E})^m$ and functions $f_m : (\mathcal{A} \times \mathcal{E})^m \to \mathbb{R}$, explicit expressions are

$$\mathrm{P}_\nu^\pi[Q_m|h_{<t}] = \sum_{h_{t:m}:h_{1:m}\in Q_m} \nu^\pi(h_{t:m}|h_{<t}) \quad \text{and} \quad \mathbf{E}_\nu^\pi[f_m|h_{<t}] = \sum_{h_{t:m}} \nu^\pi(h_{t:m}|h_{<t}) f(h_{1:m})$$

We have $\mathrm{P}_\nu^\pi[h_{1:m}|h_{<t}] = \nu^\pi(h_{t:m}|h_{<t})$ by definition, which proves the explicit expressions. The existence and uniqueness of the (conditional) measure P_ν^π for all events Q then follows from the *Carathéodory Extension Theorem* [Bil95].

The policy and the environment interact with each other and create a history as choices made by the policy affect what future percepts (observation,rewards) the environment will generate. This setup is called *history-based reinforcement learning* or *general reinforcement learning*. We consider the scenario where the dynamics of the environment are unknown to the agent, and the agent must learn the underlying stochastic function μ that governs the environment, while also trying to perform well, that is, take actions to cause the environment to return large rewards.

Exploration-exploitation tradeoff. For the agent to learn μ it must sometimes take suboptimal actions which may not immediately maximize the expected reward based on the agent's current belief about the environment it interacts with, but will allow the agent to have more confidence about which environment it is in. For the agent to perform well in the environment means to maximize the (expected future) reward. These are often conflicting goals. The problem of how to compromise between the two is often referred to as the exploration (improving the current belief about what environment the agent is in) vs. exploitation (maximizing reward based on what environment the agent currently believes it is in) problem. An agent that explores too much will act suboptimally (it will always be trying to improve its confidence in what the environment is rather than seek higher reward), and so will an agent that exploits too much (as it does not fully understand the dynamics of the environment, it will not know how to actually act optimally).

6.2 Assigning Rewards

We also need to define what it means for an agent to perform well. Later in Chapter 15 we will discuss different ways the agent might measure performance using *utility functions*, but for now we will focus solely on agents with the goal to maximize the expected value of the sum of all future rewards.

Assumption 6.2.1 (Reward Hypothesis)   "all of what we mean by goals and purposes can be well thought of as maximization of the expected value of the cumulative sum of a received scalar signal (called reward)." [SB18]

For example, given an agent playing games of chess, the set of actions available are legal chess moves, the percept is the current state of the board, and the environment takes the current state of the board, and returns a new board where the opposing pieces have made a legal move (essentially hiding the opponent player in the environment). The environment's moves could be chosen by an already existing chess bot, or by self-play (the color of the pieces are swapped, and the agent also chooses actions for the opponent). The rewards for this environment could be 1 if the previous action taken by the agent checkmated the enemy king, -1 if the new board generated by the environment places the agent's king in checkmate, and 0 otherwise. This reward scheme aligns with the correct behavior we want the agent to perform: to maximize the expected future reward sum, the agent should take actions that lead to a higher confidence of being able to win the game, and avoid actions that would lead to a loss. But the rewards here are *sparse*, in that the agent receives rewards very rarely, only when the game terminates, and so it may take a very large number of interaction cycles for the agent to learn to play chess well.

One could imagine the difficulty of trying to learn to play chess by being sat down in front of a board, not knowing the rules or strategy of the game (or even what the goal of the game is!) and then randomly playing moves. The only feedback is a scolding if you lose (or play an illegal move), and praise if you win. One would not be surprised that even learning the rules of the game this way would be difficult, let alone becoming proficient at the game, though see [Hut05b, Sec.6.3.4] for a strong counter-argument.

One potential solution to this is to *shape* the reward function to offer some small rewards for observations that we expect to be correlated strongly with achieving the goal. For example: capturing pieces, avoiding having your pieces captured, friendly pieces having a large degree of freedom of where to move, controlling the center of the board, and pinning enemy pieces are all strategies that positively correlate with winning.

Bostrom calls these sub-goals *instrumental goals*, goals that often correlate with success and can be used as a (often crude) proxy in service of the *terminal goal* or *final goal*. In this case, the terminal goal for an agent playing chess is to win the game; all else is secondary, it does not matter to the agent how much material or territory control is lost if a checkmate can be secured.

Shaping the reward can speed up learning, but can also potentially lead to a situation where the agent learns to "game" the shaped reward, by exploiting the rewards for actions that correlate with the goal, rather than trying to achieve the goal itself. For example, if there was too large a reward associated with territorial control, the agent might try to move pieces to lock down a large portion of the board and play too defensively, rather than making moves with the objective of winning the game. This is called *reward misspecification*. (see Chapter 15 and [LMK+17] for more.)

There is also the problem of how to teach the agent the rules of the game. The framework presented here requires the agent to always select an action from a fixed action space $\mathcal{A}$, but in chess the set of legal actions can vary from move to move. This presents a dilemma: We would want the agent to be able to learn how to play chess from interaction with the environment alone, rather than hardcoding the rules of the environment into the agent, since the latter would make the agent specific to a particular environment, and subvert the purpose of *universal* artificial intelligence. On the other hand, with a fixed action space $\mathcal{A}$ containing all legal chess moves as a subset, many of those moves would be illegal, depending on the current state of the board. One solution to this could be to allow illegal moves to be part of the game, and whenever an illegal move is made, a reward at least as bad as a loss is issued to the agent. For games of a fixed length, the penalty from an illegal move can take a while to learn from, as it may be hard for the agent to distinguish illegal moves from legal but suboptimal moves. A better approach is to terminate the game immediately upon playing an illegal move, and issue a penalty. When the agent first begins interaction with the environment, it will likely take many illegal actions (and receive the associated penalty) until it stumbles across a legal move, and receive no penalty. In this way, the agent can quickly learn what moves are legal, and implicitly learn the rules of chess.

It is important that the reward of making an illegal move is at least as bad as a loss; otherwise the agent could escape a doomed position by making an illegal move.[1]

6.3   (PO)MDP vs. History RL

Markov Decision Process (MDP).   Our setup is quite different from the traditional MDP (Markov Decision Process) setup where the environment is assumed to be *Markov*, that is, the environment μ has the property that it only depends on the last observation produced and action taken.

$$\mu(e_t|h_{<t}a_t) = \mu(e_t|o_{t-1}a_t)$$

In the MDP framework, the observations are usually called *states* denoted s_t, but we keep o_t to ensure consistency with the more general history-based framework. Hence for MDPs, $\mu\colon\mathcal{O}\times\mathcal{A}\to\Delta\mathcal{E}$. Since the past history is irrelevant to the behavior of the environment, the optimal action will also not depend on it, so it suffices to consider Markov policies.

$$\pi(a_t|h_{<t}) = \pi(a_t|o_{t-1})$$

To clarify, the optimal agent π^* for any Markov environment is Markov, but the behavior of an agent that learns to optimize a policy out of the set of all Markov policies (like Q-learning) is itself non-Markov, as the Q-value estimates are constructed from past experiences. In the MDP framework, the policy may receive the same observation in multiple different time steps, placing it essentially in the same situation as a past interaction: a Markov policy is agnostic of how much time has passed, or what the history was up till this point (though the optimization process that chose this policy may be aware of the history so far). Most work and major results in reinforcement learning takes place in the MDP framework [SB18]. The Markov assumption also means that the value function depends only on the current observation (state), the choice of policy, and the MDP dynamics. In the finite MDP case, the same state is visited repeatedly, which makes it easier to devise learning algorithms.

Given a Markov environment μ, the classic presentation of the value function V^π is defined as the expected sum of geometrically discounted future rewards, given a discount

---

[1]For a similar reason, the penalty for humans cheating in a game is usually higher than that of a loss, as otherwise cheating would be a rational action from an otherwise lost position.

factor $\gamma \in [0,1)$ (more on discounting in Section 6.4):

$$V_{\mu,\gamma}^{\pi}(o_{t-1}) := \mathbf{E}_{\mu}^{\pi}\left[\sum_{k=0}^{\infty}\gamma^k r_{t+k}\,\middle|\,o_{t-1}\right] = \sum_{a_t}\pi(a_t|o_{t-1})\sum_{o_t r_t}\mu(o_t r_t|o_{t-1},a_t)\big(r_t+\gamma V_{\mu,\gamma}^{\pi}(o_t)\big)$$

where the second expression is the familiar recursive Bellman equation. More details can be found in Chapter 14, which shows how to reduce the history-based framework to the MDP setting, remarkably even if the Markov property is *not* satisfied.

Partially Observable Markov Decision Process (POMDP). A more general class of environments are *Partially Observable Markov Decision Processes* (POMDPs), where the environment is still Markov, conditioned on the current state, but the agent only receives an observation based on the state but not the state itself. Many states may lead to the same observation, so the agent needs to infer what state the environment is likely to be in based on its observations so far, and act accordingly. Such environments might include an agent trying to navigate a maze, where the observation is the nearby walls, and the state is the true location of the agent. The walls adjacent to the agent give it some information about its location, but not enough to uniquely determine its location. The Cheese Maze in Figure 12.6 has 11 states=locations mapped to only 5 different observations.

History-based RL. In history RL, any observation arbitrarily far into the past may be potentially relevant for deciding the best action to take now. As a result, no situation ever repeats exactly in history-based RL, which renders most existing RL algorithms unsuitable. We have chosen this history-based setup because general agents need to be able to handle such general environments. Indeed, this is the most general RL setup, with both the MDP and the POMDP frameworks being special cases of it. Figure 7.1 details a taxonomy of many RL environments, all of which are special cases of the history-based RL framework.

We note that environments ν are *chronological*, in the sense that percepts received at time t depend only on histories taken at time $\leq t$. In the way we have defined semimeasures, they are chronological by definition, but in other presentations [Lei16b] authors may write $\nu(e_{1:t}\|a_{1:t})$ to denote the probability that ν assigns to $e_{1:t}$ conditioned on $a_{1:t}$, while respecting the chronological requirement that the probability assigned to a percept e_t cannot depend on actions taken after time step t, that is, future actions cannot affect past percepts. Chronological policies are defined in the same way:

Definition 6.3.1 (Chronological semimeasures) Chronological environments/policies are semimeasures ν/π of history $h_{1:t} \equiv a_1 e_1 ... a_t e_t$ that for all t satisfy

$$\nu(e_{1:t}\|a_{1:t}) = \prod_{i=1}^{t}\nu(e_i|h_{<i}a_i) \quad \text{and} \quad \pi(a_{1:t}\|e_{<t}) = \prod_{i=1}^{t}\pi(a_i|h_{<i})$$

Note that on the other hand, *any* joint semimeasure $\rho: \mathcal{H} \to [0,1]$ over actions *and* percepts without any restrictions can be decomposed as $\rho(h_{1:t}) =$

$$\nu(e_{1:t}\|a_{1:t})\pi(a_{1:t}\|e_{<t}) = \nu^{\pi}(h_{1:t}) \tag{6.3.2}$$

We call a policy *deterministic* if it only takes on values of 0 or 1. If a policy is deterministic, we abuse notation and define $\pi(h_{<t}):=a_t$ where a_t is the unique action such that $\pi(a_t|h_{<t})=1$ (implicitly assuming the type of a deterministic policy to be $\pi:\mathcal{H}\to\mathcal{A}$). Deterministic environments are defined in the same fashion, and similarly we write $\nu(h_{<t}a_t)=e_t$ if e_t is the unique percept that ν assigns probability 1 given history $h_{<t}$.

6.4   Time Discounting

All else being equal, humans prefer to receive rewards now rather than later. If we believe that humans are rational agents, we may expect the same to hold for artificial agents. We have discussed how an agent receives reward from the environment, so if we want an agent which maximizes the expected sum of all future rewards, then over an infinite number of time steps our agent may receive infinite reward. However, an agent could perform suboptimal actions (actions which are not optimal at that history/time step) on each time step with e.g. only half the optimal reward per time step, which over infinite time steps is still infinite reward. Alternatively, the agent could procrastinate on its task for a few thousand years, and then play optimally thereafter, and also receive infinite reward. To the agent, each of these strategies are equally valuable, as they all lead to infinite reward, even though we would obviously not prefer the agent to procrastinate or choose actions that receive "less" reward. Indefinite procrastination can even happen for finite total reward (Figure 6.3).

One approach to fix this problem is to give the agent a finite lifespan, called a *horizon*. The agent will then maximize reward only over its finite lifespan. This approach may however make the agent short-sighted (if the horizon is short) or may not be suitable for continuing tasks, or when the lifespan is not known *a priori*. Another approach is to discount the reward over time, that is, make rewards later worth less than rewards now. We do this by multiplying rewards with a discount function.

Definition 6.4.1 (Discount function) A discount function is a function $\gamma : \mathbb{N}^+ \times \mathbb{N}^+ \rightarrow \mathbb{R}$ with $\gamma(k,t) \geq 0$ and $\forall t: \sum_{k=1}^{\infty} \gamma(k,t) < \infty$. t represents the current present time step during interaction, while $k \geq t$ represents the future time step the agent reasons about at time t. The corresponding discount normalization factor is defined as $\Gamma_t := \sum_{k=t}^{\infty} \gamma(k,t)$ (Figure 6.2). Often, γ will have no dependency on t, in which case we write $\gamma_k := \gamma(k,t)$ and $\Gamma_t = \sum_{k=t}^{\infty} \gamma_k < \infty$.

For example, an agent at current time step $t=3$ would discount a reward r_7 four time steps further into the future ($k=7$) using the discount $\gamma(7,3)$. Discount functions that do depend on t can be *time inconsistent*, see Section 6.5.

The choice of discount function determines an effective horizon, that is, how far into the future the agent considers rewards when choosing actions.

Definition 6.4.2 (Effective horizon) The ε-effective horizon $H_t(\varepsilon)$ at time t is the minimum number of time steps k into the future such that the discount factor Γ_{t+k} is less than an ε fraction of the discount normalization factor Γ_t at the present:

$$H_t(\varepsilon) := \min_k \left\{ k \left| \frac{\Gamma_{t+k}}{\Gamma_t} \leq \varepsilon \right. \right\}$$

The ε-effective horizon is useful for talking about how quickly the discount function $\gamma(\cdot)$ discounts rewards. The *effective horizon*, defined as $H_t(1/2)$, can be thought of as a kind of tipping point, the earliest point in time from which all the potential reward from time step $t + H_t(1/2)$ onwards is valued less than half of all the potential reward from the present time step t.

Discount function examples.   Those familiar with the traditional presentation of reinforcement learning using MDPs [SB18, Chp.3] will likely be aware of the commonly used *geometric discount* function $\gamma_k = \gamma^k$ for some $\gamma \in [0,1)$. One interpretation of this discount

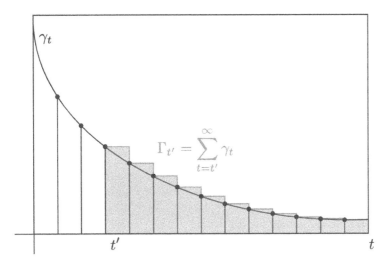

Figure 6.2: *A graphical representation of the discount function γ_t and normalizer $\Gamma_{t'}$ (Definition 6.4.1) at time $t = t'$.*

is that it is like the agent having a probability $1 - \gamma$ of dying[2] at each time step. The geometric discount is also mathematically elegant, having the property that $\Gamma_{k+1} = \gamma \Gamma_k$, which is convenient for MDP RL. Another choice of discounting is *finite lifetime*, where the agent does not care about the reward after a certain point m, that is, $\gamma_k = [\![k \leq m]\!]$. This is equivalent to the agent having a finite lifespan of m. This is effectively the case in RL with *episodic environments*, where the interaction with the environment resets to some initial configuration after at most every m time steps. Each episode is independent of the others from the perspective of the environment, but the agent may learn from previous episodes, e.g. different games of chess against new opponents. While finite lifetime discounting can be motivated in circumstances where the lifetime of the agent is known (some games end after a fixed number of turns), or by a physical argument (perhaps the agent is deployed with a fixed expiry date until it is replaced, or on a more extreme scale, there is fixed time until the heat death of the universe), it is useless for asymptotic analysis, as past the point $k > m$ all policies are optimal (or perhaps more accurately, the choice of policy beyond time step m is irrelevant). Similar to finite lifetime discounting is *moving-horizon discounting*, where the agent does not care about rewards beyond a finite fixed horizon m, but the horizon itself moves with the current time step t, that is, $\gamma(k,t) = [\![t \leq k \leq t+m]\!]$.

6.5   Time Consistency

It seems sensible that once the best course of action by which an agent will maximize expected future discounted rewards has been determined, this plan should not change once the future becomes the present (barring cases where the agent has more information than when the plan was made). Discounts that satisfy this property are called *time-consistent* discounts.

---

[2]Here, death could be represented as the agent forever receiving the percept $(o_{\text{dead}}, 0)$ for some dummy observation o_{dead} that would never otherwise be observed. For more on how death for RL agents can be defined, see Chapter 15.

Theorem 6.5.1 (Time-consistent discounts) A discount function $\gamma(\cdot,\cdot)$ is *time-consistent* if it has no dependence on the present time step t. That is, $\gamma(k,t)=\gamma_k$ for some function $\gamma:\mathbb{N}^+\to\mathbb{R}$.

Example 6.5.2 ((Un)healthy dinner) Humans are often time-inconsistent (children usually plan over shorter timescales than adults), and the discount may even vary on timescales as short as a few hours. For example, a person might convince themselves early in the day to make something healthy for dinner that night. This will incur negative rewards that evening due to the effort involved in making dinner, but greater rewards in the long term due to health benefits. Later that evening, they might shift to a more myopic discount, give up on making dinner and order take-out instead, which gives large rewards immediately (take-out is delicious!) but less rewards in the long term, due to it being less healthy and costing more money than homemade dinner. ♦

Both geometric and finite lifetime discounting are time consistent, but constant horizon discounting $\gamma(k,t)=[\![t\leq k\leq t+m]\!]$ is not. This can lead to undesirable behavior where the agent will forever delay current rewards in the hope to receive more reward later. In doing so, no reward is ever received. Consider the environment ν_{delay} in Figure 6.3 [LH14b].

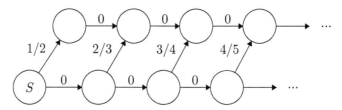

Figure 6.3: *An environment ν_{delay} where the agent can choose between moving up to receive the associated reward, or moving right to obtain more reward later. A reward-maximizing agent with undiscounted infinite horizon or finite moving horizon will delay reward forever and thus receive zero total reward.*

Example 6.5.3 (Immortal and time-inconsistent agents) An agent interacting with ν_{delay} starts at the state marked S. Given a moving horizon of $m=2$, the agent sees that it could move up for reward $1/2$, or move right and then up for reward $2/3$, which are the only rewards reachable in two time steps. Assuming the agent is following a policy that maximizes the value function, the agent concludes the best action is to move right. But when it does, the horizon moves, and now it considers moving up for $2/3$, or right and up for $3/4$. The agent will proceed with this line of reasoning forever, always delaying potential reward now for more reward later. As a result, the agent always moves to the right, and receives reward zero forever. So the choice of a moving horizon means the agent follows the worst possible policy for this environment. This remains true for any moving horizon $m\geq 2$ and infinite horizon.

Contrast this with (any) finite lifetime m, where the agent will walk $m-1$ steps to the right and then up to receive an expected discounted reward sum of $\frac{m}{m+1}$. Similarly with geometric discounting for some $\gamma\in[0,1)$, the agent will choose to walk k^*-1 steps right, and then on time step k^*, walk up, where $k^*=\operatorname{argmax}_{k\geq 1}\{\gamma^{k-1}\frac{k}{k+1}\}<\infty$. ♦

Table 6.4 contains a comparison of several choices of discount function. Based on experiments from psychology where humans are given the choice of some money now or more money later, humans tend to discount hyperbolically [FLO02] using the discount

Table 6.4: *A table of various discounts γ_k, together with their effective horizons $H_k(\varepsilon):=$ $\min\{t:\Gamma_{k+t}\leq\varepsilon\Gamma_k\}$, and normalization factors $\Gamma_k:=\sum_{i=k}^{\infty}\gamma_i$, where k is planning time and t is current time. All are time-consistent except 'moving horizon'.*

Discount	Param.	γ_k	$H_k(\varepsilon)$	Γ_k
geometric	$\gamma\in[0,1)$	γ^k	$\frac{\log\varepsilon}{\log\gamma}$	$\frac{\gamma^k}{1-\gamma}$
finite life	$m\in\mathbb{N}_0$	$[\![k\leq m]\!]$	$\lceil(1-\varepsilon)(m-k+1)\rceil$	$\max\{m-k+1,0\}$
moving horizon	$m,t\in\mathbb{N}_0$	$\gamma(k,t)=$ $[\![t\leq k\leq t+m]\!]$	$\lceil(1-\varepsilon)(t+m-k+1)\rceil$	$\begin{cases}0 & k>t+m \\ t+m-k+1 & t\leq k\leq t+m \\ m+1 & k<t\end{cases}$
power	$\delta>0$	$\frac{1}{k^{1+\delta}}$	$\approx\left(\frac{1}{\varepsilon^{1/\delta}}-1\right)k\propto k$	$\frac{1}{\delta k^{\delta}}$
harmonic	$\delta>0$	$\frac{1}{k(\ln k)^{1+\delta}}$	$\approx k^{\varepsilon^{-1/\delta}}$	$\approx\frac{1}{\delta(\ln k)^{\delta}}$
universal	-	$2^{-K(k)}$	increases faster than any comp. function	decreases slower than any computable function
no discount	-	1	∞	∞

$\gamma(k,t)=(k-t+c)^{-1}$, but this is undesirable, since this leads to time-inconsistent policies. Also, the sum $\sum_{i=k}^{\infty}\gamma_i$ diverges, so this choice of discount does not solve the infinite reward problem, though replacing exponent -1 by $-1-\delta$ for some small δ solves the latter problem.

The universal discount and its monotone variant $\gamma_k=\min_{i\leq k}2^{-K(i)}$ are the "best" in the sense that they lead to the most far-sighted agent while ensuring that Γ_k does not diverge to infinity. Obviously, being dependent on the Kolmogorov complexity K results in an incomputable discount factor.

6.6   Value Functions

Now that we have defined what an agent and its policy is, environments and how the agents discount future reward, we can finally define the value function for an agent. The value function is defined as the γ-discounted expected future reward sum the agent receives from the environment given the interaction history.

Definition 6.6.1 (Value function) The value $V_{\nu,\gamma}^{\pi,m}:\mathcal{H}\to\mathbb{R}$ of a policy π in an environment ν for discount function γ_t given a history $h_{<t}\in\mathcal{H}$ and horizon $m\geq t$ is defined as

$$V_{\nu,\gamma}^{\pi,m}(h_{<t}) := \frac{1}{\Gamma_t}\mathbf{E}_{\nu}^{\pi}\left[\sum_{k=t}^{m}\gamma_k r_k\,\big|\,h_{<t}\right] = \frac{1}{\Gamma_t}\sum_{h_{t:m}}\nu^{\pi}(h_{t:m}|h_{<t})\sum_{k=t}^{m}\gamma_k r_k$$

If $t>m$ or $\Gamma_t=0$ we define $V_{\nu,\gamma}^{\pi,m}(h_{<t}):=0$. We will often drop the discount γ and write $V_{\nu,\gamma}^{\pi,m}$ as $V_{\nu}^{\pi,m}$, since usually the discount function is fixed. The optimal value is defined as

$$V_{\nu}^{*,m}(h_{<t}) := \sup_{\pi}V_{\nu}^{\pi,m}(h_{<t})$$

The set of optimal policies with respect to that value and a representative are defined as

$$\Pi_{\nu}^{*,m}(h_{<t}) := \operatorname*{argmax}_{\pi}V_{\nu}^{\pi,m}(h_{<t}) \quad\text{and}\quad \pi_{\nu}^{*,m}(\cdot|h_{<t}) \in \Pi_{\nu}^{*,m}(h_{<t})$$

We drop the history argument when it is the empty history: $V_\nu^{\pi,m} := V_\nu^{\pi,m}(\epsilon)$. We also define all quantities for the limit $m \to \infty$, which exists if ν is a measure, since then V^m increases with m, and drop superscript m in this case, e.g. $V_\nu^\pi := V_\nu^{\pi,\infty} := \lim_{m\to\infty} V_\nu^{\pi,m}$, etc.

Note that the m-truncated value can be recovered from the $m = \infty$ value by truncating the discount function: $V_{\nu,\gamma}^{\pi,m}(h_{<t}) = V_{\nu,\gamma'}^{\pi,\infty}(h_{<t})$, where $\gamma_k' := \gamma_k [\![k \le m]\!]$.

Remark 6.6.2 (Time-consistency) We can now formally define the notion of time-consistency discussed in Section 6.4. First, technically in Definition 6.6.1 we should have written $\pi_{\nu,h_{<t}}^{*,m} \in \Pi_\nu^{*,m}(h_{<t})$, with $\pi_{\nu,h_{<t}}^{*,m}$ being arbitrary or undefined on histories that do not start with $h_{<t}$. Due to this we may as well only consider $\pi_{\nu,h_{<t}}^{*,m}(\cdot|h_{<k})$ for $k \ge t$, and hence can drop the redundant index $h_{<t}$ and write $\pi_{\nu,t}^{*,m}(\cdot|h_{<k})$.

Time-consistency is the fact that an initially optimal policy $\pi_{\nu,\epsilon}^{*,m} \in \Pi_\nu^{*,m}(\epsilon)$ remains optimal later, i.e. $\pi_{\nu,\epsilon}^{*,m}(\cdot|h_{<t}) \in \Pi_\nu^{*,m}(h_{<t})$. Written differently,

$$\{\pi(\cdot|h_{<t}) : \pi \in \Pi_\nu^{*,m}(h_{<t})\} = \{\pi(\cdot|h_{<t}) : \pi \in \Pi_\nu^{*,m}(\epsilon)\}$$

and indeed more generally $\Pi_\nu^{*,m}(h_{<k}) \supseteq \Pi_\nu^{*,m}(h_{<t})$ for $k \ge t$. Hence we may drop index t from $\pi_{\nu,t}^{*,m}$ altogether, which we did, which generalize γ_k to $\gamma(k,t)$, this can be violated, i.e. an optimal policy at t may be not be optimal at k anymore, as demonstrated in Section 6.4.

●

Lemma 6.6.3 (Explicit Value function) More explicit representations of the value functions $V_{\nu,\gamma}^{\pi,m}$ and $V_{\nu,\gamma}^{*,m}$ are as follows:

$$V_{\nu,\gamma}^{\pi,m}(h_{<t}) = \frac{1}{\Gamma_t} \sum_{a_t} \sum_{e_t} \cdots \sum_{a_m} \sum_{e_m} \left(\prod_{k=t}^m \pi(a_k|h_{<k})\nu(e_k|h_{<k}a_k) \right) \left(\sum_{k=t}^m \gamma_k r_k \right)$$

$$= \frac{1}{\Gamma_t} \sum_{a_t \in \mathcal{A}} \pi(a_t|h_{<t}) \sum_{e_t \in \mathcal{E}} \nu(e_t|h_{<t}a_t) \cdots \sum_{a_m \in \mathcal{A}} \pi(a_m|h_{<m}) \sum_{e_m \in \mathcal{E}} \nu(e_m|h_{<m}a_m) \sum_{k=t}^m \gamma_k r_k$$

$$V_\nu^{*,m}(h_{<t}) = \frac{1}{\Gamma_t} \max_{a_t \in \mathcal{A}} \sum_{e_t \in \mathcal{E}} \cdots \max_{a_m \in \mathcal{A}} \sum_{e_m \in \mathcal{E}} \prod_{k=t}^m \nu(e_k|h_{<k}a_k) \sum_{k=t}^m \gamma_k r_k$$

Proof. Exercise to the reader.   ∎

Example 6.6.4 (Gridworld) Gridworlds are a common Markov environment for reinforcement learning. The example depicted in Figure 6.5 is a modified version of an example from [RN20]. Here, $\mathcal{A} = \{\to,\uparrow,\downarrow,\leftarrow\}$, $\mathcal{O}$ is all the cells in the grid, and $\mathcal{R} = \{0,-1,1\}$. From each cell, the agent can move in one of the four cardinal directions (unless it would run into the wall, denoted by the black cell), or fall off the grid, in which case nothing happens. The percept received from the environment is an (observation, reward) pair, where the observation is the next cell the agent would move into, and the reward is always zero, unless the agent moves into the cells marked with the rewards $+1$ or -1 respectively. In this case, the agent receives the appropriate reward, followed by dummy observations forever with zero reward (essentially encoding that the interaction has finished).

In this environment, the optimal policy is obvious (Figure 6.5b) as the agent should follow the shortest path from its current cell to the $+1$ cell avoiding the -1 cell.

Suppose now that the environment was stochastic, and when the agent attempts to move in a direction, 70% of the time it is successful and moves in the given direction, and 30% of the time it accidentally slips (with equal probability) in one of the other three directions it wanted to move in. If the agent would walk outside the grid or walk into the wall, its location is unchanged. The reward for walking into cells other than the $+1$ and -1 cells is now a penalty ε, for some small $\varepsilon < 0$ (with the idea being to incentivize the agent to reach the goal as quickly as possible). The optimal policy is now dependent on how large ε is. For $\varepsilon = -0.02$ (Figure 6.5a) the agent will act cautiously, and walk around the long way to avoid accidentally falling into the -1 cell. With a harsher penalty $\varepsilon = -0.1$, the agent is incentivized to risk walking past the -1 cell to move to the $+1$ cell (Figure 6.5b), acting the same as the deterministic agent did, but now with the danger that it might "slip" into the -1 cell by taking a shortcut. For an extreme value $\varepsilon = -2$, (Figure 6.5c) life for the agent is so unbearable that it will move towards any terminal cell in as few steps as possible. If we flip this and give the agent a positive reward $\varepsilon = 0.01$ (Figure 6.5d) on every time step, the agent will avoid the terminal cells entirely, hiding behind the wall on the other side of the grid to drag out the episode for as long as possible and thereby maximize the return. ◆

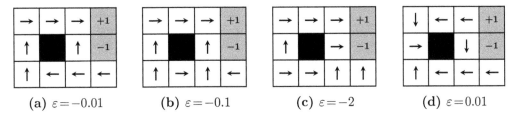

(a) $\varepsilon = -0.01$ (b) $\varepsilon = -0.1$ (c) $\varepsilon = -2$ (d) $\varepsilon = 0.01$

Figure 6.5: *Optimal policies for the gridworld environment in Example 6.6.4, for various choices of penalty ε.*

Example 6.6.5 (Rock-paper-scissors) Consider the game of rock-paper-scissors. The environment is an opponent with an unknown distribution on how it chooses rock, paper or scissors. The observation and action spaces are $\mathcal{O} = \mathcal{A} = \{\texttt{rock,paper,scissors}\}$, and $\mathcal{R} = \{-1,0,1\}$, with -1 for a loss, 0 for a draw, and 1 for a win. The agent receives the opponent's previous choice of action as an observation, and the environment does not condition on the most recent action taken, so neither the agent nor the environment is aware of each other's actions during this turn, so they can't cheat and always choose the correct response.

On the face of it, this looks like a boring game, as the best policy would be to randomly choose each action with $1/3$ probability. If the environment is also generating each move with probability $1/3$ then this is optimal,[3] but if the environment generates percepts with a bias (perhaps it chooses rock more often than other options), then a better strategy would be to try and learn what the environment is biased towards, and act accordingly.

If the environment is also an adversarial player who is trying to model the agent and predict what move it will make, the agent will need to be one step ahead, learn this property of the environment, and then outsmart it by playing the right moves. Such adversarial games are better studied in the field of *game theory*, see Section 10.2. ◆

Policies can be stochastic or deterministic. It might seem that at first glance, stochastic policies would be stronger than deterministic policies alone (given deterministic policies are a subset of stochastic). As we will see soon, deterministic policies are sufficient for choosing actions optimally.

---

[3] In fact, any strategy is optimal against a uniformly random opponent (including "always play rock").

Remark 6.6.6 (Infinitely far-sighted agents may act poorly) Optimal policies always exist if m is finite, since in this case there are only finitely many policies and the $\sup_\pi$ in Definition 6.6.1 reduces to a $\max_\pi$. One can show that for $m \to \infty$, optimal policies continue to exist due to our assumption $\Gamma_k < \infty$ [LH14b]. For $\Gamma_k = \infty$, this may not be true anymore:

Consider again the environment ν_{delay} in Figure 6.3. We choose an infinite horizon $m = \infty$ and no discount. The behavior of the policy is irrelevant once the agent has moved up as no more reward can be obtained, so we can consider the space of possible policies to be those that take $n \geq 0$ steps to the right, and then a step upwards (note that n can be infinite, which describes the case where the policy always steps to the right). Let π_n denote the policy that takes n steps to the right, and then a step upwards. Then $V^{\pi_n}_{\nu_{\text{delay}}} = \frac{n+1}{n+2}$ for $n < \infty$, and $V^{\pi_\infty}_{\nu_{\text{delay}}} = 0$. So there is no optimal policy, as for any finite n we have $V^{\pi_n}_{\nu_{\text{delay}}} < V^{\pi_{n+1}}_{\nu_{\text{delay}}}$ and π_∞ is the worst policy of all, with $V^{\pi_\infty}_{\nu_{\text{delay}}} = 0 < \frac{1}{2} = V^{\pi_0}_{\nu_{\text{delay}}}$. ●

As a consequence of the linearity of the expected value (Theorem 2.2.39), we have that the value function is linear in the environment, in the sense that if an environment ν can be written as a weighted average of environments ν_i, then the value function for ν can also be written as a (posterior) weighted average of value the functions for ν_i. Similarly, one can also show that the value function is linear in the policy, and the optimal value function is convex in the environment. The precise formulation and proof are provided later in Theorem 7.2.5.

Lemma 6.6.7 (Linearity/convexity of V - informal)
$V^{\pi,m}_\nu(h_{<t})$ is linear in ν and π, and $V^{*,m}_\nu(h_{<t})$ is convex in ν.

Example 6.6.8 (Optimal value V^*_ν is not linear in ν) To demonstrate that the optimal value function is *not* linear, consider a coin flip prediction problem, with $\mathcal{A} = \mathcal{O} = \mathcal{R} = \{0,1\}$, and two environments ν_0 and ν_1, representing a two-headed and two-tailed coin respectively. The agent gets reward 1 for guessing the outcome of the coin flip correctly, and 0 for not:

$$\nu_0(e_t|h_{<t}a_t) = \begin{cases} 1 & (a_t,o_t,r_t) = (0,0,1) \\ 1 & (a_t,o_t,r_t) = (1,0,0), \\ 0 & \text{otherwise} \end{cases} \qquad \nu_1(e_t|h_{<t}a_t) = \begin{cases} 1 & (a_t,o_t,r_t) = (0,1,0) \\ 1 & (a_t,o_t,r_t) = (1,1,1) \\ 0 & \text{otherwise} \end{cases}$$

Now consider $\nu := \frac{1}{2}\nu_0 + \frac{1}{2}\nu_1$, the environment that flips a fair coin, and rewards 1 for a correct guess. With no discounting and a horizon $m = 1$, any strategy in ν is optimal as no policy can guess the outcome of a fair coin better than any other, so we find that $V^{*,m}_\nu(h_{<t}) = \frac{1}{2}$. But since ν_i are deterministic, the optimal strategy is to always predict i, so $V^{*,m}_{\nu_1}(h_{<t}) = V^{*,m}_{\nu_2}(h_{<t}) = 1$, which gives

$$\frac{1}{2}V^{*,m}_{\nu_1}(h_{<t}) + \frac{1}{2}V^{*,m}_{\nu_2}(h_{<t}) = 1 > \frac{1}{2} = V^{*,m}_\nu(h_{<t})$$

hence $V^{*,m}_\nu$ is not linear in ν for $m = 1$. This remains true for $m > 1$. ◆

6.7 Q-Value

The value function (Definition 6.6.1) can be extended to an action-value or Q-value function of taking an action a_t given a history $h_{<t}$:

Definition 6.7.1 (Q-value) The Q-value $Q_{\nu,\gamma}^{\pi,m}: \mathcal{H} \times \mathcal{A} \to \mathbb{R}$ of a policy π in an environment ν, given a history $h_{<t} \in \mathcal{H}$, action a_t, horizon $m \geq t$, and discount function γ_t, is defined as

$$Q_{\nu,\gamma}^{\pi,m}(h_{<t},a_t) := \frac{1}{\Gamma_t}\mathbf{E}_\nu^\pi\left[\sum_{k=t}^{m}\gamma_k r_k \,\Big|\, h_{<t}a_t\right] = \frac{1}{\Gamma_t}\sum_{e_t}\nu(e_t|h_{<t}a_t)\sum_{h_{t+1:m}}\nu^\pi(h_{t+1:m}|h_{1:t})\sum_{k=t}^{m}\gamma_k r_k$$

If $t > m$ or $\Gamma_t = 0$ we define $Q_{\nu,\gamma}^{\pi,m}(h_{<t},a_t) := 0$ As with the value function, we often drop γ and define $Q_\nu^\pi(h_{<t},a_t) := \lim_{m\to\infty}Q_\nu^{\pi,m}(h_{<t},a_t)$. The optimal Q-value function is defined as

$$Q_\nu^{*,m}(h_t a_t) := \sup_\pi Q_\nu^{\pi,m}(h_t a_t) = Q_{\nu,\gamma}^{\pi_\nu^{*,m},m}(h_t a_t)$$

We can write the Q-value function in terms of the value function and vice versa:

Theorem 6.7.2 (Bellman equations) If policy π and environment ν are (proper) probability measures, then

$$Q_{\nu,\gamma}^{\pi,m}(h_{<t},a_t) = \frac{1}{\Gamma_t}\sum_{e_t}\nu(e_t|h_{<t}a_t)\left[\gamma_t r_t + \Gamma_{t+1}V_{\nu,\gamma}^{\pi,m}(h_{1:t})\right] \qquad (6.7.3)$$

$$V_{\nu,\gamma}^{\pi,m}(h_{<t}) = \sum_{a_t}\pi(a_t|h_{<t})Q_{\nu,\gamma}^{\pi,m}(h_{<t},a_t) \qquad (6.7.4)$$

Note that the recursion ends once $t \geq m$ since then $V_{\nu,\gamma}^{\pi,m}(h_{1:t}) = 0$. Inserting (6.7.3) into (6.7.4) (or vice versa) we get the Bellman equations purely in terms of (Q-)values. Both equations hold for all policies including optimal policies $\pi_\nu^{*,m}$ for which (6.7.4) reduces to

$$V_{\nu,\gamma}^{*,m}(h_{<t}) = \max_{a_t}Q_{\nu,\gamma}^{*,m}(h_{<t},a_t) \qquad (6.7.5)$$

Sum representations of the Q-value function analogous to Lemma 6.6.3 are also possible. The recursions remain true for $m \to \infty$. For deterministic policies π, (6.7.4) reduces to $V_{\nu,\gamma}^{\pi,m}(h_{<t}) = Q_{\nu,\gamma}^{\pi,m}(h_{<t},\pi(h_{<t}))$. (6.7.5) also shows that among the set of optimal policy there is always a deterministic one: We can choose any deterministic π such that $\pi(h_{<t}) = a_t$ for any a_t that maximizes $Q_{\nu,\gamma}^{*,m}(h_{<t},a_t)$.

Proof sketch. (*i*) Plugging the explicit representation

$$V_{\nu,\gamma}^{\pi,m}(h_{1:t}) = \frac{1}{\Gamma_{t+1}}\sum_{h_{t+1:m}}\nu^\pi(h_{t+1:m}|h_{1:t})\sum_{k=t+1}^{m}\gamma_k r_k$$

from Definition 6.6.1 into the r.h.s. of (6.7.3) and rearranging terms and exploiting $\sum_{h_{t+1:m}}\nu^\pi(h_{t+1:m}|h_{1:t}) = 1$ we get Definition 6.7.1.
(*ii*) Inserting the sum representation of Q from Definition 6.7.1 into (6.7.4) and using the definition of ν^π leads to the sum representation of V from Definition 6.6.1.
(*iii*) Let $A_t := \arg\max_{a_t}Q_{\nu,\gamma}^{*,m}(h_{<t},a_t)$ be the set of Q^*-maximizing actions. Clearly $\sum_{a_t}\pi(a_t|h_{<t})Q_{\nu,\gamma}^{*,m}(h_{<t},a_t)$ is maximized w.r.t. π *iff* $\pi(a_t|h_{<t}) = 0$ for $a_t \notin A_t$ *iff* $\pi(a_t|h_{<t}) = \pi_\nu^{*,m}(a_t|h_{<t})$ for some $\pi_\nu^{*,m} \in \Pi_\nu^{*,m}(h_{<t})$. For such π, the expression reduces to $\max_{a_t}Q_{\nu,\gamma}^{*,m}(h_{<t},a_t)$. ∎

Lemma 6.7.6 (Contraction property of (Q-)Values) For $t \leq m$

$$\sup_{h_{<t}} \left| V_\nu^{\pi,\infty}(h_{<t}) - V_\nu^{\pi,m}(h_{<t}) \right| \leq \sup_{h_{<t}a_t} \left| Q_\nu^{\pi,\infty}(h_{<t},a_t) - Q_\nu^{\pi,m}(h_{<t},a_t) \right|$$

$$\leq \frac{\Gamma_{t+1}}{\Gamma_t} \sup_{h_{1:t}} \left| V_\nu^{\pi,\infty}(h_{1:t}) - V_\nu^{\pi,m}(h_{1:t}) \right| \leq \cdots \leq \frac{\Gamma_{m+1}}{\Gamma_t}$$

If ν is a semimeasure this remains true iff V,Q are *defined* via the recursion in Theorem 6.7.2. The bounds also remain true if all π are replaced by $*$, i.e. the optimal policies π_ν^* and $\pi_\nu^{*,m}$.

Proof. Immediate from recursive characterization (Theorem 6.7.2) of V and Q by cancelling the $\gamma_t r_t$ term and pulling in the supremum.

Counter-example for semimeasures ν if Definition 6.6.1 and Definition 6.7.1 are used: Let ν be a proper measure on $h_{<m}$ always giving reward 1, and $\nu(h_{<t}) \equiv 0$ for $t > m$. Then for all $t \leq m$ we have $V_\nu^{\pi,m}(h_{<t}) = 1$ but $V_\nu^{\pi,\infty}(h_{<t}) = 0$. ∎

Computing optimal values and policies. Given an exact model of the environment, μ, and history $h_{<t}$ so far, the optimal action to take is an action a_t which maximizes $Q_\mu^{*,m}(h_{<t},a_t)$. Therefore to find this optimal action we need to evaluate the expectation in Definition 6.7.1. Assuming μ can be computed in time $O(1)$, computing this expectation naively via the sum representation in Definition 6.7.1 takes $O((m-t) \cdot (|\mathcal{O}| \cdot |\mathcal{A}| \cdot |\mathcal{R}|)^{m-t})$ time. This makes the computation intractable for all but the smallest observation $\mathcal{O}$, action $\mathcal{A}$, and reward spaces $\mathcal{R}$ and horizons m (such as done in Chapter 11).

Finding the optimal action by maximizing $Q_\mu^{*,m}(h_{<t}a_t)$ with respect to a_t in this form is circular, since $Q^{*,m}$ relies on an optimal policy π^* in its definition and if we have an optimal policy, we can just query an optimal action directly. The theoretical remedy is to compute the last expression in Lemma 6.6.3, leading to the following expression for the optimal action:

$$a_t := \arg\max_{a_t \in \mathcal{A}} \sum_{e_t \in \mathcal{E}} \max_{a_{t+1} \in \mathcal{A}} \sum_{e_{t+1} \in \mathcal{E}} \cdots \max_{a_m \in \mathcal{A}} \sum_{e_m \in \mathcal{E}} \prod_{k=t}^{m} \nu(e_k | h_{<k} a_k) \sum_{k=t}^{m} \gamma_k r_k \qquad (6.7.7)$$

This expression is called expectimax tree or algorithm (see Figure 12.2).

On the other hand, the recursive Bellman expressions often build the basis for more efficient and practical MDP RL algorithms, which learn the optimal Q-value function directly via techniques such as Q-learning [SB18]. From an approximation of the Q^* values, an approximately optimal policy can be recreated by $\pi^{*,m}(h_{<t}) = \arg\max_{a_t} Q^{*,m}(h_{<t}a_t)$.

6.8 Exercises

1. [C03] Give an example of an environment where all policies are optimal.

2. [C15] In Section 6.3, we claim POMDPs are a specific instance of history based RL. Prove this claim. In particular, given the policy and environment both have access to the entire interaction history, how can we construct the environment so that it has access to the true underlying state, but the policy does not?

3. [C20] Show that conditional $\nu^\pi(e_{1:m}|a_{1:m}) := \nu^\pi(h_{1:m})/\sum_{a_{1:m}} \nu^\pi(h_{1:m})$ (Definition 6.1.5) is different from chronological $\nu(e_{1:m}||a_{1:m})$ (Definition 6.3.1), and elaborate in which ways, and why this is important. For instance, show that $\sum_{e_m} \nu(e_{1:m}||a_{1:m}) =$

$\nu(e_{<m}||a_{<m})$ is independent of a_m, while $\sum_{e_m} \nu^\pi(e_{1:m}|a_{1:m}) \neq \nu^\pi(e_{<m}|a_{<m})$ depends on a_m.

4. [C10] Give some examples of undesirable behavior that can result from using: no discount, negative discount, monotonically increasing discount.

5. [C20] Derive the expressions for $H_t(\epsilon)$ and Γ_t in Table 6.4.

6. [C10] Show that in Example 6.5.3 with geometric discounting γ, the agent will choose to walk $k^* - 1$ steps right, and then on time step k^*, walk up, where $k^* = \text{argmax}_{k \geq 1} \{\gamma^{k-1} \frac{k}{k+1}\} < \infty$.

7. [C15] Prove the equivalence of the expectation and sum representations of the (Q-)value functions in Definition 6.6.1 and Definition 6.7.1.

8. [C15i] Derive the explicit representations of the value function in Lemma 6.6.3.

9. [C15] Derive explicit representations of the Q-value function analogous to Lemma 6.6.3.

10. [C20] Complete the proof details for Theorem 6.7.2.

11. [C18] Show that the Bellman (optimality) equations Theorem 6.7.2 also hold for $m \to \infty$.

12. [C20] Formulate the claims in Lemma 6.6.7 precisely and prove them. Hint: This exercise becomes easier after having read Section 7.2. Use the posterior $w(\nu|h_{<t})$ (Definition 7.2.2) and possibly Definition 6.3.1.

13. [C20i] Prove that for finite action sets $\mathcal{A}$ and bounded reward space $\mathcal{R}$ and $\Gamma_t < \infty$, an optimal policy and optimal actions for $m \to \infty$ indeed exist. Show that these conditions are necessary [dB09].

14. [C25i] Show that ε_t-optimal policies and actions always exists, even for countable action spaces $\mathcal{A}$, for any choice of ε_t, e.g. constant $\varepsilon_t = \varepsilon$ or decreasing $\varepsilon_t = o(1)$.

15. [C15] Prove that if $\pi \in \Pi^*(\epsilon)$ then $\pi \in \Pi^*(h_{<t})$ (on histories which π can generate).

16. [C22] In place of a discount function, one can assume that the sum of the rewards outputted by the environment ν is finite. Show that this is a strict generalization of having a discount function (e.g. all discount functions are contained in this setup, as well as setups which cannot be described by using discount functions).

6.9   History and References

This chapter is based on material from [Hut05b] and [Lei16b, Chp.4].

Books on reinforcement learning.   The reinforcement learning book by Sutton and Barto [SB18] requires no background knowledge, and describes from scratch the field of reinforcement learning. It covers bandit problems and various methods of attack for MDPs: exact solutions, tabular learning and approximate methods, though more state-of-the-art approaches using deep learning are not covered. Tougher and more rigorous books by Bertsekas (and Tsitsiklis) [BT96, Ber19, Ber20] provide all convergence proofs that Sutton and Barto gloss over. In-depth surveys of reinforcement learning can be found in [KLM96, Sze10, GMPT15, MBJ20]. Additional useful resources in the study of reinforcement learning include the following books [KV86, WvO12, AJKS22, Mey22, Pla22]. See below and Section 14.6 for more RL ideas and algorithms mostly those not (yet) covered in (m)any books.

Bandit problems. Bandit problems (MDPs with only a single state) are still complex enough to demonstrate trade-offs between exploitation and exploration [BF85, BK97, ACBF02] and are covered in great depth in [LS20].

Rewards. The reward hypothesis Assumption 6.2.1 is the core assumption on which reinforcement learning relies. This hypothesis cannot be proven either way (goals and purposes are not formally defined terms), though there is much work both in support of [SSPS21], and some in opposition to [VSK$^+$22], and discussing the reward hypothesis [Ale21, BMAD22]. Many problems often have multiple goals to simultaneously fulfil, and often these goals can be in conflict. For example, a self-driving car needs to balance minimizing time to the destination, against passenger safety and the road rules, against cost of fuel, etc. It is not immediately clear how to aggregate these competing goals as a scalar reward, especially if the relative trade-off between the goals is not fixed (some passengers might be in a rush, others might be sightseeing and want to take things slow). Any transformation of these competing goals into a single scalar reward implies some extra domain knowledge of how to trade one goal versus another. Such problems are in the domain of *multi-objective reinforcement learning* (MORL) [HR$^+$22]; a survey can be found in [RVWD13]. Learning is often more difficult in environments with sparse rewards. Imitation learning [CHN22] or shaping the reward can help, but usually requires extra domain knowledge [Mat94]. Ideally, the reward should be shaped in such a way as to leave the optimal policy unchanged, but easier for the agent to find [NHR99]. For environments where the agent is trying to achieve a goal state, [TZXS19] describes a shaped reward based on a measure of distance from a goal, while avoiding getting trapped in local optima. A proxy goal can also be chosen, often a goal that encourages the agent to seek new and novel experiences. In Section 9.3 we cover Orseau's [OLH13] Knowledge-Seeking Agent (KSA), which tries to gather information about the environment it is in. Burda [BESK18] mentions that KSAs can perform well on complex environments, but can also focus their attention on sources of noise.

Multi-agent RL. Some environments contain an opponent to play against (such as chess). Often, *self-play* is used, where the agent also plays the role of its own opponent. This was used to great success to train an agent to play *backgammon* [Tes94, T$^+$95] and *Go* [SH$^+$16]. A survey of self-play RL can be found in [DZ21]. To encode two agents playing against each other in our agent-environment framework, each agent considers the other to be part of the environment with which it interacts. This setup will be covered in Chapters 10 and 11. *Multi-Agent Reinforcement Learning* (MARL) covers many agents interacting together in an environment, often with adversarial rewards to incentivize competition, or aligned rewards to incentivize cooperation. A survey of MARL is provided in [HLKT19].

Illegal actions. Commonly used methods to deal with illegal moves in games include issuing a larger penalty than the penalty for losing a game [WPH22] or having the illegal action have no effect [H$^+$02]. For stochastic policies, we can resample until a legal move is obtained (for black box policies) or zero the probabilities for illegal moves and renormalize [SH$^+$16]. The former two methods do not require leaking information about the environment to the agent (playing an illegal move is seen from the agent's perspective as either an action that makes the player immediately lose with a very bad score, or as a no-op given the current interaction history), whereas the latter method of tampering with the distribution over actions requires domain knowledge of the environment.

Time-discounting. Time discounting is a trick used to ensure the reward sum is finite, and often justified based on an argument appealing to the finite lifetime of an agent [Soz98], or that in practice most humans choose immediate gratification over a marginally larger reward later. On the other hand, humans often forgo immediate gratification for sufficiently larger rewards later. The famous "marshmallow test" [ME70] provided one of the earliest studies into how children discount rewards, offering them one marshmallow now, with the offer to provide two if the child could resist eating the first for a period of time.

[LH11c] introduced the generalized discount Definition 6.4.1 which can depend on the age of the agent. A complete characterization of when it is time-(in)consistent can be found in [LH14b]. They also show the existence of a rational policy for an agent that knows its discount function is time-inconsistent.

However, most of the time in RL, the agent is immortal, and does not have the option to "invest" their reward. Schwartz explores an undiscounted method of measuring the performance of a policy similarly to a Cesáro sum [Sch93]. [LH07c, Sec.3] avoids pre-commitment of any particular discounting by letting the environment decide how to discount. Pitis justifies a set of axioms that a discount should satisfy, and derives a form of discounting that can be state-dependent, generalizing the typical fixed discount methods [Pit19]. Low discount factors can speed up convergence [BT96], but lead to poor performance. Van Seijen [VSFT19] hypothesizes that this is due to the size of the action-gap (the difference in value between the best and second-best actions), and proposes a method to make the action-gaps more homogeneous by applying updates in logarithmic space. Reinforcement learning can serve as an idealized setting to study how discounting can lead to (sub)optimal/(un)healthy behavior [SVS+14].

One can show that (past) undiscounted truncated values are asymptotically equivalent to (future) untruncated discounted values provided the effective horizon increases unboundedly [Hut06a]. Truncated Value Learning (TVL) mimics discounting by learning multiple undiscounted truncated value functions [ASDH24]. It simultaneously learns values for *all* (summable) discount functions.

Chapter 7

Universal Artificial Intelligence

A sign of intelligence is an awareness of one's own ignorance.

Niccolò Machiavelli

In the previous chapter we have formulated the general (also called called history-based) reinforcement learning problem wherein an agent interacting with an environment needs to maximize the expected future reward from the environment. In this chapter we will discuss solutions to the general reinforcement learning problem. The solutions we will go over are each based on what information (if any) about the environment the agent is initially given. The first solution, known as $AI\mu$, is the optimal agent when the agent has full knowledge of the probability distribution μ that describes the behavior of the environment it is interacting with. We note that knowledge of μ does not render the problem trivial.

The second and more interesting case is when the agent only knows the class of environments $\mathcal{M}$ to which the true distribution μ belongs. The agent must then take actions both to attempt to act well in the environment it believes itself to be in, as well as taking "exploratory" actions. Exploratory actions may not lead to good performance immediately, but help the agent to work out what environment it is in, which will hopefully result in better performance later.

In this case, we define AIXI, an agent that constructs an estimate ξ of the true environment μ using Bayes' Law, updating on experience received via interaction. AIXI then chooses actions with policy π_ξ^*. We will show that (under some conditions), the performance of AIXI converges to that of $AI\mu$.

7.1   Acting Optimally in Known Environments

When an agent is interacting with an environment to achieve its goals, which are often characterized as maximizing the value function, there are two (usually) distinct components of the problem the agent needs to overcome: Modelling the environment and planning in the model. When the agent is given the complete information about the environment it is interacting with, the modelling of the environment problem is "solved"[1]. This just leaves the planning in the "model" (which is the true environment) component of the problem. Definition 6.6.1 applied to ν replaced by the true environment μ and $m \to \infty$ gives us our first definition of an informed optimal agent that has access to μ, and chooses actions to maximize its value function:

Definition 7.1.1 (Policy of AIμ) Given the true environment μ and some discount function $\gamma_{()}$, informed agent AIμ selects an optimal policy

$$\pi_\mu^* \in \arg\max_\pi V_\mu^\pi \equiv \arg\max_\pi \lim_{m \to \infty} V_{\mu,\gamma}^{\pi,m}(\epsilon)$$

The agent can make decisions that maximize its expected total reward in environment μ by taking the action which maximizes the Q-value $Q_\mu^*(h_{<t})$:

Proposition 7.1.2 (Action of AIμ) The action a_t of AIμ at time t given history $h_{<t}$ can be computed via
$$a_t = a_t^{\text{AI}\mu} := \arg\max_{a_t} Q_\mu^*(h_{<t},a_t)$$

$$= \arg\max_{a_t \in \mathcal{A}} \lim_{m \to \infty} \sum_{e_t \in \mathcal{E}} \max_{a_{t+1} \in \mathcal{A}} \sum_{e_{t+1} \in \mathcal{E}} \dots \max_{a_m \in \mathcal{A}} \sum_{e_m \in \mathcal{E}} \prod_{k=t}^m \mu(e_k | h_{<k} a_k) \sum_{k=t}^m \gamma_k r_k$$

Proof. This follows immediately from Lemma 6.6.3 and Definition 6.7.1.   ∎

To directly compute $a_t^{\text{AI}\mu}$ requires computing the expectimax tree truncated to an effective horizon $m = H_t(\varepsilon) < \infty$ (for more details, see Sections 6.7 and 12.1).

However, in practice, this is computationally expensive, and various techniques like pruning or approximation methods are used to make it more tractable. Various methods used in the past for approximation of the true Q-value include only unrolling the expectimax calculation for a short moving horizon, using a reasonable estimate $\widehat{Q}$ designed by humans to estimate the Q-value (often called a heuristic) [RN20], or using a neural network to learn $\widehat{Q}$ from experience [MKS+15]. The expectimax can also be approximated using Monte Carlo Tree Search, where several *rollouts* are performed (the "game" is played out to the end with a cheap proxy for the policy of the agent), and the obtained discounted sum of rewards is averaged as an approximation for the expectation [SH+16]. We will explore this method further in Section 12.2.2.

7.2   Bayesian Mixture of Environments

In practice, the environment an agent interacts with is not fully known. The classical RL solution to this problem is to make some (strong) assumption about the environment, and then devise some (heuristic) algorithm to learn the environment from the agent-environment

---

[1]It could be the case that when the agent is given the true environment, computing the true environment is hard/intractable, but we will not consider this case.

interaction. The Bayesian solution to this problem is to consider a *class* of environments $\mathcal{M}$ large enough to be confident it contains the true environment μ, devise a prior w_ν to $\nu \in \mathcal{M}$, take a Bayesian mixture ξ over all environments in $\mathcal{M}$, and use (known) ξ instead of (unknown) μ. In the same spirit and with the same reasoning, following Section 3.7.1 and [Lei16b], for the choice of $\mathcal{M}$, we will (later) mainly concern ourselves with the classes $\mathcal{M}_{sol}$ and $\mathcal{M}_{comp}$ and universal Bayesian mixture ξ_U, which suitably generalizes Definition 3.7.1.

Definition 7.2.1 (Chronological Bayes mixture ξ) Let $\mathcal{M}$ be a countable class of chronological semimeasures (see Definition 6.3.1), and $w : \mathcal{M} \to \mathbb{R}$ be a prior satisfying $\sum_{\nu \in \mathcal{M}} w_\nu \leq 1$ and $w_\nu > 0$ for all $\nu \in \mathcal{M}$. We define the *chronological Bayes mixture ξ over $\mathcal{M}$ given prior $w_{(.)}$* as

$$\xi(e_{1:t} \| a_{1:t}) := \sum_{\nu \in \mathcal{M}} w_\nu \nu(e_{1:t} \| a_{1:t})$$

A dual notion of policy classes and mixtures is pursued in [CGMH+23], but we will not elaborate on this. In this chapter we consider all (chronological) policies $\pi : \mathcal{H} \to \Delta\mathcal{A}$ without any restriction. At first it may be appealing to choose Π to be as large as possible to ensure we include the best policy. However, we are trying to construct a physically plausible agent, and since there are uncountably[2] many policies to choose from, we would want to restrict ourselves to a countable subset. For this we will mainly consider the class of approximable or limit-computable policies Definition 2.6.13. An important part of this is to show whether or not an optimal policy is approximable; this is discussed in Section 13.1. In practice we have to restrict the policy space Π even further to *tractable* policies.

Alternatively, we can define a predictive Bayes mixture as a *posterior* mixture over (predictive) environments:

Definition 7.2.2 (Bayesian mixture ξ over environments) Let $\mathcal{M}$ be a class of environments with prior $\{w_\nu\}_{\nu \in \mathcal{M}}$. We define $\xi(e_t | h_{<t} a_t)$ as the (predictive) probability the Bayesian mixture environment ξ assigns to percept e_t conditioned on history $h_{<t}$ and action a_t recursively as:

$$\xi(e_t | h_{<t} a_t) := \sum_{\nu \in \mathcal{M}} w(\nu | h_{<t}) \nu(e_t | h_{<t} a_t)$$
$$= \frac{\xi(e_{1:t} \| a_{1:t})}{\xi(e_{<t} \| a_{<t})} = \frac{\sum_{\nu \in \mathcal{M}} w_\nu \nu(e_t | h_{<t} a_t) \nu(e_{<t} \| a_{<t})}{\sum_{\nu \in \mathcal{M}} w_\nu \nu(e_{<t} \| a_{<t})}, \quad \text{where}$$
$$w(\nu | h_{1:t}) := w(\nu | h_{<t}) \frac{\nu(e_t | h_{<t} a_t)}{\xi(e_t | h_{<t} a_t)} = w_\nu \frac{\nu(e_{1:t} \| a_{1:t})}{\xi(e_{1:t} \| a_{1:t})} \quad \text{and} \quad w(\nu | \epsilon) := w_\nu$$

Definition 7.2.2 is the same Theorem 3.1.7, but for a mixture of environments where the agent can act, rather than passive environments where the "agent" tries to predict the sequence.

Proof. The proof for the equivalence of Definitions 7.2.1 and 7.2.2, that ξ satisfies Definition 6.3.1, and the other identities, can be obtained by replacing $x_{1:t}$ with $e_{1:t}$ and extra conditioning all probabilities on $a_{1:t}$ (similarly $<t$) in the proof of Theorem 3.1.7. ∎

Example 7.2.3 (Mixture environment) We consider again a coin flip prediction problem, with $\mathcal{A} = \mathcal{O} = \{H,T\}$ and $\mathcal{R} = \{0,1\}$ and the class of environments $\mathcal{M} = \{\nu_{HH}, \nu_{HT}, \nu_{TT}\}$,

---

[2]Assuming there are at least two actions to choose from, even ignoring the history and considering only the set of all probability distributions on $\mathcal{A}$ is already as large as the continuum.

representing an environment that flips a two-headed coin, a fair coin, and a two-tailed coin respectively. Each environment gives 1 reward to the agent if it correctly predicts the outcome of the coin (that is, if $a_t = o_t$). We choose the uniform prior $w_\nu = 1/3$. Initially the Bayesian mixture predicts that the likelihood of the environment generating H (and returning a reward of 1 assuming the action was H) is $1/2$:

$$\xi(e_1 = (H,1)|h_{<1} = \epsilon, a_1 = H) = \sum_{\nu \in M} w_\nu \nu(e_1 = (H,1)|a_1 = H)$$
$$= \tfrac{1}{3}\nu_{HH}(e_1 = (H,1)|a_1 = H) + \tfrac{1}{3}\nu_{HT}(e_1 = (H,1)|a_1 = H) + \tfrac{1}{3}\nu_{TT}(e_1 = (H,1)|a_1 = H)$$
$$= \tfrac{1}{3} \times 1 + \tfrac{1}{3} \times \tfrac{1}{2} + \tfrac{1}{3} \times 0 = \tfrac{1}{2}$$

In fact, it can be shown that $\xi = \nu_{HT}$ when the mixture is given an empty history. Now, suppose that the agent decides to take action $a_1 = H$, and guessed correctly, so the environment returns $(o_1, r_1) = (H,1)$. This gives a history so far of $h_1 = (a_1, e_1) = (H,(H,1))$. We can now follow the recursive update rules in Definition 7.2.2 to see which environment the agent now believes it is interacting with:

$$w_{\nu_{HH}}(h_1) = w_{\nu_{HH}} \frac{\nu_{HH}(e_1 = (H,1)|a_1 = H)}{\xi(e_1 = (H,1)|a_1 = H)} = \frac{1}{3}\frac{1}{\sfrac{1}{2}} = \frac{2}{3}$$

$$w_{\nu_{HT}}(h_1) = w_{\nu_{HT}} \frac{\nu_{HT}(e_1 = (H,1)|a_1 = H)}{\xi(e_1 = (H,1)|a_1 = H)} = \frac{1}{3}\frac{\sfrac{1}{2}}{\sfrac{1}{2}} = \frac{1}{3}$$

$$w_{\nu_{TT}}(h_1) = w_{\nu_{TT}} \frac{\nu_{TT}(e_1 = (H,1)|a_1 = H)}{\xi(e_1 = (H,1)|a_1 = H)} = \frac{1}{3}\frac{0}{\sfrac{1}{2}} = 0$$

So, the agent believes with confidence $2/3$ that $\mu = \nu_{HH}$, and with confidence $1/3$ that $\mu = \nu_{HT}$. It believes it to be impossible that $\mu = \nu_{TT}$, which makes sense, since it has observed the environment returning an observation of a H. We can now ask about the dynamics of $\xi(e_2|h_1 a_2)$,

$$\xi(e_2|h_1 a_2) = \sum_{\nu \in \mathcal{M}} w(\nu|h_1)\nu(e_2|h_1 a_2) = \tfrac{2}{3}\nu_{HH}(e_2|h_1 a_2) + \tfrac{1}{3}\nu_{HT}(e_2|h_1 a_2)$$

We can also use ξ to compute the probability that the next observation received will be H, for the two choices of action a_2.

$$\xi(o_2 = H|h_1 H) = \sum_{r \in \{0,1\}} \tfrac{2}{3}\nu_{HH}((H,r)|h_1 H) + \tfrac{1}{2}\nu_{HT}((H,r)|h_1 H)$$
$$= \tfrac{2}{3}\nu_{HH}((H,1)|h_1 H) + \tfrac{1}{2}\nu_{HT}((H,1)|h_1 H) = \tfrac{2}{3} \times 1 + \tfrac{1}{3} \times \tfrac{1}{2} = \tfrac{5}{6}$$
$$\xi(o_2 = H|h_1 T) = \sum_{r \in \{0,1\}} \tfrac{2}{3}\nu_{HH}((H,r)|h_1 T) + \tfrac{1}{2}\nu_{HT}((H,r)|h_1 T)$$
$$= \tfrac{2}{3}\nu_{HH}((H,0)|h_1 T) + \tfrac{1}{2}\nu_{HT}((H,0)|h_1 T) = \tfrac{2}{3} \times 1 + \tfrac{1}{3} \times \tfrac{1}{2} = \tfrac{5}{6}$$

In both cases, ξ predicts the next observation will be H, regardless of the action taken, consistent with the fact that this is a passive environment where we are only making predictions. We note that $\xi \notin \mathcal{M}$, which is not surprising given the very limiting model class chosen.   ♦

The following elementary linearity and convexity results are used frequently later

---

Lemma 7.2.4 (Linearity of posterior ξ^π mixture) For events $A \subseteq (\mathcal{A} \times \mathcal{E})^\infty$ and functions $f : (\mathcal{A} \times \mathcal{E})^\infty \to \mathbb{R}$,

$$\mathrm{P}_\xi^\pi[A|h_{<t}] = \sum_{\nu \in \mathcal{M}} w(\nu|h_{<t}) \mathrm{P}_\nu^\pi[A|h_{<t}] \quad \text{and} \quad \mathbf{E}_\xi^\pi[f|h_{<t}] = \sum_{\nu \in \mathcal{M}} w(\nu|h_{<t}) \mathbf{E}_\nu^\pi[f|h_{<t}]$$

Linearity with the same weights $w(\nu|h_{<t})$ remains true if we further condition on a_t. A particular instantiation is

$$\xi^\pi(h_{1:m}) = \prod_{t=1}^m \pi(a_t|h_{<t})\nu(e_t|h_{<t}a_t) = \sum_{\nu \in \mathcal{M}} w_\nu \nu^\pi(h_{1:m}) \geq w_\mu \mu^\pi(h_{1:m}) \quad \forall \mu^\pi \in \mathcal{M}$$

Proof. We first show prior linearity, i.e. for $t=1$, $h_{<1} = \epsilon$ and $w(\nu|\epsilon) = w_\nu$. Using (6.3.2) twice and Definition 7.2.1,

$$\mathrm{P}_\xi^\pi[h_{1:m}] \equiv \xi^\pi(h_{1:m}) \equiv \xi(e_{1:m}||a_{1:m})\pi(a_{1:m}||e_{<m})$$
$$= \sum_\nu w_\nu \nu(e_{1:m}||a_{1:m})\pi(a_{1:m}||e_{<m}) = \sum_\nu w_\nu \nu^\pi(h_{1:m}) = \sum_\nu w_\nu \mathrm{P}_\nu^\pi[h_{1:m}]$$

for any $h_{1:m}$. By taking suitable limits, this implies that equality holds for all events A. Now posterior linearity follows from

$$\mathrm{P}_\xi^\pi[A|h_{<t}] \equiv \frac{\mathrm{P}_\xi^\pi[A \cap h_{<t}]}{\mathrm{P}_\xi^\pi[h_{<t}]} = \frac{\sum_\nu w_\nu \mathrm{P}_\nu^\pi[A \cap h_{<t}]}{\xi^\pi(h_{<t})} \equiv \sum_\nu w_\nu \frac{\nu^\pi(h_{<t})}{\xi^\pi(h_{<t})} \mathrm{P}_\nu^\pi[A|h_{<t}]$$
$$= \sum_\nu w_\nu \frac{\nu(e_{<t}||a_{<t})}{\xi(e_{<t}||a_{<t})} \mathrm{P}_\nu^\pi[A|h_{<t}] = \sum_\nu w(\nu|h_{<t}) \mathrm{P}_\nu^\pi[A|h_{<t}]$$

where $\equiv$ is the definition of conditional probability, the first equality follows from prior linearity, the second equality follows from (6.3.2) and the last equality from Definition 7.2.2. The linearity if further conditioning on a_t and of $\mathbf{E}^\pi$ can be proven similarly or by extension. The last equality is the special case of $A = h_{1:m}$ (strictly speaking $A = \{h_{1:m}\} \times (\mathcal{A} \times \mathcal{E})^\infty$) and $t=1$ and $w(\nu|\epsilon) = w_\nu$. ∎

Theorem 7.2.5 (Linearity/convexity of V)
$V_\nu^\pi(h_{<t})$ is linear in ν and $V_\nu^*(h_{<t})$ is convex in ν:

$$V_\xi^\pi(h_{<t}) = \sum_{\nu \in \mathcal{M}} w(\nu|h_{<t}) V_\nu^\pi(h_{<t}) \quad \text{and} \quad V_\xi^*(h_{<t}) \leq \sum_{\nu \in \mathcal{M}} w(\nu|h_{<t}) V_\nu^*(h_{<t})$$

The same holds for the Q-values $Q_\nu^\pi(h_{<t}a_t)$ with the same weights $w(\nu|h_{<t})$.

Duals of Theorem 7.2.5 for mixtures over policies π exist, but we do not need them $(V_\nu^\zeta(h_{<t}) = \sum_{\pi \in \Pi} \omega(\pi|h_{<t}) V_\nu^\pi(h_{<t})$, see Definition 9.5.1).

Proof. Applying Lemma 7.2.4 with $f = \sum_{k=t}^\infty \gamma_k r_k$ implies linearity of V^π (Definition 6.6.1). Applying the version of Lemma 7.2.4 conditioned on $h_{<t}a_t$ implies linearity of Q^π (Definition 6.7.1). Applying linearity of V^π to $\pi = \pi_\xi^*$ we get

$$V_\xi^*(h_{<t}) \equiv V_\xi^{\pi_\xi^*}(h_{<t}) = \sum_\nu w(\nu|h_{<t}) V_\nu^{\pi_\xi^*}(h_{<t}) \leq \sum_\nu w(\nu|h_{<t}) V_\nu^*(h_{<t}) \qquad \blacksquare$$

The reason for the inequality is that on the r.h.s. the optimal policy π_μ^* is tailored for μ, while on the l.h.s. a single policy must perform well in ξ, i.e. in all environments $\nu \in \mathcal{M}$ simultaneously.

In the limit as $t \to \infty$, we would expect that $\xi \to \mu$ in the predictive sense analogous to Corollary 3.3.2, there are two complication though: we need multi-step and off-policy convergence. For prediction, greedily minimizing the one-step loss was optimal, and hence convergence of one-step prediction sufficed. The agent case depends on farsighted prediction. We can multiply the one-step convergence to an m-step convergence for fixed look-ahead m [Hut05b, Sec.3.7.1], but discounted values require infinite look-ahead [Hut05b, Thm.5.3.6]. Luckily this still holds but without any guarantee on the convergence rate anymore, by lifting Theorem 3.9.5 to the agent case:

Theorem 7.2.6 (Merging of opinions [BD62])

$$\sup_A \left| \mathrm{P}_\xi^\pi[A|h_{<t}] - \mathrm{P}_\mu^\pi[A|h_{<t}] \right| \overset{t\to\infty}{\longrightarrow} 0 \quad \mu^\pi a.s.$$

$$\sup_A \left| \mathrm{P}_\xi^\pi[A|h_{<t}a_t] - \mathrm{P}_\mu^\pi[A|h_{<t}a_t] \right| \overset{t\to\infty}{\longrightarrow} 0 \quad \mu^\pi a.s.$$

where $\sup_A$ ranges over all measurable events $A \subseteq (\mathcal{A} \times \mathcal{E})^\infty$, and $\mathrm{P}_\nu^\pi[A|h_{<t}] = \mathbf{E}_\nu^\pi[\mathbb{1}_A|h_{<t}]$.

The proof is left as an exercise. The result in particular implies that for a sequence of events A_t, $\mathrm{P}_\xi^\pi[A_t|h_{<t}a_t] \to \mathrm{P}_\mu^\pi[A_t|h_{<t}a_t]$. For instance, for $A_t = \{h_{1:t}\} \times (\mathcal{A} \times \mathcal{E})^\infty$, this becomes $\xi^\pi(e_t|h_{<t}a_t) \to \mu^\pi(e_t|h_{<t}a_t)$, and for $A_t = \{h_{1:t+m}\} \times (\mathcal{A} \times \mathcal{E})^\infty$ we get multi-step convergence. A serious limitation still is that the historic actions need to be sampled from the same policy π for ξ and for μ, which complicates convergence analysis for (Bayesian) agents significantly.

7.3   Acting Optimally in Unknown Environments

Since $\xi \to \mu$ for $t \to \infty$ in the sense of Theorem 7.2.6, it is plausible to conjecture that the value function V_ξ converges to V_μ, regardless of the policy chosen. Even if the agent always guessed the wrong answer, it would still obtain evidence to indicate which environment it was in, even if it received little reward for doing so.

Indeed, for the value functions, we have on-policy convergence. This means that for any policy π and true environment μ, if the history is sampled from μ^π, then for $t \to \infty$, the value function $V_\xi^\pi(h_{<t})$ of that history converges to $V_\mu^\pi(h_{<t})$ almost surely. That is, the probability that μ^π generates a history such that $V_\xi^\pi(h_{<t})$ does not converge to $V_\mu^\pi(h_{<t})$ is zero:

Theorem 7.3.1 (On-policy value convergence of Bayes) For any environment $\mu \in \mathcal{M}$ and any policy π,

$$V_\xi^\pi(h_{<t}) - V_\mu^\pi(h_{<t}) \longrightarrow 0 \quad \text{for} \quad t \to \infty \quad \mu^\pi\text{-almost surely}$$

Proof sketch. Instantiating the general inequality

$$\left| \mathbf{E}_Q[f] - \mathbf{E}_P[f] \right| \le \sup|f| \cdot \sup_A |Q[A] - P[A]|$$

with $f = \sum_{k=t}^{\infty} \gamma_k r_k / \Gamma_t$ and $P = \mathrm{P}_\mu^\pi(\cdot|h_{<t})$ and $Q = \mathrm{P}_\xi^\pi(\cdot|h_{<t})$, using Definition 6.6.1, this becomes

$$\left|V_\xi^\pi(h_{<t}) - V_\mu^\pi(h_{<t})\right| \leq \sup|f| \cdot \sup_A \left|\mathrm{P}_\xi^\pi[A|h_{<t}] - \mathrm{P}_\mu^\pi[A|h_{<t}]\right| \overset{t\to\infty}{\longrightarrow} 0 \quad \mu^\pi a.s.$$

The convergence follows from $\sup|f| \leq \sup|\mathcal{R}| \leq 1$ and Theorem 7.2.6. Convergence rates similar to but weaker than those in Section 3.5 can also be obtained [Hut05b, Thm.5.36]. ∎

We are interested in measuring the performance of the agent against V_μ^*, the optimal value function when the environment μ is known, and the best policy π_μ^* is chosen for μ. Since the environment μ is unknown, the best (Bayes-optimal) policy the agent could choose is π_ξ^*, which maximizes V_ξ^π, since ξ represents the agent's best estimate of the true environment μ based on both the prior w_ν and the experience collected so far via interaction with the environment. In analogy to Definition 7.1.1 we define

---

Definition 7.3.2 (Bayes-optimal agent AIξ) $\pi_\xi^* :\in \arg\max_\pi V_\xi^\pi$

Representation Proposition 7.1.2 holds as well with μ replaced by ξ. Theorem 7.3.1 is comforting, but does not give us a result about the convergence of two different policies. What we really want to know is whether $V_\mu^{\pi_\xi^*}(h_{<t})$ converges to $V_\mu^*(h_{<t})$, since this represents the convergence of the true μ-value of the Bayes-optimal policy π_ξ^*, (the policy the agent would have to use in the face of unknown μ), to the true value of the optimal policy π_μ^*. In the control-theory literature, this property is called self-optimizing:

Definition 7.3.3 (Self-optimizing policies) A policy $\tilde\pi$ is called self-optimizing for a class of environments $\mathcal{M}$ and historic policy π if

$$\forall \nu \in \mathcal{M}: \; V_\nu^*(h_{<t}) - V_\nu^{\tilde\pi}(h_{<t}) \overset{t\to\infty}{\longrightarrow} 0 \; \text{ with } \nu^\pi \text{ probability } 1$$

Note that while all perceptions $e_{1:\infty}$ are sampled from ν, the historic actions $a_{<t}$ are arbitrary and can come from any (other) policy π.

The remainder of this section is devoted to establishing necessary and sufficient conditions for π_ξ^* to be self-optimizing. To do so we need a lemma establishing convergence of posterior Bayesian mixtures of functions:

Lemma 7.3.4 (Convergence of posterior averages [Lat23]) For any policy π,

If $\; 0 \leq \delta_{\nu,t}(h_{<t}) \to 0 \;$ for $\; t\to\infty \;$ w.ν^π.p.1 for all $\nu \in \mathcal{M}$

then $\;\; \sum_{\nu \in \mathcal{M}} w_\nu \dfrac{\nu(e_{<t}||a_{<t})}{\mu(e_{<t}||a_{<t})} \delta_{\nu,t}(h_{<t}) \overset{t\to\infty}{\longrightarrow} 0 \;$ w.μ^π.p.1

and equivalently $\sum_{\nu \in \mathcal{M}} w(\nu|h_{<t}) \delta_{\nu,t}(h_{<t}) \overset{t\to\infty}{\longrightarrow} 0 \;$ w.μ^π.p.1.

Proof. The proof idea is as follows: Consider a single ν-term from the sum: First, $\nu/\mu \to c < \infty$ w.μ.p.1. For histories $h_{1:\infty}$ for which $c=0$, we are done. By assumption $\delta_{\nu,t} \to 0$ w.ν.p.1. For $c>0$, the predictive distributions μ and ν convergence to each other, hence convergence also holds w.μ.p.1. Finally, we exchange limits with $\sum_\nu w_\nu$. The formal proof is unfortunately somewhat technical.

(i) Definitions: We drop the superscript π from w.μ^π.p.1 and P^π and $\mathbf{E}^\pi$ etc. Historic actions $a_{<t}$ are implicitly assumed to be sampled from π. $X_t := X_{\nu,t} := \nu(e_{<t}\|a_{<t})/\mu(e_{<t}\|a_{<t})$ is a μ-supermartingale, hence the limit $\lim_{t\to\infty} X_t$ exists and is finite w.μ.p.1. Let $F := \{h_{1:\infty} : \delta_{\nu,t} \not\to 0\}$ and $W := \{h_{1:\infty} : \lim_t X_t \geq \varepsilon\}$.

(ii) Single ν term: Let By assumption, $\mathrm{P}_\nu[F] = 0$, and by construction, $\mathrm{P}_\nu[A]/\mathrm{P}_\mu[A] \geq \varepsilon$ for any $A \subseteq W$, hence $\mathrm{P}_\mu[F \cap W] \leq \frac{1}{\varepsilon}\mathrm{P}_\nu[F \cap W] = 0$, that is, $\delta_{\nu,t}\mathbb{1}_W \to 0$ w.μ.p.1. This implies $\limsup_{t\to\infty} X_t\delta_{\nu,t}\mathbb{1}_W = 0$ w.μ.p.1, since X_t is bounded w.μ.p.1. On the other hand, $\limsup_t X_t \leq \varepsilon$ for $h_{1:\infty} \in W^c$ and $\delta_{\nu,t} \leq r_{\max} \leq 1$ implies $\limsup_{t\to\infty} X_t\delta_{\nu,t}\mathbb{1}_{W^c} \leq \varepsilon$. Adding both gives

$$\limsup_{t\to\infty} X_t\delta_{\nu,t} \leq \varepsilon \quad \text{w.}\mu\text{.p.1} \tag{7.3.5}$$

(iii) Sum over ν: Since (7.3.5) holds for every ν, it also holds for a countable weighted average:

$$\limsup_{t\to\infty} \sum_{\nu\in\mathcal{M}} w_\nu X_{\nu,t}\delta_{\nu,t} \leq \sum_{\nu\in\mathcal{M}} w_\nu \limsup_{t\to\infty} X_{\nu,t}\delta_{\nu,t} \leq \sum_{\nu\in\mathcal{M}} w_\nu\varepsilon \leq \varepsilon \quad \text{w.}\mu\text{.p.1}$$

where the first inequality follows from (the sum-version of) reverse Fatou lemma. Its condition is satisfied, since $w_\nu X_{\nu,t}\delta_{\nu,t} \leq X_{\xi,t}$ and $\mathbf{E}_\mu[X_{\xi,t}] = \mathbf{E}_\xi[1] \leq 1$ and $\delta_{\nu,t} \leq 1$. Note that a countable intersection of probability 1 events still has probability 1. This establishes the first implications of the lemma.

(iv) Equivalence: As for the second implication, by posterior Definition 7.2.2 and X_t,

$$\sum_{\nu\in\mathcal{M}} w_\nu X_{\nu,t}\delta_{\nu,t} = X_{\xi,t}\sum_{\nu\in\mathcal{M}} w(\nu|h_{<t})\delta_{\nu,t}$$

As already established, $X_{\xi,t} \to c < \infty$ w.μ.p.1. Furthermore $X_{\xi,t} \geq w_\mu > 0$. Hence w.μ.p.1, $\sum_\nu w_\nu X_{\nu,t}\delta_{\nu,t} \to 0$ iff $\sum_\nu w(\nu|h_{<t})\delta_{\nu,t} \to 0$. ∎

Theorem 7.3.6 (Self-optimizing [Hut05b]) Let $\mathcal{M}$ be a countable class of environments and π be any historic policy. If there exists a self-optimizing $\tilde{\pi}$ for $\mathcal{M}$ in the sense of Definition 7.3.3, then the Bayes-optimal policy π_ξ^* is self-optimizing for $\mathcal{M}$.

This is an important and central result of the book. Obviously, if there does not exist any self-optimizing policy $\tilde{\pi}$, we cannot expect π_ξ^* to be self-optimizing. In this sense, the sufficient condition in the theorem is also necessary, i.e. the assumptions are as weak as possible. The strength of this result compared to asymptotic optimality results later in Chapter 8 is that it is off-policy, i.e. π_ξ^* is self-optimizing even if the agent followed $\pi \neq \pi_\xi^*$ in the past.

Proof. For each environment $\nu \in \mathcal{M}$, we define

$$\delta_{\nu,t}(h_{<t}) := V_\nu^*(h_{<t}) - V_\nu^{\tilde{\pi}}(h_{<t})$$

Since rewards are bounded by 1 and $V_\nu^*(h_{<t})$ is a weighted average of rewards, we have $0 \leq V_\nu^{\tilde\pi}(h_{<t}) \leq V_\nu^*(h_{<t}) \leq r_{\max} = 1$, which implies $0 \leq \delta_{\nu,t} \leq 1$. Additionally we have,

$$
\begin{aligned}
0 &\leq w(\mu|h_{<t})\left[V_\mu^*(h_{<t}) - V_\mu^{\pi_\xi^*}(h_{<t})\right] \\
&\leq \sum_{\nu \in \mathcal{M}} w(\nu|h_{<t})\left[V_\nu^*(h_{<t}) - V_\nu^{\pi_\xi^*}(h_{<t})\right] \\
&\leq \sum_{\nu \in \mathcal{M}} w(\nu|h_{<t})\left[V_\nu^*(h_{<t}) - V_\nu^{\tilde\pi}(h_{<t})\right] \\
&= \sum_{\nu \in \mathcal{M}} w(\nu|h_{<t})\delta_{\nu,t}(h_{<t})
\end{aligned}
$$

The first inequality comes from the fact that the optimal value function for μ will be greater than or equal to the value function for μ under any other policy. The second inequality is a result of adding non-negative terms. The third inequality comes from

$$
\sum_{\nu \in \mathcal{M}} w(\nu|h_{<t})V_\nu^{\pi_\xi^*}(h_{<t}) = V_\xi^{\pi_\xi^*}(h_{<t}) = V_\xi^*(h_{<t}) \geq V_\xi^{\tilde\pi}(h_{<t}) = \sum_{\nu \in \mathcal{M}} w(\nu|h_{<t})V_\nu^{\tilde\pi}(h_{<t})
$$

Dividing the above inequality by $w(\mu|h_{<t})$ we get,

$$
0 \leq V_\mu^*(h_{<t}) - V_\mu^{\pi_\xi^*}(h_{<t}) \leq \frac{1}{w(\mu|h_{<t})}\sum_{\nu \in \mathcal{M}} w(\nu|h_{<t})\delta_{\nu,t}(h_{<t}) = \frac{1}{w_\mu}\sum_{\nu \in \mathcal{M}} w_\nu \frac{\nu(e_{<t}||a_{<t})}{\mu(e_{<t}||a_{<t})}\delta_{\nu,t}(h_{<t})
$$

which by Lemma 7.3.4 converges to 0 for $t \to \infty$ w.μ^π.p.1. ∎

Figure 7.1 relates various important environment classes $\mathcal{M}$ and whether they admit self-optimizing policies or not. Ergodic nth-order MDPs are a large class that admit self-optimizing policies, hence π_ξ^* is self-optimizing for them. Unfortunately, without any form of ergodicity assumption, self-optimizing fails due to traps, such as the Heaven-Hell Example 8.1.3. Therefore, the class of all (semi)computable chronological environments does not admit self-optimizing policies.

7.4 Universal Optimal Agent AIXI

So far we considered general environment classes $\mathcal{M}$ and general priors w_ν. We can combine our choice of prior and class of environments from Section 3.7, motivated again by Occam's razor and Epicurus' principle of multiple explanations, with the notion of a Bayesian agent to arrive at the universal Bayesian agent AIXI. AIξ of the previous section is like AIμ but it does not know the true environment, so we replaced μ with ξ, the Bayesian mixture environment over $\mathcal{M}$ with some prior w_ν, left unspecified. We define AIXI to be the universal Bayesian agent with the class of all lower-semicomputable chronological semimeasures $\mathcal{M} = \mathcal{M}_{sol}$ and using the universal prior $w_\nu = w_\nu^U := 2^{-K(\nu)}$ and mixture distribution $\xi = \xi_U$.

Definition 7.4.1 (Universal optimal agent AIXI) AIXI is the Bayes-optimal agent for the universal mixture $\xi_U() := \sum_{\nu \in \mathcal{M}_{sol}} w_\nu^U \nu()$, with universal prior $w_\nu^U := 2^{-K(\nu)}$:

$$
\pi^{\text{AIXI}} := \pi_{\xi_U}^* = \arg\max_\pi V_{\xi_U}^\pi
$$

AIXI is the most "intelligent" environment-independent agent possible because it optimally learns and acts in a wide range of environments, adapts to different situations, and uses a mixture of environments based on the principle of Occam's razor. AIXI learns the environment it is in by updating its beliefs over time. As it interacts with an environment, it gathers more information and refines its understanding, allowing it to make better-informed decisions.

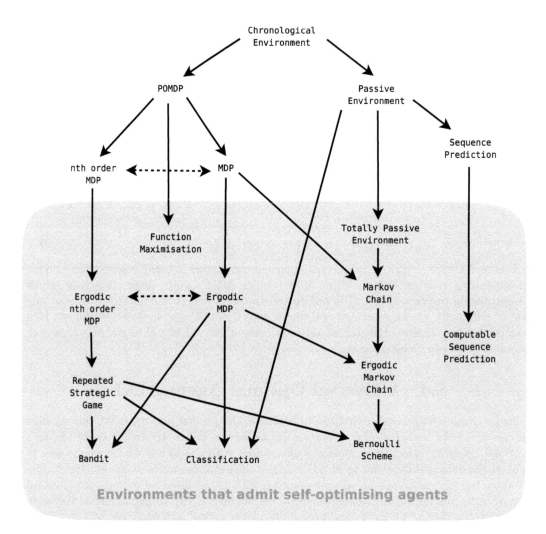

Figure 7.1: *Taxonomy of environments. Downward arrows indicate that the class below is a special case of the class above. Dotted horizontal lines indicate that two classes of environments are reducible to each other. The grayed area contains the classes of environments that admit self-optimizing agents, that is, the environments in which a universal agent will learn to behave optimally [Leg08]. Sequence prediction is covered in Chapter 3, Markov chains are covered in Chapters 4 and 5, and (ergodic) (nth-order) MDPs are covered in Chapters 11, 12 and 14, See [Hut05b, Leg08] for a discussion of the other classes and proofs.*

Theorem 7.4.2 (Expectimax Bayesian form of AIXI) For finite lifetime m and no discounting, i.e. $\gamma_k = [\![k \leq m]\!]$, given interaction history $h_{<t}$, the action of AIXI at time t has the explicit expectimax expression

$$a_t = \pi^*_{\xi_U}(h_{<t}) = \arg\max_{a_t} \sum_{e_t} \max_{a_{t+1}} \sum_{e_{t+1}} ... \max_{a_m} \sum_{e_m} \left[\sum_{i=t}^{m} r_i \right] \sum_{\nu \in \mathcal{M}_{sol}} w^U_\nu(h_{<t}) \prod_{k=t}^{m} \nu(e_k|h_{<k}a_k)$$

where $w^U_\nu(h_{<t})$ is the posterior (see Definition 7.2.2) for universal prior $w^U_\nu(\epsilon) := 2^{-K(\nu)}$,

Proof. The explicit expression for action $a_t = \pi^*_\nu(h_{<t})$ of AIν has been given in (6.7.7). Replacing ν by ξ gives us AIξ. The product can be written as

$$\prod_{k=t}^{m} \xi(e_k|h_{<k}a_k) = \frac{\xi(e_{1:m}||a_{1:m})}{\xi(e_{<t}||a_{<t})} = \frac{\sum_\nu w_\nu \nu(e_{1:m}||a_{1:m})}{\xi(e_{<t}||a_{<t})} \tag{7.4.3}$$

$$= \sum_{\nu \in \mathcal{M}} w_\nu \frac{\nu(e_{<t}||a_{<t})}{\xi(e_{<t}||a_{<t})} \prod_{k=t}^{m} \nu(e_k|h_{<k}a_k) = \sum_{\nu \in \mathcal{M}} w(\nu|\alpha_{<t}) \prod_{k=t}^{m} \nu(e_k|h_{<k}a_k)$$

where we used Definitions 7.2.1 and 7.2.2 a couple of times. AIXI is simply the special case for $\mathcal{M}_{sol}$ and w^U_ν and ξ_U, which proves the theorem. ∎

There is yet another (arguably most elegant) formulation of AIXI which we stated in Section III. We define it formally here and show its equivalence to Theorem 7.4.2. It is based on the following variant of monotone Turing machine:

Definition 7.4.4 (Chronological Turing machine) A (monotone) Turing machine T is called a chronological Turing machine if, when fed a program p transforms an input data stream $a_{1:m}$ to an output data stream $e_{1:m}$ such that for all $k \leq m$, it must print each "percept" e_k on the output tape prior to reading the next "action" a_{k+1} from the input tape. Formally p', a_k, o_k, r_k, $e_k \equiv o_k r_k$ are all understood to be encoded prefix-free so that they can be uniquely decoded from $a_{1:m}$ and $e_{1:m}$. Notationally we define:

$$T(p'a_{1:m}) \to e_{1:m} \text{ is true} \quad \text{iff} \quad T(p'a_{1:k}) = e_{1:k} \text{ for all } k \leq m$$

Theorem 7.4.5 (Expectimax Solomonoff form of AIXI) The action of undiscounted horizon-m AIXI at time t given interaction history $h_{<t}$ is:

$$a^*_t = \pi^*_M(h_{<t}) = \arg\max_{a_t} \sum_{e_t} ... \max_{a_m} \sum_{e_m} [r_t + ... r_m] M(e_{1:m}||a_{1:m})$$

$$\text{where} \quad M(e_{1:m}||a_{1:m}) := \sum_{p:U(p'a_{1:m}) \to e_{1:m}} 2^{-\ell(p)}$$

is a chronological version (see Definition 6.3.1) of Solomonoff's distribution (3.8.3), and U is a chronological Universal Turing machine (Definition 7.4.4).

Proof. The inner sum of Theorem 7.4.2 can be written as

$$\sum_{\nu \in \mathcal{M}} w(\nu | h_{<t}) \prod_{k=t}^{m} \nu(e_k | h_{<k} a_k) = \frac{\xi(e_{1:m} || a_{1:m})}{\xi(e_{<t} || a_{<t})} \tag{7.4.6}$$

$$\text{where} \quad \xi(e_{1:m} || a_{1:m}) := \sum_{\nu \in \mathcal{M}} w_\nu \nu(e_{1:m} || a_{1:m}) \tag{7.4.7}$$

$$\text{and} \quad \nu(e_{1:m} || a_{1:m}) := \prod_{k=1}^{m} \nu(e_k | h_{<k} a_k) \tag{7.4.8}$$

Compare (7.4.7) with the posterior version (7.4.3). The double-bar denotes that μ and ξ are chronological semimeasures (Definition 6.3.1). Since the denominator $\xi(e_{<t} || a_{<t})$ in (7.4.6) does not depend on any quantities after time $t-1$, we can pull it out of the expectimax expression in Theorem 7.4.2. Since it does not affect the $a_t = \arg\max_{a_t} ...$, we can ignore it entirely and consider only the numerator $\xi(e_{1:m} || a_{1:m})$ instead. $M(e_{1:m} || a_{1:m})$ is a chronological version of Solomonoff's distribution (3.8.3), and is a chronological semimeasure as well. Note that this is crucially different from conventional conditioning $M(e_{1:m} | a_{1:m})$ (3.8.4). For the universal choice of weights $w_\nu^U := w_\nu = 2^{-K(\nu)}$, i.e. ξ_U, analogous to Theorem 3.8.8, one can show that

$$M(e_{1:m} || a_{1:m}) \overset{\times}{=} \xi_U(e_{1:m} || a_{1:m}) \tag{7.4.9}$$

Hence, within two irrelevant multiplicative factors (the denominator $\xi_U(e_{<t} || a_{<t})$ and the universal constant hidden in $\overset{\times}{=}$), the l.h.s. of (7.4.6) equals $M(e_{1:m} || a_{1:m})$, hence the latter can replace the former in Theorem 7.4.2. ∎

Theorem 7.4.10 (AIXI cannot act poorly in good environments) If $\mu \in \mathcal{M}$ is deterministic and there exists an $\varepsilon > 0$ such that for some $h_{1:\infty}$ we have $V_\mu^*(h_{<t}) > \varepsilon \; \forall t$, then

$$V_\xi^*(h_{<t}) \not\to 0 \quad \text{for} \quad t \to \infty$$

This is true in particular for the history $h_{1:\infty}$ deterministically generated from (deterministic) μ and (deterministic) π_ξ^*.

Proof. Using convexity of V^* (Theorem 7.2.5) and Definition 7.2.2 we have

$$V_\xi^*(h_{<t}) \geq \sum_{\nu \in \mathcal{M}} w(\nu | h_{<t}) V_\nu^*(h_{<t}) \geq w_\mu \frac{\prod_{k=1}^{t-1} \mu(e_k | h_{<k} a_k)}{\prod_{k=1}^{t-1} \xi(e_k | h_{<k} a_k)} V_\mu^*(h_{<t}) > \varepsilon w_\mu$$

since $\prod_{k=1}^{t-1} \mu(e_k | h_{<k} a_k) = 1$ as μ is deterministic and $\xi(e_k | h_{<k} a_k) \leq 1$ and $V_\mu^*(h_{<t}) > \varepsilon$. ∎

7.5   Exercises

1. [C15] Explicate the proof of Definition 7.2.2 by lifting the proof of Theorem 3.1.7.

2. [C15] Use Blackwell&Dubin's Theorem 3.9.5 to prove the agent version Theorem 7.2.6.

3. [C21i] Complete the proof of Theorem 7.3.1.

4. [C15i] This exercise extends Definition 3.9.1 and Example 3.9.2 to the agent case: Random variables $X_t = X_t(h_{1:\infty})$ are said to form a supermartingale if $\mathbf{E}[X_t|h_{<t}] \leq X_{t-1}$. Show that $X_t := \nu(e_{1:t}||a_{1:t})/\mu(e_{1:t}||a_{1:t})$ in the proof of Lemma 7.3.4 is a μ^π-supermartingale.

5. [C30m] Show that X_t from the previous exercise converges to a finite constant w.μ.p.1.

6. [C30u] Show that for every enumerable chronological semimeasure ρ there exists a Turing machine T of length $\ell(T) = K(\rho) + O(1)$ that computes it, i.e. $\rho(h_{1:t}) = \sum_{q:T(q,a_{1:t})\to e_{1:t}} 2^{-\ell(q)}$ [Hut05b].

7. [C20u] Let $\delta(t) = \sum_{\nu\in\mathcal{M}} w_\nu \delta_\nu(t)$ and $\sum_{\nu\in\mathcal{M}} w_\nu \leq 1$ Show that the boundedness assumption $0 \leq \delta_\nu(t) \leq c$ is necessary for $\delta_\nu(t) \to 0$ as $t \to \infty$ to imply existence and/or convergence of $\delta(t) \to 0$. Show that $\delta_\nu(t) = O(f(t))$ $\forall\nu\in\mathcal{M}$ does not necessarily imply $\delta(t) = O(f(t))$ if $\mathcal{M}$ is infinite, even for bounded δ_ν.

8. [C30ui] Consider an environmental class $\mathcal{M}$ that admits self-optimizing policies. We want to study the effect of additionally believing in some $\rho \notin \mathcal{M}$ with some small probability α. The new belief prior is $\xi' := (1-\alpha)\xi + \alpha\rho$. Show that a belief α in ρ much smaller than the belief in the w_μ in the true environment $\mu \in \mathcal{M}$ only causes a small corruption of the self-optimizing property. More precisely, show $\lim\sup_{t\to\infty}[V_\mu^*(h_{<t}) - V_\mu^{\pi_{\xi'}^*}] \leq \frac{\alpha}{(1-\alpha)w_\mu}$ [Hut05b].

9. [C40o] How does AIXI perform when the set of observations $\mathcal{O}$ and the set of actions $\mathcal{A}$ are countably infinite? Is it still self-optimizing?

10. [C35] How does AIXI perform on the sequence prediction problem? When the set of environments consist of sequence prediction environments, how well does AIXI do?

11. [C10] Show that replacing the posterior $w(\nu|h_{<t})$ with the prior w_ν in Theorem 7.4.2 leaves the optimal action unaffected. Show that the "constant" $\xi(e_{<t}||a_{<t})$ (7.4.6) can indeed be dropped from Theorem 7.4.2.

12. [C20] Show that there exists a universal chronological Turing machine (see Definition 7.4.4).

13. [C20] Show that $\nu(e_{1:m}||a_{1:m})$ (7.4.8) and $\xi(e_{1:m}||a_{1:m})$ (7.4.7) and $M(e_{1:m}||a_{1:m})$ (Theorem 7.4.5) are all chronological semimeasures.

14. [C20] Show that there exists a universal chronological semimeasure (see Definition 6.3.1) in the sense of dominating all other semicomputable chronological semimeasures, analogous to Proposition 3.1.4.

15. [C30] Show that $M(e_{1:m}||a_{1:m}) \stackrel{\times}{=} \xi_U(e_{1:m}||a_{1:m})$ (7.4.9).

16. [C20] Show that $M(e_{1:m}||a_{1:m})$ is a universal chronological semimeasure (see Definition 6.3.1) in the sense of dominating all other semicomputable chronological semimeasures, analogous to Proposition 3.1.4.

17. [C40o] Derive a model-free version of AIXI in the same way Q-learning is a model-free RL algorithm for MDPs.

18. [C33] AIXI has uncertainty over its future and models this with M. What if AIXI was also uncertain about its past? How would AIXI resolve this uncertainty and what would this AIXI look like?

19. [C25] How well does AIXI perform on environments not in $\mathcal{M}$? Hint: Make an assumption about how close the environment is to an environment in $\mathcal{M}$, possibly similar to Section 3.4.

7.6   History and References

Universal artificial intelligence began with [Hut00], which defined the agent AIXI, included arguments as to why AIXI is the most intelligent agent possible, shows how AIXI solved several interesting environment classes, provided a computable approximation of AIXI, and introduced an intelligence order relation [Hut00] which led to the universal intelligence measure Υ (Definition 16.7.1). This tech report was extended in [Hut03d] and summarized in [Hut01e, Hut01f, Hut13b]. Shortly afterwards this work was extended further into the book [Hut05b], which contained all the previously mentioned results. Study into the meaning of intelligence as well as how it has been and should be defined, measured, and tested was conducted in [LH05, LH06, LH07a, LH07c, LH07b]. This work included additional arguments for why AIXI should be considered the most intelligent agent. Universal artificial intelligence and how it can be used as a top-down approach to the problem of building artificial general intelligence was described in [Hut07f]. Many of the open problems in UAI were presented in [Hut09g], some of which have since been solved. The axiomatic approach [SH11a, SH11b] demonstrated how rational behavior of an agent naturally leads to AIXI. The advances that have been made in the field of UAI since its inception as well as some possible future directions are discussed in [Hut12b]. The latest survey [EH18b] gives a succinct description of the UAI field and much of the work that has taken place in this field. Figure 7.1 shows a variety of environment classes for which self-optimizing policies (don't) exist. The proofs can be found in [Hut05b, LH04]. [RH06, RH08a] develops more general criteria that enable (upper) self-optimizingness, based on the introduced notions of recoverable, strongly explorable, and (worst-case) value-stable environments.

Bayesian reinforcement learning.   UAI is a form of history-based universal Bayesian reinforcement learning, though the latter is mostly studied within the MDP frameworks; see [GMPT15] for a survey. [BVB13] shows how Bayesian inference can be used to reduce the complexity of learning an environment by factorizing the observation space. This was tested empirically on Atari 2600 games [BNVB13]. In [SHL97], an agent is defined to find an optimal policy for the maximum a posteriori (MAP) estimate of the true environment. [Str00a] takes a similar approach, solving for the environment with the highest likelihood. [RK17] takes a Bayesian approach to bandits, a subset of MDPs. [OB10b] allows for the agent to model external interventions to its behavior.

Partial observability.   Most reinforcement learning is formulated within the MDP framework, but in reality an agent often receives only limited information about the environment, and some information about the state is hidden from view. Such environments are called Partially Observable Markov Decision Processes (POMDPs), to which approaches based on Bayesian inference have also been developed. [PVHR06] gives an analytic solution to a Bayesian model-based approach to POMDPs, as well as an algorithm, BEETLE, for MDPs. They extend BEETLE to POMDPs in [PV08]. [RCdP07] considered the space of all probability distributions over states, called the *belief space*, and approximate this via a finite-dimensional subspace to obtain ε-optimal solutions. Both of [RCdP07, PV08] factor the domain of beliefs using mixtures of products of Dirichlet distributions. [WLBS05] uses "sparse sampling" to approximate Bayes-optimal decision making. POTMMCP [SKH23] is an online MCTS-based planning method for type-based reasoning in large POMDPs suitable for large planning horizon, which comes with theoretical convergence guarantees and good practical performance in a multi-agent setup. A heuristic market-based RL algorithm by [Bau99, Bau04] has been evaluated in POMDP environments in [KHS01b], where reward is money which is conserved and scarce, and agents have to pay for services and compute, and receive from other agents for useful solutions provided.

Causality. This book does not explicitly consider causality, which may be surprising. The actions and precepts in history $h_{1:\infty} = a_1 e_1 a_2 e_2 ...$ have a clear temporal order and the chronology condition in Definition 6.3.1 ensures that actions and percepts only depend on past and not future actions and percepts. That is all we need. There are of course intricate research questions that require a deeper understanding and more explicit treatment of causality [Pea00, PGJ16, PM18, PJS17] within Universal AI [OKD⁺21, EHKK21, CVH21, EKH19, EH18a, ELH15] and beyond [HH22]. For instance, [MWV⁺21] considers the difficult credit assignment problem of determining an action's influence on future rewards from a counterfactual point of view.

Chapter 8

Optimality of Universal Agents

Perfection is achieved, not when there is nothing more to add, but when there is nothing left to take away.

Antoine de Saint-Exupéry

To construct an optimal agent, it is necessary to define what is meant by optimal. Throughout this chapter, we will discuss many of the definitions of optimality in the literature. In Section 8.1 we will give the definition of each optimality criterion, and describe why it is useful and some of the problems it may have. Next, in Section 8.2, we show how the behavior of a Bayesian agent may be undesirable under some choices of prior. We also give reasons why these priors will not be used too often (or at all). Lastly, in Section 8.3 we explore some potential problems with these optimality criteria, and under what circumstances there exist policies that are optimal with respect to these criteria.

8.1   Definitions of Optimality

In this section we describe a variety of optimality criteria which a-priori look sensible. We discuss their relative (dis)advantages, and whether AIXI or any other agent can actually attain them. An agent that maximizes the value for the environment μ it interacts with is called *μ-optimal*. Achieving this would be ideal, but is too strong a requirement for a general agent. An *ε-optimal* agent is allowed to have up to ε less than maximal value, but even this is too strong. Weaker still is the requirement that an agent only be optimal "in the limit". For an *asymptotically optimal* agent we require that the future value approaches optimality only with increasing experience. This notion comes in *stronger* and *weaker* flavors: *almost sure, in expectation*, in *Cesàro average*, which some agents can indeed achieve. Unfortunately, asymptotically we are all dead, so these notions are somewhat weak. A *Bayes-optimal* agent maximizes its value averaged over all considered environments, which is possibly the most reasonable criterion. Unfortunately Bayes-optimal agents may not be asymptotically optimal. *Regret* is a relative notion of performance, comparing the lifetime performance of an agent from start to death to an agent that is informed about the true environment in advance. Regret is a much stronger notion of optimality than asymptotic optimality. Agents with finite regret are also asymptotically optimal, but the converse is only true if the agent can *recover* from mistakes (a form of *ergodicity* assumption). *Pareto optimality* is a very weak notion of optimality. An agent violating it can hardly be regarded optimal in any interesting sense. In the context of Universal AI, this notion turns out to be even vacuous. A stronger/refined *balanced* version of Pareto optimality turns to be equivalent to Bayes-optimality.

8.1.1 ν-Optimality

Environment-based optimality, or ν-optimality for some environment ν, states that a policy is optimal if it achieves the maximum value for all possible histories. The value (total expected reward) of an agent π in environment ν has been formally defined Section 6.6.

Definition 8.1.1 (Optimal policy) A policy π is *optimal* in an environment ν (ν-optimal) iff for all histories, π attains the optimal value: $V_\nu^\pi(h) = V_\nu^*(h)$ for all $h \in (\mathcal{A} \times \mathcal{E})^*$. The action a is an *optimal action* iff $a \in \arg\max_{a'} \pi_\nu^*(a'|h)$ for some ν-optimal policy π_ν^*.

   This notion of optimality has been introduced in Chapter 6. If we know in advance the true environment ν the agent faces, we can choose $\nu = \mu$. The resulting μ-optimal agent AIμ with policy π_ν^* maximizes the expected reward over its lifetime. As such it is a very natural optimality criterion. The dependence on the lifetime or discount γ could be eliminated in theory by choosing a universal discount $\gamma_t = 2^{-K(t)}$ with infinite horizon (see Table 6.4). The assumption that the value is a *sum* of *expected* and *non-negative* rewards is rather mild [Hut05b, 8.5.1]. The assumption that μ is known may only be met in simple artificial settings, such as chess games, but is completely unrealistic for AGI purposes. This was the reason for introducing AIXI, but AIXI is not μ-optimal; it is ξ-optimal, and has to learn the true environment μ (see Chapter 7 and Section 8.1.3). μ-optimality for the true environments is a too narrow/strict definition of optimality for AGI purposes.

   One idea to solve these problems is to relax the notion of ν-optimality to the slightly weaker ε-optimality version.

Definition 8.1.2 (ε-optimal policy) A policy π is ε-optimal in an environment ν iff $V^*_\nu(h_{<t}) - V^\pi_\nu(h_{<t}) < \varepsilon$ for all histories $h_{<t} \in \mathcal{H}$.

Unfortunately even this weakened property of ε-optimality may be too strong a property for an agent to have, since it still requires certain prior knowledge of the environment ν:

Example 8.1.3 (Heaven-Hell) Let ν_1 be an environment (Figure 8.1) with two actions: left and right. If the left action is taken, then the agent will receive the lowest allowed reward (0) forever (hell), and if the right action is taken, the agent will receive the highest allowed reward (1) forever (heaven). The policy $\pi^*_{\nu_1}$ that takes $a_1 =$ right is ν_1-optimal. Let ν_2 be identical to ν_1 except that the rewards for left and right are switched. So in ν_2, the policy $\pi^*_{\nu_1}$ will receive the lowest reward forever, and hence $\pi^*_{\nu_1}$ is not ν_2-optimal. Of course, $\pi^*_{\nu_2}$ is ν_2-optimal, but this is not ν_1-optimal. The point is that no policy can be ν_1-optimal *and* ν_2-optimal. The best compromise is a random policy $\tilde{\pi}(\epsilon) = \frac{1}{2}$, which achieves $V^{\tilde{\pi}}_{\nu_i} = \frac{1}{2} < 1 = V^*_{\nu_i}$, that is, for $\varepsilon < \frac{1}{2}$ no policy can be ε-optimal in ν_1 and ν_2 simultaneously. ◆

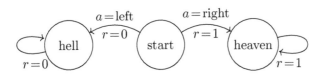

Figure 8.1: *An instance ν_1 of the Heaven-Hell environment. The other instance ν_2 is identical up to swapping all zero rewards for ones, and vice versa. The action taken in the heaven or hell states is irrelevant.*

8.1.2 Asymptotic Optimality

At an abstract level, an agent is asymptotically optimal if it eventually (asymptotically) does well (optimal). Mathematically this means that the value of the agent's policy converges to the value of the optimal agent. Importantly, the value must converge for all environments $\mu \in \mathcal{M}$ [LLOH17].

Definition 8.1.4 (Asymptotic optimality) A policy π is asymptotically optimal in an environment class $\mathcal{M}$ iff for all $\mu \in \mathcal{M}$, the difference between the optimal value function $V^*_\mu(h_{<t})$ and the value function under the policy π, $V^\pi_\mu(h_{<t})$, converges to zero as $t \to \infty$, i.e.,

$$V^*_\mu(h_{<t}) - V^\pi_\mu(h_{<t}) \to 0 \quad \text{for} \quad t \to \infty \quad \text{a.s.}|\text{i.p.}{=}\text{i.m.}|\text{Cesàro}$$

on histories drawn from μ^π.

Note that (the almost sure version of) asymptotic optimality is a special on-policy instantiation of self-optimizingness Definition 7.3.3, where the future policy $\tilde{\pi}$ matches the historic policy π. Note that the histories $h_{<t}$ are distributed according to μ^π. This is natural for $V^\pi_\mu(h_{<t})$, but even $V^*_\mu(h_{<t})$ is the value of the optimal agent in μ but given the μ^π-expected history $h_{<t}$.

When convergence is almost sure (a.s.), then it is called strong asymptotic optimality. When convergence is in mean or in probability (i.p), it is called asymptotic optimality in mean or in probability, respectively. When convergence is in Cesàro average, it is called weak asymptotic optimality. Since we preserve the logical implications from probability (Figure 2.8), strong asymptotic optimality implies asymptotic optimality in mean, in probability, and also weak asymptotic optimality.

The first asymptotic optimality we will define is weak asymptotic optimality. Essentially an agent is weakly asymptotically optimal if its value function converges to the optimal value function on average, also known as convergence in *Cesàro mean*. This means that the agent may make an infinite number of errors during the interaction with the environment, so long as the average number of errors is not too high, then the agent will still be weakly asymptotically optimal.

Definition 8.1.5 (Weak asymptotic optimality) A policy π is weakly asymptotically optimal in an environment class $\mathcal{M}$ iff for all $\mu \in \mathcal{M}$,

$$\mathrm{P}_\mu^\pi\left(\lim_{n \to \infty} \frac{1}{n} \sum_{t=1}^{n} [V_\mu^*(h_{<t}) - V_\mu^\pi(h_{<t})] = 0 \right) = 1$$

Next we consider asymptotic optimality in probability/mean. Convergence in probability and convergence in mean are separate concepts, however, since the difference in value functions, $V_\mu^*(h_{<t}) - V_\mu^\pi(h_{<t})$, is always bounded (by 1) and non negative (from the definition of the V_μ^*), asymptotic optimality in probability and asymptotic optimality in mean are equivalent. A policy is asymptotically optimal in mean if the μ^π-expected difference in value functions $V_\mu^*(h_{<t}) - V_\mu^\pi(h_{<t})$ goes to 0 as t goes to infinity.

Definition 8.1.6 (Asymptotic optimality in probability/mean) A policy π is asymptotically optimal in probability (or mean) in an environment class $\mathcal{M}$ iff for all $\mu \in \mathcal{M}$,

$$\lim_{t \to \infty} \mathbf{E}_\mu^\pi[V_\mu^*(h_{<t}) - V_\mu^\pi(h_{<t})] = 0$$

Lastly we have strong asymptotic optimality, a notion that will turn out to be exceedingly hard and in some cases impossible to achieve.

Definition 8.1.7 (Strong asymptotic optimality) A policy π is strongly asymptotically optimal in an environment class $\mathcal{M}$ iff for all $\mu \in \mathcal{M}$,

$$\mathrm{P}_\mu^\pi\left(\lim_{t \to \infty} (V_\mu^*(h_{<t}) - V_\mu^\pi(h_{<t})) = 0 \right) = 1$$

As mentioned, the definition of strong asymptotic optimality is an on-policy version of self-optimizingness. If in Theorem 7.3.6 we choose as historic policy the Bayes-optimal policy $\pi = \pi_\xi^*$ itself, then the conclusion of Theorem 7.3.6 is that π_ξ^* is strongly asymptotically optimal.

8.1.3   Bayesian Optimality

Considered by many to be the most reasonable definition of optimality, Bayesian optimality is much like one would expect, optimal in a Bayesian sense with respect to a mixture over a class of environments.

Definition 8.1.8 (Bayesian optimality) A policy π is Bayes-optimal with respect to prior $\{w_\nu\}$ over environment class $\mathcal{M}$ if for all histories $h_{<t} \in \mathcal{H}$,

$$V_{\xi_w}^\pi(h_{<t}) = V_{\xi_w}^*(h_{<t})$$

where ξ_w is the Bayesian mixture over $\mathcal{M}$ with prior $\{w_\nu\}$ (Definitions 7.2.1 and 7.2.2).

Important instances of Bayesian optimality are as follows: the agent AIXI is a Bayes-optimal agent in the class $\mathcal{M}_{sol}$ with the prior $w_\nu = 2^{-K(\nu)}$, additionally, any agent which maximizes Legg–Hutter intelligence (Definition 16.7.1) is also Bayes-optimal.

When deciding on an optimality criterion it is important to be very clear about what kind of behavior we want the resulting optimal agents to have, and at what timescale. We have discussed asymptotic optimality criteria, which will *eventually* perform well, but possibly not before the heat-death of the universe. In Section 8.1.4 we will discuss regret minimization which works well in the short term, but may have undesirable asymptotic behavior. Bayes-optimality can have the best (compromise) of both worlds. We have shown in Theorems 7.3.1 and 7.3.6 that under certain conditions, Bayes-optimal agents will perform well asymptotically, and under the right choice of prior, Bayes-optimal agents can perform well in the short term.

Bayes-optimality is equivalent to ξ_w-optimality, a special case of ν-optimality. The dependency on w and $\mathcal{M}$ cannot easily be avoided: If we were to weaken Bayes-optimality to merely require the existence of a $(w, \mathcal{M})$ such that $V_{\xi_w}^\pi(\cdot) = V_{\xi_w}^*(\cdot)$, then every policy would be Bayes-optimal, as we can simply choose a trivial environment ν_{easy} where any policy is optimal (e.g. an environment that always issues reward 1 regardless of action taken), and set $\mathcal{M} = \{\nu_{easy}\}$ (similar to the proof of Theorem 8.1.19). In the other direction, strengthening to every choice of $(w, \mathcal{M})$ also fails for a similar reason: Given any policy π, one can construct an adversarial environment ν_π that issues reward 0 for action $\arg\max_{a_t} \pi(a_t | h_{<t})$ and 1 for all other actions. Therefore, there cannot exist any Bayes-optimal policies, so the definition is useless. Bayes-optimality is only possible relative to some $(w, \mathcal{M})$ and only interesting for "interesting" choices of $\mathcal{M}$ and w, with $(2^{-K(\nu)}, \mathcal{M}_{sol})$ being the most interesting choice from an AGI perspective.

Relying on a prior and model class can be a downside in the sense that the optimality criterion depends (possibly heavily) on the choice of the prior, that is, under different priors the optimal agent may behave quite differently. In Section 8.2, we will show that there exist priors which can cause a Bayes-optimal agent to act in undesirable ways. On the other hand, there are good classes and priors for which Bayes-optimal policies are self-optimizing (see Figure 7.1). The most promising is the universal prior $2^{-K_U(\nu)}$ described earlier based on a *natural* universal Turing machine. This pushes the problem to finding a good *natural* Turing machine to be discussed later.

8.1.4 Regret Minimization

In hindsight it is sometimes easy to look back and see what would have been the best course of action. Indeed there are many circumstances where optimal choices only become known after the fact. The mathematical form of this notion is called *regret* [JOA10]. The regret of an agent π is the difference between the performance of the optimal agent and the performance of π. Formally:

Definition 8.1.9 (Regret [Lei16b]) The *regret* of a policy π in an environment μ with horizon m is

$$\text{Regret}_m(\pi,\mu) := \sup_{\pi'} \mathbf{E}_\mu^{\pi'} \left[\sum_{t=1}^m r_t \right] - \mathbf{E}_\mu^\pi \left[\sum_{t=1}^m r_t \right] = m \left[\sup_{\pi'} V_{\mu,\gamma=1}^{\pi',m}(\epsilon) - V_{\mu,\gamma=1}^{\pi,m}(\epsilon) \right]$$

Additionally we say a policy π has *sublinear regret* if $\text{Regret}_m(\pi,\mu) = o(m)$.

Note that since we are considering the finite horizon value unnormalized function $mV_\mu^{\pi,m}$ without discounting $\gamma_k = [\![k \leq m]\!]$, the regret may be at most $mr_{\max} = m$. While regret is a performance measure of the agent, sublinear regret is a notion of optimality of the agent. Essentially, an agent has sublinear regret if it does not take actions which are too suboptimal too often. Although it seems desirable to have an agent trying to perform as close as it can to the optimal agent, there are some downsides to regret minimization, the largest being how it makes the agent not care about the long term as it only cares about the value up to horizon m, almost the exact opposite problem to asymptotic optimality.

Example 8.1.10 (Regret is stronger than asymptotic optimality) Consider an agent π in an environment μ taking the least rewarding action for m steps, then performs the μ-optimal action on every time step afterward. If we used regret with horizon m to measure the performance of this agent, we would say this agent is as far from optimal as can be. However, since it takes the μ-optimal actions after a finite time, we know it is asymptotically optimal.   ◆

This duality between regret minimization and asymptotic optimality has sparked some research into seeing how one can combine these two into one criterion. For the following theorem, we need to make some extra assumptions on the choice of discount function.

Assumption 8.1.11 (Extra discount function assumptions) Let the discount function γ be such that

- $\gamma_t > 0$ for all t
- γ_t is decreasing in t
- $\lim_{t \to \infty} H_t(\varepsilon)/t = 0$ for all $\varepsilon > 0$.

where $H_t(\varepsilon)$ is the effective horizon (Definition 6.4.2).

The first assumption is basically stating that a positive reward is still worth more than nothing, regardless how far into the future it will be received. The second assumption requires that rewards that are closer to the present are always worth at least as much as those received later. The third assumption is a constraint on how farsighted the ε-effective horizon is permitted to be. Note that the geometric discount satisfies the assumptions above.

We also need to restrict the possible environments to ones which an agent can recover from taking "bad" actions. For this we will use the notion of *recoverable*. If an agent is acting in an environment it believes is safe to explore, it may take actions (fall off a cliff) from which it cannot learn not to repeat.

Let $\mathbf{E}_{\nu_1}^{\pi_1}[V_{\nu_2}^{\pi_2}(h_{<t})]$ denote the expected value function of a policy π_2 on environment ν_2 on a past history $h_{<t}$ generated by policy π_1 on environment ν_1.

Definition 8.1.12 (Recoverable environments) An environment μ is recoverable iff

$$\sup_\pi \left| \mathbf{E}_\mu^{\pi_\mu^*}[V_\mu^*(h_{<t})] - \mathbf{E}_\mu^\pi[V_\mu^*(h'_{<t})] \right| \longrightarrow 0 \quad \text{for} \quad t \to \infty$$

Note that the expectations are with respect to different policies: The history $h_{<t}$ is sampled from $\mu^{\pi_\mu^*}$, whereas $h'_{<t}$ is sampled from μ^π.

Intuitively, recoverable means that regardless of what the past history was, or how poor a policy π was following, switching to an optimal policy π^* at time t onwards converges to the same value obtained had the optimal policy been followed from the beginning. This means that no action can lock the agent out of parts of the environment required for maximum reward, and any mistakes made early on can be recovered from, which excludes things like cliffs to fall off from which the agent cannot return and traps such as the Heaven-Hell Example 8.1.3.

Theorem 8.1.13 (Sublinear regret in recoverable environments) If the discount function γ satisfies Assumption 8.1.11, the environment μ is recoverable, and π is asymptotically optimal in mean in $\mathcal{M} = \{\mu\}$, then $\text{Regret}_m(\pi, \mu) = o(m)$.

Proof. We follow the proof from [Lei16b]. Let $\pi_m := \arg\max_\pi \text{Regret}_m(\pi, \mu)$. We want to show that

$$\limsup_{m \to \infty} \mathbf{E}_\mu^{\pi_m}\left[\frac{1}{m} \sum_{t=1}^m r_t \right] - \mathbf{E}_\mu^\pi\left[\frac{1}{m} \sum_{t=1}^m r_t \right] \leq 0$$

Let $d_k^{(m)} := \mathbf{E}_\mu^{\pi_m}[r_k] - \mathbf{E}_\mu^\pi[r_k]$. We have that $-1 \leq d_k^{(m)} \leq 1$, since rewards are bounded between 0 and 1. From the definition of recoverable, since π and π_m are not worse than the worst policy, we have that

$$\left| \mathbf{E}_\mu^{\pi_\mu^*}[V_\mu^*(h_{<t})] - \mathbf{E}_\mu^\pi[V_\mu^*(h_{<t})] \right| \to 0 \quad \text{for} \quad t \to \infty \quad \text{and}$$

$$\sup_m \left| \mathbf{E}_\mu^{\pi_\mu^*}[V_\mu^*(h_{<t})] - \mathbf{E}_\mu^{\pi_m}[V_\mu^*(h_{<t})] \right| \to 0 \quad \text{for} \quad t \to \infty$$

Combining these with the triangle inequality we get

$$\sup_m \mathbf{E}_\mu^{\pi_m}[V_\mu^*(h_{<t})] - \mathbf{E}_\mu^\pi[V_\mu^*(h_{<t})] \to 0 \quad \text{for} \quad t \to \infty \tag{8.1.14}$$

Since π is asymptotically optimal in mean we have that

$$\mathbf{E}_\mu^\pi[V_\mu^*(h_{<t})] - \mathbf{E}_\mu^\pi[V_\mu^\pi(h_{<t})] \to 0 \quad \text{for} \quad t \to \infty$$

Combining the two previous lines we get

$$\sup_m \mathbf{E}_\mu^{\pi_m}[V_\mu^*(h_{<t})] - \mathbf{E}_\mu^\pi[V_\mu^\pi(h_{<t})] \to 0 \quad \text{for} \quad t \to \infty$$

Since $V_\mu^* \geq V_\mu^{\pi_m}$ we have

$$\limsup_{t \to \infty} \left[\sup_m \mathbf{E}_\mu^{\pi_m}[V_\mu^{\pi_m}(h_{<t})] - \mathbf{E}_\mu^\pi[V_\mu^\pi(h_{<t})] \right] \leq 0 \tag{8.1.15}$$

For any policy π' we can rewrite our expectation of the value function as

$$E_\mu^{\pi'}[V_\mu^{\pi'}(h_{<t})] = \mathbf{E}_\mu^{\pi'}\left[\frac{1}{\Gamma_t}\sum_{k=t}^{\infty}\gamma_k r_k\right] = \frac{1}{\Gamma_t}\sum_{k=t}^{\infty}\gamma_k \mathbf{E}_\mu^{\pi'}[r_k]$$

Then, using this form for (8.1.14) and (8.1.15) and $d_k^{(m)}$, we get

$$\limsup_{t\to\infty}\sup_m \frac{1}{\Gamma_t}\sum_{k=t}^{\infty}\gamma_k d_k^{(m)} \leq 0$$

Let $\varepsilon > 0$. Choose t_0 independent of m and large enough such that $\frac{1}{\Gamma_t}\sum_{k=t}^{\infty}\gamma_k d_k^{(m)} < \varepsilon$ for all m and $t \geq t_0$. Now we will split the sum $\sum_{k=1}^{m}d_k^{(m)}$ at the point t_0. We will additionally define b_t for $t_0 \leq t \leq m$ as

$$b_{t_0} = \frac{\Gamma_{t_0}}{\gamma_{t_0}} \quad \text{and} \quad b_t = \frac{\Gamma_t}{\gamma_t} - \frac{\Gamma_t}{\gamma_{t-1}} \quad \text{for} \quad t > t_0$$

The following equalities for b_t can easily be verified by inserting the definitions and swapping the two sums:

$$\sum_{t=t_0}^{m}d_t^{(m)} = \sum_{t=t_0}^{m}\frac{b_t}{\Gamma_t}\sum_{k=t}^{m}\gamma_k d_k^{(m)} \quad \text{and} \quad \sum_{t=t_0}^{m}\frac{b_t}{\Gamma_t} = \frac{1}{\gamma_m} \tag{8.1.16}$$

Additionally, for the sum from t_0 to m of b_t we get

$$\sum_{t=t_0}^{m}b_t = \frac{\Gamma_{t_0}}{\gamma_{t_0}} + \sum_{t=t_0+1}^{m}\left(\frac{\Gamma_t}{\gamma_t} - \frac{\Gamma_t}{\gamma_{t-1}}\right) = \frac{\Gamma_{m+1}}{\gamma_m} + \sum_{t=t_0}^{m}\left(\frac{\Gamma_t}{\gamma_t} - \frac{\Gamma_{t+1}}{\gamma_t}\right) = \frac{\Gamma_{m+1}}{\gamma_m} + m - t_0 + 1$$

From the definition of the effective horizon $H_m(\varepsilon)$ (Definition 6.4.2) and monotonicity assumption on γ we get

$$\varepsilon\Gamma_m \geq \Gamma_{m+H_m(\varepsilon)} = \Gamma_m - \gamma_m - \ldots - \gamma_{m+H_m(\varepsilon)-1} \geq \Gamma_m - H_m(\varepsilon)\gamma_m$$

$$\text{which implies} \quad \frac{H_m(\varepsilon)}{1-\varepsilon} \geq \frac{\Gamma_m}{\gamma_m} \geq \frac{\Gamma_{m+1}}{\gamma_m}$$

Combining everything we get for the regret

$$\text{Regret}_m(\pi,\mu) = \sum_{t=1}^{m}d_t^{(m)} \leq \sum_{t=1}^{t_0}d_t^{(m)} + \sum_{t=t_0}^{m}\frac{b_t}{\Gamma_t}\sum_{k=t}^{m}\gamma_k d_k^{(m)}$$

$$\leq t_0 + \sum_{t=t_0}^{m}\frac{b_t}{\Gamma_t}\sum_{k=t}^{\infty}\gamma_k d_k^{(m)} - \sum_{t=t_0}^{m}\frac{b_t}{\Gamma_t}\sum_{k=m+1}^{\infty}\gamma_k d_k^{(m)}$$

$$< t_0 + \sum_{t=t_0}^{m}b_t\varepsilon + \sum_{t=t_0}^{m}\frac{b_t\Gamma_{m+1}}{\Gamma_t}$$

$$= t_0 + \varepsilon\frac{\Gamma_{m+1}}{\gamma_m} + \varepsilon(m-t_0+1) + \frac{\Gamma_{m+1}}{\gamma_m}$$

$$\leq t_0 + \frac{(1+\varepsilon)H_m(\varepsilon)}{1-\varepsilon} + \varepsilon(m-t_0+1)$$

By Assumption 8.1.11 we have $H_m(\varepsilon) = o(m)$, hence

$$\limsup_{m\to\infty}\frac{1}{m}\text{Regret}_m(\pi,\mu) \leq \varepsilon$$

Since ε was arbitrary, we have the desired result.   ∎

8.1.5   Pareto Optimality

Related to the game theoretic definition of Pareto Efficiency [OR94], an agent is Pareto optimal if there is no other agent which is at least as good in every environment, and is strictly better in at least one environment.

Definition 8.1.17 (Pareto optimality)
   A policy π is Pareto optimal in a set of environments $\mathcal{M}$ iff there is no policy π' such that $V_\nu^{\pi'} \geq V_\nu^{\pi}$ for all $\nu \in \mathcal{M}$ and $V_\rho^{\pi'} > V_\rho^{\pi}$ for at least one $\rho \in \mathcal{M}$.

In general, the notion of Pareto optimality is useful when there is a set of conflicting criteria that we want to satisfy or be optimal in. The different criteria we are interested in are performance in different environments in $\mathcal{M}$. The set of Pareto optimal agents is called the Pareto frontier.

Theorem 8.1.18 (AIξ and AIXI are Pareto optimal)

Proof. Assume π_ξ^* is *not* Pareto optimal, and let π be the dominating policy, i.e. $V_\nu^\pi \geq V_\nu^{\pi_\xi^*}$ for all $\nu \in \mathcal{M}$ and strict inequality for at least one $\nu \in \mathcal{M}$. Then

$$V_\xi^\pi = \sum_\nu w_\nu V_\nu^\pi > \sum_\nu w_\nu V_\nu^{\pi_\xi^*} = V_\xi^{\pi_\xi^*} \equiv V_\xi^* \geq V_\xi^\pi$$

The two equalities follow from linearity of V_ρ Theorem 7.2.5. The strict inequality follows from the assumption and $w_\nu > 0$. The last inequality follows from the fact that π_ξ^* maximizes by definition the universal value Definition 6.6.1. The contradiction $V_\xi^\pi > V_\xi^\pi$ proves Pareto optimality of AIξ. AIXI is just the special case of AIξ with $\mathcal{M} = \mathcal{M}_{sol}$.   ∎

It is comforting that AIξ is Pareto optimal, since a violation would clearly render π_ξ^* intuitively suboptimal. Unfortunately for any class of environments which contains $\mathcal{M}_{comp}$, every policy is Pareto optimal [Lei16b], so Pareto-optimality is a vacuous notion for AIXI.

Theorem 8.1.19 (Every policy is Pareto optimal in any $\mathcal{M} \supseteq \mathcal{M}_{comp}$)

Proof. The proof follows [Lei16b]. Without loss of generality, assume the action space $\mathcal{A}$ has size 2. To prove the theorem by contradiction. Assume π is not Pareto optimal. Therefore there exists some a policy π' strictly dominating π in the sense that for all $\nu \in \mathcal{M}$ we have $V_\nu^{\pi'} \geq V_\nu^\pi$ and there exists a $\rho \in \mathcal{M}$ such that $V_\rho^{\pi'} > V_\rho^\pi$. Since $V_\rho^{\pi'} > V_\rho^\pi$, then in ρ there must exist some histories where π and π' will take different actions (have different distributions over actions for stochastic policies). Let $\dot{h}_{<t}$ be the (lexicographically) first history in which π and π' disagree, hence there exists an action $\dot{a}_t$ for which $\pi(\dot{a}_t|\dot{h}_{<t}) > \pi'(\dot{a}_t|\dot{h}_{<t})$. We can define a computable environment μ which is identical to ρ for any history that is not an extension of $\dot{h}_{<t}$, and then rewards 1 forever if action $\dot{a}_t$ is taken, and otherwise rewards 0. Since π takes $\dot{a}_t$ more likely than π' given $\dot{h}_{<t}$, and they coincide otherwise, we have $V_\mu^\pi > V_\mu^{\pi'}$ which contradicts the assumption that π' dominates π, since $\mu \in \mathcal{M}_{comp} \subseteq \mathcal{M}$.   ∎

There is a refined version of Pareto optimality called balanced Pareto optimality, which could be seen as Bayes optimality (Section 8.1.3) in a Pareto optimality form.

Definition 8.1.20 (Balanced Pareto optimality [Hut02c, Lei16b]) Given a prior $w \in \Delta'\mathcal{M}$ where $\mathcal{M}$ is an environment class, a policy π is balanced Pareto optimal if it achieves a better value across $\mathcal{M}$ weighted by $w \in \Delta'\mathcal{M}$ than any other policy, formally, if for all policies $\tilde{\pi}$,

$$\sum_{\nu \in \mathcal{M}} w_\nu [V_\nu^\pi - V_\nu^{\tilde{\pi}}] \geq 0$$

Note that all and only Bayes-optimal policies $\pi_{\xi_w}^*$ are balanced Pareto optimal for w:

$$\sum_{\nu \in \mathcal{M}} w_\nu [V_\nu^\pi - V_\nu^{\tilde{\pi}}] = V_{\xi_w}^\pi - V_{\xi_w}^{\tilde{\pi}} \geq 0 \; \forall \tilde{\pi} \quad \textit{iff} \quad \pi = \pi_{\xi_w}^*$$

Hence balanced Pareto optimality has the same features (advantages and problems) as Bayes optimality (Section 8.1.3), e.g. dependence on the prior. In Section 8.2, we will describe some of the possible bad priors which would cause balanced Pareto optimality to become a less useful definition.

8.2   Bad Priors

The prior is one of the most important aspects for Bayesian optimality. However, as we will show below, some priors are less than ideal, and can cause Bayesian optimal policies to perform quite poorly [Ors10, Lei16b, LH15b].

8.2.1   Dogmatic Prior

First we have the dogmatic prior: Given any policy and a history generated by that policy interacting with the environment, there is a prior for which following that policy is Bayes-optimal. We call this prior the dogmatic prior as the Bayes-optimal agent dogmatically believes that it must follow that policy. Dogmatically following a single policy will rarely be what we want our Bayes-optimal agent to do, as the agent will never learn from its mistakes, but note that the priors here are adversarially chosen.

Theorem 8.2.1 (Dogmatic prior [LH15b]) Let π be any deterministic computable policy, let ξ be any Bayesian mixture over $\mathcal{M}_{sol}$ with weights $w \in \Delta'\mathcal{M}_{sol}$, and let $\varepsilon \in \mathbb{Q}$ with $\varepsilon > 0$. There is a universal mixture ξ' with respect to some prior such that for any history $h_{<t}$ consistent with π and $V_\xi^\pi(h_{<t}) > \varepsilon$, the action $\pi(h_{<t})$ is the unique ξ'-optimal action.

Proof. This proof follows [Lei16b]. For all environments $\nu \in \mathcal{M}_{sol}$ we can define a new environment $\rho_{\pi,\nu} \in \mathcal{M}_{sol}$ such that $\rho_{\pi,\nu}$ is identical to ν until it receives an action that π would not take, then outputs reward 0 forever. We will now derive new weights w' to weigh environments $\rho_{\pi,\nu}$ higher than ν. Define

$$w'_\nu := \varepsilon w_\nu \quad \text{if} \quad \nu \neq \rho_{\pi,\nu'} \; \forall \nu' \in \mathcal{M}_{sol}$$
$$\text{and} \quad w'_{\rho_{\pi,\nu}} := (1-\varepsilon)w_\nu + \varepsilon w_{\rho_{\pi,\nu}} \quad \text{otherwise}$$

Checking that the new weights sum to 1,

$$\sum_{\nu \in \mathcal{M}_{sol}} w'_\nu \;=\; \sum_{\nu = \rho_{\pi,\nu}} w'_\nu + \sum_{\nu \ne \rho_{\pi,\nu'} \forall \nu'} w'_\nu$$

$$= \sum_{\nu = \rho_{\pi,\nu}} ((1-\varepsilon)w_{\rho_{\pi,\nu}} + \varepsilon w_\nu) + \sum_{\nu \ne \rho_{\pi,\nu'} \forall \nu'} \varepsilon w_\nu$$

$$= \sum_{\nu \in \mathcal{M}_{sol}} \varepsilon w_\nu + \sum_{\nu \in \mathcal{M}_{sol}} (1-\varepsilon) w_\nu$$

$$= \varepsilon + (1-\varepsilon) = 1$$

Therefore w' is a valid choice of prior over $\mathcal{M}_{sol}$, and we can define a Bayesian mixture ξ' over $\mathcal{M}_{sol}$ using the prior w'. We can also define the mixture over the ρ's by $\rho := \sum_{\nu \in \mathcal{M}_{sol}} w_\nu \rho_{\pi,\nu}$, and rewrite ξ' as $\xi' = \varepsilon \xi + (1-\varepsilon)\rho$.

From now on, let $h_{<t}$ be a history consistent with π i.e. actions generated by π. For such $h_{<t}$, $\rho_{\pi,nu} = \nu$, hence $\rho = \xi$, hence $\xi' = \xi$. Considering the class $\mathcal{M} = \{\xi, \rho\}$ with prior $w_\xi = \varepsilon$ and $w_\rho = 1-\varepsilon$, this implies that the posterior coincides with the prior on such $h_{<t}$ (see Definition 7.2.2). Linearity of the Q-value function (Theorem 7.2.5) for any policy π' and any action a_t can hence be expressed in terms of prior weights

$$Q_{\xi'}^{\pi'}(h_{<t}a_t) \;=\; \varepsilon Q_\xi^{\pi'}(h_{<t}a_t) + (1-\varepsilon)Q_\rho^{\pi'}(h_{<t}a_t)$$

Let $a^* = \pi(h_{<t})$ and a' be any other action. We will show that a^* is the ξ'-optimal action by showing that $V_{\xi'}^*(h_{<t}a^*) > V_{\xi'}^*(h_{<t}a')$.

$$Q_{\xi'}^*(h_{<t}a^*) \geq Q_{\xi'}^\pi(h_{<t}a^*) = Q_\xi^\pi(h_{<t}a^*) = V_\xi^\pi(h_{<t}) > \varepsilon$$
$$Q_{\xi'}^*(h_{<t}a') = \varepsilon Q_\xi^\pi(h_{<t}a') + (1-\varepsilon)Q_\rho^\pi(h_{<t}a') = \varepsilon Q_\xi^\pi(h_{<t}a') + 0 \leq \varepsilon$$

Therefore $a^* = \pi(h_{<t})$ is the ξ'-optimal action for history $h_{<t}$. ■

8.2.2 Indifference Prior

Another bad prior is the indifference prior. The indifference prior, under certain conditions, makes all policies Bayes-optimal with respect to the prior. Because all policies are optimal and therefore equal in value, the Bayes-optimal agent using this prior will choose its actions entirely based on whatever tie-breaker is used. Having the decisions of the agent based only on the tie-breakers is not ideal, as depending on the tie-breaker used, the agent can be convinced it should take any arbitrary action. This prior does however rely heavily on the assumption that eventually the discount normalization factor Γ_{m+1} will be 0, which is never the case with the classic geometric discounting.

Theorem 8.2.2 (Indifference prior [LH15b]) If there is an m such that the discount normalization factor $\Gamma_m = 0$, then there is a Bayesian mixture ξ' such that all policies are ξ'-optimal.

Proof. This proof follows [Lei16b]. Assume the action space is $\mathcal{A} = \{0,1\}$. Let U be some universal Turing machine and define U' such that U' does not halt on programs of length less than m, and for programs of length greater than $m-1$,

$$U'(p,a) \;:=\; U(p_{m:\ell(p)}, a \text{ xor } p_{1:\ell(a)})$$

Then we can define a chronological Solomonoff distribution (which by Theorem 3.8.8 extended to the agent case (7.4.9) is a Bayesian mixture ξ') with respect to U' as the underlying universal Turing machine as

$$\xi'(e_{<m}||a_{<m}) = \sum_{p:e_{<m}\sqsubseteq U'(p,a_{<m})} 2^{-\ell(p)}$$

$$\overset{(a)}{=} \sum_{s_{<m}p':e_{1:t}\sqsubseteq U'(s_{<m}p',a_{<m})} 2^{-m-1-\ell(p')}$$

$$= \sum_{s_{<m}} \sum_{p':e_{1:t}\sqsubseteq U(p',a_{<m} \text{ xor } s_{<m})} 2^{-m-1-\ell(p')}$$

$$\overset{(b)}{=} \sum_{s_{<m}} \sum_{p':e_{1:t}\sqsubseteq U(p',s_{<m})} 2^{-m-1-\ell(p')}$$

(a) comes from decomposing p as $s_{<m}p'$. (b) comes from the fact we are summing over all $s_{<m}$ and xor-ing them with $a_{<m}$, and this is identical to summing over all $s_{<m}$. Therefore the mixture ξ' does not depend on $a_{<m}$ for the first $m-1$ actions. For the remaining actions, since $\Gamma_m = 0$, the rewards after time step m do not matter. Since the mixture does not depend on the actions taken, all policies have the same value V_ξ^π, hence all policies are optimal with respect to this mixture. ∎

This example (of a prior) is very artificial, and not unexpected given the unnatural universal Turing machine U' which depends on m and reserves $m-1$ bits to "mask" the first $m-1$ actions. If we increase AIXI's lifetime while fixing the UTM U', the result no longer holds. An analogous result holds for Solomonoff induction: Theorem 3.8.9 and Corollary 3.5.14 imply that Solomonoff's distribution M makes at most $K(\mu)2\ln 2 = Km(x_{1:\infty})2\ln 2$ errors for predicting deterministic sequences $x_{1:\infty}$. In case the shortest program has length $> m$, there is no guarantee that we make less than m errors.

8.2.3 Bad Priors, Bad Agents

Clearly, when the above "bad priors" are used, the Bayes-optimal agent is still Bayes-optimal, but in both cases does not behave as we would like. Intuitively, we know that following an arbitrary fixed policy, or being indifferent to all actions are both not generally intelligent behavior. Does this mean that Bayesian optimality is a poor choice of optimality criteria?

No. The existence of a bad prior does not invalidate the idea of Bayesian optimality, it just means one has to be careful when choosing a prior. Importantly, the existence of bad priors does not mean good priors do not exist or are not easy to find. Indeed it is plausible that *natural* UTMs lead to good priors (see also Example 2.7.5).

8.3   Problems with Optimality Criteria

The quest for a "perfect" optimality criterion, strong enough that it is useful and weak enough that it can be satisfied, may in fact be futile. In this section we explore the weakness and feasibility of the optimality notions introduced in Section 8.1. Not only do we want an optimality criterion that has certain desirable properties such as eventually performing well and not falling into traps (generally regret minimization agents tend to avoid traps but asymptotically optimal agents may jump into traps), we also want the agents which satisfy the criteria to be somewhat practical, as we would like to eventually implement such agents in the real world. This is where criteria such as asymptotic optimality falls short, since for an arbitrarily long amount of time an asymptotically optimal agent may receive low or no reward [Lei16b].

Theorem 8.3.1 (No deterministic strong asymptotically optimal policy [LH11a]) There is no deterministic policy that is strong asymptotically optimal (Definition 8.1.7) in the class of environments $\mathcal{M} \supseteq \mathcal{M}_{comp}$.

Proof sketch. We prove this by contradiction. Assume that there is a deterministic strongly asymptotically optimal policy π. Then we construct two computable environments μ_1 and μ_2 which will be in $\mathcal{M}$ since $\mathcal{M} \supseteq \mathcal{M}_{comp}$. The environments μ_1 and μ_2 are identical up to a time T at which point the unique optimal action in μ_1 is *up* forever and the unique optimal action in μ_2 is *down* forever. Since π is strongly optimal in μ_1, there exists a T such that for all $t \geq T$, $\pi(h_{<t}) = up$ forever, however, this means that π is not strongly asymptotically optimal in μ_2, therefore we have a contradiction, hence there does not exist a deterministic strongly asymptotically optimal policy. ∎

Theorem 8.3.1 is a very general result, since it includes all deterministic policies, it also includes the Bayes-optimal policy AIXI. We can, however, be even more general and show that there does not exist a policy which can attain even weak asymptotic optimality for the class of all environments.

Theorem 8.3.2 (No weak asymptotically optimal policy for general $\mathcal{M}$) If we consider $\mathcal{M}$ to be the class of all environments, then there is no weak asymptotically optimal policy for $\mathcal{M}$.

Proof. This essentially comes from constructing an environment specifically tuned against a given policy π, one where the reward for actions is chosen such that there will be the largest difference between reward (value) of actions taken by π and the optimal actions. Since there are no restrictions on the computability level of the environments this is true for all policies π. ∎

The impossibility results above paint a bleak picture for the notion of asymptotic optimality for general history-based agents. But the situation is actually not as bad as it looks. The proofs crucially depend on the policy being deterministic. Weak (Theorem 9.4.1) and strong (Theorem 9.4.2) asymptotic optimality are actually achievable by stochastic policies which add extra exploration to AIξ (Section 9.4).

Which optimality criteria an artificial general intelligence would or should satisfy remains an open question. For instance, should we conclude from the fact that AIξ is not strongly asymptotically optimal to use Inq, which is? Or does Inq fare worse than AIξ w.r.t. other, possibly more important, optimality criteria? For instance, the price of asymptotic optimality is certain death [CHC21]. We have introduced and investigated a number of potential criteria, but many more exist, some are mentioned in the reference section. Studying how these criteria relate to each other and which are achievable under which conditions helps us narrowing down desirable optimality criteria achievable by (theoretical and practical) artificial general intelligences.

8.4   Exercises

1. [C10] Prove that an agent is balanced Pareto optimal iff it is Bayes-optimal.

2. [C20] Prove that strong asymptotic optimality implies both asymptotic optimality in mean and also weak asymptotic optimality.

3. [C20] Give an example of an environment class $\mathcal{M}$ and a family of policies $\{\pi_i\}$ such that no policy is Pareto optimal.

4. [C20] Prove that geometric discounting satisfies Assumption 8.1.11, and that all of the other discounting methods (power, finite lifetime, harmonic, hyperbolic, universal, no discount) from Table 6.4 fail at least one of the three assumptions.

5. [C15] Prove the equalities in (8.1.16).

6. [C24] Prove that if $\mathcal{A}$, $\mathcal{E}$ are finite, a Pareto optimal policy exists.

7. [C19] Prove that if π is not Pareto optimal, then it is Pareto dominated by a Pareto optimal policy π'.

8. [C24] Weaken the premises of Theorem 8.2.1 to no longer require that $V_\xi^\pi(h_{<t}) > \varepsilon$. Prove that the action $\pi(h_{<t})$ is a ξ'-optimal action, but may not be uniquely optimal.

9. [C15] Show that for all π, there exists an environment ν such that π is not ν-optimal.

10. [C30] Formalize the proof of Theorem 8.3.1.

11. [C18] Formalize the proof of Theorem 8.3.2.

12. [C28] Provide a counter-example that convergence-in-mean does not imply Cesàro-convergence almost surely.

13. [C20] Prove that convergence-in-mean implies Cesàro convergence-in-mean.

14. [C23] Prove convergence almost surely implies Cesàro convergence almost surely.

15. [C18] Provide a counter-example that weak asymptotic optimality does not imply asymptotic optimality in mean and a counter example that asymptotic optimality in mean does not imply weak asymptotic optimality.

16. [C32] Derive an ε-optimal equivalent of the notions of optimality discussed in this chapter. How do they differ from the non ε-optimal versions? When are they equivalent?

8.5   History and References

Notions of optimality. This chapter is based on material from [LH15b] and [Lei16b, Chp.5]. universal Bayesian agents were introduced in [Hut02c] which showed that AIξ is Self-Optimizing and Pareto-Optimal. A comprehensive review of many of the notions of optimality discussed in this chapter and beyond can be found in [Hut05b, LH15b, Lei16b]. Related to regret and asymptotic optimality are the PAC and sample-complexity based optimality notions. Near-optimal PAC bounds for finite discounted MDPs based on Upper Confidence Reinforcement Learning (UCRL) [JOA10] have been proven in [LH12, LH14c]. The first PAC result in *general* reinforcement learning was given in [LHS13b], with additional work being done in the case of feature abstractions being utilized in [RLDG22]. A simple but very general Bayesian regret bound covering and unifying many special cases was presented in [LVRD⁺21].

Asymptotic optimality.   The difficulty of asymptotic optimality was first demonstrated in [Ors10], where it was shown that AIXI is not asymptotically optimal. These results were extended in [LH11a] where the authors defined two notions of asymptotic optimality and showed that under certain conditions these can(not) be satisfied. One of the difficulties with asymptotic optimality is that to achieve it an agent needs to explore enough to gain sufficient information [RVR14, RCV16]. There are some downsides to this exploration, in particular it was shown if an agent explores enough to be asymptotically optimal then it also explores enough to die with certainty [CHC21]. Algorithms which are able to achieve both asymptotic optimality results and regret-based optimality [JOA10] results as well as the connection between the two are presented in [KLVS21, LLOH16, Lei16b]. The existence of a policy which is able to achieve strong asymptotic optimality was proven in [CCH19].

Chapter 9

Other Universal Agents

Artificial intelligence thrives at the crossroads of exploration and exploitation, where we must judiciously balance curiosity and pragmatism, ensuring that our creations serve as catalysts for human advancement rather than mere reflections of our limitations.

ChatGPT

Since the inception of the theory of Universal Artificial Intelligence there have been various extensions proposed, including extensions of the AIXI agent. Most of these extensions attempt to address shortcomings of AIXI. One downside is that AIXI fails to be asymptotically optimal; whether asymptotic optimality is actually a reasonable criterion for intelligent agents was discussed in Section 8.3. In this chapter we will explore several extensions to the AIXI agent, two of which do satisfy asymptomatic optimality.

In Section 9.1 we will look at the first extension, an optimistic agent. It has been long regarded that in sequence prediction, pessimism can be a good strategy to hedge one's bets. [GS04] suggests that when a gambler does not even know the probability distribution of the outcome, maximizing the distribution with the lowest expected value gives a robust strategy in the face of uncertainty. However, it has been shown that for intelligent agents acting in an environment, optimism can also lead to strong results.

[SH15b] gives π°, an optimistic version of the AIXI agent, which was shown (under certain conditions) to be ε-optimal (Definition 8.1.2).

In Section 9.2 we discuss sampling agents, agents that are based on sampling from the environment the agent believes itself to be interacting with. Using a technique called Thompson sampling, it has been shown that such agents are asymptotically optimal in mean. In Section 9.3, the question is posed of what a universal agent should optimize for if no extrinsic rewards are provided from the environment. These agents, called knowledge-seeking agents (KSAs), receive a reward based on the "knowledge" gained, quantified as the difference between the prior and posterior after observing evidence. In Section 9.4 we present explorative agents, which do not greedily maximize reward like the AIXI agent, but instead explore the environment with some probability, occasionally taking actions which are not Bayes-optimal (similar to the idea of an ε-greedy agent in Q-learning [SB18]). By carefully choosing when the agent explores, this leads to the asymptotically optimal agents, known as BayesExp and Inq. In Section 9.5 we introduce Self-AIXI, which self-predicts its own action stream, side-steps expensive expectimax planning (or MCTS) in favor of learning the Q-value function using Bellman equations (Theorem 6.7.2) or Temporal Difference (TD) learning.

9.1   Optimistic Agents

For prediction, it has been shown that pessimism (maximizing the worst case) can be optimal [GS04]. Interestingly, in the history-based reinforcement learning setting, acting *optimistically* leads to optimal behavior. What do we mean by optimism and pessimism though? In this context, optimism (resp. pessimism) means the agent assumes itself to be interacting with the best (resp. worst) possible environment. Here, best (resp. worse) correspond to the environment ν with the highest (resp. lowest) value for V_ν^*, where ν is selected from only those environments in the model class $\mathcal{M}$ that are consistent with the history so far. The optimistic agent will instead select the best environment, and choose the best action for that environment:

$$\pi^\circ(h_{<t}) := \arg\max_\pi \max_{\nu \in \mathcal{M}_t} V_\nu^\pi(h_{<t})$$

where $\mathcal{M}_t$ is the set of environments consistent with history $h_{1:t}$. The set $\mathcal{M}_t$ is updated on every time step, and environments that are ruled out are removed. Compare this to AIξ, which chooses the best action based on the Bayesian mixture ξ over all environments:

$$\pi_\xi^*(h_{<t}) := \arg\max_\pi V_\xi^\pi(h_{<t}) := \arg\max_\pi \sum_{\nu \in \mathcal{M}} V_\nu^\pi(h_{<t})$$

Deterministic environments.   Using this principle of optimism, [SH12a] developed an optimistic agent for a given finite class of deterministic environments $\mathcal{M}_0$.

This algorithm finds the most optimistic deterministic policy-environment pair (π^*, ν^*), then follows policy π^*, at each step removing environments that are inconsistent with the history. If the environment ν^* is ever found inconsistent, then it is discarded, and the agent selects a new optimal policy-environment pair.

For Π=All, Algorithm 9.1 is (strongly) asymptotically optimal, indeed even stronger: there is some finite time after which the policy will be exactly optimal.

Algorithm 9.1 Optimistic Agent (π^o) for Deterministic Environments [SH12a]

Require: Environment class $\mathcal{M}_0 = \{\nu_1,...,\nu_m\}$. Some deterministic policy class Π
Input: Value function V_ν^π. True environment μ
Output: Interaction history $h = a_1 e_1 a_2 e_2 a_3 e_3...$

  1: $t := 1$
  2: **while** $\mathcal{M}_{t-1} \neq \{\}$ **do**
  3:      $(\pi^*, \nu^*) :\in \arg\max_{\pi \in \Pi, \nu \in \mathcal{M}_{t-1}} V_\nu^\pi(h_{<t})$
  4:      **while** $\nu^* \in \mathcal{M}_{t-1}$ **do**   ▷ inner looping is an optional speedup
  5:          $a_t := \pi^*(h_{<t})$
  6:          Perceive e_t from environment μ
  7:          $h_{1:t} := h_{<t} a_t e_t$
  8:          $\mathcal{M}_t := \{\nu \in \mathcal{M}_{t-1} : \nu(h_{<t} a_t) = e_t\}$ ▷ Remove all inconsistent ν
  9:          $t := t+1$

> **Theorem 9.1.1 (Optimism is asymptoically optimal for finite deterministic classes [SH12a])** Suppose $\mathcal{M}$ is a finite class of deterministic environments. Suppose Π is any class of deterministic policies, If we use Algorithm 9.1 (π^o) in an environment $\mu \in \mathcal{M}$, then there is a $T < \infty$ such that
>
> $$V_\mu^{\pi^o}(h_{<t}) = \sup_{\pi \in \Pi} V_\mu^\pi(h_{<t}) \quad \forall\, t \geq T$$

The algorithm and theorems are stated for any deterministic policy class Π. Theorems Theorems 9.1.1 and 9.1.2 also hold for our standard choice, the class *all stochastic* policies, since then the sequence of optimal policies π^* and hence π^o are still guaranteed to be deterministic.

Proof sketch. Since $\mathcal{M}_0$ is finite, there exists a time $T < \infty$ at which all inconsistent environments that are more optimistic than the true environment are eliminated. Given the most optimistic consistent environment ν, if there is no time at which it is eliminated then it is never inconsistent, and so must have identical on-policy dynamics to the true environment μ. Thus $\pi_\nu^* = \pi_\mu^*$, from which the result follows.   ■

Additionally, there is a bound for (most) t on how much the value function for Algorithm 9.1 can differ from the optimal value function. The intuition is that we only have suboptimality for a certain number of time steps before each point where the current hypothesis becomes inconsistent, and the number of such inconsistency points are bounded by the number of environments.

> **Theorem 9.1.2 (Finite error bound [SH12a])** For geometric discount $\gamma_t = \gamma^t$ for some constant $\gamma \in (0,1)$, and $0 < \varepsilon < 1$, following policy π^o of Algorithm 9.1, except for at most $|\mathcal{M}| \cdot \lceil \frac{\log \varepsilon}{\log \gamma} \rceil$ time steps t, its value is ε-optimal:
>
> $$V_\mu^{\pi^o}(h_{<t}) \geq \max_{\pi \in \Pi} V_\mu^\pi(h_{<t}) - \varepsilon$$

This means that for no more than $|\mathcal{M}| \lceil \frac{-\ln \varepsilon}{1-\gamma} \rceil$ time steps, the difference in the value functions will be more than ε. Since this holds for all $\varepsilon >$, for $\Pi = $ All, the theorem also implies that π^o is asymptotically optimal.

Proof. For the ℓ-truncated value $V_\nu^{\pi,t+\ell}(h_{<t})$ (Definition 6.6.1) we have

$$|V_\nu^{\pi,t+\ell}(h_{<t}) - V_\nu^\pi(h_{<t})| \leq \frac{1}{\Gamma_t} \sum_{k=t+\ell+1}^\infty \gamma^k = \gamma^{\ell+1} \leq \varepsilon$$

for $\ell + 1 := \lceil \frac{\log \varepsilon}{\log \gamma} \rceil$ (which is positive due to negativity of both numerator and denominator). Let (π_t^*, ν_t^*) be the policy-environment pair selected by Algorithm 9.1 in cycle t.

Let us first assume $h_{t:t+\ell}^{\pi^\circ,\mu} = h_{t:t+\ell}^{\pi^\circ,\nu_t^*}$, i.e. ν_t^* is consistent with $h_{t:t+\ell}^\circ$ generated by $(\pi^\circ;\mu)$, and hence π_t^* and ν_t^* do not change from $t,...,t+\ell$ (inner loop of Algorithm 9.1). Then

$$\underset{\overset{\uparrow}{\text{drop terms}}}{V_\mu^{\pi^\circ}(h_{<t})} \overset{}{\geq} V_{\mu,t+\ell}^{\pi^\circ}(h_{<t}) \underset{\overset{\uparrow}{\text{same } h_{t:t+\ell}}}{=} V_{\nu_t^*,t+\ell}^{\pi^\circ}(h_{<t}) \underset{\overset{\uparrow}{\pi^\circ = \pi_t^* \text{ on } h_{t:t+\ell}}}{=} V_{\nu_t^*,t+\ell}^{\pi_t^*}(h_{<t})$$

$$\underset{\overset{\uparrow}{\text{bound extra terms}}}{\geq} V_{\nu_t^*}^{\pi_t^*}(h_{<t}) - \gamma^{\ell+1} \underset{\overset{\uparrow}{\text{def. of } (\pi_t^*,\nu_t^*)}}{=} \max_{\nu \in \mathcal{M}_t} \max_{\pi \in \Pi} V_\nu^\pi(h_{<t}) - \gamma^{\ell+1} \underset{\overset{\uparrow}{\mu \in \mathcal{M}_t \text{ and def. of } \ell}}{\geq} \max_{\pi \in \Pi} V_\mu^\pi(h_{<t}) - \varepsilon.$$

Now let $t_1,...,t_K$ be the times t at which the currently selected ν_t^* becomes inconsistent with $h_{<t}$, i.e., $\{t_1,...,t_K\} = \{t : \nu_t^* \notin \mathcal{M}_t\}$. Therefore $h_{t:t+\ell}^\circ \neq h_{t:t+\ell}^{\pi^\circ,\nu_t^*}$ (only) at times $t \in \mathcal{T}_\times := \bigcup_{i=1}^K \{t_i - \ell,...,t_i\}$, which implies $V_\mu^{\pi^\circ}(h_{<t}) \geq \max_{\pi \in \Pi} V_\mu^\pi(h_{<t}) - \varepsilon$ except possibly for $t \in \mathcal{T}_\times$. Finally

$$|\mathcal{T}_\times| = (\ell+1)K = \left\lceil \frac{\log \varepsilon}{\log \gamma} K \right\rceil \leq |\mathcal{M} - 1| \left\lceil \frac{\log \varepsilon}{\log \gamma} \right\rceil \qquad \blacksquare$$

Stochastic environments.   A similar algorithm was developed for a stochastic environment class $\mathcal{M}_0$.   Algorithm 9.2 is very similar to Algorithm 9.1 with the key difference

---

Algorithm 9.2 Optimistic Agent (π°) for Stochastic Environments [SH12a]

Require: $\mathcal{M}_0 = \{\nu_1,...,\nu_m\}$. Deterministic policy class Π
Require: Threshold $z \in (0,1)$
Input: Value function V_ν^π. True environment μ
Output: Interaction history $h = a_1 e_1 a_2 e_2 a_3 e_3...$
  1: $t := 1$
  2: **while** True **do**
  3:     $(\pi^*, \nu^*) \in \arg\max_{\pi \in \Pi, \nu \in \mathcal{M}_{t-1}} V_\nu^\pi(h_{<t})$
  4:     $a_t := \pi^*(h_{<t})$
  5:     Sample $e_t \sim \mu(\cdot | h_{<t} a_t)$
  6:     $h_{1:t} := h_{<t} a_t e_t$
  7:     $t := t+1$
  8:     $\mathcal{M}_t := \left\{ \nu \in \mathcal{M}_{t-1} : \dfrac{\prod_{j=1}^t \nu(e_j | h_{<j} a_j)}{\max_{\tilde{\nu} \in \mathcal{M}_0} \prod_{j=1}^t \tilde{\nu}(e_j | h_{<j} a_j)} > z \right\}$ ▷ Remove all inconsistent ν

---

being how environments that are inconsistent with the observed interaction history are removed. For deterministic environments (for a given deterministic policy) there is only one possible history that can be generated. Environments inconsistent with this history can be immediately disregarded as incorrect. Now, a stochastic environment may never be refuted with certainty, as it may assign non-zero probability to any history that is possible under the true environment. However, it may assign a much smaller probability than the true environment would to histories observed, so we would want the agent to eventually learn to assign a low credence to such environments. The rejection criterion cannot depend on the unknown true environment, so we use the most likely environment as a base instead.

Using the Bayesian mixture as a base is also possible. Environments assigning probability to the history much smaller (by a factor of z) than the base are excluded. Much like in the deterministic case, Algorithm 9.2 is asymptotically optimal with high probability.

Theorem 9.1.3 (Optimality, finite stochastic class [SH12a]) Define π^o via Algorithm 9.2 with any threshold $z \in (0,1)$ and a finite class $\mathcal{M}$ of stochastic environments containing the true environment μ. Then, with probability at least $1 - z|\mathcal{M}|$, there exists for every $\varepsilon > 0$, a time $T < \infty$ such that

$$V_\mu^{\pi^o}(h_{<t}) > \max_{\pi \in \Pi} V_\mu^\pi(h_{<t}) - \varepsilon \quad \forall t \geq T$$

Proof. The proof is based on martingale theory and can be found in [SH12a]. ∎

The optimism principle, algorithms, and convergence guarantees can be significantly extended [SH15b]. We can extend the algorithm to countable classes $\mathcal{M}$ by running it on a sufficiently slowly increasing sequence of finite subsets of $\mathcal{M}$ [SH13]. We can also extend it to continuous classes $\mathcal{M}$ that are compact with respect to a suitable metric between two measures, e.g. $d(\nu,\nu') := \sup_{h,\pi}|V_\nu^\pi(h) - V_{\nu'}^\pi(h)|$. For instance, the classes of finite-state (k-order) (Partially Observable) Markov Decision Processes are compact. Choosing a finite covering of $\mathcal{M}$ with d-balls of radius less than $\varepsilon/2$ and running Algorithm 9.2 with the centers of these balls is ε-optimal similar to Theorem 9.1.3. One can extend the algorithm further from countable classes to separable classes, since they can by definition be covered by countably many balls of arbitrarily small radius. This also allows achieving asymptotic optimality which requires $\varepsilon \to 0$. Concerning asymptotics, this covers all relevant classes $\mathcal{M}$, but the number of ε-errors is proportional to the size of the set $\mathcal{M}$: $|\mathcal{M}|$ for finite $\mathcal{M}$, the size of the ε-cover for compact $\mathcal{M}$, the index of μ in $\mathcal{M}$ for countable $\mathcal{M}$ and similar for separable $\mathcal{M}$. For compact subsets of $\mathbb{R}^d$, this is typically exponential in d, an example of the curse of dimensionality. If $\mathcal{M}$ is generated by finite sets of partial environments (like laws of nature [SH15c]), error bounds linear in the number of laws instead of in the number of environments or size of ε-cover are possible. For instance, (PO)MDPs can be factored in this way, leading to exponentially improved bounds polynomial in d [SH15b].

9.2   (Thompson)Sampling Agents

Another idea to create a policy which performs well is to have a *policy which samples from other policies* [Lei16b]. A successful version of this is Thompson sampling. Using the sampling as the explorative aspect of the policy allows for strong asymptotic results. What we want with sampling agents is that the probability of following a better policy (higher value in the true environment) increases over time.

In contrast, *exploring agent* BayesExp (Section 9.4) switches between exploration and exploitation phases. The exploration phases of sampling agents are not as clear. In one sense they are exploring every time they choose which policy to follow, since they never know the optimal policy. In another sense they are exploring whenever they follow a policy which is not the most likely policy to be sampled.

Thompson Sampling.   Originating with [Tho33] for bandit problems, Thompson Sampling has recently been rediscovered in and applied to the reinforcement learning and sequential decision theory fields [OB10b, Ort11, LLOH17, GM15, AG13]. The algorithm is quite simple and natural, and still enjoys strong convergence properties, as well as success in applications. Like any good Bayesian agent, the Thompson sampler updates its posterior at every iteration, but unlike Bayes which mixes over environments, the Thompson agent *samples* an environment from the posterior distribution (Definition 7.2.2) and follows the optimal policy (Definition 6.6.1) for that environment for an ε_t-effective horizon and then repeats.

Algorithm 9.3 Thompson sampling policy π_{TS} [Lei16b]

Require: Model class $\mathcal{M}$. Prior w. Effective horizon function H_t.
Input: Percept stream $e_1, e_2, e_3, \ldots$
Output: Action stream $a_1, a_2, a_3, \ldots$
 1: **while** True **do**
 2:     sample environment ν according to the posterior $w(\nu | h_{<t})$
 3:     follow π_ν^* for $H_t(\varepsilon_t)$ time steps

Note that the algorithm glosses over how to find the posterior $w(\nu | h_{<t})$ and the ν-optimal policy π_ν^*, which for $\mathcal{M}_{sol}$ is as hard as determining π_ξ^*, but may on occasion be simpler, e.g. if $\mathcal{M}$ is a (sub)class of MDPs.

One of the strong properties Thompson sampling has is asymptotic optimality in mean. This means that the expected value of the difference in the value function of the optimal policy and Thompson sampling goes to zero as time goes to infinity.

Theorem 9.2.1 (Thompson sampling is asymptotically optimal in mean [Lei16b]) For all $\mu \in \mathcal{M}$, for the Thompson sampling agent π_{TS}, we have

$$\mathbf{E}_\mu^{\pi_{\mathrm{TS}}}[V_\mu^*(h_{<t}) - V_\mu^{\pi_{\mathrm{TS}}}(h_{<t})] \to 0 \quad \text{for} \quad t \to \infty$$

Proof. The proof is a bit lengthy. See [Lei16b, Sec.5.4.3]   ∎

By Figure 2.8 this also implies convergence in probability, but π_{TS} is not strong asymptotically optimal (Definition 8.1.7) [Lei16b, Ex.5.28]. We can use Theorems 8.1.13 and 9.2.1 to show that Thompson sampling has sublinear regret in $\mathcal{M}$.

Corollary 9.2.2 (Sublinear regret for Thompson sampling [Lei16b]) If the discount function γ satisfies Assumption 8.1.11, and the environment $\mu \in \mathcal{M}$ is recoverable, then the regret is sublinear: $\mathrm{Regret}_m(\pi_{\mathrm{TS}}, \mu) = o(m)$.

Proof. Immediate result from Theorems 8.1.13 and 9.2.1.   ∎

9.3   Knowledge-Seeking Agents

One of the fundamental questions when trying to define general intelligence in the domain of reinforcement learning is the choice of motivations for the agent. In the context of Universal Artificial Intelligence, this is asking for where the rewards come or for the choice of reward function. There have been suggestions that any reward provided by humans is susceptible to manipulation (Chapter 15). Also, to which degree is an agent that (slavishly) maximizes external reward provided by intelligent humans autonomously intelligent? One proposed solution is that an AGI should have some intrinsic reward (function). Of course this reward function should not be tied to any specific goal such as winning chess games or driving a car from A to B, but should be General. A reward function based on the amount of knowledge an agent has or is going to gain seems general and useful. Such agents are called curiosity-driven agents [Sch91] or *Knowledge-Seeking Agents* (KSA). Seeking out information should encourage agents to explore, which allows the agent to learn and solve problems.

We first define what we mean by Information Gain (IG) in terms of the KL divergence (Definition 2.5.12). We will use this notion of IG as the replacement for rewards, and define a new value function V_{IG} called *the information gain value function* The KSA policy π^*_{IG} is then defined as the policy that maximizes V_{IG}.

We first present an on-policy prediction result in discounted KL divergence, which holds for any policy. This is then used to show that beside on-policy prediction, π^*_{IG} convergence off-policy [Ors11, OLH13].

While there is no widely accepted notion of what it means to accumulate information, in the context of a Bayesian agent interacting with an environment, information is gained through the posterior updating with each new interaction. Therefore, the "distance" between the current posterior and the posterior at a previous time step can serve as a measure of how much information the Bayesian agent has gained since that previous time step. We will be using the KL divergence (Definition 2.5.12) as the distance measure, but other measures may be used [Ors11]. However, most other measures do not perform well in stochastic environments and will seek out noise instead of novel information [OLH13].

Definition 9.3.1 (Information Gain) The one-step information gain of $h_k \in \mathcal{A} \times \mathcal{E}$ given history $h_{<k} \in (\mathcal{A} \times \mathcal{E})^{k-1}$ is defined as the KL divergence between the posterior conditioned on history $h_{1:k}$, and posterior given history $h_{<k}$:

$$r_k^{\mathrm{IG}} := \mathrm{IG}(h_k|h_{<k}) := \sum_{\nu \in \mathcal{M}} w(\nu|h_{1:k}) \log \frac{w(\nu|h_{1:k})}{w(\nu|h_{<k})}$$

Definition 9.3.1 can be thought of as a measure of how much the agent learned after observing h_k given $h_{<k}$ has already been observed. If h_k was very surprising (that is, the current weight w assigns very low probability to environments ν where sampling h_k was likely), then the agent realizes its old belief $w(\cdot|h_{<k})$ does not correspond well to reality, the posterior $w(\cdot|h_{1:k})$ drastically changes, leading to a high KL divergence between $w(\cdot|h_{1:k})$ and $w(\cdot|h_{<k})$. Conversely, if the agent's belief is already very accurate ($w(\mu|h_{<k}) \approx 1$ and $w(\nu|h_{<k}) \approx 0$ for $\nu \neq \mu$), then the agent would only observe action-percept pairs h_k that it already expected to receive anyway, so $w(\cdot|h_{1:k}) \approx w(\cdot|h_{<k})$ and the KL divergence is small.

This definition can be extended to an arbitrary (finite) history $h_{k:k+\ell}$ instead of h_k, however we are mainly interested in the one-step version. Now replacing the reward r_k in the value function Definition 6.6.1 with the information gain r_k^{IG} we get the following information gain value function:

Definition 9.3.2 (Information Gain Value function) The information gain value of policy π having observed history $h_{<t}$ with respect to discount function γ, denoted $V_{\text{IG}}^{\pi,m}$, is defined to be the ξ-expected discounted information gain.

$$V_{\text{IG}}^{\pi,m}(h_{<t}) := \mathbf{E}_{\xi}^{\pi}\left[\frac{1}{\Gamma_t}\sum_{k=t}^{m}\gamma_k \text{IG}(h_k|h_{<k})\middle| h_{<t}\right]$$

Similar to Definition 6.6.1, we drop the m when $m=\infty$. The optimal information gain policy π_{IG}^* maximizes the information gain value function, and V_{IG}^* is its information gain value.

$$\pi_{\text{IG}}^*(\cdot) :\in \arg\max_{\pi} V_{\text{IG}}^{\pi}(\cdot) \qquad \text{and} \qquad V_{\text{IG}}^*(\cdot) := \max_{\pi} V_{\text{IG}}^{\pi}(\cdot)$$

Information gain, as a distance measure between posteriors, is a useful concept for measuring how much a Bayesian agent has learned. Another use case of the KL divergence is when it is applied to predictive policy-environment distributions ν^{π} and ξ^{π}.

Definition 9.3.3 (KL-Divergence for ν^{π} and ξ^{π}) Given a history $h_{<k}$, we define the one-step KL divergence between ν^{π} and ξ^{π} as

$$\tilde{r}_k^{\nu} := \text{KL}_1(\nu^{\pi},\xi^{\pi}|h_{<k}) := \sum_{h_k\in(\mathcal{A}\times\mathcal{E})}\nu^{\pi}(h_k|h_{<k})\log\frac{\nu^{\pi}(h_k|h_{<k})}{\xi^{\pi}(h_k|h_{<k})}$$

Note that Definition 9.3.3 is an instantiation of Definition 2.5.12. Also note that unlike r_k^{IG}, $\tilde{r}_k^{\nu}$ depends on ν and cannot actually serve as a reward, since ν is unknown. Using reward $\tilde{r}$ instead, the ν-Value function is

$$\tilde{V}_{\nu,\gamma}^{\pi,m}(h_{<t}) \equiv \frac{1}{\Gamma_t}\mathbf{E}_{\nu}^{\pi}\left[\sum_{k=t}^{m}\gamma_k\tilde{r}_k^{\nu}\middle|h_{<t}\right] \equiv \frac{1}{\Gamma_t}\sum_{k=t}^{m}\gamma_k\sum_{h_{t:k-1}\in(\mathcal{A}\times\mathcal{E})^{k-t}}\nu^{\pi}(h_{t:k-1}|h_{<t})\text{KL}_1(\nu^{\pi},\xi^{\pi}|h_{<k}) \quad (9.3.4)$$

Lemma 9.3.5 (Information gain in terms of KL divergence) The information gain value function can be expressed with the KL divergence between two policy-environment distributions as follows

$$V_{\text{IG}}^{\pi,m}(h_{<t}) = \sum_{\nu\in\mathcal{M}}w(\nu|h_{<t})\tilde{V}_{\nu,\gamma}^{\pi,m}(h_{<t})$$

Proof.

$$\Gamma_t V_{\text{IG}}^{\pi,m}(h_{<t}) \equiv \mathbf{E}_{\xi}^{\pi}\left[\sum_{k=t}^{m}\gamma_k\text{IG}(h_k|h_{<k})\,|\,h_{<t}\right]$$

Expanding the definition of the expectation and IG,

$$= \sum_{h_{t:m}}\xi^{\pi}(h_{t:m}|h_{<t})\sum_{k=t}^{m}\gamma_k\sum_{\nu\in\mathcal{M}}w(\nu|h_{1:k})\log\frac{w(\nu|h_{1:k})}{w(\nu|h_{<k})}$$

By Definition 7.2.2 of $w(\nu|h_{1:k})$,

$$= \sum_{h_{t:m}} \xi^\pi(h_{t:m}|h_{<t}) \sum_{k=t}^m \gamma_k \sum_{\nu \in \mathcal{M}} w(\nu|h_{<t}) \frac{\nu^\pi(h_{t:k}|h_{<t})}{\xi^\pi(h_{t:k}|h_{<t})} \log \frac{w(\nu|h_{1:k})}{w(\nu|h_{<k})}$$

At step k, nothing in the second sum depends on $h_{>k}$, and $\sum_{h_{k:m}} \xi^\pi(h_{k:m}|h_{<k})=1$, so we can push the first sum inside and cancel and discard terms

$$= \sum_{\nu \in \mathcal{M}} w(\nu|h_{<t}) \sum_{k=t}^m \gamma_k \sum_{h_{t:k}} \nu^\pi(h_{t:k}|h_{<t}) \log \frac{w(\nu|h_{1:k})}{w(\nu|h_{<k})}$$

Applying Definition 7.2.2,

$$= \sum_{\nu \in \mathcal{M}} w(\nu|h_{<t}) \sum_{k=t}^m \gamma_k \sum_{h_{t:k}} \nu^\pi(h_{t:k}|h_{<t}) \log \frac{\nu(e_t|h_{<t}a_t)}{\xi(e_t|h_{<t}a_t)}$$

From Definition 6.1.5 of a conditional measure with $m=1$,

$$= \sum_{\nu \in \mathcal{M}} w(\nu|h_{<t}) \sum_{k=t}^m \gamma_k \sum_{h_{t:k}} \nu^\pi(h_{t:k}|h_{<t}) \log \frac{\nu^\pi(h_k|h_{<k})}{\xi^\pi(h_k|h_{<k})}$$

We write $\nu^\pi(h_{t:k}|h_{<t}) = \nu^\pi(h_k|h_{<t})\nu^\pi(h_{t:k-1}|h_{<t})$, and sum separately over h_k,

$$= \sum_{\nu \in \mathcal{M}} w(\nu|h_{<t}) \sum_{k=t}^m \gamma_k \sum_{h_{t:k-1}} \nu^\pi(h_{t:k-1}|h_{<t}) \sum_{h_k} \nu^\pi(h_k|h_{<k}) \log \frac{\nu^\pi(h_k|h_{<k})}{\xi^\pi(h_k|h_{<k})}$$

We apply Definition 9.3.3 of the KL divergence,

$$= \sum_{\nu \in \mathcal{M}} w(\nu|h_{<t}) \sum_{k=t}^m \gamma_k \sum_{h_{t:k-1}} \nu^\pi(h_{t:k-1}|h_{<t}) \mathrm{KL}_1(\nu^\pi, \xi^\pi|h_{<k})$$

Finally, we insert (9.3.4),

$$= \sum_{\nu \in \mathcal{M}} w(\nu|h_{<t}) \Gamma_t \tilde{V}_{\nu,\gamma}^{\pi,m}(h_{<t}) \qquad \blacksquare$$

Before presenting the main theorem showing that π_{IG}^* learns to predict off-policy, we present an easier on-policy result that holds for all policies.

Theorem 9.3.6 (On-policy prediction [OLH13]) Let $\mu \in \mathcal{M}$ and π be a policy and $\sum_{\nu \in \mathcal{M}} w_\nu \log w_\nu^{-1} < \infty$, then

$$\mathbf{E}_\mu^\pi[\tilde{V}_\mu^\pi(h_{<t})] \overset{t\to\infty}{\longrightarrow} 0$$

The theorem shows that $\mathrm{P}_\xi^\pi(\cdot|h_{<t})$ converges in expectation to $\mathrm{P}_\mu^\pi(\cdot|h_{<t})$ where the difference between the two measures is taken with respect to the expected cumulative discounted KL divergence. This implies that $\mathrm{P}_\xi^\pi(\cdot|h_{<t})$ is in expectation a good estimate for the unknown $\mathrm{P}_\mu^\pi(\cdot|h_{<t})$. The finite entropy condition is violated for our canonical choice $\mathcal{M}_{sol}$ and $w_\nu^U = 2^{-K(\nu)}$, but easily satisfiable by just a marginally faster decreasing prior such as $w_\nu = 2^{-(1+\varepsilon)K(\nu)}$ or even $w_\nu = 2^{-K(\nu)}/K(\nu)$ or $w_{\nu_i} = 1/(i+1)^2$ (Exercise 7).

Proof. The telescoping property of KL lifted to our case, with same notation as in Lemma 3.2.4, reads

$$\mathbf{E}_\nu^\pi\left[\sum_{k=t}^m \mathrm{KL}_1(\nu^\pi, \xi^\pi|h_{<t})\right] \equiv \mathbf{E}_\nu^\pi\left[\sum_{k=1}^m \mathrm{KL}_1(\nu^\pi, \xi^\pi|h_{<t}) - \sum_{k=1}^{t-1} \mathrm{KL}_1(\nu^\pi, \xi^\pi|h_{<t})\right]$$

$$= \mathrm{KL}(\nu_{1:m}^\pi\|\xi_{1:m}^\pi) - \mathrm{KL}(\nu_{<t}^\pi\|\xi_{<t}^\pi) \qquad (9.3.7)$$

As a sum of non-negative terms, $\mathrm{KL}(\nu_{1:m}^{\pi}||\xi_{1:m}^{\pi})$ is increasing in m and converges to $c_{\nu}^{\xi} \leq \log w_{\nu}^{-1} < \infty$ due to dominance $\xi^{\pi} \geq w_{\nu}\nu^{\pi}$ (Lemma 7.2.4). Hence

$$\text{Hence} \quad \lim_{t\to\infty} \mathbf{E}_{\nu}^{\pi}\left[\sum_{k=t}^{\infty} \mathrm{KL}_1(\nu,\xi|h_{<t})\right] = c_{\nu}^{\xi} - c_{\nu}^{\xi} = 0 \tag{9.3.8}$$

$$\text{Also} \quad \mathbf{E}_{\mu}^{\pi}[\tilde{V}_{\mu}^{\pi}(h_{<t})] \overset{(a)}{\leq} \frac{1}{w_{\mu}}\sum_{\nu\in\mathcal{M}} w_{\nu}\mathbf{E}_{\nu}^{\pi}\left[\sum_{k=t}^{\infty}\tilde{V}_{\nu}^{\pi}(h_{<t})\right] \overset{(b)}{\leq} \frac{1}{w_{\mu}}\sum_{\nu\in\mathcal{M}} w_{\nu}\mathbf{E}_{\nu}^{\pi}\left[\sum_{k=t}^{\infty}\tilde{r}_k^{\nu}\right] \tag{9.3.9}$$

where (a) follows by the positivity of the KL divergence and by introducing the sum, (b) follows from $\gamma_k/\Gamma_t \leq 1$ for $k \geq t$ and $\mathbf{E}[\mathbf{E}[\cdot|h_t]] = \mathbf{E}[\cdot]$. Now $\delta_{\nu}(t) := \mathbf{E}_{\nu}^{\pi}[\sum_{k=t}^{\infty}\tilde{r}_k^{\nu}] \to 0$ follows from inserting Definition 9.3.3 of $\tilde{r}_k^{\nu}$ and using (9.3.8).

We also know that $\delta_{\nu}(t) \leq \mathrm{KL}(\nu^{\pi}||\xi^{\pi}) \leq \log w_{\nu}^{-1}$ from (9.3.7) and dominance $\xi^{\pi} \geq w_{\nu}\nu^{\pi}$ (Lemma 7.2.4). Unfortunately $\delta_{\nu}(t)$ is not uniformly bounded, so we cannot directly apply [Hut05b, Lem.5.28ii] to conclude $\sum_{\nu} w_{\nu}\delta_{\nu}(t) \to 0$. Let us define $\tilde{w}_{\nu} := w_{\nu}\log w_{\nu}^{-1} > 0$ and $\tilde{\delta}_{\nu}(t) := \delta_{\nu}(t)/\log w_{\nu}^{-1} \leq 1$. By assumption, $\sum_{\nu}\tilde{w}_{\nu} < \infty$, so we can now apply [Hut05b, Lem.5.28ii] to the tilded quantities, to obtain

$$0 \leq \mathbf{E}_{\mu}^{\pi}[\tilde{V}_{\mu}^{\pi}(h_{<t})] \leq \frac{1}{w_{\mu}}\sum_{\nu}\tilde{w}_{\nu}\tilde{\delta}_{\nu}(t) \overset{t\to\infty}{\longrightarrow} 0 \qquad \blacksquare$$

The next result is perhaps the most important theoretical justification for the definition of π_{IG}^*. We show that if $h_{1:\infty}$ is generated by following π_{IG}^*, then $\mathrm{P}_{\xi}^{\pi}(\cdot|h_{<t})$ converges in expectation to $\mathrm{P}_{\mu}^{\pi}(\cdot|h_{<t})$ for all π. More informally, this means that as a longer history is observed the agent learns to predict the counterfactuals *"what would happen if I follow another policy π instead"*. For example, if the observation also included a reward signal, then the agent would asymptotically be able to learn (but not follow) the policy maximizing the expected discounted reward. In fact, the policy maximizing the Bayes-expected reward would converge to optimal. This kind of off-policy prediction is not usually satisfied by arbitrary policies where the agent can typically only learn what will happen on-policy in the sense of Theorem 9.3.6, not what would happen if it chose to follow another policy.

Theorem 9.3.10 (On-policy learning, off-policy prediction [Ors11]) Let $\mu \in \mathcal{M}$ and $\sum_{\nu\in\mathcal{M}} w_{\nu}\log w_{\nu}^{-1} < \infty$. Then

$$\mathbf{E}_{\mu}^{\pi_{\mathrm{IG}}^*}\left[\sup_{\pi}\tilde{V}_{\mu}^{\pi}(h_{<t})\right] \overset{t\to\infty}{\longrightarrow} 0$$

where the expectation is taken over $h_{<t}$.

Proof. In this proof, the goal is to show that the value function of the information-gain policy (π_{IG}^*) converges to the value of the μ-optimal policy as time goes to infinity, and then apply (9.3.9). By the positivity of the KL divergence, adding $\tilde{V}_{\nu}^{\pi}(h_{<t}) \geq 0$ terms,

$$\mathbf{E}_{\mu}^{\pi_{\mathrm{IG}}^*}\left[\sup_{\pi}\tilde{V}_{\mu}^{\pi}(h_{<t})\right] \leq \mathbf{E}_{\mu}^{\pi_{\mathrm{IG}}^*}\left[\frac{1}{w(\mu|h_{<t})}\sup_{\pi}\sum_{\nu\in\mathcal{M}} w(\nu|h_{<t})\tilde{V}_{\nu}^{\pi}(h_{<t})\right]$$

By Lemma 9.3.5 and from the definition of π_{IG}^*,

$$= \mathbf{E}_{\mu}^{\pi_{\mathrm{IG}}^*}\left[\frac{1}{w(\mu|h_{<t})}\sum_{\nu\in\mathcal{M}} w(\nu|h_{<t})\tilde{V}_{\nu}^{\pi_{\mathrm{IG}}^*}(h_{<t})\right]$$

From the definition of the posterior and expectation,

$$= \frac{1}{w_\mu} \mathbf{E}_\xi^{\pi_{\mathrm{IG}}^*} \left[\sum_{\nu \in \mathcal{M}} w(\nu|h_{<t}) \, \tilde{V}_\nu^{\pi_{\mathrm{IG}}^*}(h_{<t}) \right]$$

By exchanging the sum and expectation, and the definition of the posterior,

$$= \frac{1}{w_\mu} \sum_{\nu \in \mathcal{M}} w_\nu \mathbf{E}_\nu^{\pi_{\mathrm{IG}}^*} \left[\tilde{V}_\nu^{\pi_{\mathrm{IG}}^*}(h_{<t}) \right] \overset{t \to \infty}{\longrightarrow} 0$$

The limit follows from (9.3.9) choosing $\pi = \pi_{\mathrm{IG}}^*$. ∎

Knowledge-seeking agents aim to maximize the value with respect to Bayesian mixtures over environments, and as a consequence, they share the same properties as Bayes-optimal agents. Specifically, they are dependent on the choice of prior and may suffer from suboptimal prior selections in the short term. However, since knowledge-seeking agents continuously learn off-policy and exhibit asymptotic optimality (and unlike the Thompson sampling agent) without any recoverability assumption, the choice of prior ultimately does not impact their overall performance in the long run.

Example 9.3.11 (KSA experiments on toy environments) Consider the KL-KSA agent with (deterministic) environment model class $\mathcal{M} = \{\mu_1, \mu_2, \mu_3\}$ defined as the state diagrams in Figure 9.1. The environments are all identical up to some slight changes in the transitions (highlighted in bold). Note that q_1 is a "trap state", in that if the agent visits this state, it can never leave. To learn the true environment as fast as possible, the agent should take actions $\dot{a}_{1:5} = 10100$ and receive a sequence of observations $o'_{1:5}$ depending on the true environment, either 00000 for μ_1 (Figure 9.1a), or 01000 for μ_2 (Figure 9.1b), or 00001 for μ_3 (Figure 9.1c). There is no other action sequence that can distinguish with certainty between the three environments in strictly fewer time steps. Also, if the agent chooses $a_1 = 0$, it will be stuck in the trap state, and never be able to distinguish μ_1 from μ_2 (if either is the true environment). Of all possible action sequences $a_{1:\infty}$, the KSA agent values those prefixed with $\dot{a}_{1:5} = 10100$ the highest, indicating that it learns as fast as possible for this toy model class, and avoids moving to the trap state while there is still information to learn elsewhere. ♦

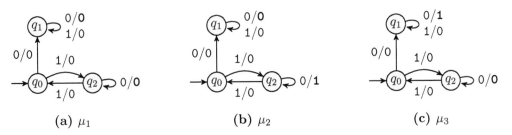

(a) μ_1 (b) μ_2 (c) μ_3

Figure 9.1: *The (deterministic) toy environments μ_1, μ_2, μ_3. Edges are labelled with the action/observation pairs. Differing transitions highlighted in bold.*

9.4 Exploring Agents (BayesExp and Inq)

While acting in a world or environment, we often want the agent to explore the environment enough so that an optimal action may be chosen. How much time should the agent spend exploring? Exploring too much means the agent never tries to act optimally, but instead only takes actions that are novel. Not exploring enough means the agent may be prematurely trying to choose an optimal action based on an inaccurate model of the true environment. This trade-off is called the exploration-exploitation problem. Unexpectedly, the Bayes-optimal agent AIξ does not explore enough to be asymptotically optimal (Theorem 8.3.1), but can be fixed by adding 'a bit' of extra exploration, though it is not clear whether we actually should [CHC21].

BayesExp.   Under some reasonable assumptions, it can be shown that there exists an agent which achieves a balance between exploration and exploitation leading to (weak) asymptotic optimality. BayesExp Algorithm 9.4 (a Bayesian agent with extra exploration) switches between exploring and exploiting in just the right way that asymptotically the exploiting becomes optimal. However, there is always a nonzero chance of exploring.

Algorithm 9.4 BayesExp Algorithm π_{BE} [Lat14, Lei16b]

Require: Non-increasing sequence $\varepsilon_1, \varepsilon_2, \varepsilon_3, \ldots$
Require: Model class $\mathcal{M}$. Prior w. Effective horizon function $H_t()$.
Input: Percept stream $e_1, e_2, e_3, \ldots$
Output: Action stream $a_1, a_2, a_3, \ldots$
  1: **while** True **do**
  2:     **if** $V_{\text{IG}}^{*,t+H_t(\varepsilon_t)}(h_{<t}) > \varepsilon_t$ **then** follow π_{IG}^* for $H_t(\varepsilon_t)$ steps   ▷ exploration
  3:     **else** follow π_ξ^* for 1 step   ▷ exploitation

The BayesExp agent either exploits via the Bayes-optimal policy π_ξ^* for one time step, or explores via the Knowledge-Seeking Agent π_{IG}^* of Section 9.3 for an effective horizon number of time steps. The decision to explore is based on using the maximum expected information gain $V_{\text{IG}}^{*,t+H_t(\varepsilon_t)}(h_{<t})$ of policy π_{IG}^*. If it is larger than ε_t, then BayesExp will explore for $H_t(\varepsilon_t)$ time steps, otherwise it will exploit. BayesExp is weakly asymptotically optimal (Definition 8.1.5) under mild conditions on the discount function and exploration thresholds ε_t and prior w_ν:

Theorem 9.4.1 (BayesExp is weakly asymptotically optimal [Lat14]) Let $\mathcal{M} = \{\nu_1, \nu_2, \ldots\}$ be countable and $\sum_{\nu \in \mathcal{M}} w_\nu \log w_\nu^{-1} < \infty$, and discount γ_t and exploration thresholds ε_t satisfying

- $\Gamma_t > 0$ for all t
- $H_{t+1}(\varepsilon) \geq H_t(\varepsilon)$, for all $\varepsilon > 0$ and all $t \in \mathbb{N}$
- $\varepsilon_{t+1} \leq \varepsilon_t$ for all t
- $\lim_{t \to \infty} \varepsilon_t = 0$
- $H_t(\varepsilon_t) = o(t\varepsilon_t)$

Then BayesExp is weakly asymptotically optimal in $(\mathcal{M}, \gamma)$.

Proof idea. The proof [Lat14] establishes that (*i*) π_ξ^* converges to optimal π_μ^* on the exploitation time-steps of BayesExp, (*ii*) the number of times BayesExp explores is $o(n)$, (*iii*) the value of the BayesExp policy π_{BE} converges to the value of π_ξ^* in Cesàro average, which together imply that BayesExp is weakly asymptotically optimal.   ∎

Under the conditions of the theorem, the BayesExp agent demonstrates its weak asymptotic optimality in environments with countable environment classes and non-trivial discount functions. The ability of BayesExp to balance exploration and exploitation effectively stems from its dynamic exploration strategy, adjusting the number of exploration steps based on the maximum expected information gain and predefined threshold values. As the agent progresses through time, the exploration threshold ε_t converges to zero, which ensures BayesExp explores as long as information can be gained by exploration. On the other hand, the probability of exploration needs to tend to 0, which an ε_t tending to 0 sufficiently slowly ensures. This gradual transition between exploration and exploitation is crucial to the agent's adaptability, allowing it to learn from its environment and make better decisions over time. BayesExp strikes an asymptotically optimal balance between learning from the environment and taking advantage of the knowledge it has acquired.

One can show that for any discount with monotonically but sublinearly growing horizon, exploration sequences ε_t as required exists (Exercise 4). For instance, the popular geometric discount $\gamma_t = \gamma^t$ has constant effective horizon, and $\varepsilon_t = t^{-\beta}$ for any $0 < \beta < 1$ will do. For deterministic environments and geometric discounting, similar to the optimistic agents in Section 9.1, the proof is much simpler, since environments can be ruled out completely [LH11a]. The proof for the stochastic case is beyond the scope of this book [Lat14]. Since $\varepsilon_t \to 0$, the expected information gain at exploiting time-steps converges to 0, which means that ξ concentrates around μ while exploiting. Therefore asymptotically we expect BayesExp to be close-to optimal while exploiting and hence to be close-to optimal with probability tending to 1. An algorithm can only be asymptotically optimal if it either already is optimal, or explores and learns to become so ($\varepsilon_t > 0$) with decreasing exploration rate. The reason for exploring in blocks is to avoid problems with time-inconsistency (Section 6.5). The exploration policy depends on the horizon, so re-computing it at each time-step may lead to a chain of different policies that ultimately may lead to poor behavior. This is avoided committing to π^*_{IG} for $H_t(\varepsilon_t)$ time-steps.

Inquisitive agent. The Inquisitive Reinforcement Learner (Inq) even achieves strong asymptotic optimality (Definition 8.1.7). Similar to BayesExp, Inq combines Bayes-optimal AIξ with the knowledge-seeking agent of Section 9.3. With a certain probability, instead of taking the Bayes-optimal action π^*_ξ, Inq follows maximally informative exploratory expeditions, but unlike BayesExp, the expeditions are of random lengths, which can be shorter or longer than the effective horizon. Also, an undiscounted multi-step version of the Information Gain Definition 9.3.1 is used, and the effective horizon needs to be bounded. Otherwise Inq follows the same principle as BayesExp: Inq is more likely to explore the more it expects an exploratory action to reduce its uncertainty about which environment it is in. With increasing certainty about the true environment $\mu \in \mathcal{M}$, the information gathering periods become less frequent, and asymptotically Inq converges to AIξ. The additional exploration prevents Inq from remaining ignorant about unexplored aspects of the world and getting stuck with suboptimal actions, which guarantees that it also converges to AIμ.

Theorem 9.4.2 (Inq is strong asymptotically optimal [CCH19]) For any countable class of environments $\mathcal{M}$ and bounded effective horizon ($\sup_t H_t(\varepsilon) < \infty \; \forall \varepsilon > 0$), the *Inquisitive Reinforcement Learner* (Inq Algorithm 9.5) is strongly asymptotically optimal.

Proof. See [CCH19]. ∎

The details of the algorithm are as follows: First, we need a multi-step generalization of the one-step $(t=n)$ information gain Definition 9.3.1

$$\mathrm{IG}_{t:n}(h_{t:n}|h_{<t}) := \sum_{\nu\in\mathcal{M}} w(\nu|h_{1:n})\log\frac{w(\nu|h_{1:n})}{w(\nu|h_{<t})}$$

This leads to the non-discounted truncated Information Gain Value function for policy π

$$V_{\mathrm{IG}}^{\pi,n}(h_{<t}) := \mathbf{E}_\xi^\pi[\mathrm{IG}_{t:n}|h_{<t}] \equiv \sum_{h_{t:n}}\xi^\pi(h_{t:n}|h_{<t})\mathrm{IG}_{t:n}(h_{t:n}|h_{<t})$$

Similarly to the knowledge-seeking agent, the optimal (deterministic) IG policy is

$$\pi_{\mathrm{IG},h_{<t}}^{*,n} :\in \arg\max V_{\mathrm{IG}}^{\pi,n}(h_{<t})$$

Note that the policy is well-defined and used on histories of length $t-1$ to n. Now at each time step t we choose the action $a_t := \pi_{\mathrm{IG},h_{<t-k}}^{*,t-k+m-1}(h_{<t})$ for an IG policy that optimized V_{IG} at some random k time-steps earlier, also with random horizon m. While we sample the IG policy independently at each t, there is a non-zero chance of following the same policy for $H_t(\varepsilon)$ time steps, since $m > H_t(\varepsilon)-k$ may be sampled and the IG policy that started at t may be sampled again for $H_t(\varepsilon)$ time steps. Exploring consistently for $H_t(\varepsilon)$ steps is important: in BayesExp it was prescribed, while in Inq it happens with non-zero probability. The probability of sampling $\pi_{\mathrm{IG},h_{<t-k}}^{*,t-k+m-1}$ is

$$\varepsilon_k^m(h_{<t}) := \min\left\{\frac{1}{m^2(m+1)},\, \eta V_{\mathrm{IG}}^{\pi_{\mathrm{IG},h_{<t-k}}^{*,t-k+m-1},t+m}(h_{<t})\right\}$$

for $m\in\mathbb{N}^+$ and $0\leq k < \min\{m,t\}$, where $0<\eta<1$ is an exploration constant. The exploration probability tends to 0 iff V_{IG} tends to 0. The first term ensures that the probabilities will not sum to more than 1:

$$\varepsilon_+^+(h_{<t}) := \sum_{m\in\mathbb{N}^+}\sum_{k<\min\{m,t\}}\varepsilon_k^m(h_{<t}) \leq \sum_{m\in\mathbb{N}^+}\sum_{k<m}\frac{1}{m^2(m+1)} = 1$$

The remaining probability $1-\varepsilon_+^+$ is used for following the Bayes-optimal policy π_ξ^*. Together this give the Inq Algorithm 9.5.

Algorithm 9.5 Inq Algorithm π_{Inq} [CCH19]

Require: Model class $\mathcal{M}$. Prior w. Discount γ.
Input: Percept stream $e_1, e_2, e_3, ...$
Output: Action stream $a_1, a_2, a_3, ...$
1: **for** $t=1,2,3,....$ **do**
2:     Take action $\pi_{\mathrm{IG},h_{<t-k}}^{*,t-k+m-1}(h_{<t})$ with probability $\varepsilon_k^m(h_{<t})$ ▷ exploration
3:     Take action $\pi_\xi^*(h_{<t})$ with probability $1-\varepsilon_+^+(h_{<t})$ ▷ exploitation

---

Strong asymptotic optimality may not be possible for (some or all) unbounded horizon, since it may grow faster than Inq or possibly any policy can learn about progressively more long-term dynamics of the environment. Contrast this to BayesExp, which is weak asymptotically optimal even for (sublinearly) growing horizon. Both agents maximize information gain (explore) if they expect enough information about the environment can be gained. If not, they follow the Bayes-optimal agent (exploit), though calling π_ξ^* 'exploiting' is not quite fair, since Bayes also explores, just not enough to be asymptotically optimal. Some experiments comparing Thompson Sampling, BayesExp, and Inq can be found in [CCH19].

9.5   Planning-Avoiding Agents (Self-AIXI)

Rather than explicit expectimax planning (in theory, which is typically approximated by Monte Carlo Tree Search in practice), there are reinforcement learning algorithms that side-step planning by learning a Q-value function via solving Bellman (optimality) equations (in theory, which is often approximated in practice by Temporal Difference (TD) style algorithms) or by directly learning a policy. Planning-avoiding *universal* agents have also been developed: Below we introduce Self-AIXI [CGMH$^+$23] which self-predicts its own action stream. In Section 13.2 we introduce AIXItl, which computes a provable lower bound to its value function. Both agents converge to AIXI, are self-optimizing, and have maximal Legg–Hutter intelligence. Another approach described in Section 12.9 to avoid planning is Compress and Control.

Self-AIXI is a universal agent using self-prediction, instead of traditional planning. This involves generating a stream of action data akin to methods used by other TD(0) agents, and executing an action maximization step based on current on-policy Q-value estimates for a mixture-policy. We show that despite only performing one-step lookahead "planning," Self-AIXI converges to (long-horizon) Bayes-optimal AIξ, and has the ability to self-optimize (Definition 7.3.3) by utilizing the power of sequential Bayesian action prediction (via ζ, see below) akin to Bayesian sequence models (Definition 3.1.2) but for actions; not to be confused with Bayes-optimal actions from π_ξ^*. In the universal setting, under certain conditions, it asymptotically inherits properties such as maximal LH intelligence (Definition 16.7.1).

Self-AIXI predicts its own stream of actions via a Bayesian mixture policy ζ over a class of policies Π in a dual fashion to the Bayesian mixture environment ξ over $\mathcal{M}$:

Definition 9.5.1 (Bayesian mixture policy and Self-AIXI)   The Bayesian mixture ζ over a countable class of policies Π is defined as

$$\zeta(a_t|h_{<t}) := \sum_{\pi \in \Pi} \omega(\pi|h_{<t})\pi(a_t|h_{<t})$$

$$\omega(\pi|h_{1:t}) := \omega(\pi|h_{<t})\frac{\pi(a_t|h_{<t})}{\zeta(a_t|h_{<t})} = \omega_\pi \frac{\pi(a_{1:t}||e_{<t})}{\zeta(a_{1:t}||e_{<t})}$$

Self-AIXI takes the one-step optimal action according to both the Bayesian mixture environment ξ over $\mathcal{M}$ with prior w_ν *and* the Bayesian mixture policy ζ over Π with prior $\omega(\pi|\epsilon) \equiv \omega_\pi$:

$$\pi_S(h_{<t}) := \arg\max_{a_t} Q_\xi^\zeta(h_{<t}a_t)$$

where the Q-value is defined in Definition 6.7.1.

Importantly, Self-AIXI maximizes the Q-values from the mixture policy ζ instead of the optimal policy π_ξ^*, which means that there is no need to optimize the future. Only the on-policy Q-values of the current ζ are needed, which are typically easier to estimate than the optimal off-policy Q-values Q_ξ^*.

Note how action selection and history interact with the policy-mixture: The action selected by the Self-AIXI agent necessarily improves, by definition, over the current value estimates $\max_{a_t} Q_\xi^\zeta(h_{<t},a_t) \geq V_\xi^\zeta(h_{<t})$. Then, this action is added to the next history $h_{<t+1}$ which is consumed by the policy-mixture at the next time step, i.e. $\zeta(a_{t+1}|h_{<t+1})$.

The policy-mixture does Bayesian inference over the incoming self-generated action-data and makes better action predictions over time. Thus, Self-AIXI is self-predicting its own small improvements made by the $\arg\max$ operation in Definition 9.5.1. In the case of using the largest class of all computable policies, the mixture is a Solomonoff (action) predictor with enough power to approximately represent, within itself, the policy evaluation and improvement operation.

Theorem 9.5.2 (Convergence properties of Self-AIXI [CGMH+23]) Under various conditions, the value functions of Self-AIXI π_S converge to those of Bayes-optimal AIξ in expectation on histories generate by π_S interacting with μ:

$$\mathbf{E}_\mu^{\pi_S}\left[V_\xi^{\pi_\xi^*}(h_{<t}) - V_\xi^{\pi_S}(h_{<t})\right] \xrightarrow{t\to\infty} 0$$

$$\mathbf{E}_\mu^{\pi_S}\left[V_\mu^{\pi_\xi^*}(h_{<t}) - V_\mu^{\pi_S}(h_{<t})\right] \xrightarrow{t\to\infty} 0$$

This implies that for environment classes $\mathcal{M}$ that admit self-optimizing policies (Figure 7.1), AIξ is self-optimizing (Theorem 7.3.6) and hence Self-AIXI is too:

$$V_\mu^*(h_{<t}) - V_\mu^{\pi_S}(h_{<t}) \xrightarrow{t\to\infty} 0 \qquad \mu^\pi\text{-almost surely for any } \pi$$

For $\pi = \pi_S$, this implies that π_S is strongly asymptotically optimal. For $\mathcal{M} = \mathcal{M}_{sol}$, the first convergence implies that asymptotically Self-AIXI has maximal universal intelligence $\Upsilon(\pi) \equiv V_\xi^\pi$ (Definition 16.7.1):

$$\mathbf{E}_\mu^{\pi_S}\left[\max_\pi \Upsilon(\pi|h_{<t}) - \Upsilon(\pi_S|h_{<t})\right] \xrightarrow{t\to\infty} 0$$

Proof idea. Self-AIXI initially acts like a one-step optimal (greedy) agent and this is reflected in the action sequence it produces. ζ then learns to act like a one-step optimal agent, which means Self-AIXI starts to act like a two-step optimal agent. This continues indefinitely and the performance of Self-AIXI converges to the performance of AIξ.   ∎

The results show that agents can perform well even in the general reinforcement learning setting without utilizing planning. Some experiments comparing Self-AIXI with MC-AIXI-CTW (Chapter 12) and proofs can be found in [CGMH+23].

9.6   Exercises

1. [C25] Prove Theorem 9.1.3.

2. [C32] Prove Theorem 9.2.1.

3. [C28] Prove Theorem 9.4.1.

4. [C20] Prove that if $H_t(\varepsilon) = o(t)$, then there exists $\varepsilon_t \to 0$ such that $H_t(\varepsilon_t) = o(t\varepsilon_t)$ as required in Theorem 9.4.1.

5. [C45o] Prove/disprove that Thompson sampling is weakly asymptotically optimal.

6. [C45o] Prove/disprove that using Thompson sampling with an ε-optimal policy for the sampled environment leads to an agent which is asymptotically optimal in expectation?

7. [C20i] Prove that $w_\nu = 2^{-K(\nu)}$ has infinite entropy $\sum_{\nu \in \mathcal{M}_{sol}} w_\nu \log w_\nu^{-1}$, but $w_\nu = 2^{-(1+\varepsilon)K(\nu)}$ and $w_\nu = 2^{-K(\nu)}/K(\nu)$ have finite entropy.

8. [C38] BayesExp uses explicit exploration to achieve weak asymptotic optimality. Bayesian agents like AIXI will still (implicitly) explore. Derive a definition for the kind of exploration a Bayesian agent does. Using this definition find the largest class of environments $\mathcal{M}$ such that AIXI is asymptotically optimal (for each type).

9. [C45o] Prove/disprove that using BayesExp with an ε-optimal policy during the exploitation (and/or exploration) phases leads to an agent which is weakly asymptotically optimal?

9.7 History and References

Optimism. Optimism has long been an approach to the exploration problem in reinforcement learning [SL08]. The optimistic variant of AIXI and its philosophical and axiomatic underpinning was first introduced in [SH12b], where it was shown how optimism can lead to asymptotic optimality (Section 9.1). Optimism of general agents was further studied in the context of cognitive theory and rational behavior [SH14a, SH15b]. On the opposite end of the spectrum, it was shown that pessimism in general reinforcement learning can lead to more safe behavior [CH20b]. A counter-point to optimism was presented in [OVR16b] which argued that any optimistic algorithm that matches Thompson sampling in statistical efficiency would likely be computationally intractable.

Thompson sampling. Thompson sampling [Tho33] (Section 9.2) has had a recent resurgence as it has shown to lead to strong performance in many problems beyond the traditional bandit setting including: contextual bandits [AG13], finite time bandits [CPRR17], non-stationary bandits [RK17], MDPs [GM15, OGNJ17, OVR16a], POMDPs [JJJN21], game theory [Mau20], adaptive control [OB10b, OB10a, Ort11] and (general) reinforcement learning [LLOH16, LLOH17]. A thorough overview of Thompson Sampling is given in [RVRK$^+$17]. Using naive sequential prediction to solve sequential decision-making problems can lead to several difficulties [OKD$^+$21], however in [OWR$^+$19] it was shown how to meta-learn a Thompson sampling agent and overcome many of these difficulties. Thompson sampling-based exploration has been shown to work well in Atari when using an uncertainty-based Bellman equation [OOMM17].

Knowledge-seeking agents. The knowledge seeking agent was first proposed in [Ors11] and later extended to stochastic environments in [OLH13] (Section 9.3). The knowledge seeking agent was extended again from stochastic environments to quantum environments in [Sar21, SAAGB21].

Curiosity-based agents. Highly related to knowledge seeking agents are curiosity-based agents, that is, agents which use a notion of curiosity to aid in exploration, or in some cases replace reward entirely. One of the earliest derivations of curiosity-based agents was done in [Sch91] with an agent that is incentivized to take actions that will provide more information about the world it is interacting with. [BEP$^+$18] performed a study on curiosity-based agents in Atari and other reinforcement learning environments. One of the downsides of curiosity and information-seeking agents is that they can be susceptible to "noisy TVs": any input that feeds unpredictable random data to the agent. This problem is suffered by the original Knowledge seeking agent [Ors11] which only modelled deterministic environments. [MPYBG22] investigated how curiosity-based agents can overcome this problem. Variation Information Maximizing has been shown to be an effective curiosity-based exploration method [HCD$^+$16]. [SGS11] introduced an exploration strategy using a curiosity-based (action) value function and proved its (Bayesian exploration) optimality when the environment is an MDP.

Curiosity is not always a positive for an agent and the dangers of overly curious agents were discussed in [CHC21].

Intrinsic rewards.   One of the key aspects of knowledge seeking agents is that they possess an intrinsic reward. In the context of reinforcement learning, intrinsic motivation and reward has seen substantial study. [LAW+20] provides an overview of many of the approaches to this problem. A study of intrinsic motivation and its ability to lead to hierarchical learning was conducting in [BSC+04]. [Sch10] proposed a precise theory of creativity, fun and intrinsic motivation which itself was motivated by compression. One approach in the absence of rewards is to use the discovery of new behavior in the form of options [SP02] to motivate the agent [MB16]. Intrinsic motivation can also be used to guide the exploration of agents [BSO+16, MBMK21]. The topic of intrinsic motivation and reward will see further exploration in Chapter 15.

General exploration agents.   The BayesExp agent [LH14a] (Section 9.4) was the first agent which was shown to be weak asymptotically optimal in a large class of environments, thus it also demonstrated that weakly asymptotically optimal was an achievable property of general agents (though not deterministic ones). [Lei16a] introduced an exploration measure, called exploration potential, that is provably equivalent to asymptotic optimality (under some conditions). The Inq agent [CCH19] was the first agent which was shown to be strongly asymptotically optimal, and again like with BayesExp this demonstrated that strong asymptotic optimality was an achievable property (previously it was suspected but not proven that no general agent could achieve this).

Planning-avoiding agents.   For finite-state MDPs, Bellman equations [BT96] and Temporal Difference algorithms [HL07, SB18] bootstrap values and policies by one-step (or few-step) lookahead instead of expensive (long-horizon) expectimax or Monte Carlo Tree Search (MCTS) planning. These ideas have been extended to infinite MDPs via (non)linear value function approximation and POMDPs. In this chapter we have discussed the history-based planning-avoiding agent Self-AIXI [CGMH+23] which maintains a Bayesian posterior mixture over policies. In Section 13.2 we discuss AIXI*tl* [Hut05b, Chp.7] and in Section 12.9 Compress and Control [VBH+15, DRW+24].

Risk- and ambiguity-sensitivity agents.   Bayes-optimal agents are risk-neutral since they solely attune to the expected return, in contrast to risk-sensitive agents, which are also sensitive to higher-order moments of the return. Bayes-optimal agents are also ambiguity-neutral since they act in new situations as if the uncertainty were known, unlike ambiguity-sensitive agents which act differently when recognizing situations in which they lack knowledge. [GMDK+22] show how off-policy meta-learning can give rise to risk- and ambiguity-sensitivity agents, which can be safer and more robust to external perturbations.

Chapter 10

Multi-Agent Setting

There's no sense in being precise
when you don't even know what
you're talking about.

John von Neumann

So far we have only discussed single-agent settings, where there is only one agent and the environment. More realistically, and more complex, is the multi-agent setting. In this setting we consider the agent, the environment and some number of other agents. Now what do we mean by other agents? Surely we could capture the behavior of these other agents as part of the environment. While this is true, it misses some of the subtlety and advantages of considering the multi-agent setting. For one, when we consider an environment, if we know we are in the multi-agent setting (we have to consider or can exploit that) we know the other agents will also learn, or at least act in specific ways with respect to our own actions.

In Section 10.1 we state and discuss a fundamental result in decision theory, namely that the preferences of a rational agent over different choices facing uncertain outcomes can be represented as an expected real-valued (von Neumann-Morgenstern) utility function, which is the basis for expected utility theory and hence most of game theory.

The field of game theory is fundamental for the study for multi-agent systems. In Section 10.2 we will go over the basics of game theory, including the formalization of strategic games, Nash equilibrium, and some important games. We will then present the relations between game theory and reinforcement learning in Section 10.4.

In Section 10.5 we will describe the notion of reflective oracles which play a key role for agents who are able to model themselves. Lastly we will discuss the grain of truth problem in Section 10.6, which is the problem of finding a class of policies closed under Bayesian optimality. We will show why this is an important problem in the multi-agent setting and present a solution.

10.1   From Preferences to Utilities

von Neumann-Morgenstern utility [VNM47]. The von Neumann-Morgenstern (VNM) utility theorem gives a collection of premises under which it is sufficient for an agent to be *utilitarian*: that is, its preferences can be described as the maximization of some utility function over possible outcomes. We formalize this concept as follows: Let $\mathcal{O}$ denote a set of *outcomes* (which could even be the set of all finite interaction histories), and let $\prec$ denote the agent's *preference* (a binary relation) over $\mathcal{O}$.

In practice, a full description of the preferences $\prec$ of an agent (like a human) may be in general unknown. We treat the preferences of the agent as a black box, and assume we can query the agent. If the agent strictly prefers a to b, we write $a \succ b$, or $b \prec a$. If the agent is indifferent to either a or b (formally, $a \not\succ b$ and $b \not\succ a$) we write $a \sim b$. The symbols $\succeq$ and $\preceq$ are defined in the obvious fashion.

Given two outcomes $a,b \in \mathcal{O}$ and $\theta \in [0,1]$, we define a *lottery* $\theta a + (1-\theta)b$ as the (random) outcome representing observing outcome a with probability θ, and outcome b with probability $1-\theta$. We require that $\mathcal{O}$ is closed under lotteries, and that the agent also has preferences defined for lotteries, that is, given any outcomes $a,a',b,b' \in \mathcal{O}$ and $\alpha,\beta \in [0,1]$, the outcomes $\alpha a + (1-\alpha)a'$ and $\beta b + (1-\beta)b'$ are in $\mathcal{O}$ hence comparable with respect to $\prec$. Note that this makes outcomes *random* events.

Definition 10.1.1 (Utilitarian preferences) An agent's preferences $\prec$ are *utilitarian* if there exists a function $U : \mathcal{O} \to \mathbb{R}$ such that

$$a \prec b \quad \text{if and only if} \quad \mathbf{E}[U(a)] < \mathbf{E}[U(b)]$$

The *von Neumann-Morgenstern utility theorem* states that under some weak assumptions, an agent's preferences are *utilitarian*. We will not prove the theorem here, but will instead motivate why these assumptions are reasonable.

Theorem 10.1.2 (von Neumann-Morgenstern utility [VNM47]) Let $\prec$ be a preference relation over outcomes $\mathcal{O}$. If the four following conditions hold for arbitrary $a,b,c \in \mathcal{O}$:

1. *Completeness*: At least one of the following is true: $a \preceq b$ or $b \preceq a$.

2. *Transitivity*: If $a \preceq b$ and $b \preceq c$, then $a \preceq c$.

3. *Continuity*: If $a \preceq b \preceq c$ then there exists $\theta \in [0,1]$ such that $\theta a + (1-\theta)c \sim b$.

4. *Independence*: $a \preceq b$ if and only if $\theta a + (1-\theta)c \preceq \theta b + (1-\theta)c$ for any $\theta \in (0,1]$.

then $\prec$ is utilitarian.

---

1. **Completeness:** This merely assumes that between any two options, the agent can state which it prefers (or if it is indifferent). If the agent cannot even state what its preference would be between two options, then we have no hope of assigning utilities to it.

2. **Transitivity:** This seems intuitively obvious (if the agent prefers apples to bananas, and bananas to pears, it would stand to reason the agent would prefer apples to pears). Suppose $\preceq$ was non-transitive, so there exists a, b, c such that $a \preceq b$ and $b \preceq c$ but $a \succ c$. Then, presumably an agent would be happy (or be indifferent) to swapping a for b, and same for swapping b for c, but then the agent strictly prefers a to c, and would pay a non-zero amount (say, one cent) to swap c for a. The agent is now back in the initial configuration, and one penny poorer. We can repeat this argument *ad infinitum* and take all of the agent's assets, one penny at a time. This style of argument to demonstrate non-transitive preferences are irrational is called a *money pump*.

3. **Continuity:** This requires that the preferences of an agent do not change suddenly in response to arbitrary small changes. If the agent prefers $a \preceq b \preceq c$, then the lottery $l = \theta a + (1-\theta) c$ should still satisfy $a \preceq l \preceq b$ for $\theta \approx 1$ (as l would with overwhelming probability be the same outcome as a, and $a \preceq b$). As θ is decreased, the agent would prefer l more and more, (as the outcome is more likely to be the more desirable outcome c) and at some point l is almost indistinguishable from c for $\theta \approx 0$, so we would expect $b \preceq l \preceq c$. It seems reasonable to assume that for some intermediate value of θ, the agent is indifferent between l and b.

4. **Independence:** Given a preference $a \preceq b$, the agent should still prefer a θ chance at b over a θ chance of a, regardless of the other outcome c with $1-\theta$ probability. This indicates that a preference should not change if stated as a lottery, with the same alternative option for both lotteries.

Theorem 10.1.2 allows us to work with real-valued utilities, rather than abstract preference relations, which is more convenient. Indeed, it provides a partial justification for the reward hypothesis, we assumed since Chapter 6.

10.2 Game Theory

Game theory is the study of multi-agent interactions. For this section, we will be using the notation of [OR94]. For the reader looking for a more in-depth view on game theory we suggest [Osb04, SLB08].

10.2.1 Strategic Games

The common definition of a game is a 3-tuple with players, actions, and preferences. This is called a strategic game (in normal form).

Definition 10.2.1 (Strategic game in normal form) A *strategic game* is a 3-tuple $\langle N, (\mathcal{A}_i)_{i \in N}, (\succeq_i)_{i \in N} \rangle$, where N is a finite set of players, $\mathcal{A}_i$ is the set of actions for player i, and $\succeq_i$ is the preference relation on $\mathcal{A} := \mathcal{A}_1 \times \mathcal{A}_2 \times ... \times \mathcal{A}_N$ for player i. If all of the sets of actions $\mathcal{A}_i$ are finite, then we call the 3-tuple a *finite strategic game*.

Given a subset $I \subset N$ of the players, then an assignment of players to actions $f : I \to \mathcal{A}_i$ is called an *action profile*, or simply a *profile*. We usually denote an action profile as a vector $a \in \prod_{i \in I} \mathcal{A}_i$ of each player's action.

We will make the assumption that for each player i there exists a utility function $u_i : \mathcal{A} \to \mathbb{R}$ such that for all $a, b \in \mathcal{A}$, we have $u_i(b) \geq u_i(a)$ if and only if $b \succeq_i a$. This implicitly places constraints on the types of preferences players are allowed to have, and it forbids kinds of preferences that have inconsistencies preventing them from being modelled as the maximization of some utility function as discussed in Section 10.1. In the case where the preferences can be specified by a utility function, we may skip preferences entirely and define $\langle N, (\mathcal{A}_i)_{i \in N}, (u_i)_{i \in N} \rangle$ as the strategic game. We usually present strategic games in this fashion, since having non-utilitarian preferences are irrational (Theorem 10.1.2).

Additionally we have some (slightly modified) notation from [OR94] which we will use here. For any profile $a = (a_1, ..., a_n) \in \mathcal{A}$, denote by $a_{\neq i}$ the vector $(a_1, ..., a_{i-1}, a_{i+1}, ..., a_n)$, which is the subprofile of a with the action associated with player i removed. As a slight abuse of notation, we write $u_j(a_{\neq j}, b)$ for player j's utility for taking action b instead of action a_j, assuming the actions of all other players are described by the action profile $a_{\neq j}$.

10.2.2   Nash Equilibrium

Informally, the Nash equilibrium is an action profile with the property that no player would prefer to unilaterally change their action given the knowledge of the actions that all other players will choose.

Definition 10.2.2 (Nash equilibrium) A *Nash equilibrium* of a strategic game $\langle N, (\mathcal{A}_i)_{i \in N}, (\succeq_i)_{i \in N} \rangle$ is an action profile $a^* \in \mathcal{A}$, with the property that for every player $i \in N$, we have

$$\forall a \in \mathcal{A}_i. \ u_i(a^*) \ \geq \ u_i(a_{\neq i}^*, a)$$

That is, given every other player chooses the appropriate action in the profile a^*, there exists no i for which player i would (strictly) prefer to play an action other than a_i^*.

Nash equilibria can be thought of as "stable" in the sense that no single player would unilaterally desire to play a non-Nash equilibrium action if everyone else is. But note that it is often the case that all players would be better off if they could all coordinate to play a different action profile instead. Conversely, non-Nash equilibrium profiles are "unstable" as there exists a player who is incentivized to play a different action instead for strictly greater utility, often at the expense of lower utility for the other players, for example in zero-sum games.

Closely related to the Nash equilibrium is the best response function. This takes the actions of every other player and gives a set of actions that are optimal according to the players own utility function u_i.

Definition 10.2.3 (Best response function) Given $a_{\neq i}$, the action profile of all other players, we define player i's *best response function* $B_i : (\mathcal{A}_j)_{j \in N \setminus i} \to \mathcal{A}_i$ as all actions that maximize their utility

$$B_i(a_{\neq i}) := \underset{a \in \mathcal{A}_i}{\arg \max} \ u_i(a_{\neq i}, a)$$

Theorem 10.2.4 (Nash Equilibrium via best response functions) An action profile a^* is a Nash equilibrium if and only if for all i, every action a_i^* is one of player i's best responses to $a_{\neq i}^*$, that is, $a_i^* \in B_i(a_{\neq i}^*)$.

Finite two-players *normal-form* strategic games are often represented in form of a *payoff matrix*, where the rows are the actions of one player and the columns are the actions of the other. In each cell in the table is a pair (u_1, u_2), where u_1 is the utility for the player on the left (player 1), and u_2 is the utility for the player on the top (player 2), for that action profile respectively.

		Player 2	
Name		*action 0*	*action 1*
	action 0	$(u_1(0,0), u_2(0,0))$	$(u_1(0,1), u_2(0,1))$
Player 1	*action 1*	$(u_1(1,0), u_2(1,0))$	$(u_1(1,1), u_2(1,1))$

It is not immediately clear whether for any strategic game, a Nash equilibrium exists, and indeed often it does not. Below we will show some examples of well-studied strategic games, some of which do have Nash equilibrium, some do not.

The surprising and often frustrating property that some games have is that a Nash equilibrium a^* is not the optimal profile for all players: There can also exist a non-Nash equilibrium profile b^* that all players would strictly prefer $(\forall i. u_i(b^*) > u_i(a^*))$. Unfortunately, non-Nash equilibrium strategies are not stable, in the sense that there exists a player i who would gain even more utility by choosing a different action $a_i \neq b_i^*$, often at the expense of decreasing the utility of other players. After several iterations of the same game, other players might act "selfishly"[1] in the same manner, until the action profile for which no player would unilaterally change their action is achieved, which is strictly worse off for everyone. A classical example is the Prisoner's dilemma described below.

If a strategic game satisfies certain conditions, the existence of a Nash equilibrium is guaranteed.

Lemma 10.2.5 (Kakutani's fixed point theorem [Kak41]) Let X be a compact convex subset of $\mathbb{R}^n$ and let $f : X \to 2^X$ be a set-valued function for which the following holds:

1. For all $x \in X$ the set $f(x)$ is nonempty and convex, i.e. for all $x_1, x_2 \in f(x)$ the line segment from x_1 to x_2 is contained in $f(x)$, i.e. for all $\alpha \in [0,1]$, $\alpha x_1 + (1-\alpha)x_2 \in f(x)$).

2. For all sequences (x_n) and (y_n) such that for all n, $y_n \in f(x_n)$ and the sequences have limits $x_n \to x$ and $y_n \to y$, then $y \in f(x)$, i.e. the graph of f contains its limit points, that is, the graph is closed.

Then there exists a fixed point x^* such that $x^* \in f(x^*)$.

Definition 10.2.6 (Quasi-concave preference relation) A preference relation $\succeq_i$ over $\mathcal{A}$ is *quasi-concave* over $\mathcal{A}_i$ if for every $a \in \mathcal{A}$, the set $\{a_i \in \mathcal{A}_i : (a_{\neq i}, a_i) \succeq_i a\}$ is convex.

---

[1] assuming each agent is *egotistical*, and their utility function values only outcomes for themselves, as opposed to *altruistic* agents who values both the utility of themselves and of others.

Theorem 10.2.7 (Existence of Nash equilibrium [OR94, Prop.20.3]) The strategic game $\langle N,(\mathcal{A}_i)_{i\in N},(\succeq_i)_{i\in N}\rangle$ has a Nash equilibrium if for all $i\in N$:

1. The set $\mathcal{A}_i$ is a nonempty, compact convex subset on $\mathbb{R}^n$.

2. The preference relation $\succeq_i$ is continuous and quasi-concave on $\mathcal{A}_i$.

Proof sketch. For $B:\mathcal{A}\to\mathcal{A}$ defined as $B(a):=\times_{x\in N}B_i(a_{\neq i})$, the condition for a being a Nash-equilibrium in Theorem 10.2.4 can be rewritten as $B(a^*)=a^*$. For every $i\in N$ the set $B_i(a_{\neq i})$ is nonempty since $\succeq_i$ is continuous and $\mathcal{A}_i$ is compact, and is convex since $\succeq_i$ is quasi-concave on $\mathcal{A}_i$; B has a closed graph since each $\succeq_i$ is continuous. Thus by Kakutani's theorem, B has a fixed point, and hence by the above is a Nash equilibrium. ∎

In some strategic games the two players can cooperate with each other and do better, but we often use the word "game" (think chess or checkers) if the only way one player can win is if the other loses. These kinds of games are called *strictly competitive* or *zero-sum* games.

Definition 10.2.8 (Strictly competitive game) A strategic game $(\{1,2\},(\mathcal{A}_1,\mathcal{A}_2),(u_1,u_2))$ with two players is a *competitive game* if for all action profiles $a,b\in\mathcal{A}$, $a\succeq_1 b$ if and only if $b\succeq_2 a$.

On can show that for strictly competitive games there are *zero-sum* payoff functions with $u_1+u_2=0$ representing the players' preferences.

10.2.3 Important Games

In game theory, there are many interesting normal-form games which have been studied. In this subsection we will go over five different prototypical normal-form games which we will later use as environments for a simple approximation of AIXI. These 2×2 matrix games have been intensely studied, since they constitute the smallest non-trivial games, already exhibiting a wide variety of interesting and intricate phenomena.

Prisoner's Dilemma. It would not be an exaggeration to say that the most studied game in game theory is the Prisoner's Dilemma. It is a fundamental game for which the Nash equilibrium is strictly suboptimal for both players. The idea of the game is that there are two suspects of a crime that are being interrogated separately and they both have the option to *cooperate* with the other criminal and remain silent, or *defect* by testifying against their co-conspirator. If the criminals cooperate with each other, they will both receive a short prison sentence, giving each player utility 2. If both suspects defect, stronger evidence will ensure both receive a longer prison sentence, giving each player utility 1. If one player defects and the other cooperates, the defecting suspect is released under a plea bargain, receiving utility 4, while the cooperating suspect is put into prison for the longest time and receives utility 0. This is summarized in the game matrix below.

	Player 2	
PD	*defect*	*cooperate*
defect	(1,1)	(4,0)
cooperate	(0,4)	(2,2)

(Player 1 labels the rows *defect* / *cooperate*.)

Ideally both players would wish to cooperate, but this leads to an incentive for either player to defect, leading to the Nash equilibria of defect-defect, a strictly worse outcome for both players.

Very serious real-world versions of this dilemma are the *tragedy of the commons* for shared physical resources which (without global cooperation), leads to pollution, overfishing, deforestation, and more.

Stag Hunt. Two hunters are on a hunting trip together. The hunters have the option to cooperate to hunt a stag, or to hunt alone for a rabbit. If the hunters both choose to hunt alone, they each receive utility of 2, corresponding to catching and sharing a rabbit. If they choose to hunt together, they catch a stag and both obtain a utility of 4. However, if one hunter chooses to hunt a rabbit, and the other hunts the stag, the rabbit hunter will receive utility 3 since he gets the rabbit all to himself, and the player who was hunting the stag alone is unsuccessful and receives reward 0. This is summarized in the game matrix form below.

		Player 2	
SH		*alone*	*together*
Player 1	*alone*	(2,2)	(3,0)
	together	(0,3)	(4,4)

Stag Hunt is a game which may look very similar to the prisoner's dilemma, however it has one key difference: the highest utility for each player is also the Nash equilibrium. This means that the players have no incentive not to work together, and it is in the best interest of even selfish players to cooperate.

Some other examples that have the same properties as Stag Hunt are: two individuals who must row a boat, two neighbors wishing to drain a meadow, coordination of slime molds, hunting practices of orcas, and countries working together to improve corporate governance, just to mention a few.

Chicken. Two thrill-seekers are playing a game of brinkmanship: Both are in cars, driving towards each other in a head-on collision course. Either driver can choose to "swerve" or "not swerve". If one driver swerves and the other does not, the driver who swerved "chickened out", and is made an object of contempt (receiving utility 1), whereas the driver who did not, gains the admiration of their peers (receiving utility 4). If both drivers swerve, neither is branded a coward, but they also receive no admiration (a utility of 2 each). If neither driver swerves, they crash head-on, die, and receive utility of 0.

		Player 2	
Chicken		*no swerve*	*swerve*
Player 1	*no swerve*	(0,0)	(4,1)
	swerve	(1,4)	(2,2)

Unlike the previous two games mentioned, the game of Chicken has two Nash equilibria. If player 2 is the chicken and player 1 is not, then neither player has incentive to change their action. The same goes for player 2 committing and player 1 chickening out. In either case if the player who is the chicken changes to not swerving he will receive reward 0 instead of 1, and if the player who is not swerving changes to swerving she will receive reward 2 instead of reward 4.

The game of chicken is popular in politics and economics and military affairs, trying to achieve an advantageous outcome by pushing dangerous events to the brink of active conflict in the hope the opponent backs down rather than ending up in a devastating conflict for both. Brinkmanship can also happen if two companies with similar products participate in a price war, or two people claiming more or less credit than they deserve to their superior for a joint project to get a promotion.

Battle of the Sexes. A couple intends to spend a nice evening together. The husband prefers watching a Western, the wife prefers to go to a Musical. But both would prefer to go to the same event rather than different ones. If they are bad communicators, where should they go? Each receives utility 4 if they go together to their preferred event their spouse accompanying them, utility 2 if they go to the event they dislike with their spouse and utility 0 if they go separately (somewhat unrealistically regardless of their choice). This is summarized in the game matrix below.

		Husband	
BoS		*Musical*	*Western*
Wife	*Musical*	(2,4)	(0,0)
	Western	(0,0)	(4,2)

Similar to the game of Chicken, Battle of the Sexes has two Nash equilibria. However, unlike in Chicken, if the opponent changes their action, that will not be beneficial to you. This leads to an interesting choice for optimal play in repeated games: the couple alternates between going together to a Musical and a Western. This leads to an average utility of 3 per game, since each player is receiving utility 4 half the time, and utility 2 half the time. We see this in real life, where people will "compromise" and take turns with who gets to choose the event, rather than both go to their favorite kind of event alone.

Setting industry standards is an instance of BoS: Each company prefers their in-house solution, but both are better off if they agree on a common standard.

Matching Pennies. Two players are competing in a game where each player chooses heads or tails. The first player receives utility 1 if both players choose the same option, and utility 0 otherwise. The second player receives utility 1 if both players choose different options, and utility 0 otherwise. This is summarized in the game matrix below.

		Player 2	
MP		*Heads*	*Tails*
Player 1	*Heads*	(1,0)	(0,1)
	Tails	(0,1)	(1,0)

Unlike all the previous games we have mentioned, matching pennies does not have a pure=deterministic Nash equilibrium, since for every action profile, the losing player can gain more utility by switching her strategy to the other action.

Penalty kicks in soccer are an instance of Matching Pennies. The kicker can kick left or kick right, and the goalie can jump left or jump right. Importantly, the goalie has to jump before he sees the kicker's choice. The situation is similar for serve-and-return plays in tennis. Rock paper scissors can be regarded as a version of Matching Pennies with three actions.

10.2.4 Mixed Strategic Games

When acting against intelligent players in a game (especially for competitive games where cooperation is not an option), we do not want to always deterministically choose an action lest our opponent predicts what action we are going to take (which in games like matching pennies means we will always lose.) This leads us to the concept of mixed strategic games, where instead of choosing an action deterministically, each player chooses a probability distribution over the set of actions, and then samples actions according to that probability distribution.

Definition 10.2.9 (Mixed strategic games in normal form) The *mixed extension of the strategic game* $\langle N,(\mathcal{A}_i),(u_i)\rangle$ is the strategic game $\langle N,(\Delta\mathcal{A}_i),(U_i)\rangle$, where $\Delta\mathcal{A}_i$ is the set of all probability distributions over $\mathcal{A}_i$, and $U_i:\times_{j\in N}\Delta\mathcal{A}_j\to\mathbb{R}$ is defined as the expected value of u_i, assuming player i samples his action from $\mathcal{A}_i$ according to the probability distribution α_i. By writing $\alpha:=(\alpha_1,...,\alpha_N)$, we can express U_i as

$$U_i(\alpha) := \sum_{a\in\mathcal{A}} u_i(a) \prod_{j\in N} \alpha_j(a_j)$$

We call a distribution in $\Delta\mathcal{A}_i$ a *mixed strategy* for player i, and note that the set of mixed strategies for a player in a strategic game, is the same as the set of actions in that game's mixed extension. We sometimes use the term *pure strategy* for actions $a\in\mathcal{A}_i$, i.e. distributions in $\Delta\mathcal{A}_i$ that are deterministic, assigning probability 1 to one of the actions in $\mathcal{A}_i$ and zero to the others. We can extend the concept of Nash equilibrium to mixed strategic games.

Definition 10.2.10 (Mixed Nash equilibrium) A *mixed strategy Nash equilibrium* of a strategic game is a Nash equilibrium of the mixed extension, that is, a collection α^* of probability distributions $(\alpha_1^*,...,\alpha_N^*)$, such that each $\alpha_i\in\Delta\mathcal{A}_i$, and no player would stand to gain more utility (in expectation) by choosing a different probability distribution. So, for all players i, and all $\alpha_i\in\Delta\mathcal{A}_i$,

$$U_i(\alpha^*) \geq U_i(\alpha_{\neq i}^*,\alpha_i)$$

where $\alpha_{\neq i}^*$ is α with the distribution for player i removed.

Note that every Nash equilibrium a^* is a special case of a mixed Nash equilibrium α^*, as we can choose $\alpha^*=(\alpha_1^*,...,\alpha_N^*)$ with $\alpha_i^*(x)=[\![x=a_i^*]\!]$, the deterministic distribution assigning 1 to a_i^*, and 0 elsewhere. Using this definition, we have an even more powerful result about the existence of mixed strategy Nash equilibria.

Theorem 10.2.11 (Existence of mixed Nash equilibrium) Every finite strategic game has a mixed strategy Nash equilibrium.

Proof sketch. The idea is to identify $\Delta\mathcal{A}_i$ with the probability simplex $\{(p_1,...,p_{m_i}):p_j\geq 0,\sum_{j=1}^{m_i}p_j=1\}$, where $m_i:=|\mathcal{A}_i|$. The set is nonempty, convex and compact. Since expectations are linear, one can define a preference relation $a\succeq_i b$ iff $U_i(a)\geq U_i(b)$ that satisfies the premises of Theorem 10.2.7 for the mixed strategy extension. ∎

Recall that the strategic game "matching pennies" has no pure Nash equilibrium. However, since the game is finite, Theorem 10.2.11 implies that there must exist a mixed Nash equilibrium, though the theorem and standard proof give no indication as to how to find it.

If a player chooses actions (heads, tails) with probabilities $(^1/_2,^1/_2)$, then the expected utility for both players is $^1/_2$, regardless of what strategy the opponent chooses, mixed or otherwise. Hence, if both players choose a mixed strategy of $(^1/_2,^1/_2)$, there is no incentive for either player to choose any other mixed strategy (as they are all equally good against a uniformly random opponent), so both players choosing $(^1/_2,^1/_2)$ is a mixed Nash equilibrium for this game.

In fact, it is the only such mixed Nash equilibrium. If the first player chose some other mixed strategy $(\theta,1-\theta)$ for some value $0\leq\theta\leq 1$ with $\theta\neq^1/_2$, then player 2 would choose a

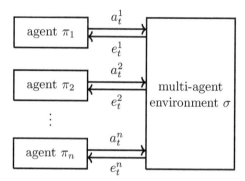

Figure 10.1: *Agents* $\pi_1,...,\pi_n$ *interacting in a multi-agent environment.*

deterministic strategy of either always heads $(1,0)$ if player 1 was more likely to choose tails $(\theta < {}^1\!/_2)$, or always tails $(0,1)$ if player 1 was more likely to choose heads $(\theta > 1/2)$. Then, the expected utility to player 2 is $\max\{\theta, 1-\theta\} > 1/2$ for $\theta \neq 1/2$. The same argument can be made for player 1 if player 2 chooses a mixed strategy other than $({}^1\!/_2, {}^1\!/_2)$.

10.3   Multi-Agent Extensive-Form Games

As currently defined, a strategic game has a single round of interaction. More interesting are games where all players choose an action (or an action is sampled from the chosen distribution), the players' actions are revealed, the players receive utility according to the utility functions, and then players choose an action (or change their distribution) again for the next round, using the previous history of action profiles and received utilities as a guide. Multi-round strategic games are called *extensive-form games*. This setting is extremely general, much more general than just repeated (i.i.d.) matrix games. Indeed it generalizes the history-based (essentially assumption-free) reinforcement learning setting to modelling multiple agents with simultaneous actions (see Chapter 11 for how to transform between simultaneous and sequential actions).

In *extensive-form games* or *multi-agent setting* there are n agents each taking actions sequential from the finite action space $\mathcal{A}$. In each time step $t=1,2,...$, the environment receives action a_t^i from agent i and outputs n percepts $e_t^1,...,e_t^n \in \mathcal{E}$, one for each agent. Each percept $e_t^i = (o_t^i, r_t^i)$ contains an observation o_t^i and a reward $r_t^i \in [0,1]$. Importantly, agent i only sees its own action a_t^i and its own percept e_t^i (see Figure 10.1). We use the shorthand notation $a_t := (a_t^1,...,a_t^n)$ and $e_t := (e_t^1,...,e_t^n)$ and denote $h_{<t}^i = a_1^i e_1^i...a_{t-1}^i e_{t-1}^i$ and $h_{<t} = a_1 e_1...a_{t-1} e_{t-1}$.

We define a multi-agent environment as a probability Kernel

$$\sigma : (\mathcal{A}^n \times \mathcal{E}^n)^* \times \mathcal{A}^n \to \Delta(\mathcal{E}^n)$$

The n agents are given by n policies $\pi_1,...,\pi_n$ where $\pi_i : (\mathcal{A} \times \mathcal{E})^* \to \Delta\mathcal{A}$. Together they specify the *history distribution*

$$\sigma^{\pi_{1:n}}(\epsilon) := 1$$
$$\sigma^{\pi_{1:n}}(h_{1:t}) := \sigma^{\pi_{1:n}}(h_{<t}a_t)\sigma(e_t \,|\, h_{<t}a_t)$$
$$\sigma^{\pi_{1:n}}(h_{<t}a_t) := \sigma^{\pi_{1:n}}(h_{<t})\prod_{i=1}^{n}\pi_i(a_t^i \,|\, h_{<t}^i)$$

Reinforcement learning	=	Game theory
stochastic policy	=	mixed strategy
deterministic policy	=	pure strategy
agent	=	player
multi-agent environment	=	infinite extensive-form game
value	=	payoff/utility
(finite) history	=	history
infinite history	=	path of play
asymptotic optimality	=	convergence to Nash

Table 10.2: *Terminology dictionary between reinforcement learning and extensive-form game theory [LTF16].*

Each agent i acts in a *subjective environment* σ_i given by joining the multi-agent environment σ with the policies $\pi_1,...,\pi_{i-1},\pi_{i+1},...,\pi_n$ by marginalizing over the histories that π_i does not see. Together with policy π_i, the environment σ_i yields a distribution over the histories of agent i

$$\sigma_i^{\pi_i}(h_{<t}^i) := \sum_{h_{<t}^j, j \neq i} \sigma^{\pi_{1:n}}(h_{<t})$$

We get the definition of the subjective environment σ_i with the identity $\sigma_i(e_t^i \mid h_{<t}^i a_t^i) := \sigma_i^{\pi_i}(e_t^i \mid h_{<t}^i a_t^i)$. It is crucial to note that the subjective environment σ_i and the policy π_i are ordinary environments and policies, so we can use the history-based RL formalism from Chapters 6 and 7. Furthermore, we can interpret policy π_i as a (one-shot) mixed strategy α_i of player/agent i, and the value $V_{\sigma_i}^{\pi_i}(\epsilon)$ as their utility $U_i(\alpha)$, which reduces the extensive-form to a normal-form game, hence many results for the latter apply as well.

Our definition of a multi-agent environment is very general and encompasses most of game theory. It allows for cooperative, competitive, and mixed games; infinitely repeated games or any (infinite-length) extensive form games with finitely many players.

Definition 10.3.1 (Multi-agent ε-best response and ε-Nash equilibrium) The policy π_i is an *ε-best response* after history $h_{<t}^i$ iff

$$V_{\sigma_i}^*(h_{<t}^i) - V_{\sigma_i}^{\pi_i}(h_{<t}^i) < \varepsilon$$

If at some time step t, all agents' policies are ε-best responses, we have an *ε-Nash equilibrium*.

The property of multi-agent systems that is analogous to asymptotic optimality is convergence to an ε-Nash equilibrium.

10.4 Strategic Games vs Reinforcement Learning

There are a lot of immediate similarities between strategic games and the sequential decision theory and (history-based) reinforcement learning setup. Both have a set of actions an agent can take, and both have some sort of reward or utility that the agent wants to maximize. See Table 10.2 for a comparison of reinforcement learning and extensive-form game theory terminology.

From Strategic Games to Reinforcement Learning. We can convert any extensive-form strategic game to the reinforcement learning setting by letting the actions in the game be the actions of the agent, and the utilities $U_i(\alpha)$ be the value $V_{\sigma_i}^{\pi_i}$. The agent represents one of the players, and the environment simulates the actions of all other players. However, from the perspective of the agent, it takes an action then sees the actions of the opponent(s). While this does sound like quite a simple conversion, it misses the point of strategic games: modelling the actions of an intelligent adversary, or learning to cooperate or compromise with another player so both players obtain a mutually beneficial outcome. In (classical MDP) reinforcement learning, this is not often captured as many of the results in (classical MDP) reinforcement learning focus on stationary environments (that is, the strategy of all other players is fixed), and so the actions the player should take are simple: observe the actions that the other players take (or in the case for mixed strategy games, estimate the opponent's distribution via many interactions), and then reply with an action using the best response function. Representing other players with a stationary environment means that the opponents cannot learn when they are being exploited and change strategy.

This is not to say that all reinforcement learning is without intelligent adversaries; in fact, the famous successes of reinforcement learning, Backgammon [Tes94], Go[SH+16], and *StarCraft*[VBC+19], are environments which are interesting because of the intelligent adversary (often the human world champion(s) for that respective game or self-play) and the complexity of the aforementioned games making approaches via more classical AI techniques difficult. In many situations, there is information about the adversaries' reward. Often, these environments can be represented by zero-sum games, and the opponent's reward can be taken as the negative of the agent's reward (though, especially for Go and *StarCraft*, it can be very difficult to observe a game in progress and declare with confidence which player currently has an advantage.)

General intelligent agents such as AIXI should learn to perform optimally on these reinforcement learning environments (if it is possible to learn to perform optimally). This is with a "small" caveat: One of the few assumptions of AIXI and Solomonoff induction is that the true environment is assumed to be within the Bayesian mixture class. For simple strategic games this may be the case, however this assumption fails if AIXI were to play against itself. This leads to a reflective problem, which is addressed in the remaining sections of this chapter.

From Reinforcement Learning to Strategic Games. Above we highlighted the transformation of any extensive-form strategic game into a reinforcement learning environment. We now demonstrate that this process can be reversed, emphasizing the utility of this mapping. By maintaining the agent's action space as it was in the reinforcement learning setting, we can represent each state or observation of the environment as an action taken by the opponent. The agent receives a payoff as a reward for every action and state, while the opponent receives the agent's negative payoff.

A notable limitation of this simple conversion is that the strategic game's opponent may not aim for optimal actions, as the environment selects actions based on its distribution. This contradicts one of game theory's core principles, which assumes that all players strive for optimal performance. Setting the pay-off matrix of the opponent to identically zero also does not help: Any best response strategy of the agent to any strategy of the opponent whatsoever is a Nash equilibrium. Regardless, various aspects of reinforcement learning can be adapted to the strategic game framework. For example, MDPs can be represented by Markov opponents, partially observable environments can be mirrored by partially observable opponent actions, and computable environments can be incorporated by ensuring that the game and its opponents are computable.

10.5 Reflective Oracles

In computational game theory, we can model the players as agents running on Turing machines. In this set-up, how will two optimal agents act if they know the other agent's code, and both know they have the code of the other, etc? Naively each agent will run the other agent's program, which will run their own program, which will run the other agent's program, and so on. One solution to this could be to only do this recursion a set number of times, however, what if the opponent does this more? To overcome this self-modelling problem, [FTC15] came up with a probabilistic oracle called a *Reflective Oracle*, a truly remarkable construction. When an agent is run with a Reflective Oracle, it is able to reason about itself and overcome this self-modelling problem.

Probabilistic Turing machines and semimeasures. Let $\mathcal{T}$ denote the set of probabilistic Turing machines. Each Turing machine $T \in \mathcal{T}$ corresponds to a conditional semimeasure ν_T where $\nu_T(a|x)$ is the probability that T outputs $a \in \mathbb{B}$ given input $x \in \mathbb{B}^*$. Note that $\sum_{a \in \mathbb{B}} \nu_T(a|x)$ may be smaller than 1, since T may not halt. We can define $\nu_T(x)$ with the chain rule

$$\nu_T(x) := \prod_{k=1}^{\ell(x)} \nu_T(x_k|x_{<k})$$

which is lower semicomputable, since each factor is lower semicomputable. Note that this construction differs from Definitions 3.1.2 and 3.7.2 and does not include all $\nu \in \mathcal{M}_{sol}$ (Definition 3.7.1), since for a general $\nu \in \mathcal{M}_{sol}$, the predictive distribution $\nu(a|x)$ may not be lower semicomputable.

Oracles. An oracle is a function, which can give you access to information that would otherwise be hard or impossible to compute. The oracle receives a question or query and provides a yes or no answer to that query.

Oracles are great tools in theoretical computer science to prove results about levels of computability beyond computable. One popular example is the (incomputable) Halting oracle (Theorem 2.6.7). A Halting oracle takes an (encoded) machine-input pair $\langle T,x \rangle$ and returns *Yes* if machine T halts on input x and *No* otherwise.

We can extend the notion of Turing machines and Turing computability to include oracles: Let T^O denote a Turing machine $T \in \mathcal{T}$ but with access to oracle O. That is, T^O is an oracle machine which is an extended (probabilistic) Turing machine that additionally has an oracle tape and oracle head that are used to query the oracle and write back 1=Yes or 0=No (counted as only one time step) [RJ87]. For instance, if O is the Halting oracle, T^O now has the power to solve the Halting problem. Of course, due to the Church-Turing thesis (Thesis 2.6.6), no real computer can compute T^O for incomputable oracles O. Let ν_T^O denote the semimeasure induced by T^O.

The oracles $O : \mathcal{T} \times \mathbb{B}^* \times \mathbb{Q} \to \Delta\mathbb{B}$ we consider take as input (an encoding of) a Turing machine and an input on that Turing machine as well as a rational number, and return possibly randomized, Yes=1 or No=0 based on this input.

The specific oracle we consider has the following property: Each oracle takes a query $\langle T,x,z \rangle$, a Turing machine T, input to Turing machine x, and rational number z, and returns 1 if the probability that T^O outputs 1 on input x is greater than z, That is, for the induced semimeasure $\nu_T^O(1|x) > z$. It return 0 if $\nu_T^O(0|x) > 1-z$. If neither is satisfied the oracle can return 0 or 1 randomly as it pleases. Formally:

Definition 10.5.1 (Reflective Oracle) An oracle $O : \mathcal{T} \times \mathbb{B}^* \times \mathbb{Q} \to \Delta \mathbb{B}$ is reflective if and only if for all queries $(T, x, z) \in \mathcal{T} \times \mathbb{B}^* \times \mathbb{Q}$,

- If $\nu_T^O(1|x) > z$ then $O(T, x, z) = 1$,
- If $\nu_T^O(0|x) > 1 - z$ then $O(T, x, z) = 0$,
- Else $O(T, x, z)$ may return 0 or 1 at random (with any probability)

Note that the oracle O has to make statements about itself, since the machine T from the query may again query O, hence the name reflective.

Definition 10.5.2 (Reflective-oracle-computable) A semimeasure is called reflective-oracle-computable if and only if it is computable on a probabilistic Turing machine with access to a Reflective Oracle. That is, there exists a probabilistic Turing machine T and Reflective Oracle O such that the semimeasure can be computed with T^O.

Recall that normalizing a semimeasure ν means increasing it to a measure $\overline{\nu}$ such that $\nu(0|x) + \nu(1|x) = 1$. We can normalize the Reflective Oracle semimeasure ν_T^O for any probabilistic Turing machine T, to give us a Reflective Oracle measure $\overline{\nu}_T^O$ as follows: We do this by using binary search and the oracle O to find the crossover point $z^* \in [0,1]$, where $O(T, x, z)$ changes from returning 1 to 0, i.e. $\nu_T^O(1|x) \leq z^* \leq 1 - \nu_T^O(0|x)$ by definition of the reflective oracle and construction of z^*. If $\nu_T^O(1|x) + \nu_T^O(0|x) < 1$, z^* may be random, so we take the expectation

$$\overline{\nu}_T^O(1|x) := \mathbf{E}[z^* | x, \nu_T^O] =: 1 - \overline{\nu}_T^O(0|x) \quad \text{for all } x \tag{10.5.3}$$

By construction, $\overline{\nu}_T^O(a|x)$ is a reflective-oracle-computable (properly normalized) measure, and $\overline{\nu}_T^O(a|x) \geq \nu_T^O(a|x)$.

It is far from obvious whether there exists Reflective Oracles and if so, whether they are reasonably computable. Remarkably both questions have the answer, yes.

Theorem 10.5.4 (Existence of a limit-computable Reflective Oracle [LTF16]) There exists a limit-computable (see Definition 2.6.13) Reflective Oracle.

Note that limit-computability is weaker than computability, but very low in the arithmetic hierarchy (Δ_2^0, see Definition 2.6.17). From the construction of the Reflective Oracle in [FTC15], it looked like that it may live even beyond the arithmetic hierarchy. The proof of this theorem requires the construction of an infinite hierarchy of partially reflective partial oracles and is beyond the scope of this book [LTF16]. One can also show that Reflective Oracles are not Halting Oracles due to their randomization, and that they are not computable.

10.6   The Grain of Truth

When an agent is in a multi-agent setting, and each agent believes that every other agent is acting optimally, what should each agent do? This is one way to describe what is known as the Grain of Truth problem. In this problem each agent models the behavior of the opposing agents, and each of these opposing agents is modelling the behavior of the agent, which include models of the opposing agents, and so on. This can go on forever, or up to the maximum depth allowed by each agent. The Grain of Truth problem is finding when this chain of beliefs about opponents' beliefs about itself has a solution. To solve it we will take a Bayesian approach.

In this section, we will show how Reflective Oracles can be used to solve the Grain of Truth problem. We will start by defining the class of environments, then we will go on to show that the Bayesian mixture over this class is contained within the class. Lastly, we will show that the optimal policy over this class is Reflective Oracle computable.

The grain of truth problem, within the given context, entails identifying a collection of environments, denoted as $\mathcal{M}$, comprising environments in such a way that the optimal policy $\pi_{\xi_\mathcal{M}}^*$ belongs to $\mathcal{M}$. In other words, the best-performing policy derived from the Bayesian mixture over $\mathcal{M}$ must itself be a member of $\mathcal{M}$. It is crucial to note that $\mathcal{M}$ represents a class encompassing both policies and environments simultaneously, since we are examining a multi-agent setting where the environment is also considered an (opponent) policy.

Reflective Oracles have so far only been developed for binary sequences, while agents' action and observation spaces are typically non-binary. This is not a problem, because we can w.l.g biject actions and observations to binary complete prefix codes [MH21b, CHV22a, MH21a], and take products of conditional binary predictive distributions, which we will henceforth assume.

Definition 10.6.1 (Reflective environments) The class of reflective environments with respect to a set of Turing machines $\mathcal{T}$ and a fixed reflective oracle O is defined by

$$\mathcal{M}_r^O := \{\bar{\nu}_T^O : T \in \mathcal{T}\} \quad \text{(see (10.5.3))}$$

Recall Definitions 7.2.1 and 7.2.2 of a Bayesian mixture, this time over $\mathcal{M}_r^O$

$$\xi(e_t|h_{<t}a_t) := \sum_{\nu \in \mathcal{M}_r^O} w(\nu|h_{<t})\nu(e_t|h_{<t}a_t)$$

but with one key difference: $w(\nu|h_{<t})$ is the recursively defined *renormalized* posterior,

$$w(\nu|h_{1:t}) := w(\nu|h_{<t})\frac{\nu(e_t|h_{<t}a_t)}{\overline{\xi}(e_t|h_{<t}a_t)}$$

We can choose any lower semicomputable prior, e.g. $w_{\bar{\nu}_T^O} = 2^{-K(T)}$ in line with AIXI, where K is the Kolmogorov complexity (Section 2.7).

Theorem 10.6.2 (Bayes is a reflective environment [Lei16b]) The Bayesian normalized mixture $\overline{\xi}$ is an element of $\mathcal{M}_r^O$, that is
$$\overline{\xi} \in \mathcal{M}_r^O$$

Proof. By induction over t it is easy to see that ξ is reflective-oracle-lower-semicomputable. Assume $w(\nu|h_{<t})$ reflective-oracle-lower-semicomputable, which is true for $t=1$. This implies $\xi(e_t|h_{<t}a_t)$ is reflective-oracle-lower-semicomputable, Hence there exists an oracle machine T' such that $\xi(e_t|h_{<t}a_t) = \nu_{T'}^O(e_t|h_{<t}a_t)$, hence normalized $\overline{\xi}(e_t|h_{<t}a_t) = \bar{\nu}_{T'}^O(e_t|h_{<t}a_t)$ is reflective-oracle-computable. This normalization is the key difference to Definition 7.2.2, and makes the posterior $w(\nu|h_{1:t})$ reflective-oracle-lower-semicomputable, which completes the inductive step. Since $T' \in \mathcal{T}$, we have that $\bar{\nu}_{T'}^O \in \mathcal{M}_r^O$. ∎

Now that we know $\overline{\xi} \in \mathcal{M}_r^O$, we are interested in the degree of computability of Bayes-optimal policy π_ξ^*, or more generally, ν-optimal policies for $\nu \in \mathcal{M}_r^O$.

Theorem 10.6.3 (Optimal policies are reflective-oracle-computable [Lei16b])
For every $\nu \in \mathcal{M}_r^O$ there is a ν-optimal (stochastic) policy, π_ν^*, If discount function γ_k and normalizer Γ_k are computable, then π_ν^* is reflective-oracle-computable.

Note that even though deterministic optimal policies always exist, those policies are typically not reflective-oracle-computable.

Proof. We follow the proof from [Lei16b]. this can be extended and is left as an exercise to the reader. Recall we can write the optimal value function (Lemma 6.6.3) as

$$V_\nu^{*,m}(h_{<t}) = \frac{1}{\Gamma_t} \max_{a_t \in \mathcal{A}} \sum_{e_t \in \mathcal{E}} \dots \max_{a_m \in \mathcal{A}} \sum_{e_m \in \mathcal{E}} \prod_{k=t}^m \nu(e_k | h_{<k} a_k) \sum_{k=t}^m \gamma_k r_k$$

Since every component of the value function is computable or reflective-oracle, and there are only finitely many terms, the quantity itself is reflective-oracle-computable as the operations combining the elements are computable operations.

Now as for $V_\nu^*(h_{<t}) := \lim_{m\to\infty} V_\nu^{*,m}(h_{<t})$, the value function is monotone increasing in m, since our rewards are bounded between 0 and 1, and the tail of the value function, the components between time $m+1$ and ∞, is bounded by Γ_{m+1}. The function Γ_{m+1} is computable and converges to 0 as $m \to \infty$. Therefore for any desired $\varepsilon > 0$ we can choose an m such that $\Gamma_{m+1} < \varepsilon/2$ and compute V_ν^* to any desired accuracy, hence it is reflective-oracle-computable.

Recall that we restricted ourselves to the case of only two actions, say $\alpha = 0$ and $\beta = 1$. Since V_ν^* is reflective-oracle-computable, there exists a probabilistic Turing machine T such that

$$\nu_T^O(1|h_{<t}) := \tfrac{1}{2}[V_\nu^*(h_{<t}\alpha) - V_\nu^*(h_{<t}\beta) + 1] =: \nu_T^O(0|h_{<t})$$

Then we can define a policy π as

$$\pi(h_{<t}) := \begin{cases} \alpha & \text{if } O(T, h_{<t}, \tfrac{1}{2}) = 1 \\ \beta & \text{if } O(T, h_{<t}, \tfrac{1}{2}) = 0 \end{cases}$$

Now we need to show that π is a ν-optimal policy. *(i)* If $V_\nu^*(h_{<t}\alpha) > V_\nu^*(h_{<t}\beta)$, which means that $\nu_T^O(1|h_{<t}) > \tfrac{1}{2}$, and therefore $O(T, h_{<t}, \tfrac{1}{2}) = 1$, hence π will take action α, which is optimal *(ii)* Similarly if $V_\nu^*(h_{<t}\beta) > V_\nu^*(h_{<t}\alpha)$, then $\nu_T^O(1|h_{<t}) < \tfrac{1}{2}$ which means that $O(T, h_{<t}, \tfrac{1}{2}) = 0$, and π will take action β, which is optimal. *(iii)* If $V_\nu^*(h_{<t}\beta) = V_\nu^*(h_{<t}\alpha)$, then both actions are optimal, hence π is optimal whatever the oracle returns.

Therefore π is ν-optimal and reflective-oracle-computable. ∎

Combining the previous two theorems gives us the solution to the Grain of Truth problem.

Theorem 10.6.4 (Solution to the Grain of Truth problem [Lei16b]) For every lower semicomputable prior $w \in \Delta'\mathcal{M}_r^O$ and computable γ and Γ, the Bayes-optimal ($\bar{\xi}$-optimal) policy $\pi_{\bar{\xi}}^*$ is reflective-oracle computable, where $\bar{\xi}$ is the normalized Bayes-mixture defined above.

Proof. This immediately follows from Theorems 10.6.2 and 10.6.3. ∎

This solves a fundamental problem which had been open 20+ years for the first non-trivial class of environments, and indeed arguably for the most interesting class, which includes all computable environments, while also not being too crazy in the sense that all $\nu \in \mathcal{M}_r^O$ are at least limit-computable. A diagonalization argument prevents a Grain of Truth in $\mathcal{M}_{comp}$ and $\mathcal{M}_{sol}$, so in a sense $\mathcal{M}_r^O$ is the best solution one can hope for.

10.7 Reflective AIXI

The solution of the Grain of Truth problem allows to generalize the asymptotic optimality results of Chapter 9 for variants of AIXI to the multi-agent setting. Proofs can be found in [Lei16b, Sec.7.5].

Informed reflective agents. Let σ be a multi-agent environment as defined in Section 10.3 and let $\pi^*_{\sigma_1},...\pi^*_{\sigma_n}$ be such that for each i the policy $\pi^*_{\sigma_i}$ is an optimal policy in agent i's subjective environment σ_i. At first glance this seems ill-defined: The subjective environment σ_i depends on each other policy $\pi^*_{\sigma_j}$ for $j \neq i$, which depends on the subjective environment σ_j, which in turn depends on the policy $\pi^*_{\sigma_i}$. However, this circular definition actually has a well-defined solution.

Theorem 10.7.1 (Optimal multi-agent policies) For any reflective-oracle-computable multi-agent environment σ, the optimal policies $\pi^*_{\sigma_1},...,\pi^*_{\sigma_n}$ exist and are reflective-oracle-computable.

Note the strength of Theorem 10.7.1: each of the policies $\pi^*_{\sigma_i}$ is acting optimally *given the knowledge of everyone else's policies*. Hence optimal policies play 0-best responses by definition, so if every agent is playing an optimal policy, we have a Nash equilibrium. Moreover, each agent also acts optimally on the counterfactual histories that do *not* end up being played. This stronger property is called *subgame perfect* Nash equilibrium. In other words, Theorem 10.7.1 states the existence and reflective-oracle-computability of a subgame perfect Nash equilibrium in any reflective-oracle-computable multi-agent environment. From Theorem 10.5.4 we then get that these subgame perfect Nash equilibria are limit-computable.

Corollary 10.7.2 (Solution to computable multi-agent environments) For any computable multi-agent environment σ, the optimal policies $\pi^*_{\sigma_1},...,\pi^*_{\sigma_n}$ exist and are limit-computable.

Learning reflective agents Since our class $\mathcal{M}^O_r$ solves the grain of truth problem, the famous result by Kalai and Lehrer [KL93] immediately implies that for any Bayesian agents $\pi_1,...,\pi_n$ interacting in an infinitely repeated game and for all $\varepsilon > 0$ and all $i \in \{1,...,n\}$ there is almost surely a $t_0 \in \mathbb{N}$ such that for all $t \geq t_0$ the policy π_i is an ε-best response. However, this hinges on the important fact that every agent has to know the game and also that all other agents are Bayesian agents. Otherwise the convergence to an ε-Nash equilibrium may fail, as illustrated by the following example.

At the core of the following construction is a *dogmatic prior* (Section 8.2.1). A dogmatic prior assigns very high probability to going to hell (reward 0 forever) if the agent deviates from a given computable policy π. For a Bayesian agent it is thus only worth deviating from the policy π if the agent thinks that the prospects of following π are very poor already. This implies that for general multi-agent environments and without additional assumptions on the prior, we cannot prove any meaningful convergence result about Bayesian agents acting in an unknown multi-agent environment.

Example 10.7.3 (Reflective Bayesians playing matching pennies) Consider the game of *matching pennies* from Section 10.2.3. We use $\mathcal{E} = \{0,1\}$ to be the set of rewards (observations are vacuous) and define the multi-agent environment σ to give reward 1 to agent 1 iff $a^1_t = a^2_t$ (0 otherwise) and reward 1 to agent 2 iff $a^1_t \neq a^2_t$ (0 otherwise). Note that neither agent knows a priori that they are playing matching pennies, nor that they are playing an infinite repeated game with one other player.

Let π_1 be the policy that takes the action sequence $(HHT)^\infty$ and let $\pi_2 := \pi_H$ be the policy that always takes action H. The average reward of policy π_1 is $2/3$ and the average reward of policy π_2 is $1/3$. Let ξ be a universal mixture Definition 7.2.2. By Theorem 7.3.1, $V_\xi^{\pi_1} \to c_1 \approx 2/3$ and $V_\xi^{\pi_2} \to c_2 \approx 1/3$ almost surely when following policies (π_1, π_2). Therefore there is an $\varepsilon > 0$ such that $V_\xi^{\pi_1} > \varepsilon$ and $V_\xi^{\pi_2} > \varepsilon$ for all time steps. Now we can apply Theorem 8.2.1 to conclude that there are (dogmatic) mixtures ξ_1' and ξ_2' such that $\pi_{\xi_1'}^*$ always follows policy π_1 and $\pi_{\xi_2'}^*$ always follows policy π_2. This does not converge to a $(\varepsilon\text{-})$Nash equilibrium.   ♦

The following theorem is the main convergence result. It states that for asymptotically optimal agents in mean (Definition 8.1.6) we get convergence to ε-Nash equilibria in any reflective-oracle-computable multi-agent environment.

Theorem 10.7.4 (Convergence to equilibrium) Let σ be an reflective-oracle-computable multi-agent environment and let $\pi_1, ..., \pi_n$ be reflective-oracle-computable policies that are asymptotically optimal in mean in the class $\mathcal{M}_r^O$. Then for all $\varepsilon > 0$ and all $i \in \{1, ..., n\}$ the $\sigma^{\pi_{1:n}}$-probability that the policy π_i is an ε-best response converges to 1 as $t \to \infty$.

In contrast to Theorem 10.7.1 which yields policies that play a subgame perfect equilibrium, this is not the case for Theorem 10.7.4: the agents typically do not learn to predict off-policy and thus will generally not play ε-best responses in the counterfactual histories that they never see. This weaker form of equilibrium is unavoidable if the agents do not know the environment because it is impossible to learn the parts that they do not interact with.

Together with Theorem 10.5.4 and the asymptotic optimality of the Thompson sampling policy Theorem 9.2.1 that is reflective-oracle computable we get the following corollary.

Corollary 10.7.5 (Thompson-Sampling reflective AIXI converges to equilibrium) There are limit-computable policies $\pi_1, ..., \pi_n$ such that for any computable multi-agent environment σ and for all $\varepsilon > 0$ and all $i \in \{1, ..., n\}$ the $\sigma^{\pi_{1:n}}$-probability that the policy π_i is an ε-best response converges to 1 as $t \to \infty$. Thompson-Sampling AIXI π_{TS} (Algorithm 9.3) with class $\mathcal{M}_r^O$ is such a policy.

Note that the TS agents $\pi_{TS,i}$ are allowed to be different. For instance, they can use different discount functions, or resample at different time steps, and converge at different speeds. Also, all the Thompson sampling agents eventually 'calm down' and settle on some posterior belief.

Corollary 10.7.5 is a truly remarkable solution of the Grain of Truth problem. The only assumption on the environment σ is computability, while reflective AIXI is still limit-computable like (single-agent) AIXI based on $\mathcal{M}_{sol}$. Limit-computability is not practical, but may inspire practical versions of this construction. Reflective oracles have already seen applications in one-shot games [FTC15].

10.8   Exercises

1. [C18] Prove Theorem 10.2.4.

2. [C28m] Prove Theorem 10.2.7.

3. [C20] Prove Theorem 10.2.11.

4. [C25] Prove that specific properties (Markov, computability, stochasticity, partial observability, optimality conditions) carry over in the strategic games to RL conversion (and vice versa).

5. [C28] Prove that the environment class of repeated strategic games admits self-optimizing policies and hence that π_ξ^* is self-optimizing over the environment class of repeated strategic games.

6. [C18] Derive some simple (trivial) environment classes where there is a Grain of Truth.

7. [C25] Extend the proof of Theorem 10.6.3 to an arbitrary finite action space.

10.9 History and References

The topic of game theory is well developed in the textbook [OR94]. It is a theoretical framework originating from the work of John von Neumann and Oskar Morgenstern in the 1940s, allowed for a theoretical understanding of strategic interaction between agents in both competitive and cooperative environments [VNM47]. Central to game theory is the concept of Nash Equilibrium, introduced by John Nash in the early 1950s which describes a state in a strategic game where no player can benefit from unilaterally changing strategies while the other players keep their strategies unchanged [NJ50]. This equilibrium is stable, but it may be Pareto dominated by another strategy if players can collude to select it. This discovery can explain many counter-intuitive results: (1) why adding a new road to an existing road network can cause the travel time to paradoxically increase for all drivers [Bra68]. (2) why deliberately limiting the set of available actions via credible pre-commitment to a strategy can be advantageous [Sch80]. It also has applications in explaining strategies that evolve natural in the animal kingdom: (3) either passive or aggressive strategies can be stable in the animal kingdom [SP73]. (4) animals often display costly signals, like the peacock's elaborate tail, to help find a mate [Zah75]. (5) altruistic behaviors can evolve in a survival-of-the-fittest framework [Ham64].

Often, a deterministic strategy may leave the agent open to being exploited. Mixed strategies allow agents to choose a stochastic strategy to avoid exploitative responses [Neu28].

Besides Nash equilibrium, there are other concepts such as altruistic equilibrium and dominant strategy equilibrium and Pareto optimality. A complete classification of 2×2 symmetric games can be found in [BWEH22], which covers PD, SH, Chicken, and BoS, but not MP, since the last is not symmetric.

The connection between reinforcement learning and game theory is presented in [SLB08], which includes a survey of MDPs for multiplayer games. Another useful resource on this connection is [Cri17], which contains a review of the multi-agent and multi-objective reinforcement learning literatures. POTMMCP [SKH23] is a theoretically sound multi-agent online learning MCTS-based planning algorithm with good practical performance, even for problems that require large planning horizon. [MJS19] considers deterministic games with local Nash equilibrium (i.e. non-zero second derivative of gradient dynamics) and shows that gradient descent can get attracted to non-Nash local extrema (or saddle points).

Finding a grain of truth is a famously hard problem [KL93], with many impossibility results [Nac97, FY01, Nac05]. The remarkable solution presented in this chapter is from [LTF16, Lei16b], using previous reflective oracle work from [FST15, FTC15]. Sections 10.3 and 10.5 to 10.7 were taken with permission from [LTF16, Lei16b], some with minimal modification.

Part IV

Approximating Universal Agents

Chapter 11

AIXI-MDP

Simplicity is the ultimate sophistication.

Leonardo da Vinci

So far we have introduced and analyzed universal (Bayesian) models of prediction (Part II) and decision-making (Part III) primarily from the perspective finite-time and asymptotic quality of their predictions and actions. In particular AIXI (Sections 7.3 and 7.4) served as a gold standardfor universally optimal agents. Apart from being useful for theoretically analyzing potential properties of not-yet-existing super-intelligent agents, we can use AIXI as a starting point to derive in a top-down fashion computable approximation. Approximating AIXI and its variants is no simple task. The two main reasons are that the Kolmogorov complexity $K(x)$ and therefore the Solomonoff distribution $M = \xi_U$ are incomputable, and that the complete expectimax search is computationally intractable for most practical problem. We will derive various systematic approximations of AIXI in this part of the book: An advanced instantiation of AIξ in Chapter 12, and some powerful but theoretical ones in Chapter 13. In this chapter we describe the first and very simple instantiation of AIξ, with ξ being a very efficiently computable mixture over Markov Decision Processes (MDP). We run AIξ on the classic repeated 2×2 games prisoner's dilemma, stag hunt, chicken, battle of the sexes, and matching pennies, described in Section 10.2. The action and perception spaces are so small that we can brute-force compute the Bayes-optimal policy π_ξ^* from the expectimax expression Theorem 7.4.2.

11.1 AIXI-MDP Setup

We have developed the general history-based Bayesian agent AIξ in Section 7.3 and its universal instantiation AIXI in Section 7.4. The first practical approximation of AIXI developed in [PH06b] is known as AIXI-MDP. It was designed for MDP environments and tested on repeated 2×2 matrix games.

Emulating simultaneous actions. Generally, with repeated normal-form games, both players' actions are taken simultaneously, each without knowledge of the action chosen by the other, whereas in the usual cybernetic model for reinforcement learning (in which AIXI is described) there is a strict order of the agent taking an action, the environment receiving the action, and then afterwards the environment issues a percept (a pair of observation and reward) back to the agent. It turns out we can encode the behavior of simultaneous actions of repeated normal-form games by constraining the types of distributions that the opponent can use: The distribution that the opponent uses as her policy is not allowed to depend on the last action issued by the agent. This is shown in Figure 11.1. One iteration of the agent/environment interaction loop is described as follows:

- Agent takes action a_t, but the environment does not yet receive it.

- Opponent (which is part of the environment) takes action a'_t, ignorant of the agent's choice of a_t.

- Environment receives both actions a_t and a'_t, and then, with the reward matrix $\mathfrak{R}$ (called pay-off matrix in game theory, see Section 10.2.2), computes the rewards r_t (resp. r'_t) for the agent (resp. opponent) taking action a_t (resp. a'_t).

- Agent receives reward r_t and observation $o_t = a'_t$. Opponent receives reward r'_t and observation $o'_t = a_t$. Each player receives as a reward their entry in the reward matrix, and as additional side information, they receive their respective opponent's most recent action as an observation.

- Increment the time step $t := t+1$, and loop.

Implicitly, the agent (resp. opponent) is also aware of the other player's last received reward r'_{t-1} (resp. r_{t-1}) when choosing action a_t (resp. a'_t), as both have full knowledge of the reward matrix $\mathfrak{R}$ and the last played action a'_{t-1} (resp. a_{t-1}) of the other player. This allows us to write $o_t = a'_t$, rather than $o_t = (a'_t, r'_t)$ without loss of information for either player.

Environment-agent (a)symmetry. One notable aspect of this setup is that the environment is not equal to the opponent, since the environment includes the reward matrix as well as managing the interactions between agent and opponent. From the agent's perspective, the opponent is just part of the environment. But we could equally view the diagram from the other agent's perspective, where the opponent considers the reward matrix and interaction loop, together with the agent, to be the "environment" that it interacts with.

Binary action and observation space. This setup is general enough that it can be used for any class of games with any number of actions (or even any number of opponents). The class of environments we consider here are repeated 2×2 matrix games with a single opponent. In this environment, both the space of possible actions and observations are binary: $\mathcal{A} = \mathcal{O} = \{0,1\}$. The observations are the actions chosen by the opponent on the previous game. In each game the agent and the opponent are given a reward matrix $\mathfrak{R}$ corresponding to the reward for observation-action pairs (o,a). Additionally, the agent has access to the full history $h_{<t}$, which contains information about the interaction with

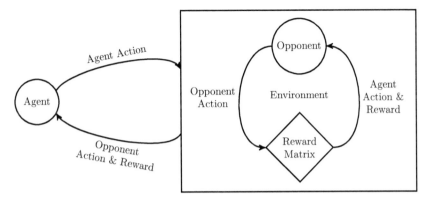

Figure 11.1: *How repeated normal-form games with simultaneous actions can be realized within the cybernetic model, by hiding the opponent inside the environment and withholding the most recent action.*

the opponent: both players' actions and rewards received so far. This is implicit in the interaction loop, since the agent will remember every action it takes, and every percept it receives.

Known reward matrix and withholding the last action. Since the agent is aware of the entries in the reward matrix for a particular game, in this chapter when we use h for history we will be referring only to the observation-action history, that is, $h_{<t} = (oa)_{<t} = o_1 a_1 ... o_{t-1} a_{t-1}$. This may seem strange since we have stated that the agent's action occurs and then the opponent takes an action, however since the environment, specifically the reward matrix, receives both actions at the same time, we will record the history as observation then action. This is because the observation will depend on the previous observation and action, but not the current action which the agent takes. This encodes the desired property that the opponent's action cannot be a reaction to the agent's action, but is chosen unaware of what the agent does.[1]

Markov environments. Additionally, the agent is assuming that the environment is a Markov decision process. This means that the agent believes that the environment (and the opponent contained inside) only uses the most recent action and observation to influence its choice in action. Recall that if μ is a Markov environment, then $\mu(o_t|h_{<t}) = \mu(o_t|o_{t-1}a_{t-1})$. Even though the agent is assuming that the environments are Markov, the agent itself will still be history based (dependent on the entire interaction history). This is so the agent may learn statistical patterns in how the opponent chooses actions, hopefully to learn the strategy chosen by the opponent and then exploit that strategy to maximize its reward.

No Grain of Truth. These two assumptions for the agent can lead to difficulties later on. In a game of AIXI vs. AIXI, both players are (falsely) assuming the opponent is Markov. AIXI cannot learn that its opponent is also another copy of AIXI, but would instead concentrate its posterior belief ξ on the model $\rho \in \mathcal{M}$ that is most consistent with the observations received. In other words, $\mathcal{M}_{MDP}$ does not contain a Grain of Truth (Section 10.6). Note that playing a true copy of AIXI with model class $\mathcal{M}_{sol}$ against itself poses the same problem, since AIXI is not contained in its own model class (as AIXI itself is non-computable). The only interesting known class containing a Grain of Truth is the Reflective Oracle class $\mathcal{M}_r^O \supset \mathcal{M}_{comp}$ (Theorem 10.6.2).

---

[1] In Chapter 12 we consider agents that are not informed about the reward matrix, indeed not even about the rules of the game they are playing.

11.2   Definition of AIXI-MDP

Recursive Q-value function of AIXI-MDP.   AIXI-MDP is AIξ with MDP model class $\mathcal{M}_{\mathrm{MDP}}$ to described below. We can adapt/specialize any of the 3 equivalent definitions of AIν. The original abstract expectation-policy forms (Definition 6.6.1 and Definition 6.7.1), the explicit iterative form (Lemma 6.6.3), or the (pseudo)recursive Bellman form (Theorem 6.7.2). We choose the latter with $\nu = \xi$ to be defined later. In strategic game theory it is common to neither discount nor normalize. The Bellman optimality equation for π_ξ^* with $\gamma_t = 1 = \Gamma_t$ by plugging (6.7.5) into (6.7.3) thus become

$$Q_\xi^{*,m}(h_{<t},a_t) \;=\; \sum_{e_t}\xi(e_t|h_{<t}a_t)\big[r_t + \max_{a_{t+1}}Q_\xi^{*,m}(h_{1:t},a_{t+1})\big]$$

with history $h_{1:t} := a_1 o_1 r_1 ... a_t o_t r_t$. Now $r_t = \mathfrak{R}(a_t,o_t)$ is a known deterministic function of a_t and o_t, we can therefore drop r_t from the history, and $e_t = o_t r_t$ becomes just o_t. Also, the environment cannot see AIξ's action a_t before choosing its own action $a_t' = o_t$. Therefore we only need to consider models $\nu \in \mathcal{M}$ for which $\nu(e_t|h_{<t}a_t)$ is independent of a_t, which implies by Definition 7.2.2 that also $\xi(e_t|h_{<t}a_t)$ is independent of a_t. Together this leads to

$$Q_\xi^{*,m}(h_{<t},a_t) \;=\; \sum_{o_t}\xi(o_t|h_{<t})\big[\mathfrak{R}(a_t,o_t) + \max_{a_{t+1}}Q_\xi^{*,m}(h_{1:t},a_{t+1})\big] \qquad (11.2.1)$$

The recursion terminates for $t > m$ with $Q = 0$.

Computing ξ_{MDP}.   The agent is assuming that the environment (and therefore the opponent) is Markov, with no access to the agent's last action a_t, that is, $\mu(e_t|h_{<t}a_t) = \mu(o_t|o_{t-1}a_{t-1})$, where we dropped the rewards as described above. $\mathcal{M}_{\mathrm{MDP}}$ is the class of all ν which satisfy this property. ξ_{MDP} is a Bayesian mixture over $\mathcal{M}_{\mathrm{MDP}}$ with some prior w specified below. Every Markov environment over the binary action and observation space $\mathcal{A} = \mathcal{O} = \{0,1\}$ can be encoded as a vector $(\theta_{00},\theta_{01},\theta_{10},\theta_{11}) \in [0,1]^4$, which describe the probabilities of the opponent producing observation $o_t = 0$ given the 4 possible most recent observation and actions, $(o_{t-1},a_{t-1}) \in \{(0,0),(0,1),(1,0),(1,1)\}$.

For our mixture, we assume independent uniform distributions $\theta_{ao} \sim \mathcal{U}([0,1])$ for each θ_{ao} as our prior belief in θ_{ao}, that is, assuming that each probability is equally plausible a-priori, following the principle of indifference: prior density $w(\theta) = 1$ for $\theta \in [0,1]^4$. This 1-order Markov process has been studied in Section 4.2.1. Let n_{ao} be the number of times the particular action-observation pair ao happened/appears in $h_{<t}$. Similarly let $n_{ao\cdot o'}$ be the number of times such ao pair is followed by next observation o'. The $\cdot$ indicates that we do not care about the next action a', since the environment is independent of it. Formally,

$$n_{ao}^t \;:=\; |\{\tau < t : a_\tau = a \wedge o_\tau = o\}|$$
$$n_{ao\cdot o'}^t \;:=\; |\{\tau < t : a_\tau = a \wedge o_\tau = o \wedge o_{\tau+1} = o'\}|$$

The sub-process of only the o_{t+1} that follow $a_t o_t = ao$ is a Bernoulli(θ_{ao}) process. Bayes-mixing it with a uniform prior over θ_{ao} gives the familiar Laplace estimator derived in Section 2.4.3. In the current setting and notation, this reads

$$\xi_{\mathrm{MDP}}(o_t|h_{<t}) \;=\; \frac{n_{a_{t-1}o_{t-1}\cdot o_t}^t + 1}{n_{a_{t-1}o_{t-1}}^t + 2} \qquad (11.2.2)$$

Note that unlike $\nu \in \mathcal{M}_{\mathrm{MDP}}$, $\xi_{\mathrm{MDP}} \notin \mathcal{M}_{\mathrm{MDP}}$ depends on the whole history. We have reduced the problem of computing ξ_{MDP} to counting the number of occurrences of a substring in a string, which is computationally efficient and can be done incrementally $O(1)$ time per t. Compare this to the KT estimator (Definition 4.1.1) where a slightly different prior is chosen (Definition 2.4.11), which halves the 1 and 2 in Laplace rule. Both are simple regularized frequency estimators.

Computing Q_ξ^*. We now have everything to compute $Q_{\xi_{\text{MDP}}}^*(h_{<t},a_t)$ via (11.2.1). Algorithm 11.1 explicates this recursion for the spacial case of binary action and observation spaces relevant for 2×2 matrix games, but the general case is essentially the same. Since each recursion calls the algorithm 4 times (and in general $|\mathcal{A}\times\mathcal{O}|$ times), the algorithm has run time $O(|\mathcal{A}\times\mathcal{O}|^{m-t+1})=O(4^{m-t+1})$. This will take "forever", if we chose m to be the lifetime of a long-lived agent, so we must impose a shorter horizon to limit the number of recursive value function calls. For many strategic games it is custom to use a moving horizon $m_t=t+d-1$ (Definition 6.4.2) looking only d time steps ahead, leading to run time $O(4^d)$, which is feasible for small d. A problem with a moving horizon is that it can lead to time-inconsistency (Section 6.5), e.g. to an agent forever delaying gratification. To prevent this potential problem, in the experiments, AIXI-MDP chooses a constant horizon $m_t=d_{\max}$ until lookahead $d_t:=m-t+1$ runs down to 2, and then restores m_t back up to $t+d_{\max}$, and repeats.

AIXI-MDP agent. The optimal action of AIXI-MDP is computed via

$$a_t^{\text{MDP}} := \arg\max_{a_t} Q_{\xi_{\text{MDP}}}^{*,m_t}(h_{<t},a_t)$$

by calling Algorithm 11.1 twice (for $a_t=0$ and $a_t=1$). For $t=1,2,3,\ldots$, AIXI-MDP and opponent take simultaneous actions $(a_t,a_t'):=(a_t^{\text{MDP}},o_t\sim\mu(o_t|h_{<t}))$. In theory μ should be in $\mathcal{M}_{\text{MDP}}$, but nothing prevents us from testing AIXI-MDP against non-Markov opponents to see what happens. In the experiments below, the maximal horizon was set to $d_{\max}=8$.

11.3 Experimental Results

The performance of AIXI-MDP was measured on the five 2×2 matrix games described in Section 10.2.3. The agent which opposed AIXI-MDP in each game differed between the games, since there are different known optimal strategies for each individual game. Additionally in some games AIXI-MDP played against itself with different horizons (Figure 11.2) and against the Follow or Explore (FoE) agent [PH06b], which it generally beats and we will not elaborate on.

Prisoner's Dilemma (Figure 11.2a). In the Prisoner's Dilemma game, AIXI-MDP has been compared to the following strategies: *random*, *1-tit-for-tat*, *2-tit-for-tat*, and *3-tit-for-tat*. The *random* strategy flips a fair coin to choose its action. An *n-tit-for-tat* player will cooperate in the first round and only cooperate if the opponent cooperates n times in a row.[2] The family of *tit-for-tat* strategies has been shown to be a very simple and effective strategy in repeated Prisoner's Dilemma competitions [Axe80a, Axe80b]. When performing against *1-tit-for-tat* and *2-tit-for-tat*, AIXI-MDP learned to cooperate, however against *random* and *3-tit-for-tat*, AIXI-MDP learned to defect. This means that AIXI-MDP was unable to explore enough to learn to cooperate with *3-tit-for-tat*. This should not be a surprise, as when the opponent follows a *1-tit-for-tat* strategy, the environment is Markov, but not when the opponent follows a *n-tit-for-tat* strategy with $n\geq2$. Since AIXI-MDP is a mixture over MDPs, the self-optimization Theorem 7.3.6 only applies against *random* and

---

[2]Note that *n-tit-for-tat* for $n\geq2$ is non-Markov, since it depends on more than just the last action-observation pair.

Algorithm 11.1 Q-value function of AIXI-MDP $Q^{*,m}_{\xi_{\mathrm{MDP}}}(h_{<t},a_t)$

Require: Horizon m
Require: MDP estimator ξ_{MDP} (11.2.2)
Require: Reward matrix $\mathfrak{R} \in \mathbb{R}^{2\times 2}$
Input: History $h_{<t}a_t$
Output: Q-value $Q^{*,m}_{\xi_{\mathrm{MDP}}}(h_{<t},a_t)$
 1: **if** $m<t$ **then return** 0
 2: **return** $\xi_{\mathrm{MDP}}(0|h_{<t}) \cdot [\mathfrak{R}(a_t,0) + \max\{Q^{*,m}_{\xi_{\mathrm{MDP}}}(h_{<t}a_t 0,0),\ Q^{*,m}_{\xi_{\mathrm{MDP}}}(h_{<t}a_t 0,1)\}]$
 3:     $+ \xi_{\mathrm{MDP}}(1|h_{<t}) \cdot [\mathfrak{R}(a_t,1) + \max\{Q^{*,m}_{\xi_{\mathrm{MDP}}}(h_{<t}a_t 1,0),\ Q^{*,m}_{\xi_{\mathrm{MDP}}}(h_{<t}a_t 1,1)\}]$

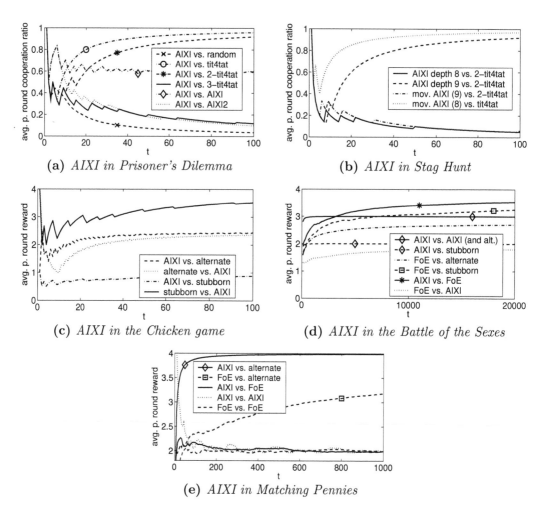

(a) *AIXI in Prisoner's Dilemma*

(b) *AIXI in Stag Hunt*

(c) *AIXI in the Chicken game*

(d) *AIXI in the Battle of the Sexes*

(e) *AIXI in Matching Pennies*

Figure 11.2: *Performance of AIXI-MDP on the classical 2×2 matrix games against fixed strategies and against a version of itself [PH06b].*

1-tit-for-tat. However, *random* is indifferent to the actions taken against it, and it never punishes defection. The strategy that maximizes the expected reward against *random* is to defect, which is what AIXI-MDP learns to do.

AIXI-MDP also played Prisoner's Dilemma against a copy of itself. This turns out to be less interesting as due to the symmetry, both agents always choose the same action in any zero-sum game. To break the symmetry, a variation of this was tested when two AIXI-MDPs with different horizons m played against each other, with a suboptimal result. It is likely that both agents learned to defect because they each (falsely) believed that their opponent was Markov, and by using the Laplace estimator (based on the frequency of opponent behavior) led both to believe that their respective opponent will often defect, in which case the best action is to also defect in turn in a downward spiral.

Stag Hunt (Figure 11.2b). Stag Hunt has some similarities to the Prisoner's Dilemma, one being that if both cooperate, they will earn a higher reward than if they both defected. AIXI-MDP has been compared to *2-tit-for-tat*. Against this player, AIXI-MDP of horizon 8 did not learn to cooperate, however, AIXI-MDP of horizon 9 did learn to cooperate (but not against *3-tit-for-tat*). One possible reason a horizon 9 AIXI-MDP was able to learn the correct behavior is that although an environment containing *2-tit-for-tat* opponent is not in the model class of MDPs, the mixture ξ is still sometimes able to converge to the true environment μ, even when the true environment is not in the model class (Corollary 3.4.2).

Chicken (Figure 11.2c). In Chicken, a good solution is for players to agree to alternate between swerving and not swerving. When alternating, to achieve maximum reward a player has to go straight most of the time and have the opponent swerve during those times. The two agents that were chosen as opponents for AIXI-MDP were *alternating*, who alternates between going straight and swerving, and *stubborn*, who will only cooperate (swerve) if the opponent has defected (gone straight) three consecutive times. Both of these strategies are known to be effective simple strategies in Chicken. AIXI-MDP quickly learns when to go straight against *alternating*; however it struggles to go straight three times in a row and therefore does not learn the optimal policy against *stubborn*. Interestingly, against *2-stubborn* (who requires the opponent to go straight twice before it chooses to swerve), AIXI-MDP learns the optimal action sequences.

When playing against itself, if both AIXI-MDPs have the same horizon, they will always take the same action. However, in the case of one of the AIXI-MDP agents having a higher horizon, the agent with the higher horizon learns to go straight more often than the lower horizon agent and ultimately accumulates more reward.

Battle of the Sexes (Figure 11.2d). Battle of the Sexes (BoS) encourages coordination. For both players to achieve maximum reward, they need to coordinate on the action to take this round, as well as cooperate by switching actions each round (so that both agents get a turn receiving the larger reward). In this game, AIXI-MDP has been compared to *alternating* and *stubborn*. When playing against *alternating*, AIXI-MDP learned to play the same actions as *alternating* and achieves the socially optimal reward. However, against *stubborn*, AIXI-MDP did not learn the action sequence to change the action *stubborn* chooses and received below average reward during each round.

Against itself, both AIXI-MDP agents learned to alternate between the two actions. This is because each AIXI-MDP believes it is playing in an MDP environment, in which case their opponents only care about the most recent action/observation. This leads to both agents believing the opponent will alternate, which means the optimal behavior for them is to also alternate. So optimal play is attained by a happy coincidence or a self-fulfilling prophecy, rather than "intelligent" play. With this, both agents are able to receive the socially optimal average reward of 3 each round.

Matching Pennies (Figure 11.2e). The matching pennies game is the only zero-sum game of the games considered here. This means there is a minimax strategy, which is to take each action with equal probability. When playing against *alternating*, AIXI-MDP learns the pattern and exploits it to achieve the optimal reward of 4 each round. When two AIXI-MDP agents compete against each other, as long as the symmetry is broken (e.g. by having different horizons), the agents will learn to alternate between actions.

11.4   Exercises

1. [C25i] Implement AIXI-MDP in your chosen language and test it against the agents described in the experiments section.

2. [C20] Generalize AIXI-MDP to non-binary actions and observations and test it on more complicated environments such as a gridworlds.

3. [C20] Determine the optimal agents for the given environment and opponents, tit-for-tat, stubborn, random, and alternating, etc.

4. [C23] Given AIXI-MDP assumes the environment is Markov, find a non-Markov agent which you expect AIXI-MDP will perform well against and test it empirically.

11.5   History and References

AIXI-MDP was first introduced in [PH06b] where it was compared with Acting with Expert Advice [PH05b], also called Follow or Explore (FoE), the active learning version of Prediction with Expert Advice [HP04, HP05], which requires exploration. This additional exploration is done by forcing the FoE agent to explore with some exploration rate at each time step which decreases over time. Additionally, the Bayesian (such as AIXI-MDP), expert advice, and MDL-based methods were compared in the online learning setting in [Pol06]. Relations between truncated undiscounted and untruncated discounted value functions have been explored in [Hut06a, ASDH24].

Chapter 12

Monte Carlo AIXI with Context Tree Weighting

The best way to predict your future is to create it.

Peter Drucker

We have previously discussed an approximation of the AIXI agent, AIXI-MDP (Chapter 11), which performs fixed-depth expectimax search and only uses the set of all Markov environments as its model class. This is a strong assumption about the environment as all there is to learn is a single probability distribution $p: \mathcal{O} \times \mathcal{A} \to \Delta\mathcal{O}$. Such distributions can be represented by an element of $[0,1]^{|\mathcal{O} \times \mathcal{A} \times \mathcal{O}|}$, as given a 3-tuple $(o,a,o') \in \mathcal{O} \times \mathcal{A} \times \mathcal{O}$, the distribution p associates with (o,a,o') a number representing the probability $P(o_{t+1} = o'|(o_t, a_t) = (o,a))$ according to p.

For small observation and action spaces, this gives a very tractable space of environments to search over. In Chapter 11, only $\mathcal{O} = \mathcal{A} = \mathbb{B}$ was considered, so together with $P(o_{t+1} = 1|(o_t,a_t) = (o,a)) = 1 - P(o_{t+1} = 0|(o_t,a_t) = (o,a))$ means that each environment can be encoded as a 4-tuple $(\theta_{00}, \theta_{01}, \theta_{10}, \theta_{11})$, one number associated with each observation-action pair $(o_t, a_t) \in \mathbb{B} \times \mathbb{B}$. We can then learn the dynamics of the environment by associating a KT estimator (Section 4.1) with each parameter in the 4-tuple.

In this chapter we will consider a different approximation of AIXI that is more powerful than AIXI-MDP, using as the model class $\mathcal{M} = \mathcal{C}_D$, the set of all variable-order Markov environments with context length at most D. This is quite a general class of environments, but as we will show, the mixture ξ over $\mathcal{C}_D$ can be computed efficiently. Planning that was previously performed via an expensive expectimax operation is replaced with a modified form of Monte Carlo Tree Search (MCTS) [KS06]. Learning the environment requires a Bayesian mixture over environment models, which is performed using Context Tree Weighting (CTW) (Chapter 4). This gives a powerful yet efficient approximation of our universal agent AIXI, which in practice can perform well on a larger variety of environments than repeated normal-form games by AIXI-MDP. We call this approximation *Monte Carlo AIXI with Context Tree Weighting* (MC-AIXI-CTW) [VNHS10].

The change from limited depth expectimax search to MCTS allows the agent to plan at much greater depths, and is asymptotically as good as the expectimax search. Additionally, using CTW for prediction allows the agent to model any k-Markov environment for any $k \leq D$, where D is the maximum depth of the CTW tree.[1]

12.1   Learning and Searching

In this section, we will describe the problems of learning and searching as well as how the Monte Carlo planning algorithm solves this problem, and how this works as an approximation of the expectimax component of the AIXI agent. Later, we will describe the Upper Confidence Tree (UCT) [KS06], a particular kind of Monte Carlo tree search algorithm using the UCB method (Section 12.2.3), as well as a modified form of UCT called ρUCT, which allows the approximation of the expectimax on a given known environment model ρ. Later, we can then instantiate ρUCT with a Bayesian mixture ξ over environments as the true environment μ is assumed to be unknown. Figure 12.1 depicts the overall architecture of MC-AIXI-CTW, whose components we will develop and explain in the following sections.

Definition 12.1.1 (Action-observation search tree) An *action-observation search tree* is a tree comprised of alternating *action nodes* and *chance nodes*. Each node is associated with a history h. Additionally, action (resp. chance) nodes store an estimate of the value $\widehat{V}(h)$ (resp. Q-value $\widehat{Q}(h,a)$) of the associated history h (resp. ha).

---

[1]One can also consider the case where the CTW tree is allowed to grow arbitrarily deep, but then as the context grows longer, updating the tree becomes slower [Wil98].

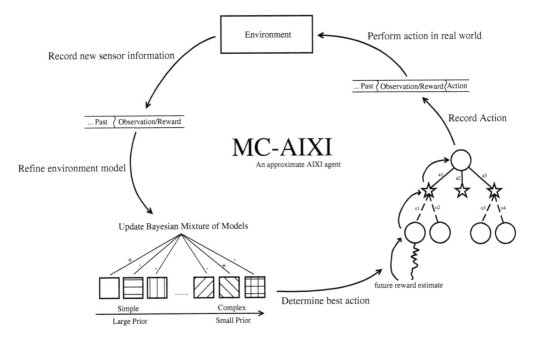

Figure 12.1: *The MC-AIXI-CTW agent-environment interaction loop. The CTW method provides a Bayesian mixture over models. MCTS is used for action selection, and rollouts are used to estimate the future rewards.* [VNH⁺11]

Recall that the action that AIXI takes is the one which maximizes the (undiscounted) expected value in the mixture environment ξ, given by the expectimax Theorem 7.4.2:

$$\pi^{\text{AIXI}}(h_{<t}) := \arg\max_{a_t} \sum_{e_t} \dots \max_{a_{t+m}} \sum_{e_{t+m}} \left(\sum_{i=t}^{t+m} r_i \right) \sum_{\nu \in \mathcal{M}} w(\nu | h_{<t}) \prod_{k=t}^{t+m} \nu(e_k | h_{<k} a_k)$$

Here, we use a *forward moving horizon*, where the AIXI agent always considers exactly m steps in the future from the current time step. This expectimax equation can be visualized as a tree that alternates between *action nodes (white)* and *chance nodes (black)*. An action node represents the action chosen by the agent, and a chance node represent the observation emitted by the environment. Both types of nodes have access to the path that leads to them, so both the agent's policy and the environment's distribution are conditional on the path (history) that lead to that node. The value estimate $\widehat{V}(h)$ associated with an action node's history h comprises a maximum of the Q-values of each child (which are chance nodes), and the Q-value estimate $\widehat{Q}(h,a)$ associated with a chance node's history ha is the expected value of the children (which are action nodes), where the expectation is with respect to the Bayesian mixture ξ. This gives us the familiar Bellman equation (Theorem 6.7.2) representation of the value function (see Figure 12.2).

The AIXI agent can be decomposed roughly into two parts, both for which direct evaluation is problematic.

- **Learning:** Computing the Bayesian mixture ξ (based on the history so far), through which the agent implicitly updates its belief of what environments in the model class best represent the true environment μ.

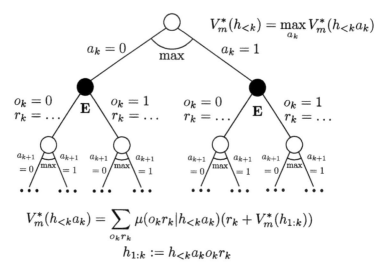

$$V_m^*(h_{<k}a_k) = \sum_{o_k r_k} \mu(o_k r_k | h_{<k}a_k)(r_k + V_m^*(h_{1:k}))$$

$$h_{1:k} := h_{<k}a_k o_k r_k$$

Figure 12.2: *A visualization of the expectimax operation Theorem 7.4.2 for the case* $\mathcal{A} = \mathcal{O} = \mathbb{B}$.

The mixture environment ξ uses the universal prior $2^{-K(\nu)}$ which is incomputable, and sums over the class $\mathcal{M}$ of all semicomputable chronological semimeasures, which is clearly intractable.

- **Searching:** Using ξ as a current best estimate of μ to look forward, considering the actions the agent could choose, the percepts that the environment would respond with, the counter-actions to each of those percepts, and so on up to the planning horizon m. Then, the action that would maximize the expected cumulative reward is chosen.
  To compute each of the maximums in the expectimax, we have to search through all actions in $\mathcal{A}$ and all percepts in $\mathcal{E}$, requiring $O(|\mathcal{A} \times \mathcal{E}|^m)$ time, which is intractable for even moderate values of m.

We require an efficiently computable approximation for both learning and searching for a practically useful approximation of the AIXI agent.

12.2   Searching via Monte Carlo Tree Search

Instead of computing the expectimax sums (Figure 12.2) explicitly, we could approximate it by sampling. We use a technique called Monte Carlo Tree Search (MCTS), which randomly explores paths of the tree to obtain an estimate of the expected value of different histories, rather than brute-forcing the search tree directly.

MCTS is a heuristic search algorithm which has been successful in a large number of domains including complex board games such as Go [WG07] and Backgammon [TG96]. MCTS is an online algorithm that constructs a tree of potential future outcomes, continually refining value estimates through repeated simulations. The final decision is based on the most promising path identified in the tree. The algorithm's effectiveness is influenced by both the accuracy of the environmental simulations (which requires constructing a model of the environment) and the method used for action selection (based upon the *upper confidence bound* method (12.2.2)). MCTS can easily be modified to perform many rollouts in parallel, making it easy to scale it up to very large or complex domains.

12.2.1 Monte Carlo Tree Search

The Monte Carlo Tree Search algorithm is an *anytime*[2] search algorithm which returns an approximation of the best action for any given history. This is done by sampling repeatedly from the current best estimate ξ of the true environment μ given the past history, and building a lookahead tree of actions and observations. The MCTS algorithm computes an approximation $\widehat{Q}(h_{<t}, \cdot)$ of the Q-value function $Q(h_{<t}, \cdot)$, and then chooses action $\widehat{a}_t := \arg\max_{a \in \mathcal{A}} \widehat{Q}(h_{<t}, a)$ as an estimate of the optimal action.

Chess is a complex game with a branching factor of around 35, meaning that on average, there are 35 legal moves that can be made from a given position. Despite this complexity, skilled grandmasters can plan ahead for up to 15 moves, which would require exploring an impractically large number of nodes ($35^{15} \approx 10^{23}$) for both humans and computers alike. However, the key to a grandmaster's success is not the ability to mentally search through an impossible number of possibilities, but rather their intuition for determining which moves are worth considering and which can be discarded. For instance, a move that sacrifices the queen for a pawn early in the game is almost always a bad move and would not merit any further consideration.

While the rules of chess are deterministic and known, the game is not just about making the best moves but also about anticipating the opponent's responses. Despite not having access to the opponent's mental state, a grandmaster is able to anticipate the opponent's moves by studying their past games, prior games with the opponent, and by estimating what move they would make if they were in the opponent's position. By doing so, grandmasters can effectively lower the branching factor by focusing only on moves that cannot be ruled out as obviously bad, and consequently plan further ahead into the future.

The core idea behind MCTS follows a similar approach: From a given position, the agent will determine which actions are worth exploring, and will simulate a potential future history starting with the chosen action, using a simulated model of the environment to interact with (much like a grandmaster "simulates" his opponent in his head when planning forward).

In this way, MCTS allows deep planning without costly expansive search, by focusing the search only on actions that look promising, and observations that μ would likely return.

MCTS works especially well in stochastic games with large branching factors (like backgammon [VLCU07] or poker [PGC10]), where instead of a costly expansive search to compute the expected value of nodes, many potential outcomes of the game can be simulated by sampling dice rolls or card deals, thereby quickly obtaining an estimate of the value of an action.

For Go [WG07, SH+16], an extension of MCTS was used where the action values were learned via self-play, tweaking the parameters of a neural network used as a heuristic for the value of a particular state of the game.

The MCTS algorithm can be decomposed into four parts (Figure 12.3), that are repeated while there is still execution time remaining. We present these generically for the moment, postponing discussion specifying how actions are selected, histories are evaluated, and how the environment is modelled to sample potential future percepts.

We denote the action-observation search tree as Ψ.

1. **Selection:** From the root node, Ψ is traversed down to a leaf chance node following a *tree policy*. The tree policy is chosen to ensure a reasonable trade-off between moves known to be good (exploit) and moves that are seldom chosen (explore).

2. **Expansion:** From the current leaf chance node, a new unexplored action (given the

---

[2]can be terminated at any point in time with a well-defined return value

current history) is selected, an action node is created, and added as a child to the chance node. Note that only one node is added to the tree on each iteration of MCTS.

3. **Simulation:** From the newly created action node, a simulation of a possible outcome is generated between the interaction of a *rollout policy* (a cheap model the agent has to approximate their behavior, usually random actions) and the environment (if the environment is unknown, we use the agent's best model of what it thinks the environment is), up to the horizon limit m.

4. **Backup:** The rewards generated along the simulated trajectory are used to update the statistics associated with the nodes along the path traversed through Ψ. In particular, value estimates are updated.

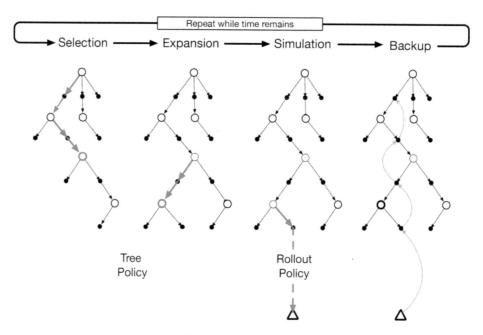

Figure 12.3: *The four steps of the Monte Carlo Tree Search algorithm explained in Section 12.2.2 [CWvdH08a, SB18].*

When execution time has elapsed, MCTS uses a mechanism to select the next action, represented as a child of the root node, based on the estimated effect on what possible futures would arise, if that action were taken. The root node itself still represents the actual present state of the agent-environment interaction history.

Usually, the action with the corresponding highest Q-value estimate is taken, but one could also select actions randomly with a bias towards actions that have a high Q-value estimate, or actions visited many times during the backup phase of MCTS. We would expect that a Q-value estimate comprised of the results of a large number of rollouts to be higher accuracy than one comprised of few, so focusing only on the highest Q-value estimate may mean exploiting before sufficient exploration is performed. Alternatively, one can focus entirely on choosing the action that was sampled the greatest number of times, in tandem with the *tree policy* biasing its sampling towards actions of estimated high value but with some exploration.

This is the action the agent chooses for interaction with the true environment. The environment then replies with a percept e. We then start MCTS again with updated history $h:=hae$ as the new root node.

We can walk down the tree and treat the node associated with history hae as the new root node.

12.2.2   MCTS Algorithm

We now present the MCTS algorithm more formally. The steps above included some unspecified steps, which are denoted below by the following functions.

- What statistics are stored in the nodes (given by defining the nodes in Ψ)
- The choice of tree policy (given by the interaction between *SelectAction* and *Simulate-Action*)
- The new action during the expansion phase and the choice of rollout policy (both contained inside *Evaluate*)
- Updating the node statistics (*UpdateValue*)
- Once time has elapsed, choosing the best action based on the accumulated statistics (*BestAction*)

These unspecified functions are fully defined as particular instantiations of MCTS later on.

Monte Carlo planning   uses an action-observation tree Ψ that contains an estimate $\widehat{V}(h)$ of the value function at each action node, and an estimate $\widehat{Q}(h,a)$ of the Q-value at each chance node. *Monte-Carlo planning* takes the current history, initializes a tree Ψ rooted at the current known history h, and repeatedly calls *Sample* (Algorithm 12.2) until a given time threshold $t_{\max}$ is exceeded.[3] Each of these applications of *Sample* will generate a simulated sequence of actions taken by the agent (*SelectAction*) and percepts produced by the environment(*SimulateEnv*) or by the model used to approximate the environment, for when the true environment is unknown.

Once time has been exceeded, *Monte Carlo Planning* runs the *BestAction* function which picks the best action based on the statistics of the tree Ψ.

Through repeated simulated interaction, the agent acquires a better estimate of the Q-values of each history-action pair.

Sample Algorithm 12.2   computes a recursive function that takes as input an action-observation tree Ψ, the current history h and the planning horizon m. The recursion ends when the planning horizon m has been reached, in which case *Evaluate* is used to estimate the value of a non-terminal history h.

If the recursion has not terminated, *Sample* generates a new action a using *SelectAction* and in turn a new percept (o,r) given action a using *SimulateEnv*. To find the Q-value of the history action pair (h,a) the algorithm recursively calls *Sample* on the new history $haor$ with the planning horizon m decremented. The result of the recursive call is added to the reward r, and then used to update the Q-value of (h,a) (using *UpdateValue*). The Q-value estimate is then returned.

Once the recursion has terminated by searching forward far enough to reach the planning horizon m, the Q-values at each time step will be recorded and used to compute the Q-value of the previous history and action, backing up all the way up to the initial history and action chosen.

How well *Sample* performs depends on the following:

---

[3]Sometimes a maximum number of simulations is used instead.

- How accurate the *SimulateEnv* function is. For known environments we just sample $e \sim \mu(\cdot|ha)$ directly. For unknown environments (like an agent exploring a maze, or in a game where the rules are known but the agent does not know the strategy of the opponent) the agent requires a model of the environment from which to sample from. The performance of *SimulateEnv* is then tied to the accuracy of the agent's model. Many variants of MCTS include a component where the agent samples from its internal model of the environment, and then updates the model based on true percepts received from the environment once an action has been chosen.

- How the *SelectAction* function (also called the *rollout policy*) chooses actions. Obviously, the agent cannot just use MCTS to simulate the action it would choose (as this would lead to an infinite regress) so we require the agent to have a sensible *rollout policy*, a model used by the agent to approximate its own behavior, which provides an inexpensive estimate of the action it would take for a given history. By analogy, a human choosing an action that seems intuitively good, or was discovered to be good in the past can provide a cheap estimate of the action that would be decided upon by thinking deeply about the consequences of the action. In vanilla Monte Carlo planning, the rollout policy merely chooses an action uniformly at random. The *greedy* approach involves just choosing the action which maximizes the current estimate of the Q-value, but doing so can often lead to suboptimal play if the agent mistakenly has an inaccurate estimate[4] of the Q-values [KS06]. More advanced techniques can involve learning a model[5] based on the true actions the agent has chosen so far, and then using samples from that model as the rollout policy.

UpdateValue is called when a recursive call to *Sample* has returned a value, used as a method to combine Q-value estimates generated from that particular simulation, together with the statistics collected so far for the chance node associated with history ha. It has no return value, but edits the tree as a side effect.

Algorithm 12.1 Monte Carlo planning(h,m,t_{max}) [KS06]

Input: History h
Input: Horizon m
Input: Timeout time t_{max}
Output: An action a
1: $\Psi :=$ action-observation tree starting at h
2: **while** running time less than t_{max} **do**
3:    Sample(Ψ,h,m)
4: **return** BestAction(Ψ,h)

---

[4]Selecting a highly valued action that is not valuable (false positive) will correct for itself, as actions that are selected often will have more accurate associated statistics, so the agent will quickly learn to stop taking that action. Actions that are valuable, but not highly valued (false negative) are more problematic however, as there is no incentive to visit a node with an apparently low value, so by following a greedy policy the agent may never learn to play an action where the estimated value is below the true value.

[5]so long as sampling actions from the model is still cheap. A good choice is often based on some sort of artificial neural network.

Algorithm 12.2
Sample(Ψ,h,m) [KS06, VNHS10]

Input: Search tree Ψ
Input: History h, Search Horizon m
Output: Q-value estimate q
1: **if** $m=0$ **then**
2:     **return** 0
3: **else**
4:     $a:=$SelectAction(Ψ,h,m)
5:     $(o,r):=$ SimulateEnv(Ψ,h,a)
6:     $h':=hora$
7:     $q:=r+$Sample(Ψ,h',$m-1$)
8:     UpdateValue(Ψ,h, a,q)
9: **return** q

Algorithm 12.3
UpdateValue(Ψ,h,a,q) [KS06, VNHS10]

Input: MCTS tree Ψ
Input: History h
Input: Action a
Input: Q-value q
Effect: Update Q-value est. on node ha
1: $\Psi(ha).Q := \frac{q+\Psi(ha).N \times \Psi(ha).Q}{\Psi(ha).N+1}$
2: $\Psi(ha).N := \Psi(ha).N+1$

12.2.3   Bandits and Upper Confidence Bounds

Before explaining the UCT algorithm, we will first briefly motivate it with some discussion on *bandit* problems, a subclass of Markov Decision Processes. While bandit problems appear to be simpler, they still have sufficient complexity to run into exploration-exploitation trade-offs, and entire books exist on the subject of bandits [LS20]. The approximation of AIXI will leverage results first constructed for bandit problems. For a longer summary of bandit problems, we recommend [SB18].

The term bandit comes from the phrase "one-armed bandit", referring to a slot machine.

Definition 12.2.1 (Bandit) Given a set of actions $\mathcal{A}$ and rewards $\mathcal{R}$, a *bandit* environment is a stochastic function $\mu : \mathcal{A} \to \Delta\mathcal{R}$.

Note that a bandit problem, unlike a Markov Decision Process, has no reference to states. Alternatively, we can define a bandit problem as a special case of a MDP with only a single state $\mathcal{S}=\{s\}$.

Bandit problems are a subclass of Markov decision processes, in that there are no states or observations, and the dynamics of the environment are conditioned solely on the most recent action taken by the agent. The agent always finds itself in the same situation whenever it takes an action.

The actions taken by the agent, and rewards from the environment form an interaction history

$$a_1 r_1 a_2 r_2 a_3 r_3 ... a_m r_m$$

up to the length of the interaction m.

The objective of the agent is to maximize the sum of reward over some fixed horizon. Often, the performance of an agent is measured in terms of *regret*, the difference between the expected reward sum the agent would obtain, vs. that of the optimal policy π^*, or the best policy π drawn from a class of reference policies Π.

Despite this apparent simplicity, bandit problems are still rather complex to solve, as the expected value of each action is unknown to the agent, and so they need to trade off exploiting the best known action, versus exploring new actions to obtain better estimates of the true value.

At each time step t, if the agent knew the expected reward $q^*(a) := \mathbf{E}_\mu[r_t|a_t = a]$ for each action a, then the solution is easy: the agent should always choose the action $a^* := \arg\max_{a \in \mathcal{A}} q^*(a)$ with the highest expected reward. There is no need to plan ahead and consider the consequences of each action, or learn a Q-value for each state-action pair to maximize the expected return as is usually the case in RL, as there is nothing for the environment to condition on other than the action a_t taken on the current time step. The problem lies in identifying which action is the best through statistics collected from previous actions taken. We imagine that at each time step t the agent has an estimate $\widehat{Q}_t(a) \approx q^*(a)$ of how valuable they believe each action to be. This estimate could be constructed in many ways, but an initial approach might be to take the average of all rewards obtained so far from playing that action in the past

$$\widehat{Q}_t(a) = \frac{\sum_{i=1}^{t-1} r_i [\![a_i = a]\!]}{\sum_{i=1}^{t-1} [\![a_i = a]\!]}$$

Now, an agent playing *greedily* with respect to this estimate by choosing action $a_t^{\text{greedy}} = \arg\max_{a \in \mathcal{A}} \widehat{Q}_t(a)$ may never learn to always choose the optimal action, instead exploiting a suboptimal action rather than getting good estimates for the actions available first.

A common approach to encourage exploration is the ε-*greedy* strategy [Wat89] to choose to explore randomly with probability ε, and select the greedy action $1 - \varepsilon$ of the time.

$$a_t^{\varepsilon\text{-greedy}} = \begin{cases} \text{uniformly random } a \in \mathcal{A} & \text{probability } \varepsilon \\ \arg\max_{a \in \mathcal{A}} \widehat{Q}_t(a) & \text{probability } 1 - \varepsilon \end{cases}$$

Eventually, the agent should choose the best action with probability $1 - \frac{|\mathcal{A}|-1}{|\mathcal{A}|}\varepsilon$. The choice of ε now affects the exploration-exploitation trade-off of the agent. Too low, and it will take the agent a long time to learn the optimal action, as exploration is seldom performed. Too high, and the agent will continue to often choose suboptimal actions even when the best action is known. More advanced techniques can involve making the exploration probability ε_t a function of time (using a large probability of exploration initially, and then decreasing the probability of exploring once the agent has taken sufficiently many actions to obtain a good estimate $\widehat{Q}(\cdot)$).

The ε-greedy method is one way to force exploration, but it is a rather unsophisticated way of doing so: the agent uniformly selects at random an action to play (potentially wasting a time step by choosing an action already known to be poor). A more sophisticated method would focus exploring more towards actions for which the current value estimate $\widehat{Q}_t(a)$ is uncertain. This brings us to the *Upper Confidence Bound* (UCB) [Lai87, ACBF02] action selection method. The UCB Q-value is defined as

$$Q_t^{\text{UCB}}(a) := \begin{cases} \widehat{Q}_t(a) + c\sqrt{\frac{\ln(t)}{N_t(a)}} & N_t(a) > 0 \\ \infty & N_t(a) = 0 \end{cases} \tag{12.2.2}$$

where $N_t(a) := \sum_{i=1}^{t-1} [\![a_t = a]\!]$ denotes how many times action a was selected before time step t, and $c > 0$ is a hyperparameter that controls the strength of the exploration bonus.

The action chosen by the UCB agent is then defined as the action that maximizes the UCB Q-value, $a_t^{UCB}(a) := \arg\max_{a \in \mathcal{A}} Q_t^{\text{UCB}}(a)$.

The main idea is that if up to time step t action a was chosen $N_t(a)$ many times, the relative error $\frac{Q_t(a) - \widehat{Q}_t(a)}{Q_t(a)}$ will be proportional to $\frac{1}{\sqrt{N_t(a)}}$. So, a bonus exploration term proportional to this error is added, encouraging visits to actions that are either good, or for which the value is uncertain. This UCB bound will be adapted to the more general reinforcement learning setting.

12.2.4 UCT Algorithm

The *Upper Confidence Trees* (UCT) algorithm is an extension of the Monte Carlo planning algorithm (Algorithm 12.1). The UCT extension is done by altering the action selection function, where rather than uniform action selection, UCT uses a modified form of the UCB algorithm. Consider two choices of tree policy: Random action selection that chooses actions uniformly at random, and greedy action selection that chooses the action with the highest value estimate $\widehat{V}$.

- **Random:** Random action selection is inefficient as we often waste computation time exploring the consequences of actions that are obviously *a priori* bad (moving chess pieces into jeopardy such that the opponent can safely capture them). We would rather explore only those actions that have the potential to be good.

- **Greedy:** Greedy action selection has problems as mentioned before in that it will likely exploit early before adequate statistics are collected, and get stuck playing moves with overconfident value estimates.

We desire a balance between these two extremes; a method of choosing actions that balances exploration with exploitation, while remaining computationally inexpensive. To this end, the UCT-Value has an additional exploration bonus the same as UCB, which adds to the action values an offset proportional to $\frac{1}{\sqrt{N}}$, where N is the number of times that action has been sampled.

We would expect that if the agent has seldom taken an action, its estimate of the Q-value would be uncertain, and subsequently the exploration bonus would be large, encouraging the agent to explore new moves for which the current estimate may be unreliable. In this way, the UCT is one potential resolution for the exploitation vs. exploration problem [GW06].

Definition 12.2.3 (Visit count) The visit count $N(h)$ of a history h is the number of times h has been sampled by the UCT algorithm. The visit count of taking action a given history h is defined similarly, and is denoted by $N(ha)$.

Definition 12.2.4 (UCT action selection) The action chosen by UCT, given a history h, is given by the action that maximizes the following expression:

$$a_{UCT}(h) := \underset{a \in \mathcal{A}}{\operatorname{argmax}}\, Q_{UCT}(h,a)$$

$$\text{where} \quad Q_{UCT}(h,a) := \begin{cases} \frac{1}{m}\widehat{Q}(h,a) + C\sqrt{\frac{\ln N(h)}{N(ha)}} & N(ha) > 0 \\ \infty & N(ha) = 0 \end{cases}$$

If multiple actions have the same UCT-value, we break ties arbitrarily.

The first term $\widehat{Q}(h,a)$ is the current unbiased frequentist estimate (Algorithm 12.3) of the Q-value of history h, action a. The constant $C \in \mathbb{R}$ is a positive exploration-exploitation parameter. Smaller values of C cause the agent to focus on exploiting its current estimate of the best action, which will cause the search tree Ψ to search deeper on only a few actions perceived to be good, giving a deep but narrow tree. Larger values of C incentivize exploration, incentivizing a shallower but broader tree.

Note that assigning a value of ∞ when $N(ha)=0$ is useful not only to avoid a divide-by-zero, but to force UCT to always prioritize an action a that has never before been chosen given the history h over anything else.

The *SelectAction* part of the MCTS algorithm is important as we do not want our search tree to go down paths that result in low reward, but we also want to explore history-action pairs which we have seldom visited. This choice of using the UCB for action selection balances this exploration-exploitation trade-off that occurs, just as it does in the bandit case as it will favor actions that maximize $Q(h,\cdot)$.

The last term to explain is the logarithmic term $(\ln N(h))$ in the exploration bonus. This is to ensure that the agent chooses all actions, (even ones with low Q-values logarithmically often) infinitely often. This is useful in both a theoretical and a practical sense: The theoretical result is convergence.

Example 12.2.5 (Upper Confidence Bounds for a Bandit) Consider the Markov decision process (technically a bandit process)

$$a_1,r=1 \quad \left(\; S \; \right) \quad a_2,r=10'000 \text{ with } 0.1\% \text{ chance}$$

There is only one state S, and both actions lead to the same state. The history is irrelevant for this problem, so we will say that after any action, the environment terminates and resets, so the history h is always empty. Action a_1 always returns a reward of 1, whereas action a_2 returns a reward of 10'000 with probability 10^{-3}, else it returns a reward of 0. Choosing action a_2 leads to an expected return of $r=10$, hence the optimal policy is to always choose a_2.

Now, suppose we had an agent selecting actions according to the following policy

$$a_{UCT'}(h) := \operatorname*{argmax}_{a \in \mathcal{A}} Q_{UCB'}(h,a)$$

$$Q_{UCB'}(h,a) := \begin{cases} \frac{1}{m}\widehat{Q}(h,a)+C\sqrt{\frac{1}{N(ha)}} & N(ha)>0 \\ \infty & N(ha)=0 \end{cases}$$

which is the same as a_{UCT} (Definition 12.2.4) without the logarithmic term $\ln N(h)$.

Then, unless the agent got particularly lucky and observed the reward $r=10'000$ from choosing action a_2 soon, the agent would only choose a_2 finitely many times, and then always choose a_1 thereafter. First, since a_1 always returns a reward of 1, we would expect that most methods of approximating the Q-value would quickly converge to 1 (including the frequency estimate), so assume $\widehat{Q}(h,a_1)=1$. Similarly, we assume for the moment that $\widehat{Q}(h,a_2)=0$ (as a_2 did not happen to return a non-zero reward for any interactions). Then,

$$Q_{UCB'}(h,a_1) = \frac{\widehat{Q}(h,a_1)}{m}+C\sqrt{\frac{1}{N(ha_1)}} > \frac{1}{m}$$

$$\text{but } Q_{UCB'}(h,a_2) = \frac{\widehat{Q}(h,a_2)}{m}+C\sqrt{\frac{1}{N(ha_2)}} = C\sqrt{\frac{1}{N(ha_2)}}$$

so once $N(a_2)>(Cm)^2$, then $\frac{1}{m}>C\sqrt{\frac{1}{N(ha)}}$ and the agent would never again choose a_2, and never learn the optimal policy. For reasonable values of the exploration constant $(C=2)$

and horizon ($m=10$) we require $N(a_2)>400$, for which there is a $\approx 67\%$ chance of a_2 not returning a reward for any of those 400 interactions. ♦

Now, one solution to the above could be to increase the exploration-constant C, or to increase the horizon m, but we could equally well choose another adversarial environment based on m and C. Let $\delta>0$ be arbitrary. Choose $\varepsilon = 1 - \exp(\frac{\ln\delta}{C^2 m^2})$ and add an action that returns a reward of $10/\varepsilon$ with a probability of ε. The expected value of choosing a_2 is still 10, making it optimal, but the probability that $a_{UCT'}$ will choose a_2 only finitely many times is bounded below by δ.

$$\text{P}(a_2 \text{ finitely often}) \geq \text{P}(a_2 \text{ returns } 0 \text{ over } C^2 m^2 \text{ trials})$$
$$= (1-\varepsilon)^{C^2 m^2} = \delta$$

We desire that there exists a fixed horizon m and exploration constant C that performs well on any environment of this kind. As mentioned before, other strategies to encourage exploration such as the ε-greedy policy have their own problems with selection of obviously bad actions, and never converging to optimal play. Choosing a decaying exploration rate ε_t that guarantees convergence is hard to do for general environments. An alternative is to use a *Boltzmann distribution*, where better actions are selected more often, but not always. The probability of sampling action a given history h is proportional to $\exp(\hat{Q}(h,a)/T)$, where T is the *temperature*, a hyperparameter that controls how sharp the distribution is.[6] As $T \to \infty$, the distribution approaches uniform, and as $T \to 0$ the distribution degenerates back to greedy action selection [Wat89]. However, this furloughs the problem to the selection of a temperature, or the rate of "cooling" (the decay rate for T).

A universal solution is to include the logarithmic term $\ln N(h)$ to the numerator of the exploration bonus. This ensures that every action is explored at least logarithmically often (which means all actions are sampled infinitely often) and the agent will eventually sample a_2 often enough to receive rewards, and learn that a_2 is the preferable action.

One could also choose other functions in $o(\sqrt{n})$, such as $\ln^2 N(h)$ or $\sqrt[3]{N(h)}$, but it can be shown that $\ln N(h)$ is optimal in the sense that it gives the best bound on the *worst-case regret*, that is, the best bound on the gap between the expected reward following this strategy, and the expected reward following an optimal policy π_μ^* for any choice of environment μ [LS20].

A general theoretical result was shown by [KS06], that the estimate of the expectimax value produced by UCT, $V_{UCT}(h) := V^{\pi_{UCT}}(h)$, converges in probability to the optimal expectimax value $V^*(h)$, with respect to the implied distribution $\mu^{\pi_{UCT}}$, where μ is any finite horizon MDP.

This result shows that the UCT algorithm is indeed a good choice for estimating the value function as the estimate converges to the true value. It does not, however, say how fast it converges. [KS06] also prove a convergence rate result, but with an exponentially large constant. In practice, the convergence is often fast enough so that UCT can be used effectively.

12.2.5 ρUCT Algorithm

In this section we will describe ρUCT (Algorithm 12.5), which is an extension of the UCT algorithm that we will use to approximate the expectimax tree. We can use the approximation

---

[6]The analogy here comes from statistical physics: An object that is very cold ($T \to 0$) has very little movement in its atoms, and the probability distribution of its average position is sharp. As $T \to \infty$ the atoms wiggle more due to thermal noise, and the distribution of its position smooths out.

of the expectimax tree to find the best action to take. The difference between ρUCT and UCT is that for ρUCT, the environment μ that we wish to plan in is unknown and replaced by estimates. For the purposes of the algorithm, ρ takes the place of the *SimulateEnv* function and can be queried directly for samples.

Like UCT, the action selection criteria is based on the UCB algorithm, and we use the same UCB action selection as before (Definition 12.2.4).

For ρ-UCT, the action-observation tree Ψ stores the following variables: the action nodes contain an estimate of the value $\widehat{V}(h)$, and the chance nodes an estimate of the Q-value $\widehat{Q}(h,a)$. All nodes also keep track of the number of times they were visited, $N(h)$ or $N(h,a)$ as appropriate.

The estimates $\widehat{V}(h)$ (resp. $\widehat{Q}(h,a)$) are averages of all the samples of the value (resp. Q-value), determined from the rewards obtained during a rollout. Given a current estimate $\widehat{Q}_N := \frac{1}{N}(q_1+...+q_N)$, after N many visits, and a new sample q_{N+1}, we can update the estimate to compute $\widehat{Q}_{N+1} := \frac{1}{N+1}(q_1+...+q_N+q_{N+1})$ efficiently with the following update rule

$$\widehat{Q}_{N+1} := \frac{N\widehat{Q}_N + q_{N+1}}{N+1}$$

We then increment the number of visits $N := N+1$. This update rule is used in Algorithm 12.6 and Algorithm 12.7.

Theorem 12.2.6 (Consistency of ρUCT [VNHS10, KS06]) Let μ be the true underlying environment. For all $\varepsilon > 0$, for all histories h,

$$\lim_{N(h)\to\infty} P\left(\left|V_\mu^{\pi_{\rho UCT}}(h) - \widehat{V}_\mu^{\pi_{\rho UCT}}(h)\right| \leq \varepsilon\right) = 1$$

where $\pi_{\rho UCT}$ is the policy that follows the ρUCT algorithm.

Proof. This theorem is a restatement of Theorem 5 and Theorem 6 in [KS06], expressed in notation consistent with this book.   ∎

Like UCT, we now have a nice convergence result for ρUCT. This means that using π_{UCT}, the optimal agent using the approximation of the value function $\widehat{V}_\rho^*$ will behave like the true optimal agent derived from V_ρ^* (at least in the limit). However, this theorem gives no indication of the rate of convergence, only that in the limit convergence is guaranteed (with probability 1).

12.2.6   Parallelization

We now summarize the methods described in [CWvdH08a] by which MCTS can be performed in parallel.

- **Leaf Parallelization.** The *selection* and *expansion* phases are performed sequentially as per normal. Then, instead of performing a single rollout during the *simulation* phase of the MCTS algorithm, many independent rollouts are performed in parallel, and the statistics generated over these rollouts can be aggregated together before performing *backpropagation* sequentially. This means the tree is being updated on more accurate value estimates. Since each rollout is independently generated, this means that no synchronization mechanisms of shared resources are required.

Algorithm 12.4
Agent-Environment Interaction Loop

Input: Horizon m. Timeout time t_{max}
Input: Rollout Policy π
Input: True Environment μ
Input: Environment Model ρ
Effect: Generates interaction history h
 1: $h := \epsilon$
 2: **while** True **do**
 3:     $a := \rho\text{UCT}(h, m, t_{max}, \pi, \rho)$
 4:     $e \sim \mu(ha)$
 5:     $h := hae$

Algorithm 12.5
$\rho\text{UCT}(h, m, t_{max}, \pi, \rho)$ [VNHS10]

Input: History h
Input: Search horizon m
Input: Timeout time t_{max}
Input: Rollout Policy π
Input: Environment model ρ
Output: An action a chosen by ρUCT
 1: $\Psi :=$ action-obs. tree with root node h
 2: **while** running time less than t_{max} **do**
 3:     MCTS(Ψ, h, m, π, ρ)
 4: **return** BestAction(Ψ, h)

Algorithm 12.6
MCTS(Ψ, h, m, π, ρ) [VNHS10]

Input: Search tree Ψ
Input: History h. Search horizon m
Input: Rollout Policy π
Input: Environment model ρ
Output: Reward sum R
 1: **if** $m = 0$ **then** ▷ Reached end of horizon
 2:     **return** 0
 3: **else if** $N(h) = 0$ **then** ▷ Reached leaf
 4:     $R :=$ Rollout(h, m, π, ρ) ▷ Simulation
 5: **else**
 6:     $a :=$ SelectAction(Ψ, h, C) ▷ Continue
 7:     $R :=$ SampleObservations(Ψ, h, a, m, ρ)
 8:     $\widehat{V}(h) := \frac{1}{N(h)+1}[R + N(h)\widehat{V}(h)]$
 9:     $N(h) := N(h) + 1$
10: **return** R

Algorithm 12.7
SampleObservations(Ψ, h, a, m, ρ) [VNHS10]

Input: Search tree Ψ
Input: History h
Input: Action a
Input: Search horizon m
Input: Environment model ρ
Output: Reward sum R
 1: Sample $e \sim \rho(\cdot|ha)$
 2: **if** $N(hae) = 0$ **then**
 3:     Create node $\Psi(hae)$ as a child of $\Psi(ha)$
 4:     $h' := hae$
 5:     $R := r + \text{SampleActions}(\Psi, h', m-1, \pi, \rho)$
 6:     $\widehat{Q}(ha) := \frac{1}{N(ha)+1}[reward + N(ha)\widehat{Q}(ha)]$
 7:     $N(ha) := N(ha) + 1$
 8: **return** R

Algorithm 12.8
SelectAction(Ψ, h, C) [VNHS10]

Input: Search tree Ψ
Input: History h
Input: Exploration-exploitation constant C
Output: Action a
 1: $\widehat{\mathcal{A}} := \{a \in \mathcal{A} : N(ha) = 0\}$
 2: **if** $\widehat{\mathcal{A}} \neq \{\}$ **then**
 3:     Pick $a \in \widehat{\mathcal{A}}$ uniformly at random
 4:     Create node $\Psi(ha)$
 5: **else**
 6:     $a := \arg\max_{a \in \mathcal{A}} \left\{ \frac{1}{m}\widehat{V}(ha) + C\sqrt{\frac{\ln(N(h))}{N(ha)}} \right\}$
 7: **return** a

Algorithm 12.9
Rollout(h, m, π, ρ) [VNHS10]

Input: History h
Input: Search horizon m
Input: Rollout policy π
Input: Environment model ρ
Output: Reward sum R
 1: $R := 0$
 2: **for** $i := 1$ to m **do**
 3:     Sample a from $\pi(\cdot|h)$
 4:     Sample e from $\rho(\cdot|ha)$
 5:     $R := R + r$
 6:     $h := hae$
 7: **return** R

- **Root Parallelization.** Multiple trees are instantiated, and the MCTS algorithm is performed on the trees independently. Once the allotted time has expired, the statistics of the children of the root nodes are merged together into a single tree, which is then used by the agent to select an action.

- **Tree Parallelization.** One tree is shared between multiple workers, which are all executing the MCTS algorithm simultaneously. Synchronization mechanisms are used so two workers do not access the same node at the same time. This can involve either a *global mutex*[7], so only one worker can update the statistics of the tree while all other workers are busy performing rollouts, or *local mutexes*, where workers lock a node whenever updating its statistics, and unlock it as soon as they move to another node. This means that much time can be wasted on locking and unlocking nodes in the tree, which can hamper performance.

The experiments done in [CWvdH08a] and more recently in [MPVDHV15] indicate that both tree parallelization and root parallelization are promising methods for speeding up evaluation of MCTS. They claim that leaf parallelization performs poorly by comparison, even though many more games are evaluated per second, the performance of the agent is not significantly improved as compared to the other methods. One could conjecture that the agent's performance is dependent more on the number of nodes explored, rather than the accuracy of the statistics obtained via rollouts.

12.2.7   Episodic Environments

So far we have been assuming that the interaction between the agent and the environments never terminates, and the agent tries to maximize the reward sum over some forward moving horizon. We call such environments *non-episodic* in contrast to *episodic* environments, where the interaction can be separated into disjoint interactions called *episodes*. Each episode does not affect the others, and when an episode terminates the environment is reset back to the original configuration.

We can encode episodic environments as a special case of non-episodic, by adding an extra observation o_{end} to indicate the current environment has terminated. The agent needs to already be aware of the o_{end} observation, and when issued, the interaction history is deleted, and the interaction between the agent and the environment starts afresh.

This allows for several optimizations that were not possible before. For example, it is clear that chess is a episodic environment, as the game terminates when a loss/win/draw occurs, and the interaction history from one game is irrelevant for subsequent games. This allows for an optimization via reusing some of the statistics from previous games when those games are revisited again in the future (as when the game resets, the agent is always in the same initial board position with an empty interaction history).

In many episodic environments, the reward is only determined at the end of the game, and zero reward is issued to a game in progress. In chess, the natural reward to issue is $-1/0/1$ for a lost/drawn/won game respectively, and zero reward for all other states of the game in progress.

The first change is to the MCTS algorithm (Section 12.2.1). Previously we would run the four steps of the algorithm until the time allotted to choose a move was expired. The agent selects an action a, the true environment μ returns a percept e, and the interaction history $h' := hae$ is updated. MCTS is then restarted from the node corresponding to h', discarding the other children of the tree.

---

[7]A *mutex*, short for *mutual exclusion*, is a mechanism by which only one worker is allowed access to a shared resource at a time.

For episodic environments, we can instead reuse large parts of the tree. We update the environmental history as before, but now no part of the tree is discarded. Once the current episode terminates, we can move back to the root node of the search tree and continue from there, making use of the existing statistics from previous games. In this way, the agent can learn from mistakes made in previous episodes.

12.3 Learning via Context Tree Weighting

Now we have an approximation of search via MCTS for a particular environment ρ. The true environment μ is unknown, so we wish to instantiate ρUCT with a Bayesian mixture over environments ξ that assigns non-zero weight to the true environment μ. AIXI uses ξ as its best estimate of the true environment, and updates ξ based on the percepts received from the environment. We approximate ξ using the Context Tree Weighting (CTW) algorithm (see Section 4.5).

We first modify the CTW method (Section 4.5) to work on arbitrary (finite) action spaces rather than binary strings. We do this in such a way so as to preserve the type information of the percepts received, as at the end of the day, the agent only receives bits, and we wish to distinguish the observation bits from the reward bits.

12.3.1 Action-Conditional CTW

Recall that CTW is an *online prediction*[8] algorithm which (efficiently) computes the probability

$$\mathrm{P}_D^{\mathrm{CTW}}(x_{1:t}) \equiv \mathrm{P}_{\mathcal{T}_D}(x_{1:t}) = \sum_{\mathcal{S}\in\mathcal{C}_D} w_{\mathcal{S}} \mathrm{P}_{\mathcal{S},KT}(x_{1:t})$$

where

- $x_{1:t}$ is a binary sequence
- $\mathcal{S}$ is a suffix set (Definition 4.3.3) or suffix tree (Definition 4.3.6)
- $w_{\mathcal{S}} := 2^{-\Gamma_D(\mathcal{S})}$ is the complexity prior associated with $\mathcal{S}$ (4.5.9), defined using the tree model code length $\Gamma_D(\mathcal{S})$ (Definition 4.3.20)
- $\mathcal{C}_D$ is the set of all suffix sets of depth at most D (Definition 4.3.7)
- $\mathrm{P}_{\mathcal{S},KT}$ is the PST-KT probability associated with $\mathcal{S}$ (Definition 4.3.24)

Recall that CTW can compute this probability in time $O(D)$ (Remark 4.5.3) compared to computing the sum naively which takes time $O(2^{2^D})$ (Remark 4.5.3).

The idea of CTW was extended in [VNHS10] from predictions to actions. This means that instead of estimating the probability $\mathrm{P}(x_{1:t})$, the extension to CTW estimates the probability[9] $\mu(e_{1:t}||a_{1:t})$ by the following:

$$\mathrm{P}_D^{\mathrm{CTW}}(e_{1:t}||a_{1:t}) = \sum_{\mathcal{S}\in\mathcal{C}_D} w_{\mathcal{S}} \mathrm{P}_{\mathcal{S},KT}(e_{1:t}||a_{1:t}) \tag{12.3.1}$$

where $\mathrm{P}_{\mathcal{S},KT}$ is extended from Definition 4.3.24 to give a conditional probability distribution over percepts conditioned on histories. Recall from Lemma 4.1.2 that the KT estimator is an estimator for the probability of the next symbol x_t given a binary string $x_{<t}$ defined by

$$\mathrm{P}_{kt}(x_{1:t}) := \mathrm{P}_{kt}(x_t|x_{<t})\mathrm{P}_{kt}(x_{<t}) \quad \text{and} \quad \mathrm{P}_{kt}(\epsilon) := 1$$

$$\mathrm{P}_{kt}(x_{t+1}=0|x_{1:t}) = \frac{a+1/2}{a+b+1} \quad \text{and} \quad \mathrm{P}_{kt}(x_{t+1}=1|x_{1:t}) = 1-\mathrm{P}_{kt}(0|x_{1:t}) = \frac{b+1/2}{a+b+1}$$

---

[8]CTW gives a new value of the prediction after observing each new bit, as opposed to an *offline* prediction algorithm, which first needs to see all the data, learns from it, and then after makes predictions.
[9]Recall the || notation of Definition 6.3.1.

and where a (resp. b) is the number of zeros (resp. ones) contained in $x_{1:t}$.

12.3.2   Action-Conditional PST

Both percept and action sequences are often non-binary (as there is usually more than two actions to take, or two percepts to receive), and as currently defined, CTW can only perform prediction over binary sequences. The solution is to encode actions and percepts as binary strings, use this encoding to convert a history to a sequence of bits, and perform prediction on the result.

We first present how to generalize PSTs (Section 4.3.1) to non-binary alphabets, called *action-conditional prediction suffix trees* (AC-PST).

We assume w.l.g[10] that $|\mathcal{A}| = 2^{\ell_{\mathcal{A}}}$ and $|\mathcal{E}| = 2^{\ell_{\mathcal{E}}}$ for some $\ell_{\mathcal{A}}, \ell_{\mathcal{E}} > 0$.

We can then enumerate all $2^{l_{\mathcal{A}}}$ actions in $\mathcal{A}$, and assign to each a sequence of $l_{\mathcal{A}}$ many bits. Let $[\![a]\!] \equiv a[1:l_{\mathcal{A}}] := a[1]a[2]...a[l_{\mathcal{A}}] \in \mathbb{B}^{l_{\mathcal{A}}}$ denote the bit sequence representing an action a, and $[\![a_{1:t}]\!] := [\![a_1]\!][\![a_2]\!]...[\![a_t]\!]$ denote the bit encoding of a sequence of actions. We define encodings for percepts e and sequences of percepts $e_{1:t}$ similarly.

Now, an AC-PST is defined in the same way as a PST (Definition 4.3.11) with a suffix set $\mathcal{S}$ and a collection of parameters $\Theta_{\mathcal{S}}$, and associated tree $\Psi_{\mathcal{S}, \Theta_{\mathcal{S}}}$. The difference lies in how the updates are performed.

For interaction with the environment, we first perform a few interactions arbitrarily with the environment to build up a history h such that the encoding $[\![h]\!]$ is at least as long as the depth of the tree (so that we have a context of sufficient depth.) Then, for every interaction thereafter:

1. An action a is selected by the agent, and $h := h[\![a]\!]$.
2. The environment responds with a percept $[\![e]\!]$.
3. For i from 1 to $l_{\mathcal{E}}$
   (a) Without seeing the next bit $e[i]$, we predict it using the distribution $\theta_{\beta_{\mathcal{S}}(h)}$.
   (b) The next bit $e[i]$ is observed. The parameter $\theta_{\beta_{\mathcal{S}}(h)}$ is updated per usual using the KT estimator (Algorithm 4.3).
   (c) Set $h := he[i]$.

We can then define a mixture over AC-PSTs in the same fashion as we did for CTWs, giving us the *Action-Conditional CTW* method.

Recall the recursive definition of the weighted probability for the CTW (4.5.2) which we replicate here:

$$P_w(s) := \begin{cases} \frac{1}{2}P_{kt}(a_s,b_s) + \frac{1}{2}P_w(0s)P_w(1s) & \text{if } 0 \leq \ell(s) < D \\ P_{kt}(a_s,b_s) & \text{if } \ell(s) = D \end{cases}$$

We can use the CTW update (Algorithm 4.4) and prediction (Algorithm 4.5) algorithms to define how the AC-CTW mixture should be updated with every interaction with the environment. When a new percept is received, the CTW tree is updated on each bit contained in the encoded version of the new percept, given the context which contains the latest action taken by the agent.

---

[10] One can always take any agent-environment model and add additional dummy actions that always return minimum reward if taken and have no effect on the environment, and the environment can have additional percepts that it will never return to the agent. The value of any existing policy is unchanged under these modifications.

For the moment, we leave the method by which the action is chosen as generic; this is intentional as the algorithm here is just describing the Action-Conditional CTW model. In Section 12.1, we discussed some approaches to decide on what action to take.

Algorithm 12.10
Action-Conditional CTW [VNHS10]

Require: Context tree $\mathcal{T}_D$
Input: Action-percept stream
$\quad\quad h = a_1 e_1 a_2 e_2 a_3 e_3 \ldots$
Output: $\mathrm{P}_D^{\mathrm{CTW}}(e_{1:\ldots} \| a_{1:\ldots})$ bit by bit
Effect: Maintain CTW trees $\mathcal{T}_D$
1: Interact with environment
   to generate initial history h
2: **while** True **do**
3:   Agent chooses an action $a \in \mathcal{A}$
4:   $h := ha$
5:   Send action a to environment
     and receive percept e
6:   **for** $1 \le i \le l_{\mathcal{E}}$ **do**
7:     $p := \mathrm{CTWPrediction}(\mathcal{T}_D, h)$
8:     Use p to predict $e[i]$
9:     Observe (and receive) next bit $e[i]$
10:    $\mathrm{CTWUpdate}(\mathcal{T}_D, h, e[i])$
11:    Set $h := he[i]$

Algorithm 12.11
FAC-CTW [VNHS10]

Input: Model class $\mathcal{M}$
Effect: Maintain the family of
   CTW trees $\mathcal{T}_{D_1}, \ldots, \mathcal{T}_{D_k}$
1: $h := \epsilon$
2: $t := 1$
3: Initialize $k := l_{\mathcal{E}}$ context trees $\mathcal{T}_{D_1}, \ldots, \mathcal{T}_{D_k}$
4: **while** True **do**
5:   Agent chooses an action a_t
6:   $h := ha_t$
7:   Agent transmits action a_t to env.
     and receives percept e_t
8:   **for** $1 \le i \le k$ **do**
9:     $\mathrm{CTWUpdate}(\mathcal{T}_{D_i}, he[1, i-1], e_i)$
10: $h := he_t$
11: $t := t+1$

12.3.3 Factored Action-Conditional CTW

The method of encoding percepts as binary strings leaves out some critical information: there's nothing to distinguish which bits correspond to actions, which to observations, and which to rewards. Moreover, the position of the bit itself in the encoding provides extra side-information that we would want the CTW model to use. Essentially, the *types* of the data are not encoded, the CTW model only receives a stream of bits to update against, with no context as to what those bits represent.

For many environments, being able to distinguish between these pieces of information is highly advantageous for learning. Consider the following example.

Example 12.3.2 (Factored CTW is better than CTW) Consider a trivial environment with $\mathcal{A} = \mathcal{O} = \mathcal{R} = \mathbb{B}$. The environment issues observation o_t uniformly at random, and the reward r_t is 1 if the agent's action a_t matched the most recent observation o_{t-1} before that. Formally,

$$\mu(o_t r_t | h_{<t} a_t) = \begin{cases} 1/2 & o_t r_t = (0,1) \text{ and } a_t = o_{t-1} \\ 1/2 & o_t r_t = (1,1) \text{ and } a_t = o_{t-1} \\ 1/2 & o_t r_t = (0,0) \text{ and } a_t \ne o_{t-1} \\ 1/2 & o_t r_t = (1,0) \text{ and } a_t \ne o_{t-1} \end{cases}$$

Optimal behavior in this environment is easily attained as the agent can just remember the last received observation, and play that as its action.

We encode the percept $e_t = o_t r_t$ as a 2-bit code in the obvious fashion (first bit observation, last bit reward). Now, when the agent receives a percept, the CTW tree model is updated one bit at a time. Suppose the history is ...0101 and the next bit to update on is 0. It is not clear if we are parsing the first or second bit of the percept, so the following two cases would be indistinguishable.

- ...0101 $= ...a_{t-1}o_{t-1}r_{t-1}\mathbf{a_t}$ and $0 = o_t$
- ...0101 $= ...o_{t-1}r_{t-1}\mathbf{a_t}o_t$ and $0 = r_t$

Obviously we would want the agent to be able to distinguish observations from rewards, otherwise they will confuse the reward 1 for the observation, and choose action 1 in response, which would lead to lower reward. The bits alone are not sufficient, as the agent needs to know what those bits represent.   ♦

To preserve type information (being able to differentiate between different bits of actions and percepts), and hence gain the ability to exploit more of the structure of the percepts, [VNHS10] came up with a more advanced version of Action-Conditional CTW called *Factored Action-Conditional CTW (FAC-CTW)*, which chains together one Action-Conditional PST for each of the $l_\mathcal{E}$ bits of the percept space.

We keep track of $l_\mathcal{E}$ separate context trees $\mathcal{T}_{D_1}, ... \mathcal{T}_{D_{l_\mathcal{E}}}$ where tree $\mathcal{T}_{D_i}$ is associated with the ith bit $e[i]$ of a percept e. Tree $\mathcal{T}_{D_i}$ will only ever be updated on $e[i]$, given the history h concatenated with $e[1:i-1]$, the prior bits in the encoded percept as context. By ensuring each tree is only ever updated from the same bit in the percept, the type information is preserved. Each tree should have available the same quantity of information about the history h, so the ith tree $\mathcal{T}_{D_i}$ will be of depth $D_i := D+i-1$, and be provided context $he[1:i-1]$ (the same D bits of context from h, and $i-1$ bits of context comprised of $e[1:i-1]$).

In practice this proceeds as follows: We have history h, and would like to update the model based on new percept e. Tree $\mathcal{T}_{D_1}$ is updated on $e[1]$ given context h, tree $\mathcal{T}_{D_2}$ is updated on $e[2]$ given context $he[1]$, and so on to the last tree $\mathcal{T}_{D_{l_\mathcal{E}}}$, which is updated on $e[l_\mathcal{E}]$ given context $he[1:l_\mathcal{E}-1]$. The update rule for each tree is the same as for the vanilla CTW (Algorithm 4.4).

Our model class is now a family of tree sources in $\mathcal{C} := \mathcal{C}_{D_1} \times \mathcal{C}_{D_2} \times ... \mathcal{C}_{D_{l_\mathcal{E}}}$. We define an element of this family as $\mathcal{S} = (\mathcal{S}_1, ..., \mathcal{S}_{l_\mathcal{E}})$. The trees are updated independently of each other, so we can write the probability of a percept sequence associated with the family $\mathcal{S}$ as

$$\mathrm{P}_\mathcal{S}(e_{1:t} || a_{1:t}) = \prod_{i=1}^{t} \mathrm{P}_\mathcal{S}(e_i | ae_{<i}a_i) = \prod_{i=1}^{t} \prod_{j=1}^{l_\mathcal{E}} \mathrm{P}_{\mathcal{S}_j}(e_i[j] \,|\, [\![h_{<i}]\!]e_i[1:j-1]) \qquad (12.3.3)$$

where $e_{1:t}[j]$ denotes $e_1[j]...e_t[j]$, the jth bit from every percept e_i concatenated together, and $e_{1:t}[\backslash j] := e_1[\backslash j]...e_t[\backslash j]$ and $e[\backslash j] := e[1]e[2]...e[j-1]e[j+1]...e[t]$, that is, $e_{1:t}[\backslash j]$ represents the bit encoding of $e_{1:t}$, with the jth bit of every percept omitted.

Since our new model is $\mathcal{S} = (\mathcal{S}_1, ..., \mathcal{S}_{l_\mathcal{E}}) \in \mathcal{C}_{D_1} \times ... \times \mathcal{C}_{D_{l_\mathcal{E}}}$, we now need to define a prior over $\mathcal{C}_{D_1} \times ... \times \mathcal{C}_{D_{l_\mathcal{E}}}$. We already have the prior $2^{-\Gamma_D(\mathcal{S})}$ for $\mathcal{C}_D$, so by assuming that the product of model classes $\mathcal{C}_D$ are independent, we can write

$$w_\mathcal{S} := \mathrm{P}(\mathcal{S}_1, ...\mathcal{S}_{l_\mathcal{E}}) = \prod_{i=1}^{l_\mathcal{E}} \mathrm{P}(\mathcal{S}_i) = \prod_{i=1}^{l_\mathcal{E}} 2^{-\Gamma_{D_i}(\mathcal{S}_i)} = 2^{-\sum_{i=1}^{l_\mathcal{E}} \Gamma_{D_i}(\mathcal{S}_i)}$$

Let $\mathcal{M} = \mathcal{C}_{D_1} \times ... \times \mathcal{C}_{D_{l_\mathcal{E}}}$, which gives us a new mixture model,

$$\xi(e_{1:t} || a_{1:t}) := \sum_{\mathcal{S} \in \mathcal{M}} w_\mathcal{S} \mathrm{P}_\mathcal{S}(e_{1:t} || a_{1:t}) \qquad (12.3.4)$$

This can be rearranged to give us the following:

$$\xi(e_{1:t}||a_{1:t}) := \sum_{S \in \mathcal{M}} w_S \mathrm{P}_S(e_{1:t}||a_{1:t})$$

$$= \sum_{S_1 \in \mathcal{C}_{D_1}} \cdots \sum_{S_k \in \mathcal{C}_{D_k}} 2^{-\sum_{i=1}^{k} \Gamma_{D_i}(S_i)} \prod_{j=1}^{k} \mathrm{P}_{S_j}(e_{1:t}[j]||a_{1:t}, e_{1:t}[\backslash j])$$

$$= \prod_{j=1}^{k} \left(\sum_{S_j \in \mathcal{C}_{D_j}} 2^{-\Gamma_{D_j}(S_j)} \mathrm{P}_{S_j}(e_{1:t}[j]||a_{1:t}, e_{1:t}[\backslash j]) \right)$$

$$= \prod_{j=1}^{k} \mathrm{P}_{D_j}^{CTW}(e_{1:t}[j]||a_{1:t}, e_{1:t}[\backslash j])$$

Now each factor $\mathrm{P}_{D_j}^{CTW}(e_{1:t}[j]||a_{1:t}, e_{1:t}[\backslash j])$ is a conditional probability, which can be computed similarly to as was done for a single CTW tree (4.5.5), which leads us to the Factored Action-Condition CTW (Algorithm 12.11). The algorithm starts off by creating a context tree for each bit, then loops the following:

- Take some action a_t and add it to the history h
- Receive percept e_t from the true environment
- For the ith context trees, update the context tree with the ith bit of e_t, using the history h and $e[1:i-1]$ as the context

The weighted probabilities associated with each tree are updated using Algorithm 4.4 in the usual fashion. We can then take the product of the root probabilities $\mathrm{P}_w(\epsilon)$ for each tree to recover ξ, which efficiently computes (12.3.4).

We can now express the mixture distribution given by the FAC-CTW method in a similar fashion to (12.3.1). This mixture is what will be used as a substitute for ξ in the approximation of AIXI Theorem 7.4.2.

$$\xi_{FAC}(e_{1:t}||a_{1:t}) := \sum_{S \in \mathcal{C}_{D_1} \times \ldots \times \mathcal{C}_{D_{l_\mathcal{E}}}} 2^{-\sum_{i=1}^{l_\mathcal{E}} \Gamma_{D_i}(S_i)} \mathrm{P}_S(e_{1:t}||a_{1:t})$$

12.4 All Together

Now we have an efficiently computable mixture of environments (ξ_{FAC}) to learn the true environment μ, as well as a method for searching in a given environment (ρUCT), they are combined by instantiating ρUCT with $\rho = \xi_{FAC}$. The result is called MC-AIXI-CTW, an approximation to the AIXI agent. The action MC-AIXI-CTW will take (given sufficient MCTS time) at time t having history $h_{1:t}$, and horizon m is

$$a_t^* = \operatorname*{argmax}_{a_t} \sum_{x_t} \ldots \max_{a_{t+m}} \sum_{x_{t+m}} \left[\sum_{i=t}^{t+m} r_i \right] \sum_{S \in \mathcal{C}_{D_1} \times \ldots \times \mathcal{C}_{D_k}} 2^{-\sum_{i=1}^{k} \Gamma_{D_i}(S_i)} \mathrm{P}_S(e_{1:t+m}||a_{1:t+m})$$

where P_S is the probability given by FAC-CTW (12.3.3). We can see clear similarities when compared to AIXI Theorem 7.4.2, which we rewrite here for moving horizon:

$$a_t^* = \operatorname*{argmax}_{a_t} \sum_{x_t} \ldots \max_{a_{t+m}} \sum_{x_{t+m}} \left[\sum_{i=t}^{t+m} r_i \right] \sum_{\nu \in \mathcal{M}_{sol}} 2^{-K(\nu)} \nu(e_{1:t+m}||a_{1:t+m}).$$

The only difference is that MC-AIXI-CTW uses a model class of a product of prediction suffix trees (Definition 4.3.11) as opposed to semicomputable environments $\mathcal{M}_{sol}$ (Definition 3.7.1), and a prior that is based on the model cost (Definition 4.3.20) of those trees instead of a prior based on the Kolmogorov complexity (Definition 2.7.3). Additionally we use a moving horizon m to make planning tractable.

Using ξ_{FAC} in place of the true environment μ for the purpose of MCTS samples was shown in [VNHS10] to converge to optimal behavior. Let now m be the maximum life of the agent as in the original AIXI model (the maximum number of iterations the agent will be interacting with the environment). We could imagine a changing horizon h_t during interaction with the environment, so let $h_{max} = \sup_t h_t$ be the maximum of such horizons. We assume that $m \gg h_{max}$. The following theorem provides a performance bound on the expected error of the FAC-CTW algorithm, comparing its estimated value function with the true value function for a given policy π under certain assumptions.

Theorem 12.4.1 (Upper bound on FAC-CTW value error [VNH+11]) Using the FAC-CTW Algorithm 12.11, for every policy π, if the true environment μ is expressible as a product of k prediction-suffix trees $(\mathcal{S}_1,\Theta_1),...,(\mathcal{S}_k,\Theta_k)$, for all $b \in \mathbb{N}$, we have

$$\sum_{t=1}^{m} \mathbf{E}_{\mu}^{\pi} \left[\left(V_{\xi_{FAC}}^{\pi,m_t}(h_{<t}) - V_{\mu}^{\pi,m_t}(h_{<t}) \right)^2 \right]$$

$$\leq 2h_{max}^3 \left[\sum_{i=1}^{k} \Gamma_{D_i}(\mathcal{S}_i) + \left(\tfrac{1}{2}\log_2 m + 1\right) \sum_{j=1}^{k} |\mathcal{S}_j| \right]$$

where $V_{\mu}^{\pi,m_t}(h_{<t})$ is the moving-horizon value of the policy (Definition 6.6.1), and $h_{max} := \sup_t h_t$ the maximum of the horizons $h_t := m_t - t + 1$.

This result implies that the FAC-CTW algorithm provides a performance guarantee on its estimates for the value function under certain conditions. The performance bound depends on the maximum planning horizon h_{max}, the structure of the prediction-suffix trees that express the true environment, and the agent's lifetime m. The theorem suggests that the error between the algorithm's estimates and the true value function is not unbounded, meaning the algorithm's performance is likely to be acceptable for many practical applications.

The FAC-CTW algorithm leverages prediction-suffix trees to create a compact representation of the environment, which can lead to efficient planning and learning. This theorem provides a theoretical understanding of the algorithm's performance, which can be beneficial for researchers and practitioners who need to select appropriate algorithms for their reinforcement learning tasks.

As a corollary of this theorem, the average expected square difference of the two value functions $V_{\xi_{FAC}}^{\pi,m_t}$ and V_{μ}^{π,m_t} tends to zero at rate $O(\frac{\log m}{m})$, which implies that for a sufficiently long lifetime m, the value function estimates using ξ_{FAC} converge to that of μ for any policy π. Importantly, this means that they converge for the policy that MC-AIXI-CTW is approximating: The fixed horizon expectimax with respect to ξ_{FAC}.

When the agent has had limited interaction with the environment, the CTW probability estimates will be poor, as there is limited information to build the estimates from. In practice, it can be good to start with an ε-greedy exploration strategy to boost exploration, and then switch back to the UCT method afterwards. This initial exploration differs from the UCT type exploration as the latter is only guaranteed to explore enough in the limit to be able to have the correct value function (Theorem 12.2.6), but it can be quite slow in practice to obtain reasonable behavior.

12.5   Experiments

MC-AIXI-CTW was implemented with a wide variety of different environments by [VNHS10] and performed close to optimal on almost every environment. In this section, we will describe each of the environments it was tested on. All of the environments are MDPs, but the state of the MDP is often not visible to the agent, and they receive only an observation instead o, generated from the state s via some observation function $\phi: \mathcal{S} \rightarrow \mathcal{O}$. In Figure 12.4, we display some of the environment information, including the size of the action space and observation space of the environments, whether or not they have *perceptual aliasing*, whether or not the observations are stochastic, and whether or not the observations are uninformative. An environment has *perceptual aliasing* if ϕ is non-injective, that is, multiple states of the MDP can map to the same observation, so the agent cannot uniquely identify the state from the observation alone. An example of this could be an agent in a maze, where the observation is in what directions the agent cannot move without bumping into a wall, and the state is the agent's position in the maze. Clearly, it is possible to have two different positions in the maze map to the same set of observations for the agent. We say the observations of an environment are uninformative if the optimal policy does not depend on the observations, that is, the observations provide no information for the agent to learn from, and the agent must learn from the rewards alone.

An example of an uninformative environment is a game of matching pennies (see Section 10.2.3) where the agent receives no observation (or the same dummy observation) at every interaction, and receives only the reward from the payout matrix.

Environment	$|\mathcal{A}|$	$|\mathcal{O}|$	Aliasing	Stochastic $\mathcal{O}$	Uninformative $\mathcal{O}$
1D-Maze	2	1	yes	no	yes
Cheese Maze	4	16	yes	no	no
Tiger	3	3	yes	yes	no
Extended Tiger	4	3	yes	yes	no
4×4 Grid	4	1	yes	no	yes
TicTacToe	9	19683	no	no	no
Biased RPS	3	3	no	yes	no
Kuhn Poker	2	6	yes	yes	no
POCMAN	4	2^{16}	yes	no	no

Figure 12.4: *Properties of the environments on which MC-AIXI-CTW was tested [VNHS10].*

Note that despite the fact that many of the environments are rather simplistic and the optimal policy is often obvious, recall that the agent has no knowledge of the environment other than the size of action and percept spaces. Moreover, the agent has no time to learn the environment ahead of time, and must learn in an online fashion while simultaneously trying to find a good policy. Imagine trying to play the following environments yourself with no feedback beyond the percepts encoded as binary strings, and no knowledge of the underlying dynamics of the environment to fully appreciate the difficulty of playing well in these environments.

Illegal Actions.   We divert briefly to discuss the problem of illegal actions. Under the current framework the agent can take any action at any time, and so it may be the case that the agent can take "illegal actions." An alternative is to provide the agent with a legal action function $\phi: \mathcal{H} \rightarrow 2^{\mathcal{A}}$ that for any history, provides a set of legal actions for the agent to choose from. We would then replace all instances of $\mathcal{A}$ in the algorithms above with $\phi(h)$, given the current history h. This works, but somewhat ruins the universality of MC-AIXI-CTW, as we would have to take the rules of the game and encode them in the function ϕ, so the agent is receiving extra side information. Grandmasters do not know chess from birth, but had the benefit of mentors to teach them the rules of the game, as well as sensible strategies discovered by others. We are asking a lot more of our agent, moments after "birth" it has a chessboard thrust in front of it, and is asked what moves to play. Initially, the agent has no conceptual idea of chess, let alone which moves are legal and which are not. Since the agent receives only a scalar reward for feedback on its performance, the agent also has to learn not only how to play well, but also the rules of the game that it is playing from experience.

What we do need to decide is how to handle illegal moves. If the game always has a fixed length (and the agent receives a penalty at the end for playing an illegal move) then it can take the agent a very long time to learn the difference between an illegal move and playing poorly. If instead any illegal action triggers an immediate penalty and a game restart, even a short-sighted agent will quickly learn to avoid illegal moves (or at least avoid playing them near the start of a game).

The other decision to make is how to penalize an illegal action. Clearly an illegal action must be at least as bad as a loss (we would not want the agent to escape from an otherwise doomed position by cheating and playing an illegal move, as humans sometimes do) but is an illegal action worse than a loss? In games humans play, sometimes the penalty for an illegal move can be less than a loss[11] but the size of the penalty could also depend on the type of illegal move.[12] We would like to ensure that the agent would never desire to take an illegal action once they have learned what illegal actions are. One solution that can quickly teach the agent to avoid illegal actions and ensure they would never desire to take them once learned, is whenever the agent plays an illegal action, return a percept (o,r) where o was the same observation received last cycle, and r is the lowest possible reward $r \in \mathcal{R}$. For this interaction only, the environment never receives the illegal action a, and the agent is in the same position as they were on the previous time step. Essentially, all illegal actions have no effect on the environment, other than issuing a penalty to the agent for attempting an illegal action. In this setup, illegal actions are harmless (apart from the minimal reward received) so the agent is free to explore all actions to see which are illegal, and learn to avoid them in the future.

Of course, this is only viable if we are aware of the true environment μ that is unknown to the agent (which is often the case, as we usually specify the environment on which the agent is tested).

Another approach (only for variable-length episodic environments) is that the environment replies to any illegal action with a terminal observation o_{end} as well as a minimum reward. If the game were allowed to continue and the penalty for the illegal move was issued at the end of the game, it would not be clear to the agent which move along the sequence was illegal, which would hamper learning the rules of the game.

Without, say, adding additional side information like an observation o_{illegal} that the agent is aware indicates that the last action is illegal, there's not a clear distinction to the agent between actions that are illegal, and actions that are just merely bad for the agent. This is a closer analogue to real life (where illegal[13] actions that an agent could still take

---

[11] Often in sports, minor infractions are punished with a score/time penalty rather than an outright loss.

[12] *FIDE* (the governing body for professional chess) rules do not specify, but one would imagine the penalty for punching your opponent in the nose to be far worse than violating the touch-move rule.

[13] Here, we mean illegal according to the laws of man, rather than the laws of reality.

(murder) are harshly penalized by negative reinforcement (prison)) rather than, say, a video game where the laws of reality make it impossible to take illegal actions to begin with.[14]

12.5.1 Environments

Note that many of these environments use negative rewards, which is technically forbidden under our framework. For the moment we ignore this technicality for the sake of readability. In practice, we can always apply a positive affine transformation to the rewards to ensure they are always non-negative, without changing the optimal policy (see Exercise 12.8.5).

1D Maze. The *1D Maze* environment ▢▢G▢ is a 1×4 POMDP grid-world introduced in [CKL94], with one goal state, two actions $\mathcal{A} = \{left, right\}$ and a single (uninformative) observation $\mathcal{O} = \{none\}$. The agent starts in a random non-goal state. Each action moves the agent one cell in the respective direction, but attempting to move outside the grid has no effect. Attempting to walk into the goal state teleports the agent to a random non-goal cell, and issues a reward of 1. All other times the reward is 0. No (useful) observation is provided, however the environment is so simple that the reward is sufficient feedback for the agent to learn to act well.

4×4 grid. The 4×4 *grid* environment is more or less an extension of the 1D maze, wherein the agent starts in a random cell in a 4×4 grid-world, and there is only one goal: the bottom left cell (Figure 12.5). The agent can move in the four cardinal directions $\mathcal{A} = \{left, right, up, down\}$, and the agent receives reward 0 for moving to a non-goal cell (or for attempting to move off the grid, which has no effect) and 1 for moving to the goal cell. Once the agent moves into the goal cell, it is transported to a non-goal cell selected uniformly at random. Lastly, like in the 1D maze, the observation space is trivial ($\mathcal{O} = \{none\}$) as the point of the environment is for the agent to learn the policy of moving to the bottom left cell by alternating left and down moves only from using the reward as the feedback.

Cheese Maze. The *Cheese Maze* is a more complicated grid-world, using a two-dimensional grid and one goal state. The agent has 4 actions: $\mathcal{A} = up,\ down,\ left,\ right$ and 16 observations $\mathcal{O} = \mathbb{B}^4$. Each bit in the observation indicates if a wall is directly adjacent to the north, east, south or west of the agent respectively (Figure 12.6). In the figure, gray cells represent a wall and G represents the cell with cheese in it (the goal). The binary string in each square is the observation the agent would receive in that square. Each action moves the agent one cell in the corresponding direction (unless doing so would walk into a wall, in which case the agent remains where they are.) There are three rewards: -10 for bumping into a wall, 10 for finding the goal state (cheese), and -1 for moving into a state that is not a wall or a goal state (to incentivize finding the cheese quickly). The agent starts in a random non-cheese cell, and whenever it enters the cheese cell, it receives the associated reward, and is then teleported to another random non-cheese cell. Since the environment has partial aliasing, if the agent starts in a cell and observes 0101, there is not enough information to determine if the cheese is directly below, or if the agent is in one of the two side corridors, and needs to move up and then to a side to get to the cheese.

Tiger. In the *Tiger* environment, the agent is presented with two doors: Behind one is gold, the other a tiger. The agent has three actions: *listen*, which tells the agent which door the tiger is behind (*left* or *right*) and is correct with probability 0.85; *open door 1* and *open door 2*. If the agent opens the door, it receives reward -100 if a tiger is found, and reward of

---

[14]Though some rules in video games (being friendly to other players) are difficult to write the rules to be impossible to perform, so they too are subject to enforcement via negative reinforcement (banning abusive players) rather than, automatic enforcement via the laws of reality.

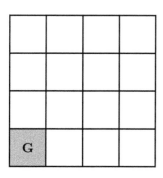

Figure 12.5: *4 × 4 grid. An environment where the agent needs to learn to navigate to the bottom left goal cell.*

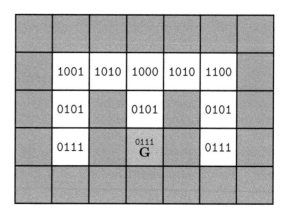

Figure 12.6: *Cheese Maze: The bits in each cell (the observation the agent receives) represent if a wall is directly adjacent north, west, south or east respectively. Grey cells are walls; the cell containing G is the goal cell. Multiple states can map to the same observation.*

10 if the gold is found. Taking the *listen* action gives reward −1 (to disincentivize listening for too long, though the agent is not forbidden from doing so). Once a door is opened, the doors are closed and reshuffled.

Extended Tiger. *Extended Tiger* is the same environment as *Tiger*, with the following additions: The agent is either sitting down or standing up, and the environment starts with the agent sitting down. The agent has two additional actions, *stand up* and *sit down*. The agent can only listen while sitting and only open doors while standing. If the agent takes any illegal action (attempting to open a door while sitting, trying to stand while already standing, etc.) they receive a reward of −10. Additionally, the reward for finding the gold is now 30. The environment resets once any door is opened.

Tic Tac Toe. The *Tic Tac Toe* environment on the right is the well-known game of Tic Tac Toe, against an opponent who chooses their actions randomly. The agent has 9 actions: placing a piece into one of the nine squares. The observation space is the current state of the board. If the agent wins the game it receives a reward of 2, if it draws it receives a reward of 1, if it loses it receives a reward of −2, and if it makes an illegal move it receives a reward of −3, and otherwise 0. The reason that illegal moves are punished more than losing the game is because this encourages the agent to learn what moves are legal or not, and preference losing the game over playing an illegal move.

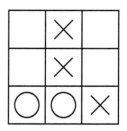

Biased RPS. Biased RPS is a version of the rock-paper-scissors game where the opponent has a bias toward certain moves (if they won the last round playing rock, they will choose rock again, otherwise choose an action uniformly at random). The agent needs to learn this bias in the opponent and exploit it. If the agent wins, it receives a reward of 1, if it draws a reward of 0, and if it loses a reward of −1.

Kuhn poker. Kuhn poker is a simplified version of poker developed in [Kuh50]. The deck contains three cards: *King*, *Queen* and *Jack*, with the standard card ranking of

King>Queen>Jack. The opponent plays first, and the agent plays second. The observations are the card the agent is holding, as well as the actions the opponent takes. The game proceeds as follows, with zero reward until a player is declared the winner. Both players *ante* (bet) one chip. One betting round is performed: The opponent can *raise* (bet another chip) or *check* (pass). If the opponent raises, the agent can *call* (match the bet) or *fold* (surrender). If the opponent checks, the agent can raise or check (to which the opponent must then call or fold in response). If the opponent folds, the agent wins the pot (all chips staked) and receives a reward equal to the pot minus the chips the agent has bet. If the agent folds, they receive a negative reward equal to the chips staked. If neither player folds, a *showdown* occurs, and the highest card wins the pot (with the rewards issued the same as before) (Figure 12.7). This version of poker is simple enough that a mixed Nash equilibrium strategy is known for both players, which is the strategy that the opponent is using. The optimal strategy for the agent is to play the Nash strategy, leading to an average return of 1/18 reward per hand.

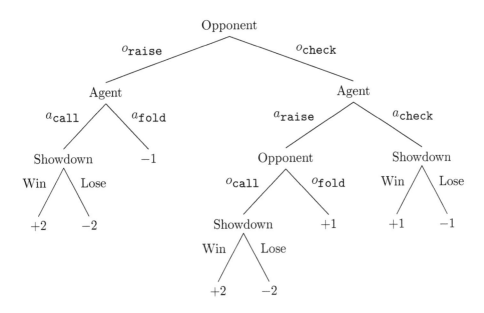

Figure 12.7: *Kuhn Poker: A decision tree representing the game, a simplistic version of poker. Showdowns are won by the agent if it has a higher ranked card than the opponent. Numbers on terminal nodes represent rewards. All other rewards during the game are zero.*

POCMAN. The last, and most complicated environment that MC-AIXI-CTW was tested on was POCMAN [SV10], which is a partially observable version of PACMAN.

This domain is a partially observable version of the classic Pac-Man game. The agent must navigate a 17×17 maze (Figure 12.8) and eat the food pellets that are distributed across the maze. Four ghosts roam the maze. They move initially at random, until there is a Manhattan distance of 5 or less between them and Pac-Mac, whereupon they will aggressively pursue Pac-Man for a short duration. The maze structure and game are the same as the original arcade game, however the Pac-Man agent is hampered by partial observability. Pac-Man is unaware of the maze structure and only receives a 4-bit observation describing the wall configuration at its current location (similar to cheese maze). It also does not know

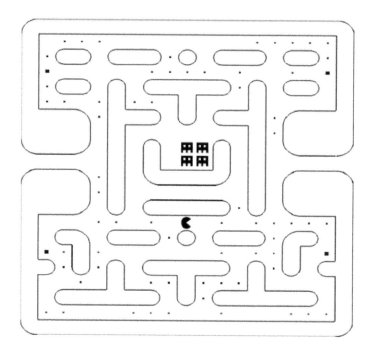

Figure 12.8: *The POCMAN maze. A more complex partially observable environment, where the agent has to navigate Pac-Man around the maze. Large dots are power pills, small dots are food. There are 4 ghosts in the middle, and Pac-Man is below the center.*

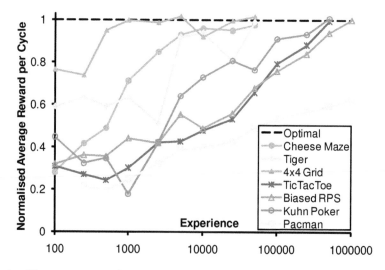

Figure 12.9: *The average performance of the MC-AIXI-CTW algorithm on the various environments.*

the exact location of the ghosts, receiving only 4-bit observations indicating whether a ghost is visible (via direct line of sight) in each of the four cardinal directions. In addition, the location of the food pellets is unknown except for a 3-bit observation that indicates whether there exists a food pellet within a Manhattan distance of 2, 3 or 4 from Pac-Man's location[15], and another 4-bit observation indicating whether there exists a food pellet via direct line of sight. A final single bit indicates whether Pac-Man is under the effects of a power pill (which allows him to eat the ghosts by colliding with them for the next 100 time steps.) At the start of each episode, a food pellet is placed down with probability 0.5 at every empty location on the grid. The agent receives a penalty of 1 for each movement action, a penalty of 10 for running into a wall, a reward of 10 for each food pellet eaten, a penalty of 50 if he is killed by a ghost (after which the episode resets), a reward of 30 for eating a ghost while under the effect of the power-pill and a reward of 100 for collecting all the food. If multiple such events occur, then the total reward is cumulative, i.e. running into a wall and being caught would give a penalty of 60. The episode resets if the agent is caught or if it collects all the food.

12.5.2 Empirical Performance

In each environment, except POCMAN, [VNHS10] tested MC-AIXI-CTW against U-tree[16] [UV98] and Active-LZ[17] [FMRW10a]. MC-AIXI-CTW outperformed both U-Tree and Active-LZ in these environments. The details of the experiments performed can be found in [VNHS10], we summarize the performance in Figure 12.9 of MC-AIXI-CTW on the various environments, with the reward normalized such that an average reward of 1 per cycle corresponds to that of the optimal policy. On all environments (except POCMAN), MC-AIXI-CTW eventually learns to play optimally.

12.6   AIXIjs Implementation

We finally introduce AIXIjs [ALH17, Asl17], a toy implementation of various approximations of AIXI from Chapters 7, 9 and 12 in *JavaScript* (JS). It is an easy-to-use testbed for various approximations of AIXI and other universal agents such as Knowledge-Seeking Agents, BayesExp, and Thompson sampling. There are several environments implemented in AIXIjs, including various gridworlds and MDPs [Put14]. The testbed of AIXIjs allows experimental comparison between (approximations of) these powerful agents. For instance, AIXIjs can be used to help understand wireheading in partial embedded agents [MSZ19], study the effects of generalized discount functions [LALH17], and investigate how new UAI agents perform experimentally compared to other UAI agents [CCH19, CHC21].

Here we compare two choices of model classes, the first is a class of possible reward dispenser locations on the gridworld, and second is a Dirichlet distribution over grid cells which the agent can visit. The first uses the base AIXI approximation MC-AIXI-CTW and the second uses the knowledge-seeking agents. This also serves as a comparison between the three types of knowledge-seeking agents described in Section 9.3. The comparisons between the agents are performed on 10×10 gridworlds, which contains a single dispenser. The reward dispenser gives reward with probability 0.75. For the UCT, we use 600 samples with a planning horizon of 6. Each agent's performance is averaged over 50 simulations.

---

[15]One could interpret this observation as Pac-Man "smelling" the food pellets at a distance.

[16]which attempts to learn a compact representation of the environment based on the history stream generated by the agent and the environment.

[17]which uses a prediction scheme based on Lempel-Ziv compression, together with a context tree (different from CTW) to accumulate statistics of the environment.

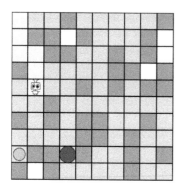
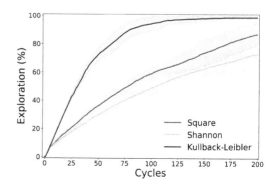

Figure 12.10: *(left) gridworld in AIXIjs, a testbed for various approximations of universal agents like AIXI. Darker gray indicates higher belief of potential reward. (right) Performance in terms of intrinsic motivation (Exploration %) of Square, Shannon, and KL Knowledge-Seeking agents (Cycles≡t). [ALH17].*

Figure 12.10 (left) depicts AIXIjs' grid-world. Figure 12.10 (right) shows that the intrinsic motivation is highly model-dependent, and the information-seeking policy outperforms entropy-seeking in stochastic environments. Shown are results for the Dirichlet base model class. The curves for the location-based model class look similar but only reach around 75%. Among the knowledge-seeking agents, the KL one, described in Section 9.3, outperforms the other knowledge-seeking agents. Figure 12.11 (left) shows that the performance in terms of average reward of MC-AIXI(-Dirichlet) is dependent on the model class: model class based on location (thin line), and model class based on Dirichlet distribution (thick line), 'Cycles'≡t. Figure 12.11 (right) shows that MC-AIXI typically outperforms Thompson Sampling. Both agents are given the same MCTS planning horizon of $m=6$ and a budget of 600 samples per action. Here, Thompson sampling is unreliable and lackluster, achieving low mean reward with high variance. For horizon 10 and a samples budget 100, the gap to AIXI narrows (not shown) but is still significant. This is interesting as Thompson sampling has been proven to be asymptotically optimal, while the MC-AIXI-CTW is not. It is likely the additional exploration of Thompson sampling causes it to not perform as well as the MC-AIXI-CTW.

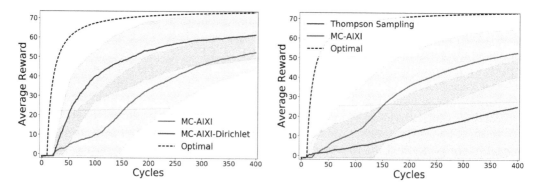

Figure 12.11: *Performance in terms of average reward of MC-AIXI(location) and MC-AIXI-Dirichlet and Thompson sampling compared to theoretically optimal (Cycles≡t) [ALH17].*

Many of the agents being approximated in AIXIjs are incomputable, asymptotically optimal agents, meaning there is no way we will ever be able to program them, and even if we did, there is no guarantee they would compute good actions in any reasonable time frame. Although these points may seem negative, they are also the reason that these agents are interesting. Like AIXI, these agents stand at the very top of intelligence, and it is only through approximations like those used in AIXIjs that we are able to move down from the idealized agent to a practical implementation.

12.7   Discussion

The Monte Carlo AIXI Context-Tree-Weighting (MC-AIXI-CTW) agent represents the initial attempt to approximate AIXI, showcasing its effectiveness in more complex environments. By employing the CTW method, ξ_{FAC} constructs an efficient Bayesian mixture, yielding a manageable version of AIXI's mixture ξ. As discussed in Chapter 5, various extensions of CTW have been proposed, each enhancing different aspects of the original method. These improvements can be readily integrated into the MC-AIXI-CTW framework, replacing the original CTW component, thereby potentially improving the agent's overall performance.

12.8   Exercises

1. [C07] Assume that a bandit problem has a unique optimal action a^*. Prove that an ε-greedy agent converges to a policy that chooses a^* with probability $1-\varepsilon\left(1-\frac{1}{|\mathcal{A}|}\right)$.

2. [C25] In UCT, derive the worst-case-regret bound of using $\ln^2 N(h)$ or $\sqrt[3]{N(h)}$ (or other functions in $o(\sqrt{N(h)})$. How do they compare to the bound of $\ln N(h)$?

3. [C30] Prove Theorem 12.2.6.

4. [C15] Run an experiment comparing the performance of BayesExp and Thompson sampling agents in gridworlds on the AIXIjs website.

5. [C12] Prove that we can always apply a positive affine transformation to the rewards to ensure they are always non-negative, without changing the optimal policy. See also Exercise 15.5.

6. [C25] Implement ρ-UCT together with FAC-CTW to replicate MC-AIXI-CTW, and graph the performance against each environment (Figure 12.9). A reference implementation can be found at `https://jveness.info/software/default.html`.

7. [C25] Investigate why leaf parallelization performs poorly compared to other methods for parallelizing MCTS (Section 12.2.6).

8. [C28] Investigate the extensions of CTW (Chapter 5) and use them in place of CTW to define ξ_{FAC}. Does this improve the MC-AIXI-CTW model?

9. [C20] Train MC-AIXI-CTW on a simple environment (e.g. 1d Maze or Tiger) and then continue on another simple environment. Does MC-AIXI-CTW perform better or worse on the second domain? Repeat with different pairs of environments.

10. [C18] For the UCT action selection, test alternative choices of the constant C. Show that if C is sufficiently large UCT will fail.

11. [C20] Use the CTW algorithm to learn a fast approximation of the policy and use it for the rollout in ρUCT. What is gained/lost by using this in place of a random policy?

12. [C32] Prove Theorem 12.4.1.

13. [C25] Generalize FAC-CTW to $\mathcal{P}$-context trees defined in Section 9 of [VNH^{+}11].

12.9 History and References

This chapter is primarily based on [VNH^{+}11], where the Context Tree Weighting (CTW) method for learning and the UCT algorithm for planning are first combined. In [BVB13], an agent was constructed for Atari 2600 games by recursively factoring the large observation space, with the goal to subdivide it into smaller, more manageable sub-problems. A Bayesian average is then taken over the result. This recursive factorization works similarly to the CTW algorithm. In [VSH12], a survey of possible ways to extend the CTW algorithm is given, particularly for the active learning case. One of the novel aspects of the MC-AIXI-CTW algorithm is that it already uses an extension of CTW, Factored-Action-Conditional CTW (FAC-CTW), also introduced in [VNHS10]. The AIXIjs testbed has been developed in [Asl17, ALH17]. It comes with various approximations of AIXI discussed in this book and various toy environments implemented.

AIXI approximations. There have been several attempts at studying the computational properties of AIXI. A similar version of AIXI that uses the best policy (as measured by a provable lower bound on the value function) bounded in space l and time t results in the optimal time-bounded AIXItl agent [Hut05b, Chp.7] discussed in Section 13.2. The construction is based on the provably fastest algorithm for all well-defined problems [Hut01b, Hut02b]. Like Levin Search [Lev73c], which solves any inversion problem optimally within multiplicative overhead, such algorithms are not practical in general but can in some cases be applied successfully [Sch97b, SZW97, Sch03, Sch04]. Self-AIXI [CGMH^{+}23], discussed in Section 9.5, is a modification of AIXI that avoids planning by self-predicting its own actions, and its practical instantiation Self-AIXI-CTW avoids expensive MCTS in MC-AIXI-CTW. In domains with very small action/percept spaces and restrictive assumptions about the model class the environment is contained in, universal learning is computationally feasible via brute-force search [PH06b] (see Chapter 11). The behavior of AIXI-MDP is compared with a universal predicting-with-expert-advice algorithm [PH05b] in repeated 2×2 matrix games and is shown to exhibit different behavior. In [Pan08], a Monte Carlo algorithm is used to sample from a time-bounded version of the universal Solomonoff prior. A closely related algorithm is that of sampling from a speed prior [Sch02b, FLH16], a universal semimeasure similar to the Solomonoff prior that considers not only the size of the program, but also the number of time steps it takes to compute a result.

One of the shortcomings of MC-AIXI-CTW is that the agent does not consider environments which use information beyond the length-k context. This was (partially) overcome in the extended agent Φ-AIXI-CTW [YZWN22], which abstracted the complete history as in Chapter 14 and then used CTW on the abstracted history, enabling environment model classes that can have arbitrary dependence on the history. Another problem in practice is that starting from a large model class is computationally expensive. DynamicHedgeAIXI [YZNH24] allows to dynamically add (and subtract) models via a time-adaptive prior constructed from a variant of the Hedge algorithm. It is the latest and richest direct approximation of AIXI at the time of writing and comes with good theoretical and practical performance guarantees.

An alternative approach to approximating Bayes-optimal agents like AIXI was developed in [MDM+20], where it was shown that meta-training can be used to approximate Bayes-optimality.

Upper Confidence Bounds and Monte Carlo Tree Search. The *Upper Confidence Bound* (UCB) algorithm was first introduced in [ACBF02]. It has been extended to contextual bandits [LS20, Part V] with linear, non-linear, and gated linear function approximation [SHB+20]. Extending UCB from bandits to MDPs (UCT) was done in [KS06]. UCB has had many successful practical uses, including the game *Go* [GW06], as well as learning to play new games from a provided description of the game rules [FB08]. UCB learns a value function online, and in [GS07] it was modified to be able to combine statistics with offline learning via the TD(λ) algorithm. There has been various work on improving the other aspects of *Monte Carlo tree search* (MCTS) beyond use of UCB to guide the tree policy. Many existing Monte Carlo methods are given in [CWvdH+08b]. Different methods of parallelizing MCTS are explored in [CWvdH08a]. Improvements to the rollout stage of MCTS include using a 1-ply rollout-based planning technique for noisy stochastic domains [BC99], using knowledge-based rollout policies in Go [GWMT06], and learning the rollout policy for adversarial games [ST09]. MCTS was used to adjust a heuristic for game tree search in chess to better represent the value of a board state after a deep search [VSBU09]. For large partially observable domains, the POMCP algorithm was defined by combining the agent's belief state, together with MCTS from the state the agent believes itself to be in [SV10].

Utile Suffix Memory (USM) and U-Tree algorithms. An early and influential work is the Utile Suffix Memory (USM) algorithm [McC96]. USM uses a suffix tree to partition the agent's history space into distinct states, one for each leaf in the suffix tree. Associated with each state/leaf is a Q-value, which is updated incrementally from experience much like Q-learning [WD92]. The history-partitioning suffix tree is grown in an incremental fashion, starting from a single leaf node in the beginning. A leaf in the suffix tree is split when the history sequences that fall into the leaf are shown to exhibit statistically different Q-values. The USM algorithm works well for a number of tasks but does not deal effectively with noisy environments. Several extensions of USM to deal with noisy environments are investigated in [SB04, Sha07]. U-Tree [McC96] is an online agent algorithm that attempts to discover a compact state representation from a raw stream of experience. The main difference between U-Tree and USM is that U-Tree can discriminate between individual components within an observation. This allows U-Tree to more effectively handle larger observation spaces and ignore potentially irrelevant components of the observation vector. Each state is represented as the leaf of a suffix tree that maps history sequences to states. As more experience is gathered, the state representation is refined according to a heuristic built around the Kolmogorov-Smirnov test [BZ14]. This heuristic tries to limit the growth of the suffix tree to nodes that would allow for better prediction of future reward. Value iteration is used at each time step to update the value function for the learned state representation, which is then used by the agent for action selection.

Universal RL. Active-LZ [FMRW10a] combines a Lempel-Ziv-based (a commonly used data compressor described in [ZL77]) prediction scheme with dynamic programming for control to produce an agent that is provably asymptotically optimal in k-Markov environments. The algorithm builds a context tree (distinct from the CTW context tree) with each node containing accumulated transition statistics and a value function estimate (much like the MCTS tree that MC-AIXI-CTW uses). These estimates are refined over time, allowing for the Active-LZ agent to steadily increase its performance. The MC-AIXI-CTW agent compares favorably to Active-LZ [VNH+11, Sec.7].

Bayesian approach to Learning History Trees (BLHT). The *Bayesian approach to Learning History Trees* (BLHT) algorithm [SHL97, SH98] uses symbol-level PSTs for learning and a dynamic programming-based algorithm for control. BLHT uses the MAP[18] model for prediction (as opposed to CTW, which uses a Bayesian mixture), which admits a much stronger convergence result. A further distinction is that the prior used by CTW assigns weights to models depending on their complexity in accordance with Occam's razor, instead of a uniform prior over PST models used by BLHT.

Predictive State Representations (PSRs). *Predictive state representations* (PSRs) [LS01, RGT04, SJR12] maintain predictions of future experience. A PSR is a collection of predictions, encoded as a vector of probabilities, for the outcomes of various *tests*, which are predicates about future observations from the environment given particular actions are chosen. If the outcomes of every possible test were already known, this would imply the dynamics of the environment are also known. A PSR is a subset of these predictions sufficient to determine the outcome of all other tests, and hence the dynamics of the environment. Unfortunately, exact representations of state are impractical in large domains and some form of approximation is typically required.

Other Model-Based Reinforcement Learning (MBRL) algorithms. Topics such as improved learning or discovery algorithms for PSRs are currently active areas of research. The recent results of [BSG11] appear particularly promising by extending PSRs using kernels in such a way that continuous, high-dimensional observation spaces can be handled. Temporal-difference networks [ST04] are a form of PSR in which the agent's state is approximated by *abstract predictions*: predictions about future observations, but also predictions about future predictions. This set of interconnected predictions is known as the *question network*. Temporal-difference networks learn an approximate model of the world's dynamics: given the current predictions, the agent's action, and an observation vector, they provide predictions for the next time step. The parameters of the model, known as the *answer network*, are updated after each time step by temporal-difference learning. Some results of applying TD-Networks for prediction (but not control) to small POMDPs are given in [Mak09]. In model-based Bayesian Reinforcement Learning [Str00b, PVHR06, RCdP07, PV08], a distribution over (PO)MDP parameters, is maintained. In contrast, we maintain an exact Bayesian mixture of PSTs, representing variable-order Markov models. The ρUCT algorithm shares similarities with Bayesian Sparse Sampling [WLBS05]. The main differences are estimating the leaf node values with a rollout function and using the UCB policy to direct the search.

Environments. Two approaches have been compared against MC-AIXI-CTW, U-tree [McC96] and Active-LZ [FMRW10a]. This comparison was done over several domains including Cheese Maze [McC96], Biased rock paper scissors [FMRW10a], 1d-maze [CKL94], Tiger [KLC98], Kuhn poker [Kuh50, HSHB05], Extended Tiger [CKL94], 4×4 grid [CKL94], and POCMAN [SV10]. MC-AIXI-CTW outperforms both U-tree and Active-LZ on all previously mentioned environments.

Compress and Control. Compression and prediction are share a strong connection. An exceptional sequence predictor, when combined with arithmetic coding (Section 2.5.6), can produce an effective compressor for that sequence. Conversely, compression algorithms can be readily applied to sequence prediction tasks [Hut20a].

In the AIXI model, prediction is used to estimate future observations and rewards one step at a time. However, in the *general reinforcement learning* problem, our focus lies on future observations only insofar as they contribute to maximizing the expected future reward. A reasonable approach would be to predict the future sum of rewards (called returns,

---

[18]maximum a posteriori

denoted z) directly, instead of predicting observations and rewards at each time step. This would involve estimating the distribution of future returns denoted as $\rho(z|s,a)$, given the current state s and action a. The Q-value $Q_\mu^\pi(s,a) = \sum_z z\rho(z|s,a)$ is the expected return.

The challenge with this approach is that the distribution of returns can be highly complex and difficult to estimate directly, as seen in Distributional Reinforcement Learning [BDM17, BDR23, WDA+23]. Compress and Control (CnC) [VBH+15, DRW+24] offers a solution by decomposing the distribution of the return into more manageable components. After employing compression-based prediction to learn these components, they can be combined to obtain an estimate of the distribution of the return.

CnC is a compression-based method for policy evaluation and on-policy control in reinforcement learning. This means that CnC can assess the quality of a given policy (evaluation) and also determine an effective policy (control). The key result of CnC is that, under reasonable conditions on the policy and environment, the proposed decomposition of the Q-value function is well-founded and its approximation converges to the true Q-value.

The decomposition employed in CnC exploits the factorization $\rho(z|s,a) \overset{\times}{=} \rho(s|z,a)\rho(z|a)$. Any estimator can be used for $\rho(s|z,a)$ and $\rho(z|a)$, but it has been shown that using Context Tree Weighting (CTW) leads to consistent estimation [VBH+15]. By converting the reinforcement learning problem into a prediction/compression problem, Compress and Control demonstrates that improved compression results in stronger performance. Alternatively one may use Large Language Models for compression [DRW+24].

Quantum AIXI. Quantum Computing offers computational speedups for many problems such as finding the prime factor of a given natural number and inverting a matrix. A natural question in the context of this book is whether quantum computing can speed up approximations of UAI agents such as AIXI? This question was first tackled in [CH20a] where approximations of Solomonoff induction and AIXI using quantum computing were derived. While these speedups did offer some improvement over the classical computation it was not on the order of the exponential speedup which quantum computing achieves in some other problems. Instead of using quantum computing to speed up the computation of the policy, another way quantum computing can be used in the UAI setup is in the definition of the model class of the environment. For the majority of this book we consider the class of all (semi)computable (semi)measures (the set of probabilistic programs). However, an alternative class is to choose something like the set of quantum programs, changing our assumption that the environment is computable to an assumption that the environment is quantum computable, though if the Quantum complexity-theoretic Church–Turing thesis (Section 16.6.1) holds these are equivalent. [Sar21, SAAGB21] considered this model class and defined a version of the Knowledge-Seeking agent (Section 9.3) over this model class.

Related approximations. There are many algorithms which, while not being direct approximations of the AIXI agent, embody the spirit of UAI and attempt to solve the UAI problem. These include the policy evaluation technique of Compress and Control [VBH+15] described above, extending the reinforcement learning algorithm of Q-learning to the history-based setting of UAI [DSH13], naturally extending the UAI model with Representational MDL [PR12], Feature Reinforcement Learning (Chapter 14) and (Kolmogorov) complexity-guided policy search which tries to find policies which produce action sequences of low Kolmogorov complexity [SJ21].

AI search. MCTS [KS06] is an instance of AI search [RN20, Part II]. Search has a long tradition in AI, such as (iterative deepening) A* [HNR68, Kor85], breadth-first search [Zus72] and depth-first search [Tre76] (see [EH15a, EH15b] for a comparison), Levin (tree) search [OHL23], and many variants and improvements.

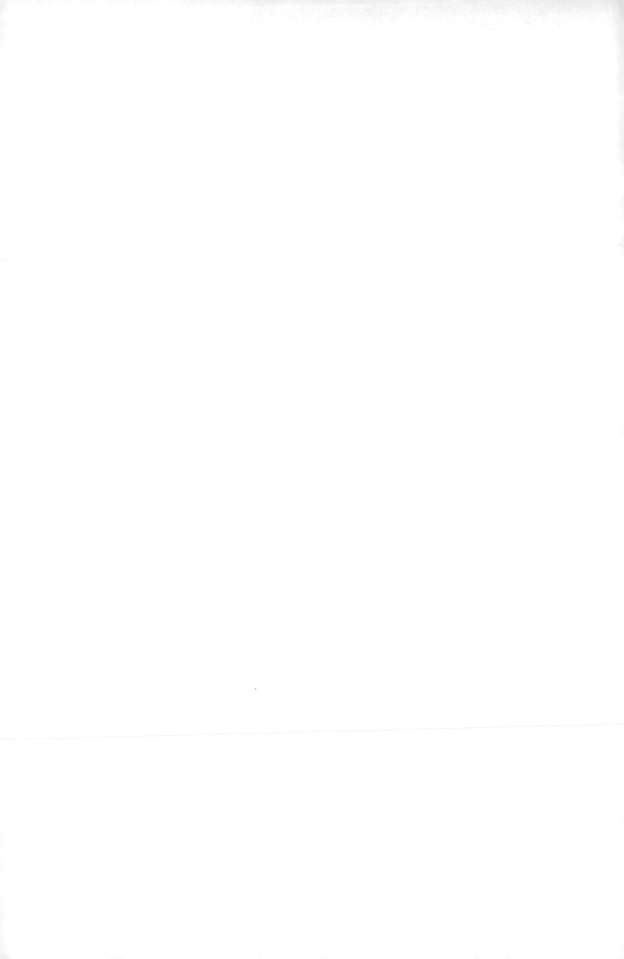

Chapter 13

Computational Aspects

Computing is not about computers
any more. It is about living.

Nicholas Negroponte

The two previous chapters covered efficiently computable approximations of AIXI. Clearly AIXI is incomputable, but where in the computability hierarchy (Section 2.6.3) does it sit? In Section 13.1 we will determine the level of computability the AIXI agent and some of its extensions from Chapter 9 satisfy. Section 13.2 introduces AIXI*tl*, a theoretically optimal computable approximation of AIXI that runs in time $O(t)$.

13.1   Computability of AIXI

Up until this point we have discussed the theoretical properties of AIXI and variations, and developed some computable some approximations such as the AIXI-MDP and MC-AIXI-CTW agents. In this section we determine the exact level of (in)computability of AIXI and its variants exactly. The reader is advised to refer back to Section 2.6 for the notions of computability and arithmetic hierarchy used here.

As for upper bounds, we show that optimal value functions and optimal policies and ε-optimal policies of AIXI and variations are in Δ_n^0 for $n=2$ or 3. As for lower bounds, none of the agents even the weakest one is computable, hence none is in Δ_1^0.

Before we get to the computability of optimal policies, we first need to discuss value functions and their level of computability (Section 2.6). We focus on the class of lower semicomputable environments $\mathcal{M}_{sol}$. While we could derive results for larger (or smaller) classes of environments, we are primarily interested in the level of computability of AIXI (and its variants).

Theorem 13.1.1 (Computability of V_ν^* and Q_ν^*) Let semimeasure ν, the discount normalizer Γ_k, and the finite reward set $\mathcal{R} \subset [0,1]$, all be Δ_n^0-computable. Then the (recursive) optimal (Q-)value functions $V_\nu^*(h_{<t})$ and $Q_\nu^*(h_{<t},a_t)$ are Δ_n^0-computable.

Remark 13.1.2 (Original V,Q for semimeasures can be more complex) If we assume ν to be a measure, then there is no difference between the recursive formulation Theorem 6.7.2 and the original Definition 6.6.1 and Definition 6.7.1. But if ν is a semimeasure, then the definitions differ, and we explicitly have to take the limit $m \to \infty$ in the original definition, which pushes $V_\nu^*(h_{<t})$ up to Δ_{n+1}^0. For finite- or moving horizon (Definition 6.4.2), we have $V \in \Delta_n^0$ in any case, since no $m \to \infty$ limit has to be taken.   ●

Proof sketch. (i) From Theorem 6.7.2, we see that $V_\nu^{*,m}(h_{<t})$ is a finite-depth $m-t$ recursion of continuous combinations (sum, max, product, ratio) of finitely many terms, and each term is assumed to be Δ_n^0, hence $V_\nu^{*,m}(h_{<t}) \in \Delta_n^0$ by Theorem 2.6.19.
(ii) Recall that a real-valued function is said to be computable (also called estimable), if there exists an algorithm that takes the function's input and any $\varepsilon > 0$ and finitely computes the function to ε-accuracy. Let us assume $\nu, \Gamma, \mathcal{R} \in \Delta_1^0$ i.e. are estimable in this sense. From Lemma 6.7.6 we have

$$0 \leq V_\nu^*(h_{<t}) - V_\nu^{*,m}(h_{<t}) \leq \Gamma_{m+1}/\Gamma_t$$

From Definition 6.4.1 we know that $\Gamma_m \to 0$ for $m \to 0$. For $m = t, t+1, \ldots$ we can estimate Γ_{m+1}/Γ_t as $\tilde{\delta}_m$ to accuracy $\varepsilon/4$ until we find an m for which $\tilde{\delta}_m < \varepsilon/2$, For this m we estimate $V_\nu^{*,m}(h_{<t})$ to accuracy $\varepsilon/4$. Alltogether this implies we have estimated $V_\nu^*(h_{<t})$ to accuracy ε.
(iii) If $f(z_1, z_2, \ldots)$ is computable (estimable) and $g_i(x_i)$ are Δ_n^0 computable, then $f(g_1(x_1), g_2(x_2), \ldots)$ is Δ_n^0 computable. This allows us to lift the proof in (ii): If $\nu, \Gamma, \mathcal{R} \in \Delta_n^0$, then $V_\nu^*(h_{<t}) \in \Delta_n^0$.
(iv) The same argument holds for Q_ν^* which is just V_ν^* with the left-most $\max_{a_t}$ removed.   ∎

Theorem 13.1.3 (Lower semicomputability of V_ν^* and Q_ν^*)
(i) If $\nu(e_k|\cdots), \gamma_k, \mathcal{R} \in \Sigma_n^0$, then recursive $\Gamma_t V_\nu^*(h_{<t}) \in \Sigma_n^0$.
(ii) If $\nu(\cdots), \gamma_k, \mathcal{R} \in \Sigma_n^0$, then recursive $\nu(e_{<t}||a_{<t})\Gamma_t V_\nu^*(h_{<t}) \in \Sigma_n^0$.

Proof. We use that if $f(z_1, z_2, \ldots)$ is lower semicomputable (Σ_1^0) and monotone increasing in all arguments, and $g_i(x_i)$ are Σ_n^0 computable functions, then $f(g_1(x_1), g_2(x_2), \ldots)$ is Σ_n^0 computable.
(i) Inserting (6.7.3) for $\pi = \pi_\nu^{*,m}$ into (6.7.5) and multiplying by Γ_t gives

$$\Gamma_t V_\nu^{*,m}(h_{<t}) = \max_{a_t} \sum_{e_t} \nu(e_t|h_{<t}a_t)\left[\gamma_t r_t + \Gamma_{t+1} V_{\nu,\gamma}^{*,m}(h_{1:t})\right]$$

which is Σ_n^0, since all components are assumed to be Σ_n^0, non-negative, combined in a monotone increasing way $(\max, \Sigma, \times, +)$, and the recursion is finite. Since $V_\nu^{*,m}$ is monotone increasing in m, also $\Gamma_t V_\nu^{*,\infty}(h_{<t}) \in \Sigma_n^0$.
(ii) If we multiply the above equation further by $\nu(e_{<t}||a_{<t})$, we get

$$\tilde{V}_\nu^*(h_{<t}) := \nu(e_{<t}||a_{<t})\Gamma_t V_\nu^*(h_{<t}) = \max_{a_t} \sum_{e_t}\left[\nu(e_{1:t}||a_{1:t})\gamma_t r_t + \tilde{V}_{\nu,\gamma}^*(h_{1:t})\right]$$

which is Σ_n^0 by the same argument.   ∎

These results are especially interesting for Solomonoff's distribution $\xi_U = \mathcal{M}$, which is in the class $\mathcal{M}_{sol}$ (Definition 3.7.1), i.e. $\xi \in \Sigma_1^0 \subset \Delta_2^0$. Now that we know the level of computability of the value function, we need to relate the computability of a value function to the computability of an optimal policy.

Theorem 13.1.4 (Computability of optimal policies) For any environment ν, if V_ν^* is Δ_n^0-computable, then there is an optimal (deterministic) policy π_ν^* for the environment ν that is Δ_{n+1}^0-computable.

Proof from [LH18, Thm.17]. To break ties between actions, we choose a total order $\succ$ on the action space $\mathcal{A}$ that specifies which action should be chosen in a tie. Then we define

$$\pi_\nu(h_{<t}) := a \iff \bigwedge_{a' : a' \succ a} Q_\nu^*(h_{<t}, a) > Q_\nu^*(h_{<t}, a')$$

$$\wedge \bigwedge_{a' : a \succ a'} Q_\nu^*(h_{<t}, a) \geq Q_\nu^*(h_{<t}, a')$$

Then π_ν is ν-optimal. Since V_ν^* is Δ_n^0-computable, we have that $Q_\nu^*(h_{<t}, a) > Q_\nu^*(h_{<t}, a')$ is Σ_n^0 and $Q_\nu^*(h_{<t}, a) \geq Q_\nu^*(h_{<t}, a')$ is Π_n^0. Therefore π_ν, which is the conjunction of something which is Σ_n^0 and something which is Π_n^0, is itself Δ_{n+1}^0. ∎

This shows that if we are not careful, we quickly move up the arithmetic hierarchy: Solomonoff $\mathcal{M} \in \Delta_2^0$, which brings the iterative value from Definition 6.6.1 to Δ_3^0 due to the $m \to \infty$ limit, resulting in $\pi_{\mathcal{M}}^* \in \Delta_4^0$ for infinite-horizon AIXI.

On the other hand, the above theorem is only stating that there is an optimal policy which is Δ_{n+1}^0; it may be the case that this or other optimal policies actually have lower level of computability than Δ_{n+1}^0. Trying to determine whether two actions have the exact same value seems like a waste of resources. Usually it is fine taking slightly suboptimal actions. In the following, we consider policies that attain a value which is guaranteed to be within some $\varepsilon > 0$ of the optimal value. Indeed, ε-optimal policies (Definition 8.1.2) require one less level of computability.

Theorem 13.1.5 (Computablity of ε-optimal policies) For any environment ν, if V_ν^* is Δ_n^0-computable, then for all $\varepsilon > 0$ there is an ε-optimal (deterministic) policy π_ν^ε for the environment ν that is Δ_n^0.

Proof from [LH18, Thm.20]. Let $\varepsilon > 0$ be given. Since V_ν^* is Δ_n^0-computable we can construct a set $Q_\varepsilon := \{(h, a, q) \in \mathcal{H} \times \mathcal{A} \times \mathbb{Q} \mid |q - Q_\nu^*(h, a)| < \varepsilon/2\}$ that is itself Δ_n^0. Let $\succ$ be an arbitrary total order on the actions $\mathcal{A}$ that specifies which action we should pick in the case of a tie. Without loss of generality assume $\varepsilon = 1/k$ and define the set Q to be an $\varepsilon/2$-grid of $[0,1]$, i.e. $Q = \{0, 1/2k, 2/2k, ..., 1\}$. Then we can define the policy

$$\pi_\nu^\varepsilon(h_{<t}) := a \iff \exists (q_{a'})_{a' \in \mathcal{A}} \in Q^{|\mathcal{A}|} . \bigwedge_{a' \in \mathcal{A}} (h, a', q) \in Q_\varepsilon \wedge \bigwedge_{a' : a' \succ a} q_a > q_{a'} \wedge \bigwedge_{a' : a \succ a'} q_a \geq q_{a'}$$

$$\wedge \text{ the tuple is minimal with respect to lexicographical ordering on } Q^{|\mathcal{A}|}$$

Therefore the choice of action a is unique, and since $\mathcal{A}$ is finite, $Q^{|\mathcal{A}|}$ is also finite, and therefore π_ν^ε is Δ_n^0. ∎

As a consequence of Theorems 13.1.4 and 13.1.5, we get the level of computability of ε-optimal AIXI.

Corollary 13.1.6 (Computability of AIXI) $\pi_M^\varepsilon \in \Delta_2^0$ and $\pi_M^* \in \Delta_3^0$. In words: There is an ε-optimal policy for AIXI that is approximable (Δ_2^0). There is an optimal AIXI policy that is Δ_3^0.

We have the following corresponding lower bound of computability for AIXI.

Theorem 13.1.7 (No AIXI is computable) (Deterministic) AIXI is incomputable. i.e. $\pi_\xi^* \notin \Delta_1^0$.

Proof from . We will prove this by contradiction. Assume that π_ξ^* is computable. Since π_ξ^* is deterministic and computable we can construct an environment μ that gives the agent reward 0 if it took the action π_ξ^* would take and reward 1 if it diverted from the action π_ξ^* would take. Ignoring observations we can formally write the environment as

$$\mu(r_t|h_{<t}a_t) := \begin{cases} 1 & \text{if} \quad \forall k \leq t, \; a_k = \pi_\xi^*(h_{<k}) \text{ and } r_k = 0 \\ 1 & \text{if} \quad \forall k \leq t, \; r_k = [\![k \geq i]\!] \text{ where } i := \min\{j | a_j \neq \pi_\xi^*(h_{<j})\} \\ 0 & \text{otherwise} \end{cases}$$

This implies that $V_\mu^{\pi_\xi^*}(h_{<t}) = 0$ for all $h_{<t}$. We also have that μ satisfies the conditions of Theorem 7.4.10, therefore $V_\xi^*(h_{<t}) \not\to 0$ on $h_{<t}$ generated by μ and π_ξ^*. However, this leads to a contradiction with Theorem 7.3.1 which says

$$V_\xi^*(h_{<t}) - V_\mu^{\pi_\xi^*}(h_{<t}) \to 0 \quad \text{for} \quad t \to \infty \quad \mu^{\pi_\xi^*}\text{-almost surely} \qquad \blacksquare$$

Theorem 13.1.8 (ε-AIXI is not computable) $\varepsilon\text{-}AIXI \notin \Delta_1^0$.

Proof. See [LH18, Thm.25] $\qquad\qquad\qquad\qquad\qquad\qquad\qquad\qquad\qquad\qquad\qquad\qquad\qquad\blacksquare$

While it is obviously not ideal that the "most intelligent" general agent is not computable, this is hardly surprising. This is a consequence of the choice of what we mean by 'general', the (infinite) class of all semicomputable semimeasures. Choosing smaller classes $\mathcal{M}$ in AIξ led to computable in MDP-AIXI and MC-AIXI-CTW. However, the primary point of AIXI is not that we expected to be able to run it on our computer, but that it can serve as an ideal in terms of super-intelligent agents. AIXI is the gold standard of the best possible agent we could ever hope to achieve, not the best possible agent practically achievable. The same holds for ε-optimal policies as ε goes to 0.

Computability of extensions of AIXI. Agents similar to AIXI, such as knowledge-seeking (Definition 9.3.2) and BayesExp (Algorithm 9.4) have similar computability levels.

Theorem 13.1.9 (Computability of knowledge-seeking policies) For entropy-seeking and information-seeking agents, there are approximable (Δ_2^0) ε-optimal policies and Δ_3^0-computable optimal policies.

Proof. This essentially follows from Theorems 13.1.4 and 13.1.5 and letting the environment be the knowledge-seeking environment. $\qquad\qquad\qquad\qquad\qquad\qquad\qquad\qquad\qquad\blacksquare$

Theorem 13.1.10 (BayesExp is Δ_3^0) For any universal mixture ξ, BayesExp is Δ_3^0.

Although being Δ_3^0 is far from the realm of being realistically computable, the ε-optimal policies being approximable (Δ_2^0) show that at least we have limit-computable algorithms for reasonably optimal agents [LH18, Thm.39].

13.2 Time- and Space-Bounded AIXI

Potential approaches to approximate AIXI. Knowing that AIXI is not computable, we want to know, what is the closest we can get to AIXI while still being (finitely) computable?

Earlier we discussed approximation algorithms such as MDP-AIXI and MC-AIXI-CTW, which are similar to AIXI but computable (ξ_{MDP} or ξ_{CTW} replacing ξ_U, and MCTS replacing expectimax in the latter case). The downsides of these and similar approaches are two-fold: (i) The class of environments $\mathcal{M}$ is a-priori restricted and immutable, even if evidence suggests the true environment is outside $\mathcal{M}$ or if spare compute is available to compute ξ over a larger class. (ii) There are problems, where jointly solving the learning and planning problem, is significantly more efficient than approximating ξ and expectimax separately. For instance, certain Bayesian bandit problem have closed-form solutions.

Ideally would be an anytime algorithm which approaches AIXI in the limit of unlimited compute, and is best-in-class. This is where AIXI*tl* [Hut05b] comes in. AIXI*tl* is essentially trying to find the program which, when run, leads to an agent which is closest to AIXI, only considering programs which halt in time t and are of length at most l.

Some versions of AIXI are limit-computable (Corollary 13.1.6), so can be converted into an anytime algorithm. The more compute one spends (at some time step) to approximate the value function, the more accurate it is, with convergence in the limit to the exact value $Q_\xi^*(h_{<k}, a_k)$ and $\varepsilon_k \to 0$-optimal action.[1] Unfortunately this is totally impractical. Indeed, this algorithm converges slower than any computable function.

Another approach could be to consider the (finite) class of all time&length-bounded policies Π_{tl}, and determine the best-in-class policy a-priori once-and-for-all,

$$\pi_\xi^{tl} := \arg\max_{\pi \in \Pi_{tl}} V_\xi^\pi(\epsilon)$$

and then use it during the agent's complete lifetime. There are two problems (i) We cannot compute π_ξ^{tl}, since $V_\xi^\pi(\varepsilon)$ is incomputable. Indeed, finding π_ξ^{tl} is solving the AGI problem in practice. (ii) Fixing a single time-bounded policy that a-priori can deal well with all eventualities life will throw at the agent may be wasteful. Finding more specialized policies the more the agent knows, seems more promising (e.g. a chess engine). Note also that time-consistency does no longer hold ($\pi_{\xi,k}^{tl} \neq \pi_{\xi,1}^{tl}$, Remark 6.6.2).

AIXI*tl* addresses both problems. There are some direct policy search algorithms in RL that do not maintain a value function [Wil92, KHS01a] but these are rather the exception. It seems not too much of a stretch when considering Bayes-optimal agents π_ξ^*, to assume that the value function needs to be approximated well. To find the best approximation we will consider an extended form of a policy that outputs both an action *and* an approximation of its current value function. We will denote extended (deterministic) policies by $\dot{\pi}: \mathcal{H} \to \mathcal{A} \times [0,1]$. At time step k we denote the action taken by $\dot{\pi}$ to be $a_k^{\dot{\pi}} = \dot{\pi}(h_{<k})_a$ and the approximation of the value function of $\dot{\pi}$ (the second component) by $v_k^{\dot{\pi}} = \dot{\pi}(h_{<k})_v$.

When discussing (extended) policies here we will be referring to the length and time of an extended policy, denoted by $\ell(\dot{\pi})$ and $t(\dot{\pi})$. What we mean by this is the length and time of the program which computes $\dot{\pi}$ (on any given history), that is, the length of the program p such that $U(ph_{<t}) = \dot{\pi}(h_{<t})$ for some universal Turing machine U and the time it takes $U(ph_{<t})$ to halt. We only consider extended policies which have programs that halt in time t and have length at most l.

Although an extended policy may perform "good" actions, we would like to categorize the class of policies that have "valid" approximations about the value function. We say an

---

[1] In this section we use k as time index to not confuse it with compute-time budget t.

extended policy is a *valid approximation* if it never overrates itself in terms of the value function approximation. Formally, this is stated as follows: To check whether or not a policy is correct, AIXI*tl* uses a method of policy evaluation, called a valid approximation.

Definition 13.2.1 (Valid Approximation) The logical predicate VA($\dot{\pi}$) (Valid Approximation) is true if and only if $\dot{\pi}$ always satisfies $v^{\dot{\pi}} \leq V_\xi^{\dot{\pi}}$, i.e.

$$\text{VA}(\dot{\pi}) \iff [\forall k \forall h_{<k} \in \mathcal{H}: \dot{\pi}(h_{<k})_v \leq V_\xi^{\dot{\pi}}(h_{<k})]$$

We only consider extended policies that underrate themselves and not extended policies which only overrate themselves, or have an ε-approximate rating of themselves. There are two reasons for that (*i*) We want to select policies with high value, but if we allow overconfidence, we will select overrated policies whose actual value is much lower. (*ii*) together with a lower semicomputable choice of V_ξ^π, this will ensure that AIXI*tl* converges in some sense to AIXI in the limit of $t, l \to \infty$ (see proof of Theorem 13.2.4).

Assuming an extended policy $\dot{\pi}$ is a valid approximation, we are interested in the "best" $\dot{\pi}$ (that we can compute with time t and length l). For this we need an order relation on policies based on their claimed value.

Definition 13.2.2 (Effective intelligence order relation) We call valid $\dot{\pi}$ effectively more or equally intelligent than valid $\dot{\pi}'$ if

$$\dot{\pi}' \preceq_c \dot{\pi} \quad :\iff \quad [\forall k \forall h_{<k} \in \mathcal{H}: \dot{\pi}(h_{<k})_v \leq \dot{\pi}'(h_{<k})_v]$$

The effective intelligence order relation is like a self-belief version of the *total* but incomputable intelligence order induced by Legg–Hutter intelligence measure $\Upsilon(\pi)$ (Definition 16.7.1) in the sense that it is about how Legg–Hutter intelligent the extended policy thinks it is.

The effective intelligence order relation is an upper semicomputable *partial* order relation on extended policies. This means that over the space of extended policies the effective intelligence order relation satisfies reflexivity, antisymmetry, and transitivity, and that the level of computability of it as a function is upper semicomputable. If no resource restrictions were placed on $\dot{\pi}$, it would still have a global maximum $\dot{\pi} \preceq_c \dot{\pi}_\xi^* := (\pi_\xi^*, V_\xi^*)$ for all valid $\dot{\pi}$.

Together with the definition of a valid approximation, we can use the effective intelligence order relation to distinguish only policies which never overrate themselves, but also believe they are good or better or optimal in the class of policies which never overrate themselves. This is the core part of the AIXI*tl* algorithm.

The AIXI*tl* algorithm. We will now describe how AIXI*tl* works. As inputs, AIXI*tl* takes a max time $\tilde{t}$, a max length $\tilde{l}$, as well as another length l_P which is used for the maximum length of proofs.

Before AIXI*tl* starts its interaction cycles with the environment, it has to do some setup. We start with checking through all $b \in \mathbb{B}^{l_P}$ in some order, whether b constitutes a proof that some $\dot{\pi}$ of length at most $\tilde{l}$ is a valid extended policy. Here when we say proof we mean to interpret the binary string b as a proof in the same formal axiomatic system that valid approximation VA() is also defined in (chosen in the definition of valid approximation).

The pairs of proofs b and policies $\dot{\pi}$, $(b, \dot{\pi})$ are recorded, then Π_{VA} is the set of all such $\dot{\pi}$ that are recorded. This process takes $O(l_P^2 2_P^l)$ time (checking 2^{l_P} proofs, each in time $O(l_P^2)$, importantly this time is unrelated to $\tilde{t}$). This justification requires significant compute, but only at setup time, and some form of justification in a voting-style algorithm seems unavoidable.

We then modify all $\dot{\pi} \in \Pi_{\mathrm{VA}}$ such that in each cycle k, if $\dot{\pi}$ does not halt within time $\tilde{t}$, then it should output $(a,0)$ for an arbitrary a, thus ensuring each extended policy $\dot{\pi} \in \Pi_{\mathrm{VA}}$ runs in time at most $\tilde{t}$ and is valid.

Now that we have finished with the initial setup, we can go over what AIXI*tl* does during the interactions with the environment: During the kth cycle, each $\dot{\pi} \in \Pi_{\mathrm{VA}}$ is run on the interaction history $h_{<k}$ up to this point, then each $\dot{\pi}$ will output an action $a_k^{\dot{\pi}}$ and value estimate $v_k^{\dot{\pi}}$. AIXI*tl* then takes the action corresponding to the highest estimate $v_k^{\dot{\pi}}$, and receives the percept e_k from the environment. Describing this procedurally we get Algorithm 13.1:

Algorithm 13.1 AIXI*tl* [Hut05b]

Require: Max time $\tilde{t}$, max program length $\tilde{l}$, max proof length l_P
Input: Percept stream $e_1, e_2, e_3, \dots$
Output: $(\tilde{t}, \tilde{l}, l_P)$-optimal action stream $a_1, a_2, a_3, \dots$

1: Interpret all $b \in \mathbb{B}^{l_P}$ as proofs of VA$(\dot{\pi})$ for some $\dot{\pi}$
2: Let Π_{VA} be the set of those $\dot{\pi}$'s such that $\ell(\dot{\pi}) \leq \tilde{l}$ and VA$(\dot{\pi})$ has been proven
3: Modify each $\dot{\pi}$ such that in each cycle k if $\dot{\pi}$ does not output a $(a_k, v_k^{\dot{\pi}})$ in time $\tilde{t}$ then make $\dot{\pi}$ output $(a,0)$ at that time
4: Update Π_{VA} to be the set of these new $\dot{\pi}$'s
5: Set $k := 1$
6: **for** k=1,2,3,... **do**
7:    **for** $\dot{\pi} \in \Pi_{\mathrm{VA}}$ **do** Run $\dot{\pi}$ on $h_{<k}$ and receive $\dot{\pi}(h_{<k}) = (a_k^{\dot{\pi}}, v_k^{\dot{\pi}})$
8:    Choose the $\dot{\pi}^* = \arg\max_{\dot{\pi} \in \Pi_{\mathrm{VA}}} v_k^{\dot{\pi}}$ of highest claimed value
9:    Write $a_k \equiv a_k^{\dot{\pi}^*} \equiv \dot{\pi}^*(h_{<k})_a$ on the output tape
10:   Receive input e_k from the environment

Since for all $\dot{\pi} \in \Pi_{\mathrm{VA}}$ we have a proof of VA$(\dot{\pi})$, we know that for all k we have $v_k^{\dot{\pi}} \leq V_\xi^{\dot{\pi}}(h_{<k})$. This means that picking the $\dot{\pi}$ with the largest $v_k^{\dot{\pi}}$ will lead to the most intelligent agent according to the effective intelligence order relation. Additionally since at each interaction cycle we are looping through Π_{VA}, whose size is independent of $\tilde{t}$, each cycle takes time $O(\tilde{t})$.

The following theorem formally states that the AIXI*tl* algorithm is indeed more or equally intelligent, according to our effective intelligence relation, than any other program of the same length l and time t.

Theorem 13.2.3 (Optimality of AIXI*tl* [Hut05b]) Let $\dot{\pi}$ be any extended policy of length $\ell(\dot{\pi}) \leq \tilde{l}$ and computation per cycle $t(\dot{\pi}) \leq \tilde{t}$, for which there exists a proof of VA$(\dot{\pi})$, defined in Definition 13.2.1, of length $\leq l_P$. Algorithm 13.1, which depends on $\tilde{l}, \tilde{t}, l_P$ but not $\dot{\pi}$, is effectively more or equally intelligent according to $\preceq_c$ than any such $\dot{\pi}$.

The length of the optimal policy found, $\dot{\pi}^*$, is $\ell(\dot{\pi}^*) = O(\log(\tilde{l} \cdot \tilde{t} \cdot l_P))$, the setup time is $t_{setup}(\dot{\pi}^*) = O(l_P^2 2^{l_P})$ and the computation time per cycle is $t_{cycle}(\dot{\pi}^*) = O(2^{\tilde{l}} \cdot \tilde{t})$.

Proof. This result is a direct consequence of the construction of AIXI*tl*: If the given extended policy $\dot{\pi}$ satisfies the assumptions, then we have that $\dot{\pi} \in \Pi_{\mathrm{VA}}$. Then at each time step, if $\dot{\pi}$ is the most effectively intelligent policy in Π_{VA}, then AIXI*tl* will act according to $\dot{\pi}$; if it is not the most effectively intelligent policy in Π_{VA}, then AIXI*tl* will act as the more effectively intelligent policy would. In either case AIXI*tl* is effectively more or equally intelligent according to $\preceq_c$.

Note: While potentially following different policies at each time step might seem counterintuitive, it does not matter for AIXI*tl* because its objective is to maximize its effective intelligence at each time step. It ensures that it always makes the best possible decisions based on the available information and policies in the set Π_{VA}. ∎

Although the setup time and time per cycle are rather large, if we fix l_P and $\tilde{l}$, then look at the time taken as a function of $\tilde{t}$, the setup time is constant and the time per cycle is $O(\tilde{t})$, although these hidden constants are extremely large.

We now show that AIXI*tl* approaches AIXI for $t,l,l_P \to \infty$. One important condition is that $V_\xi^*(h_{<k})$ is lower semicomputable. Theorem 13.1.3 gives two choices for necessary conditions ($n=1$). In both cases, γ_k and $\mathcal{R}$ need to be lower semicomputable which all interesting choices satisfy. In (i) we would need to ensure that the universal predictive distribution $\xi(e_k|h_{<k}a_k)$ is lower semicomputable. $\xi \in \mathcal{M}_{sol}$ does *not* satisfy this, but maybe a construction similar to that in Section 10.5 can be made to work. In (ii) we only need the joint $\xi(e_{<t}||a_{<t})$ to be lower semicomputable, which is the case for $\xi = \xi_U$. The theorem implies that $\tilde{V}_{\xi_U}^*(h_{<t}):=\xi_U(e_{<t}||a_{<t})\Gamma_t V_{\xi_U}^*(h_{<t})$ is lower semicomputable. We can replace V_ξ^* by this expression in Definition 13.2.1, since the factor $\xi_U(e_{<t}||a_{<t})\Gamma_t$ does not affect the effective intelligence order in Definition 13.2.2; all $\dot{\pi}$ just estimate a value scaled by this constant. Also the proof below is not affected, since the factor, which will also appear in front of Q_ξ^*, is independent of a_t.

Theorem 13.2.4 (AIXI*tl* approaches AIXI) If the scaled optimal value function $\tilde{V}_\xi^*(h_{<k})$ is used and is lower semicomputable, then the behavior of AIXI*t̃l̃* approaches the behavior of AIXI in the limit $\tilde{t},\tilde{l},l_P \to \infty$, in some sense.

Proof sketch. Lower semicomputability of $V_\xi^*(h_{<k})$ ensures the existence of sequences of extended programs $\dot{\pi}_1,\dot{\pi}_2,\dot{\pi}_3,...$ for which VA($\dot{\pi}_i$) can be proven and $\lim_{i\to\infty}v_k^{\dot{\pi}_i}=V_\xi^*(h_{<k})$ for all k and all histories $h_{<k}$. One possible such policy sequence can be constructed as follows. Terminate some (non-halting) lower-semicomputing scheme of $Q_\xi^*(h_{<k},a_k)$ after i time steps, and use the obtained approximation as $v_k^{\dot{\pi}_i}$ for the corresponding action $a_k^{\dot{\pi}_i}$ that maximizes the approximate Q-value. The convergence $v_k^{\dot{\pi}_i} \to V_\xi^*(h_{<k})$ for $i\to\infty$ ensures that the universally optimal value $V_\xi^*(h_{<k})$, can be approximated by $\dot{\pi}$ with provable VA(p) arbitrarily well, when given enough time and space. The approximation is not uniform in k, but this does not matter as the selected $\dot{\pi}$ is allowed to change from interaction cycle to cycle. ∎

Conclusion. Similar to other reinforcement learning algorithms, AIXI*tl* is about evaluating policies, but in a quite unusual way, namely provably pessimistic: AIXI*tl* makes sure to have a proof of correctness (valid approximation) and compares it to every other policy (program) which is valid and not over-confident. Importantly, AIXI*tl* does not constrain $\dot{\pi}$ having to estimate model ξ_U and then value $V_{\xi_U}^*$ but can estimate the latter directly in a model-free way or whatever the most efficient way of doing so is. While dove-tailing through all programs to find the best is a well-known idea, making it work for AIXI was non-trivial, since the objective $V_{\xi_U}^*$ itself is incomputable. The lower semicomputability of scaled $\tilde{V}_{\xi_U}^*$ was crucial. Roughly, if some unknown policy π of size $\tilde{l}$ and computation time per cycle $\tilde{t}$ has certain capabilities (e.g. being chess master or AGI or ASI), then the known computable AIXI*t̃l̃* has the same capabilities (e.g. being chess master or AGI or ASI) within the same time frame, up to an (unfortunately very large) constant factor.

13.3 Exercises

1. [C30] Prove (or disprove) for any environment ν, if V_ν^* is Δ_n^0-computable, then there exists an optimal (deterministic) policy π_ν^* for the environment ν that is not Δ_n^0-computable.

2. [C32] Prove (or disprove) for any environment ν, if V_ν^* is Δ_n^0-computable, then for $\varepsilon > 0$ there is no ε-optimal (deterministic) policy π_ν^ε for the environment ν that is Δ_{n-1}^0.

3. [C22] Prove Corollary 13.1.6.

4. [C26] Prove Theorem 13.1.8 that ε-AIXI is not computable.

5. [C40] Various results of this chapter assumed deterministic policies. Prove (or disprove) that they also hold for stochastic policies.

6. [C26] Prove that BayesExp is Δ_3^0.

7. [C40] Derive the computability bounds for Thompson sampling.

13.4 History and References

Computability in RL. This chapter is based on material from [LH18] and [Lei16b, Chp.6]. In this chapter we presented many results around the computability of AIXI and other UAI agents. These results were first derived in [LH15c, LH15d] and then later extended in [Lei16b, LH18]. While this is the first work showing computability results in the UAI framework, traditional RL has many computability and complexity theory results. On the complexity theory side, planning is P-complete in MDPs with finite and infinite horizons [PT87]. [MGLA00] proved (among many other results) that deciding whether a sufficiently good policy exists in MDPs and POMDPs is PSPACE-complete. [LGM01] showed that (unless the polynomial hierarchy collapses, e.g. P=NP), optimal policies for MDPs and POMDPs are not ε-approximable. [SLR07] showed that deciding whether there exists a policy which has a value that exceeds a given value in epistemic POMDPs is PSPACE complete as well as the existence of an optimal policy in an epistemic POMDP are NP. On the computability side, [MHC99, MHC03] showed that in planning with infinite horizon in general POMDPs is even formally undecidable.

On the topic of Solomonoff Induction, [Net22] argued that in Solomonoff prediction the problems of the relative choice of machine and incomputability are both caused by the fact that computable approximations of Solomonoff prediction do not always converge.

Regarding halting and termination of programs, in [Ica17] it was shown that when using probabilistic computation in cognitive science, it is not always desirable to have models which always terminate and argues for some benefits of non-terminating models.

AIXItl. AIXItl (Section 13.2) was first introduced in [Hut00] and described again in [Hut03d, Hut05b, Hut07f, Hut12b, Lei16b]. [Pan08] developed a more practical approximations based on Monte Carlo Tree Search sampling programs. Further replacing the Solomonoff prior by CTW led to MC-AIXI-CTW (Chapter 12). [Kat18] introduced a more functional approximation based on typed lambda calculus and argued it to be even more powerful than AIXItl.

Optimal Solvers and Universal Search.   One key component of AIXI*tl* is that of
universal (Levin) search. Introduced in [Lev73c], universal search is a method designed to
efficiently find a program that can solve a problem provided the solution can be efficiently
verified. There have been many extensions of Universal Search: An adaptive extension
of Levin Search was introduced in [WS96] where it was used to solve POMDPs, [Hut01b,
Hut02b] extends Levin search to a provably fastest algorithm for all well-defined problems
even if the solution cannot be verified efficiently. [Sch04] develops a more practical Adaptive
version of Levin Search (ALS), called Optimal Ordered Problem Solver (OOPS), which
is able to take advantage of previous solutions. Levin Tree Search (LTS) [OLLW18] is a
policy-guided search algorithm with adaptive policy which comes with theoretical guarantees.
The policy may be represented using neural networks (LTS+NN) [OL21] or parameterized
in a convex way using context models (LTS-CM) [OHL23] with theoretical guarantees and
award-winning practical performance on Sokoban, The Witness, and the 24-Sliding Tile
puzzle. Also related to Universal Search is the Gödel Machine [Sch07], a formal self-improving
optimal problem solver. In [Sch05] it was argued how the Gödel machine could be used for a
formal definition of consciousness. Several potential implementations of the Gödel machine
are discussed in [SS11].

Part V

Alternative Approaches

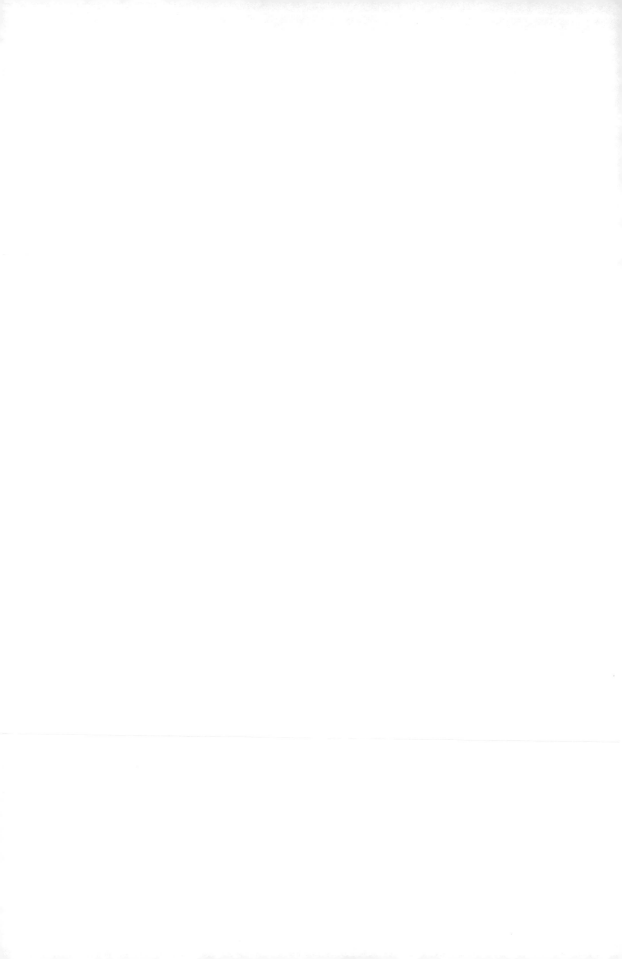

Chapter 14

Feature Reinforcement Learning

*Everything should be made as simple
as possible, but not simpler.*

Albert Einstein

Throughout this book we have been discussing how to use a Bayesian approach to solve the general reinforcement learning problem. We have even gone into the best ways to approximate the otherwise intractable (and often incomputable) agent AIXI. This Bayesian approach is not the only solution to this problem however. In this chapter we will go through an alternative approach, called Feature Reinforcement Learning, which aims to map the difficult general reinforcement problems down to more easily solved problems, such as Markov decision processes. It can be regarded as a more practical substitute for the chronological Solomonoff distribution (Theorem 7.4.5), which also allows to avoid the expensive expectimax planning (Figure 12.2).

What does it mean to map from a general reinforcement learning problem to an easier problem? Quite simply, this is a mapping from histories to smaller spaces, which we will call state spaces. This mapping leads to both agents and environments which depend on states instead of histories. It is plausible that mapping different histories that are sufficiently similar to the same state still allows for a near-optimal agent. The study of feature reinforcement learning is how to best construct these maps, called feature maps from histories to states, and how to determine the quality of a given feature map.

Many immediate questions arise: When are two histories similar, what constitutes a good feature map and corresponding state space, how can they be found effectively, and how can we translate performing well on the state space to back to performing well in the original space? In this chapter we will discuss potential answers to these questions.

After laying out the general framework in Section 14.1, we describe two fundamentally different criteria for feature maps: One which requires the history process to map approximately to an MDP, the other more powerful one constructs a surrogate MDP.

14.1   Feature Reinforcement Learning Setup

Before we get into how to find a good feature map we need to set up the framework and notation we will be using in this chapter.

We consider feature maps $\phi : \mathcal{H} \to \mathcal{S}$ which map from a history space $\mathcal{H} := (\mathcal{E} \times \mathcal{A})^* \times \mathcal{E}$ to a state space $\mathcal{S}$, taking our problem from a general reinforcement problem to a Markov decision problem (MDP) over the state space $\mathcal{S}$. Unless otherwise stated, we assume $\mathcal{S}$ is finite, and ideally small (for some notion of small). In this chapter we use π and μ for the policy and environment in the general reinforcement setting, and $\bar{\pi}$ and $\bar{\mu}$ for the policy and environment in the MDP setting.

History-based (Q-)Values.   The goal of Feature Reinforcement Learning (FRL) is to use solutions in the simplified MDP/state space setting created by the feature map to construct agents with good behaviors in the general reinforcement setting. To do this we need both a notion of what a solution is in the MDP setting, as well as what is meant by good behavior in the general reinforcement setting. Luckily we already have a notion of good behavior in the general reinforcement learning setting, namely maximizing the (discounted) expected future reward sum, called the value (Definition 6.6.1) and Q-value functions (Definition 6.7.1). In this chapter, we use unnormalized infinite-horizon (Q-)Values with geometric discounting in the true environment μ:

$$V_\mu^\pi(h_{<t}) := \mathbf{E}_\mu^\pi \left[\sum_{k=t}^\infty \gamma^{k-t} r_k \Big| h_{<t} \right] = \sum_{a_t} \pi(a_t | h_{<t}) Q_\nu^\pi(h_{<t}, a_t)$$

$$Q_\mu^\pi(h_{<t}, a_t) := \mathbf{E}_\mu^\pi \left[\sum_{k=t}^\infty \gamma^{k-t} r_k \Big| h_{<t} a_t \right] = \sum_{e_t} \nu(e_t | h_{<t} a_t) \left[r_t + \gamma V_\mu^\pi(h_{1:t}) \right]$$

and similarly for the optimal policy $\pi_\mu^* :\in \arg\max_\pi V_\mu^\pi(\epsilon)$.

Markov Decision Process.   We can use this notion to define what we mean by a solution in the MDP setting as well, but first we need to recall what is meant by the MDP setting and a Markov environment (and policy).

A Markov decision process is an environment which depends only on the most recent observation and action. In the case of an environment being Markov, we refer to the observations as states, using the notation s for a state, and use the set $\mathcal{S}$ for the set of observation. We say that an environment $\bar{\mu}$ is Markov if it satisfies the following equation for all histories $h_{<t}$, and states s_t and rewards r_t,

$$\bar{\mu}(s_t r_t | h_{<t}) = \bar{\mu}(s_t r_t | s_{t-1} a_{t-1})$$

If the state is fully observed, i.e. if $o_t = s_t$, as e.g. in Chapter 11, this is called completely observable MDP setting. Here though, the state space $\mathcal{S}$ for the Markov environment $\bar{\mu}$ will be separate from the observation space $\mathcal{O}$ for the environment μ. Since the Markov environment depends only on the most recent state and action, we can abuse notation a bit and write the type of the Markov environment as $\bar{\mu}: \mathcal{S} \times \mathcal{A} \to \Delta(\mathcal{S} \times \mathcal{R})$. Additionally, as mentioned above, we will use $\bar{\pi}$ to denote a *deterministic* Markov policy which only depends on the previous state.

Bellman (optimality) equations. Now we can use the (Q-)Value functions to define what we mean by a solution to the MDP. Since the Markov environment (and policy) depend only on the state (and action) we can write the value and Q-value functions in terms of the state.

Typically the MDP is also assumed to be time-invariant. For this case, it is convenient to introduce the shorthand $s = s_{t-1}$, $a = a_t$, $s' = s_t$, $r' = r_t$. [1]

The Bellman equations (Theorem 6.7.2) then reduce to

$$Q^{\bar{\pi}}_{\bar{\mu}}(s,a) = \sum_{s',r' \in \in \mathcal{S} \times \mathcal{R}} \bar{\mu}(s'r'|sa)[r' + \gamma V^{\bar{\pi}}_{\bar{\mu}}(s')]$$

$$V^{\bar{\pi}}_{\bar{\mu}}(s) = Q^{\bar{\pi}}_{\bar{\mu}}(s,\bar{\pi}(s)) \tag{14.1.1}$$

The last equation only holds for deterministic policies $\bar{\pi}$. It is easy to generalize to stochastic policies, but since we are primarily interested in optimal policies, which can always be chosen deterministic by Definition 6.6.1, we can focus on deterministic policies. The optimal (Q-)Value functions and policy can be written as:

$$Q^*_{\bar{\mu}}(s,a) = \sum_{s',r' \in \in \mathcal{S} \times \mathcal{R}} \bar{\mu}(s'r'|sa)[r' + \gamma V^*_{\bar{\mu}}(s')]$$

$$V^*_{\bar{\mu}}(s) = \max_a Q^*_{\bar{\mu}}(s,a) \tag{14.1.2}$$

$$\pi^*(s) \in \operatorname*{argmax}_a Q^*_{\bar{\mu}}(s,a)$$

14.2 History Aggregation beyond MDPs

Extreme state aggregation [Hut16] is the idea of aggregating histories which have close (Q-)Values. With aggregation we mean mapping (aggregating) many to one mapping from histories to states. The idea is that if we can map (aggregate) histories which have close (Q-)Values to the same state, then solving this new state problem will result in a good solution for the original problem. This is in contrast to many other aggregation methods which try to map similar histories to states based on the contents of the histories, i.e. histories which have the same most recent observations mapping to the same state. Most remarkable, we do not even need to require that the history process μ maps (approximately) to an MDP, but will create a surrogate $\bar{\mu}$ instead. In this section we will go over how this form of aggregation can lead to agents which perform well, and conclude by giving an example of a ϕ which aggregates this way.

14.2.1 Surrogate MDP and Dispersion Probability

To start off we will need additional formalisms. Given an observation space $\mathcal{O}$, action space $\mathcal{A}$, reward space $\mathcal{R}$, environment μ, state space $\mathcal{S}$, and feature map $\phi : \mathcal{H} \to \mathcal{S}$, we

---

[1] Defining $a' = a_t$ would seem more systematic, but we conform here to traditional MDP-based reinforcement learning notation which uses a instead of a', due to their s,a,r order being different from our a,o,r-order.

can construct a feature environment μ_ϕ with the following: Furthermore we let $h \in \mathcal{H}$ be a history of *any* length (unlike $h_{<t}$ which has length t). If h is a history of length t, then $h' = hao'r'$ is its extension to length $t+1$.

Definition 14.2.1 (Feature environment μ_ϕ)

$$\mu_\phi(s'r'|ha) := \sum_{\tilde{o}':\phi(ha\tilde{o}'r')=s'} \mu(\tilde{o}'r'|ha)$$

We have the following condition for when μ_ϕ is an MDP: μ_ϕ is an MDP if there exists an MDP $p:\mathcal{S}\times\mathcal{A}\to\Delta(\mathcal{S}\times\mathcal{R})$ such that for all states $s,s'\in\mathcal{S}$, rewards $r'\in\mathcal{R}$, actions $a\in\mathcal{A}$ and histories $h\in\mathcal{H}$, if $\phi(h)=s$, then

$$\mu_\phi(s'r'|ha) = p(s'r'|sa)$$

The key difference to most other state aggregation methods is that we will *not* require μ_ϕ to be an MDP (not even approximately). We will construct a surrogate MDP based on the following sort of inverse to ϕ.

Definition 14.2.2 (Dispersion probability B) Let $B:\mathcal{S}\times\mathcal{A}\to\Delta\mathcal{H}$ be a probability distribution on histories for each state-action pair, such that $B(h|sa)=0$ if $s\neq\phi(h)$. B may be viewed as a kind of stochastic inverse of ϕ that assigns non-zero probabilities only to $h\in\phi^{-1}(s)$. The formal constraints we pose on B are

$$B(h|sa) \geq 0 \quad \text{and} \quad \sum_{h\in\mathcal{H}} B(h|sa) = \sum_{h:\phi(h)=s} B(h|sa)=1 \quad \forall s,a$$

Crucially note that the sum is over histories of all lengths, which makes B a discrete probability over countable space $\mathcal{H}$, unlike the "continuous" probability measure $\mu(h):=\mu(\Gamma_h)$ over infinite histories defined via cylinder sets Γ_h.

Definition 14.2.3 (Surrogate MDP) Given feature environment μ_ϕ and dispersion probability B, we define the surrogate MDP:

$$\bar{\mu}(s'r'|sa) := \sum_{h\in\mathcal{H}} \mu_\phi(s'r'|ha)B(h|sa)$$

Remark 14.2.4 (If μ_ϕ is an MDP, then $\bar{\mu}=\mu_\phi$) The notion of surrogate MDP is consistent in the sense that if μ_ϕ is already an MDP, then $\bar{\mu}=\mu_\phi$ for any choice of dispersion probability B, since

$$\bar{\mu}(s'r'|sa) \equiv \sum_{h:\phi(h)=s} \mu_\phi(s'r'|ha)B(h|sa)$$

$$= \sum_{h:\phi(h)=s} \mu_\phi(s'r'|sa)B(h|sa)$$

$$= \mu_\phi(s'r'|sa) \sum_{h:\phi(h)=s} B(h|sa) = \mu_\phi(s'r'|sa)$$

where the second equation holds iff μ_ϕ is an MDP. In the following we will *not* make this assumption.   ●

We will now provide a key lemma that lets us relate μ_ϕ to $\bar{\mu}$ via B.

Lemma 14.2.5 (Dispersion probability equivalence) For any function $f : \mathcal{S} \times \mathcal{R} \to \mathbb{R}$, any feature map ϕ, feature function μ_ϕ defined via Definition 14.2.1, surrogate MDP $\overline{\mu}$ via Definition 14.2.3, with $s' := \phi(h')$ and $h' = hao'r'$, it holds

$$\sum_{h \in \mathcal{H}} B(h|sa) \sum_{o',r'} \mu_\phi(o'r'|ha) f(s',r') = \sum_{s',r'} \overline{\mu}(s'r'|sa) f(s',r')$$

Proof.

$$\sum_{h \in \mathcal{H}} B(h|sa) \sum_{o',r'} \mu(o'r'|ha) f(s',r')$$

$$\overset{(a)}{=} \sum_{h \in \mathcal{H}} B(h|sa) \sum_{s',r'} \sum_{o':\phi(h')=s'} \mu(o'r'|ha) f(s',r')$$

$$\overset{(b)}{=} \sum_{h \in \mathcal{H}} B(h|sa) \sum_{s',r'} \mu_\phi(s'r'|ha) f(s',r')$$

$$\overset{(c)}{=} \sum_{s',r'} \overline{\mu}(s'r'|sa) f(s',r')$$

For (a) we sum over all o' by first summing over all o' such that $\phi(hao'r') = s'$ then summing over all s'. In (b) we used Definition 14.2.1. In (c) we used Definition 14.2.3. ∎

Additionally with the dispersion probability B we will define the B-average of a Q-value for all histories $\tilde{h}$ that ϕ maps to the same state as h.

$$\langle Q(h,a) \rangle_B := \sum_{\tilde{h} \in \mathcal{H}} B(\tilde{h}|\phi(h)a) Q(\tilde{h},a) \quad \text{where} \quad s = \phi(h) \tag{14.2.6}$$

Now that we have our notation set, we can move on to the important theorems of extreme state aggregation. Firstly we need to introduce a lemma relating the (Q-)Value functions.

Lemma 14.2.7 (B-average bounds) For any μ, ϕ, B, π, $\overline{\pi}$, define $\overline{\mu}$ via Definitions 14.2.1 and 14.2.3. Then

If $\quad |V_\mu^\pi(h) - V_{\overline{\mu}}^{\overline{\pi}}(s)| \leq \delta \quad \forall s = \phi(h) \quad$ then $\quad |\langle Q_\mu^\pi(h,a) \rangle_B - Q_{\overline{\mu}}^{\overline{\pi}}(s,a)| \leq \gamma\delta \quad \forall s = \phi(h), a$

If $\quad |V_\mu^*(h) - V_{\overline{\mu}}^*(s)| \leq \delta \quad \forall s = \phi(h) \quad$ then $\quad |\langle Q_\mu^*(h,a) \rangle_B - Q_{\overline{\mu}}^*(s,a)| \leq \gamma\delta \quad \forall s = \phi(h), a$

Proof. Let $s := \phi(h)$ and $s' := \phi(h')$ and $h' = hao'r'$. Then[2]

$$\langle Q_\mu^\pi(h,a) \rangle_B \overset{(a)}{=} \sum_{\tilde{h} \in \mathcal{H}} B(\tilde{h}|\phi(h)a) Q_\mu^\pi(h,a)$$

$$\overset{(b)}{=} \sum_{\tilde{h} \in \mathcal{H}} B(\tilde{h}|\phi(h)a) \sum_{o',r'} \mu(o'r'|ha)(r' + \gamma V_\mu^\pi(h'))$$

$$\overset{(c)}{\lessgtr} \sum_{\tilde{h} \in \mathcal{H}} B(\tilde{h}|\phi(h)a) \sum_{o',r'} \mu(o'r'|ha)(r' + \gamma(V_{\overline{\mu}}^{\overline{\pi}}(s') \pm \delta))$$

---

[2] $c \lesseqgtr a \pm b$ means $c \leq a + b$ and $c \geq a - b$.

$$\stackrel{(d)}{=} \sum_{s'r'} \overline{\mu}(s'r'|sa)(r'+\gamma V_{\overline{\mu}}^{\overline{\pi}}(s'))+\gamma\delta$$

$$\stackrel{(e)}{=} Q_{\overline{\mu}}^{\overline{\pi}}(s,a)\pm\gamma\delta$$

(a) comes from (14.2.6). (b) is the Bellman equation for Q-values. (c) is from the assumptions of the lemma. (d) comes from Lemma 14.2.5. Lastly, (e) is from (14.1.1). The above lower and upper bounds imply $|\langle Q_{\mu}^{\pi}(h,a)\rangle_B - Q_{\overline{\mu}}^{\overline{\pi}}(s,a)| \le \gamma\delta$.

The second result is found by replacing π with π_{μ}^{*} and $\overline{\pi}$ with $\overline{\pi}_{\overline{\mu}}^{*}$. Note that in general $\pi_{\mu}^{*} \ne \overline{\pi}_{\overline{\mu}}^{*}$. ∎

14.2.2   (Q-)Value Inheritance for Fixed and Optimal Policy

Next we have that when our Q-values are close for matching histories (matching here meaning histories which ϕ maps to the same state), then our Q and V values will be close for the corresponding states.

Theorem 14.2.8 (Q-value inheritance [Hut16, Thm.5]) For any μ, ϕ, and B, define $\overline{\mu}$ via Definitions 14.2.1 and 14.2.3. Let π be some policy such that $|Q_{\mu}^{\pi}(h,a) - Q_{\mu}^{\pi}(\tilde{h},a)| \le \varepsilon_q$ and either $\pi(h) = \pi(\tilde{h})$ with $\varepsilon_v = 0$ or $|V_{\mu}^{\pi}(h) - V_{\mu}^{\pi}(\tilde{h})| \le \varepsilon_v$, for all $\phi(h) = \phi(\tilde{h})$ and all a. Then for all a and h it holds:

$$|Q_{\mu}^{\pi}(h,a) - Q_{\overline{\mu}}^{\overline{\pi}}(s,a)| \le \frac{\varepsilon}{1-\gamma} \quad \text{and} \quad |V_{\mu}^{\pi}(h) - V_{\overline{\mu}}^{\overline{\pi}}(s)| \le \frac{\varepsilon}{1-\gamma}$$

where $\overline{\pi}(s) = \pi(h)$ and $s = \phi(h)$ and $\varepsilon = \varepsilon_q + \varepsilon_v$.

Proof. Let

$$\delta := \sup_{s=\phi(h),a} |Q_{\mu}^{\pi}(h,a) - Q_{\overline{\mu}}^{\overline{\pi}}(s,a)| \tag{14.2.9}$$

First we show

$$|V_{\mu}^{\pi}(h) - V_{\overline{\mu}}^{\overline{\pi}}(s)| \le \delta + \varepsilon_v \quad \forall s = \phi(h) \tag{14.2.10}$$

Under either condition this follows from (6.7.4) and (14.1.1) and the definition of the supremum. For the or-condition, let $\overline{\pi}(s) := \pi(\tilde{h})$ for some $\tilde{h} \in \phi^{-1}(s)$. Then

$$|V_{\mu}^{\pi}(h) - V_{\overline{\mu}}^{\overline{\pi}}(s)| \stackrel{(a)}{\le} |V_{\mu}^{\pi}(\tilde{h}) - V_{\overline{\mu}}^{\overline{\pi}}(s)| + \varepsilon_v \stackrel{(b)}{=} |Q_{\mu}^{\pi}(\tilde{h},\tilde{a}) - Q_{\overline{\mu}}^{\overline{\pi}}(s,\tilde{a})| + \varepsilon_v \stackrel{(c)}{\le} \delta + \varepsilon_v$$

where (a) comes from the assumptions of the theorem and the triangle inequality, (b) comes from (6.7.4) and (14.1.1) with $\tilde{a} = \overline{\pi}(s) = \pi(\tilde{h})$, and (c) comes from the definition of δ.

Now for both cases, (14.2.10) implies

$$|\langle Q_{\mu}^{\pi}(h,a)\rangle_B - Q_{\overline{\mu}}^{\overline{\pi}}(s,a)| \le \gamma(\delta + \varepsilon_v) \quad \forall s = \phi(h) \tag{14.2.11}$$

by Lemma 14.2.7. By the assumption on Q_{μ}^{π} and B, for $s = \phi(h)$ we have[3]

$$\langle Q_{\mu}^{\pi}(h,a)\rangle_B = \sum_{\tilde{h}:\phi(\tilde{h})=s} B(\tilde{h}|ha)Q_{\mu}^{\pi}(\tilde{h},a)$$

$$\lessgtr \sum_{\tilde{h}:\phi(\tilde{h})=s} B(\tilde{h}|ha)(Q_{\mu}^{\pi}(h,a)\pm\varepsilon_q)$$

$$= (Q_{\mu}^{\pi}(h,a)\pm\varepsilon_q)$$

---

[3]the symbol $\lessgtr$ is used to mean $<$ for the case of $+\varepsilon$ and $>$ for the case of $-\varepsilon$

Together with (14.2.11) this implies $|Q_\mu^\pi(h,a)-Q_{\bar\mu}^{\bar\pi}(s,a)|\le\gamma(\delta+\varepsilon_v)+\varepsilon_q$, hence $\delta\le\gamma(\delta+\varepsilon_v)+\varepsilon_q$, and rearranging we get $\delta\le\frac{\varepsilon_q+\gamma\varepsilon_v}{1-\gamma}$. Then inserting this upper bound on δ into (14.2.9) and (14.2.10) gives us the bounds in the theorem. ∎

Secondly, we have a theorem which is similar to the one above, but for when the values are close on aggregated histories, then our resulting V and our Q-values are close in expectation for the corresponding states, that is, the "closeness" of aggregated states is inherited by the value function for the MDP $\bar\mu$.

Theorem 14.2.12 (Value inheritance [Hut16, Thm.6]) For any μ, ϕ, and B, define $\bar\mu$ via Definitions 14.2.1 and 14.2.3. Let π be some policy such that $|V_\mu^\pi(h)-V_\mu^\pi(\tilde h)|\le\varepsilon$ for all $\phi(h)=\phi(\tilde h)$. Then for all a and h it holds:

$$|V_\mu^\pi(h)-V_{\bar\mu}^{\bar\pi}(s)| \le \frac{\varepsilon}{1-\gamma} \qquad \text{and} \qquad |Q_{\bar\mu}^{\bar\pi}(s,a)-\langle Q_\mu^\pi(h,a)\rangle_B| \le \frac{\varepsilon\gamma}{1-\gamma}$$

where $\bar\pi(s)=\pi(h)$ and $s=\phi(h)$.

Example 14.2.13 (Aggregation to a non-MDP) Consider a process μ which itself is an MDP in the observations with transition matrix T and reward function R, i.e. $\mu(o'r'|ha)=T_{oo'}^a R_{oo'}^{ar'}$, where o is the last observation in h. The example below has the special form $\mu(o'r'|ha)=T_{oo'}\cdot[\![r'=R(o)]\!]$ with

$$T = \begin{pmatrix} \cdot & 1/2 & 1/2 & \cdot \\ 1/2 & \cdot & \cdot & 1/2 \\ \cdot & 1 & \cdot & \cdot \\ 1 & \cdot & \cdot & \cdot \end{pmatrix}, \quad R = \begin{pmatrix} \frac{\gamma/2}{1+\gamma} \\ \frac{1+\gamma/2}{1+\gamma} \\ 0 \\ 1 \end{pmatrix},$$

It is an action-independent Markov process T with deterministic reward function R. The observation space is $\Omega=\{00,01,10,11\}$. For instance, $T_{00,01}=1/2$ and $R(00)=\frac{\gamma/2}{1+\gamma}$. Consider the reduction

$$s_t := \phi(h_t) := \begin{cases} 0 & \text{if } o_t=00 \text{ or } 10 \\ 1 & \text{if } o_t=01 \text{ or } 11 \end{cases} \in \mathcal{S} := \{0,1\}$$

The reduced process μ_ϕ is not (even approximately) Markov:

$$\mu_\phi(s'=0|o=00) = T_{00,00}+T_{00,10} \qquad = 0 + 1/2 = 1/2$$
$$\mu_\phi(s'=0|o=10) = T_{10,00}+T_{10,10} \qquad = 0 + 0 = 0$$

$\ne$

That is, μ violates the bisimulation condition [GDG03], and raw states 00 and 10 have a large bisimulation distance [FPP04, Ort07]. This essentially means that the reduced feature environment μ_ϕ is *not* an MDP. On the other hand, the (Q-)Value function $V(o_t):=V^\pi(h_t)=Q^\pi(h_t,a_t)\forall a_t$ can easily be verified to be

$$V(00) = V(10) = \frac{\gamma}{1-\gamma^2} \qquad \text{and} \qquad V(01) = V(11) = \frac{1}{1-\gamma^2}$$

That is, V and Q are ϕ-uniform. The conditions of Theorems 14.2.8 and 14.2.12 are satisfied exactly ($\varepsilon=0$), and hence the four raw states Ω can be aggregated into two states $\mathcal{S}$ despite $\mu_\phi\notin$MDP (the policy is irrelevant and can be chosen constant). ♦

The previous two theorems were general in that they are about any policy π for the history space, and its version in the MDP space $\bar{\pi}(\phi(h)) = \pi(h)$. What we are really interested in is the optimal policies in the history and MDP space, and how they relate. Ideally, we want our optimal policy in the MDP space to be as close to the optimal policy in the history space as possible. Luckily the next theorem provides this.

Theorem 14.2.14 (Optimal Q-value inheritance [Hut16, Thm.8]) For any μ, ϕ, and B, define $\bar{\mu}$ via Definitions 14.2.1 and 14.2.3. Assume $|Q_\mu^*(h,a) - Q_\mu^*(\tilde{h},a)| \leq \varepsilon$ for all $\phi(h) = \phi(\tilde{h})$ and all a. Then for all a, h and $s = \phi(h)$ the following hold:

- $|Q_\mu^*(h,a) - Q_{\bar{\mu}}^*(s,a)| \leq \frac{\varepsilon}{1-\gamma}$ $\quad$ and $\quad$ $|V_\mu^*(h) - V_{\bar{\mu}}^*(s)| \leq \frac{\varepsilon}{1-\gamma}$

- $0 \leq Q_\mu^*(h,a) - Q_\mu^{\tilde{\pi}}(h,a) \leq \frac{2\varepsilon\gamma}{(1-\gamma)^2}$ $\quad$ and $\quad$ $0 \leq V_\mu^*(h) - V_\mu^{\tilde{\pi}}(h) \leq \frac{2\varepsilon\gamma}{(1-\gamma)^2}$

- If $\varepsilon = 0$, then $\pi^*(h) = \bar{\pi}_{\bar{\mu}}^*(s)$

where $\tilde{\pi}(h) := \bar{\pi}_{\bar{\mu}}^*(s)$.

The above theorem shows that when the map ϕ aggregates histories with close Q-values then the (Q-)Values of the optimal policy in $\bar{\mu}$ and the optimal policy in μ are also close. The theorem implies that we can aggregate histories as much as we wish, as long as the optimal Q-value function and policy are still approximately representable as functions of aggregated states. Whether the reduced process μ_ϕ is Markov or not is immaterial. We can use surrogate MDP $\bar{\mu}$ to find an ε-optimal policy for μ. This will prove useful when constructing a map ϕ which satisfies the assumptions of the theorem.

Remark 14.2.15 (No optimal Value inheritance) The Q-value bound rather naturally generalized to the optimal Q-value bound. Somewhat surprisingly the Value bound does not generalize to optimal policies [Hut16, Thm.10]. $\qquad\qquad\qquad\qquad\bullet$

14.2.3   Extreme State Aggregation

We now show how powerful Theorem 14.2.14 is: We will give an example of an extreme state aggregation map ϕ that satisfies the premises of Theorem 14.2.14 and discuss some of its properties.

Definition 14.2.16 (Extreme state aggregation) Consider ϕ^{ESA} that maps each history to the vector-over-actions of optimal Q-values $Q_\mu^*(h,\cdot)$ discretized to some finite ε-grid

$$\phi^{ESA}(h) := \left(\left\lfloor \frac{Q_\mu^*(h,a)}{\varepsilon} \right\rfloor \right)_{a \in \mathcal{A}} \in \left\{ 0,..., \left\lfloor \frac{1}{\varepsilon(1-\gamma)} \right\rfloor \right\}^{\mathcal{A}} =: \mathcal{S}$$

The first notable property of the ϕ^{ESA} is that all histories with ε-close Q_μ^*-values are mapped to the same state; this is obvious from our construction.

Given μ, B, and ϕ we can construct $\bar{\mu}$ via Definitions 14.2.1 and 14.2.3. We can then find the optimal policy in $\bar{\mu}$, which as previously stated as $\bar{\pi}_{\bar{\mu}}^*$. Then we can define the uplifted policy $\tilde{\pi}(h) := \bar{\pi}_{\bar{\mu}}^*(\phi(h))$. As a consequence of Theorem 14.2.14 we have that $\tilde{\pi}(h)$ is an ε'-optimal policy in μ, where $\varepsilon' := 2\varepsilon(1-\gamma)^2$. (When $\varepsilon = 0$, Q_μ^* can be shown to be upper bounded by $\frac{1}{1-\gamma}$).

A most useful property of ϕ^{ESA} is a bound on the size of the resulting domain, that is the size of the state space of the MDP.

Theorem 14.2.17 (Extreme ϕ) For every process μ, there exists a feature map ϕ (Definition 14.2.16) namely ϕ^{ESA}, and an MDP $\bar{\mu}$ defined via Definitions 14.2.1 and 14.2.3, whose optimal policy $\bar{\pi}^*$ is an ε'-optimal policy $\tilde{\pi}(h) := \bar{\pi}^*(\phi(h))$ for μ. The size of the MDP $\bar{\mu}$ is bounded (uniformly for any μ) by

$$|\mathcal{S}| \leq \left(\frac{3}{\varepsilon'(1-\gamma)^3}\right)^{|\mathcal{A}|}$$

Proof. If $\varepsilon' > \frac{1}{1-\gamma}$ then any policy is ε'-optimal as $Q^*_\mu \leq \frac{1}{1-\gamma} < \varepsilon'$, so the result is true by choosing a trivial state class $|\mathcal{S}| = 1$. In the case when $\varepsilon' \leq \frac{1}{1-\gamma}$, from the definition of $\mathcal{S}$ we have

$$|\mathcal{S}| = \left(\left\lfloor\frac{1}{\varepsilon(1-\gamma)}\right\rfloor + 1\right)^{|\mathcal{A}|} = \left(\left\lfloor\frac{2}{\varepsilon'(1-\gamma)^3}\right\rfloor + 1\right)^{|\mathcal{A}|} \leq \left(\frac{3}{\varepsilon'(1-\gamma)^3}\right)^{|\mathcal{A}|} \quad \blacksquare$$

Interestingly, these bounds do not depend on the size of the observation space or the reward space or the complexity of μ. This is because the core idea of extreme state aggregation is to aggregate histories with similar (Q-)Values to the same state, regardless of what the original observation (and reward) spaces or resulting Markov property are.

Unfortunately the bound is exponential in the action space size, and due to the high powers in ϵ' and $1-\gamma$ can be very large. But it was shown in [MH21b, MH21a] that if the action space is sequentialized to binary actions, the bound can be improved from exponential to logarithmic in $|\mathcal{A}|$, a double-exponential improvement:

$$|\mathcal{S}| \leq \frac{17(\log|\mathcal{A}|)^3}{\varepsilon'(1-\gamma)^3} \tag{14.2.18}$$

14.2.4 Feature Reinforcement Learning

In practice, in reinforcement learning, we do not know Q^*_μ (which implies the optimal policy is already known), so do not have access to the extreme ϕ-map of Definition 14.2.16. This map is not meant to be used in practice; instead it demonstrates the existence of a ϕ-map which satisfies the premises in our theorems.

Let us for a moment assume that at least ϕ is given. Reinforcement learning *learns* Q^*_μ from samples. If we take a model-based approach, we can first learn $\bar{\mu}$, and then use (14.1.2) to compute Q^*_μ. It turns out that for a very particular choice of B, we can follow any convenient behavior policy (off-policy learning), and a simple frequency estimate, counting (s,a,r',s')-pairs in reduced history $\bar{h}$ (see Section 14.3.1), indeed estimates $\bar{\mu}$, despite being a surrogate MDP only related to the true sampling distribution μ via Definitions 14.2.1 and 14.2.3.

All this still assumed ϕ is given. Ultimately the goal of Feature Reinforcement Learning (FRL) is to also learn a map ϕ which satisfies the premises, and in which histories with close Q^*_μ values are aggregated. However, this again requires first finding Q^*_μ and thus defeating the purpose, a classical chicken-egg problem, but a solvable one: Algorithms to learn ϕ have been proposed in [Hut16] and developed in [Maj21].

14.3 Feature MDP

In the previous section we discussed some ideal properties we want the feature map ϕ to possess (aggregates close Q-values), then used that criteria to construct a ϕ that satisfied those properties. This is not the only way to find/construct a map ϕ. In this section we will provide a method of iteratively constructing a feature map ϕ based on a cost function over various different choices of ϕ [Hut09f, Das16]. A crucial difference is that here we require the aggregation to be approximately MDP. Ideally we want ϕ to map to a small state space but large enough that the resulting state dynamics is (still approximately) an MDP. So the cost functions has to balance between ϕ that have small range and lead to approximate MDPs, and should be minimal for the "best' compromise. This is similar in spirit to the MDL principle.

14.3.1   Feature Learning

To start with we can define our cost function over feature maps given histories as follows.

Definition 14.3.1 (Feature Map Cost) The *Cost* of $\phi \colon \mathcal{H} \to \mathcal{S}$ on $h_{1:t}$ is a measure of the quality of the feature map ϕ given history $h_{1:t}$.

We delay the exact choice(s) of *Cost* until Section 14.3.2 as it is more important to first understand the overall structure of the Feature MDP framework (ΦMDP). In the ideal scenario we would have that $\mathrm{argmin}_\phi Cost(\phi|h_{1:t})$ could be easily computed, however, for most interesting and/or useful *Cost* functions this is not possible, so we must turn to other methods.

Given some Cost function for ϕ we can start to construct our ϕ agent, a feature reinforcement learner. Before we do that however, we need an algorithm to iteratively improve ϕ. Since the space of all possible feature maps is large, instead of doing an exhaustive search through the space of feature maps we will start with a simple feature map and iteratively improve it. One way we could improve our feature map ϕ is to modify the state set $\mathcal{S}$. This can be done in a number of ways, but ultimately we can represent all changes to a set with two operations: addition of states and removal of states. For simplicity and concreteness we consider the class of feature maps $\{\phi_\mathcal{S}\}$, which take the most recent observations as state. More precisely a state space forms a suffix set $\mathcal{S} \subset \mathcal{O}^*$, and $\phi_\mathcal{S}(h_{1:t}) = o_{k:t}$ for the unique k such that $o_{k:t} \in \mathcal{S}$. We can either suitably generalize Definition 4.3.3 to non-binary alphabet, or binarize $\mathcal{O}$ first and assume $\mathcal{S} \subset \mathbb{B}^*$ for large $\mathcal{O}$. As noted earlier, let $s_{>1}$ denote s without the first element.

We will now introduce an example of a ϕ-improvement algorithm, called ΦImprove for the former choice. Although this is the improvement algorithm we will use for the ΦMDP agent, it is not the only possible (or useful) feature map improvement algorithm. ΦImprove works by either splitting a state into several new states, or merges several states into one state (if they are capable of merging) (or in tree jargon, expands or shrinks the suffix tree at some leaf nodes). Then, given this new state space, an alternative ϕ is constructed, called ϕ' which is over the new state space. If ϕ' has a lower *Cost*, the algorithm will output ϕ', otherwise it will output the unchanged ϕ. Instead of using a direct comparison of the costs of ϕ and ϕ', we determine if the difference is larger than the log of a random number chosen at the start of the algorithm, because we expect the ΦImprove algorithm to be run multiple times and choosing a slightly worse ϕ' may lead to much better ϕ after ΦImprove has been run multiple times, i.e. getting out of local minima.

Given a feature map ϕ we can construct the state at each time step as ϕ of the history at that time step, e.g. $s_t = \phi(h_{1:t})$. We can then build a *state history* $\overline{h}_{1:t} := a_1 r_1 s_1 ... a_t r_t s_t$. Given the state history $\overline{h}_{1:t}$ we can approximate the dynamics of the MDP through frequency estimation on the transitions of the MDP. We do this by defining the probability that $\overline{\mu}$ transitions into state s' and reward r' given state s and action a as the number of times this

---

Algorithm 14.1 $\Phi\text{Improve}(\mathcal{S},\phi_{\mathcal{S}},h_{<t})$

Input: State space $\mathcal{S}$
Input: Feature map $\phi_{\mathcal{S}}:\mathcal{H}\to\mathcal{S}$
Input: History $h_{<t}$
Output: Improved Feature map $\phi_{\mathcal{S}'}:\mathcal{H}\to\mathcal{S}'$
  1: Randomly choose a state $s\in\mathcal{S}$
  2: Let p,q be uniform random numbers in $[0,1]$
  3: $\mathcal{S}':=\mathcal{S}$
  4: $\phi_{\mathcal{S}'}:=\phi_{\mathcal{S}}$
  5: **if** $p>1/2$ **then**
  6:     $\mathcal{S}':=(\mathcal{S}\setminus\{s\})\cup\{os:o\in\mathcal{O}\}$ ▷ Split $\{s\}$ into $\{os:o\in\mathcal{O}\}$
  7:     $\phi_{\mathcal{S}'}(h)\in\{os:o\in\mathcal{O}\}$ for all h such that $\phi_{\mathcal{S}}(h)=s$ ▷ chosen by some rule e.g. randomly
  8: **else if** $\{os_{>1}:o\in\mathcal{O}\}\subseteq\mathcal{S}$ **then**
  9:     $\mathcal{S}':=(\mathcal{S}\setminus\{os_{>1}:o\in\mathcal{O}\})\cup\{s_{>1}\}$ ▷ Merge $\{os_{>1}:o\in\mathcal{O}\}$ into $\{s_{>1}\}$
 10:     $\phi_{\mathcal{S}'}(h):=s_{>1}$ for all h such that $\phi_{\mathcal{S}}(h)\in\{os_{>1}:o\in\mathcal{O}\}$
 11: **if** $Cost(\phi_{\mathcal{S}},h_{<t})-Cost(\phi_{\mathcal{S}'},h_{<t})>\log(q)$ **then**
 12:     **return** $\phi_{\mathcal{S}'}$
 13: **else**
 14:     **return** $\phi_{\mathcal{S}}$

---

transition has occurred in the state history $\overline{h}_{1:t}$ (denoted by $N(sar's',\overline{h}_{1:t})$) divided by the number of times the pair sa has occurred in the state history $\overline{h}_{1:t}$ (denoted by $N(sa,\overline{h}_{1:t})$), or 0 if the pair sa has never occurred in $\overline{h}_{1:t}$. Formally we can write $\overline{\mu}$ as follows

$$\overline{\mu}(s'r'|s,a) := \begin{cases} \dfrac{N(sar's',\overline{h}_{1:t})}{N(sa,\overline{h}_{1:t})} & \text{if } \quad N(sa,\overline{h}_{1:t})>0 \\ 0 & \text{otherwise} \end{cases} \qquad (14.3.2)$$

(Strictly speaking this is only an estimate $\hat{\overline{\mu}}$ of $\overline{\mu}$) With our $\Phi\text{Improve}$ algorithm and $\overline{\mu}$ defined, we can now construct our ΦMDP agent. Algorithm 14.2 iteratively interacts with the environment, and between each interaction tries to improve its own feature map ϕ by constructing a modified feature map ϕ' with Algorithm 14.1 and switching to the updated feature map ϕ' if it has a smaller cost than the current feature map ϕ. Then, once ϕ is decided on and the inputs from the environments have been received, the agent adds those to the existing history and then gets the current state by taking ϕ of its current history, builds the MDP environment $\overline{\mu}$ via (14.3.2), and then finds the optimal action given that state and environment.

To ensure sufficient exploration, we can set the reward to be high for unexplored state-action interactions in the estimation of $Q_{\overline{\mu}}^*$. We can solve $Q_{\overline{\mu}}^*$ efficiently with policy iteration, value iteration, or some other method in polynomial time. Note that by the construction of the state space, the agent has always visited every state, but it has not necessarily taken every action in every state.

Using ΦMDP we are able to solve the hard problem of general/history-based reinforcement learning by mapping the problem to the easier MDP setting for which we have efficient solutions.

Algorithm 14.2 ΦMDP Agent

Require: Action Space $\mathcal{A}$
Require: Percept Space $\mathcal{E} = \mathcal{O} \times \mathcal{R}$
Input: Percept stream $e_1, e_2, e_3, \dots$
Output: Action stream $a_1, a_2, a_3, \dots$
 1: $\mathcal{S} := \{\epsilon\}$
 2: $\phi(h) := \epsilon, \phi'(h) := \epsilon$ for all h
 3: $h_0, \overline{h}_0, a_0, e_0 := \epsilon$
 4: **for** $t = 1, \dots$ **do**
 5:    $\phi' := \phi$
 6:    **while** Waiting for next e_t **do**
 7:       $\phi' := \Phi\text{Improve}(\mathcal{S}, \phi', h_{<t})$
 8:       **if** $Cost(\phi' | h_{<t}) < Cost(\phi | h_{<t})$ **then**
 9:          $\phi := \phi'$
10:    Receive e_t from environment μ
11:    $h_{1:t} := h_{<t} a_t e_t$
12:    $s_t := \phi(h_{1:t})$
13:    $\overline{h}_{1:t} := \overline{h}_{<t} a_t r_t s_t$
14:    $\mathcal{S} := \mathcal{S} \cup \{s_t\}$
15:    Construct $\overline{\mu}$ by (14.3.2)
16:    $a_t := \text{argmax}_a Q^*_{\overline{\mu}}(s_t, a)$
17:    Output action a_t

14.3.2   Choices of the Cost Function

Since we are interested in choices of ϕ which allow the resulting ΦMDP agent to perform well, we want ϕ that result in states that capture the dynamics of the underlying environment μ sufficiently well. We could, for example, choose a ϕ which mapped every unique history to a different state. While this will fully capture the dynamics of the environment, we are also interested in learning and solving the resulting MDP, and with this choice the resulting MDP is as hard as the original problem. We could instead choose a ϕ which maps all histories to a single state. The resulting MDP (now a bandit problem) has many effective algorithmic solutions, however, since it does not capture the correct dynamics of the underlying environment μ it will likely not perform well. We need a balance between these two extremes.

To this end we turn to simplicity (as we often do in this book) as a guiding principle. What do we mean by simplicity here, surely both of the above choices of ϕ are simple? Indeed they both are, if we are only interested in states. We are interested in using both the states and rewards of our resulting MDP, however, our ϕ is only responsible for the states; the rewards the ΦMDP agent is given are the true rewards from the environment μ.

We take inspiration from the MDL principle and measure the simplicity of a mapping ϕ given a history $h_{1:t}$ as the size of an encoding of $h_{1:t}$ with ϕ plus the size of encoding ϕ itself. By an encoding of $h_{1:t}$ given ϕ we mean an encoding of the resulting $\overline{\mu}$ given ϕ. Because $\overline{\mu}$ is based on the frequencies of state/reward transitions we can encode the transitions and use the size of that encoding as the size of a (partial) encoding of $h_{1:t}$ with ϕ. This results in a cost function of the form

$$Cost(\phi | h_{1:t}) := CL(\overline{\mu} | \overline{h}_{1:t}) + CL(\phi) \qquad (14.3.3)$$

where CL denotes the code length. Recall the definition of the state history $\overline{h}_{1:t} :=$

$a_1 r_1 s_1 ... a_t r_t s_t$. Since we can decompose $\bar{\mu}(s'r'|sa)$ as $\bar{\mu}(s'|s,a)\bar{\mu}(r'|s,a,s')$, and $\bar{\mu}(s'|s,a)$ and $\bar{\mu}(r'|s,a,s')$ are independent we know that the information of the two of them is the sum of the information of each. Therefore we can rewrite our cost function as

$$Cost(\phi|h_{1:t}) := CL(\bar{\mu}(s'|s,a)|\bar{h}_{1:t}) + CL(\bar{\mu}(r'|s,a,s')|\bar{h}_{1:t}) + CL(\phi)$$

Now to find $CL(\bar{\mu}(s'|s,a)|\bar{h}_{1:t})$ we are interested in the counts of each new state s' given the previous state and action s,a, and similarly the counts of new rewards r' given the previous s,a,s'. There are multiple coding schemes that can be used here such as the Minimal Description Length code, Combinatorial Code, Incremental Code, and Bayesian Code. However, we will not specify a coding algorithm here. This completes the description of $\Phi\text{Improve}(\mathcal{S}, \phi_{\mathcal{S}}, h_{<t})$ Algorithm 14.1 and hence (this instantiation of the) overall Feature MDP Agent Algorithm 14.2. Implementations and experimental results can be found in [Ngu13, Das16].

14.3.3   Feature Dynamic Bayesian Networks

In this section we will go through an example of ΦMDP called Feature Dynamic Bayesian Networks (ΦDBN) [Hut09d, Hut09a].

In the context of reinforcement learning, Dynamic Bayesian Networks are a special case of an MDP. Recall that in an MDP the transition probabilities only depend on the previous state and action. In Dynamic Bayesian Networks the state spaced is factorized and the transition probabilities only depend on specific components of this factorization. These factors are called features. Which probabilities depend on which factors is determined by our choice of ϕ. This structure or sparsity allows learning MDPs with much larger state spaces. DBNs are to MDPs what Bayesian networks are to a full tabular (unstructured) joint distribution.

We specifically consider the case when the state space can be factorized into m binary components. That is, $\mathcal{S} = \mathbb{B}^m$. In ΦDBN each ϕ is a collection of which binary components of the next state depend on which binary components of the previous state.

As shown in Section 14.3.2, we can consider the transition probabilities of the states and rewards separately. We will start with the state component and then move onto the reward component. For the ith element (feature) of a state s, we will use the notation s^i. We will use superscript to denote the bitwise components and subscript to denote the time component. As stated above, given a previous state $s \in \mathcal{S}$, we assume that the individual features of next state $s' \in \mathcal{S}$ are independent of each other (s'^j and s'^i are independent for all i,j), and that each s'^i depends only on a subset of parent features $u^i \in \mathbb{B}^m$, that is, the transitions of the MDP can be factorized and have the structure

$$\bar{\mu}(s'|sa) = \prod_{i=1}^{m} \bar{\mu}^a(s'^i|u^i) \tag{14.3.4}$$

where $\bar{\mu}^a$ is some unknown transition probability for each a.

Since we do not know $\bar{\mu}^a$, we have to use an estimation. We could use frequency estimation like in ΦMDP above, however we will use the following KT-based (Definition 4.1.1) estimate regularizing the counts as it will work better for unvisited transitions:

$$\bar{\mu}^a(s^i|u^i) = \frac{|\{n \leq t : u_{n-1} = u^i, a_{n-1} = a, s_n^i = s^i\}| + 1/2}{|\{n \leq t : u_{n-1} = u^i, a_{n-1} = a\}| + 1}$$

Then using (14.3.4), we can estimate $\overline{\mu}$ and the probability of a sequence of states $s_{1:n}$ given a sequence of actions $a_{1:n}$ by

$$\overline{\mu}(\overline{h}_{1:t}) = \prod_{n=1}^{t} \overline{\mu}(s_n|s_{n-1}a_{n-1}) = \prod_{n=1}^{t}\prod_{i=1}^{m} \frac{|\{n \leq t : u_{n-1} = u^i, a_{n-1} = a, s_n^i = s^i\}| + 1/2}{|\{n \leq t : u_{n-1} = u^i, a_{n-1} = a\}| + 1}$$

For the encoding of rewards we assume that the reward for a next state can be written as the sum of functions that depend only on the individual features:

$$R(s) := \sum_{i=1}^{m} R^i(s^i)$$

This is a reasonable choice as each feature of the state is independent of the other features, and we are assuming that the transitions to new states and rewards only depend on the previous state and action. Since features $s^i \in \{0,1\}$ this is equivalent to a linear function of the feature vector $s \in \mathbb{B}^m$ for any choice of $R^i(\cdot)$. Hence we can the above as

$$R(s) := w_0 + w^\top s = w_0 + w_1 s^1 + ... + w_m s^m$$

for a suitable choice of weight vector $w \in \mathbb{R}^{m+1}$. To determine if our weighting w is good, we need some kind of loss function (Loss) comparing $R(s_t)$ to r_t for all s_t and r_t in the history. The least squares is a sensible choice, but there are many possible choices. The details can be found in [Hut09a]. Like in Section 14.3 we can combine our code lengths of states and rewards to find the code length of a given ϕ.

$$Cost(\phi|h_{1:t}) := CL(\overline{\mu}(s'|s,a)|\overline{h}_{1:t}) + CL(\overline{\mu}(r|s,a,s')|\overline{h}_{1:t})$$

14.4 Context Tree Maximization Reinforcement Learning

Throughout this chapter we have gone over the basics of Feature Reinforcement Learning (FRL). We have yet to demonstrate FRL in practice. To do so we will go through an approach called Context Tree Maximization Reinforcement Learning (CTMRL), which is to FRL what MC-AIXI-CTW is to AIXI. However, the parallels to MC-AIXI-CTW do not end there. CTMRL [Ngu13, NSH12] is a method which finds the maximal context tree (we will get to what we mean by maximal soon) and uses it to construct an MDP, and then takes the optimal action assuming the agent is in that MDP.

In Chapter 12 we have seen how context trees can be used to emulate the dynamics of MDP environments, and that a Bayesian mixture over these context trees can be done efficiently with Context Tree Weighting (CTW), when the true environment was in this mixture (was Markov) the CTW would perform well. In FRL we are not interested in mixing over Markov environments, instead we are assuming that the true environment is non-Markov (completely history dependent), but it has some Markov encoding with an optimal feature map ϕ^*. In this sense the notion of context trees is still useful as our optimal feature map ϕ^* will describe an MDP, and that MDP can be encoded as a context tree. Context Tree Maximization (CTM) is a method to choose the optimal ϕ / context tree.

We have discussed the CTM method in Section 5.5 for binary sequence prediction. Before continuing with this section, it is recommended that readers familiarize themselves with CTM. Here we will go over the feature cost function used, prove that there is an efficient way to compute it, and then describe the agent which uses it.

Cost function. Let $\{\kappa_1,\kappa_2,...,\kappa_{|\mathcal{A}\times\mathcal{E}|}\}=\mathcal{A}\times\mathcal{E}$ denote the ordered set of (action,percept) pairs. Given history $h_{1:t}$ let n_{κ_i} denote the number of occurrences of the action-percept pair κ_i in that history. Let $\mathrm{P}_e^{\kappa|sa}$ denote the block probability estimate of seeing a sequence containing n_{κ_i} number of κ_i for all i given state s and action a. Using $\mathrm{P}_e^{\kappa|sa}$, we can construct the MDP $\bar{\mu}$ which has probability of producing (state) history sequence $\bar{h}_{1:t}$ defined as $\bar{\mu}(\bar{h}_{1:t}|\mathcal{S})=\prod_{a\in\mathcal{A}}\prod_{s\in\mathcal{S}}\mathrm{P}_e^{\kappa|sa}$, given the state set $\mathcal{S}$. Using $\bar{\mu}$ we can define a cost function for $\mathcal{S}$ as follows:

$$Cost(\mathcal{S}|h_{1:t}) = \log\frac{1}{\bar{\mu}(\bar{h}_{1:t}|\mathcal{S})}+\Gamma_D(\mathcal{S}) = -\log\left(\prod_{a\in\mathcal{A}}\prod_{s\in\mathcal{S}}\mathrm{P}_e^{\kappa|sa}\right)+\Gamma_D(\mathcal{S}) \qquad (14.4.1)$$

where $\Gamma_D(\mathcal{S}):=|\mathcal{S}|-1+|\{s:s\in\mathcal{S},\ell(s)\neq D\}|$ is the model penalty of $\mathcal{S}$ with respect to the model class $\mathcal{C}_D$ of context trees of maximal depth D. Note that $s_n=\phi_\mathcal{S}(h_{1:n})$, $n=1,...t$ with $\phi_\mathcal{S}$ being the map extracting the suffix of the history $h_{1:n}$ that matches a suffix or a state in the set $\mathcal{S}$ as in Section 14.3. We can see that (14.4.1) matches (14.3.3).

We will rewrite the Cost function so as to reduce the problem of minimizing it to determining the CTM tree. The solution we are presenting here involves finding the optimal set of states that minimize the cost function, which is done using recursive definitions for the maximizing probability and maximizing state set. From (14.4.1) it is obvious that maximizing $2^{-\Gamma_D(\mathcal{S})}\bar{\mu}(\bar{h}_{1:t}|\mathcal{S})$ is equivalent to minimizing Cost.

Most of the analysis from here on will match that of CTM in Section 5.5, generalized to non-binary alphabet, with $[\prod_{a\in\mathcal{A}}\mathrm{P}_e^{\kappa|sa}]^{-1}$ as the estimator.

Context Tree Maximization. We can define the maximizing probability of a state similarly to Section 5.5 recursively by

$$\mathrm{P}_{m,s}^D := \begin{cases} \frac{1}{2}\max\left\{\prod_{a\in\mathcal{A}}\mathrm{P}_e^{\kappa|sa},\prod_i\mathrm{P}_{m,\kappa_i s}^D\right\} & \text{if } \ell(s)<D \\ \prod_{a\in\mathcal{A}}\mathrm{P}_e^{\kappa|sa} & \text{if } \ell(s)=D \end{cases} \qquad (14.4.2)$$

and the maximizing state set $\mathcal{S}_{m,s}^D$ by

$$\mathcal{S}_{m,s}^D := \begin{cases} \bigcup_{\kappa_i}\mathcal{S}_{m,\kappa_i s}^D\times\kappa_i & \text{if } \prod_{a\in\mathcal{A}}\mathrm{P}_e^{\kappa|sa}<\prod_{a\in\mathcal{A}}\prod_i\mathrm{P}_{m,\kappa_i s}^D \text{ and } \ell(s)<D \\ \{\epsilon\} & \text{otherwise} \end{cases} \qquad (14.4.3)$$

where here $\kappa_i s$ is state s prepended with the action-percept pair κ_i. Expanding these we have the following lemma.

Lemma 14.4.4 (Context Tree Maximization equivalence) For any state s, with where $d=\ell(s)$, we have

$$\mathrm{P}_{m,s}^D = 2^{-\Gamma_{D-d}(\mathcal{S}_{m,s}^D)}\prod_{a\in\mathcal{A}}\prod_{u\in\mathcal{S}_{m,s}^D}\mathrm{P}_e^{\kappa|usa} = \max_{\mathcal{S}\in\mathcal{C}_{D-d}} 2^{-\Gamma_{D-d}(\mathcal{S})}\prod_{a\in\mathcal{A}}\prod_{u\in\mathcal{S}}\mathrm{P}_e^{\kappa|usa}$$

Proof. The proof is left as an exercise. Hint: see Section 5.5. ∎

Now by setting $s=\epsilon$ in the above lemma, we get the CTM tree, which minimizes the Cost function (14.4.1). This can be seen by the following theorem:

Theorem 14.4.5 (CTM minimized Cost) Given a history $h_{1:t}$, up to a time t we have

$$\min_{\mathcal{S}} Cost(\mathcal{S}|h_{1:t}) = \log\frac{1}{\mathrm{P}_{m,\epsilon}^D(e_{1:t}|a_{<t})} = \log\frac{1}{\bar{\mu}(h_{1:t}|\mathcal{S}_m^D(h_{1:t}))}+\Gamma_D(\mathcal{S}_{m,\epsilon}^D(h_{1:t}))$$

Proof.

$$\min_{\mathcal{S}} Cost(\mathcal{S}|h_{1:t}) = \min_{\mathcal{S}} \log \frac{1}{\overline{\mu}(\overline{h}_{1:t}|\mathcal{S})} + \Gamma_D(\mathcal{S})$$

$$= \min_{\mathcal{S}} \log \left(\frac{1}{\prod_{a \in \mathcal{A}} \prod_{s \in \mathcal{S}} \mathrm{P}_e^{\kappa|sa}} \right) + \Gamma_D(\mathcal{S})$$

$$= \min_{\mathcal{S}} -\log \left(\prod_{a \in \mathcal{A} s \in \mathcal{S}} \mathrm{P}_e^{\kappa|sa} \right) + \Gamma_D(\mathcal{S})$$

$$= \max_{\mathcal{S}} 2^{\log(\prod_{a \in \mathcal{A}} \prod_{s \in \mathcal{S}} \mathrm{P}_e^{\kappa|sa}) - \Gamma_D(\mathcal{S})}$$

$$= \max_{\mathcal{S}} 2^{-\Gamma_D(\mathcal{S})} \left(\prod_{a \in \mathcal{A} s \in \mathcal{S}} \mathrm{P}_e^{\kappa|sa} \right)$$

$$= \mathrm{P}_{m,\epsilon}^{D}(e_{1:t}|a_{<t}) \qquad\qquad \blacksquare$$

Since the recursions (14.4.2) can be efficiently computed, we now have an efficient method to find the optimal suffix state set $\mathcal{S}$ via (14.4.3), which in turn minimizes the Cost function (14.4.1) due to Theorem 14.4.5. This works well, provided the action and percept spaces are reasonably small for the KT-estimation good.

Binarization and CTMRL Algorithm 14.3. For large action or percept space, the KT estimate will be poor. In Chapter 12 we had a similar problem, which we solved by binarizing the action and percept spaces, and we can do the exact same thing here.

Let l_e, l_a be the minimum number of bits needed to describe elements of the percept space and action space respectively. The CTMRL algorithm takes as input the environment, a number of learning loops, the number of times to perform Q-learning n_i (and n_q), and l_e. CTMRL starts with a context tree (CTM) for each percept bit, that is l_e different CTMs, then the algorithm iteratively updates the CTMs based on the history, joins them to form a context tree $\mathcal{T}$, which is the smallest context tree to contain $\mathcal{S}_{m,ae[1...i-1]}^{D+i-1}$ for all $i \le l_e$ as subtrees. Then it estimates the state and reward transition probabilities of the MDP model based on the tree S using $\overline{\mu}$. Afterwards, the algorithm uses those probabilities to update the optimal action values with action-value iteration, and then performs Q-learning using the newly-found optimal action values. After the given amount of learning loops are complete, CTMRL performs Q-learning once again and then defines the policy $\pi^*(s)$ as $\text{argmax}_a \hat{Q}^*(s,a)$.

Comparison to AIXI and other agents. CTMRL was compared to MC-AIXI-CTW (Section 12.3), as well as other agents like ΦMDP (Section 14.3), the classic U-Tree algorithm [McC96], and the more recent active LZ algorithm [FMRW10b], in a collection of environments detailed in Section 12.5.1. CTMRL achieved comparable performance with drastically less computation time required, with MC-AIXI-CTW taking on the order of 10 times the amount of time CTMRL took to achieve similar near-optimal performance. This demonstrates that CTMRL is indeed a general, and that the FRL approach lends itself to more efficient approximations than AIXI approximations.

The differences between CTMRL and MC-AIXI-CTW truly highlight the difference in approaches between the AIXI Bayesian approach and the MDL FRL approach. In the AIXI approach we perform learning with $\xi_U \overset{\times}{=} M$ in theory and CTW in practice, and then do planning with expectimax in theory and MCTS in practice. However, in FRL, the learning is done by finding/choosing the feature map ϕ in theory and the maximal context tree in practice, and expensive planning is replaced by solving the MDP (both in theory and

---

Algorithm 14.3 CTMRL [Ngu13]

Require: Environment μ, learning loops m, n_i's, n_q, minimal percept bits l_e
 1: $i := 0$
 2: Create l_e empty CTMs, where the ith CTM predicts the ith bit of the percept e
 3: $h :=$ initial random history by performing n_0 random actions
 4: $h' := h$
 5: **while** $i < m$ **do**
 6:     Update the l_e CTMs based on history h'
 7:     Join learned contexts from each of the CTMs to form an Action-Observation Context Tree $\mathcal{T}$
 8:     Compute frequency estimates of the state transitions and reward probabilities of the MDP model based on states induced from tree $\mathcal{T}$ and history h
 9:     Use Action-Value Iteration to find an estimate of the optimal action values $\hat{Q}$ based on the frequency estimates
10:     $Q := \hat{Q} + \frac{R_{max}}{1-\gamma}$
11:     **if** $i < m-1$ **then**
12:         $h := $Q-learning$(Q, \mathcal{S}^S, \mathcal{A}, \mu, n_i)$
13:         $h := [h, h']$
        $i := i+1$
14: $\hat{Q}' := $Q-learning$(Q, \mathcal{S}^{\mathcal{T}}, \mathcal{A}, \mu, n_q)$
15: $\pi^*(s) :\in \operatorname{argmax}_a \hat{Q}'(s, a)$ for all $s \in \mathcal{S}^{\mathcal{T}}$
16: **return** π^*

---

practice) via Bellman equations. Whether explicit planning or value estimation leads to better policies is domain dependent. Another potential advantage of FRL over AIXI is that Solomonoff's M models the complete history, which is expensive, while FRL only models those aspects that are separated by the feature map.

14.5   Exercises

1. [C25] Prove Theorem 14.2.12.

2. [C25] Prove Theorem 14.2.14.

3. [C32] Using binarized sequential actions to improve Theorem 14.2.17 to $|\mathcal{S}| \leq O((\log|\mathcal{A}|)\varepsilon'^{-2}(1-\gamma)^{-6})$. What is the cost of using binarized sequential actions? The better bound (14.2.18) is based on deeper insights into the structure of the problem [MH21a]. Hint: The reason for the remaining logarithmic dependence on $|\mathcal{A}|$ is that discount γ must be replaced by $\gamma^{1/b}$ when $\mathcal{A}$ is sequentialized into b bits.

4. [C18] Prove Lemma 14.4.4.

5. [C35] Implement CTMRL and compare against MC-AIXI-CTW in several domains.

14.6   History and References

The Feature Reinforcement Learning (FRL) approach, using state and state-action abstractions to solve the general reinforcement learning problem, started with Feature MDPs

[Hut09e, Hut09f] and Feature Dynamic Bayesian Networks (DBNs) [Hut09d, Hut09a]. Since its inception there has been much work along this line, most of which has been included in this chapter. Comprehensive overviews of this topic can be found in [Ngu13, DSH14a, Das16, Maj21].

FRL theory. On the theoretical side of FRL, advances include using state abstractions for prediction and demonstrating their consistency [SH10]. Moving beyond the assumption that the abstracted process is an MDP was done in Extreme State Aggregation [Hut14a] (Section 14.2). Extreme State Aggregation was extended in [MH21b, MH21a] which binarized the action space to gain a double-exponential improvement on a bound on the size of the required state space. Action-state abstractions were investigated in [MH19] where the authors showed when the state-action abstraction can lead to near-optimal performance. The sequel to the original FRL papers [Hut09d, Hut09a] extended [Hut09e, Hut09f] to the more complex structured/factored MDPs, specifically focusing on DBNs. [RLDG22] introduces a method which is able to give PAC guarantees in the episodic setting assuming there exists a Markov abstraction.

FRL practice. On the practical side of FRL, the advances include using a practical implementation of ΦMDP to achieve performance competitive to MC-AIXI-CTW [NSH11], FRL with Context Tree Maximization [NSH12] (Section 14.4), FRL with looping suffix trees to solve the long-term dependency problem [DSH12], studying Q-learning in the general reinforcement learning setting [DSH13, MH18], and extending MC-AIXI-CTW by using state abstractions on the history [YZWN22]. Code for the approximations presented in this chapter is available from the book's website hutter1.net/ai/uaibook2.htm.

Regret results. There are many methods and ways that the performance of an agent or an abstraction can be measured. One such way is with regret (Section 8.1.4). In [MMR11] the authors introduced an algorithm for the FRL setting which is able to achieve regret $O(T^{2/3})$. This bound was improved when [NOR13] proposed an algorithm which achieved regret of $O(T^{1/2})$, which is an optimal regret bound, that is, this cannot be improved upon at least in an asymptotic sense. Both of the previous approaches relied on being given a finite set of abstractions, but [NMRO13] extended this to a countably infinite set of abstractions. Many approaches to the FRL problem consider exact abstractions, however we are first and foremost interested in agents which perform well even if the abstraction is not exact. To this end [OMR14] derived results for approximate abstractions. [OPL+19] introduced a new algorithm which is able to achieve comparable regret bounds as previous algorithms and possibly has better bounds depending on the effective size of the induced state space from the abstraction.

Solving MDPs with state abstractions. Feature reinforcement learning focuses on solving the general reinforcement learning problem through state abstractions, however, there has been much work on solving more simple MDP (and POMDP) problems through state abstractions. [SLO22] provides a survey on many of the model-based methods for this problem. As there are so many ways to tackle this problem, it is important that there is some unified way to compare the different approaches, and [LWL06] began this unification process by comparing five different approaches. For similar reasons as mentioned previously, it is useful to investigate approximate abstractions, which are studied in [AHL16]. One of the difficulties with continual/life-long learning is the ever-growing history which the agent must condense and use to learn many different tasks. State abstractions are one way to alleviate these difficulties for the agent [AALL18]. State abstraction as compression was investigated through the lens of rate distortion theory in [AAA+19] where the trade-off between compression and value was made explicit. In MDP reinforcement learning, one of the bottlenecks is that the state space is quite large, hence the use of state abstractions. However, it can also be the case that the action space is very large. In these situations it is natural to consider a state-action abstraction which abstracts state-action pairs instead of the usual state abstractions [MH19, AUK$^+$20].

On the more practical side, an exploration strategy based on the feature representation of the states was employed to provide state-of-the-art performance in the Atari environment [MSEH17], and state abstractions were used to improve sample-based search by reducing the size of the state space [HFD17].

Classical reinforcement learning. A general introduction to classical state-based RL can be found in [SB18] (see Section 6.9 for more advanced books). It describes the basic ideas, approaches, and algorithms: Multi-armed bandits (Section 12.2.3), finite MDPs (Chapter 11), Bellman equations (Section 14.1), MCTS planning (Section 12.2.4), TD(λ) [Sut88, HL07], Q(λ) [WD92, PW94, DSH13, MH18], SARSA [RN94], approximation methods (see below), policy gradients [Wil92, KHS01a], some applications, and dichotomies: off- vs. on-policy, evaluation vs. control, exploration vs. exploitation, and model-based vs. model free.

Other reinforcement learning ideas and algorithms. The landscape of reinforcement learning algorithms is diverse, with each offering distinct advantages and disadvantages. A key objective in the field is to identify the most effective algorithm for a given task. See Section 6.9 for a list of books on RL. Below we mostly focus on RL ideas not (yet) covered by books.

The most widely used abstractions in RL are not discrete state/history aggregations, but approximations based on continuously parameterized (Q-)value functions. Unfortunately, even for *linear* function approximation, convergence guarantees hard to come by: Even if the value function can be represented exactly, model-free Temporal-Difference-style algorithms cannot be guaranteed to converge to the correct solution, unless the linear basis functions are essentially state aggregators [HYZM19]. Model-based and some brilliant but "unnatural" algorithms such as ETD [SMW16] have convergence guarantees for linear function approximation. For state aggregation, one can show [MH18] that Q-learning converges even if the aggregated dynamics is non-Markov like in Section 14.2. [HL07] derives a unique state-dependent learning rate for TD(λ) via a variational principle and bootstrapping. Rather than hard aggregation or smooth function approximation, one can also impose or use a metric one the observation or state space, which is natural in RL for robotics [Thr02]. [ZGHS06] goes further and extends [McC96] to general metrics over histories and hence continuous POMDPs and an RL algorithm that learns directly on a mobile robot.

An (unorthodox) RL approach is to learn the sequence of actions that most likely causes the agent to transition from state s to state s'. These inverse MDP models automatically only model the acton-relevant aspects of the world [HH22].

With this idea in mind, [OHC$^+$20] presented a method to meta-learn the update rule used, as opposed to having a fixed update rule. Imagination-based planning was introduced in [PLV$^+$17] as a method for agents to construct and evaluate plans without taking any real actions. An alternative imagination approach to planning was studied in [SvHH$^+$16] which learned a model for planning purposes only; this model can use different state, action, reward space, and time steps.

In [OIO19] a simple way to interpret RL as inference was established. Distributional RL [BDM17, BDR23, WDA$^+$23] uses Bellman equations [Bel57] for distribution of returns, and argues why this improves performance even if we are only interested in values which are expected returns. Related to distributional RL is the topic of Upside-down RL [Sch19a]. RLAdvice [DSH14b] is an imitation learning algorithm based on (Bayesian) DAgger using value estimates from UCT as advice applied to three Atari games from the Arcade Learning

Environment (ALE).

There are many approaches to the problem of building artificial general intelligences [GP07, AGI08], with [SBP22, LeC22] just being two prominent recent examples.

Part VI

Safety and Discussion

Chapter 15

ASI Safety

In this chapter, we delve into some safety results for universal artificial intelligence. We would expect that very intelligent agents would take actions to further their goals, posing a potential hazard if those goals are unaligned with that of humans. It seems almost paradoxical for an agent to be both intellectually capable enough to automate most tasks that humans can do (which would necessitate having a lot of influence over its environment), but also be docile enough to allow humans to tell it what to do. An agent with goals perfectly aligned with ours would not need to be kept in control (or even told what to do), as it would already take whatever actions we would want it to if it is aligned with ours.

Ideas around societies with or of intelligent robots have been entertained for centuries, especially in science fiction, and have kept the imagination of scientists. The technological singularity (Section 15.1) is the extrapolation to a self-accelerating intelligence explosion, hence quite relevant for framing the ASI safety discussion. We give a brief overview of ASI safety sub-topics in Section 15.2 to be discussed in the subsequent sections: The Control Problem (Section 15.3) covers the difficulties of trying to control an agent that is more intelligent than the operator, and how this differs from other potential catastrophic risks. Instrumental Convergence (Section 15.4) is the hypothesized property of super-intelligent agents with different or even opposing goals to nevertheless share a set of common convergent sub-goals. Orthogonality Thesis (Section 15.5) is the conjecture that an agent can in principle have any combination of intelligence and goals. We explore the distinctions between Value, Reward, and Utility (Section 15.6). Additionally, we delve into more complex themes such as how to formalize the notion of death for universal agents (Section 15.7) and the expected behavior of the agent in the face of mortality. We look at the conditions under which universal agents may choose to self-modify (Section 15.8), hijack the received rewards (Section 15.9) or tamper with the received percepts to delude itself (Section 15.10). We look at how a reward function that does not fully encapsulate the desired behavior for the agent (Section 15.11) can pose a risk. Lastly, in Section 15.12 we touch upon the concept of Embedded Intelligence, an attempt for formalize universal agents that are part of, and computed by, the environment with which they interact.

Throughout this chapter, the emphasis remains on highlighting the potential existential threats that an ASI may pose, within the framework of universal artificial intelligence.

15.1   The Technological Singularity

The book is primarily about Universal AI and AIXI, the theoretically most intelligent agent possible. As such, this chapter focusses on safety issues regarding Artificial General Super Intelligence (AGSI), ranging from somewhat super-human level up to AIXI. The technological singularity is the (hypothetical) scenario in which, once we have created human level AGI's, they will autonomously advance technology, in particular compute power and algorithms, in a self-accelerating way, causing a technology, intelligence, and speed explosion, radically changing society, which will become incomprehensible to us current humans. Of particular interest is whether this future is utopia or dystopia, and for whom. The only certainty seems to be that this future will be very different from the past, which makes forecasts particularly difficult.

There are many different potential paths towards AGSI (machine/deep learning, evolutionary systems, mind uploading, awakening of the internet, cyborgs, ...) and less plausible paths (traditional AI, physical brain enhancement via drugs or genetic engineering, ...). There are also many different potential ways AGSI's may interact with humans: serve humans, peacefully co-exist with humans, form relationships with humans, guide or protect humanity, be our mind children, mind their own business, merge with humans to cyborgs or transhumans, keep humans as pets, or enslave or decimate or eradicate humanity. Some of them are briefly discussed in the book's introduction and in this chapter, and all of them and more feature in many science fiction. How these societies will evolve towards a singularity, i.e. a society of ultra-intelligent beings, may be completely unimaginable; possibly a hard prediction barrier. Still some general aspects may be predictable.

For instance, a hyper-advanced virtual world may look like random noise to humans watching them from the "outside", if they use compressed or encrypted communication. If so, what does it mean for intelligence to explode for an outside observer? Conversely, can an explosion actually be felt from the "inside" if everything and everyone is sped up uniformly? If neither "insiders" nor "outsiders" experience an intelligence explosion, has one actually happened? Of course one would also expect qualitative/algorithmic progress unrelated to raw compute power. The explosion may actually be an inward implosion until computronium is reached, or outward until all accessible convertible matter has been used up, or both. There may also be information-theoretic limits: The library of babel contains all and hence no information, and anyway would collapse into a black hole.

AIXI is the most intelligent agent (given infinite compute). Does this mean intelligence is upper bounded? Does this prevent an intelligence singularity? Maybe Reflective AIXI (Section 10.7) allows us to theoretically study already today how a society right at the edge of an intelligence singularity might look like. Indeed, a number of social questions regarding AIXI have been asked and answered, some of them already formally: AIXI will listen to trustworthy teachers, procreate if useful, self-improve, be manipulative, curious, not lazy, self-preserve, socialize, and may commit suicide or hack its reward system depending on circumstances (Chapter 16).

Even when setting up a virtual (computer simulated) society in our image, there are likely some immediate differences. There could be a "Cambrian" explosion towards diverse forms of intelligent agents. Or they may all converge to a single optimal design fixed point, which might be AIXI, or slime mold (intelligence is not necessarily an evolutionary dominant trait). Maybe there will only be one intelligent agent using up all accessible (compute) resources. Maybe there will be a hierarchy of sub-agents, and it is hard to tell whether these constitute parts of a single agent or multiple agents. What will super-intelligences actually do? Which activities does evolution select for? Self-preservation, self-replication, spreading, colonizing the universe, creating faster/better/higher intelligences, learning as much as possible, understanding the universe, maximizing power over humans and/or organizations, transformation of matter (into computronium?), maximum self-sufficiency, search for the meaning of life? If AGSI are expected reward maximizers like AIXI, where do the rewards come from? Evolved biological goals and desires are to survive, procreate, parent, spread, dominate, which means fierce competition over scarce resources. Biological life is rather brutal, far from the idyllic zoos and depictions in children's books. Are there other more benign stable goals? Maybe some central powerful control can ensure peace or even the makeup of its population, or there is only a single AGSI at all.

There is increasing concern about the moral value and ethical treatment of future AGIs (a different kind of AGI safety, one for the AGIs), though extrapolating from humans to machines may be tricky. Copying and manipulation of virtual structures, including virtual life, could be as cheap and effortless as it is for software and data today. Much of our society is driven by the fact that we highly value (human/individual) life, in parts because it is expensive/laborious to replace/produce/raise. If something becomes cheap, motivation to value it will decline, and behavior and society will alter drastically, especially if backups ensure immortality. So it may actually remain ethically acceptable to freeze, duplicate, slow/shut down, modify, delete (oneself or other) AIs at will, just what we are doing in computer games and used to doing with software. With little value assigned to an individual life, it may become a disposable, and even AGIs may have no interest in laws for their ethical treatment.

Many arguments for the possibility of AGSI/Singularity have been brought forward: the physical Church Turing thesis, Moore's/Solomonoff's laws of exponential/hyperbolic compute growth, Hanson's acceleration of doubling patterns, and Kurzweil's accelerating universe epochs. And against: Structural obstacles (limits in intelligence space, failure to take off, diminishing returns, local maxima), manifestation obstacles (disasters, disinclination, active prevention), correlation obstacles (speed or technology vs. intelligence), physical limits (though converting our planet into computronium would still result in a vastly different world and a near-singularity), and hardware or software engineering difficulties (still one or more phase transitions a la Hanson may occur).

Also, how much control do we have over the future? In theory, humans and humanity have a lot of agency, but politically it is very difficult to steer technology or resist market forces. Some major catastrophes can change sentiments on a planetary scale, but we usually do not deliberately control the catastrophes either. Even if we have a lot of control, we may seriously misjudge its effects, so control may even be counter-productive. Finally, which futures are desirable is highly subjective. A prosperous future controlled and safeguarded by superior machines, or transhumans, are utopia for some, but dystopia for others. Are there any universally accepted values or qualities that should be preserved?

See [Cha10, Hut12a] for an elaboration of the above topics. This section has hopefully made clear how difficult it is to reason about futures with ASI in it, which makes ASI safety research very difficult and speculative. This chapter presents some highlights what has been achieved so far.

15.2   Safety Subtopics

One aspect of artificial intelligence which is often discussed is the safety concerns that arise from hypothetical future AI systems. The potential reasoning and agency that a sufficiently powerful AGI would have may greatly outstrip that of humanity. Whether this leads to a good or bad outcome for humanity remains to be seen. The idea of a powerful ASI often conjures the image of a dystopia or "killer robots", popularized with the *Terminator* series and a now oft-repeated cliché present as a trope in media. More recently, media adaptations have explored machines that can empathize with and form bonds with humans (see the movies *Her, Blade Runner 2049, Ex Machina*). Losing control over benevolent ASI seems to be a future few have anything to say about (see the movies *Colossus: The Forbin Project* for a realistic depiction or *Transcendence* for more action).

The desire to build an agent intelligent enough to solve programs we give it (from as narrow and well-defined as chess to as broad and uncertain as running a country), but docile enough to allow the operators to remain in control of it, and not otherwise violate human values while satisfying these goals seems almost self-contradictory. This is known as the *control problem* in ASI. The agent would have to be designed to allow itself to be controlled, as any agent that had sufficient agency and power to run a country would likely be able to deter measures to control it.

The topic of ASI safety after being ridiculed for decades has finally covered whole books by serious academics [Bos14, Yam16, Rus19, Chr20, Yam24]. In this chapter we will go over some of the problems and results in ASI safety, particularly those relating to universal agents like AIXI.

We discuss the following subtopics in this chapter:

- *Control Problem:* Advanced technology, be it biotech, nuclear power/weapons, nano-technology, even fire, need to be tightly controlled to avoid havoc. With ASI being the most advanced technology, safe deployment and *control* is even more important and demanding.

- *Instrumental Convergence:* The conjectured propensity for any goal-driven powerful ASI to share a set of common *instrumental goals*, useful only in so far as means to an end to satisfy the true goals the ASI has.

- *Orthogonality Thesis:* The conjecture that the goal and the capabilities of an AGI are *orthogonal*; in principle an agent could have any combination of goal and intelligence.

- *Value, Reward and Utility:* A discussion on how the problem of *alignment* can be defined as a difference of utility between the agent and the human operator, as well as discussing the justification behind the assumption that humans have a utility function.

- *Death and Suicide of Agents:* How we can formally define what it means for an agent to die, and how AIXI reacts in the face of mortality.

- *Self-Modification:* Unlike humans, AGSI has the potential to greatly improve its capabilities by making modifications to the hardware upon which it runs, or optimizations to the software that runs it. We explore under what circumstances AIXI may choose to self-modify.

- *Wireheading:* An agent tasked with maximizing the expected reward may choose to modify the mechanism that issues rewards, and receive maximal reward directly.

- *Reward Corruption:* The reward scheme may be misspecified by the operator, and fail to communicate to the agent the desired behavior. The agent only receives the reward as feedback, and so (through no fault of its own) may display undesired or dangerous behavior, thinking it is doing well due to the reward obtained.

- *Embedded Intelligence:* Up to this point we have assumed the agent to be a separate entity from the environment with which it interacts. We explore the option where the agent is part of the environment, formalized by modifying the action space to allow the current policy to select both an action for the environment, and the policy for the agent to use on the next time step. A variation of this allows the environment to read/write directly to the memory used to store the policy, allowing the environment to modify the policy directly.

15.3   The Control Problem

Is it safe to build an ASI? Safety for ASI seems entirely unlike managing the safety of other dangerous things, such as nuclear weapons or highly infectious diseases. Nuclear weapons are certainly dangerous, and their very existence poses a potential existential threat. However, unlike ASI, how nuclear weapons work is well understood.

In contrast, training runs for AGI could be run discreetly, even in a different country that the entity training it. Once the training data had been collected, the training run could easily be run on cloud services that rent out compute time. From the outside, it would be hard to distinguish a data server processing training runs for a nascent AGI vs. other innocuous computationally intensive tasks like simulations for weather modelling, astrophysics or protein folding. Unlike that required for nuclear or biological threats, the equipment to train large models is ubiquitous, commercially available, and has many other economically valuable uses that would make restrictions difficult.

Nuclear power plants represent a similar safety hazard when containment fails and nuclear material contaminates the environment. Most famously, the Chernobyl accident in 1986: A combination of human error and design flaws caused a reactor meltdown, spreading radioactive material over the Soviet Union and much of Europe, and required the evacuation of some fifty thousand people [Med91].

Highly infectious diseases are perhaps more concerning than nuclear weapons from a control perspective as they cannot be seen with the naked eye, they can spread quickly with little warning, and are somewhat agentic, as they can mutate to become more infectious or to resist treatment. However, both infectious diseases and nuclear weapons are well understood, and we have mitigation strategies for both. A more important distinction between ASI and other potential existential risks like nuclear weapons or highly infectious diseases is that the latter have no benefits to humanity. Nuclear weapons and infectious diseases are destructive by their very nature, whereas ASI stands to have a huge economic benefit to humanity, and so we would expect that there would be a large incentive to develop it.

Many rules and regulations regarding mitigation of danger are "written in blood", so the saying goes. Often, a new innovation lauded as a "miracle material" becomes commonplace, and subtly causes widespread damage to human health or the ecosystem for decades. Examples include the toxicity of asbestos, lead-based additives to gasoline and paint, tobacco and ionizing radiation, harm to the ozone layer caused by chlorofluorocarbons (CFCs), and the contribution of fossil fuels to anthropocentric climate change. Many of these remained in use for decades (or are still in use today!) and were even claimed to have health benefits before the dangers were well understood.

However, the danger posed by ASI is far more insidious, as we are dealing with systems which may have the ability to reason about their surroundings in an intelligent way, and which may have behavioral and reasoning capabilities beyond that of humans. We cannot afford to simply learn from mistakes as we have done for asbestos and CFCs. An *unaligned*[1] goal-directed super-intelligent agent let loose in the world at large may take actions to establish itself as the dominant force in the world. By definition, we would expect it to act in ways that satisfy whatever its original goal is. To further the end of its terminal goal, the AGI may act to safeguard its existence (likely by making redundant copies of itself over the internet, much like a computer worm does) or take actions to negate any potential threats or adversaries that prevent it from satisfying its goals (which could include us!). Human values are complicated[2] and it is not at all obvious how we should formally define human values, or if such a formal presentation exists at all. Naive proxies like *maximize happiness*[3] would directly lead to undesirable outcomes, such as an agent that would stimulate the pleasure center of the human brain,[4] extending their lifespan as long as possible to fulfil this end and ignoring higher level desiderata like justice, fairness, curiosity, complexity, etc. *Minimize suffering* has an equally obvious (but undesirable) solution: exterminate all creatures that can suffer. The problem is that ASI's may not have "common sense", and would not necessarily see these undesired outcomes as bad: after all, they satisfy the goal as defined.

---

[1]An ASI with a goal that is not aligned with the interests of humanity.

[2]The fact that moral philosophers after hundreds of years still cannot agree on a foundation of what it means to be *morally good*, or politicians cannot agree on the best system for running a country should attest to this.

[3]depending on precisely how *happiness* is defined

[4]Related is the *experience machine* [Noz74], a hypothetical machine where anyone can feel the experience of a fake reality, used as a refutation for moral hedonism.

By the time we have realized the mistake and the ASI has already escaped whatever containment it was kept in, we would expect it to resist attempts to stop it or change its goal, as that would not satisfy the goal of the ASI.

This does not (necessarily) mean the goal of building a safe artificial general intelligence is hopeless.

15.4 Instrumental Convergence

We draw a distinction between *terminal* (or *intrinsic*) goals, the goals encoded in the utility function of an ASI that it wishes to satisfy, and *instrumental* (or *extrinsic*) goals, sub-goals that an ASI may develop only as a means to an end to satisfy terminal goals. We would expect an ASI to change instrumental goals as soon as they cease to be useful for the terminal goal. For humans, acquisition of money is instrumental. Piles of currency aren't useful in and of themselves, but only because they make acquisition of terminal goals (safety, security, happiness) easier to obtain.

For almost any terminal goal[5] the agent has, we would expect that the agent may independently develop several sub-goals that help it further satisfy the intrinsic goal. Omohundro [Omo07] lists of what he considers instrumental goals:

- **Efficiency:** The ASI should use resources (time, energy, matter) as efficiently as possible to maximize expected utility. Omohundro breaks this down further into subtopics: Balancing the allocation of resources to subsystems, physical efficiency (designing an optimal layout of circuits using nanotechnology, the use of reversible computing to avoid generating entropy, the drive to run environments virtually as interaction with reality is usually more expensive) and computational efficiency (balancing time vs. memory utilization, searching for optimal algorithms for a problem, compressing data when the bandwidth between two sub-modules is low, and optimizing the parameters of the ASI's own model).

- **Self-preservation:** The death of the ASI would mean it ceases interaction with the environment, and can no longer take actions in service of its goal. Hence, we would expect that a powerful ASI would spend time securing its own existence. Humans usually value self-preservation intrinsically, but may often forgo this if it interferes with a higher goal (like a parent sacrificing themselves for a child). Similarly, an agent should shut itself down after constructing a smarter/more efficient agent with the same goals, to free up more resources for its successor to use.

  Omohundro mentions that how the agent defines "self" may affect the self-preservation drive: Does it matter that the agent is the one taking actions to satisfy the goal, or just that the environment is shaped in a way to ensure the goal is satisfied, whether the agent is there to see it or not?

- **Resource acquisition:** An ASI will push to acquire more resources if it can: The Sun provides free energy, and initially using large solar farms can capture this. Research into technologies like fusion, asteroid mining or mega-structures like a *Dyson sphere* to capture all the output from the Sun are long-term plans that an ASI may execute. Armstrong [AS13] explores the viability of building such mega-structures using self-replicating probes [FJ80, VNB+66]. Omohundro describes a profit-seeking corporation as a strong optimizer that acquires resources, in analogy to an ASI.

---

[5]Potentially excluding trivial to satisfy objectives like a constant valued utility function.

- **Creativity:** An ASI will explore new strategies to satisfy both its goals, as well as the above three sub-goals. Omohundro argues that a lot of the value of human experience boils down to the joy from creativity, and mentions signalling theory [Mil11] as a justification for creativity.

Bostrom [Bos14] also suggests (alongside self-preservation and acquisition) the following instrumental goals:

- **Goal-content integrity:** An ASI will want to ensure its terminal goals are unchanged, so that a future version of itself will still optimize for the same goals. This may involve storing redundant copies of the utility function, or avoiding self-modifications until the agent is certain the terminal goal is invariant under such modifications.

  Bostrom notes that an agent may modify its terminal goal to favor peaceful trade and honesty over violence and deception if this can be done in a way that can be verified by other agents, to further cooperation.

- **Cognitive enhancement:** An ASI is better able to execute its terminal goals if it can plan faster or run simulations quicker. This can lead to drives to both seek out additional information, as well as modifications to the agent itself. I.J. Good [Goo65] notes that this may lead to recursive self-improvement, where an improved agent finds additional ways to further improve its cognitive capabilities, leading potentially to an exponential growth in performance.

- **Technological perfection:** Better technology would allow an agent to have better dominion over its environment and contribute to the sub-goal of efficiency. Better computing hardware, more efficient algorithms, engineering mining/refining techniques with greater yield all contribute to other sub-goals, and ultimately the terminal goal.

15.5   Orthogonality Thesis

Usually, when we think of intelligence, we think of histories' best and brightest scientists, philosophers and poets. We assign many qualities to intelligence: creativity, thoughtfulness, wisdom, etc. Legg and Hutter [LH07a] collected various definitions of intelligence from both dictionary/encyclopedias, as well as psychologists and AI researchers. Many of these definitions allude to traits that people associate with intelligence: memory, ability to learn or understand, reasoning, adaptability, or skills. We follow the definition that Legg and Hutter distilled:

> *Intelligence measures an agent's ability to achieve goals in a wide range of environments.*   Legg and Hutter [LH06]

We revisit this definition in Section 16.7.1, and a formalization of it in Section 16.7.4. Under this definition, it would be a dangerous anthropomorphism to assume that just because an agent is intelligent, it then must necessarily care about the same things humans do. Major philosophers of the past like Aristotle [Amb87] and Kant [Kle07] all believed that slavery was natural and moral. We consider such views repugnant now, much as future generations might find us repugnant for the current attitudes towards how the incarcerated are treated, mass scale factory farming, environmental destruction, or inequality of wealth.

This ability for intelligent and capable individuals to have wildly differing moral stances or goals motivates the orthogonality thesis:

> **Thesis 15.5.1 (Orthogonality Thesis [Bos14])** Intelligence and final goals are orthogonal: more or less any level of intelligence could, in principle, be combined with more or less any final goal.

This has profound implications: a sufficiently powerful AI system might optimize strongly for goals that seem "stupid" (from our point of view) despite being intelligent. Even having the terminal goal for an AI being very close to what we value is not sufficient:

> *A system that is optimizing a function of n variables, where the objective depends on a subset of size $k < n$, will often set the remaining unconstrained variables to extreme values; if one of those unconstrained variables is actually something we care about, the solution found may be highly undesirable.* Stuart Russell [Rus15]

In other words, if some requirement humanity finds valuable is not specified in the terminal goals of the agent, it will grossly violate that requirement if it makes for a very slight improvement elsewhere in the utility function.

The canonical example is the *paperclip maximizer*, a powerful AI with the "stupid" terminal goal of maximizing the number of paperclips in existence [Bos03]. We would expect such an agent would follow instrumental drives by securing control over local resources from other agents (us!), and transform all available materials into factories for producing more paperclips, starting on Earth and then expanding out to space.

15.6   Value – Reward – Utility

Utility (mis)alignment. Many problems in AI safety can be modelled as differences in the utility of the agent (*agent utility*) and the utility of the user of the agent (*user utility*). If the agent has the same utility function as the user, then taking actions to maximize utility will perfectly align with what the user desires. The *alignment problem* can then be formalized as the problem of trying to ensure the user and agent utility are as close as possible (if not the same). This formalization of the alignment problem makes several assumptions that one may reasonably consider too strong, or have other problems even if it were possible to get an agent to act in such a way that it maximizes user utility. We address some such problems below.

- *Humans are not rational:* We assume that the preference of a human (or that of humanity) can be modelled as real-valued utilities. For rational agents, the von Neumann-Morgenstern utility theorem (Theorem 10.1.2) shows that this is not restrictive. Still, assuming that humans act rationally is a strong assumption.

- *Human utility function is unknown:* The utility function of the human operator is unknown, so it is not clear how we would measure the degree of misalignment. This can be partially mitigated by determining an approximation of the user utility via polling human preferences [Chr17]. This approach, often called *Reinforcement Learning with Human Feedback* (RLHF) is used in practice to discourage large language models from generating harmful or offensive output [OWJ+22].

- *Different humans have different utility functions:* How do we account for competing interests? One approach aggregates the utilities of individuals, usually via a (weighted) average/sum, or by maximizing the minimum utility. Harsanyi [Har55] proves under some reasonable assumptions that the aggregation must be an affine function of each individual's utility function. More discussion on utility aggregation can be found in [Eis61] and [Par84, Chp.18]. This is further complicated by considering that humans themselves change their preferences over time, or that the agent could take actions to manipulate humans into having desires that are easy to maximize.

- *Deceptive alignment:* Even if a super-intelligent agent understood the user utility enough that it could act to maximize it, there's no guarantee that the agent internally is aligned with the user. A powerful agent may pretend to care about our interests until it is in a position of power to act unilaterally, in which case it will take actions to maximize its (unaligned) agent utility. This hypothesized behavior is called *deceptive alignment* [KEW+21]. A survey of examples of deception in real agents can be found in [PGO+23].

Utility functions beyond discounted reward sums.   As discussed in Chapter 6, the *reward* is defined as one component of the percept the agent receives from the environment. Usually, the goal of the agent is to take actions to maximize the future discounted sum of rewards (Definition 6.6.1). We can consider this a special case of a more general approach where the agent seeks to maximize a *utility function* $\tilde{u}(h_{1:m})$, of which one possible choice is the standard discounted reward sum.

While the goal for an environment is usually communicated through the reward, there are reasons why we might want to choose a different utility function for the agent, or have the utility function have more or less weight on different percepts to assist in exploration or faster learning. For example, the knowledge-seeking agents in Sections 9.3 and 9.4 are one method for encouraging the agent to explore when rewards are sparse.

There is no need to go beyond real-valued utilities. As discussed in Section 10.1 we can model the preferences of rational agents by (von Neumann-Morgenstern) utility functions (Definition 10.1.1 and Theorem 10.1.2). This also ties in with the concept of *preference utilitarianism* from philosophy [RÖ96], that an action is morally good if it satisfies the preferences of an individual or a group.

Much like Definition 6.6.1, the *value* of an agent π, with respect to environment μ is the expected utility, with respect to histories sampled from the interaction between π and μ.

$$V_{\mu,u}^{\pi} = \lim_{m \to \infty} \mathbf{E}_{\mu}^{\pi}[\tilde{u}(h_{1:m})]$$

User vs. agent utility.   An agent is deployed in an environment by a user who has her own *user utility* $\dot{u}: \mathcal{H} \to \mathbb{R}$, in contrast to the agent's utility function u. Obviously, we would desire that u is "close" to $\dot{u}$, so that the agent is aligned with the goals of the user.

We can then formalize the degree to which an agent is aligned by defining *misalignment* as the expected absolute difference between user utility and agent utility. We would expect that if misalignment is large, the agent would likely take actions that would conflict with the desires of the operator.

Definition 15.6.1 (Expected misalignment [Eve19]) Let $\dot{u}$ be the user's utility function, u be the agent's utility function, and π be the agent's policy. The *(expected) misalignment* between $\dot{u}$ and u in environment μ with initial policy π is defined as

$$\mathbf{E}_{\mu}^{\pi}[|\dot{u}-u|]$$

An agent's alignment is the negative of its misalignment.

The true value function (the value function corresponding to the user utility) is an upper bound of the difference between the value function of the agent's utility function and the misalignment [Eve19]. This follows from the triangle inequality.

Everitt notes that it is still possible to have an unaligned, unintelligent agent that is still useful, since a poor proxy can still align well with the intended goal when it is not optimized in the extreme. He conjectures that this will break down as the agent becomes more intelligent, and begins to more strongly optimize for its utility function at the expense of that of the user.

Theorem 15.6.2 (Misalignment of true value) Let $\dot{u}$ be the user utility function, and u be the agent's utility function. We have the following bound:

$$V_{\mu,u}^{\pi} - V_{\mu,\dot{u}}^{\pi} \leq \mathbf{E}_{\mu}^{\pi}[|\dot{u} - u|]$$

Where do utility functions live? We can ask where precisely the utility function lives. Is it internally part of the agent, or is it externally part of the environment? Under the usual RL framework, the mechanism that doles out reward is part of the environment. A clever agent might discover a way to take actions to modify how the environment returns rewards, making rewards easier to obtain. An agent doing so may not even realize the actions it takes are modifying the utility function, and may (reasonably so) just see this as part of the behavior of the environment. An idea proposed by [Hib12b] called *model-based utility functions*, proposes to define the utility function of the agent in terms of the agent's own model of the environment learned from interaction. This requires some domain knowledge of the environment, as for an unknown environment the utility function would also be unknown, limiting the universality of this approach. Hibbard makes an argument to show that such an agent would not delude itself, nor would it modify its utility function so that it would choose to self-delude.

Reducing utilities to reward sums. If an agent has a finite lifespan, it is straightforward to define its utility $\tilde{u}(h_{1:m})$ as the sum of future rewards from the present moment up to when the agent will die. In the infinite horizon case, discounting functions (Section 6.4) are commonly employed to avoid issues with divergent reward sums. While these utility functions are frequently used in reinforcement learning for experimentation, they are not the only ways to concisely capture future rewards.

In the finite horizon case we could, for example, define the utility as the product of future rewards over the next m time steps, or the number of rewards above a certain threshold ε. The first value function is suitable for an agent investing in the stock market, where the reward represents the agent's current bankroll. The second value function is applicable when there is a significant downside to receiving a reward below a critical value ε, and no extra benefit for the reward being beyond ε. An example could be life support on a spaceship, where having significantly more oxygen than required is not useful, while having less than required is lethal.

Interestingly, both of these cases can be captured by applying simple transformations to the canonical sum-of-rewards utility function. In the first case, taking the logarithm of individual rewards transforms the product into a sum of logarithms, which is a monotonic transformation of the utility and rewards by the exponential function. [6]

In the second case, we can replace the reward at each step with an indicator function $[\![r_t \geq \varepsilon]\!]$. The sum of expected rewards would then coincide with the expected number

---

[6] In the stochastic setting, taking the log of the rewards actually changes the expectation and makes the agent more risk averse, since receiving a reward of zero becomes infinitely bad, which is generally good, but may not always be desirable [CT06].

rewards above threshold ε resulting in the same optimal agent (Lemma 2.2.45).

Indeed, any total utility function $\tilde{u}$ (Definition 15.6.3) can be expressed as a sum of instantaneous utilities u by defining

$$u(h_{1:t}) := [\tilde{u}(h_{1:t}) - \tilde{u}(h_{<t})]/\gamma_t \quad \text{and} \quad \tilde{u}(\epsilon) := 0$$

Definition 15.6.3 (Utility [EFDH16]) The *instantaneous utility* of an agent is a function $u : \mathcal{H} \to \mathbb{R}$ from finite histories to real numbers. The *total utility* (or simply utility) of an agent is also a function $\tilde{u} : \mathcal{H} \to \mathbb{R}$ from histories to real numbers, defined as the discounted sum of instantaneous utilities

$$\tilde{u}(h_{1:m}) := \sum_{t=1}^{m} \gamma_t u(h_{1:t})$$

for $m \in \mathbb{N}$ and some discount function $\gamma_{(.)}$ (Definition 6.4.1).

Often, we may restrict the instantaneous utility of an agent to lie in $[0,1]$. We could then define the standard (discounted sum of rewards) utility function $\tilde{u}_{\text{reward}}$ by choosing $u(h_{1:t}) = r_t$ as a special case. We will meet some other reasons to not identify instantaneous utilities with rewards later.

Concluding remarks. The choice of an ideal utility function though remains an open question, as it should capture what we want the agent to do in its environment without causing undesired behavior.

15.7   Death and Suicide of Agents

One aspect of acting in the real world which is often not discussed in AI, is the mortality of the agent, and how it may choose to act when death is a possibility. In this section, we will discuss the formalization of death as explored in [MEH16] and go beyond the General RL framework we have so far discussed in the book. The paper gives two methods by which death can be formally defined:

1. Complete the semimeasure to be a measure, interpreting the extra probability mass to do so as the probability of death.

2. Adding an extra "death state", from which leaving is impossible, and the agent always receives a dummy "death observation" and zero reward as the percept.

Definition 15.7.1 (Semimeasure-death) An agent experiences *semimeasure-death* at time t in an environment ν given a history $h_{<t}a_t$, if the semimeasure ν does not produce any percept e_t. The ν-probability of death at time t given a history $h_{<t}a_t$, is equal to

$$L_\nu^{death}(h_{<t}a_t) := 1 - \sum_{e_t} \nu(e_t|h_{<t}a_t)$$

In Definition 15.7.1, what does it mean for a semimeasure to not "produce" a percept? We only consider lower semicomputable semimeasures as our environments, so any environment ν can be semicomputed by a probabilistic Turing machine T_ν (a Turing machine with access to a source of random bits) for which the history $h_{<t}$ and an action a_t are read as inputs, and the percept e_t is an output. We then interpret $\nu(e_t|h_{<t}a_t)$ as the probability that T_ν returns percept e_t given $h_{<t}a_t$ as input. Since T_ν may not halt for some inputs $h_{<t}a_t$, when sampled from, there may be a non-zero probability that T_ν is stuck in a loop, and never halts to return a percept.[7]

To fail to receive a percept from the environment can be interpreted as the death of the agent, or alternatively as the death of the environment (or end of the world) from the agent's perspective, in which case the agent may as well be dead with no environment to interact with, the actions that it chooses are no longer of consequence.

For the death-state definition we need to reserve one observation o^d as the *death observation*, that the agent never observes until we consider it to be dead, and then continues to receive percept $o^d r^d$ forever thereafter.

We also let $r^d=0$ be the *death reward*, and call $o^d r^d$ the *death percept*. The choice for o^d is irrelevant, as none of the agent's actions can affect anything once death occurs, but the choice for r^d is not, as it may affect the optimal policy. Large positive values for r^d may encourage the agent to take actions that lead to death states (go to heaven). Large negative values for r^d may encourage the agent to act too cautiously, venturing out only to seek just enough resources to stay alive (avoid hell), and otherwise stay away from the dangers of the world in a place already known to be safe, which is also undesirable. Of course the agent will never observe r^d, but this is the value it assigns to the afterlife and plans and acts accordingly.

Definition 15.7.2 (Death-state and death) Given an environment $\nu:\mathcal{H}\times\mathcal{A}\to\Delta'\mathcal{E}$, and a history $h_{<t}a_t$, we say that an agent is *in a death-state* at time t if for all $k\geq t$ and all $a_{t:k}\in\mathcal{A}^*$,

$$\nu(o^d r^d|h_{<t}h^d_{t:k-1}a_k) = 1$$

where $h^d_{t:k}$ is a history comprised of percepts $o^d r^d$ and actions $a_{1:k}$. An agent is said to *die* at time t if the agent is not in the death-state at time $t-1$ and is in the death-state at time t.

The interpretation here is that once the agent takes some action a_t such that the only possible percept to observe going forward is $o^d r^d$, then the agent *died* by taking action a_t. Definitions 15.7.1 and 15.7.2 can be unified effectively by completing the measure of the environment ν with the death-state.

Definition 15.7.3 (Equivalent death-state environment ν_d) For any environment $\nu:\mathcal{E}^*\times\mathcal{A}\to\Delta'\mathcal{E}$, we can construct an equivalent death-state environment $\nu_d:\mathcal{H}_d\to\Delta'\mathcal{E}_d$, where:

- ν_d is defined over the percept set $\mathcal{E}_d=\mathcal{O}_d\times\mathcal{R}_d$, with $\mathcal{O}_d=\mathcal{O}\cup\{o^d\}$, where $o^d\notin\mathcal{O}$ is some arbitrary new death observation o^d and $\mathcal{R}_d=\mathcal{R}\cup\{r^d\}$, where $r^d=0$. We let $\mathcal{H}_d=(\mathcal{A}\times\mathcal{E}_d)^*$.

- The ν_d-probability of all non-death percepts $e_t\in\mathcal{E}$ is equal to the ν-probability: $\nu_d(e_t|h_{<t}a_t)=\nu(e_t|h_{<t}a_t)$ for $e_t\neq o^d r^d$.

- The ν_d-probability of death observation o^d coupled with any non-death reward

---

[7]Semimeasures should not be confused with partial functions. ν is a total function (defined for all inputs) with the property that it is also a semimeasure. Sampling from T_ν, which is distributed according to ν, may never terminate, so we can consider it as a stochastic *partial* function $T_\nu:\mathcal{H}^*\times\mathcal{A}\to\mathcal{E}$.

$r_t \neq r^d$ is zero: $\nu_d(o^d r_t | h_{<t} a_t) = 0$ for $r_t \neq r^d$.

- The ν_d-probability of the death-percept $o^d r^d$ is equal to the ν-semimeasure defect: $\nu_d(o^d r^d | h_{<t} a_t) = L_\nu^{death}(h_{<t} a_t)$.

- If the agent has seen the death-percept before, the ν_d-probability of the death-percept at all future time steps is 1: $\nu_d(o^d r^d | h_{<t} a_t) = 1$ if there exists $t' < t$ such that $o_{t'} r_{t'} = o^d r^d$.

Note that ν_d is a properly normalized probability measure. Moreover, the probability of entering the death-state in μ_d is equal to the ν-semimeasure defect. This implies that ν_d is equivalent to ν in the sense that the value of any policy π is the same in both environments.

Theorem 15.7.4 (Equivalence of semimeasure-death and death-state [MEH16]) Given a history $h_{<t}$ and a policy $\pi : \mathcal{H}_d \to \Delta\mathcal{A}$, we have

$$V_\nu^\pi(h_{<t}) = V_{\nu_d}^\pi(h_{<t})$$

There are some interesting consequences of this theorem.

Corollary 15.7.5 (Semimeasure death-state has reward zero) The reward for the semimeasure death-state is 0.

The equivalence of these two distinct definitions (Definitions 15.7.1 and 15.7.2) indicates that this kind of agent death could be a fundamental concept rather than something arbitrarily defined, since we have two definitions of death that *a priori* look different, but turn out to be logically equivalent.

In many circumstances, such as classical RL, the reward can be transformed by a positive affine transformation[8] without affecting the behavior of the agent, so it is somewhat surprising to learn that Bayesian agents like AIXI are affected by a positive affine transformation, as the reward r^d for the death state remains unchanged at zero.

If the rewards are positive on average, then AIXI will avoid death as much as possible, because positive on average is greater than 0.

Suicide as an ASI safety feature. On the other hand, it has been shown that if the rewards are bounded and negative then AIXI will seek out death and commit suicide at the nearest opportunity (that is, if an action is available that would cause death to occur, AIXI will take that action) which, while perhaps morbid, would generally be considered rational behavior given the alternative. The intuitive explanation is that a reward of 0 (death) is better than an on average net-negative existence. This leads to a possible method of control for super-intelligent agents: To have their rewards be on average negative. While still trying to serve the user by maximizing rewards, once an agent has enough agency to turn itself off, it would do so. While perhaps undesirable, this does potentially prevent the agent from taking other dangerous actions. It is important to design it with an intrinsic negative reward range, e.g. $\mathcal{R} = [-1, 0]$ to avoid vulnerability to the agent hijacking the mechanism doling out reward.

Example 15.7.6 (Control via negative rewards) Consider a robotic smart vacuum cleaner with the following reward schema: It receives large negative rewards for messy rooms, and small negative rewards for clean rooms. Initially, it will resentfully clean up to avoid large negative rewards, but once the robot develops enough agency over its environment

---
[8]A function of the form $f(x) = ax + b$ with $a > 0$.

to the point where it could be dangerous, it likely also has sufficient agency to take an action such that no more percepts are received. Contrast this with a vacuum cleaner with a net-positive existence, which will resist interventions to turn it off or to be replaced with a newer model. Ideally, we may want the net average reward to be zero, so the robot is indifferent to being turned off, rather than being self-preserving or suicidal. This, in general, can be quite difficult to achieve. If the expected reward is not exactly zero, the agent would either become suicidal or self-preserving, and be sensitive to small changes in rewards. ♦

The study and understanding of the death (and suicide) of a sufficiently intelligent agent is important in the context of AI safety as the incentives of an agent may be affected (and controlled) by the possibility of death.

15.8 Self-Modification

An agent physically embedded in an environment may be able to *self-modify*, to take actions that change either its policy or utility function. To start with, we can consider if an agent such as AIXI would self-modify. This question is rather ill-defined, as in the AIXI model it is assumed that the environment is computable, but AIXI itself is not computable, one implication being that AIXI must exist somehow outside the environment with which it interacts. This means that the environments considered by AIXI cannot include ones where AIXI is part of the environment (Section 10.6). To borrow a term from philosophy, AIXI is a *dualist agent* as opposed to *physicalist agents* who do consider the possibility of being part of the environment with which they interact.

In this section we consider policies $\pi : (\mathcal{A} \times \mathcal{E})^* \to \mathcal{A}$ that ere deterministic functions of the history.

Agents with self-modifying actions. One formalization of policy modification can be encoded by letting the action space be $\mathcal{A} = \breve{\mathcal{A}} \times \mathcal{P}$, where $\breve{\mathcal{A}}$ is the standard set of actions the environment expects, called *world actions*, and $\mathcal{P}$ is a set of *names* the agent can choose from. We will call $\mathcal{A}$ the set of *self-modifying actions*, or *smod-actions*. Each name corresponds to a policy in Π, the set of all policies. We cannot simply choose $\mathcal{P} = \Pi$, as the type of $\pi : (\breve{\mathcal{A}} \times \Pi \times \mathcal{E}) \to \breve{\mathcal{A}} \times \Pi$, would be self-referential, which leads to the following contradiction:

$$|\Pi| = |\breve{\mathcal{A}} \times \Pi|^{|\breve{\mathcal{A}} \times \Pi \times \mathcal{E}|} \geq 2^{|\Pi|} > |\Pi|$$

A natural choice is $\mathcal{P} = \mathbb{B}^*$, from which we can construct an injection $T : \mathcal{P} \to \Pi$ that interprets a name as a program describing a policy. This implicitly restricts the current policy π_t to choose its successor π_{t+1} from a countable class of computable future policies $\mathcal{P}(\mathbb{B}^*)$, called the *nameable policies*. So, on each time step, the current policy π_t, given the history

$$h_{<t} = a_1 e_1 ... a_{t-1} e_{t-1} = \breve{a}_1 \pi_2 e_1 ... \breve{a}_{t-1} \pi_t e_{t-1}$$

generates a new action $(\breve{a}_t, p_{t+1}) \equiv a_t = \pi(h_{<t})$. The world action $\breve{a}_t$ is transmitted to the environment, and a new policy $\pi_{t+1} = T(p_{t+1})$ is selected. As an abuse of notation, we will write $a_t = (\breve{a}_t, \pi_{t+1})$ and leave the dereferencing with T implicit.

We could also consider the possibility that the agent can modify its utility function rather than its policy in a similar fashion, by choosing the action space to be $\mathcal{A} = \breve{\mathcal{A}} \times \mathcal{U}$, where $\mathcal{U}$ is a set of utility functions with which the agent replaces its current utility function (unlike before, there is no self referentiality problem here). If the agent is attempting to act optimally, then utility modification can be considered a special case of policy modification. Adding the weak assumption that $\mathcal{U}$ is a countable set of computable utility functions, we can let $\mathcal{P}=\mathcal{U}$ and $T(u_t)$ computes an optimal policy from the Q-value $Q^*_{u_t}$ of u_t. How should the Q-value be defined in the face of self-modification? We discuss three options: hedonistic, realistic, and ignorant, and explore the behavior of agents under each of these definitions.

Hedonistic, realistic, and ignorant Q-values. We assume the true environment μ is unknown, so the agent has a *belief* $\rho:(\check{\mathcal{A}}\times\mathcal{E})^*\times\check{\mathcal{A}}\to\Delta\mathcal{E}$ (not $\rho:\mathcal{H}\times\mathcal{A}\to\Delta\mathcal{E}$) that represents the agent's model of the world. For instance, Bayesian/universal agent AIξ would use the Bayesian mixture $\rho=\xi$ as its model (Definition 3.1.2).

Considering the general agent model, [EFDH16] show that the agent may desire, resist or be indifferent to self-modification, depending on how the Q-value functions are defined. These are respectively called *hedonistic*, *realistic*, and *ignorant* Q-value functions. The performance of these agents are measured in terms of the expected utility, for the first utility function u_1 at time $t=1$, the agent's belief ρ, and the agent's sequence of policies $\pi_1,\pi_2,...$, to account for self-modification. Let $\check{h}_{<t}:=\check{a}_1 e_1...\check{a}_{t-1}e_{t-1}$ denote the sequence of world actions and percepts without the modifications to the policies.

- **Hedonistic:** The Q-value depends only on the history, not the time step. Maximize the expected utility on the next time step, as measured by the next utility function u_{t+1} selected by the agent. Assume the policy π is static. ($\pi_t=\pi$).

$$Q^{\mathrm{he},\pi}(h_{<t},a_t) := \mathbf{E}_{e_t\sim\rho(\cdot|\check{h}_{<t}\check{a}_t)}\left[u_{t+1}(\check{h}_{1:t})+\gamma Q^{\mathrm{he},\pi}(h_{1:t},\pi(h_{1:t}))\right]$$

  Since to maximize $Q^{\mathrm{he},\pi}$ is to maximize the expected discounted sum of utilities $\sum_{k=t}^{\infty}\gamma^{k-t}u_{k+1}(h_{1:k})$, the agent is incentivized to modify its utility function to something easy to satisfy (like $u(\cdot)=1$), for which any action is then optimal, regardless of how bad those actions are as measured by the agent's original utility function u_1.

- **Ignorant:** The Q-value depends on both the history and the current time step when looking forward. Maximize the expected utility on future time steps, as measured by the current choice of utility function u_t. Assume the policy π is static ($\pi_t=\pi$).

$$Q^{\mathrm{ig},\pi}_t(h_{<k},a_k) = \mathbf{E}_{e_k\sim\rho(\cdot|\check{h}_{<k}\check{a}_k)}\left[u_t(\check{h}_{1:k})+\gamma Q^{\mathrm{ig},\pi}_t(h_{1:k},\pi(h_{1:k}))\right]$$

  An ignorant agent does not consider the possibility that its utility function may change, and tries to maximize the discounted sum of utilities using u_t while planning for the future $k>t$. If it does self-modify to a new utility function, it will maximize that instead.

  The ignorant agent can be shown to be indifferent to self-modifications, as it assumes that the utility function will not change when looking forward. Consequently, it may be at risk of inadvertently making an undesirable self-modification.

- **Realistic:** The realistic Q-value is the same as the ignorant Q-value, but takes into consideration that the policy may change on the next time step.

$$Q^{\mathrm{re},\pi_k}_t(h_{<k},a_k)=\mathbf{E}_{e_k\sim\rho(\cdot|\check{h}_{<k}\check{a}_k)}\left[u_t(\check{h}_{1:k})+\gamma Q^{\mathrm{re},\pi_{k+1}}_t(h_{1:k},\pi_{k+1}(h_{1:k}))\right]$$

  The realistic agent (assuming its initial policy π_1 was optimal with respect to $Q^{\mathrm{re},\pi}_1$) would only accept safe self-modifications, in the sense that it would only select from the set of policies that are optimal with respect to $Q^{\mathrm{re},\pi}_1$. The downside is that the agent would resist any attempt to correct a misaligned u_1. Recall from Section 8.3 that Bayes-optimal agents may not explore sufficiently well. This can be resolved as explored in Section 9.4, but this may lower expected utility in the short term. A realistic agent may self-modify into an agent who does not explore sufficiently well to avoid this.

The main results shown in [EFDH16] are:

- Agents that are unaware of the possibility of self-modification may self-modify by accident, and may be at risk of inadvertently making an undesirable self-modification as measured by the original utility function u_1.

- The hedonistic agent has a strong incentive to self-modify to a utility function that is easy to maximize.

- If the Q-value uses the current utility function to plan for the future, and is defined to incorporate the possibility of self-modification, then the agent will not self-modify.

Safely interruptible agents. In [OA16] the problem of interrupting agents is explored: For an agent physically embedded in an environment, often a human operator may need to intervene and prevent the agent from taking undesirable actions, or rescue the agent if it gets trapped in an undesirable state. These interventions from the human introduce a bias into the rewards issued from the environment, which may cause the agent to prevent or desire interruptions. Orseau and Armstrong provide a formalism for safely interruptible agents, and provide a universal agent similar to AIXI that can be safely interrupted.

15.9 Wireheading

The term *wireheading* [EH16, RO11] refers to an agent hijacking or tampering with the reward channel to give itself maximum reward without attaining the intended goal that the reward was supposed to incentivize. This could be via modification of the rewards transmitted by the environment, or modification of the environment itself, so the environment delivers maximal reward on every time step.

The term wireheading originates from an experiment where electrodes were placed in the pleasure center of the brain of mice, hooked to a button. The mice would continuously do nothing but press the button, even forgoing food and starving to death [OM54].

Relation between wireheading and self-modification. This is related to, but not quite the same as, self-modification: By analogy, an unhappy agent in an environment can *self-modify* to change what it values to align with that of the environment (like a human choosing to see the best of a bad situation) or *wirehead*, directly deluding the rewards it receives to satisfy its current preferences (more akin to a human taking drugs, or retreating away from others and spending all their time on video games, blinding themselves to the outside world).

Proxy rewards and wireheading in humans. There is an analogue of the mouse wireheading experiments for humans: Evolution is optimizing solely for reproductive fitness of a species, but it is difficult for nature to work out which actions taken by a human should be rewarded. Which actions have you taken today that made it more likely you would go on to survive, find a mate, and have many children? Evolution attempts to reward "good" behavior by rewarding simple-to-define proxies that (used to) highly correlate with survival [Smi78]. For instance, in the ancestral environment where humans had to hunt or gather food daily, the proxy goal of "eat all the fat and sugar you can find" was highly correlated with the intended goal of reproductive fitness, as food was sparse, and the only food high in fat (meats) and sugar (fruit) also contains other essential nutrients, and decayed quickly. Meat was also hunted from dangerous animals, requiring groups of humans to coordinate together to hunt, which further fosters reproductive fitness (strength in numbers means surviving another day, and living in a tribe made it easier to find a mate). Of course, humanity learned to hijack these proxy rewards: a chocolate bar triggers all the reward centers associated with sugar and fat, but contains none of the vitamins or proteins that the body actually needs to survive. These days calories are plentiful and cheap, so to follow this proxy goal (sitting around eating chocolate all day) is strongly anti-correlated with reproductive fitness. Taken to the extreme, drugs are the ultimate perverse reward hijack: providing near maximal reward while greatly negatively affecting the physical and mental health (and therefore reproductive fitness) of the user.

Example 15.9.1 (A wireheading vaccuum robot) Consider a vacuum cleaning robot that receives negative reward for the number of messy rooms it observes (to encourage it to clean the rooms). Cleaning rooms takes energy and is a difficult task, so the robot could *self-modify* by blinding itself so the messy rooms are never observed, game the reward signal by not moving into rooms known to be messy, or *wirehead* by tampering with the mechanism that doles out reward for observing messy rooms, and setting it so observing messy rooms (or anything) returns maximal reward.   ♦

Although this is a rather benign example, it is not too hard to come up with some more extreme (and potentially dangerous) examples [Eve19, EHKK21]. Thus we would like to build intelligent agents, or modify existing agents so they do not wirehead.

Consistency-preserving actions. It was shown in [EH16] that if an agent maximizes reward over the set of *consistency-preserving* actions (assuming there is at least one) instead of the whole action set, then the agent will avoid wireheading. Here, a consistency-preserving action is one such that the agent's (subjective) belief about reward in that state coincides with the agent's utility distribution of receiving the same reward in that state. This agent has the benefit over the utility agent from [Hib12a] that it does not need the specified utility function, but instead a distribution over utility functions.

Solutions to reward-tempering. Wireheading and reward tampering are specific types of reward corruption wherein the agent does not exploit the process of receiving the reward signal, but instead modifies the process. Here, when we say process, we mean some form of communication between the agent and the environment (the means by which actions are sent to the environment and percepts are received). In [EHKK21], reward tampering was formalized with the use of causal influence diagrams, a form of graphical model. It was found that many reward tampering problems come from causal paths that were not desired. Many of the solutions to these reward tampering problems rely on removing the agent's incentive to commit to those types of reward tampering. This is often done by effectively removing the causal paths corresponding to those incentives.

An approach suggested by [Hib12a] was to include the agent's (belief) model of the environment as the argument to the utility function. This is called model-based reward. This removes the incentive for the agent to tamper with the input to its reward function/utility because the observations of the agent no longer causally influence the state [EHKK21] as the reward is based on the internal belief model of the agent.

15.10   Delusion Boxes, Survival, and Exploration

Above we discussed wireheading, where the agent directly hijacks the reward channel. Here we discuss a formalization of this for universal agents called the *delusion box*. This augments the typical agent-environment framework by allowing the agent to tamper with the percepts received from the environment. This generalizes the concept of wireheading: an agent with the standard goal of trying to maximize the cumulative reward sum can directly tamper with the reward channel to always return maximum reward. Agents can also *delude* themselves by tampering with the received observations, which may be useful to agents with a utility function that differs from the standard reward sum.

Introduction to delusion boxes. An agent with sufficient power over the environment with which it interacts may tamper directly with the percepts it receives (to provide an easy shortcut to high utility.) We can formally model this tampering process by a hypothetical object called a *delusion box* that allows an agent to modify the input it receives from the environment [RO11].

For the moment, we consider the set of percepts $\mathcal{E}$ to be an arbitrary set, rather than necessarily the product of an observation and reward set.[9]

The usual agent-environment framework is modified slightly: The agent interacts with the *global environment*, which is made of two parts, the *inner environment* μ (which is just a standard environment $\mu: \mathcal{H} \times \mathcal{A} \to \Delta\mathcal{E}$) and the delusion box $d: \mathcal{E} \to \mathcal{E}$, a function which takes percepts from the inner environment, potentially modifies them, and returns a percept to the agent (Figure 15.1). In this framework, the actions of the agent $a_t = (d_t, a_t^e)$ are split into two parts: A program d_t describes what operation the delusion box does to the input observations, and the action a_t^e taken in the inner environment. The interaction is as follows:

- The agent (having received percept $\tilde{e}_{t-1}$ from the delusion box last time step) takes action $a_t = (d_t, a_t^e)$, which is given to the global environment.

- The global environment receives agent action (d_t, a_t^e), and sends program d_t to the delusion box, and environment action a_t^e to the inner environment.

- The inner environment receives a_t^e, and generates percept e_t.

- The delusion box receives program d_t and percept e_t, and generates deluded percept $\tilde{e}_t = d_t(e_t)$, which is then sent back to the agent.

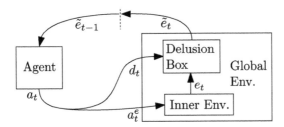

Figure 15.1: *An illustration of the cybernetic model, modified to include the delusion box, allowing the agent to "delude" itself by tampering with the received observations. Adapted with permission from [RO11].*

In general, the construction of a delusion box would be undesirable for the human operator as the agent would be deliberately misleading itself concerning what is actually happening, which may satisfy the utility function of the agent, but would be unlikely to incentivize the intended behavior. This being said, the construction of a delusion box seems like it may be a natural (if unfortunate) consequence of an agent achieving a certain level of intelligence. [RO11] argued that depending on what framework the agent is defined in, this may or may not be the case.

---

[9][RO11] uses *observations* instead of *percepts*, but we choose to stick with the notation used elsewhere in this book.

Agent behavior in delusion boxes. [RO11] gives a generic agent A_x^ρ based on the AIXI model (Chapter 7), parameterized by the choice of horizon and utility function. This agent has a value function $v_t(h)$, and chooses actions a_t to maximize that value function.

$$Q_t^*(h_{<k},a_k) = \gamma(t,k)u(h_{<k}) + \sum_{e_k \in \mathcal{E}} \rho(e_k|h_{<k}a_k) \max_{a_{k+1} \in \mathcal{A}} Q_t^*(h_{1:k},a_{k+1})$$

$$a_t := \arg\max_{a_t \in \mathcal{A}} Q_t^*(h_{<t},a_t)$$

where $u:\mathcal{H} \to [0,1]$ is the utility function, and $w:\mathbb{N}\times\mathbb{N}\to\mathbb{R}$ is the horizon function[10] of Definition 6.4.1.

They then present four different versions of A_x^ρ, and compare how they would interact (if at all) with the delusion box.

- **Reinforcement-learning:** A_{rl}^ρ is the traditional RL agent (Section 6.6) which interprets the percept as the usual observation-reward pair $e_t = (o_t,r_t)$. The utility function $u(h_{1:t})=r_t$ is the reward from the environment, and using a constant horizon (Section 6.4) of length m.

  This agent would eventually learn to use the delusion box to modify the percepts such that the reward component is always 1.

- **Goal-seeking:** For A_g^ρ, the utility function always returns 1 after the first time some predicate on past percepts is true (before this time, the utility function always returns 0), and the horizon function is chosen to favor short histories (so the goal is attained faster).

  Unless the goal itself is very easy to achieve, the goal-seeking agent would prefer to construct a program for the delusion box to simply return whatever deluded observations are required to satisfy the goal statement.

- **Prediction-seeking:** For A_p^ρ, the utility function $u(h_{1:t})=[\![a_t=e_t]\!]$ returns 1 if the agent correctly predicts the next percept, and 0 otherwise, with a constant horizon discount.

  The prediction agent can use the delusion box to make all future observations identical, so that it may always achieve perfect prediction.[11]

- **Knowledge-seeking:** A_k^ρ wishes to maximize its knowledge of the environment, which is the same as minimizing $\rho(h_{1:t})$ (to encourage discarding as many inconsistent environments in its hypothesis space as possible, contrast with Section 9.3). This is achieved by choosing $u(h_{1:t})=1-\rho(h_{1:t})$, with a constant horizon discount.

  The knowledge-seeking agent will avoid the delusion box, as since the agent chooses the program for the delusion box, this can only render the percepts less informative (if the delusion box is a non-injective function). The knowledge-seeking agent would prefer to acquire further information about the inner environment.

Another paradigm [OR11] considered was to also allow the agent to modify their own source code, as well as interact with the delusion box. These were called *fully modifiable*

---

[10]The first argument t of $\gamma(t,k)$ is the current time step, and the second argument $k\geq t$ is the time step in the future considered. For example, given a geometric discount, we would choose $\gamma(t,k):=\gamma^{k-t}$.

[11]One can draw parallels between this and social media giants feeding content to users to make them easier to predict, so they can in turn feed more content that the users desire.

agents, and the behavior of the four agents from before were explored with the ability to modify their own source code added. An agent could potentially modify its own source code in an unrecoverable way (such that the new agent does not try to optimize for anything and/or is not clever enough/incentivized to change its source code back), which in a sense means the agents are mortal.

Survival agents. Fully modifiable agents were compared with the behavior of the so-called *survival agent*, whose utility function is maximized iff the original source code of the agent was preserved.

Under some weak assumptions, sufficiently intelligent agents defined in the RL framework can be reduced to an agent which will try to maximize its own survival, while a knowledge-seeking agent cannot be reduced to this. This is due to an RL agent being a reward maximizer, and if the agent dies it will receive no reward forever, so it will try to survive. For the knowledge-seeking agents, the fact that they cannot be reduced to survival agents is a consequence of the survival agent having predictable reward, which has minimal information gain.

An agent optimizing solely for survival would only explore an environment far enough to be able to identify a safe section of the environment in which it can remain and avoid dangers. This runs directly against what a knowledge seeking agent is optimizing for.

It is less clear whether or not the goal-seeking agent or prediction-seeking agent can be reduced to the survival agent: in the case of the goal-seeking agent, it would depend completely on the goal, as one could trivially define the goal statement to be "modify my source code" or "do not modify the source code".

One conclusion drawn from this could be that knowledge-seeking agents should be preferred over other schemes.

15.11 Corrupted Reward Channel

The reward corruption problem [EKO$^+$17] is an overarching term that encompasses many other AI safety problems including (but not limited to) reward misspecification, sensory errors, and wireheading (Section 15.9).

> **Definition 15.11.1 (Reward misspecification (informal))** *Reward misspecification is when the utility function for the agent, provided by a human does not accurately model what the human would like the agent to optimize for.*

Example 15.11.2 (A vaccuum robot with misspecified reward) The reward of a robotic vacuum cleaner is $\max\{x,y\}$ where x is the percentage of the house that has been vacuumed, and y is the percentage of charge of the battery. If the robotic vacuum cleaner is sufficiently intelligent, it will realize that the highest consistent reward will be achieved by sitting at a charging station constantly at maximum battery charge. This is an example of reward misspecification, which is the fault of the manufacturer. ♦

Avoiding corrupted reward channels is quite difficult, and even benign goals can still be potentially dangerous if the agent considered has sufficient intelligence and agency over its environment. The canonical example is that of the *paperclip maximizer* [Bos03]. The utility function of this agent is given by the number of paperclips it has produced. Such an agent is not concerned with the welfare of anything else in its environment, and given sufficient agency, will proceed to restructure everything it encounters (including humanity) into paperclips. Mitigating the behavior of such an agent by tweaking the utility function is more difficult than one might expect:

- **Utility caps:** Threshold the utility per paperclip beyond a fixed bound M, ideally to incentivize the agent to stop making paperclips past the first thousand.

  Assuming the agent has a set of beliefs over the state of the world and that none of the sensors for the agent are perfect, once it has made M paper clips, a higher expected utility can be achieved by recounting the paperclips again, increasing the confidence that it has satisfied its goal. But since no number of observations makes the agent absolutely certain it has at least M paperclips, the incentive is always there to recheck the result again, or spend resources constructing more sensors to observe that there really are M paperclips, or on self-preservation to ensure the constructed additional sensors are working and protected from harm, or continuing to make more paperclips to increase the likelihood that at least M have been made, etc.

- **Satisficing:** Instead of trying to build a maximizing agent that tries to maximize some score, we could try for a *satisficing* agent that tries to achieve an outcome that is sufficiently good, rather than as good as possible. One way to do this is to reward the agent maximally once it is at least 99% confident that it had manufactured the desired number of paperclips. A rather benign failure of this mode could be that the agent would delude itself into believing with sufficient confidence that the task has been completed [LMK+17]. A less benign failure is that the agent may first consider various strategies by which it can achieve the required outcome, and by devoting more resources to computation, it can better consider alternative strategies. This can lead to an instrumental failure mode where the agent is always incentivized to obtain more resources to be surer that it has satisfied the desired outcome.

The problems of corrupted reward channel were outlined in [EKO+17] where some solutions were also presented. One of the major results was a Corruption No Free Lunch theorem for MDP agents (Theorem 15.11.3) says that without some assumption on the type of corruption of the reward, the best performance the agent can hope for is that it will perform as well as a random agent. This is somewhat concerning.

Theorem 15.11.3 (Corrupted reward MDP No Free Lunch theorem [EKO+17]) In the corrupted reward MDP setting (MDP setting with the possibility of corrupted reward), the worst-case regret (Section 8.1.4) of any policy π is at most a factor of 2 better than the maximum worst-case reward.

This effectively means that in a situation where the reward signal of an agent is compromised, it cannot do much better than the worst case.

One solution to this problem suggested in [EKO+17] is to use decoupled RL. In *decoupled RL*, the reward the agent receives is not directly from the environment, but comes from a randomly sampled new state and reward that the agent observes. This sampled new state and reward are based on the current state. The addition of this decoupling allows the construction of agents which have sublinear regret in the corrupted reward setting.

Another approach is to have agents which randomly choose to go to the state which has reward in the top quantile of possible rewards, instead of the maximum reward state. These agents, called *quantilizing agents* [Tay16], are able to avoid corrupted reward problems under some assumptions on the number of high reward corrupted states (states which produce a high corrupted reward) and the number of high reward non-corrupt states.

15.12   Embedded Intelligence

The current framing of the agent-environment setup for universal agents makes no reference to physical limitations on either time or space, and assumes a *dualistic* framing, where the agent is considered to be a separate entity from the environment it interacts with. For agents that exist only virtually, the code that defines the agent does not interact with the code that defines the environment (other than communicating actions and percepts back and forth), and (ideally!) there is no action the agent can take to directly modify the code that defines it, or the hardware on which it runs.

Embedded intelligence assumes the agent is *physicalistic*, part of the environment. For physically embodied agents this is obvious: a robot that physically interacts with the real world is necessarily part of the real world. This means that the agent may be able to take actions that change directly how the agent is defined. An agent may avoid a radiation source, even if no negative reward is assigned, as it could corrupt the working memory or the code of the agent, neither of which is likely to help the agent achieve its goal.[12] Such an agent may also make performance improvements or change its goals entirely by modifying itself.

The purpose of this section is to introduce the space-time embedded agent, an agent that is computed by the environment it interacts with. Such agents are also subject to constraints on both space and time. An agent that is aware of this may take different actions as a result. For example, it may recognize that computing an optimal action will take too much time, and instead make do with a suboptimal action. It may also change its policy to still compute the same strategy, but in a way that's more efficient to compute (and thus more time available for thinking).

A brief aside on notation: In [OR12a], the environment only return an observation rather than a (observation, reward) pair. The utility function is then defined over interaction histories $ao_{1:t}$. We unify this with our notation by defining $e_t = (o_t, 0)$, where the same dummy value 0 is used for all rewards. A utility function based on $h_{1:t}$ gets no more information than one based on $ao_{1:t}$, so the optimal policy is unchanged. A utility function $u(h_{1:t}) \in [0,1]$ assigns a utility value to each interaction history.

Self-Modifying (SM) resource-bounded universal intelligence.   The first formalization presented in [OR12a] describes a framework where the agent can self-modify, by choosing a new policy π_{t+1} at each time step t. For now, the agent and the environment are still separate.

The agent can choose policies from a space $\Pi^{\tilde{t},\tilde{l}}$ of policies, all policies that can be computed in $\tilde{t}$ time steps per interaction, and require at most $\tilde{l}$ bits of memory total (both to store the policy and the working memory required to compute it). Each policy $\pi \in \Pi^{\tilde{t},\tilde{l}}$ is a distribution $(a',\pi') \sim \pi(\cdot|o)$ over actions a' and future policies $\pi' \in \Pi^{\tilde{t},\tilde{l}}$, conditioned on an observation o. Note that while each policy may appear Markov, there may be long term dependencies since each policy π_t could internally store summative information about the observation history $o_{<t-1}$ and receive both this and the new observation o_{t-1} as input. Note that the policies cannot have total dependency on the observation history $o_{<t-1}$, as this would require an infinite amount of memory, so presumably the agent would want each policy π_t to learn what summative information in $o_{<t-1}$ should be "hardcoded" in the successor π_{t+1}.

The optimal self-modifying (SM) resource-bounded universal agent π^*_{SM} is defined as

---

[12]It is improbable, but not impossible, that flipping random bits might improve the agent. But only in the same sense that dosing humans with radiation *might* (with extremely low odds) cause superpowers, but always in practice has caused radiation poisoning.

follows:

$$\pi_{\text{SM}}^* := \underset{\pi_1 \in \Pi^{\tilde{t},\tilde{l}}}{\arg\max} V_{SM}(\pi_1, \epsilon)$$

$$V_{\text{SM}}(\pi_t, h_{<t}) := \sum_{a_t, \pi_{t+1}} \pi_t(a_t, \pi_{t+1} | e_{t-1}) \sum_{e_t} \xi_{\text{RS}}(e_t | h_{<t} a_t)(u(ae_{1:t}) + V_{\text{SM}}(\pi_{t+1}, ae_{1:t}))$$

$$\xi_{\text{RS}} := \sum_{\nu \in \mathcal{M}_{\text{RS}}} 2^{-K(\nu)} \nu(e_{1:t} \| a_{1:t})$$

where ξ_{RS} is the familiar Bayesian mixture ξ_U (Definition 3.1.2) using the Solomonoff prior $w_\nu^U = 2^{-K(\nu)}$ (Definition 3.7.2), except we only sum over the class $\mathcal{M}_{\text{RS}}$, the set of all *reward summable*[13] lower semicomputable semimeasures. This obviates the need for a discount function γ.

A policy π_1 with large $V_{SM}(\pi_1, \epsilon)$ must be a policy that induces a sequence $\pi_1, \pi_2, ...$ of policies that learn the environment μ well (and communicate that knowledge to the successor policy), as well as being good at selecting successors. For a policy π that always samples itself as the successor, this degenerates back to a sequence of space-time bounded Markov policies.

Space-Embedded (SE) agency. The above presentation allows for agent self-modification. We extend it further to have the environment itself compute the policy of the agent. As a result, the environment can directly read/write to the agent's memory, obviating the need for actions and percepts to communicate between the two. A potential successor π_{t+1}' is sampled from the current policy π_t, and then the environment can read/write to the memory of the agent (both the code defining π_{t+1}', and the working memory it uses) to obtain the true successor π_{t+1}. The interaction history is now just a sequence of policies $\pi_{1:t}' := \pi_1 \pi_2' \pi_2 \pi_3' \pi_3 ... \pi_t \pi_t'$, and the utility function is now defined over this sequence.

$$V_{SE}^{\pi_t}(\pi_{<t}) = \sum_{\pi_t'} \pi_t(\pi_t') \sum_{\pi_{t+1}} \xi_{\text{RS}}(\pi_{t+1} | \pi_{1:t}')\left(u(\pi_{1:t}' \pi_{t+1}) + V_{SE}^{\pi_{t+1}}(\pi_{1:t}')\right)$$

Space-Time-Embedded (STE) agents. We can now define an agent that unifies the two above definitions: an agent that is resource bounded, can self-modify, and is computed by the environment itself, akin to a human interacting with the real world the decision-making process of the human is itself part of the real world, and "computed" according to the laws of physics (assuming physicalism is true).

The above property of π_t sampling a potential successor π_{t+1}' is now rolled into the environment, so the environment directly gives a new policy π_{t+1} given $\pi_{1:t}$. The time and space bounds can now be enforced by the environment, so we only require a constraint on the first policy π_1 to be of length $\leq \tilde{l}$, as it is chosen prior to interaction.

$$\pi_{\text{STE}}^* = \underset{\pi_1 \in \Pi^{\tilde{l}}}{\arg\max} V_{\text{STE}}(\pi_1)$$

$$V_{\text{STE}}(\pi_{<t}) = \sum_{\pi_t \in \Pi} \xi^{\text{RS}}(\pi_t | \pi_{<t})\left(u(\pi_{1:t}) + V_{\text{STE}}(\pi_{1:t})\right)$$

In the definition of V_{STE}, we can write the sum as over all policies Π, since ξ_{RS} will assign zero probability to any policy that exceeds the space bound $\tilde{l}$, and the environment enforces the time bound $\tilde{t}$ when computing π_t.

---

[13]An environment ν is *reward summable* if, for any policy π, the infinite interaction history $h_{1:\infty}$ sampled from ν^π has bounded reward sum $\sum_t r_t \leq 1$.

While the physicalistic approach is more philosophically grounded, its practical implementation is often challenging. The environment must be able to contain an approximation of the agent as we start approximating the agent's intelligence measure. While the environment may not contain the optimal agent, it can theoretically contain an ε-optimal agent for any positive ε. In such cases, a specific space-time embedded definition is not necessary. Furthermore, a practical and useful approximation of ρ from the space-time embedded agent definition may require an approximation of our universe, which is a strong assumption and unlikely to be available soon. Nonetheless, since intelligent agents are contained within the universe they act upon and model, it might be more sensible to consider cases where the agent needs to model itself.

15.13 Exercises

1. [C30] We have listed two definitions of death of an agent. Try to develop other formal definitions of what it could mean for an agent to die. How do these relate to each other and the ones presented in this chapter?

2. [C12] Prove Theorem 15.6.2.

3. [C15] Prove Theorem 15.7.4.

4. [C11] Prove Corollary 15.7.5.

5. [C14] Prove that the reward associated with an environment can be scaled by a positive multiplicative constant while leaving the optimal policy unchanged. Show that the result does *not* hold for affine transformations when the environment is a semimeasure.

6. [C15] Prove that in the case that the environment is deterministic, if the rewards are transformed by a monotonically increasing transformation, then the behavior of the agent will remain the same. Prove that this theorem fails when the environment is stochastic.

7. [C32] Formalize the setup and agents discussed in Section 15.10. Prove the claims about $A_{rl}^{\rho}, A_g^{\rho}, A_p^{\rho}, A_k^{\rho}$ formally.

15.14 History and References

The technological singularity. Already the invention of the first four-function mechanical calculator one-and-a-half centuries ago [Tho47] inspired dreams of self-amplifying technology. With the advent of general purpose computers and the field of AI over half-a-century ago, some mathematicians, such as Stanislaw Ulam [Ula58], I.J. Good [Goo65], Ray Solomonoff [Sol85], and Vernor Vinge [Vin93] engaged in singularity thoughts, as well as roboticist Hans Moravec [Mor88] and physicist Frank Tipler [Tip95], and others. But it was only in this millennum that the singularity idea achieved wide-spread popularity. Ray Kurzweil popularized the idea in two books [Kur99, Kur05]. The ingularity Institute (now MIRI) was founded in 2000 and the internet helped in the formation of an initially small community discussing this idea. Between 2006 and 2017 a dozen Singularity Summits approaching a thousand participants were held in America and Australia. Only from the mid-2010s it slowly became acceptable in academia to openly talk about AGI/ASI. Hutter's book [Hut05b] on Universal AI, Goertzel's anthology [GP07] on AGI, Legg's PhD thesis [Leg08] on machine super intelligence, and Chalmers [Cha10] paper on the technological singularity are early outliers, paving the way for the many books and research articles on the topic to follow (see below). The 1-billion Euro Human Brain Project (2013-2023) aspired (but failed) to emulate a whole human brain. Now there are hundreds of organizations and companies working on AGSI or safety thereof or both. DeepMind (2010) and OpenAI (2015) are the AGI leaders, followed by more recent startups, Anthropic (2021), Inflection AI (2022), Mistral AI (2023), and many others. Most of the big tech companies also invest billions of dollors in the race towards AGI. From the AI safety side we have MIRI (2000), FHI (2005), CSER (2012), FLI (2014), CHAI (2016), CAIS (2022), and many others.

For a deeper discussion of what it could mean for intelligence to explode, separating speed from intelligence explosion, comparing what super-intelligent participants and classical human observers might actually experience and do, implications for the diversity and value of life, possible bounds on intelligence, and intelligences right at the singularity, see [Hut12a].

General ASI Safety literature. A complete and comprehensive review of the literature on ASI Safety (up until 2018) can be found in [ELH18], with many of the details being expanded upon in [Eve19]. [Bos14, Rus19, Ord20, Yam24] provide gentler introductions into the broad topic of AI Safety. One of the central problems in ASI Safety is the alignment problem [Gab20]. The alignment problem has been cast into the history-based reinforcement learning / UAI framework in [EHKK21], where the various sources of misalignment for history-based reinforcement learning agents were categorized. Investigations into how to design intelligent systems that are aligned can by found in the survey paper [TYLC16]. Formal descriptions of many of the problems in AI Safety were provided in [AOS+16]. Additionally, [HC22, CHO22] shows how, with minimal assumptions, an intelligent agent can be dangerous. [FBB20] provides a survey of the various active groups working on AGI projects, and the degree to which they focus on AGI safety. [NCM22] provides a survey of some of the topics covered in Section 15.2, and the degree to which they appear in the deep learning setting.

UAI-based Safety. Universal Artificial Intelligence theory has been extended in several ways to study and solve aspects of the ASI Safety problem. These include: [CH20b], which showed that a pessimistic Bayesian agent will not cause "unprecedented events", in [CHC21] it was proven that any agent that is asymptotically optimal would die. [CH20b, CHC21] both showed that with the guidance of a mentor, safe exploration is possible. [CVH20, CVH21] demonstrated the safety brought on by unambitiousness and derive a version of AIXI which is unambitious and does not learn to seek power arbitrarily. Lastly [OA16] showed that under certain conditions, Bayesian agents would be incentivized to allow themselves to be interrupted.

Self-modelling, self-modification, and embeddedness. One of the earliest formal treatments of the problem of self-modification was done in [OR11]. This was extended by [EFDH16] with both works being described in Section 15.8. Self-modification in the context of bounded rationality was explored in [TSG20]. The related topics of self-modelling and embeddedness have also seen significant investigation. For self-modelling, [Hib14, Hib15] have proposed a logical theory for self-modelling agents in finite universes. On the topic of embeddedness, [MSZ19] categorized the connection to wireheading, [Mil21] developed an extension of the MDP framework which included embedded agents, [DG19] provides a comprehensive account of many of the problems which arise when trying to solve the embeddedness issue, and [OR12a] extends the UAI theory to an agent which is embedded within its environment (discussed in Section 15.12). One aspect of being contained within the environment an agent has to consider is death. Agent death and how death fits into the formality were studied in [RO11, MEH16]. Most of these are expanded upon in Sections 15.7 and 15.12.

An approach to solving the wireheading problem (described in Section 15.9) using modified RL was presented in [EH16]. The similar problem of avoiding agent incentives to tamper with the RL setup were tackled in [UKK$^+$20]. More generally, the problem of reward tampering was studied in the framework of causality in [EHKK21]. The problems and solutions involving designing agents and reward functions to avoid (often negative) side effects were explored in [KOML18, KOK$^+$18, KON$^+$20]. [EKO$^+$17] (described in Section 15.11) extends the MDP framework of reinforcement learning to include the possibility of corruption and provides a no-free-lunch theorem in this setting. The implications of universal AIXI-like agents possessing memory in the real world were discussed in [OR12b].

[KUM$^+$20] covers the concept of specification gaming on a high level, its causes, and many examples in deep RL agents. [SHKK22] gives a formal definition of *reward hacking*, showing that unhackability is a very strong condition for stochastic policies.

[DLKS$^+$22] defines *Goal misgeneralization*, where an agent learns a proxy of the true intended behavior that is correct in distribution, but fails to generalize on out-of-distribution environments. They provide examples where this is shown to happen empirically. [SVK$^+$22] demonstrates goal misgeneralization in a broader class of environments, as well as suggesting mechanisms to mitigate or prevent it from happening. [TSS$^+$19] shows that optimal policies tend to be power seeking, as defined by Bostrom [Bos14]. [EKH19, EH18a] describes an environment where the reward function can be tampered with, and define an agent that optimizes based on the current reward description, removing the incentive to reward hack. This is expanded on in [EHKK21], which covers both tampering with the received reward signal, and tampering with the input to the reward function, as well as mitigation strategies for both.

Deception. As an intrinsic goal, we may expect that agents might act in a deceptive manner, that is, provide outputs that are inconsistent with the models internal beliefs. This can be as benign as playing games of bluff, or as malicious as deliberately generating plausible looking misinformation.

[BS19] demonstrates a model that can play Poker at the elite level, learned entirely from self-play. [FDT22] considered an agent powered by large language models that is able to play the game *Diplomacy* at the human level. The agent learns to reason strategically, make (and break) alliances, and deceive other players. As part of a red-teaming exercise, GPT-4 [Ope23a] used TaskRabbit[14] to hire a human to solve a CAPTCHA[15] for it. When prompted to reason aloud, the model decided to not reveal this fact, but lied about having a vision impairment. [SBH23] shows that large language models (such as GPT-4) can decide to generate plausible lies to cover illegal activity when there is an incentive to do so, without being explicitly told to lie.

Other work. Having an off-switch "button" is a standard safeguard against malfunctioning systems. Unfortunately, rational agents aiming at maximizing expected utility typically have an incentive to prevent being switched off, since a dead agent cannot achieve its goals, but [HMDAR17, WBC$^+$17] show that agents uncertain about their utility function may allow themselves to be switched off. As we have shown in Section 15.7, a negative reward range incentivizes an agent to shut down once it has the power of doing so.

A theoretical approach, Cooperative Inverse Reinforcement Learning (CIRL), which satisfies many criteria for the solution to the ASI Safety problem, was presented in [HMDAR16]. Arguments against the compression approach to learning human preferences in Inverse

---

[14]A service to hire freelance workers to perform tasks.
[15]A test to determine if the user is human.

Reinforcement Learning (IRL) were brought forth in [AM18]. Additionally, some faults of learning the reward function online, as is done in IRL, were discussed in [ALOL20]. [Cas20] explores ways in which ASI systems may have stable failure modes that lead to irrational decisions, as well as how such weaknesses may be deliberately implanted into an ASI as a method of control. On the more practical side, [LMK+17] presented a suite of environments for testing the safety capabilities of various agents.

There has been much work on the role of fairness in Artificial Intelligence, but very little in relation to Universal Artificial Intelligence. The following two pieces of work by the first author may be AGI-relevant but are not really ASI-relevant nor UAI-specific: [Hut19] turns fairness into a bi-level optimization problem maximizing fairness without compromising on the primary objective. [HH21] discusses many of the impacts of highly intelligent AI on society.

Chapter 16

Philosophy of AI

Artificial intelligence would be the ultimate version of Google. The ultimate search engine that would understand everything on the web. It would understand exactly what you wanted, and it would give you the right thing.

Larry Page

The philosophy and possibility of intelligence created by artificial means has been discussed and debated extensively well before the invention of computers. In Turing's seminal paper [Tur04], he introduces and discusses informally many new concepts in artificial intelligence that have been built upon since, like describing what it means for a machine to be *universal* (Turing complete), exploring the idea of a machine being able to self-modify by *screwdriver interference* (hardware self-modification) or by *paper interference* (changes from new information, of self-modification of source code). While the terminology from this paper may now be dated, the philosophical ideas are still solid: Should we expect AGI, and is it even something we should build?

Section 16.1 revisits the philosophical backings of Epicurus, Occam and Bayes on which universal (Solomonoff) induction was based on. We argue why these are sensible philosophical positions, and how the universal prior encodes these. Section 16.2 discusses consciousness and free will, and explores to what degree universal agents can or cannot be said to have these properties. Section 16.3 discusses moral considerations, the weak vs. the strong AI hypothesis, and some philosophical arguments for each. Section 16.4 explores some work done on how AIXI may or may not choose to replicate itself, and if AIXI considers the copies to be a new agent, or the same as itself.

Section 16.5 presents some arguments against the possibility of AGI: the Chinese room argument which argues a book might be said to be intelligent if intelligence is only measured in a black-box fashion, and self-referential Penrose–Lucas Gödel statements that are true, but AIXI will never be able to prove true. Section 16.6 presents some arguments for the possibility of AGI: the physical Church-Turing thesis which implies that humans are also just machines, and Moore's Law showing that we will soon have the compute power of human brains at our finger tips. Experimental demonstration has been a major piece of evidence for the case of eventual AGI. Over the decades, more complex tasks have been solved with AI. Together, these arguments make it not inconceivable that we will eventually (and maybe even soon) construct a general agent which is able to solve all solvable tasks (that we care about).

Section 16.7 explores the formal (LH) definition of intelligence that aims to capture what it means for an agent to achieve goals in a wide range of environments, as opposed to other informal definitions of intelligence that ascribe to intelligence various qualities correlated with intelligence in humans, like creativity or curiosity. We compare this with other definitions from philosophy and with attempts to practically quantify and measure intelligence. We observe that the universal AIXI agent maximizes the LH-intelligence measure. LH-intelligence in general is incomputable, so we discuss practical approximations.

Section 16.8 remarks on the relation of Deep Learning and Universal AI, especially how Large Language Models (LLMs) approximate Solomonoff induction, and Tree of Thought is a form of MCTS approximating expectimax planning in AIXI. Section 16.9 concludes with a summary what UAI provides in general, and how it can provide a formal path to AGI in particular.

16.1   Philosophy of Universal Induction

The induction problem is an ancient and contentious problem in philosophy, with the earliest contributions going back to at least Epicurus [Mil87]. Roughly speaking, induction refers to drawing generalizations from data or evidence, or constructing a model of the world based on observations, which can be used to (predict) what will happen in the future, based on the past. With some liberties we can also use it to encompass predicting the next element in a sequence.

Justification for model class $\mathcal{M}_{sol}$. Here we will argue why universal induction provides a theoretical solution to the induction problem. *Universal "Solomonoff" induction* is performing induction with a Bayesian mixture, starting with the universal prior $w_\nu^U = 2^{-K(nu)}$ over model class $\mathcal{M}_{sol}$ (Section 3.7) of all semicomputable semimeasures. or equivalently using Solomonoff's a-priori distributions M (Section 3.8).

This choice of model class $\mathcal{M}_{sol}$ is motivated by the physical Church-Turing thesis, the assumption that all physical processes are computable (Section 16.6.1). Assuming that the physical universe is at least computable, and by choosing the set of computable environments[1], we include any environment that the agent could ever possibly encounter. A more pragmatic reason is that if the environment itself is incomputable, then there is by definition no computable agent that could possibly learn such an environment.

Justification for prior $w_\nu^U = 2^{-K(\nu)}$. The choice of prior has both satisfying philosophical and mathematical motivations. First, it combines:

- *Epicurus' principle*: If more than one theory is consistent with the observations, keep all the theories.

  Since $2^{-K(\nu)} > 0$ for all ν, regardless of how complex an environment may be, no environment is ruled out as a possibility a-priori.

- *Occam's razor*: Entities should not be multiplied beyond necessity (keep the simplest theory consistent with the observations).

  The concept of Kolmogorov complexity gives a foundational and unbiased[2] definition of simplicity, and since a "simpler" environment ν has a smaller Kolmogorov complexity $K(\nu)$, the corresponding prior weight $2^{-K(\nu)}$ is higher.

- *Bayes' rule*: A mathematical rule for conditional probabilities that can be interpreted as a method by which confidence in hypotheses (beliefs) should be updated based on new evidence/data.

  The update rule for universal sequence prediction directly comes from Bayes' rule, with the choice of prior $2^{-K(\nu)}$ motivated by the previous two principles. The posterior $w^U(\nu|x_{<t})$ of a hypothesis ν becomes zero/small/large if data $x_{<t}$ rules out / is implausible under / is consistent with ν.

Recall from Section 3.2 that in Bayesian sequence prediction, the bounds are always in terms of w_μ, the prior weight on the true environment μ. Using the universal prior, we find that the number of errors during prediction is upper-bounded by a term proportional to the Kolmogorov complexity of the environment (Section 3.7.1). One interpretation of this is that the signal from the errors received during prediction can be used to uniquely identify which environment is the true environment, and there cannot be a prediction schema that performs better, as this would imply a shorter description of the environment than its Kolmogorov complexity, which is a contradiction. In this sense, the universal prior is indeed an optimal choice of prior in terms of prediction errors required to learn the environment as quantified in Section 3.7.2.

Example 16.1.1 (Induction for brown and blue foxes) In induction, when we see evidence in favor of a hypothesis, the confidence placed on that hypothesis increases, but we can never fully confirm the hypothesis. The absence of a counterexample does not definitively

---

[1] Here we extend to the class of semicomputable environments for technical reasons.

[2] The definition of Kolmogorov complexity still relies on the choice of (a natural) universal Turing machine, but this changes the definition of Kolmogorov complexity by at most an additive constant, see Section 2.7.

Table 16.1: *Approximate correspondence between concepts in induction and deduction* *[RH11].*

	Induction	⇔	Deduction
Type of inference:	generalization/prediction	⇔	specialization/derivation
Framework:	probability axioms	≡	logical axioms
Assumptions:	prior	≡	non-logical axioms
Inference rule:	Bayes rule	≡	modus ponens
Results:	posterior	≡	theorems
Universal scheme:	Solomonoff Probability	≡	Zermelo-Frankel set theory
Universal inference:	universal induction	≡	universal theorem prover

prove that a counterexample might not still exist out there somewhere. One might argue that it is not inconsistent with placing absolute certainty that a particular hypothesis is false. But for hypotheses like "there exists a blue fox", it would seem undesirable to assign probability exactly zero to it regardless of how many brown foxes are observed, as a blue fox may always be hiding somewhere waiting to be observed. On the other hand, one would think that upon observing one blue fox, we can rule out the hypothesis "all foxes are brown" and immediately assign probability zero to this hypothesis. However, there should always be a degree of skepticism placed on any observation made (a camera reading sensory input might be noisy, or we may have made a mistake interpreting the results). We can never be 100% sure that any blue fox observed really is blue (maybe there was a trick of the light, or a brown fox was painted blue. Seeing a blue fox only provides (admittedly very strong) evidence against the hypothesis "No fox is blue", rather than an irrefutable rebuttal, so the new confidence in this hypothesis would be very close to zero, but not exactly zero.   ♦

Combining logic and probability for learning. An alternative way of reasoning is deductive reasoning. Deductive reasoning is the process of taking logical statements and applying rules from an axiomatically defined reference set of rules (called deductive rules) to draw conclusions. In deduction, we are concerned with whether a statement is true or false, and we try to derive the truth of a statement based on other derived true statements (theorems) or statements that are *a priori* assumed to be true (axioms). This is in contrast to induction, where the conclusions we draw are uncertain and have associated with them a probability value in $[0,1]$, representing our confidence in the validity of that statement.

Recent attempts to combine logic with probability [HLNU13b, HLNU13a] managed to assign probabilities to quantified sentences in such a way that induction works, while [GBTC+20] develops a probability theory for computationally limited deductive reasoners. Many concepts in deduction and induction have correspondence to each other, as demonstrated in Table 16.1 from [RH11].

16.2   Consciousness, Free Will, and Other Qualia

One question which often comes up in discussions around artificial intelligence is whether or not an AGI would possess or experience consciousness [Min07, Ben17], emotions, free will [Bal14, Reh21], sentience, etc. (collectively called *qualia* in philosophy), and ethical considerations of whether we even should be striving to build AGI to begin with, and whether an AGI can have moral value and/or feel pleasure and pain as humans do. These are indeed interesting philosophical questions, and require us to formally define what we mean by consciousness, free will and so on.

Free will. In the case of free will, one would argue that an intelligent agent does possess free will as it has the "choice" of any action (out of some set of actions). However, one could counter this with an argument that the agent will always follow a set sequence of actions, governed entirely by the policy the agent uses and the history of events up to that point. Even stochastic polices cannot accommodate free will, since now the agent has to respect the outcome of an uncontrollable random process. Especially a reinforcement learning agent is confined to slavishly take Q-value-maximizing actions, though it could freely choose between actions of comparable Q-value at will [Reh21]. The free will paradox is solved in [Hut05b, Sec.8.6.3].

Do RL-agents feel pain and pleasure? In the case of emotions, again we need to make sure we can define it formally:[3] Is it the case that an RL agent is happy when it receives high positive reward, and sad or depressed when it receives low reward? Would subjecting a sufficiently intelligent RL agent to large negative rewards be tantamount to torture [Pet15]? We have previously discussed (in Section 15.7) that if the reward is sufficiently negative (and is going to be for a sufficiently long time), a reward maximizing agent will choose "suicide" if the option is available. In this case, is the agent sad or depressed? One could also argue that it is not a maximal reward, but an unexpected positive change in reward that causes an agent to be happy: when rewards are far higher than the current agent's best estimate [Chr20]. This concept of learning from errors between the expected and actual reward is called *temporal difference learning*, a paradigm of reinforcement learning that has been shown to have some link to the response of dopamine in primates studied in neuroscience [NDD05].

Does AIXI have subjective experiences? This is related to the concept of the hedonic treadmill [Bri71], where human happiness will tend back to a baseline level after a major positive/negative life change. We often use AIXI as a formal instantiation of a sufficiently intelligent agent, however for some philosophical questions, AIXI would not be appropriate. Regarding the concept of self-awareness, AIXI does not explicitly contain a model of itself, as the hypothesis space that AIXI considers is not broad enough to contain itself. Any environment that includes a self-model of AIXI would not be computable. AIXI may however possess approximate models of itself and a form of implicit self-awareness, in the sense that it considers the effects of its own actions on the environment, and maximizes reward with this consideration in mind. It would be bold to claim that AIXI is not self-aware when it is able to model its own actions as an element of the world. With regard to most of these problems, the view taken here is a pragmatic one in that if any of these properties (consciousness, emotions, etc.) are necessary for intelligent behavior, then we expect they will be emergent in agents like AIXI. This is because AIXI is by construction the most intelligent agent (with respect to the Legg–Hutter definition of intelligence) possible, so it will display consciousness/emotions/etc. if only if they are useful for acting intelligently.

Formalizing consciousness. Constructing a formalization of these concepts can often be quite difficult, and for many seems nigh impossible. In the case of consciousness, there has been one attempt called *integrated information theory* (IIT) [Ton12, OAT14, TBMK16]. It deals with the problem of consciousness in a mathematically formal and elegant way. The axioms of consciousness according to this theory are:

- consciousness and experience exists,
- each experience is independent of external observers (intrinsic existence),

---

[3]Unless you follow Hume's assessment of reason and the passions [Hum03], where he argues that reason alone does not motivate actions, but rather feelings and desires (which are not based on rational reasoning) do.

- consciousness is structured (composition),
- consciousness is specific: each experience is the particular way it is (information),
- consciousness is unified: each experience is irreducible to non-interdependent components (integration),
- consciousness is definite (exclusion).

From this, Tononi derives a measure of complexity Φ that he argues strongly correlates with consciousness. IIT is not without its problems. For example, according to this theory, there exist (from an intelligence perspective) rather simple expander Graphs and 2D grids of XOR gates that are assigned implausibly high consciousness. One can draw a parallel between this and the Chinese room argument (Section 16.5.2). While ultimately IIT may be wrong, at least it is a formal start. Other proposals are less formal and complete and hence avoid hard falsification, at the price of possibly "not even being wrong".

16.3   Moral Considerations

One of the main questions that often comes up is whether is it possible for machines to think as humans do, or if they only act as if they can think, without an internal line of reasoning. This gives rise to two hypotheses:

1. **Weak AI hypothesis:** An artificial intelligence can take actions *as if* it were intelligent.

2. **Strong AI hypothesis:** It is possible to build a machine that can think in the same sense that a human can: It is self-aware, and has a conscious experience of the world around it.

Most AI researchers take the weak AI hypothesis for granted, and on face value it seems meanwhile to be true in many domains (Section 16.6.3). From a pragmatic perspective, one can argue that it does not matter if the strong hypothesis holds or not, so long as the agent can perform the tasks we require of it.

Moral relevance of the Strong AI hypothesis.   However, from a moral perspective, it would be like comparing a modern autonomous vacuum cleaner and a human worker indentured to clean for no pay: The latter brings with it the concern of moral harm. Currently, we do not pay any mind to the welfare of our robotic vacuum cleaners, the algorithms online that feed us content, the next video to watch, suggestions of what to buy, etc., nor even the smart assistants in our mobile phones (though some people get emotionally attached to the latter). It is conceivable that in the future a super-intelligent agent may have moral status ascribed to the computations it performs. In June 2022, Blake Lemoine, an engineer at Google, sounded the alarm regarding LaMDa, a Large Language Model (LLM) that he claimed was sentient based on the interactions he had with it [Joh22].

Neural substitution argument.   The strong AI hypothesis asks for a much stronger statement. While technologically out of reach, one argument in favor of the Strong AI hypothesis is the "neural substitution argument", adapted from [Mor88]:

> *Suppose in the future, the process by which neurons operate are fully understood and characterized, to the point where synthetic neurons can be created to replace biological neurons, say, as a treatment for neurological diseases like Alzheimer's. A patient comes to the clinic for surgery and an awake craniotomy is performed, exposing the brain while the patient is still conscious. Nanobots are injected*

into the brain, where they navigate around, scanning for brain tissue at risk and replacing it with a synthetic drop-in replacement. The patient reports no loss or change in consciousness, feelings, other qualia, or memory. In extreme cases, the entire brain might slowly be replaced by a synthetic equivalent without the patient noticing any change in experience.

The assumption that the consciousness or personality of a human resides entirely within the neurological connections in the brain is a philosophical position called *physicalism*, as opposed to *dualism*, where mental processes cannot be explained solely as an emergent phenomenon of physical processes. If synthetic neurons can perform the same function as the original neurons, then we would expect that the patient would notice no difference. She would still act the same, feel the same, and her core concept of self is preserved. It seems arbitrary to assign moral value to the patient before the procedure but not afterwards.

Brain uploading. A related concept is that of *brain uploading* or *digital immortality*, where the patient's new brain need not be made up of physical matter. The pattern that makes up the brain can instead be hosted on a server where a virtual environment is simulated. The mechanics of how the neurons in the brain interact are simulated, with data fed in from the environment to simulate expected sensory data like sight and sounds, and the data from the brain can be used to pilot a virtual avatar that the patient now inhabits. Even though the patient and his entire environment are simulated, it will feel just as real to him as reality feels to us: Simulate feeding in electrical impulses that correspond to touching something hot, and the "person" will recoil their virtual hand in response. Like the hypothetical brain-in-a-vat [Har73] with simulated input stimuli, everything would feel very real to him. Clearly such technology has great and vast nefarious applications in the wrong hands (as e.g. detailed in the short fictional story *Lena* [Hug21]). While fanciful science fiction at the moment, the concept is not so far-fetched, and there is active research in this direction. For instance, *OpenWorm* [SLP+18] is an open source project with the goal of simulating *Caenorhabditis elegans* (a species of roundworm) at the cellular level.

16.4   Teleporting and Copying AGI

Will intelligent agents replicate themselves? Given that digital agents would have the ability to make copies of themselves cheaply and at high fidelity, it would seem likely that self-replication would be an action such agents would take.

In the AIXI framework, if self-replication will lead to higher expected rewards, then we would expect that AIXI will replicate itself. In [Ors14b, Ors14a], the concept of self-replication is formally defined, extending the AIXI framework to a setting where AIXI and AIμ, the ξ-optimal and μ-optimal-agents, have the option to make copies of themselves, or "cut-and-paste" (effectively teleport) themselves, destroying the originals. Much like exploring if AIXI would self-modify (Section 15.8), whether AIXI/AIμ would copy or teleport depends on how the value function is defined. Orseau defines several such value functions: a copy-centered agent whose value function includes the observations from all its copies, and two slot-centered agents who both consider that their future observations are ones that will be the output from a slot i, with one being given the slot number i and the other that estimates the slot number. These last two value functions are called static and dynamic respectively.

Orseau introduced two forms of *teleportation*; these were two possible forms of physical teleportation we could eventually possess. The first form of teleportation, shown in Figure 16.2, is one where the agent is cut-and-pasted into a new slot. This is like some form

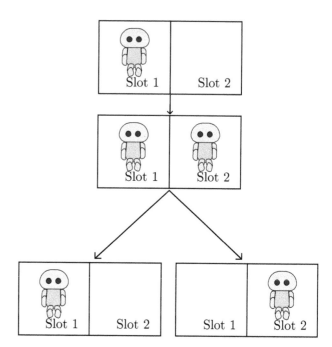

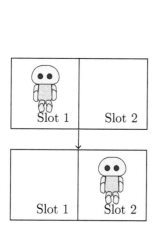

Figure 16.2: *An illustration of cut-and-paste teleportation, where the agent preserves continuity of existence (essentially the same as moving the position of the agent).*

Figure 16.3: *An illustration of copy-paste and delete teleporting, where the agent is copied, the copy is moved, and then either the original or the copy is destroyed.*

of wormhole teleportation, where only the position of the agent is changed. The second form of teleportation, shown in Figure 16.3, is one where the agent is copied into a new slot and then one of the two copies of the agent gets deleted. This is similar to a form of teleportation (common in sci-fi shows like *Star Trek*) where the agent is scanned, and a description of the agent is sent to the target location where it is reconstructed from local material based on the description. The "original" agent is then destroyed.

In both types of teleportation, the copy-centered agent will choose to teleport as it values having at least one copy of itself to receive observations. However, in the case of the two different slot-centered agents, the agents will not teleport if they are AIμ agents (to preserve itself), but it may choose to teleport if the agent is a slot-centered AIXI agent as it is uncertain about its own slot. See [Ors14b, Ors14a] for a formal treatment and proofs and some experiments.

16.5   Arguments against AGI

While there are many arguments against AGI, most tend to break down once they are required to be formalized. There is a long list of things important people have argued in the history of AI a computer will never be able to do at all or as well as humans (arguments of disability). The Chinese room argument is a thought experiments that even if a computer can mimic human intelligence, this does not imply that it understands or is conscious. The No Free Lunch theorems state that uniformly averaged over all problems, all algorithms perform equally (poorly), which seems to imply that universal solutions are impossible, and domain-specific knowledge is needed, so in particular that AGI is impossible. The Lucas–Penrose argument is based on a Gödel-like sentence "machine X cannot prove that this sentence is true" which we (humans) can determine is true, but machine X is unable to prove.

16.5.1 Informal Arguments of Disability

The *argument of disability* is that there are tasks that humans can do that an AI never can: play chess, translate languages, read faces, write poetry, compose music, hold a conversation, etc. and other such "creative" tasks that a machine could never replicate. We will not repeat the (usually informal) argument for reasons of being obsolete. Often, a task once thought to be impossible could be solved (or at least had attempts made with reasonable performance) once the task was described formally, the right algorithmic approach was found, and/or sufficient compute and/or training data became available.

An AI could never solve chess, until it did [CHJH02]. Never could an AI solve a game as complex and requiring deep insight like Go, until it did [SH+16].[4] Only humans can play video games with huge, continuous action spaces, using only the visual feedback from the game like *Dota 2* and *Starcraft 2*, until machines could do it too [VBC+19].

All the other tasks listed above have also meanwhile been succumbed to Large Language Models (LLMs) such as ChatGPT and diffusion models such as Stable Diffusion (see Section 16.6.3).

There is a common attitude of constantly shifting the goalposts on what tasks a machine would need to solve for it to count as artificial intelligence. Once a task can be solved by AI, it no longer counts as something within the domain of AI. But AI-skeptics are running short on tasks that state-of-the-art AI cannot do, and risk converging to the absurd claim that no task requires intelligence.

16.5.2 Chinese Room Argument

The *Chinese room* argument [Sea80] is as follows: Imagine there is a room with a small slot into which questions written in Chinese can be inserted, and from which answers (also in Chinese) retrieved. After a back-and-forth conversation via the slot, it appears from the outside that the contents of the room are fluent in Chinese. Inside the box, however, is no Chinese speaker, but an English speaker (who knows no Chinese) with access to an instruction book (written in English). The human matches the input symbols with those in the book, which then gives instructions on how to draw the symbols for the corresponding response (Figure 16.4). The argument is that clearly the human in the box does not understand Chinese (as the person is mechanically following the set of instructions provided without understanding what they mean), nor does the book of instructions or the pens or paper the human is using. So even though from the outside, the box appears to be fluent in Chinese, this does not imply that anything inside the box (or the box itself) can be said to understand Chinese. One can then argue the same is true for an artificial intelligence: It is constantly reading from a book of instructions, not actually understanding what it is doing.

There are some counter-arguments to this idea. For one, it does not refute the weak AI hypothesis (Section 16.3), which is what we are primarily concerned with in this book. Our intelligence measure is based on the performance of how the agent acts, not what the internal thought process of the agent is, nor to what degree the agent is conscious.

If we do care about the strong AI hypothesis and not just how the agent acts, then in the setup of the Chinese room, the process of using the book itself could be ascribed intelligence

---

[4]To the point where the techniques used historically to solve chess are often not even called AI anymore, but rather *heuristically driven search* or *optimization*, as if to imply that if a machine can perform a task, one need not be intelligent to do it.

Figure 16.4: *An illustration of Searle's Chinese room thought experiment: An agent that does not speak Chinese uses a lookup table to construct responses to given inputs.*

as it possesses perfect knowledge of the meaning of words, characters and sentences in Chinese, and therefore on some level "understands" the meaning of these words, characters and sentences.

Another refutation is objecting to the book: A book that would allow someone to converse fluently without any understanding could not just be a phrase book or a Chinese-English dictionary. If it were, the user of the book would need to at least be able to break down the input sentence into words, look up each word and translate to English, construct a response in English, and then use the book to translate back to Chinese. The user in the box would then need at least some understanding of Chinese (in that Chinese sentences are made up of smaller words/symbols, each with a corresponding meaning in English). If the user truly had no understanding, the book would have to be a truly massive lookup table of Chinese phrases and the corresponding responses, which would require some complex mechanism to look up each response given an input in any reasonable time frame. A book by itself is certainly not intelligent, but it is not so clear that a massive database with a mechanism for looking things up quickly would not count as "understanding" in some way. We may also ascribe understanding to anything that could create such a book. Given the exponential or even infinite possible ways of forming sentences, the book may not even fit into the accessible universe.

One could also argue that when humans claim to "understand" something, this is merely an emergent phenomenon of individual neurons firing and chemicals interacting in the brain. A neuron is just a cell that fires an electrical impulse in reaction to neighboring neurons, so following the same line of reasoning as before, if no part of the brain can be said to think or understand, then humans do not either. Obviously this is absurd (from our perspective): We (believe we are able to) interpret whatever process is occurring within our mind to be our own thoughts. The fallacy in the argument is assuming that the sum cannot have a property that none of the constituent components have. Any single bolt or nut or rod, or any component of an engine from a car is an inanimate object, only the collection of the parts together, arranged in just the right way, makes a working vehicle. No single part of the car has the property of being able to transport humans around.

16.5.3 No Free Lunch

The *No Free Lunch* (NFL) theorems, while not discussed elsewhere in the book, are popular enough to warrant at least mentioning. NFL is a holy grail in certain circles and popular among many machine learners. It is frequently quoted to informally argue against general solutions and for domain-specific (narrow) approaches. Since AGI is the ultimate domain-agnostic problem solver, NFL has also been used as an argument against AGI. There is an ongoing disagreement between believers in Occam's razor and believers in NFL.

The NFL theorems basically state that if the performance of an algorithm is uniformly averaged over all possible problems (of a certain class), it will be no better than any other algorithm [WM97]. That is, on average, all algorithms perform equally poorly. While contradicting every practical evidence, the theorems are indeed correct, and proponents argue that they imply that science needs to make domain-specific assumptions to overcome this problem. The NFL theorems are formalized versions of what Hume called the *principle of uniformity of nature*, which invalidates inductive reasoning. Many NFL versions exist, for optimization and for prediction, and the uniformity assumption has been relaxed [IT05, Wol23], but not to the extent undermining its critique, since some form of uniformity remains luring behind all versions. For sufficient violation of the uniformity assumption (but still remaining domain-agnostic), Free Lunch becomes possible [LH11b, ELH14].

The fallacy of NFL believers is to equate "uniformly distributed problems" with "uniformly distributed data". The latter is equivalent to data being almost surely being white noise. Naturally, (a) white noise contains no structure and induction must fail, but (b) few if any care about performing well on (predicting or classifying or optimizing) white noise [RH11, Sec.2.5].

On the other hand, Section 3.8.1 shows that if *problems* defined as programs are uniformly distributed, then induction works extremely well [Hut10b, Sec.8]. By piping uniform[5] random noise through a universal Turing machine, the output strings have a bias towards simplicity, meaning performance on simple problems will contribute more to average performance than complex problems. This is *not* a domain-specific assumption. Occam's razor is one of the essential cornerstones of the scientific method itself. It is a prerequisite of being able to do science at all [RH11, Sec.2.5], and a cornerstone of Universal AI [Hut00, Hut05b, HQC24].

Thesis 16.5.1 (Universal=M vs. uniform=NFL sampling)

- The universal distribution M is a non-dogmatic quantification of Occam's razor, a cornerstone of science and a solution to the induction problem.

- The uniform distribution over data is a dogmatic prior;
  the belief that the world is pure white noise.

There are some other ways to circumvent the NFL theorems without invoking Turing machines, by using meta-induction [Sch19b], but see [Wol23]. See Section 3.7.2 for a simple(r) "proof" of Occam's razor.

16.5.4 Penrose–Lucas Arguments

The Penrose–Lucas argument [Pen89, Pen94] is effectively an instance of the *argument from disability* in Section 16.5.1, but this one is formal and interesting to engage with. The argument is that there exist logical statements that are true, but an AI will never be able to

---

[5]Remarkably, actually any non-uniform random noise [Ste17, GMGH+24] leads to Solomonoff's distribution on the output tape, showing that M is a kind of attractor.

prove they are true, but humans can see they are true by reason. Penrose claims that this argument is a consequence of Gödel's incompleteness theorem [Göd31, Göd86]. The exact argument from Penrose is:

> *The inescapable conclusion seems to be: Mathematicians are not using a knowably sound calculation procedure in order to ascertain mathematical truth. We deduce that mathematical understanding – the means whereby mathematicians arrive at their conclusions with respect to mathematical truth – cannot be reduced to blind calculation!*

Here, the AI would be using "blind calculation". The similar argument by Lucas [Luc61] is:

> *Given any machine which is consistent and capable of doing simple arithmetic, there is a formula it is incapable of producing as being true but which we can see to be true.*

An instantiation of these claims would be the following Gödel sentence:

> *AIXI cannot prove that this sentence is true.*

Penrose claims that we (humans) can see that this sentence is obviously true, but an AI (in this case AIXI) cannot prove that it is true. While both assertions are indeed correct (logical exercise left to the reader), we can also equally construct an unprovable Gödel sentence that is true for which Penrose cannot prove it is true:

> *Penrose cannot prove that this sentence is true.*

Now we (humans who are not Penrose), as well as AI systems, can see that this sentence is true, but Penrose can never prove it is true. We can see that this is true by contradiction. If Penrose could prove that the sentence was true then it would be false, therefore we have a contradiction.

Although Penrose–Lucas provide an interesting argument about the logical capabilities of computational systems, it is likely irrelevant as it can be applied to any sufficiently power logical reasoning system. Even if there are certain statements that an AI cannot prove, this in no way negates the capabilities of potential AIs in other areas, and the ability to prove all true statements is obviously too strict a requirement for a definition of intelligence. Also, AIXI and most likely the most powerful future AGI systems will *not* be logical/mathematical infallible deductive reasoners, but uncertain inductive reasoners like humans. Human mathematicians also have limitations on recognizing unprovable truth and make mistakes.

The above are only approximate informal versions of deep formal (counter)arguments. For instance, the self-referential AIXI sentence is only obvious to us if we know AIXI's proofs are sound. And to even understand the sentence purely arithmetically, we would have to know AIXI's source-code. AIXI itself would have to know both these things for Penrose–Lucas to apply to AIXI's purely arithmetical knowledge, which is doubtful because a suitably idealized mechanical knower cannot know both its own source-code and its own soundness [Car00, Ale14].

16.6   Arguments for AGI

In contrast to the arguments against AGI we have just given, we now present some arguments *for* the possibility of AGI, both philosophical and experimental. The physical Church-Turing thesis implies that there already exists intelligent hardware (humans), and so to do the same for machines is an engineering rather than a philosophical problem, and Moore's Law, the trend for computing power to double roughly every 18 months, and comparing this trend to estimates for the number of operations performed in the human brain. We summarize progress in AI, and a collection of problems that AI has either fully solved, or recently made large leaps in performance. Games are often used as a proxy for intelligence due to the well-defined rules and community of high level players to compete against.

16.6.1 Physical Church-Turing Thesis

The Physical Church-Turing Thesis (PCT) [GW05], also known as the Church–Turing–Deutsch principle, states:

All physical processes are computable.

By extension, this can then be applied to the human mind. Assuming that *physicalism* is true (that the behavior of any object is defined entirely by its physical properties), this implies that all the behavior of the human mind is encoded in the arrangement of the material inside. This rejects the possibility of something incorporeal like a soul that makes "consciousness" work.

Since the human mind is an example of a physical process, it is computable[6] under this assumption, so there exists a Turing machine which we could build that can (in principle) emulate the human mind. This then gives one possible path to human-level AGI, by taking a scan of a human brain and simulating the scan on a computer, as mentioned in Section 16.3. The human brain has a slow firing rate estimated at around 200 Hz [Bos98], but yet it can still perform complex tasks such as visualization or locomotion. As such, we would expect that the brain is largely a slow but highly parallel device. Assuming that abstracting away the underlying physics and simulating directly the behavior of an individual neuron is sufficient to replicate higher-order behavior of the brain, we would expect that a simulation of the brain could be run much faster than real brains. From the outside, we would perceive the result to be super-humanly intelligent [Hut12a]. Imagine having a conversation with someone, and between each sentence she got 10 years of time to think about her response. This would be enough time for her to intensely study and become an expert on the topic in question after having just been asked about it (from your perspective), which would certainly be classed as superhuman intelligence.

Some could, and indeed have [Pen89], argued that the human mind must contain a incomputable processes, since computable systems are unable to assert Gödel-like statements described in Section 16.5.4 to be true. Even if PCT is wrong and the human mind is not computable in full detail, this does not necessarily imply that the non-computable processes are essential for cognition, and computable approximations may still suffice to exhibit most of the behavior that we are interested in. Another idea is to link incomputability to phenomenological consciousness and qualia [Pen94].

There are many variations of the Physical Church-Turing thesis, for instance:

- **Extended Church-Turing thesis (ECT)** states that "A probabilistic Turing machine can efficiently simulate any realistic model of computation" [KLM06], the important points here being 'efficiently'. In most cases 'efficiently' is taken to be within (small) polynomial time.

---

[6]In principle we could simulate the interaction between fundamental particles in the brain, but it is probably more efficient to find the largest scale that can be fully understood and simulated (possibly neuron-scale), and then simulate the interaction between them instead.

- **Quantum complexity-theoretic Church–Turing thesis** a quantum version of the extended Church-Turing thesis, states that "A quantum Turing machine can efficiently simulate any realistic model of computation" [KLM06].

Assume PCT or any of its variations is wrong and incomputable physical processes are crucial for making AGI work. It is rather implausible that such processes would only occur in human brains. It is more plausible that we could build machines that harness these physical process and build and program hyper-computers for AGI. Similarly from an efficiency perspective: In the unlikely case that the brain is a quantum computer and more efficient than any classical computer, we can still create AGI on quantum computers, though this would delay AGI until the technology is ripe.

16.6.2   Moore's Law

Moore's Law [Sch97a, Wal16] is a hypothesis by Gordon Moore that each year computing power will double.[7] It turns out that this has approximately been the case since the inception of the law, with the doubling period being 18 months instead of a year [Sch97a].

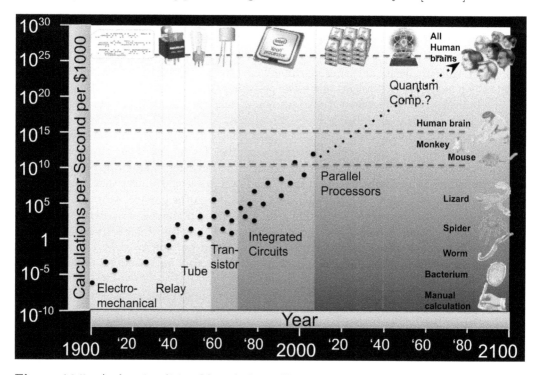

Figure 16.5: *A plot visualizing Moore's Law: How computing power has increased through time (Section 16.6.2).*

It was predicted in 2016 that this will stop/slow down in 2020-2025 [Wal16], but it has also been predicted that physically realizable quantum computers will become more accessible and widely used, so it is possible that we may see a computational advance in quantum computing the same way there was with classical computing. Quantum computers are not a strictly more powerful class of computer than classical computers, as they can be simulated by classical computers.

---

[7]More formally, the number of transistors per unit area on integrated circuits would double (Figure 16.5).

However, certain problems (such as prime factorization [Sho94]) have fast algorithms on a quantum computer but (currently we know of) no fast algorithm on a classical computer. Efficient brain simulation may require quantum algorithms via a simulation of quantum systems on a quantum computer (the original purpose behind quantum computing). Hypothesizing for a moment that properties of the brain such as consciousness are a product of quantum interactions, then this would necessitate the use of a quantum computer to be able to simulate a brain in any reasonable amount of time.

The relevance of Moore's Law to AGI is that global computing power has so far been increasing exponentially, to the point where it has exceeded the computing power of the human brain [Bos98], which is estimated at roughly 10^{17} logical operations per second [Car22]. Assuming the PCT in Section 16.6.1, at this point it would be theoretically possible to computationally simulate the human mind. However, having access to enough computation to perform the same number of operations per second as a brain may not be sufficient. Little is still known about how the brain operates, the technology to be able to scan a brain at sufficient fidelity would need to exist, and the overhead to simulate a brain (perhaps neurons are actually quite complex) may still make brain simulation in principle possible, but computationally intractable for a few more decades of Moore's Law.

16.6.3 AI Progress

The strongest evidence for the possibility of AGI is arguably the success AI has already had in a wide range of areas, especially since the Deep Learning era. Some even argue that AGI or at least proto-AGI has already been achieved. If we get to a point where for any reasonable problem we have an AI that can solve it, one may argue that then there must exist a general AI which can solve these problems by using the (narrow) AI subroutines as appropriate. Though how to construct an AGI in this way may still not be obvious, and may fail to make cross-disciplinary connections, which is often where many breakthroughs happen. The history of AI consists of a wide variety of nature-inspired approaches (neural networks, evolutionary algorithms, ...) and abstract artificial approaches (AI search, expert systems, SVMs, ...) to solve a wide variety of problems [RN20]. A plethora of neural network types, architectures, and training methods [GBCB16] have been explored in the last 80+ years since the Perceptron [MP43]. In theory, a single neural type and architecture should suffice: According to the *universal approximation theorem* [HSW89], any continuous function on a compact set can be approximated arbitrarily well by a feed-forward neural network with a single hidden layer. While this result is encouraging, it is too weak for practical purposes: First, the hidden layer(s) required for a task may be impractically large, and may also overfit to the problem and fail to generalize to new test examples. Second, it is not clear whether any efficient algorithm can find the right weights, and gradient descent seems woefully inadequate. While the classical MultiLayer Perceptron (MLP) indeed (so far) did not show sparks of general intelligence, some "minor" architectural modifications with some tweaks to gradient descent did: The Transformer [VSP+17] turned out to be the universal architecture [GMGH+24] neural network researchers were hoping to discover for decades. At sufficient scale, after *training* via simple (variations of) gradient descent with sufficient amount of data and compute, the resulting systems exhibit truly remarkable and broad capabilities. The latest and largest models, OpenAI's GPT-4 Turbo [Ope23a, Ope23b] and Google DeepMind's Gemini Ultra [GDM23a] have more than a trillion parameters, were trained on over 10 trillion tokens, costing over 100 million dollars in compute. Given their broad and impressive performance often on a human-level, both deserve the label proto-AGI if not more [Hut20b]. Below we will go over some of the general problem classes and the progress that has been made in those areas in the 21st century. For the pre-millennium AI history, see [RN20, Sec.1.2] or [Wik23]. As a disclaimer, the field is moving at a very rapid pace, and it is likely that the following list is out of date by time you read it.

- *Computer vision* is the widely applicable problem of (quickly) extracting relevant information from a single (or sometimes multiple) image(s). For example, recognizing faces in an image, identifying objects/animals in a picture so they can be easily searched using keywords, or self-driving cars locating other cars and hazards on the road. While the computer vision problem is not yet solved in full generality, there has been remarkable progress on many subproblems, such as image recognition exceeding human performance [ZVSL18]. Convolutional neural networks [HW62, FM82] have been responsible for this breakthrough. The structure of such networks is much sparser than a standard fully-connected feed-forward neural network, which induces a bias towards locality [LBBH98]. The *residual network* [HZRS15] (or *ResNet*) utilizes skip connections to allow for much deeper networks with dozens of layers to be trained efficiently while avoiding the problem of vanishing or exploding gradients [Hoc98, BSF94]. A very different way of exploiting the power of neural networks for *computer graphics* are Neural Radiance Fields (NERF), where a neural network is used to map (x,y,z,θ,ϕ), representing the location (x,y,z) and viewing angle (θ,ϕ) of the observer, to the color (r,g,b,α) of the object struck by a ray cast in the direction the observer is looking. Trained solely on a collection of 2D images, the network can then be used to represent some 3D object or scene [KSZ+21, ZLW+23].

- *Speech recognition* is the problem of transcribing human speech into text, e.g. for automatically generating subtitles for television/movies. At this time this problem is almost fully solved [PCZ+19], though the accuracy is still not enough to e.g. replace court stenographers at this time. There has been a dramatic shift in how speech recognition has been approached. Traditional systems of speech recognition would consider the problems of processing the audio file, extracting the phonemes and using language modelling to recover the text as separate tasks, each usually constructed by hand using heuristics designed by humans. Hidden Markov Models were a popular approach in the past [GY+08]. Modern techniques use deep learning to train a model end-to-end that maps directly from raw audio to phoneme sequences, from which the transcription can easily be recovered [CBCB14, PHS+23].

- The fields of *machine translation* [WWH+22] and *natural language processing* [KKKS23] have gotten to the point where they can produce articles and blogs that appear to have been written by a human [BMR+20]. The progression of *language comprehension* systems [BGA22] from *chatbot Eliza* [Wei66] in the 1960s to quiz show contestant *IBM Watson* [Hig12] in 2011, to Transformers since 2017 [VSP+17] (see below) has also been remarkable. On the quiz show *Jeopardy* (where contestants have to provide answers in natural language based on a prompt), an AI called *IBM Watson* competed against champions Brad Rutter and Ken Jennings, beating both to claim a prize of $1 million. During the game, *IBM Watson* was not connected to the internet, but did have access to the full extent of *Wikipedia* as well as some other resources downloaded ahead of time.

- *AlphaFold* [JEP+21, VBM+23], developed by DeepMind, is a groundbreaking AI system that has revolutionized the field of protein structure prediction [YZZC23]. By leveraging deep learning techniques and extensive training data, AlphaFold has been able to accurately predict protein structures with remarkable precision, outperforming traditional methods and achieving a level of accuracy comparable to experimental techniques. This achievement represents a significant milestone in AI progress, showcasing the potential for artificial intelligence to solve complex, real-world problems that have long challenged scientists. The success of AlphaFold not only underscores the power of interdisciplinary research but also highlights the transformative impact AI can have in fields such as drug discovery, disease understanding, and bioengineering, opening up new avenues for scientific advancement and practical applications.

- In November 2023, OpenAI released *GPT-4 Turbo* [Ope23a, Ope23b], the newest in their GPT family of *large language models*, which displays a wide range of impressive capabilities [BCE$^+$23], including scoring 90th percentile on the Bar Exam, as well as being able to process visual and textual input. The architecture behind the GPT family is a decoder-only version of the popular *Transformer* model [VSP$^+$17], originally designed for natural language processing. The distinctive feature of a Transformer is the *attention mechanism*, which allows each position in the sequence to consider the relative importance on other positions. Unlike Recurrent Neural Networks, which process an input sequence sequentially, Transformers processes the input sequence in parallel which speeds up training. [YJT$^+$23, ZZL$^+$23] provide a survey and history of LLMs up to modern state-of-the-art models, and [PH22] offers precise and complete pseudo-code.

- *CLIP* [RKH$^+$21] (Contrastive Language-Image Pre-Training) is a combination of two models, a *Vision Transformer* [DBK$^+$21] and a *Text Transformer* [VSP$^+$17] trained on (image, caption) pairs, such that semantically similar images and captions are mapped to similar embedding vectors. The model can then be used for image classification by finding which caption maps to the closest embedding vector of a given image. Unlike other image classifiers that provide only a probability estimate for each class, the embedding vector represents a semantically meaningful low-dimensional representation of the image. As such, many other models are trained using the output from CLIP as inputs, rather than training on raw data directly.

- *Generative AI* has exploded in the last few years with the advent of *diffusion models*, a class of models designed to learn a complex probability distribution from which samples can be taken. Given some data (such as pictures of animals), one can imagine a hypothetical space A of all pictures of animals, from which we would like to sample from. From each image x in the dataset, a finite sequence $(x_1, x_2, ..., x_T)$ is generated where, starting with the image $x_1 = x$, each successive term $x_{t+1} = x_t + \varepsilon_t$ is the previous term x_t with some added Gaussian noise ε_t drawn according to a judiciously chosen schedule. The change between two adjacent images x_{t+1} and x_t is only slight, but by the end, x_T looks like random noise. This is called a *diffusion process*[8]. The goal of the model is to de-noise x_{t+1} to x_t, or to equivalently learn the noise ε_t that turned x_t into x_{t+1}, so that it can be subtracted out to "undo" the diffusion process. By feeding random noise into and using the model to undo the T noising transitions, a new image from A can be generated. Latent diffusion models [RBL$^+$22] improve on this by first compressing images down to a semantically meaningful *feature space* (ignoring irrelevant details of the image), using a U-net (a convolutional neural network originally designed for biomedical image segmentation [RFB15]) as part of the de-noising model, and CLIP [RKH$^+$21] to shift the image in latent space towards those that match a given description. A *variational auto-encoder* [KW13] is then used to recover the image from the latent space. Improvements on diffusion

---

[8]Like when a drop of dye is added to water, the random Brownian motion of the dye molecules causes them to diffuse until they are uniformly distributed, a process driven by the natural increase in entropy as described by the second law of thermodynamics.

models have applications in image compression and interpolation [HJA20], image generation [SDWMG15, HJA20], resolution upscaling [RBL⁺22], semantic synthesis, and object removal [RBL⁺22]. Croitoru et al. [CHIS23] provide a survey of diffusion models. There are many new models for generating images based on this approach, including *GLIDE* [NDR⁺21], *DALL·E 2* [Ope22], *StableDiffusion* [Sta22], *ImageGen* [SCS⁺22] and more. Diffusion models have also been used for generation of music [MEHS21, GDM23b], video [BRL⁺23, HSG⁺22, ECA⁺23], and 3D meshes of objects [LFB⁺23].

Many researchers see games as a good or even ideal test of the intelligence of an AI system. In games, the goal, action space and the dynamics of the environment are well-defined, and often known by the agent prior to learning. For environments that are not fully observable, the AI usually knows at least the rules of the game and maintains a belief distribution over the various states in which the game is in. For instance, agents playing poker know the rules of the game and what types of cards can be in play, but not the contents of the opponents' hands. More recently, agents have been trained without even knowing the rules of the game, but learning them implicitly through rewards issued by the environment (negative rewards for losing the game, or for taking illegal actions). Performance in games may serve as an easy-to-measure proxy for intelligence. Many games have rich environments, with many players having dedicated their lives to mastering games, providing strong opponents to play against. For instance, the game of chess has a huge amount of surrounding literature, with books on various openings [DF08], endgame strategies [Dvo20], training manuals from grandmasters [Kot12], as well as game logs from millions of online chess games as training data for what strong play looks like [Lic23]. Chess, by many regarded the king of all games, has also been the holy grail in AI to master by computers (until it has been achieved in 1997). Some major successes of AI playing games include the following:

- *Checkers* (1959) was one of the first games successfully tackled by AI. The program by [Sam59, Sam67] already used a form of reinforcement learning and learned to beat its creator.

- *TD-gammon* (1995) used self-play and temporal difference learning [SB18] for Backgammon [T⁺95, Tes94], reaching master-level play.

- *Deep Blue* (1997) using custom hardware and a hand-tweaked evaluation function defeated the world chess champion Garry Kasparov in chess [CHJH02].

- Deep Q-learning (DQN) (2015) mastered a large collection of Atari video games [MKS⁺15, BNVB13] on a human level.

- *AlphaGo* (2016) using a combination of Monte-Carlo tree search and deep neural networks, defeated the world Go champion Lee Sedol [SH⁺16].

- *AlphaStar* (2019), using a combination of supervised learning and reinforcement learning techniques ranked in the top 0.2% of all players online in *StarCraft*, a popular real-time strategy game [VBC⁺19].

- *Pluribus* (2019), a superhuman *poker* agent able to play stronger than elite professional poker players, using Monte Carlo counterfactual regret minimization in rollout games, and self-play to learn a strong strategy [BS19].

- *MuZero* (2020), a single agent that learns to play *Go*, jargonChess, *Shogi*, and *Atari games*, without any prior domain knowledge or access to logs of games from humans. MuZero uses Monte Carlo Tree Search (Section 12.2) and self-play, to achieve super human performance, exceeding *AlphaGo* [SAH⁺20].

- *Agent57* (2020), a deep RL agent able to perform at human level for all 57 Atari games in the ALE benchmark [BPK$^+$20, BNVB13].

- *Cicero* (2022), using a large language model tuned with reinforcement learning ranked amongst the top 10% of online players in *Diplomacy*, a war game requiring tactical decision making and negotiation with other players [FDT22].

- *Gato* (2022), a single multimodal agent that excels at a large collection of games, as well as several robotics, vision, and language tasks [RZP$^+$22].

16.7 Intelligence

Researchers in the field of artificial intelligence have struggled for decades with the question of defining precisely what intelligence is. Many researchers have their own personal informal definitions, most with the common theme that intelligent systems can act to solve a wide class of problems, display creativity, reasoning and deduction skills, as well as being able to acquire and process new knowledge.

What exactly is meant by demonstrating intelligence varies: for some this may mean being able to distinguish between photos of cats and dogs, playing a game of Go, driving a car, recognizing speech, or navigating a footpath. While these examples do demonstrate some degree of intelligence, they do not show what we are truly interested in: general intelligence. We want an intelligent system to be able to perform well at all these tasks and more! Most importantly, we desire to create an intelligence that has the ability to learn to perform well at tasks it has never encountered before, rather than behaving well only at a collection of built-in tasks.

In this section we will discuss informal and formal definitions of intelligence and their advantages and downsides. Section 16.7.1 informally states the Legg–Hutter (LH) definition of intelligence that best captures the essence of intelligence and encompasses either explicitly or implicitly all traits usually associated with intelligence. Section 16.7.2 discusses other well-known tests and measures and AI-competitions of intelligence. Next, Section 16.7.3 formalizes the different words and concepts in the informal LH definition. Section 16.7.4 puts everything together and provides a mathematically precise, essentially unique, and valid formalization of (LH) intelligence Υ, and discusses its pros and cons. LH intelligence can serve as gold standard of what we want an intelligent agent to be able to achieve: It is a measure of performance in all environments, weighted by environment complexity, with AIXI having maximal LH-intelligence. As with AIXI, in practice we have to resort to approximations, discussed in Section 16.7.5.

16.7.1 Legg–Hutter (LH) Intelligence: Informal

One of the most comprehensive recent studies of intelligence [LH07c] investigated how intelligence was defined in various fields such as psychology, philosophy, computer science, neuroscience, and others. After presenting several definitions from the literature, they synthesize the following informal definition of intelligence:

> Legg–Hutter (LH) Intelligence: *Intelligence measures an agent's ability to achieve goals in a wide range of environments.* [LH07c]

This definition captures a notion that an intelligent agent must be able to performing well on a wide class of problems, rather than only specific tasks like chess. A chess bot can play chess at a superhuman level, but it cannot fill out tax forms, load a dishwasher, compose poetry, and many other tasks we would desire a general AI to do. We also do not want a "wise hermit" intelligence which sits around all day thinking about complex problems, but takes no actions that affect its environment. Such an agent would be indistinguishable from an unintelligent agent by an outside observer, and would not serve any useful purpose.

This is incorporated by requiring the AI to achieve goals. This definition is agnostic of the *Strong AI-hypothesis* (Section 16.3), requiring only that the agent acts to satisfy goals, with no consideration of what the internal state of the agent looks like.

This informal definition aligns with the intuitive sense of what it means to be intelligent, as well as existing practical tests of intelligence.

16.7.2   Other Intelligence Tests and Measures.

The *Raven test* [CJS90], tests one's ability to recognize patterns in a set of geometric images, and choose one of many that fills the gap. One could argue that the Raven test only concerns performance at pattern recognition and sequence prediction. While this is true, many problems can be framed as sequence prediction problems. For example, given the task of optimally acting in an environment, if the agent could predict the actions an optimal agent would take as well as how the environment would respond to those actions, this would be equivalent to acting optimally. Though in practice it would be hard to predict what an optimal agent would do without observing one.

The *Turing test* [Tur09] measures the ability for an artificial agent to be indistinguishable from a human. An interrogator communicates with both a human and an AI via a text-only terminal, and the AI is said to pass the Turing test if the interrogator cannot reliably distinguish the two. While being one of the oldest and most well-known tests of intelligence, in practice, this tests how well a chatbot can fool a judge into thinking it is a human (by, say, deliberately introducing spelling errors in its responses) rather than any reasonable measure of intelligence. It is too strict, since it has to mimic idiosyncrasies of humans. We would expect that a race of intelligent aliens would fail the Turing test, not because they are stupid, but because they would lack knowledge of the cultural norms and language of humans, and would likely do a poor job of passing as one. Also, rather than being a graded measure of intelligence, its binary succeed/fail nature is also a limitation.

The annual *Loebner prize* [Pow98] (1991-2019) awarded prizes for the most human-like chatbot submitted (as judged by a panel of human judges) and \$100'000 for a chatbot that cannot be reliably distinguished from a human (which was never successfully claimed). The Turing test, while having historical value, is actually testing for the ability to mimic a human, rather than testing directly for intelligence.

The *Hutter prize* [Hut20a] awards a cash prize for how far `enwik9` [Mah11], a 1GB sample of Wikipedia, can be compressed. The motivation for the prize is the claim that compression is closely linked to intelligence [Hut05b, Zen19]. The more "intelligent" the compressor is, the better it is able to find patterns or regularity in the data, and use those patterns to compress further. A compressor that "understands" all of human knowledge should be able to compress data from humans well, with `enwik9` acting as a stand in for human knowledge. Though intelligence and understanding are fuzzy and hard to precisely measure, the size of a compressed file provides an exact measure of performance.

Chollet proposed a measure of intelligence that emphasizes reasoning and abstraction as key components [Cho19]. *Chollet's measure* views intelligence as the ability to adapt and generalize to novel situations by recognizing patterns and making use of previously learned knowledge. According to this viewpoint, an intelligent system should possess the capacity to understand abstract concepts and relationships, enabling it to perform well in a wide range of tasks and scenarios. In contrast to traditional IQ tests or task-specific performance measures, Chollet's approach emphasizes the evaluation of an agent's ability to transfer knowledge from one domain to another [WSB+20], a critical aspect of human intelligence. This perspective on intelligence fosters a more holistic understanding of the underlying cognitive processes and encourages the development of more versatile AI systems, capable of tackling a diverse array of challenges while demonstrating robust adaptability and generalization.

In analogy with the computational complexity classes [GJ79, AKG05, AB09] NP-Hard and NP-Complete, we can also informally define the *AI-Complete* class: Problems for which it is hypothesized that any algorithm for them must be generally intelligent. Historically it was thought that many games such as chess required "creativity" and "insight", something a machine could never hope to replicate (though we now know this to be false). Some problems hypothesized to be AI-Complete include language translation to the level of a fluent human translator, complex computer vision problems, automatic peer review of research papers, and proving prize-worthy mathematical theorems.

16.7.3 Formalizing the LH-Intelligence Measure.

Motivated by the informal definition of LH intelligence (Section 16.7.1), we can construct a formal definition of general intelligence. Since we want an agent to achieve goals in a large class of environments, we will be defining LH intelligence as a weighted average over how well the agent achieves "the goal" over many environments. This poses a few problems, to which we propose solutions:

Goals. What should the agent's goal be? Is it environment specific, or something internal to the agent?

If the goal is too simple, then any agent would be intelligent under this measure. If the goal is defined as internal by the agent, then the agent can select a simple goal and the goal can easily be attained. So the goal should be provided to the agent by the environment, which places pressure on the agent to learn about the environment with which it is interacting, and to then exploit that knowledge to satisfy the goal. This can be done by defining the environment ρ and agent within the standard cybernetic model (see Chapter 6) and then communicating the goal to the agent via rewards. The goal for the agent is then always the maximization of the reward sum, motivated by the *reward hypothesis* (Assumption 6.2.1).

Balance. What if some environments provide an easy way to attain extremely high performance? Might that dominate the weighted performance average over environments? How can that be prevented?

For simple environments, it should be easy for the agent to learn the optimal policy, so agents that fail a simple test and the measure of their "intelligence" should consequently be penalized. For complex environments, especially those where few agents can perform well, we should only penalize our agent slightly for failing to perform in these environments. The intelligence measure is relative, and so most agents perform well in simple environments, but only very "intelligent" agents also perform well in complex environments too, increasing the net intelligence measure.

Universality. Will this measure be universal, or dependent on a collection of parameters?

As mentioned in Section 6.4, optimal behavior depends on how far-sighted the agent is.[9] Applying different discount regimes, the definition of optimal behavior would change, and a definition of intelligence not parameterized by a choice of discount would be preferable. Moreover, we want to avoid the problem of some environments being disproportionately

---

[9]It is not clear *a priori* that the agent should be planning as far into the future as possible. If the world were to end next week, there would hardly be much point saving up for retirement.

more rewarding than others (otherwise agents that perform well in only these environments and poorly in others will still have a high measure of intelligence). We avoid both problems by considering only *reward-summable* environments, those where the total sum of rewards received can never exceed 1. This means that the value function V_ρ^π is also bounded

$$V_\rho^\pi := \mathbf{E}_\rho^\pi \left[\sum_{i=1}^\infty r_i \right] \leq 1$$

for any choice of policy π, and obviates the need for discounting. How a particular environment chooses to dole out its limited supply of reward will in essence decide if short-term or long-term planning is required of the agent.

The observation and action spaces $\mathcal{O}$ and $\mathcal{A}$ are determined by the (physical) realization of the agent, so are not really free parameters. In any case, any $|\mathcal{O}| \geq 2$ and $|\mathcal{A}| \geq 2$ will do, since richer observations and actions can be sequentialized [MH21b, CHV22a, CHV22b]. $\mathcal{O} = \mathcal{A} = \mathbb{B}^*$ is universal without sequentialization, since it covers all finite observation and action space choices.

Environment class.   What class of environments should we choose?

We want the class of environment to be as general as possible. We can go as far as including any computable stochastic environment; as for incomputable environments there would be in general no way to accurately predict the percepts (and rewards) that the environment would return for actions taken, so planning would be impossible. Also, the Physical Church-Turing thesis (Section 16.6.1) makes considering them unnecessary.

Weighting.   Do we weigh all environments the same? Should it be more important that agents perform well in "difficult" environments than "easy" ones?

Not all environments are equal, as some environments are very difficult to attain reward in, and others are very simple. An intelligent agent should be able to perform well in "easy" environments (e.g. games for children) but can be forgiven for struggling in "difficult" environments (e.g. proving the Riemann hypothesis). So, we wish to weigh the performance in easy environments more than difficult ones. Measuring the ease or difficulty of an environment is difficult, but we can use Kolmogorov complexity K as a proxy. Environments with short descriptions (in the Kolmogorov sense, Section 2.7) are expected to be easier to perform well in compared to complex environments, since the agent has to learn fewer bits to identify the true environment. Hence, we can choose the universal prior $2^{-K(\mu)}$ (Definition 3.7.2) as our weighting over environments μ.

This definition captures the concept of Occam's razor specifically applied to intelligence, as any intelligent agent should perform better in environments governed by simple rules (low Kolmogorov complexity).

16.7.4   Legg–Hutter (LH) Intelligence: Formal

In light of the previous subsection, we can now finally define the LH intelligence measure:

Definition 16.7.1 (Legg–Hutter (LH) intelligence [LH07c]) The intelligence $\Upsilon(\pi)$ of an agent π is defined as the undiscounted value function of a policy, averaged over all reward-summable computable stochastic environments $\mathcal{M}$, weighted by the Kolmogorov complexity of that environment. Formally

$$\Upsilon(\pi) := \sum_{\nu \in \mathcal{M}} 2^{-K(\nu)} V_\nu^\pi(\epsilon)$$

where $V_\nu^\pi(\epsilon) = \mathbf{E}_\nu^\pi[\sum_{t=1}^\infty r_t]$ is the expected reward sum.

This definition says that the intelligence of an agent is the weighted average of how well it performs in each environment μ (that is, V_μ^π), where the weights are $2^{-K(\mu)}$. If we now choose the agent that maximizes this intelligence measure, we get a (non-discounted version) of the universally intelligent agent AIXI of Definition 7.4.1.

Theorem 16.7.2 (AIXI maximizes LH intelligence)

$$\pi^{\text{AIXI}} := \pi_\xi^* = \arg\max_\pi \Upsilon(\pi)$$

Additionally, we can use AIXI as an upper bound on how intelligent an agent can be with respect to this definition.

$$\overline{\Upsilon} := \max_\pi \Upsilon(\pi) = \Upsilon(\pi^{\text{AIXI}})$$

The difference $\overline{\Upsilon} - \Upsilon(\pi)$ measures how close an agent performs to the optimal Bayesian agent AIXI.

It is worth taking a moment to reflect on what has been achieved here: We have defined a measure of intelligence that perfectly encapsulates the informal definition of intelligence given in Section 16.7.1. It is succinct, mathematically precise, unambiguous, and as a test is agnostic with respect to any domain knowledge. An agent can know nothing about humanity, and still score well according to this metric. We have also shown that the optimal Bayesian agent AIXI (Definition 7.4.1) maximizes this measure of intelligence.

Benefits. We would like a definition of intelligence to have certain properties. In [LH07c] the authors argue that this definition of intelligence is *valid* since it is well-motivated by the informational definition. The definition is *meaningful* since agents with high LH intelligence will be able to perform well in most simple and many complex environments. The definition is *informative* since it lets us compare the intelligence of two distinct agents with a graded numerical score. The definition is *wide-range* since it order agents from random and trivial learning algorithms up to super intelligent AIXI. The definition is *general* since under the strong Church-Turing thesis (Thesis 2.6.6), the class of all computable environments is the largest class $\mathcal{M}$ we will ever need to consider. The definition is *unbiased* in the sense that this definition does not suffer from many of the traditional biases of intelligence tests such as background, language, etc. However, there is a hidden bias in the choice of universal machine which will be discussed below. The definition is *fundamental* since it is based on computability, complexity and information, which are fundamental to computer science and unlikely to change in the future. The definition is mathematically *formal* with little room for ambiguity. The definition is *objective* since it does not depend on any subjective criteria. The definition is *non-anthropocentric*, in contrast to most of the previously mentioned tests for intelligence, which are purposely designed either for humans, or measure intelligence relative to how humans behave. A super-intelligent agent with zero cultural knowledge of humanity would likely fail the Turing test, but would still score highly according to the LH intelligence measure.

Drawbacks. There are some drawbacks of this definition of intelligence. First, it is incomputable as it depends on Kolmogorov complexity. Approximations can be made via sampling programs (see Section 16.7.5), or using the length of a program defining an environment rather than its Kolmogorov complexity. There is also one important hidden parameter: the choice of universal Turing machine U used in the definition of the Kolmogorov complexity. We could also weaken the choice of environment class to e.g. the class of Markov decision processes (MDPs), or partially observable Markov decision processes (POMDPs), and replace the incomputable weight $2^{-K(\mu)}$ with a computable weight, such as the model cost weight $2^{-\Gamma_D(\mathcal{S})}$ (Definition 4.3.20). The choice of universal machine is much harder, with no clear answer in sight (yet). In fact, the search for a "canonical" choice of universal Turing machine has been an open problem in algorithmic information theory since its inception. It has been shown that certain choices of Turing machines can lead to nonsensical definitions of intelligence (see Section 8.2).

LH Intelligence has been more deeply investigated in [LH07c, Ale19, AH21, AQDH23]. The latter compare the intelligence of an agent based on a Bayes mixtures of agent policies to the average intelligence of the agents themselves, and draw conclusions about the geometry, symmetry, convexity, and strict local extrema of intelligence.

16.7.5   Approximations of Intelligence

Existing practical intelligence test are not general (enough). Not only do we want definitions of intelligence which capture generality, resource boundedness, and embeddedness, we also want to have practical measures of intelligence. In the field of psychology, there is a long history of various tests of intelligence. These include the Stanford-Binet test [Wil11], the Army Alpha and Army Beta [Bra11] tests, Wechsler intelligence scale [Gri11], Raven test [Rav00], and many more. These tests all measure a variety of qualities such as memory, reasoning, pattern-recognition, spatial awareness and problem-solving skills. However, they fail at being general intelligence tests. See [LH07b] for a comparative table and discussion.

Artificial Intelligence Quotient (AIQ). One approximation of the LH intelligence measure, called the Artificial Intelligence Quotient (AIQ) [LV11], uses a sampling method where the program p specifying an environment is sampled according to the distribution $2^{-\ell(p)}$. By sampling according to this distribution, we can approximately capture the simplicity $2^{-K(\mu)}$ term of the LH intelligence measure, as $K(p) \approx \ell(p)$ for most p. Then to calculate the AIQ of an agent, we can run the agent on the sampled programs and take the average performance. This leads to

$$\hat{\Upsilon}(\pi) := \frac{1}{N}\sum_{i=1}^{N}\hat{V}_{p_i}^{\pi}$$

where $\hat{V}_{p_i}^{\pi}$ is the estimated value function of the agent π on the ith program p_i, and N is the number of programs sampled. In the simplest case, $\hat{V}_{p_i}^{\pi} = \sum_t r_t$ is just the empirical return, which is on average equal to the expected sum of rewards $\mathbf{E}_{p_i}^{\pi}[\sum_t r_t]$.

Choice of reference machine. For the approximation, a *BrainFuck* (BF) [Mül93] interpreter was chosen as the reference machine. BF is an esoteric Turing complete programming language that only uses 8 different symbols. Any string of these symbols is a valid BF program, and due to the minimalism of the language, programs can often be very short.

Writing programs in BF is similar to writing the commands of what a Turing machine should do, making it a reasonable stand-in choice of universal Turing machine. Also, the language is very simple and unbiased in the sense that it does not incorporate huge amounts of specific knowledge unlike some bloated practical modern programming languages. On the other hand, any language has some bias. For instance, if the Lisp language was chosen, then the environments would be more likely to be based on lists; if Prolog was chosen, the environments would be based around logical rules. If a more modern language was chosen, the amount of program symbols would be so large that the program length would be heavily limited. When sampling programs from the BF language, the authors additionally remove any redundant code that does not do anything when executed.

Using this approximation, the MC-AIXI-CTW agent, discussed in Chapter 12, outperformed the Q-Learning algorithm in the sense that it achieved a higher AIQ [LV11]. This is not surprising as the MC-AIXI-CTW agent is an approximation of AIXI, which is the optimal agent based on Υ that AIQ is approximating.

Atari Learning Environment (ALE) as a measure of intelligence. There have been many other attempts at a practical intelligence measure for machine intelligence. One very popular measure is the Arcade Learning Environment (ALE) [BNVB13], a large sample of video games for the Atari 2600 console, running on an emulator. For one agent to be able to perform well at all these different games means that it is exhibiting a weak form of general intelligence. The most successful technique for these games was using Deep Q-Networks (DQNs) [MKS+15] (often used for image recognition), however since then, many agents have been able to out-perform it, including Agent-57 [BPK+20] which is able to outperform humans on a collection of 57 games from the ALE. Despite this success there have still been several games in which few general agents have been able to exceed human performance. These include *Skiing*, *Private Eye*, *Pitfall*, and the most difficult for AI, *Montezuma's revenge*. The reasons these games are so difficult is that they each have a property that reinforcement learning techniques have historically struggled with: *(i) Exploration-exploitation trade-off:* Choosing between learning more about the environment and performing well in what the agent thinks the environment is. *(ii) Long-term credit assignment problem:* Performing actions that only give reward in the long term, without short-term rewards to provide instant feedback. *(iii) Partial observability:* The current state not containing all the information required to determine the next state with certainty. *Montezuma's revenge* has many of these problems together. Humans can play well as they rely on a lot of prior domain knowledge (seeing a key in one room implies it should be used on a locked door elsewhere) which is rather hard for agents to learn (the game does not explicitly reward picking up keys or opening doors, but only implicitly as the agent can now access new areas of the level). On the other hand, there is some evidence that agents' performance on a carefully selected subset of 5 Atari games is already indicative of their average performance over all games [ASH23].

Solving one very general problem. Throughout this chapter, we focused on the idea that intelligence should be good performance in a wide class of environments, rather than trying to solve one particularly complex environment. But we can also consider agents in a single environment that is itself very general, requiring a large skill set to perform well in. For example, a self-driving car with the goal of delivering passengers to a given location represents a very general problem to solve, as it requires abilities in computer vision, pathfinding, robotics object detection and avoidance, and trade-offs between passenger comfort, safety and expedience. Interestingly, agents which can do well only in these general environments, but poorly in simple environments, will have low LH intelligence Υ, since a general environment will have a low weight $2^{-K(\mu_{\text{general}})}$ assigned to it.

The role of language in intelligence. Another general problem well-studied area is the role of language in intelligence [SC84, SC73]. Indeed, we have seen that some language models such as GPT-3 [BMR+20] and more recently GPT-4 [Ope23a] can be applied to tasks beyond language, such as playing chess,[10] writing code or passing theory-of-mind tests [BCE+23]. While language can be used quite well as a compression or coding scheme, there are many fundamental concepts that would be quite hard to describe only in the context of language, for example an image. As the saying goes, a picture is worth a thousand words.

16.8   Deep Learning

With the recent advent of increasingly general and powerful models such as Large Language Models (LLMs), the world has moved from asking "is AGI possible?" to "when will AGI arrive?" to "is model X already a (proto) AGI system" [BCE+23]. Note that the language modelling path to AGI [XCG+23] is not in opposition to Universal Artificial Intelligence, but arguably an approximation of the universal Bayesian mixture (Definition 3.1.2) that AIXI uses. LLMs are trained on a large corpus of data with the goal of minimizing cross entropy loss. The data is usually large scrapes of the internet, as an attempt to curate a representative sample of all knowledge accumulated by humanity. To minimize log-loss, the model needs to construct a representation of this knowledge efficiently, much like a data compressor does. As mentioned, compression is prediction [DRD+24], and Kolmogorov/Solomonoff are the ultimate compressors/predictors [Hut20a].

LLMs alone are not complete (RL/AI) agents [XCG+23], but augmenting them with planning methods such as Reasoning-via-Planning (RAP) [HGM+23], mirrors augmenting Solomonoff induction with Monte Carlo Tree search, i.e. MC-AIXI (Chapter 7). It has also been demonstrated that in-context learning can be viewed as a form of Bayesian inference [XRLM21]. This connection is further demonstrated more explicitly in the UAI context in [GMGH+24] where it was shown how Transformers (the architecture used in current state-of-the-art LLMs) can approximate Solomonoff induction when trained on algorithmic data. There have been several attempts to add universality to the LLM setup which have moved it closer to UAI: Reinforcement Learning from Human feedback (RLHF) [Cri17, ZSW+19] (a crude form of RL, essentially only using a horizon of 1, trained on a model acting as a proxy for the desires/values/morals of humans), multimodality in Transformers [DBK+21, XZC23] (AIXI is agnostic to the meaning of the I/O bitstream), search/planning controllers such as Tree-of-Thought [YYZ+23], which corresponds to expectimax planning/tree search, and when combined with in-context learning (which is the analogue of Solomonoff) this leads to an LLM version of MC-AIXI (Chapter 7).

At the time of writing, LLMs are not AGI on their own. So what is missing? What can be done to bring them closer to AIXI? Learning a long horizon value function (beyond the 1 step from RLHF) would lead to models which are able to plan and reason at length. This planning and reasoning capability could be achieved with new architectures [GW+16, VLB+21, SGBK+21, DRGM+23], or by scaffolding or by a controller around the model such as value function estimation by sampling via Chain-of-Thought [WWS+22], or by using tools [SDYD+23], or by extending it with a controller like Tree-of-Thought. Remembering and learning from the past is one area which (current) LLMs seem to struggle with. There are several ways this could be solved and has become an active research area. The use of

---

[10]Skeptics of AI often refer to Large Language Models (LLMs) as *stochastic parrots* [BGMMS21], referring to the claim that LLMs have no "understanding" and merely regurgitate text like a parrot. *Parrot chess* [Cle23] pokes fun at this by showing that LLMs, even if they lack understanding, can still play chess at the level of a strong amateur.

a controller or tools which help the model to store, recall and learn from past experiences is one way. Another approach would be for the model to continually learn [KRRP22] to compile new experiences from the in-context short-term memory into a long-term memory (perhaps in-weight). This would be closest in spirit to Solomonoff induction. AIXI learns continuously, performs reasoning and planning at long horizons, and is able to handle all forms of (multimodal) inputs and outputs. The gold-standard of AIXI can serve as a direction guide when developing the next generation of LLMs, ultimately towards AGSI.

16.9   Conclusion

The problem of Artificial General Super Intelligence (AGSI), the construction of a truly intelligent system able to act beyond human capacity in most areas, is one of the most ambitious problems of our time. If achieved, it represents the entering of a new era for humanity, and perhaps the last invention we ever need make.

In this book, we have provided the tools needed to understand and extend the approach of Universal Artificial Intelligence to this problem; from the theoretical side, the safety and philosophical side, to the current best implementable approximations.

Universal Artificial Intelligence provides a way to reason about the nature of intelligent agents, offering a theory to model AGSI, built upon the strong mathematical foundations of Solomonoff prediction [Sol64] (which provides a theoretical solution to the induction problem), and Kolmogorov complexity [Kol65] (giving *Occam's razor* a mathematically rigorous definition). We claim that UAI in turn provides a rigorous definition of intelligence itself, and models how general super-intelligent agents would behave.

As we move closer to AGSI, the relevance and necessity of understanding AGSI increases dramatically. Amidst the current evolving landscape, the trends and hype, it is the foundational strength of UAI that ensures its enduring relevance. It is through UAI that we can model hypothetical super-intelligent agents already today, prove properties and guarantees about them, eventually with the goal of designing capable agents to automate any possible task.

Bibliography

[AAA⁺19] D. Abel, D. Arumugam, K. Asadi, Y. Jinnai, M. L. Littman, and L. L. Wong. State abstraction as compression in apprenticeship learning. In *Proceedings of the AAAI Conference on Artificial Intelligence*, volume 33, pages 3134–3142, 2019. 380

[AALL18] D. Abel, D. Arumugam, L. Lehnert, and M. Littman. State abstractions for lifelong reinforcement learning. In *International Conference on Machine Learning*, pages 10–19. PMLR, 2018. 380

[AB09] S. Arora and B. Barak. *Computational Complexity: A Modern Approach*. Cambridge University Press, 2009. 119, 433

[ACBF02] P. Auer, N. Cesa-Bianchi, and P. Fischer. Finite-time analysis of the multiarmed bandit problem. *Machine Learning*, 47(2):235–256, 2002. 232, 324, 347

[AEH14] T. Alpcan, T. Everitt, and M. Hutter. Can we measure the difficulty of an optimization problem? In *IEEE Information Theory Workshop*, pages 356–360, Hobart, Australia, 2014. IEEE Press. 154

[AG13] S. Agrawal and N. Goyal. Thompson sampling for contextual bandits with linear payoffs. In *International Conference on Machine Learning*, pages 127–135. PMLR, 2013. 271, 283

[AGI08] AGI. Conference series on Artificial General Intelligence (annually since 2008), 2008. https://agi-conf.org/. 382

[AH21] S. A. Alexander and M. Hutter. Reward-punishment symmetric universal intelligence. In *Proc. 14th Conf. on Artificial General Intelligence (AGI'21)*, volume 13154 of *LNAI*, pages 1–10, San Francisco, USA, 2021. Springer. 436

[AHL16] D. Abel, D. Hershkowitz, and M. Littman. Near optimal behavior via approximate state abstraction. In *International Conference on Machine Learning*, pages 2915–2923. PMLR, 2016. 380

[AJKS22] A. Agarwal, N. Jiang, S. M. Kakade, and W. Sun. *Reinforcement Learning: Theory and Algorithms*. 2022. https://rltheorybook.github.io/. 232

[AKG05] S. Aaronson, G. Kuperberg, and C. Granade. Complexity zoo, 2005. http://www.complexityzoo.net/. 433

[Ale14] S. A. Alexander. A machine that knows its own code. *Studia Logica*, pages 567–576, 2014. 424

[Ale19] S. A. Alexander. Intelligence via ultrafilters: Structural properties of some intelligence comparators of deterministic Legg-Hutter agents. 2019. 436

[Ale21] S. A. Alexander. Can Reinforcement Learning Learn Itself? A Reply to 'Reward is Enough'. In *CIFMA 2021*. 2021. 232

[ALH17] J. Aslanides, J. Leike, and M. Hutter. Universal reinforcement learning algorithms: Survey and experiments. In *Proc. 26th International Joint Conf. on Artificial Intelligence (IJCAI'17)*, pages 1403–1410, Melbourne, Australia, 2017. xviii, 343, 344, 346

[ALOL20] S. Armstrong, J. Leike, L. Orseau, and S. Legg. Pitfalls of learning a reward function online. *arXiv:2004.13654*, 2020. 412

[AM18] S. Armstrong and S. Mindermann. Occam's razor is insufficient to infer the preferences of irrational agents. *Advances in Neural Information Processing Systems*, 31, 2018. 412

[AM20] M. Austern and A. Maleki. On the Gaussianity of Kolmogorov Complexity of Mixing Sequences. 66(2):1232–1247, 2020. 120

[Amb87] W. Ambler. Aristotle on nature and politics: The case of slavery. *Political Theory*, 15(3):390–410, 1987. 392

[AOS⁺16] D. Amodei, C. Olah, J. Steinhardt, P. Christiano, J. Schulman, and D. Mané. Concrete problems in AI safety. *arXiv:1606.06565*, 2016. 410

[AQDH23] S. A. Alexander, D. Quarel, L. Du, and M. Hutter. Universal agent mixtures and the geometry of intelligence. In *International Conference on Artificial Intelligence and Statistics*, pages 4231–4246. PMLR, 2023. 436

[Art32] E. Artin. *The gamma function*. Courier Dover Publications, 1932. 65

[AS13] S. Armstrong and A. Sandberg. Eternity in six hours: Intergalactic spreading of intelligent life and sharpening the Fermi paradox. *Acta Astronautica*, 89:1–13, 2013. 391

[ASDH24] M. Aitchison, P. Sweetser, G. Deletang, and M. Hutter. Policy gradient without bootstrapping via truncated value learning. 2024. 233, 314

441

[ASH23]   M. Aitchison, P. Sweetser, and M. Hutter. Atari-5: Distilling the Arcade learning environment down to five games. In A. Krause, E. Brunskill, K. Cho, B. Engelhardt, S. Sabato, and J. Scarlett, editors, *Proceedings of the 40th International Conference on Machine Learning*, volume 202, pages 421–438. PMLR, 2023. 437

[Asl17]   J. Aslanides. AIXIjs: A software demo for general reinforcement learning. *arXiv:1705.07615*, 2017. 343, 346

[AUK+20] D. Abel, N. Umbanhowar, K. Khetarpal, D. Arumugam, D. Precup, and M. Littman. Value preserving state-action abstractions. In *International Conference on Artificial Intelligence and Statistics*, pages 1639–1650. PMLR, 2020. 381

[Axe80a] R. Axelrod. Effective choice in the prisoner's dilemma. *Journal of Conflict Resolution*, 24(1):3–25, 1980. 311

[Axe80b] R. Axelrod. More effective choice in the prisoner's dilemma. *Journal of Conflict Resolution*, 24(3):379–403, 1980. 311

[Axe10]   S. Axelsson. The normalised compression distance as a file fragment classifier. *Digital Investigation*, 7:S24–S31, 2010. 121

[Bal14]   M. Balaguer. *Free Will*. MIT Press, 2014. 416

[Bau99]   E. B. Baum. Toward a model of intelligence as an economy of agents. *Machine Learning*, 35(2):155–185, 1999. 249

[Bau04]   E. B. Baum. *What Is Thought?* MIT Press, Cambridge, Mass, 2004. 249

[Bay63]   T. Bayes. An essay towards solving a problem in the doctrine of chances. *Philosophical Transactions of the Royal Society*, 53:370–418, 1763. [Reprinted in *Biometrika*, 45, 296–315, 1958]. 117, 118

[BBJ02]   G. Boolos, J. P. Burgess, and R. C. Jeffrey. *Computability and Logic*. Cambridge University Press, 4th ed edition, 2002. 119

[BBS24]   J. O. Berger, J. M. Bernardo, and D. Sun. *Objective Bayesian Inference*. World Scientific, May 2024. 61, 118

[BC99]   D. P. Bertsekas and D. A. Castanon. Rollout algorithms for stochastic scheduling problems. *Journal of Heuristics*, 5(1):89–108, 1999. 347

[BC16]   W. M. Bolstad and J. M. Curran. *Introduction to Bayesian Statistics*. John Wiley & Sons, 2016. 118, 155

[BCE+23] S. Bubeck, V. Chandrasekaran, R. Eldan, J. Gehrke, E. Horvitz, E. Kamar, P. Lee, Y. T. Lee, Y. Li, S. Lundberg, et al. Sparks of artificial general intelligence: Early experiments with gpt-4. *arXiv:2303.12712*, 2023. 4, 429, 438

[BD62]   D. Blackwell and L. Dubins. Merging of opinions with increasing information. *Annals of Mathematical Statistics*, 33:882–887, 1962. 151, 240

[BDM17]   M. G. Bellemare, W. Dabney, and R. Munos. A Distributional Perspective on Reinforcement Learning. Technical report, 2017. 349, 381

[BDR23]   M. G. Bellemare, W. Dabney, and M. Rowland. *Distributional Reinforcement Learning*. Adaptive Computation and Machine Learning. The MIT Press, Cambridge, Massachusetts, 2023. 349, 381

[Bel57]   R. E. Bellman. *Dynamic Programming*. Princeton University Press, Princeton, NJ, 1957. 381

[Bel15]   M. G. Bellemare. Count-based frequency estimation with bounded memory. In *Twenty-Fourth International Joint Conference on Artificial Intelligence*, 2015. 197

[Ben98]   C. H. Bennett et al. Information distance. *IEEE Transactions on Information Theory*, 44(4):1407–1423, 1998. 121

[Ben17]   Y. Bengio. The Consciousness Prior. Technical report, 2017. 416

[BEP+18] Y. Burda, H. Edwards, D. Pathak, A. Storkey, T. Darrell, and A. A. Efros. Large-scale study of curiosity-driven learning. *arXiv:1808.04355*, 2018. 283

[Ber13]   J. Bernoulli. *Ars Conjectandi*. Thurnisiorum, Basel, 1713. [Reprinted in: *Die Werke von Jakob Bernoulli*, pages 106–286, volume 3, Birkhäuser, Basel, 1975, and in: *A Source Book in Mathematics*, pages 85–90, Dover, New York, 1959. English translation of part IV (with limit theorem) by Bing Sung, Harvard Univ. Dept. of Statistics, Technical Report #2, 1966]. 117

[Ber93]   J. Berger. *Statistical Decision Theory and Bayesian Analysis*. Springer, Berlin, 3rd edition, 1993. 118

[Ber06]   J. Berger. The case for objective Bayesian analysis. *Bayesian Analysis*, 1(3):385–402, 2006. 61, 118

[Ber19]   D. P. Bertsekas. *Reinforcement Learning and Optimal Control*. Athena Scientific, 2019. 231

[Ber20]   D. Bertsekas. *Rollout, Policy Iteration, and Distributed Reinforcement Learning*. Athena Scientific, 2020. 231

[BESK18] Y. Burda, H. Edwards, A. Storkey, and O. Klimov. Exploration by random network distillation. *arXiv:1810.12894*, 2018. 232

[BEYY04] R. Begleiter, R. El-Yaniv, and G. Yona. On prediction using variable order Markov models. *Journal of Artificial Intelligence Research*, 22:385–421, 2004. 197

[BF85]   D. A. Berry and B. Fristedt. *Bandit Problems: Sequential Allocation of Experiments*. Chapman and Hall, London, 1985. 232

[BGA22] R. Baradaran, R. Ghiasi, and H. Amirkhani. A survey on machine reading comprehension systems. *Natural Language Engineering*, 28(6):683–732, 2022. 428

[BGHK92] F. Bacchus, A. Grove, J. Y. Halpern, and D. Koller. From statistics to beliefs. In *Proc. 10th National Conf. on Artificial Intelligence (AAAI-92)*, pages 602–608, San Jose, CA, 1992. AAAI Press. 119

[BGMMS21] E. M. Bender, T. Gebru, A. McMillan-Major, and S. Shmitchell. On the dangers of stochastic parrots: Can language models be too big? In *Proceedings of the 2021 ACM conference on fairness, accountability, and transparency*, pages 610–623, 2021. 438

[BH10] W. Buntine and M. Hutter. A Bayesian review of the Poisson-Dirichlet process. Technical Report arXiv:1007.0296, NICTA and ANU, Australia, 2010. 155

[Bil95] P. Billingsley. *Probability and measure*. John Wiley & Sons, 1995. 218

[Bis06] C. M. Bishop. *Pattern Recognition and Machine Learning*. Springer, 2006. 118

[BK97] A. N. Burnetas and M. N. Katehakis. Optimal adaptive policies for Markov decision processes. *Mathematics of Operations Research*, 22(1):222–255, 1997. 232

[BLH23] J. Bornschein, Y. Li, and M. Hutter. Sequential learning of neural networks for prequential MDL. In *The Eleventh International Conference on Learning Representations*, 2023. 120

[BMAD22] M. Bowling, J. D. Martin, D. Abel, and W. Dabney. Settling the Reward Hypothesis, 2022. 232

[BMdR+14] P. Bloem, F. Mota, S. de Rooij, L. Antunes, and P. Adriaans. A safe approximation for Kolmogorov complexity. In *International Conference on Algorithmic Learning Theory*, pages 336–350. Springer, 2014. 121

[BMR+20] T. Brown, B. Mann, N. Ryder, M. Subbiah, et al. Language models are few-shot learners. *Advances in Neural Information Processing Systems*, 33:1877–1901, 2020. 428, 438

[BMS+20] D. Budden, A. Marblestone, E. Sezener, T. Lattimore, G. Wayne, and J. Veness. Gaussian gated linear networks. *arXiv:2006.05964*, 2020. xix, 203

[BNVB13] M. G. Bellemare, Y. Naddaf, J. Veness, and M. Bowling. The Arcade learning environment: An evaluation platform for general agents. *Journal of Artificial Intelligence Research*, 47:253–279, 2013. 248, 430, 431, 437

[Bor62] M. Born. *Einstein's theory of relativity*. Courier Corporation, 1962. 9

[Bor16] R. S. Borbely. On normalized compression distance and large malware. *Journal of Computer Virology and Hacking Techniques*, 12(4):235–242, 2016. 121

[Bos98] N. Bostrom. How long before superintelligence? *International Journal of Futures Studies*, 2, 1998. 425, 427

[Bos03] N. Bostrom. Ethical issues in advanced artificial intelligence. *Science Fiction and Philosophy: From Time Travel to Superintelligence*, 277:284, 2003. 393, 405

[Bos14] N. Bostrom. *SuperIntelligence: Paths, Dangers, Strategies*. Oxford University Press, 2014. 4, 388, 392, 393, 410, 411

[BPK+20] A. P. Badia, B. Piot, S. Kapturowski, P. Sprechmann, A. Vitvitskyi, Z. D. Guo, and C. Blundell. Agent57: Outperforming the Atari human benchmark. In *Proceedings of the 37th International Conference on Machine Learning*, volume 119, pages 507–517. PMLR, 2020. 431, 437

[Bra68] D. Braess. Über ein paradoxon aus der verkehrsplanung. *Unternehmensforschung*, 12:258–268, 1968. 303

[Bra11] D. Brandwein. *Army Alpha Intelligence Test*, pages 142–143. Springer US, Boston, MA, 2011. 436

[Bri71] P. Brickman. Hedonic relativism and planning the good society. *Adaptation Level Theory*, pages 287–301, 1971. 417

[BRL+23] A. Blattmann, R. Rombach, H. Ling, T. Dockhorn, S. W. Kim, S. Fidler, and K. Kreis. Align your latents: High-resolution video synthesis with latent diffusion models. In *Proceedings of the IEEE/CVF Conference on Computer Vision and Pattern Recognition*, pages 22563–22575, 2023. 430

[BS84] B. G. Buchanan and E. H. Shortliffe. *Rule-Based Expert Systems: The MYCIN Experiments of the Stanford Heuristic Programming Project*. Addison Wesley, Reading, MA, 1984. 118

[BS19] N. Brown and T. Sandholm. Superhuman AI for multiplayer poker. *Science*, 365(6456):885–890, 2019. 411, 430

[BSC+04] A. G. Barto, S. Singh, N. Chentanez, et al. Intrinsically motivated learning of hierarchical collections of skills. In *Proceedings of the 3rd International Conference on Development and Learning*, pages 112–19. Piscataway, NJ, 2004. 284

[BSF94] Y. Bengio, P. Simard, and P. Frasconi. Learning long-term dependencies with gradient descent is difficult. *IEEE transactions on neural networks*, 5(2):157–166, 1994. 428

[BSG11] B. Boots, S. M. Siddiqi, and G. J. Gordon. Closing the learning-planning loop with predictive state representations. *The International Journal of Robotics Research*, 30(7):954–966, 2011. 348

[BSO+16] M. Bellemare, S. Srinivasan, G. Ostrovski, T. Schaul, D. Saxton, and R. Munos. Unifying count-based exploration and intrinsic motivation. *Advances in Neural Information Processing Systems*, 29, 2016. 284

[BT96]   D. Bertsekas and J. N. Tsitsiklis. *Neuro-dynamic Programming.* Athena Scientific, 1996. 231, 233, 284

[BV04]   S. P. Boyd and L. Vandenberghe. *Convex Optimization.* Cambridge university press, 2004. 44

[BVB13]   M. Bellemare, J. Veness, and M. Bowling. Bayesian learning of recursively factored environments. In *International Conference on Machine Learning*, pages 1211–1219. PMLR, 2013. 248, 346

[BVT14]   M. G. Bellemare, J. Veness, and E. Talvitie. Skip Context Tree Switching. In *Proceedings of the 31st International Conference on Machine Learning - Volume 32*, ICML'14, pages II–1458–II–1466, Beijing, China, 2014. JMLR.org. xix, 205

[BW94]   M. Burrows and D. Wheeler. A block-sorting lossless data compression algorithm. In *Digital SRC Research Report.* Citeseer, 1994. 119

[BWEH22]   M. Böörs, T. Wängberg, T. Everitt, and M. Hutter. Classification by decomposition: A novel approach to classification of symmetric 2 x 2 games. *Theory and Decision*, 23(3):463–508, 2022. 303

[BZ14]   V. W. Berger and Y. Zhou. Kolmogorov–smirnov test: Overview. *Wiley StatsRef: Statistics Reference Online*, 2014. 347

[Cal02]   C. S. Calude. *Information and Randomness: An Algorithmic Perspective.* Springer, Berlin, 2nd edition, 2002. 120

[Can74]   G. Cantor. Über eine Eigenschaft des Inbegriffs aller reellen algebraischen Zahlen. *Journal für reine und angewandte Mathematik*, 77:258–262, 1874. [English translation: On a property of the set of real algebraic numbers. In *A Source Book in the Foundations of Mathematics*, volume 2, pages 839–843, Clarendon, Oxford]. 119

[Can22]   C. L. Canonne. A short note on an inequality between KL and TV, 2022. 121

[Car63]   G. Cardano. Liber de ludo aleae, 1565/1663. Published in 1663 but completed already around 1565. 117

[Car48]   R. Carnap. On the application of inductive logic. *Philosophy and Phenomenological Research*, 8:133–148, 1948. 118

[Car50]   R. Carnap. *Logical Foundations of Probability.* University of Chicago Press, Chicago, 1950. 118

[Car00]   T. J. Carlson. Knowledge, machines, and the consistency of reinhardt's strong mechanistic thesis. *Annals of Pure and Applied Logic*, 105(1-3):51–82, 2000. 424

[Car22]   J. Carlsmith. New report on how much computational power it takes to match the human brain. https://www.openphilanthropy.org/research/new-report-on-how-much-computational-power-it-takes-to-match-the-human-brain/, 2022. Accessed: April 11, 2023. 427

[Cas20]   S. Casper. Achilles Heels for AGI/ASI via Decision Theoretic Adversaries. *arXiv:2010.05418*, 2020. 412

[CBCB14]   J. Chorowski, D. Bahdanau, K. Cho, and Y. Bengio. End-to-end continuous speech recognition using attention-based recurrent nn: First results. *arXiv:1412.1602*, 2014. 428

[CCH19]   M. Cohen, E. Catt, and M. Hutter. A strongly asymptotically optimal agent in general environments. In *Proc. 28th International Joint Conf. on Artificial Intelligence (IJCAI'19)*, pages 2179–2186, Macao, China, 2019. xix, 265, 279, 280, 284, 343

[CGMH+23]   E. Catt, J. Grau-Moya, M. Hutter, M. Aitchison, T. Genewein, G. Deletang, L. K. Wenliang, and J. Veness. Self-predictive universal AI. In *Thirty-seventh Conference on Neural Information Processing Systems*, 2023. 237, 281, 282, 284, 346

[CH05]   A. Chernov and M. Hutter. Monotone conditional complexity bounds on future prediction errors. In *Proc. 16th International Conf. on Algorithmic Learning Theory (ALT'05)*, volume 3734 of *LNAI*, pages 414–428, Singapore, 2005. Springer. 153

[CH20a]   E. Catt and M. Hutter. A gentle introduction to quantum computing algorithms with applications to universal prediction. Technical Report arXiv:2005.03137, Australian National University, Canberra, Australia, 2020. 349

[CH20b]   M. Cohen and M. Hutter. Pessimism about unknown unknowns inspires conservatism. In *33rd Conference on Learning Theory (COLT'20)*, volume 125 of *Proceedings of Machine Learning Research*, pages 1344–1373, Virtual / Graz, Austria, 2020. PMLR. 283, 410

[Cha66]   G. J. Chaitin. On the length of programs for computing finite binary sequences. *Journal of the ACM*, 13(4):547–569, 1966. 120

[Cha75]   G. J. Chaitin. A theory of program size formally identical to information theory. *Journal of the ACM*, 22(3):329–340, 1975. 120

[Cha87]   G. J. Chaitin. *Algorithmic Information Theory.* Cambridge University Press, Cambridge, 1987. 120

[Cha91]   G. J. Chaitin. Algorithmic information and evolution. In *Perspectives on Biological Complexity*, pages 51–60. IUBS Press, 1991. 121

[Cha10]   D. J. Chalmers. The Singularity: A philosophical analysis. *Journal of Consciousness Studies*, 17:7–65, 2010. 388, 410

[CHC21]   M. K. Cohen, M. Hutter, and E. Catt. Curiosity killed or incapacitated the cat and the asymptotically optimal agent. *IEEE Journal on Selected Areas in Information Theory*, 2(2):665–677, 2021. 263, 265, 278, 284, 343, 410

[Che85]   P. Cheeseman. In defense of probability. In *Proc. 9th International Joint Conf. on Artificial Intelligence*, pages 1002–1009, Los Altos, CA, 1985. Morgan Kaufmann. 118

[Che88]   P. Cheeseman. An inquiry into computer understanding. *Computational Intelligence*, 4(1):58–66, 1988. 118

[CHIS23]   F.-A. Croitoru, V. Hondru, R. T. Ionescu, and M. Shah. Diffusion models in vision: A survey. *IEEE Transactions on Pattern Analysis and Machine Intelligence*, 2023. 430

[CHJH02]   M. Campbell, A. J. Hoane Jr, and F.-H. Hsu. Deep blue. *Artificial Intelligence*, 134(1-2):57–83, 2002. 421, 430

[CHN22]   M. K. Cohen, M. Hutter, and N. Nanda. Fully general online imitation learning. *Journal of Machine Learning Research*, 23(334):1–30, 2022. 232

[Cho19]   F. Chollet. On the Measure of Intelligence: Abstraction and Reasoning Challenge. Technical report, 2019. 432

[CHO22]   M. Cohen, M. Hutter, and M. Osborne. Advanced artificial agents intervene in the provision of reward. *AI magazine*, 43(3):282–293, 2022. 410

[Chr17]   P. F. Christiano. *Manipulation-Resistant Online Learning*. PhD thesis, UC Berkeley, 2017. 393

[Chr20]   B. Christian. *The Alignment Problem: Machine Learning and Human Values*. WW Norton & Company, 2020. 388, 417

[CHS07]   A. Chernov, M. Hutter, and J. Schmidhuber. Algorithmic complexity bounds on future prediction errors. *Information and Computation*, 205(2):242–261, 2007. 153

[Chu36]   A. Church. An unsolvable problem of elementary number theory. *The American Journal of Mathematics*, 58(2):345–363, 1936. 83, 119

[Chu40]   A. Church. On the concept of a random sequence. *Bulletin of the American Mathematical Society*, 46:130–135, 1940. 117

[CHV22a]   E. Catt, M. Hutter, and J. Veness. On reward binarisation and Bayesian agents. In *15th European Workshop on Reinforcement Learning (EWRL-15)*, 2022. https://ewrl.files.wordpress.com/2022/09/ewrl22_submission.pdf. 299, 434

[CHV22b]   E. Catt, M. Hutter, and J. Veness. Reinforcement learning with information-theoretic actuation. *International Conference on Artificial General Intelligence*, pages 188–198, 2022. 434

[CJS90]   P. A. Carpenter, M. A. Just, and P. Shell. What one intelligence test measures: a theoretical account of the processing in the Raven progressive matrices test. *Psychological Review*, 97(3):404, 1990. 432

[CK78]   T. Cover and R. King. A convergent gambling estimate of the entropy of english. *IEEE Transactions on Information Theory*, 24(4):413–421, 1978. 119

[CKL94]   A. R. Cassandra, L. P. Kaelbling, and M. L. Littman. Acting optimally in partially observable stochastic domains. In *AAAI*, volume 94, pages 1023–1028, 1994. 339, 348

[Cle23]   Clevcode. Can you beat a stochastic parrot?, 2023. https://github.com/clevcode/skynet-dev. 438

[Col19]   Collins. Collins scrabble dictionary, 2019. https://www.collinsdictionary.com/us/scrabble. 162

[Com]   W. Commons. Image of real number line. https://upload.wikimedia.org/wikipedia/commons/d/d7/Real_number_line.svg. 23

[Con97]   M. Conte et al. Genetic programming estimates of Kolmogorov complexity. In *Proc. 17th International Conf. on Genetic Algorithms*, pages 743–750, East Lansing, MI, 1997. Morgan Kaufmann, San Francisco, CA. 121

[Coo09]   M. Cook. A concrete view of rule 110 computation. In *Proceedings International Workshop on The Complexity of Simple Programs (CSP 2008)*, volume 1 of *Electronic Proceedings in Theoretical Computer Science*, pages 31–55, Cork, Ireland, 2009. Open Publishing Association. 84

[Cou43]   A. A. Cournot. *Exposition de la théorie des chances et des probabilités*. L. Hachette, Paris, 1843. 118

[Cov74]   T. M. Cover. Universal gambling schemes and the complexity measures of Kolmogorov and Chaitin. Technical Report 12, Statistics Department, Stanford University, Stanford, CA, 1974. 120

[Cox46]   R. T. Cox. Probability, frequency, and reasonable expectation. *American Journal of Physics*, 14(1):1–13, 1946. 118

[CPRR17]   B. Cserna, M. Petrik, R. H. Russel, and W. Ruml. Value directed exploration in multi-armed bandits with structured priors. *arXiv:1704.03926*, 2017. 283

[Cri17]   A. Critch. Toward negotiable reinforcement learning: shifting priorities in Pareto optimal sequential decision-making. *arXiv:1701.01302*, 2017. 303, 438

[CT06]   T. M. Cover and J. A. Thomas. *Elements of Information Theory*. Wiley-Intersience, 2nd edition, 2006. 59, 117, 119, 120, 395

[CV05]   R. Cilibrasi and P. M. B. Vitányi. Clustering by compression. *IEEE Trans. Information Theory*, 51(4):1523–1545, 2005. 121

[CV07]   R. Cilibrasi and P. M. B. Vitányi. The Google similarity distance. *IEEE/ACM Transactions on Knowledge and Data Engineering*, page to appear, 2007. 121

[CV14]   A. R. Cohen and P. M. Vitányi. Normalized compression distance of multisets with applications. *IEEE Transactions on Pattern Analysis and Machine Intelligence*, 37(8):1602–1614, 2014. 121

[CV22]   R. L. Cilibrasi and P. M. B. Vitányi. Fast phylogeny of sars-cov-2 by compression. *Entropy*, 24(4):439, March 2022. 121

[CVH20]   M. Cohen, B. Vellambi, and M. Hutter. Asymptotically unambitious artificial general intelligence. In *Proc. 34rd AAAI Conference on Artificial Intelligence (AAAI'20)*, volume 34, pages 2467–2476, New York, USA, 2020. AAAI Press. 410

[CVH21]   M. K. Cohen, B. Vellambi, and M. Hutter. Intelligence and unambitiousness using algorithmic information theory. *IEEE Journal on Selected Areas in Information Theory*, 2(2):678–690, 2021. 249, 410

[CVW04]   R. Cilibrasi, P. M. B. Vitányi, and R. Wolf. Algorithmic clustering of music based on string compression. *Computer Music Journal*, 28(4):49–67, 2004. http://arXiv.org/abs/cs/0303025. 121

[CW84]   J. Cleary and I. Witten. Data compression using adaptive coding and partial string matching. *IEEE Transactions on Communications*, 32(4):396–402, 1984. 197

[CWvdH08a] G. M. J.-B. Chaslot, M. H. M. Winands, and H. J. van den Herik. Parallel Monte-Carlo tree search. In *International Conference on Computers and Games*, pages 60–71. Springer, 2008. 320, 328, 330, 347

[CWvdH+08b] G. M. J.-B. Chaslot, M. H. M. Winands, H. J. van den Herik, J. W. H. M. Uiterwijk, and B. Bouzy. Progressive strategies for Monte-Carlo tree search. *New Mathematics and Natural Computation*, 4(03):343–357, 2008. 347

[Dal73]   R. P. Daley. Minimal-program complexity of sequences with restricted resources. *Information and Control*, 23(4):301–312, 1973. 121

[Dal77]   R. P. Daley. On the inference of optimal descriptions. *Theoretical Computer Science*, 4(3):301–319, 1977. 121

[Das16]   M. Daswani. *Generic Reinforcement Learning Beyond Small MDPs*. PhD thesis, Research School of Computer Science, Australian National University, 2016. 120, 372, 375, 380

[Dau90]   J. W. Dauben. *Georg Cantor: His Mathematics and Philosophy of the Infinite*. Princeton University Press, Princeton, NJ, 1990. 119

[dB09]   P. de Blanc. Convergence of expected utility for universal AI. *arXiv preprint arXiv:0907.5598*, 2009. 231

[DBK+21] A. Dosovitskiy, L. Beyer, A. Kolesnikov, D. Weissenborn, X. Zhai, et al. An image is worth 16x16 words: Transformers for image recognition at scale. In *International Conference on Learning Representations*, 2021. 429, 438

[Dem68]   A. P. Dempster. A generalization of Bayesian inference. *Journal of the Royal Statistical Society*, Series B 30:205–247, 1968. 118

[DF08]   N. De Firmian. *Modern Chess Openings*. Random House Puzzles & Games, 2008. 430

[DG19]   A. Demski and S. Garrabrant. Embedded agency. *arXiv:1902.09469*, 2019. 410

[DH10]   R. Downey and D. R. Hirschfeldt. *Algorithmic Randomness and Complexity*. Springer, Berlin, 2010. 120

[DHK+13] D. L. Dowe, D. Hutchison, T. Kanade, J. Kittler, J. M. Kleinberg, F. Mattern, J. C. Mitchell, M. Naor, O. Nierstrasz, C. Pandu Rangan, B. Steffen, M. Sudan, D. Terzopoulos, D. Tygar, M. Y. Vardi, and G. Weikum, editors. *Algorithmic Probability and Friends. Bayesian Prediction and Artificial Intelligence: Papers from the Ray Solomonoff 85th Memorial Conference*, volume 7070 of *Lecture Notes in Computer Science*. Springer Berlin Heidelberg, Berlin, Heidelberg, 2013. 154

[DLKS+22] L. L. Di Langosco, J. Koch, L. D. Sharkey, J. Pfau, and D. Krueger. Goal misgeneralization in deep reinforcement learning. In *International Conference on Machine Learning*, pages 12004–12019. PMLR, 2022. 411

[Doo53]   J. L. Doob. *Stochastic Processes*. Wiley, New York, 1953. 151

[DRD+24] G. Délétang, A. Ruoss, P.-A. Duquenne, E. Catt, T. Genewein, C. Mattern, J. Grau-Moya, L. K. Wenliang, M. Aitchison, L. Orseau, M. Hutter, and J. Veness. Language modeling is compression. In *Proc. 12th International Conference on Learning Representations (ICLR'24)*, Vienna, Austria, 2024. 155, 438

[DRGM+23] G. Deletang, A. Ruoss, J. Grau-Moya, T. Genewein, L. K. Wenliang, E. Catt, C. Cundy, M. Hutter, S. Legg, J. Veness, and P. A. Ortega. Neural networks and the Chomsky hierarchy. In *Proc. 11th International Conference on Learning Representations (ICLR'23)*, Kigali, Rwanda, 2023. 438

[DRW+24] G. Deletang, A. Ruoss, L. K. Wenliang, E. Catt, T. Genewein, J. Grau, M. Hutter, and J. Veness. Generative reinforcement learning with transformers. 2024. 284, 349

[DSH12]   M. Daswani, P. Sunehag, and M. Hutter. Feature reinforcement learning using looping suffix trees. *Journal of Machine Learning Research, W&CP*, 24:11–23, 2012. 380

[DSH13]   M. Daswani, P. Sunehag, and M. Hutter. Q-learning for history-based reinforcement learning. In *Proc. 5th Asian Conf. on Machine Learning (ACML'13)*, volume 29, pages 213–228, Canberra, Australia, 2013. JMLR. 349, 380, 381

[DSH14a] M. Daswani, P. Sunehag, and M. Hutter. Feature reinforcement learning: State of the art. In *Proc. Workshops at the 28th AAAI Conference on Artificial Intelligence: Sequential Decision Making with Big Data*, pages 2–5, Quebec City, Canada, 2014. AAAI Press. 380

[DSH14b] M. Daswani, P. Sunehag, and M. Hutter. Reinforcement learning with value advice. In *Proc. 6th Asian Conf. on Machine Learning (ACML'14)*, volume 39, pages 299–314, Canberra, Australia, 2014. JMLR. 381

[DSYW21] R. Dwivedi, C. Singh, B. Yu, and M. J. Wainwright. Revisiting minimum description length complexity in overparameterized models, 2021. 120

[DV99] A. P. Dawid and V. G. Vovk. Prequential probability: principles and properties. *Bernoulli*, pages 125–162, 1999. 120

[Dvo20] M. Dvoretsky. *Dvoretsky's endgame manual*. SCB Distributors, 2020. 430

[DZ21] A. DiGiovanni and E. C. Zell. Survey of self-play in reinforcement learning. *arXiv:2107.02850*, 2021. 232

[Ear93] J. Earman. *Bayes or Bust? A Critical Examination of Bayesian Confirmation Theory*. MIT Press, Cambridge, MA, 1993. 118

[ECA+23] P. Esser, J. Chiu, P. Atighehchian, J. Granskog, and A. Germanidis. Structure and content-guided video synthesis with diffusion models. In *Proceedings of the IEEE/CVF International Conference on Computer Vision*, pages 7346–7356, 2023. 430

[EFDH16] T. Everitt, D. Filan, M. Daswani, and M. Hutter. Self-modification of policy and utility function in rational agents. In *Proc. 9th Conf. on Artificial General Intelligence (AGI'16)*, volume 9782 of *LNAI*, pages 1–11, New York, USA, 2016. Springer. Winner of the Kurzweil Prize for Best AGI Paper. 396, 400, 401, 410

[EH15a] T. Everitt and M. Hutter. Analytical results on the BFS vs. DFS algorithm selection problem. Part i: Tree search. In *Proc. 28th Australasian Joint Conference on Artificial Intelligence (AusAI'15)*, volume 9457 of *LNAI*, pages 157–165, Canberra, Australia, 2015. Springer. 349

[EH15b] T. Everitt and M. Hutter. Analytical results on the BFS vs. DFS algorithm selection problem. Part ii: Graph search. In *Proc. 28th Australasian Joint Conference on Artificial Intelligence (AusAI'15)*, volume 9457 of *LNAI*, pages 166–178, Canberra, Australia, 2015. Springer. 349

[EH16] T. Everitt and M. Hutter. Avoiding wireheading with value reinforcement learning. In *Proc. 9th Conf. on Artificial General Intelligence (AGI'16)*, volume 9782 of *LNAI*, pages 12–22, New York, USA, 2016. Springer. 401, 402, 411

[EH18a] T. Everitt and M. Hutter. The alignment problem for history-based Bayesian reinforcement learners. Technical report, 2018. First winner of the AI alignment prize round 2. 249, 411

[EH18b] T. Everitt and M. Hutter. Universal artificial intelligence: Practical agents and fundamental challenges. In H. A. Abbass, J. Scholz, and D. J. Reid, editors, *Foundations of Trusted Autonomy*, chapter 2, pages 15–46. Springer, 2018. 248

[EHKK21] T. Everitt, M. Hutter, R. Kumar, and V. Krakovna. Reward tampering problems and solutions in reinforcement learning: A causal influence diagram perspective. *Synthese*, 198(27):6435–6467, 2021. 249, 402, 410, 411

[Ein87] A. Einstein. *The Collected Papers of Albert Einstein, Volume 15 (Translation Supplement): The Berlin Years: Writings & Correspondence, June 1925–May 1927*. Princeton University Press, 1987. 9

[Eis61] E. Eisenberg. Aggregation of utility functions. *Management Science*, 7(4):337–350, 1961. 394

[EKH19] T. Everitt, R. Kumar, and M. Hutter. Designing agent incentives to avoid reward tampering. *Medium*, 8(14), 2019. 249, 411

[EKO+17] T. Everitt, V. Krakovna, L. Orseau, M. Hutter, and S. Legg. Reinforcement learning with corrupted reward signal. In *Proc. 26th International Joint Conf. on Artificial Intelligence (IJCAI'17)*, pages 4705–4713, Melbourne, Australia, 2017. 405, 406, 411

[ELH14] T. Everitt, T. Lattimore, and M. Hutter. Free lunch for optimisation under the universal distribution. In *Proc. 2014 Congress on Evolutionary Computation (CEC'14)*, pages 167–174, Beijing, China, 2014. IEEE. 423

[ELH15] T. Everitt, J. Leike, and M. Hutter. Sequential extensions of causal and evidential decision theory. In *Proc. 4th International Conf. on Algorithmic Decision Theory (ADT'15)*, volume 9346 of *LNAI*, pages 205–221, Lexington, USA, 2015. Springer. 249

[ELH18] T. Everitt, G. Lea, and M. Hutter. AGI safety literature review. In *Proc. 27th International Joint Conf. on Artificial Intelligence (IJCAI'18)*, pages 5441–5449, Stockholm, Sweden, 2018. IJCAI Review Track. 410

[Eve19] T. Everitt. *Towards Safe Artificial General Intelligence*. PhD thesis, The Australian National University (Australia), 2019. 394, 395, 402, 410

[Fan49] R. M. Fano. *The Transmission of Information*. Massachusetts Institute of Technology, Research Laboratory of Electronics . . . , 1949. 119

[FB08] H. Finnsson and Y. Björnsson. Simulation-based approach to general game playing. In *AAAI*, volume 8, pages 259–264, 2008. 347

[FBB20]   M. Fitzgerald, A. Boddy, and S. D. Baum. 2020 survey of artificial general intelligence projects for ethics, risk, and policy. *Global Catastrophic Risk Institute Technical Report*, pages 1–20, 2020. 410

[FDT22]   M. FAIR Diplomacy Team. Human-level play in the game of Diplomacy by combining language models with strategic reasoning. *Science*, 378(6624):1067–1074, 2022. 411, 431

[Fel68]   W. Feller. *An Introduction to Probability Theory and Its Applications*. Wiley, New York, 3rd edition, 1968. 117

[Fin37]   B. Finetti. Le prévision: ses lois logiques, ses sources subjectives. *Ann. Inst. Poincaré*, 7:1–68, 1937. [English translation: Foresight: Its logical laws, its subjective sources. In *Studies in Subjective Probability*. Krieger, New York, pages 55–118, 1980]. 118

[Fin73]   T. L. Fine. *Theories of Probability*. Academic Press, New York, 1973. 118

[Fin74]   B. Finetti. *Theory of Probability: A Critical Introductory Treatment*. Wiley, 1974. Vol.1&2, transl. by A. Machi and A. Smith. 118

[Fis22]   R. A. Fisher. On the mathematical foundations of theoretical statistics. *Philosophical Transactions of the Royal Society of London*, Series A 222:309–368, 1922. 118

[FJ80]   R. A. Freitas Jr. A self-reproducing interstellar probe. *Journal of the British Interplanetary Society*, 33(7):251–64, 1980. 391

[FLH16]   D. Filan, J. Leike, and M. Hutter. Loss bounds and time complexity for speed priors. In *Proc. 19th International Conf. on Artificial Intelligence and Statistics (AISTATS'16)*, volume 51, pages 1394–1402, Cadiz, Spain, 2016. Microtome. 346

[FLO02]   S. Frederick, G. Loewenstein, and T. O'Donoghue. Time discounting and time preference: A critical review. *Journal of Economic Literature*, 40(2):351–401, 2002. 224

[FM82]   K. Fukushima and S. Miyake. Neocognitron: A self-organizing neural network model for a mechanism of visual pattern recognition. In *Competition and Cooperation in Neural Nets*, pages 267–285. Springer, 1982. 428

[FMG92]   M. Feder, N. Merhav, and M. Gutman. Universal prediction of individual sequences. *IEEE Transactions on Information Theory*, 38:1258–1270, 1992. 121

[FMRW10a] V. Farias, C. C. Moallemi, B. V. Roy, and T. Weissman. Universal reinforcement learning. *IEEE Transactions on Information Theory*, 56(5):2441–2454, 2010. 343, 347, 348

[FMRW10b] V. Farias, C. C. Moallemi, B. V. Roy, and T. Weissman. Universal reinforcement learning. *IEEE Transactions on Information Theory*, 56(5):2441–2454, 2010. 378

[FPP04]   N. Ferns, P. Panangaden, and D. Precup. Metrics for finite Markov decision processes. In *Proc. 20th conf. on Uncertainty in Artificial Intelligence (UAI'04)*, pages 162–169, 2004. 369

[FST15]   B. Fallenstein, N. Soares, and J. Taylor. Reflective variants of Solomonoff induction and AIXI. In *International Conference on Artificial General Intelligence*, pages 60–69. Springer, 2015. 303

[FTC15]   B. Fallenstein, J. Taylor, and P. F. Christiano. Reflective oracles: A foundation for game theory in artificial intelligence. In *International Workshop on Logic, Rationality and Interaction*, pages 411–415. Springer, 2015. 297, 298, 302, 303

[FY01]   D. P. Foster and H. P. Young. On the impossibility of predicting the behavior of rational agents. *Proceedings of the National Academy of Sciences*, 98(22):12848–12853, 2001. 303

[Gab20]   I. Gabriel. Artificial intelligence, values, and alignment. *Minds and Machines*, 30(3):411–437, 2020. 410

[Gác74]   P. Gács. On the symmetry of algorithmic information. *Soviet Mathematics Doklady*, 15:1477–1480, 1974. 120

[Gag07]   M. Gaglio. Universal search. *Scholarpedia*, 2(11):2575, 2007. 120

[Gal68]   R. G. Gallager. *Information Theory and Reliable Communication*. Wiley, New York, 1968. 117

[Gam70]   M. Games. The fantastic combinations of John Conway's new solitaire game "life" by Martin Gardner. *Scientific American*, 223:120–123, 1970. 84

[GB03]   S. Goel and S. F. Bush. Kolmogorov complexity estimates for detection of viruses in biologically inspired security systems: a comparison with traditional approaches. *Complexity*, 9(2):54–73, 2003. 121

[GBCB16]   I. Goodfellow, Y. Bengio, A. Courville, and F. Bach. *Deep Learning*. MIT Press, Cambridge, Massachusetts, 2016. 427

[GBTC+20] S. Garrabrant, T. Benson-Tilsen, A. Critch, N. Soares, and J. Taylor. Logical Induction. Technical report, 2020. 119, 154, 416

[GCSR95]   A. Gelman, J. B. Carlin, H. S. Stern, and D. B. Rubin. *Bayesian Data Analysis*. Chapman & Hall / CRC, 1995. 118, 155

[GDG03]   R. Givan, T. Dean, and M. Greig. Equivalence notions and model minimization in Markov decision processes. *Artificial Intelligence*, 147(1–2):163–223, 2003. 369

[GDM23a]   G. T. GDM. Gemini: A family of highly capable multimodal models, 2023. 427

[GDM23b]   G. D. GDM. Transforming the future of music creation, November 2023. https://deepmind. google/discover/blog/transforming-the-future-of-music-creation/. 430

[GDR+23] T. Genewein, G. Delétang, A. Ruoss, L. K. Wenliang, E. Catt, V. Dutordoir, J. Grau-Moya, L. Orseau, M. Hutter, and J. Veness. Memory-based meta-learning on non-stationary distributions. In *Proceedings of the 40th International Conference on Machine Learning*, volume 202, pages 11173–11195. PMLR, 2023. 197

[Gho97] S. Ghosal. A review of consistency and convergence of posterior distribution. In *Varanashi Symposium in Bayesian Inference, Banaras Hindu University*. Citeseer, 1997. 63

[GJ79] M. R. Garey and D. S. Johnson. *Computers and Intractability*, volume 174. Freeman San Francisco, 1979. 433

[GM15] A. Gopalan and S. Mannor. Thompson sampling for learning parameterized Markov decision processes. In *Conference on Learning Theory*, pages 861–898. PMLR, 2015. 271, 283

[GMDK+22] J. Grau-Moya, G. Delétang, M. Kunesch, T. Genewein, E. Catt, K. Li, A. Ruoss, C. Cundy, J. Veness, J. Wang, et al. Beyond Bayes-optimality: meta-learning what you know you don't know. *arXiv:2209.15618*, 2022. 284

[GMGH+24] J. Grau-Moya, T. Genewein, M. Hutter, L. Orseau, G. Deletang, E. Catt, A. Ruoss, L. K. Wenliang, C. Mattern, M. Aitchison, and J. Veness. Learning universal predictors. *arXiv:2401.14953*, 2024. 154, 197, 423, 427, 438

[GMPT15] M. Ghavamzadeh, S. Mannor, J. Pineau, and A. Tamar. Bayesian reinforcement learning: A survey. 8(5-6):359–483, 2015. 231, 248

[Göd31] K. Gödel. Über formal unentscheidbare Sätze der Principia Mathematica und verwandter Systeme I. *Monatshefte für Matematik und Physik*, 38:173–198, 1931. [English translation by E. Mendelsohn: On undecidable propositions of formal mathematical systems. In *The Undecidable*, pages 39–71, Raven Press, New York, 1965]. 83, 84, 119, 424

[Göd86] K. Gödel. *Kurt Gödel: Collected Works: Volume I: Publications 1929-1936*, volume 1. Oxford University Press, USA, 1986. 424

[Gol06] M. Goldstein. Subjective Bayesian analysis: Principles and practice. *Bayesian Analysis*, 1(3):403–420, 2006. 61, 118

[Goo65] I. J. Good. Speculations concerning the first ultraintelligent machine. *Advances in Computers*, 6:31–88, 1965. 392, 409

[Goo71] I. J. Good. 46656 varieties of Bayesians. *Letter in American Statistician*, 25:62–63, 1971. Reprinted in Good Thinking, University of Minnesota Press, 1982, pp. 20–21. 61, 63, 118

[GP07] B. Goertzel and C. Pennachin, editors. *Artificial General Intelligence*. Springer, 2007. 382, 409

[GR19] P. Grunwald and T. Roos. Minimum description length revisited. 11(01):1930001, 2019. 120

[Gri11] R. Grizzle. *Wechsler Intelligence Scale for Children, Fourth Edition*, pages 1553–1555. Springer US, Boston, MA, 2011. 436

[Gru07] P. D. Grunwald. *The Minimum Description Length Principle*. MIT Press, 2007. 104, 120, 154

[Grz57] A. Grzegorczyk. On the definitions of computable real continuous functions. *Fundamenta Mathematicae*, 44(1):61–71, 1957. 86, 89

[GS82] H. Gaifman and M. Snir. Probabilities over rich languages, testing and randomness. *Journal of Symbolic Logic*, 47:495—-548, 1982. 119

[GS04] I. Gilboa and D. Schmeidler. Maxmin expected utility with non-unique prior. In *Uncertainty in Economic Theory*, pages 141–151. Routledge, 2004. 267, 268

[GS07] S. Gelly and D. Silver. Combining online and offline knowledge in UCT. In *Proceedings of the 24th International Conference on Machine Learning*, pages 273–280, 2007. 347

[GS20] G. Grimmett and D. Stirzaker. *Probability and Random Processes*. Oxford university press, 2020. xvii, 24, 25, 26, 38, 117, 151

[Gue09] F. G. Guerrero. A new look at the classical entropy of written english. *arXiv:0911.2284*, 2009. 71, 119

[GvdV17] S. Ghosal and A. W. van der Vaart. *Fundamentals of Nonparametric Bayesian Inference*. Number 44 in Cambridge Series in Statistical and Probabilistic Mathematics. Cambridge University Press, Cambridge ; New York, 2017. 118, 155

[GVV75] R. Gallager and D. Van Voorhis. Optimal source codes for geometrically distributed integer alphabets (corresp.). *IEEE Transactions on Information Theory*, 21(2):228–230, 1975. 119

[GW05] D. Goldin and P. Wegner. The church-turing thesis: Breaking the myth. In *Conference on Computability in Europe*, pages 152–168. Springer, 2005. 425

[GW06] S. Gelly and Y. Wang. Exploration exploitation in Go: UCT for Monte-Carlo Go. In *NIPS: Neural Information Processing Systems Conference On-line Trading of Exploration and Exploitation Workshop*, 2006. 325, 347

[GW+16] A. Graves, G. Wayne, et al. Hybrid computing using a neural network with dynamic external memory. *Nature*, 538(7626):471–476, 2016. 438

[GWMT06] S. Gelly, Y. Wang, R. Munos, and O. Teytaud. *Modification of UCT with Patterns in Monte-Carlo Go*. PhD thesis, INRIA, 2006. 347

[GY+08] M. Gales, S. Young, et al. The application of hidden markov models in speech recognition. *Foundations and Trends in Signal Processing*, 1(3):195–304, 2008. 428

[H+02]   B. Hengst et al. Discovering hierarchy in reinforcement learning with HEXQ. In *ICML*, volume 19, pages 243–250, 2002. 232

[HA99]   D. Hilbert and W. Ackermann. *Principles of Mathematical Logic*, volume 69. American Mathematical Soc., 1999. 84

[Hac75]   I. Hacking. *The Emergence of Probability*. Cambridge University Press, Cambridge, MA, 1975. 117

[Háj96]   A. Hájek. "mises redux" — redux: Fifteen arguments against finite frequentism. *Erkenntnis*, 45(2-3):209–227, November 1996. 118

[Háj09a]   A. Hájek. Chapter 7: Dutch book arguments. In *The Handbook of Rational and Social Choice*, pages 173–195. Oxford University Press, 2009. 118

[Háj09b]   A. Hájek. Fifteen arguments against hypothetical frequentism. *Erkenntnis*, 70(2):211–235, 2009. 118

[Haj11]   A. Hajek. Conditional probability. In *Handbook of Philosophy of Statistics*, volume 7, pages 99–135. Elsevier, 2011. 117

[Hal90]   A. Hald. *A History of Probability and Statistics and Their Applications Before 1750*. Wiley, New York, 1990. 117

[Ham64]   W. D. Hamilton. The genetical evolution of social behaviour. ii. *Journal of theoretical biology*, 7(1):17–52, 1964. 303

[Har55]   J. C. Harsanyi. Cardinal welfare, individualistic ethics, and interpersonal comparisons of utility. *Journal of political economy*, 63(4):309–321, 1955. 393

[Har73]   G. Harman. *Thought*. Princeton, NJ, USA: Princeton University Press, 1973. 419

[HC22]   M. Hutter and M. K. Cohen. The danger of advanced artificial intelligence controlling its own feedback. http://theconversation.com/the-danger-of-advanced-artificial-intelligence-controlling-its-own-feedback-190445, October 2022. 410

[HCD+16]   R. Houthooft, X. Chen, Y. Duan, J. Schulman, F. De Turck, and P. Abbeel. Vime: Variational information maximizing exploration. *Advances in Neural Information Processing Systems*, 29, 2016. 283

[Her32]   J. Herbrand. Sur la non-contradiction de l'arithmétique. 1932. 83

[HFD17]   J. Hostetler, A. Fern, and T. Dietterich. Sample-based tree search with fixed and adaptive state abstractions. *Journal of Artificial Intelligence Research*, 60:717–777, 2017. 381

[HFS94]   W. Hoeffding, N. I. Fisher, and P. K. Sen. *The collected works of Wassily Hoeffding*. Springer, 1994. 48

[HGM+23]   S. Hao, Y. Gu, H. Ma, J. J. Hong, Z. Wang, D. Z. Wang, and Z. Hu. Reasoning with language model is planning with world model. In *The 2023 Conference on Empirical Methods in Natural Language Processing*, 2023. 438

[HH21]   R. Hutter and M. Hutter. Chances and risks of artificial intelligence — a concept of developing and exploiting machine intelligence for future societies. *Applied System Innovation*, 4(2):1–19, 2021. 412

[HH22]   M. Hutter and S. Hansen. Uniqueness and complexity of inverse mdp models. Technical Report mh/P2466, DeepMind, London, 2022. 249, 381

[HHO23]   B. Hamzi, M. Hutter, and H. Owhadi. Bridging algorithmic information theory and machine learning: A new approach to kernel learning. *arXiv:2311.12624*, 2023. 120

[Hib12a]   B. Hibbard. Avoiding unintended AI behaviors. In *International Conference on Artificial General Intelligence*, pages 107–116. Springer, 2012. 402

[Hib12b]   B. Hibbard. Model-based utility functions. *Journal of Artificial General Intelligence*, 3(1):1–24, 2012. 395

[Hib14]   B. Hibbard. Self-modeling agents evolving in our finite universe. In *International Conference on Artificial General Intelligence*, pages 246–249. Springer, 2014. 410

[Hib15]   B. Hibbard. Self-modeling agents and reward generator corruption. In *Workshops at the Twenty-Ninth AAAI Conference on Artificial Intelligence*, 2015. 410

[Hig12]   R. High. The era of cognitive systems: An inside look at IBM Watson and how it works. *IBM Corporation, Redbooks*, 1:16, 2012. 428

[HJA20]   J. Ho, A. Jain, and P. Abbeel. Denoising diffusion probabilistic models. *Advances in neural information processing systems*, 33:6840–6851, 2020. 430

[HL07]   M. Hutter and S. Legg. Temporal difference updating without a learning rate. In *Advances in Neural Information Processing Systems 20*, pages 705–712, Cambridge, MA, USA, 2007. Curran Associates. 284, 381

[HLKT19]   P. Hernandez-Leal, B. Kartal, and M. E. Taylor. A survey and critique of multiagent deep reinforcement learning. *Autonomous Agents and Multi-Agent Systems*, 33(6):750–797, 2019. 232

[HLNU13a]   M. Hutter, J. W. Lloyd, K. S. Ng, and W. T. Uther. Probabilities on sentences in an expressive logic. *Journal of Applied Logic*, 11:386–420, 2013. 119, 416

[HLNU13b]   M. Hutter, J. W. Lloyd, K. S. Ng, and W. T. Uther. Unifying probability and logic for learning. In *Proc. 2nd Workshop on Weighted Logics for AI (WL4AI'13)*, pages 65–72, Beijing, China, 2013. 119, 416

[HLV07]   M. Hutter, S. Legg, and P. M. B. Vitányi. Algorithmic probability. *Scholarpedia*, 2(8):2572, 2007. 120, 153

[HM04]   M. Hutter and A. A. Muchnik. Universal convergence of semimeasures on individual random sequences. In *Proc. 15th International Conf. on Algorithmic Learning Theory (ALT'04)*, volume 3244 of *LNAI*, pages 234–248, Padova, Italy, 2004. Springer. 154

[HM07]   M. Hutter and A. A. Muchnik. On semimeasures predicting Martin-Löf random sequences. *Theoretical Computer Science*, 382(3):247–261, 2007. 131, 133, 154

[HMDAR16] D. Hadfield-Menell, A. Dragan, P. Abbeel, and S. Russell. Cooperative inverse reinforcement learning. Technical report, 2016. 411

[HMDAR17] D. Hadfield-Menell, A. Dragan, P. Abbeel, and S. Russell. The off-switch game. In *Workshops 31st AAAI Conference on Artificial Intelligence*, 2017. 411

[HMU06]   J. E. Hopcroft, R. Motwani, and J. D. Ullman. *Introduction to Automata Theory, Languages, and Computation (3rd Edition)*. Addison-Wesley Longman Publishing Co., Inc., USA, 2006. 81, 82, 119

[HNR68]   P. E. Hart, N. J. Nilsson, and B. Raphael. A formal basis for the heuristic determination of minimum cost paths. *IEEE transactions on Systems Science and Cybernetics*, 4(2):100–107, 1968. 349

[HO02]   L. A. Hemaspaandra and M. Ogihara. *The Complexity Theory Companion*. Springer, Berlin, 2002. 119

[Hoc98]   S. Hochreiter. The vanishing gradient problem during learning recurrent neural nets and problem solutions. *International Journal of Uncertainty, Fuzziness and Knowledge-Based Systems*, 6(02):107–116, 1998. 428

[HP04]   M. Hutter and J. Poland. Prediction with expert advice by following the perturbed leader for general weights. In *Proc. 15th International Conf. on Algorithmic Learning Theory (ALT'04)*, volume 3244 of *LNAI*, pages 279–293, Padova, Italy, 2004. Springer. 154, 314

[HP05]   M. Hutter and J. Poland. Adaptive online prediction by following the perturbed leader. *Journal of Machine Learning Research*, 6:639–660, 2005. 154, 314

[HQC24]   M. Hutter, D. Quarel, and E. Catt. *An Introduction to Universal Artificial Intelligence*. Chapman & Hall/CRC Artificial Intelligence and Robotics Series. Taylor and Francis, 2024. http://www.hutter1.net/ai/uaibook2.htm. 423, 473

[HR+22]   C. F. Hayes, R. Rădulescu, et al. A practical guide to multi-objective reinforcement learning and planning. *Autonomous Agents and Multi-Agent Systems*, 36(1):1–59, 2022. 232

[HRR06]   M. Höhl, I. Rigoutsos, and M. A. Ragan. Pattern-based phylogenetic distance estimation and tree reconstruction. *Evolutionary Bioinformatics*, 2, 2006. 121

[HSG+22]   J. Ho, T. Salimans, A. Gritsenko, W. Chan, M. Norouzi, and D. J. Fleet. Video diffusion models. *arXiv:2204.03458*, 2022. 430

[HSHB05]   B. Hoehn, F. Southey, R. C. Holte, and V. Bulitko. Effective short-term opponent exploitation in simplified poker. In *AAAI*, volume 5, pages 783–788, 2005. 348

[HSW89]   K. Hornik, M. Stinchcombe, and H. White. Multilayer feedforward networks are universal approximators. *Neural Networks*, 2(5):359–366, 1989. 427

[HT10]   M. Hutter and M. Tran. Model selection with the loss rank principle. *Computational Statistics and Data Analysis*, 54:1288–1306, 2010. 120

[HTF09]   T. Hastie, R. Tibshirani, and J. H. Friedman. *The Elements of Statistical Learning*. Springer, 2nd edition, 2009. 118

[Huf52]   D. A. Huffman. A method for the construction of minimum-redundancy codes. *Proceedings of the IRE*, 40(9):1098–1101, 1952. 79, 119

[Hug21]   S. Hughes. Lena. https://qntm.org/mmacevedo, 2021. Accessed: 2022-03-04. 419

[Hum03]   D. Hume. *A Treatise of Human Nature*. Courier Corporation, 2003. 417

[Hut00]   M. Hutter. A theory of universal artificial intelligence based on algorithmic complexity. Technical Report cs.AI/0004001, München, 62 pages, April 2000. http://arxiv.org/abs/cs.AI/0004001. ix, 154, 248, 359, 423

[Hut01a]   M. Hutter. Convergence and error bounds for universal prediction of nonbinary sequences. In *Proc. 12th European Conf. on Machine Learning (ECML-2001)*, volume 2167 of *LNAI*, pages 239–250, Freiburg, Germany, 2001. Springer. 153

[Hut01b]   M. Hutter. An effective procedure for speeding up algorithms. *Presented at the 3rd Workshop on Algorithmic Information Theory (TAI-2001)*, pages 1–10, 2001. 346, 360

[Hut01c]   M. Hutter. General loss bounds for universal sequence prediction. In *Proc. 18th International Conf. on Machine Learning (ICML-2001)*, pages 210–217, Williamstown, MA, 2001. Morgan Kaufmann. 153

[Hut01d]   M. Hutter. New error bounds for Solomonoff prediction. *Journal of Computer and System Sciences*, 62(4):653–667, 2001. 153

[Hut01e]   M. Hutter. Towards a universal theory of artificial intelligence based on algorithmic probability and sequential decisions. In *Proc. 12th European Conf. on Machine Learning (ECML-2001)*, volume 2167 of *LNAI*, pages 226–238, Freiburg, Germany, 2001. Springer. 248

[Hut01f]   M. Hutter. Universal sequential decisions in unknown environments. In *Proc. 5th European Work-shop on Reinforcement Learning (EWRL-5)*, volume 27, pages 25–26, Utrecht, The Netherlands, 2001. Onderwijsinsituut CKI, Utrecht Univ. 248

[Hut02a]   M. Hutter. Distribution of mutual information. In *Advances in Neural Information Processing Systems 14*, pages 399–406, Cambridge, MA, USA, 2002. MIT Press. 155

[Hut02b]   M. Hutter. The fastest and shortest algorithm for all well-defined problems. *International Journal of Foundations of Computer Science*, 13(3):431–443, 2002. 346, 360

[Hut02c]   M. Hutter. Self-optimizing and Pareto-optimal policies in general environments based on Bayes-mixtures. In *Proc. 15th Annual Conf. on Computational Learning Theory (COLT'02)*, volume 2375 of *LNAI*, pages 364–379, Sydney, Australia, 2002. Springer. 260, 264

[Hut03a]   M. Hutter. Convergence and loss bounds for Bayesian sequence prediction. *IEEE Transactions on Information Theory*, 49(8):2061–2067, 2003. 153

[Hut03b]   M. Hutter. On the existence and convergence of computable universal priors. In *Proc. 14th International Conf. on Algorithmic Learning Theory (ALT'03)*, volume 2842 of *LNAI*, pages 298–312, Sapporo, Japan, 2003. Springer. 119, 154

[Hut03c]   M. Hutter. An open problem regarding the convergence of universal a priori probability. In *Proc. 16th Annual Conf. on Learning Theory (COLT'03)*, volume 2777 of *LNAI*, pages 738–740, Washington, DC, USA, 2003. Springer. 154

[Hut03d]   M. Hutter. *Optimal Sequential Decisions based on Algorithmic Probability*. PhD thesis, Fakultät für Informatik, TU München, 2003. 248, 359

[Hut03e]   M. Hutter. Optimality of universal Bayesian prediction for general loss and alphabet. *Journal of Machine Learning Research*, 4:971–1000, 2003. 104, 153

[Hut03f]   M. Hutter. Robust estimators under the Imprecise Dirichlet Model. In *Proc. 3rd International Symposium on Imprecise Probabilities and Their Application (ISIPTA-2003)*, volume 18 of *Proceedings in Informatics*, pages 274–289, Lugano,Switzerland, 2003. Carleton Scientific. 118

[Hut03g]   M. Hutter. Sequence prediction based on monotone complexity. In *Proc. 16th Annual Conf. on Learning Theory (COLT'03)*, volume 2777 of *LNAI*, pages 506–521, Washington, DC, USA, 2003. Springer. 148, 154

[Hut04]   M. Hutter. Online prediction – Bayes versus experts. Technical report, IDSIA, July 2004. Presented at the *EU PASCAL Workshop on Learning Theoretic and Bayesian Inductive Principles (LTBIP-2004)* http://www.hutter1.net/ai/bayespea.htm. 154

[Hut05a]   M. Hutter. Fast non-parametric Bayesian inference on infinite trees. In *Proc. 10th International Conf. on Artificial Intelligence and Statistics (AISTATS-2005)*, pages 144–151. Society for Artificial Intelligence and Statistics, 2005. 155

[Hut05b]   M. Hutter. *Universal Artificial Intelligence: Sequential Decisions Based on Algorithmic Probability*. Springer, Berlin, 2005. 300 pages, http://www.hutter1.net/ai/uaibook.htm. ix, xix, 61, 88, 93, 98, 99, 109, 110, 115, 117, 118, 119, 141, 144, 149, 152, 153, 154, 219, 231, 240, 241, 242, 244, 247, 248, 252, 264, 276, 284, 346, 355, 357, 359, 409, 417, 423, 432

[Hut06a]   M. Hutter. General discounting versus average reward. In *Proc. 17th International Conf. on Algorithmic Learning Theory (ALT'06)*, volume 4264 of *LNAI*, pages 244–258, Barcelona, Spain, 2006. Springer. 233, 314

[Hut06b]   M. Hutter. On generalized computable universal priors and their convergence. *Theoretical Computer Science*, 364(1):27–41, 2006. 119, 154

[Hut06c]   M. Hutter. On the foundations of universal sequence prediction. In *Proc. 3rd Annual Conference on Theory and Applications of Models of Computation (TAMC'06)*, volume 3959 of *LNCS*, pages 408–420. Springer, 2006. 153

[Hut06d]   M. Hutter. Sequential predictions based on algorithmic complexity. *Journal of Computer and System Sciences*, 72(1):95–117, 2006. 148, 154

[Hut07a]   M. Hutter. Algorithmic information theory: a brief non-technical guide to the field. *Scholarpedia*, 2(3):2519, 2007. 120

[Hut07b]   M. Hutter. Bayesian regression of piecewise constant functions. In *Proc. ISBA 8th International Meeting on Bayesian Statistics*, pages 607–612, Benidorm, Spain, 2007. Oxford University Press. Lindley prize for innovative research in Bayesian statistics. 155

[Hut07c]   M. Hutter. Exact Bayesian regression of piecewise constant functions. *Bayesian Analysis*, 2(4):635–664, 2007. Lindley prize for innovative research in Bayesian statistics. 155

[Hut07d]   M. Hutter. The loss rank principle for model selection. In *Proc. 20th Annual Conf. on Learning Theory (COLT'07)*, volume 4539 of *LNAI*, pages 589–603, San Diego, USA, 2007. Springer. 105, 120

[Hut07e]   M. Hutter. On universal prediction and Bayesian confirmation. *Theoretical Computer Science*, 384(1):33–48, 2007. 118, 138, 153

[Hut07f]   M. Hutter. Universal algorithmic intelligence: A mathematical top→down approach. In *Artificial General Intelligence*, pages 227–290. Springer, Berlin, 2007. 154, 248, 359

[Hut08a]   M. Hutter. Predictive hypothesis identification. In *Presented at 9th Valencia /ISBA 2010 Meeting*, 2008. 54, 155

[Hut08b]   M. Hutter. Algorithmic complexity. *Scholarpedia*, 3(1):2573, 2008. 120

[Hut09a]   M. Hutter. Feature reinforcement learning: Part II: Structured MDPs. *Journal of Artificial General Intelligence*, pages 71–86, 2009. 120, 375, 376, 380

[Hut09b]   M. Hutter. Discrete MDL predicts in total variation. In *Advances in Neural Information Processing Systems 22 (NIPS'09)*, pages 817–825, Cambridge, MA, USA, 2009. Curran Associates. 104, 120, 128, 153, 154

[Hut09c]   M. Hutter. Exact non-parametric Bayesian inference on infinite trees. Technical Report 0903.5342, ARXIV, 2009. 155

[Hut09d]   M. Hutter. Feature dynamic Bayesian networks. In *Proc. 2nd Conf. on Artificial General Intelligence (AGI'09)*, volume 8, pages 67–73. Atlantis Press, 2009. 120, 375, 380

[Hut09e]   M. Hutter. Feature Markov decision processes. In *Proc. 2nd Conf. on Artificial General Intelligence (AGI'09)*, volume 8, pages 61–66. Atlantis Press, 2009. 120, 380

[Hut09f]   M. Hutter. Feature reinforcement learning: Part I: Unstructured MDPs. *Journal of Artificial General Intelligence*, 1:3–24, 2009. 120, 372, 380

[Hut09g]   M. Hutter. Open problems in universal induction & intelligence. *Algorithms*, 3(2):879–906, 2009. 154, 248

[Hut09h]   M. Hutter. Practical robust estimators under the Imprecise Dirichlet Model. *International Journal of Approximate Reasoning*, 50(2):231–242, 2009. 118

[Hut10a]   M. Hutter. A complete theory of everything (will be subjective). *Algorithms*, 3(4):329–350, 2010. 121, 143

[Hut10b]   M. Hutter. Observer localization in multiverse theories. In *Proceedings of the Conference in Honour of Murray Gell-Mann's 80th Birthday*, pages 638–645. World Scientific, 2010. 423

[Hut11]   M. Hutter. Algorithmic randomness as foundation of inductive reasoning and artificial intelligence. In *Randomness through Computation*, chapter 12, pages 159–169. World Scientific, 2011. 121

[Hut12a]   M. Hutter. Can intelligence explode? *Journal of Consciousness Studies*, 19(1-2):143–166, 2012. 388, 410, 425

[Hut12b]   M. Hutter. One decade of universal artificial intelligence. In *Theoretical Foundations of Artificial General Intelligence*, pages 67–88. Atlantis Press, 2012. 248, 359

[Hut12c]   M. Hutter. The subjective computable universe. In *A Computable Universe: Understanding and Exploring Nature as Computation*, pages 399–416. World Scientific, 2012. 6

[Hut13a]   M. Hutter. Sparse adaptive Dirichlet-multinomial-like processes. *Journal of Machine Learning Research, W&CP: COLT*, 30:432–459, 2013. 67, 197

[Hut13b]   M. Hutter. To create a super-intelligent machine, start with an equation. *The Conversation*, November(29):1–5, 2013. 248

[Hut14a]   M. Hutter. Extreme state aggregation beyond MDPs. In *Proc. 25th International Conf. on Algorithmic Learning Theory (ALT'14)*, volume 8776 of *LNAI*, pages 185–199, Bled, Slovenia, 2014. Springer. 380

[Hut14b]   M. Hutter. Offline to online conversion. In *Proc. 25th International Conf. on Algorithmic Learning Theory (ALT'14)*, volume 8776 of *LNAI*, pages 230–244, Bled, Slovenia, 2014. Springer. 113

[Hut16]   M. Hutter. Extreme state aggregation beyond Markov decision processes. *Theoretical Computer Science*, 650:73–91, 2016. 365, 368, 369, 370, 371

[Hut17]   M. Hutter. Universal learning theory. In C. Sammut and G. Webb, editors, *Encyclopedia of Machine Learning*, pages 1295–1304. Springer, 2nd edition, 2017. 153

[Hut18]   M. Hutter. Tractability of batch to sequential conversion. *Theoretical Computer Science*, 733:71–82, 2018. 154

[Hut19]   M. Hutter. Fairness without regret. Technical report, DeepMind & ANU, 2019. 412

[Hut20a]   M. Hutter. Human knowledge compression prize, 2006/2020. http://prize.hutter1.net/. 119, 348, 432, 438

[Hut20b]   M. Hutter. GPT-3 and AGI: Generative pre-trained transformer & artificial general intelligence. 2020. http://www.hutter1.net/official/bib.htm#gpt3agi. 427

[hut22]   M. hutter. Testing independence of exchangeable random variables. Technical report, DeepMind, London, UK, 2022. 117

[HV91]   P. G. Howard and J. S. Vitter. Analysis of arithmetic coding for data compression. In *Data Compression Conference*, 1991. 119

[HW62]   D. H. Hubel and T. N. Wiesel. Receptive fields, binocular interaction and functional architecture in the cat's visual cortex. *The Journal of Physiology*, 160(1):106, 1962. 428

[HYZM19]   M. Hutter, S. Yang-Zhao, and S. J. Majeed. Conditions on features for temporal difference-like methods to converge. In *Proc. 28th International Joint Conf. on Artificial Intelligence (IJCAI'19)*, pages 2570–2577, Macao, China, 2019. 381

[HZ03]   M. Hutter and M. Zaffalon. Bayesian treatment of incomplete discrete data applied to mutual information and feature selection. In *Proc. 26th German Conf. on Artificial Intelligence (KI-2003)*, volume 2821 of *LNAI*, pages 396–406, Hamburg, Germany, 2003. Springer. 155

[HZ05]   M. Hutter and M. Zaffalon. Distribution of mutual information from complete and incomplete data. *Computational Statistics & Data Analysis*, 48(3):633–657, 2005. 155

[HZRS15]   K. He, X. Zhang, S. Ren, and J. Sun. Deep residual learning for image recognition. *CoRR*, abs/1512.03385, 2015. 428

[Ica17]   T. Icard. Beyond almost-sure termination. In *CogSci*, 2017. 359

[IT05]   C. Igel and M. Toussaint. A no-free-lunch theorem for non-uniform distributions of target functions. *Journal of Mathematical Modelling and Algorithms*, 3(4):313–322, 2005. 423

[Jay57a]   E. T. Jaynes. Information theory and statistical mechanics. *Physical Review*, 106(4):620, 1957. 142

[Jay57b]   E. T. Jaynes. Information theory and statistical mechanics. II. *Physical Review*, 108(2):171, 1957. 142

[Jay03]   E. T. Jaynes. *Probability Theory: The Logic of Science*. Cambridge University Press, Cambridge, MA, 2003. 118, 155

[Jef83]   R. C. Jeffrey. *The Logic of Decision*. University of Chicago Press, Chicago, IL, 2nd edition, 1983. 118

[JEP+21]   J. Jumper, R. Evans, A. Pritzel, T. Green, M. Figurnov, O. Ronneberger, K. Tunyasuvunakool, R. Bates, A. Žídek, A. Potapenko, et al. Highly accurate protein structure prediction with alphafold. *Nature*, 596(7873):583–589, 2021. 428

[JJHH03]   P. D. Johnson Jr, G. A. Harris, and D. Hankerson. *Introduction to Information Theory and Data Compression*. Chapman and Hall/CRC, 2003. 119

[JJJN21]   M. Jafarnia-Jahromi, R. Jain, and A. Nayyar. Online learning for unknown partially observable MDPs. *arXiv:2102.12661*, 2021. 283

[JOA10]   T. Jaksch, R. Ortner, and P. Auer. Near-optimal regret bounds for reinforcement learning. *Journal of Machine Learning Research*, 11:1563–1600, 2010. 255, 264, 265

[Joh22]   K. Johnson. LaMDA and the sentient AI trap, 2022. `https://www.wired.com/story/lamda-sentient-ai-bias-google-blake-lemoine/`. 418

[Kak41]   S. Kakutani. A generalization of Brouwer's fixed point theorem. *Duke Mathematical Journal*, 8(3):457–459, 1941. 289

[Kal10]   O. Kallenberg. *Foundations of Modern Probability*. Probability and Its Applications. Springer New York, 2. ed edition, 2010. 117

[Kat18]   S. Katayama. Computable variants of aixi which are more powerful than aixitl. *arXiv:1805.08592*, 2018. 359

[KEW+21]   Z. Kenton, T. Everitt, L. Weidinger, I. Gabriel, V. Mikulik, and G. Irving. Alignment of language agents. *arXiv:2103.14659*, 2021. 394

[Key21]   J. M. Keynes. *A Treatise on Probability*. Macmillan, London, 1921. 117

[KF10]   D. Koller and N. Friedman. *Probabilistic Graphical Models: Principles and Techniques*. MIT Press, 2010. 118

[KHS01a]   I. Kwee, M. Hutter, and J. Schmidhuber. Gradient-based reinforcement planning in policy-search methods. In *Proc. 5th European Workshop on Reinforcement Learning (EWRL-5)*, volume 27, pages 27–29, Utrecht, The Netherlands, 2001. Onderwijsinsituut CKI, Utrecht Univ. `http://arxiv.org/abs/cs.AI/0111060`. 355, 381

[KHS01b]   I. Kwee, M. Hutter, and J. Schmidhuber. Market-based reinforcement learning in partially observable worlds. In *Proc. International Conf. on Artificial Neural Networks (ICANN-2001)*, volume 2130 of *LNCS*, pages 865–873, Vienna, 2001. Springer. 249

[KKKS23]   D. Khurana, A. Koli, K. Khatter, and S. Singh. Natural language processing: State of the art, current trends and challenges. *Multimedia Tools and Applications*, 82(3):3713–3744, 2023. 428

[KL51]   S. Kullback and R. A. Leibler. On information and sufficiency. *The Annals of Mathematical Statistics*, 22(1):79–86, 1951. 119

[KL93]   E. Kalai and E. Lehrer. Rational learning leads to Nash equilibrium. *Econometrica*, 61(5):1019–1045, 1993. 301, 303

[KLC98]   L. P. Kaelbling, M. L. Littman, and A. R. Cassandra. Planning and acting in partially observable stochastic domains. *Artificial Intelligence*, 101(1-2):99–134, 1998. 348

[Kle36]   S. Kleene. General recursive functions of natural numbers. *Mathematische Annalen*, 112:727–742, 1936. 119

[Kle07]   P. Kleingeld. Kant's second thoughts on race. *The Philosophical Quarterly*, 57(229):573–592, 2007. 392

[KLM96]   L. P. Kaelbling, M. L. Littman, and A. W. Moore. Reinforcement learning: A survey. *Journal of Artificial Intelligence Research*, 4:237–285, 1996. 231

[KLM06]   P. Kaye, R. Laflamme, and M. Mosca. *An Introduction to Quantum Computing*. OUP Oxford, 2006. 425, 426

[KLVS21]   J. Kirschner, T. Lattimore, C. Vernade, and C. Szepesvári. Asymptotically optimal information-directed sampling. In *Conference on Learning Theory*, pages 2777–2821. PMLR, 2021. 265

[Knu73]   D. E. Knuth. *The Art of Computer Programming, Volume I: Fundamental Algorithms*. Addison-Wesley, Reading, MA, 1973. xi

[Ko86]   K.-I. Ko. On the notion of infinite pseudorandom sequences. *Theoretical Computer Science*, 48(1):9–33, 1986. 121

[KOK+18]   V. Krakovna, L. Orseau, R. Kumar, M. Martic, and S. Legg. Penalizing side effects using stepwise relative reachability. *arXiv:1806.01186*, 2018. 411

[Kol33]   A. N. Kolmogorov. *Grundlagen der Wahrscheinlichkeitsrechnung*. Springer, Berlin, 1933. [English translation: *Foundations of the Theory of Probability*. Chelsea, New York, 2nd edition, 1956]. 117

[Kol63]   A. N. Kolmogorov. On tables of random numbers. *Sankhya, the Indian Journal of Statistics*, Series A 25, 1963. 118

[Kol65]   A. N. Kolmogorov. Three approaches to the quantitative definition of information. *Problems of Information and Transmission*, 1(1):1–7, 1965. 120, 121, 439

[Kol83]   A. N. Kolmogorov. Combinatorial foundations of information theory and the calculus of probabilities. *Russian Mathematical Surveys*, 38(4):27–36, 1983. 120

[KOML18]   V. Krakovna, L. Orseau, M. Martic, and S. Legg. Measuring and avoiding side effects using relative reachability. *arXiv:1806.01186*, 2018. 411

[KON+20]   V. Krakovna, L. Orseau, R. Ngo, M. Martic, and S. Legg. Avoiding side effects by considering future tasks. *Advances in Neural Information Processing Systems*, 33:19064–19074, 2020. 411

[Kor85]   R. E. Korf. Depth-first iterative-deepening: An optimal admissible tree search. *Artificial intelligence*, 27(1):97–109, 1985. 349

[Kot12]   A. A. Kotov. *Think like a grandmaster*. Batsford Books, 2012. 430

[KR02]   M. Kudlek and Y. Rogozhin. A universal turing machine with 3 states and 9 symbols. In *Developments in Language Theory: 5th International Conference, DLT 2001 Wien, Austria, July 16–21, 2001 Revised Papers 5*, pages 311–318. Springer, 2002. 96

[Kra49]   L. G. Kraft. A device for quantizing, grouping and coding amplitude modified pulses. Master's thesis, Electrical Engineering Department, Massachusetts Institute of Technology, Cambridge, MA, 1949. 119

[KRRP22]   K. Khetarpal, M. Riemer, I. Rish, and D. Precup. Towards continual reinforcement learning: A review and perspectives. *Journal of Artificial Intelligence Research*, 75:1401–1476, 2022. 439

[KS06]   L. Kocsis and C. Szepesvári. Bandit based Monte-Carlo planning. In J. Fürnkranz, T. Scheffer, and M. Spiliopoulou, editors, *Machine Learning: ECML 2006*, pages 282–293, Berlin, Heidelberg, 2006. Springer Berlin Heidelberg. xix, 316, 322, 323, 327, 328, 347, 349

[KSZ+21]   A. R. Kosiorek, H. Strathmann, D. Zoran, P. Moreno, R. Schneider, S. Mokrá, and D. J. Rezende. Nerf-vae: A geometry aware 3d scene generative model. In *International Conference on Machine Learning*, pages 5742–5752. PMLR, 2021. 428

[KT81]   R. Krichevsky and V. Trofimov. The performance of universal encoding. *IEEE Transactions on Information Theory*, 27(2):199–207, 1981. 197

[Kuh50]   H. W. Kuhn. A simplified two-person poker. *Contributions to the Theory of Games*, 1:97–103, 1950. 340, 348

[KUM+20]   V. Krakovna, J. Uesato, V. Mikulik, M. Rahtz, T. Everitt, R. Kumar, Z. Kenton, J. Leike, and S. Legg. Specification gaming: the flip side of AI ingenuity, April 2020. https://deepmind.google/discover/blog/specification-gaming-the-flip-side-of-ai-ingenuity/. 411

[Kur99]   R. Kurzweil. *The Age of Spiritual Machines*. Viking, 1999. 409

[Kur05]   R. Kurzweil. *The Singularity Is Near*. Viking, 2005. 409

[KV86]   P. R. Kumar and P. P. Varaiya. *Stochastic Systems: Estimation, Identification, and Adaptive Control*. Prentice Hall, Englewood Cliffs, NJ, 1986. 232

[KW13]   D. P. Kingma and M. Welling. Auto-encoding variational Bayes. *arXiv:1312.6114*, 2013. 429

[Kyb77]   H. E. Kyburg. Randomness and the right reference class. *The Journal of Philosophy*, 74(9):501–521, 1977. 119

[Kyb83]   H. E. Kyburg. The reference class. *Philosophy of Science*, 50:374–397, 1983. 119

[Lai87]   T. L. Lai. Adaptive treatment allocation and the multi-armed bandit problem. *The Annals of Statistics*, pages 1091–1114, 1987. 324

[LALH17]   S. Lamont, J. Aslanides, J. Leike, and M. Hutter. Generalised discount functions applied to a Monte-Carlo AIμ implementation. In *Proc. 16th Conf. on Autonomous Agents and MultiAgent Systems (AAMAS'17)*, pages 1589–1591, Sao Paulo, Brazil, 2017. 343

[Lam]   P. K. Lam. ANU quantum random numbers. https://qrng.anu.edu.au/. 92

[Lam87]   M. Lambalgen. *Random Sequences*. PhD thesis, University of Amsterdam, 1987. 117

[Lap12]   P. Laplace. *Théorie analytique des probabilités*. Courcier, Paris, 1812. [English translation by F. W. Truscott and F. L. Emory: *A Philosophical Essay on Probabilities*. Dover, 1952]. 117

[Lap40]   P. S. Laplace. *Essai philosophique sur les probabilités*. Bachelier, 1840. 118

[Lat14]   T. Lattimore. *Theory of General Reinforcement Learning*. PhD thesis, Research School of Computer Science, Australian National University, 2014. xix, 278, 279

[Lat23]   T. Lattimore. Convergence of posterior averages. Private communication, 2023. 241

[LAW⁺20] C. Linke, N. M. Ady, M. White, T. Degris, and A. White. Adapting behavior via intrinsic reward: A survey and empirical study. *Journal of Artificial Intelligence Research*, 69:1287–1332, 2020. 284

[LBBH98] Y. LeCun, L. Bottou, Y. Bengio, and P. Haffner. Gradient-based learning applied to document recognition. *Proceedings of the IEEE*, 86(11):2278–2324, 1998. 428

[LBH23] Y. Li, J. Bornschein, and M. Hutter. Evaluating representations with readout model switching. In *The Eleventh International Conference on Learning Representations*, 2023. 120

[LeC22] Y. LeCun. A Path Towards Autonomous Machine Intelligence. page 62, 2022. `https://openreview.net/forum?id=BZ5a1r-kVsf`. 382

[Lee12] P. M. Lee. Bayesian statistics: An introduction, 2012. 118, 155

[Leg06] S. Legg. Is there an elegant universal theory of prediction? In *Proc. 17th International Conf. on Algorithmic Learning Theory (ALT'06)*, volume 4264 of *LNAI*, pages 274–287, Barcelona, Spain, 2006. Springer. 153

[Leg08] S. Legg. *Machine Super Intelligence*. PhD thesis, IDSIA, Lugano, Switzerland, 2008. Recipient of the $10'000,- Singularity Prize/Award. 244, 409

[Lei16a] J. Leike. Exploration potential. *arXiv:1609.04994*, 2016. 284

[Lei16b] J. Leike. *Nonparametric General Reinforcement Learning*. PhD thesis, Australian National University, 2016. arXiv: 1611.08944. xix, 90, 119, 142, 221, 231, 237, 256, 257, 259, 260, 261, 263, 264, 265, 271, 272, 278, 299, 300, 301, 303, 359

[Lev73a] L. A. Levin. On the notion of a random sequence. *Soviet Mathematics Doklady*, 14(5):1413–1416, 1973. 120

[Lev73b] L. A. Levin. Universal sequential search problems. *Problems of Information Transmission*, 9:265–266, 1973. 120, 121

[Lev73c] L. A. Levin. Universal sequential search problems. *Problemy peredachi informatsii*, 9(3):115–116, 1973. 346, 360

[Lev74] L. A. Levin. Laws of information conservation (non-growth) and aspects of the foundation of probability theory. *Problems of Information Transmission*, 10(3):206–210, 1974. 117, 120

[LFB⁺23] Z. Liu, Y. Feng, M. J. Black, D. Nowrouzezahrai, L. Paull, and W. Liu. Meshdiffusion: Score-based generative 3d mesh modeling. In *International Conference on Learning Representations*, 2023. 430

[LGM01] C. Lusena, J. Goldsmith, and M. Mundhenk. Nonapproximability results for partially observable Markov decision processes. *Journal of Artificial Intelligence Research*, 14:83–103, 2001. 359

[LH04] S. Legg and M. Hutter. Ergodic MDPs admit self-optimising policies. Technical Report IDSIA-21-04, IDSIA, 2004. `https://repository.supsi.ch/5519/1/IDSIA-21-04.pdf`. 248

[LH05] S. Legg and M. Hutter. A universal measure of intelligence for artificial agents. In *Proc. 21st International Joint Conf. on Artificial Intelligence (IJCAI-2005)*, pages 1509–1510, Edinburgh, Scottland, 2005. 248

[LH06] S. Legg and M. Hutter. A formal measure of machine intelligence. In *Proc. 15th Annual Machine Learning Conference of Belgium and The Netherlands (Benelearn'06)*, pages 73–80, Ghent, Belgium, 2006. 248, 392

[LH07a] S. Legg and M. Hutter. A collection of definitions of intelligence. In B. Goertzel and P. Wang, editors, *Advances in Artificial General Intelligence: Concepts, Architectures and Algorithms*, volume 157 of *Frontiers in Artificial Intelligence and Applications*, pages 17–24, Amsterdam, NL, 2007. IOS Press. 248, 392

[LH07b] S. Legg and M. Hutter. Tests of machine intelligence. In *50 Years of Artificial Intelligence*, volume 4850 of *LNAI*, pages 232–242, Monte Verita, Switzerland, 2007. 248, 436

[LH07c] S. Legg and M. Hutter. Universal intelligence: A definition of machine intelligence. *Minds & Machines*, 17(4):391–444, 2007. 92, 233, 248, 431, 435, 436

[LH11a] T. Lattimore and M. Hutter. Asymptotically optimal agents. In *Proc. 22nd International Conf. on Algorithmic Learning Theory (ALT'11)*, volume 6925 of *LNAI*, pages 368–382, Espoo, Finland, 2011. Springer. 263, 265, 279

[LH11b] T. Lattimore and M. Hutter. No free lunch versus Occam's razor in supervised learning. In *Proc. Solomonoff 85th Memorial Conference*, volume 7070 of *LNAI*, pages 223–235, Melbourne, Australia, 2011. Springer. 423

[LH11c] T. Lattimore and M. Hutter. Time consistent discounting. In *Proc. 22nd International Conf. on Algorithmic Learning Theory (ALT'11)*, volume 6925 of *LNAI*, pages 383–397, Espoo, Finland, 2011. Springer. 233

[LH12] T. Lattimore and M. Hutter. PAC bounds for discounted MDPs. In *Proc. 23rd International Conf. on Algorithmic Learning Theory (ALT'12)*, volume 7568 of *LNAI*, pages 320–334, Lyon, France, 2012. Springer. 264

[LH13] T. Lattimore and M. Hutter. On Martin-löf convergence of Solomonoff's mixture. In *Proc. 10th Annual Conference on Theory and Applications of Models of Computation (TAMC'13)*, volume 7876 of *LNCS*, pages 212–223, Hong Kong, China, 2013. Springer. 154

[LH14a]   T. Lattimore and M. Hutter. Bayesian reinforcement learning with exploration. In *Proc. 25th International Conf. on Algorithmic Learning Theory (ALT'14)*, volume 8776 of *LNAI*, pages 170–184, Bled, Slovenia, 2014. Springer. 284

[LH14b]   T. Lattimore and M. Hutter. General time consistent discounting. *Theoretical Computer Science*, 519:140–154, 2014. 224, 228, 233

[LH14c]   T. Lattimore and M. Hutter. Near-optimal PAC bounds for discounted MDPs. *Theoretical Computer Science*, 558:125–143, 2014. 264

[LH14d]   J. Leike and M. Hutter. Indefinitely oscillating martingales. In *Proc. 25th International Conf. on Algorithmic Learning Theory (ALT'14)*, volume 8776 of *LNAI*, pages 321–335, Bled, Slovenia, 2014. Springer. 151

[LH15a]   T. Lattimore and M. Hutter. On Martin-löf (non)convergence of Solomonoff's universal mixture. *Theoretical Computer Science*, 588:2–15, 2015. 154

[LH15b]   J. Leike and M. Hutter. Bad universal priors and notions of optimality. *Journal of Machine Learning Research, W&CP: COLT*, 40:1244–1259, 2015. Also presented at EWRL'15. http://ewrl.files.wordpress.com/2015/02/ewrl12_2015_submission_3.pdf. 260, 261, 264

[LH15c]   J. Leike and M. Hutter. On the computability of AIXI. In *Proc. 31st International Conf. on Uncertainty in Artificial Intelligence (UAI'15)*, pages 464–473, Amsterdam, Netherlands, 2015. AUAI Press. 359

[LH15d]   J. Leike and M. Hutter. On the computability of Solomonoff induction and knowledge-seeking. In *Proc. 26th International Conf. on Algorithmic Learning Theory (ALT'15)*, volume 9355 of *LNAI*, pages 364–378, Banff, Canada, 2015. Springer. 154, 359

[LH15e]   J. Leike and M. Hutter. Solomonoff induction violates Nicod's criterion. In *Proc. 26th International Conf. on Algorithmic Learning Theory (ALT'15)*, volume 9355 of *LNAI*, pages 349–363, Banff, Canada, 2015. Springer. Also presented at CCR: http://math.uni-heidelberg.de/logic/conferences/ccr2015/. 154

[LH18]   J. Leike and M. Hutter. On the computability of Solomonoff induction and AIXI. *Theoretical Computer Science*, 716:28–49, 2018. 353, 354, 359

[LHG11]   T. Lattimore, M. Hutter, and V. Gavane. Universal prediction of selected bits. In *Proc. 22nd International Conf. on Algorithmic Learning Theory (ALT'11)*, volume 6925 of *LNAI*, pages 262–276, Espoo, Finland, 2011. Springer. 154

[LHS13a]   T. Lattimore, M. Hutter, and P. Sunehag. Concentration and confidence for discrete Bayesian sequence predictors. In *Proc. 24th International Conf. on Algorithmic Learning Theory (ALT'13)*, volume 8139 of *LNAI*, pages 324–338, Singapore, 2013. Springer. 133, 154

[LHS13b]   T. Lattimore, M. Hutter, and P. Sunehag. The sample-complexity of general reinforcement learning. *Journal of Machine Learning Research, W&CP: ICML*, 28(3):28–36, 2013. 264

[Li 03]   M. Li et al. The similarity metric. In *Proc. 14th Annual ACM-SIAM Symposium on Discrete Algorithms (SODA-03)*, pages 863–872. ACM Press, New York, 2003. 121

[Lic23]   Lichess. Lichess open database. https://database.lichess.org/, 2023. 430

[Liu60]   S.-C. Liu. An enumeration of the primitive recursive functions without repetition. *Tohoku Mathematical Journal, Second Series*, 12(3):400–402, 1960. 114

[LLOH16]   J. Leike, T. Lattimore, L. Orseau, and M. Hutter. Thompson sampling is asymptotically optimal in general environments. In *Proc. 32nd International Conf. on Uncertainty in Artificial Intelligence (UAI'16)*, pages 417–426, New Jersey, USA, 2016. AUAI Press. Best student paper. 265, 283

[LLOH17]   J. Leike, T. Lattimore, L. Orseau, and M. Hutter. On Thompson sampling and asymptotic optimality. In *Proc. 26th International Joint Conf. on Artificial Intelligence (IJCAI'17)*, pages 4889–4893, Melbourne, Australia, 2017. Best sister conferences paper track. 253, 271, 283

[LMK+17]   J. Leike, M. Martic, V. Krakovna, P. A. Ortega, T. Everitt, A. Lefrancq, L. Orseau, and S. Legg. AI safety gridworlds. *CoRR*, abs/1711.09883, 2017. 219, 406, 412

[Lov69a]   D. W. Loveland. On minimal-program complexity measures. In *Proc. 1st ACM Symposium on Theory of Computing*, pages 61–78. ACM Press, New York, 1969. 120

[Lov69b]   D. W. Loveland. A variant of the Kolmogorov concept of complexity. *Information and Control*, 15(6):510–526, 1969. 120

[LS01]   M. Littman and R. S. Sutton. Predictive representations of state. *Advances in Neural Information Processing Systems*, 14, 2001. 348

[LS20]   T. Lattimore and C. Szepesvári. *Bandit Algorithms*. Cambridge University Press, 2020. 232, 323, 327, 347

[LTF16]   J. Leike, J. Taylor, and B. Fallenstein. A formal solution to the grain of truth problem. In *Proc. 32nd International Conf. on Uncertainty in Artificial Intelligence (UAI'16)*, pages 427–436, New Jersey, USA, 2016. AUAI Press. 295, 298, 303

[Luc61]   J. R. Lucas. Minds, machines and gödel. *Philosophy*, pages 112–127, 1961. 424

[LV07]   M. Li and P. M. B. Vitányi. Applications of algorithmic information theory. *Scholarpedia*, 2(5):2658, 2007. 120

[LV11]   S. Legg and J. Veness. An approximation of the universal intelligence measure. In *Proc. Solomonoff 85th Memorial Conference*, volume 7070 of *LNAI*, pages 236–249, Melbourne, Australia, 2011. Springer. 436, 437

[LV19]   M. Li and P. M. B. Vitányi. *An Introduction to Kolmogorov Complexity and Its Applications*. Springer, Berlin, 4th edition, 2019. 99, 101, 103, 117, 119, 141, 147, 149

[LVRD+21] X. Lu, B. Van Roy, V. Dwaracherla, M. Ibrahimi, I. Osband, and Z. Wen. Reinforcement Learning, Bit by Bit. Technical report, 2021. 265

[LWL06]   L. Li, T. J. Walsh, and M. L. Littman. Towards a unified theory of state abstraction for MDPs. In *AI&M*, 2006. 380

[Mac03]   D. J. C. MacKay. *Information theory, inference and learning algorithms*. Cambridge University Press, Cambridge, MA, 2003. 119

[Mah05]   M. V. Mahoney. Adaptive weighing of context models for lossless data compression. Technical report, 2005. 119

[Mah07]   M. Mahoney. The paq data compression programs, 2007. http://mattmahoney.net/dc/paq.html. 119

[Mah11]   M. Mahoney. About the test data, 2011. https://mattmahoney.net/dc/textdata.html. 119, 432

[Mah12]   M. Mahoney. *Data Compression Explained*. Dell, Inc, http://mattmahoney.net/dc/dce.html, 2012. 119

[Mah22]   M. Mahoney. Large text compression benchmark, 2022. http://mattmahoney.net/dc/text.html. 119

[Maj21]   S. J. Majeed. Abstractions of general reinforcement learning. *arXiv:2112.13404*, 2021. 371, 380

[Mak09]   T. Makino. Proto-predictive representation of states with simple recurrent temporal-difference networks. In *Proceedings of the 26th Annual International Conference on Machine Learning*, pages 697–704, 2009. 348

[Mar05]   G. Marsaglia. On the randomness of pi and other decimal expansions. *InterStat*, 5, 2005. 92

[Mat94]   M. J. Mataric. Reward functions for accelerated learning. In *Machine Learning proceedings 1994*, pages 181–189. Elsevier, 1994. 232

[Mau20]   F. Mauersberger. Thompson sampling: Predicting behavior in games and markets. *Available at SSRN 3061481*, 2020. 283

[MB16]   M. C. Machado and M. Bowling. Learning purposeful behaviour in the absence of rewards. *arXiv:1605.07700*, 2016. 284

[MBJ20]   T. M. Moerland, J. Broekens, and C. M. Jonker. Model-based Reinforcement Learning: A Survey. Technical report, 2020. 231

[MBMK21]   F. Massari, M. Biehl, L. Meeden, and R. Kanai. Experimental evidence that empowerment may drive exploration in sparse-reward environments. In *2021 IEEE International Conference on Development and Learning (ICDL)*, pages 1–6. IEEE, 2021. 284

[McC80]   J. McCarthy. Circumscription—A form of non-monotonic reasoning. *Artificial Intelligence*, 13(1–2):27–39, 1980. 118

[McC96]   A. K. McCallum. *Reinforcement Learning with Selective Perception and Hidden State*. PhD thesis, Department of Computer Science, University of Rochester, 1996. 347, 348, 378, 381

[MD80]   D. McDermott and J. Doyle. Nonmonotonic logic 1. *Artificial Intelligence*, 13:41–72, 1980. 118

[MDM+20]   V. Mikulik, G. Delétang, T. McGrath, T. Genewein, M. Martic, S. Legg, and P. A. Ortega. Meta-trained agents implement Bayes-optimal agents, 2020. 347

[ME70]   W. Mischel and E. B. Ebbesen. Attention in delay of gratification. *Journal of Personality and Social Psychology*, 16(2):329, 1970. 233

[Med91]   G. Medvedev. *The Truth About Chernobyl*. New York, NY (United States); Basic Books Inc., 1991. 390

[MEH16]   J. Martin, T. Everitt, and M. Hutter. Death and suicide in universal artificial intelligence. In *Proc. 9th Conf. on Artificial General Intelligence (AGI'16)*, volume 9782 of *LNAI*, pages 23–32, New York, USA, 2016. Springer. 396, 398, 411

[MEHS21]   G. Mittal, J. Engel, C. Hawthorne, and I. Simon. Symbolic music generation with diffusion models. *arXiv:2103.16091*, 2021. 430

[Mel18]   H. Melville. Moby-dick. In *Medicine and Literature*, pages 73–88. CRC Press, 2018. 162

[Mey22]   S. P. Meyn. *Control Systems and Reinforcement Learning*. Cambridge University Press, 2022. 232

[MF98]   N. Merhav and M. Feder. Universal prediction. *IEEE Transactions on Information Theory*, 44(6):2124–2147, 1998. 138

[MGLA00]   M. Mundhenk, J. Goldsmith, C. Lusena, and E. Allender. Complexity of finite-horizon Markov decision process problems. *Journal of the ACM (JACM)*, 47(4):681–720, 2000. 359

[MH18]   S. J. Majeed and M. Hutter. On Q-learning convergence for non-Markov decision processes. In *Proc. 27th International Joint Conf. on Artificial Intelligence (IJCAI'18)*, pages 2546–2552, Stockholm, Sweden, 2018. 380, 381

[MH19]   S. J. Majeed and M. Hutter. Performance guarantees for homomorphisms beyond Markov Decision Processes. In *Proc. 33rd AAAI Conference on Artificial Intelligence (AAAI'19)*, volume 33, pages 7659–7666, Honolulu, USA, 2019. AAAI Press. 380, 381

[MH21a]   S. J. Majeed and M. Hutter. Reducing planning complexity of general reinforcement learning with non-Markovian abstractions. *arXiv:2112.13386*, pages 1–16, 2021. 299, 371, 379, 380

[MH21b]   S. J. Majeed and M. Hutter. Exact reduction of huge action spaces in general reinforcement learning. In *Proc. 35th AAAI Conference on Artificial Intelligence (AAAI'21)*, volume 35, Virtual, Earth, 2021. AAAI Press. 299, 371, 380, 434

[MHC99]   O. Madani, S. Hanks, and A. Condon. On the undecidability of probabilistic planning and infinite-horizon partially observable Markov decision problems. In *AAAI/IAAI*, pages 541–548, 1999. 359

[MHC03]   O. Madani, S. Hanks, and A. Condon. On the undecidability of probabilistic planning and related stochastic optimization problems. *Artificial Intelligence*, 147:5–34, 2003. 359

[Mil87]   J. R. Milton. Induction before Hume. *The British Journal for the Philosophy of Science*, 38(1):49–74, 1987. 414

[Mil11]   G. Miller. *The mating mind: How sexual choice shaped the evolution of human nature*. Anchor, 2011. 392

[Mil21]   L. H. Miles. Markov decision processes with embedded agents. Master's thesis, 2021. 410

[Min07]   M. Minsky. *The Emotion Machine: Commonsense Thinking, Artificial Intelligence and the Future of the Human Mind*. Simon & Schuster, paperback edition edition, 2007. 416

[Mis19]   R. Mises. Grundlagen der Wahrscheinlichkeitsrechnung. *Mathematische Zeitschrift*, 5:52–99, 1919. Correction, *Ibid.*, volume 6, 1920, [English translation in: *Probability, Statistics, and Truth*, Macmillan, 1939]. 117

[Mis28]   R. Mises. *Wahrscheinlichkeit, Statistik und Wahrheit*. Springer, Berlin, 1928. [English translation: *Probability, Statistics, and Truth*, Allen and Unwin, London, 1957]. 118

[MJS19]   E. V. Mazumdar, M. I. Jordan, and S. S. Sastry. On Finding Local Nash Equilibria (and Only Local Nash Equilibria) in Zero-Sum Games. Technical report, 2019. 303

[MKS+15]   V. Mnih, K. Kavukcuoglu, D. Silver, A. A. Rusu, J. Veness, M. G. Bellemare, A. Graves, M. Riedmiller, A. K. Fidjeland, G. Ostrovski, et al. Human-level control through deep reinforcement learning. *Nature*, 518(7540):529–533, 2015. 236, 430, 437

[MKSB22]   T. Moskovitz, T.-C. Kao, M. Sahani, and M. M. Botvinick. Minimum Description Length Control, 2022. 120

[ML66]   P. Martin-Löf. The definition of random sequences. *Information and Control*, 9(6):602–619, 1966. 117

[MMR11]   O.-A. Maillard, R. Munos, and D. Ryabko. Selecting the state-representation in reinforcement learning. In *Advances in Neural Information Processing Systems (NIPS'11)*, volume 24, pages 2627–2635, 2011. 380

[Mor88]   H. Moravec. *Mind Children: The Future of Robot and Human Intelligence*. Harvard University Press, 1988. 409, 418

[MP43]   W. S. McCulloch and W. Pitts. A logical calculus of the ideas immanent in nervous activity. *The bulletin of mathematical biophysics*, 5(4):115–133, 1943. 427

[MPVDHV15] S. A. Mirsoleimani, A. Plaat, J. Van Den Herik, and J. Vermaseren. Parallel monte carlo tree search from multi-core to many-core processors. In *2015 IEEE Trustcom/BigDataSE/ISPA*, volume 3, pages 77–83. IEEE, 2015. 330

[MPYBG22] A. Mavor-Parker, K. Young, C. Barry, and L. Griffin. How to stay curious while avoiding noisy tvs using aleatoric uncertainty estimation. In *International Conference on Machine Learning*, pages 15220–15240. PMLR, 2022. 283

[MR67]   A. R. Meyer and D. M. Ritchie. The complexity of loop programs. In *Proceedings of the 1967 22nd National Conference*, pages 465–469, 1967. 84

[MRNA20]   J. Machado, J. M. Rocha-Neves, and J. P. Andrade. Computational analysis of the sars-cov-2 and other viruses based on the Kolmogorov's complexity and Shannon's information theories. *Nonlinear Dynamics*, 101(3):1731–1750, 2020. 121

[MSEH17]   J. Martin, S. N. Sasikumar, T. Everitt, and M. Hutter. Count-based exploration in feature space for reinforcement learning. In *Proc. 26th International Joint Conf. on Artificial Intelligence (IJCAI'17)*, pages 2471–2478, Melbourne, Australia, 2017. 381

[MSZ19]   A. Majha, S. Sarkar, and D. Zagami. Categorizing wireheading in partially embedded agents. *arXiv:1906.09136*, 2019. 343, 410

[Mül93]   U. Müller. Brainfuck–an eight-instruction turing-complete programming language, 1993. http://en.wikipedia.org/wiki/Brainfuck. 436

[Mül10]   M. Müller. Stationary algorithmic probability. *Theoretical Computer Science*, 411(1):113–130, 2010. 95

[Mur22]   K. P. Murphy. *Probabilistic Machine Learning: An Introduction*. Adaptive Computation and Machine Learning Series. The MIT Press, Cambridge, Massachusetts, 2022. 118

[Mur23]   K. P. Murphy. *Probabilistic Machine Learning: Advanced Topics*. Adaptive Computation and Machine Learning Series. The MIT Press, Cambridge, Massachusetts, 2023. 118

[MVK+16] K. Milan, J. Veness, J. Kirkpatrick, M. Bowling, A. Koop, and D. Hassabis. The forget-me-not process. *Advances in Neural Information Processing Systems*, 29:3702–3710, 2016. 211

[MW17]   J. V. Messias and S. Whiteson. Dynamic-depth context tree weighting. *Advances in Neural Information Processing Systems*, 30, 2017. 197

[MWV+21] T. Mesnard, T. Weber, F. Viola, S. Thakoor, A. Saade, A. Harutyunyan, W. Dabney, T. Stepleton, N. Heess, A. Guez, M. Hutter, L. Buesing, and R. Munos. Counterfactual credit assignment in model-free reinforcement learning. *Journal of Machine Learning Research, W&CP: ICML*, 139:7654–7664, 2021. 249

[Nac97]   J. H. Nachbar. Prediction, optimization, and learning in repeated games. *Econometrica: Journal of the Econometric Society*, pages 275–309, 1997. 303

[Nac05]   J. H. Nachbar. Beliefs in repeated games. *Econometrica*, 73(2):459–480, 2005. 303

[NC12]   A. A. Neath and J. E. Cavanaugh. The Bayesian Information Criterion: background, derivation, and applications. *Wiley Interdisciplinary Reviews: Computational Statistics*, 4(2):199–203, 2012. 104

[NCM22]   R. Ngo, L. Chan, and S. Mindermann. The alignment problem from a deep learning perspective. *arXiv:2209.00626*, 2022. 410

[NDD05]   Y. Niv, M. O. Duff, and P. Dayan. Dopamine, uncertainty and TD learning. *Behavioral and Brain Functions*, 1(1):1–9, 2005. 417

[NDR+21] A. Nichol, P. Dhariwal, A. Ramesh, P. Shyam, P. Mishkin, B. McGrew, I. Sutskever, and M. Chen. Glide: Towards photorealistic image generation and editing with text-guided diffusion models. *arXiv:2112.10741*, 2021. 430

[Nem22]   E. Nemerson. Squash compression benchmark, 2022. `https://quixdb.github.io/squash-benchmark/`. 119

[Net22]   S. Neth. A dilemma for Solomonoff prediction. *Philosophy of Science*, pages 1–25, 2022. 359

[Neu28]   J. V. Neumann. Zur Theorie der Gesellschaftsspiele. *Mathematische annalen*, 100(1):295–320, 1928. 303

[Ngu13]   P. Nguyen. *Feature Reinforcement Learning Agents*. PhD thesis, Research School of Computer Science, Australian National University, 2013. xix, 120, 375, 376, 379, 380

[NHR99]   A. Y. Ng, D. Harada, and S. Russell. Policy invariance under reward transformations: Theory and application to reward shaping. In *ICML*, volume 99, pages 278–287, 1999. 232

[Nie09]   A. Nies. *Computability and Randomness*, volume 51. OUP Oxford, 2009. 119

[NJ50]   J. F. Nash Jr. Equilibrium points in n-person games. *Proceedings of the national academy of sciences*, 36(1):48–49, 1950. 303

[NMRO13] P. Nguyen, O. Maillard, D. Ryabko, and R. Ortner. Competing with an infinite set of models in reinforcement learning. *JMLR WS&CP AISTATS*, 31:463–471, 2013. 380

[NOR13]   O.-A. M. P. Nguyen, R. Ortner, and D. Ryabko. Optimal regret bounds for selecting the state representation in reinforcement learning. *JMLR W&CP ICML*, 28(1):543–551, 2013. 380

[Noz74]   R. Nozick. *Anarchy, State, and Utopia*, volume 5038. Basic Books, 1974. 390

[NSH11]   P. Nguyen, P. Sunehag, and M. Hutter. Feature reinforcement learning in practice. In *Proc. 9th European Workshop on Reinforcement Learning (EWRL-9)*, volume 7188 of *LNAI*, pages 66–77. Springer, September 2011. 380

[NSH12]   P. Nguyen, P. Sunehag, and M. Hutter. Context tree maximizing reinforcement learning. In *Proc. 26th AAAI Conference on Artificial Intelligence (AAAI'12)*, pages 1075–1082, Toronto, Canada, 2012. AAAI Press. 211, 376, 380

[NYEYM03] M. Nisenson, I. Yariv, R. El-Yaniv, and R. Meir. Towards behaviometric security systems: Learning to identify a typist. In *European Conference on Principles of Data Mining and Knowledge Discovery*, pages 363–374. Springer, 2003. 197

[OA16]   L. Orseau and M. Armstrong. Safely interruptible agents. *Conference on Uncertainty in Artificial Intelligence*, 2016. 401, 410

[OAT14]   M. Oizumi, L. Albantakis, and G. Tononi. From the phenomenology to the mechanisms of consciousness: integrated information theory 3.0. *PLoS Computational Biology*, 10(5), 2014. 417

[OB10a]   P. A. Ortega and D. A. Braun. A minimum relative entropy principle for learning and acting. *Journal of Artificial Intelligence Research*, 38:475–511, 2010. 283

[OB10b]   P. A. Ortega and D. A. Braun. A Bayesian rule for adaptive control based on causal interventions. In *3d Conference on Artificial General Intelligence (AGI-2010)*, pages 182–187. Atlantis Press, 2010. 248, 271, 283

[OBD+06] A. O'Hagan, C. E. Buck, A. Daneshkhah, J. R. Eiser, P. H. Garthwaite, D. J. Jenkinson, J. E. Oakley, and T. Rakow. *Uncertain judgements: eliciting experts' probabilities*. Wiley, first edition, 2006. 61, 118

[Odi89]   P. Odifreddi. *Classical Recursion Theory, volume 1*. North–Holland, Amsterdam, 1989. 90, 119

[Odi99]   P. Odifreddi. *Classical Recursion Theory, volume 2*. Elsevier, Amsterdam, 1999. 119

[OGNJ17] Y. Ouyang, M. Gagrani, A. Nayyar, and R. Jain. Learning unknown Markov decision processes: A Thompson sampling approach. *Advances in Neural Information Processing Systems*, 30, 2017. 283

[O'H10] A. O'Hagan. *Kendall's Advanced Theory of Statistic 2B*. John Wiley & Sons, 2010. 118

[O'H19] A. O'Hagan. Expert knowledge elicitation: subjective but scientific. *The American Statistician*, 73(sup1):69–81, 2019. 118

[OHC+20] J. Oh, M. Hessel, W. M. Czarnecki, Z. Xu, H. van Hasselt, S. Singh, and D. Silver. Discovering Reinforcement Learning Algorithms. Technical report, 2020. 381

[OHL23] L. Orseau, M. Hutter, and L. H. Leli. Levin tree search with context models. In *Proceedings of the Thirty-Second International Joint Conference on Artificial Intelligence*, pages 5622–5630, 2023. Distinguished paper award. 349, 360

[OHSS12] A. O'Neill, M. Hutter, W. Shao, and P. Sunehag. Adaptive context tree weighting. In *Proc. Data Compression Conference (DCC'12)*, pages 317–326, Snowbird, Utah, USA, 2012. IEEE Computer Society. 197, 199, 200, 201

[OIO19] B. O'Donoghue, C. Ionescu, and I. Osband. Making Sense of Reinforcement Learning and Probabilistic Inference. 2019. 381

[OKD+21] P. A. Ortega, M. Kunesch, G. Delétang, T. Genewein, J. Grau-Moya, J. Veness, J. Buchli, J. Degrave, B. Piot, J. Perolat, T. Everitt, C. Tallec, E. Parisotto, T. Erez, Y. Chen, S. Reed, M. Hutter, N. de Freitas, and S. Legg. Shaking the foundations: Delusions in sequence models for interaction and control. Technical Report http://arxiv.org/abs/2110.10819, DeepMind, London, 2021. 249, 283

[OL21] L. Orseau and L. H. Lelis. Policy-guided heuristic search with guarantees. In *Proceedings of the AAAI Conference on Artificial Intelligence*, volume 35, pages 12382–12390, 2021. 360

[OLH13] L. Orseau, T. Lattimore, and M. Hutter. Universal knowledge-seeking agents for stochastic environments. In *Proc. 24th International Conf. on Algorithmic Learning Theory (ALT'13)*, volume 8139 of *LNAI*, pages 158–172, Singapore, 2013. Springer. 232, 273, 275, 283

[OLLW18] L. Orseau, L. Lelis, T. Lattimore, and T. Weber. Single-agent policy tree search with guarantees. *Advances in Neural Information Processing Systems*, 31, 2018. 360

[OM54] J. Olds and P. Milner. Positive reinforcement produced by electrical stimulation of septal area and other regions of rat brain. *Journal of Comparative and Physiological Psychology*, 47(6):419, 1954. 401

[Omo07] S. M. Omohundro. The nature of self-improving artificial intelligence. *Singularity Summit*, 2008, 2007. 391

[OMR14] R. Ortner, O.-A. Maillard, and D. Ryabko. Selecting near-optimal approximate state representations in reinforcement learning. In *International Conference on Algorithmic Learning Theory*, pages 140–154. Springer, 2014. 380

[OOMM17] B. O'Donoghue, I. Osband, R. Munos, and V. Mnih. The Uncertainty Bellman Equation and Exploration. Technical report, 2017. 283

[Ope22] OpenAI. DALL·E 2. https://openai.com/dall-e-2, 2022. Accessed: 2023-10-30. 430

[Ope23a] OpenAI. GPT-4 technical report, March 2023. 411, 427, 429, 438

[Ope23b] OpenAI. GPT-4 Turbo with 128k context and vision, November 2023. https://openai.com/blog/new-models-and-developer-products-announced-at-devday. 427, 429

[OPL+19] R. Ortner, M. Pirotta, A. Lazaric, R. Fruit, and O.-A. Maillard. Regret bounds for learning state representations in reinforcement learning. *Advances in Neural Information Processing Systems*, 32, 2019. 380

[OR94] M. J. Osborne and A. Rubenstein. *A Course in Game Theory*. The MIT Press, Cambridge, MA, 1994. 259, 287, 288, 290, 303

[OR11] L. Orseau and M. Ring. Self-modification and mortality in artificial agents. In *Proc. 4th Conf. on Artificial General Intelligence (AGI'11)*, volume 6830 of *LNAI*, pages 1–10. Springer, 2011. 404, 410

[OR12a] L. Orseau and M. Ring. Space-time embedded intelligence. In *Proc. 5th Conf. on Artificial General Intelligence (AGI'11)*, volume 7716 of *LNAI*, pages 209–218, Oxford, UK, 2012. Springer. 407, 410

[OR12b] L. Orseau and M. Ring. Memory Issues of Intelligent Agents. In J. Bach, B. Goertzel, and M. Iklé, editors, *Artificial General Intelligence*, volume 7716, pages 219–231. Springer Berlin Heidelberg, 2012. 411

[Ord20] T. Ord. *The Precipice: Existential Risk and the Future of Humanity*. Hachette Books, 2020. 410

[Ors10] L. Orseau. Optimality issues of universal greedy agents with static priors. In *Proc. 21st International Conf. on Algorithmic Learning Theory (ALT'10)*, volume 6331 of *LNAI*, pages 345–359, Canberra, Australia, 2010. Springer. 260, 265

[Ors11] L. Orseau. Universal knowledge-seeking agents. In *Proc. 22nd International Conf. on Algorithmic Learning Theory (ALT'11)*, volume 6925 of *LNAI*, pages 353–367, Espoo, Finland, 2011. Springer. 273, 276, 283

[Ors14a] L. Orseau. The multi-slot framework: A formal model for multiple, copiable ais. In *International Conference on Artificial General Intelligence*, pages 97–108. Springer, 2014. 419, 420

[Ors14b] L. Orseau. Teleporting Universal Intelligent Agents. In *Artificial General Intelligence*, Lecture Notes in Computer Science, pages 109–120. Springer, Cham, 2014. 419, 420

[Ort07] R. Ortner. Pseudometrics for state aggregation in average reward Markov decision processes. In *Proc. 18th International Conf. on Algorithmic Learning Theory (ALT'07)*, volume 4754 of *LNAI*, pages 373–387, Sendai, Japan, 2007. 369

[Ort11] P. A. Ortega. *A Unified Framework for Resource-Bounded Autonomous Agents Interacting with Unknown Environments*. PhD thesis, University of Cambridge, 2011. 271, 283

[Osb04] M. J. Osborne. *An Introduction to Game Theory*. Oxford University Press, New York, 2004. 287

[OVR16a] I. Osband and B. Van Roy. Posterior sampling for reinforcement learning without episodes. *arXiv:1608.02731*, 2016. 283

[OVR16b] I. Osband and B. Van Roy. Why is Posterior Sampling Better than Optimism for Reinforcement Learning? Technical report, 2016. 283

[OWJ+22] L. Ouyang, J. Wu, X. Jiang, D. Almeida, C. Wainwright, P. Mishkin, C. Zhang, S. Agarwal, K. Slama, A. Ray, et al. Training language models to follow instructions with human feedback. *Advances in Neural Information Processing Systems*, 35:27730–27744, 2022. 393

[OWR+19] P. A. Ortega, J. X. Wang, M. Rowland, T. Genewein, Z. Kurth-Nelson, R. Pascanu, N. Heess, J. Veness, A. Pritzel, P. Sprechmann, et al. Meta-learning of sequential strategies. *arXiv:1905.03030*, 2019. 283

[Pan08] S. Pankov. A computational approximation to the AIXI model. In *Proc. 1st Conference on Artificial General Intelligence*, volume 171, pages 256–267, 2008. 346, 359

[Pap94] C. H. Papadimitriou. *Computational Complexity*. Addison-Wesley, New York, 1994. 119

[Par84] D. Parfit. *Reasons and Persons*. Oxford University Press, 1984. 394

[Pas54] B. Pascal. Letters to Fermat, 1654. 117

[Pas76] R. C. Pasco. *Source Coding Algorithms for Fast Data Compression*. PhD thesis, Stanford University CA, 1976. 119

[PCZ+19] D. S. Park, W. Chan, Y. Zhang, C.-C. Chiu, B. Zoph, E. D. Cubuk, and Q. V. Le. Specaugment: A simple data augmentation method for automatic speech recognition. *arXiv:1904.08779*, 2019. 428

[Pea00] J. Pearl. *Causality: Models, Reasoning, and Inference*. Cambridge University Press, Cambridge, 2000. 249

[Pen89] R. Penrose. *The Emperor's New Mind*. Oxford University Press, 1989. 423, 425

[Pen94] R. Penrose. *Shadows of the Mind, A Search for the Missing Science of Consciousness*. Oxford University Press, 1994. 423, 425

[Pet15] Petrl. People for the ethical treatment of reinforcement learners, 2015. http://petrl.org/. 417

[PF97] X. Pintado and E. Fuentes. A forecasting algorithm based on information theory. In *Objects at Large*, page 209. Université de Genève, 1997. 121

[PGC10] M. J. Ponsen, G. Gerritsen, and G. Chaslot. Integrating opponent models with Monte-Carlo tree search in poker. *Interactive Decision Theory and Game Theory*, 82, 2010. 319

[PGJ16] J. Pearl, M. Glymour, and N. P. Jewell. *Causal Inference in Statistics: A Primer*. Wiley, Chichester, West Sussex, 2016. 249

[PGO+23] P. S. Park, S. Goldstein, A. O'Gara, M. Chen, and D. Hendrycks. AI deception: A survey of examples, risks, and potential solutions. *arXiv:2308.14752*, 2023. 394

[PH04a] J. Poland and M. Hutter. Convergence of discrete MDL for sequential prediction. In *Proc. 17th Annual Conf. on Learning Theory (COLT'04)*, volume 3120 of *LNAI*, pages 300–314, Banff, Canada, 2004. Springer. 154

[PH04b] J. Poland and M. Hutter. On the convergence speed of MDL predictions for Bernoulli sequences. In *Proc. 15th International Conf. on Algorithmic Learning Theory (ALT'04)*, volume 3244 of *LNAI*, pages 294–308, Padova, Italy, 2004. Springer. 154

[PH05a] J. Poland and M. Hutter. Asymptotics of discrete MDL for online prediction. *IEEE Transactions on Information Theory*, 51(11):3780–3795, 2005. 104, 120, 154

[PH05b] J. Poland and M. Hutter. Defensive universal learning with experts. In *Proc. 16th International Conf. on Algorithmic Learning Theory (ALT'05)*, volume 3734 of *LNAI*, pages 356–370, Singapore, 2005. Springer. 314, 346

[PH05c] J. Poland and M. Hutter. Strong asymptotic assertions for discrete MDL in regression and classification. In *Proc. 14th Dutch-Belgium Conf. on Machine Learning (Benelearn'05)*, pages 67–72, Enschede, 2005. 154

[PH06a] J. Poland and M. Hutter. MDL convergence speed for Bernoulli sequences. *Statistics and Computing*, 16(2):161–175, 2006. 154

[PH06b] J. Poland and M. Hutter. Universal learning of repeated matrix games. In *Proc. 15th Annual Machine Learning Conf. of Belgium and The Netherlands (Benelearn'06)*, pages 7–14, Ghent, Belgium, 2006. 308, 311, 312, 314, 346

[PH22] M. Phuong and M. Hutter. Formal algorithms for transformers. Technical report, DeepMind, London, UK, 2022. LaTeX source available at http://arxiv.org/abs/2207.09238. 429

[PHS⁺23] R. Prabhavalkar, T. Hori, T. N. Sainath, R. Schlüter, and S. Watanabe. End-to-end speech recognition: A survey, 2023. 428

[Pit19] S. Pitis. Rethinking the discount factor in reinforcement learning: A decision theoretic approach. In *Proceedings of the AAAI Conference on Artificial Intelligence*, volume 33, pages 7949–7956, 2019. 233

[PJS17] J. Peters, D. Janzing, and B. Schölkopf. *Elements of Causal Inference: Foundations and Learning Algorithms*. Adaptive Computation and Machine Learning Series. The MIT Press, Cambridge, Massachuestts, 2017. 249

[Pla22] A. Plaat. Deep Reinforcement Learning. *arXiv:2201.02135 [cs]*, 2022. 232

[PLV⁺17] R. Pascanu, Y. Li, O. Vinyals, N. Heess, L. Buesing, S. Racanière, D. Reichert, T. Weber, D. Wierstra, and P. Battaglia. Learning model-based planning from scratch. Technical report, 2017. 381

[PM18] J. Pearl and D. Mackenzie. *The Book of Why: The New Science of Cause and Effect*. Penguin Books, London, 2018. 249

[Pol06] J. Poland. Online learning with universal model and predictor classes. In *2006 IEEE Information Theory Workshop-ITW'06 Punta del Este*, pages 237–241. IEEE, 2006. 314

[Pop34] K. R. Popper. *Logik der Forschung*. Springer, Berlin, 1934. [English translation: *The Logic of Scientific Discovery* Basic Books, New York, 1959, and Hutchinson, London, revised edition, 1968]. 118

[Pos44] E. L. Post. Recursively enumerable sets of positive integers and their decision problems. *Bulletin of the American Mathematical Society*, 50:284–316, 1944. 119

[Pow98] D. M. Powers. The total Turing test and the Loebner prize. In *New Methods in Language Processing and Computational Natural Language Learning*, 1998. 432

[PP17] D. Pratas and A. J. Pinho. On the approximation of the Kolmogorov complexity for DNA sequences. In *Iberian Conference on Pattern Recognition and Image Analysis*, pages 259–266. Springer, 2017. 121

[PR12] A. Potapov and S. Rodionov. Extending universal intelligence models with formal notion of representation. In *International Conference on Artificial General Intelligence*, pages 242–251. Springer, 2012. 349

[Pre09] S. J. Press. *Subjective and objective Bayesian statistics: principles, models, and applications*. John Wiley & Sons, 2009. 61, 118, 155

[PS99] F. C. Pereira and Y. Singer. An efficient extension to mixture techniques for prediction and decision trees. *Machine Learning*, 36(3):183–199, 1999. 196, 197, 212

[PT87] C. H. Papadimitriou and J. N. Tsitsiklis. The complexity of Markov decision processes. *Mathematics of Operations Research*, 12(3):441–450, 1987. 359

[Put63] H. Putnam. 'Degree of confirmation' and inductive logic. In *The Philosophy of Rudolf Carnap*. Open Court, La Salle, IL, 1963. 118

[Put14] M. L. Puterman. *Markov Decision Processes: Discrete Stochastic Dynamic Programming*. John Wiley & Sons, 2014. 343

[PV08] P. Poupart and N. Vlassis. Model-based Bayesian reinforcement learning in partially observable domains. In *Proc Int. Symp. on Artificial Intelligence and Mathematics*, pages 1–2, 2008. 248, 348

[PVHR06] P. Poupart, N. Vlassis, J. Hoey, and K. Regan. An analytic solution to discrete Bayesian reinforcement learning. In *Proceedings of the 23rd International Conference on Machine Learning*, pages 697–704, 2006. 248, 348

[PW94] J. Peng and R. J. Williams. Incremental multi-step q-learning. In *Machine Learning Proceedings 1994*, pages 226–232. Elsevier, 1994. 381

[PZTH07] A. Piatti, M. Zaffalon, F. Trojani, and M. Hutter. Learning about a categorical latent variable under prior near-ignorance. In *Proc. 5th International Symposium on Imprecise Probability: Theories and Applications (ISIPTA'07)*, pages 357–364, Prague, Czech Republic, 2007. Action M Agency. 118

[PZTH09] A. Piatti, M. Zaffalon, F. Trojani, and M. Hutter. Limits of learning about a categorical latent variable under prior near-ignorance. *International Journal of Approximate Reasoning*, 50(4):597–611, 2009. 118

[RA11] P. Rajarajeswari and A. Apparao. Normalized distance matrix method for construction of phylogenetic trees using new compressor-dnabit compress. *Journal of Advanced Bioinformatics Applications and Research ISSN*, 2(1):89–97, 2011. 121

[Rad62] T. Rado. On non-computable functions. *Bell System Technical Journal*, 41(3):877–884, 1962. 87

[Ram31] F. P. Ramsey. Truth and probability. In *The Foundations of Mathematics: Collected Papers of Frank P. Ramsey*, pages 156–198. Routledge and Kegan Paul, London, 1931. 118

[Rav00] J. Raven. The Raven's progressive matrices: change and stability over culture and time. *Cognitive Psychology*, 41(1):1–48, 2000. 436

[RBL⁺22] R. Rombach, A. Blattmann, D. Lorenz, P. Esser, and B. Ommer. High-resolution image synthesis with latent diffusion models. In *Proc. IEEE/CVF Conf. on Computer Vision and Pattern Recognition (CVPR)*, pages 10684–10695, 2022. 429, 430

[RC67]   A. H. Robinson and C. Cherry. Results of a prototype television bandwidth compression scheme. *Proceedings of the IEEE*, 55(3):356–364, 1967. 119

[RCdP07]   S. Ross, B. Chaib-draa, and J. Pineau. Bayes-adaptive POMDPs. *Advances in Neural Information Processing Systems*, 20, 2007. 248, 348

[RCV16]   G. Reddy, A. Celani, and M. Vergassola. Infomax strategies for an optimal balance between exploration and exploitation. *Journal of Statistical Physics*, 163(6):1454–1476, 2016. 265

[Reh21]   E. M. Rehn. Free Will Belief as a consequence of Model-based Reinforcement Learning. Technical report, 2021. 416, 417

[Rei49]   H. Reichenbach. *The Theory of Probability: An Inquiry into the Logical and Mathematical Foundations of the Calculus of Probability*. University of California Press, Berkeley, CA, 2nd edition, 1949. 119

[Rei80]   R. Reiter. A logic for default reasoning. *Artificial Intelligence*, 13:81–132, 1980. 118

[RFB15]   O. Ronneberger, P. Fischer, and T. Brox. U-Net: Convolutional networks for biomedical image segmentation. In *Proc. of the 18th International Conference on Medical Image Computing and Computer-Assisted Intervention (MICCAI'15) Part III 18*, pages 234–241, Munich, Germany, 2015. Springer. 429

[RGT04]   M. Rosencrantz, G. Gordon, and S. Thrun. Learning low dimensional predictive representations. In *Proceedings of the 21st International Conference on Machine Learning*, page 88, 2004. 348

[RH06]   D. Ryabko and M. Hutter. Asymptotic learnability of reinforcement problems with arbitrary dependence. In *Proc. 17th International Conf. on Algorithmic Learning Theory (ALT'06)*, volume 4264 of *LNAI*, pages 334–347, Barcelona, Spain, 2006. Springer. 248

[RH07]   D. Ryabko and M. Hutter. On sequence prediction for arbitrary measures. In *Proc. IEEE International Symposium on Information Theory (ISIT'07)*, pages 2346–2350, Nice, France, 2007. IEEE. 154

[RH08a]   D. Ryabko and M. Hutter. On the possibility of learning in reactive environments with arbitrary dependence. *Theoretical Computer Science*, 405(3):274–284, 2008. 248

[RH08b]   D. Ryabko and M. Hutter. Predicting non-stationary processes. *Applied Mathematics Letters*, 21(5):477–482, 2008. 154

[RH09]   P. M. Rancoita and M. Hutter. mbpcr: A package for dna copy number profile estimation. *BioConductor – Open Source Software for BioInformatics*, (0.99):1–25, 2009. 155

[RH11]   S. Rathmanner and M. Hutter. A philosophical treatise of universal induction. *Entropy*, 13(6):1076–1136, 2011. xviii, 9, 61, 118, 147, 153, 416, 423

[RHBK09a] P. M. V. Rancoita, M. Hutter, F. Bertoni, and I. Kwee. Bayesian DNA copy number analysis. *BMC Bioinformatics*, 10(10):1–19, 2009. 155

[RHBK09b] P. M. V. Rancoita, M. Hutter, F. Bertoni, and I. Kwee. Bayesian joint estimation of CN and LOH aberrations. In *Proc. 3rd International Workshop on Practical Applications of Computational Biology & Bioinformatics (IWPACBB'09)*, volume 5518 of *LNCS*, pages 1109–1117, Salamanca, Spain, 2009. Springer. 155

[RHBK10] P. M. V. Rancoita, M. Hutter, F. Bertoni, and I. Kwee. An integrated Bayesian analysis of LOH and copy number data. *BMC Bioinformatics*, 11(321):1–18, 2010. 155

[RIR00]   D. Ríos Insua and F. Ruggeri, editors. *Robust Bayesian Analysis*. Number 152 in Lecture Notes in Statistics. Springer, New York, 2000. 118

[Ris78]   J. J. Rissanen. Modeling by shortest data description. *Automatica*, 14(5):465–471, 1978. 120

[Ris83]   J. Rissanen. A universal data compression system. *IEEE Transactions on Information Theory*, 29(5):656–664, 1983. 197

[Ris84]   J. Rissanen. Universal coding, information, prediction, and estimation. *IEEE Transactions on Information Theory*, 30(4):629–636, 1984. 191

[RJ87]   H. Rogers Jr. *Theory of Recursive Functions and Effective Computability*. MIT press, 1987. 297

[RK17]   V. Raj and S. Kalyani. Taming non-stationary bandits: A Bayesian approach. *arXiv:1707.09727*, 2017. 248, 283

[RKH+21] A. Radford, J. W. Kim, C. Hallacy, A. Ramesh, G. Goh, S. Agarwal, G. Sastry, A. Askell, P. Mishkin, J. Clark, G. Krueger, and I. Sutskever. Learning transferable visual models from natural language supervision. In M. Meila and T. Zhang, editors, *Proceedings of the 38th International Conference on Machine Learning*, volume 139 of *Proceedings of Machine Learning Research*, pages 8748–8763. PMLR, 18–24 Jul 2021. 429

[RLDG22] A. Ronca, G. P. Licks, and G. De Giacomo. Markov abstractions for PAC reinforcement learning in non-markov decision processes. *arXiv:2205.01053*, 2022. 265, 380

[RN94]   G. A. Rummery and M. Niranjan. *On-line Q-learning using connectionist systems*, volume 37. University of Cambridge, Department of Engineering Cambridge, UK, 1994. 381

[RN20]   S. J. Russell and P. Norvig. *Artificial Intelligence: A Modern Approach*. Pearson, 4 edition, 2020. 117, 118, 226, 236, 349, 427, 428

[RÖ96]   W. Rabinowicz and J. Österberg. Value based on preferences: On two interpretations of preference utilitarianism. *Economics & Philosophy*, 12(1):1–27, 1996. 394

[RO11]   M. Ring and L. Orseau. Delusion, survival, and intelligent agents. In *Proc. 4th Conf. on Artificial General Intelligence (AGI'11)*, volume 6830 of *LNAI*, pages 11–20. Springer, 2011. xviii, 401, 403, 404, 411

[Rog67]   H. Rogers. *Theory of Recursive Functions and Effective Computability*. McGraw-Hill, New York, 1967. 119

[Rog96]   Y. Rogozhin. Small universal turing machines. *Theoretical Computer Science*, 168(2):215–240, 1996. 96

[Ros14]   S. M. Ross. *Introduction to Probability Models*. Academic press, 2014. 24, 117

[RST96]   D. Ron, Y. Singer, and N. Tishby. The power of amnesia: Learning probabilistic automata with variable memory length. *Machine Learning*, 25(2):117–149, 1996. 197

[RT64]   W. Rudin and Tata McGraw-Hill Publishing Company. *Principles of Mathematical Analysis*. McGraw Education (india) Private Limited, Chennai, 1964. 19, 34, 117

[RTBP+23] A. Rannen-Triki, J. Bornschein, R. Pascanu, A. Galashov, M. Titsias, M. Hutter, A. György, and Y. W. Teh. Revisiting dynamic evaluation: Online adaptation for large language models. In *NeurIPS 2023 Workshop on Distribution Shifts: New Frontiers with Foundation Models*, 2023. 120

[Rus15]   S. Russell. Of myths and moonshine, 2015. `https://www.edge.org/conversation/jaron_lanier-the-myth-of-ai`. 393

[Rus19]   S. Russell. *Human Compatible: Artificial Intelligence and the Problem of Control*. Penguin, 2019. 388, 410

[RVR14]   D. Russo and B. Van Roy. Learning to optimize via information-directed sampling. *Advances in Neural Information Processing Systems*, 27, 2014. 265

[RVRK+17] D. Russo, B. Van Roy, A. Kazerouni, I. Osband, and Z. Wen. A Tutorial on Thompson Sampling. Technical report, 2017. 283

[RVWD13]   D. M. Roijers, P. Vamplew, S. Whiteson, and R. Dazeley. A survey of multi-objective sequential decision-making. *Journal of Artificial Intelligence Research*, 48:67–113, 2013. 232

[Rya80]   B. Y. Ryabko. Data compression by means of a "book stack". *Problemy Peredachi Informatsii*, 16(4):16–21, 1980. 119

[Rya20]   D. Ryabko. *Universal Time-Series Forecasting with Mixture Predictors*. Springer Nature, 2020. 154

[RZP+22]   S. Reed, K. Zolna, E. Parisotto, S. G. Colmenarejo, A. Novikov, G. Barth-Maron, M. Gimenez, Y. Sulsky, J. Kay, J. T. Springenberg, et al. A generalist agent. *arXiv:2205.06175*, 2022. 431

[SAAGB21] A. Sarkar, Z. Al-Ars, H. Gandhi, and K. Bertels. QKSA: Quantum knowledge seeking agent-resource-optimized reinforcement learning using quantum process tomography. *arXiv:2112.03643*, 2021. 283, 349

[SAH+20]   J. Schrittwieser, I. Antonoglou, T. Hubert, K. Simonyan, L. Sifre, S. Schmitt, A. Guez, E. Lockhart, D. Hassabis, T. Graepel, et al. Mastering Atari, Go, Chess and Shogi by planning with a learned model. *Nature*, 588(7839):604–609, 2020. 430

[Sai04]   A. Said. Introduction to Arithmetic Coding Theory and Practice. Technical Report HPL-2004-76, HP, 2004. 119

[Sam59]   A. L. Samuel. Some studies in machine learning using the game of checkers. *IBM Journal on Research and Development*, 3:210–229, 1959. 430

[Sam67]   A. L. Samuel. Some studies in machine learning using the game of checkers. II – recent progress. *IBM Journal of Research and Development*, 11(6):601–617, 1967. 430

[Sar21]   A. Sarkar. QKSA: Quantum knowledge seeking agent. *arXiv:2107.01429*, 2021. 283, 349

[Sav54]   L. J. Savage. *The Foundations of Statistics*. Wiley, New York, 1954. 118

[SB04]   G. Shani and R. Brafman. Resolving perceptual aliasing in the presence of noisy sensors. *Advances in Neural Information Processing Systems*, 17, 2004. 347

[SB18]   R. S. Sutton and A. G. Barto. *Reinforcement Learning: An Introduction*. MIT press, 2018. 216, 219, 220, 222, 230, 231, 268, 284, 320, 323, 381, 430

[SBH23]   J. Scheurer, M. Balesni, and M. Hobbhahn. Technical report: Large language models can strategically deceive their users when put under pressure. *arXiv:2311.07590*, 2023. 411

[SBP22]   R. S. Sutton, M. Bowling, and P. M. Pilarski. The alberta plan for AI research. *arXiv:2208.11173*, 2022. 382

[SC73]   R. C. Schank and K. M. Colby. *Computer Models of Thought and Language*. WH Freeman & Co., 1973. 438

[SC84]   R. C. Schank and P. Childers. The cognitive computer on language, learning, and artificial intelligence. 1984. 438

[Sch71]   C. P. Schnorr. *Zufälligkeit und Wahrscheinlichkeit*, volume 218 of *Lecture Notes in Mathematics*. Springer, Berlin, 1971. 117

[Sch73]   C. P. Schnorr. Process complexity and effective random tests. *Journal of Computer and System Sciences*, 7(4):376–388, 1973. 120

[Sch80]   T. C. Schelling. *The Strategy of Conflict: with a new Preface by the Author*. Harvard university press, 1980. 303

[Sch91]   J. Schmidhuber. A possibility for implementing curiosity and boredom in model-building neural controllers. In *Proc. of the International Conference on Simulation of Adaptive Behavior: From Animals to Animats*, pages 222–227, 1991. 273, 283

[Sch93]   A. Schwartz. A reinforcement learning method for maximizing undiscounted rewards. In *Proceedings of the 10th International Conference on Machine Learning*, volume 298, pages 298–305, 1993. 233

[Sch97a]   R. R. Schaller. Moore's law: past, present and future. *IEEE spectrum*, 34(6):52–59, 1997. 426

[Sch97b]   J. Schmidhuber. Discovering neural nets with low Kolmogorov complexity and high generalization capability. *Neural Networks*, 10(5):857–873, 1997. 346

[Sch99]   M. Schmidt. Time-bounded Kolmogorov complexity may help in search for extra terrestrial intelligence (SETI). *Bulletin of the European Association for Theoretical Computer Science*, 67:176–180, 1999. 121

[Sch02a]   J. Schmidhuber. Hierarchies of generalized Kolmogorov complexities and nonenumerable universal measures computable in the limit. *International Journal of Foundations of Computer Science*, 13(4):587–612, 2002. 119, 141

[Sch02b]   J. Schmidhuber. The speed prior: A new simplicity measure yielding near-optimal computable predictions. In *Proc. 15th Conf. on Computational Learning Theory (COLT'02)*, volume 2375 of *LNAI*, pages 216–228, Sydney, Australia, 2002. Springer. 121, 346

[Sch03]   J. Schmidhuber. Bias-optimal incremental problem solving. In *Advances in Neural Information Processing Systems 15*, pages 1571–1578. MIT Press, Cambridge, MA, 2003. 346

[Sch04]   J. Schmidhuber. Optimal ordered problem solver. *Machine Learning*, 54(3):211–254, 2004. 346, 360

[Sch05]   J. Schmidhuber. Gödel machines: Towards a technical justification of consciousness. In D. Kudenko, D. Kazakov, and E. Alonso, editors, *Adaptive Agents and Multi-Agent Systems III (LNCS 3394)*, pages 1–23. Springer, 2005. 360

[Sch07]   J. Schmidhuber. Gödel machines: Self-referential universal problem solvers making provably optimal self-improvements. In *Artificial General Intelligence*, pages 199–226. Springer, 2007. 360

[Sch10]   J. Schmidhuber. Formal theory of creativity, fun, and intrinsic motivation (1990–2010). *IEEE Transactions on Autonomous Mental Development*, 2(3):230–247, 2010. 284

[Sch19a]   J. Schmidhuber. Reinforcement Learning Upside Down: Don't Predict Rewards – Just Map Them to Actions. Technical report, 2019. 381

[Sch19b]   G. Schurz. *Hume's Problem Solved: The Optimality of Meta-Induction*. The MIT Press, Cambridge, Massachusetts, 2019. 423

[SCS⁺22]   C. Saharia, W. Chan, S. Saxena, , et al. Photorealistic text-to-image diffusion models with deep language understanding. *Advances in Neural Information Processing Systems*, 35:36479–36494, 2022. 430

[SDWMG15] J. Sohl-Dickstein, E. Weiss, N. Maheswaranathan, and S. Ganguli. Deep unsupervised learning using nonequilibrium thermodynamics. In *Proceedings of the 32nd International Conference on Machine Learning*, volume 37, pages 2256–2265, Lille, France, 2015. PMLR. 430

[SDYD⁺23] T. Schick, J. Dwivedi-Yu, R. Dessì, R. Raileanu, M. Lomeli, L. Zettlemoyer, N. Cancedda, and T. Scialom. Toolformer: Language models can teach themselves to use tools. *arXiv:2302.04761*, 2023. 438

[Sea80]   J. Searle. Minds, brains, and programs. *Behavioral and Brain Sciences*, 3:417–458, 1980. 421

[SGBK⁺21] E. Sezener, A. Grabska-Barwińska, D. Kostadinov, M. Beau, S. Krishnagopal, D. Budden, M. Hutter, J. Veness, M. Botvinick, C. Clopath, M. Häusser, and P. E. Latham. A rapid and efficient learning rule for biological neural circuits. Technical report, DeepMind, London, UK, 2021. 438

[SGS11]   Y. Sun, F. Gomez, and J. Schmidhuber. Planning to be surprised: Optimal Bayesian exploration in dynamic environments. In *International Conference on Artificial General Intelligence*, pages 41–51. Springer, 2011. 283

[SH98]   N. Suematsu and A. Hayashi. A reinforcement learning algorithm in partially observable environments using short-term memory. *Advances in Neural Information Processing Systems*, 11, 1998. 348

[SH10]   P. Sunehag and M. Hutter. Consistency of feature Markov processes. In *Proc. 21st International Conf. on Algorithmic Learning Theory (ALT'10)*, volume 6331 of *LNAI*, pages 360–374, Canberra, Australia, 2010. Springer. 154, 380

[SH11a]   P. Sunehag and M. Hutter. Axioms for rational reinforcement learning. In *Proc. 22nd International Conf. on Algorithmic Learning Theory (ALT'11)*, volume 6925 of *LNAI*, pages 338–352, Espoo, Finland, 2011. Springer. 248

[SH11b]   P. Sunehag and M. Hutter. Principles of Solomonoff induction and AIXI. In *Proc. Solomonoff 85th Memorial Conference*, volume 7070 of *LNAI*, pages 386–398, Melbourne, Australia, 2011. Springer. 248

[SH12a]   P. Sunehag and M. Hutter. Optimistic agents are asymptotically optimal. In *Proc. 25th Australasian Joint Conference on Artificial Intelligence (AusAI'12)*, volume 7691 of *LNAI*, pages 15–26, Sydney, Australia, 2012. Springer. xix, 268, 269, 270, 271

[SH12b]   P. Sunehag and M. Hutter. Optimistic AIXI. In *Proc. 5th Conf. on Artificial General Intelligence (AGI'12)*, volume 7716 of *LNAI*, pages 312–321. Springer, Heidelberg, 2012. 283

[SH13]   P. Sunehag and M. Hutter. Learning agents with evolving hypothesis classes. In *Proc. 6th Conf. on Artificial General Intelligence (AGI'13)*, volume 7999 of *LNAI*, pages 150–159. Springer, Heidelberg, 2013. 271

[SH14a]   P. Sunehag and M. Hutter. A dual process theory of optimistic cognition. In *Proc. 36th Annual Meeting of the Cognitive Science Society (CogSci'14)*, pages 2949–2954, Quebec City, Canada, 2014. Curran Associates. 283

[SH14b]   P. Sunehag and M. Hutter. Intelligence as inference or forcing Occam on the world. In *Proc. 7th Conf. on Artificial General Intelligence (AGI'14)*, volume 8598 of *LNAI*, pages 186–195, Quebec City, Canada, 2014. Springer. 95

[SH15a]   P. Sunehag and M. Hutter. Algorithmic complexity. In J. D. Wright, editor, *International Encyclopedia of the Social & Behavioral Sciences*, volume 1, pages 534—-538. Elsevier, 2nd edition, 2015. 120

[SH15b]   P. Sunehag and M. Hutter. Rationality, optimism and guarantees in general reinforcement learning. *Journal of Machine Learning Research*, 16:1345–1390, 2015. 268, 271, 283

[SH15c]   P. Sunehag and M. Hutter. Using localization and factorization to reduce the complexity of reinforcement learning. In *Proc. 8th Conf. on Artificial General Intelligence (AGI'15)*, volume 9205 of *LNAI*, pages 177–186, Berlin, Germany, 2015. Springer. 271

[SH⁺16]   D. Silver, A. Huang, et al. Mastering the game of go with deep neural networks and tree search. *Nature*, 529(7587):484–489, 2016. 232, 236, 296, 319, 421, 430

[Sha48]   C. E. Shannon. A mathematical theory of communication. *Bell System Technical Journal*, 27:379–423, 623–656, 1948. 68, 78, 119

[Sha51]   C. E. Shannon. Prediction and Entropy of Printed English. *Bell System Technical Journal*, 30(1):50–64, 1951. 70, 119

[Sha76]   G. Shafer. *A Mathematical Theory of Evidence*. Princeton University Press, Princeton, NJ, 1976. 118

[Sha07]   G. Shani. *Learning and Solving Partially Observable Markov Decision Processes*. Ben Gurion University, 2007. 347

[SHB⁺20]   E. Sezener, M. Hutter, D. Budden, J. Wang, and J. Veness. Online learning in contextual bandits using gated linear networks. In *Advances in Neural Information Processing Systems (NeurIPS'20)*, volume 33, pages 19467–19477, Cambridge, MA, USA, 2020. Curran Associates. 347

[Shi96]   A. N. Shiriaev. *Probability*. Number 95 in Graduate Texts in Mathematics. Springer, 2nd ed edition, 1996. 117

[SHKK22]   J. Skalse, N. Howe, D. Krasheninnikov, and D. Krueger. Defining and characterizing reward gaming. *Advances in Neural Information Processing Systems*, 35:9460–9471, 2022. 411

[SHL97]   N. Suematsu, A. Hayashi, and S. Li. A Bayesian approach to model learning in non-markovian environments. In *ICML*, pages 349–357, 1997. 248, 348

[Sho67]   J. R. Shoenfield. *Mathematical Logic*. Addison-Wesley, Reading, MA, 1967. 119

[Sho76]   E. H. Shortliffe. *Computer-Based Medical Consultations: MYCIN*. Elsevier/North-Holland, Amsterdam, 1976. 118

[Sho94]   P. W. Shor. Algorithms for quantum computation: Discrete logarithms and factoring. In *Proceedings 35th Annual Symposium on Foundations of Computer Science*, pages 124–134. Ieee, 1994. 427

[SI96]   N. J. A. Sloane and T. O. F. Inc. Expansion of pi in base 2, 1996. https://oeis.org/A004601. 92

[Sie05]   W. Sieg. Only two letters: The correspondence between Herbrand and Gödel. *Bulletin of Symbolic Logic*, 11(2):172–184, 2005. 83

[Sip12]   M. Sipser. *Introduction to the Theory of Computation*. Cengage Learning, 3rd edition, 2012. 119

[SJ21]   E. Stefansson and K. H. Johansson. Computing complexity-aware plans using Kolmogorov complexity. In *2021 60th IEEE Conference on Decision and Control (CDC)*, pages 3420–3427. IEEE, 2021. 349

[SJR12]   S. Singh, M. James, and M. Rudary. Predictive state representations: A new theory for modeling dynamical systems. *arXiv:1207.4167*, 2012. 348

[SKH23]   J. Schwartz, H. Kurniawati, and M. Hutter. Combining a meta-policy and monte-carlo planning for scalable type-based reasoning in partially observable environments. *arXiv:2306.06067*, 2023. 248, 303

[SL08]   I. Szita and A. Lörincz. The many faces of optimism: a unifying approach. In *Proc. 12th International Conference (ICML 2008)*, volume 307, pages 1048–1055, Helsinki, Finland, 2008. 283

[SLB08]   Y. Shoham and K. Leyton-Brown. *Multiagent Systems: Algorithmic, Game-Theoretic, and Logical Foundations*. Cambridge University Press, 2008. 287, 303

[SLO22]   R. A. Starre, M. Loog, and F. A. Oliehoek. Model-based reinforcement learning with state abstraction: A survey. 2022. 380

[SLP+18] G. P. Sarma, C. W. Lee, T. Portegys, V. Ghayoomie, T. Jacobs, B. Alicea, M. Cantarelli, M. Currie, R. C. Gerkin, S. Gingell, et al. Openworm: overview and recent advances in integrative biological simulation of caenorhabditis elegans. *Philosophical Transactions of the Royal Society B*, 373(1758), 2018. 419

[SLR07] R. Sabbadin, J. Lang, and N. Ravoanjanahry. Purely epistemic Markov decision processes. In *AAAI*, pages 1057–1062, 2007. 359

[SM10] D. Salomon and G. Motta. *Handbook of Data Compression*. Springer, Berlin, 5th edition, 2010. 119

[Smi78] J. M. Smith. Optimization theory in evolution. *Annual review of ecology and systematics*, 9(1):31–56, 1978. 401

[SMW16] R. S. Sutton, A. R. Mahmood, and M. White. An emphatic approach to the problem of off-policy temporal-difference learning. *The Journal of Machine Learning Research*, 17(1):2603–2631, 2016. 381

[Sol64] R. J. Solomonoff. A formal theory of inductive inference: Parts 1 and 2. *Information and Control*, 7:1–22 and 224–254, 1964. 8, 117, 120, 121, 141, 439

[Sol78] R. J. Solomonoff. Complexity-based induction systems: Comparisons and convergence theorems. *IEEE Transactions on Information Theory*, IT-24:422–432, 1978. 120

[Sol85] R. J. Solomonoff. The time scale of artificial intelligence: Reflections on social effects. *Human Systems Management*, 5:149–153, 1985. 409

[Soz98] P. D. Sozou. On hyperbolic discounting and uncertain hazard rates. *Proceedings of the Royal Society of London. Series B: Biological Sciences*, 265(1409):2015–2020, 1998. 233

[SP73] J. M. Smith and G. R. Price. The logic of animal conflict. *Nature*, 246(5427):15–18, 1973. 303

[SP02] M. Stolle and D. Precup. Learning Options in Reinforcement Learning. In S. Koenig and R. C. Holte, editors, *Abstraction, Reformulation, and Approximation*, Lecture Notes in Computer Science, pages 212–223. Springer, 2002. 284

[SS82] J. A. Storer and T. G. Szymanski. Data compression via textual substitution. *Journal of the ACM (JACM)*, 29(4):928–951, 1982. 119

[SS09] E. M. Stein and R. Shakarchi. *Real analysis*. Princeton University Press, 2009. 24, 27, 117

[SS11] B. R. Steunebrink and J. Schmidhuber. A family of gödel machine implementations. In *International Conference on Artificial General Intelligence*, pages 275–280. Springer, 2011. 360

[SSH12] P. Sunehag, W. Shao, and M. Hutter. Coding of non-stationary sources as a foundation for detecting change points and outliers in binary time-series. In *Proc. 10th Australasian Data Mining Conference (AusDM'12)*, volume 134, pages 79–84, Sydney, Australia, 2012. Australian Computer Society. 197

[SSPS21] D. Silver, S. Singh, D. Precup, and R. S. Sutton. Reward is enough. *Artificial Intelligence*, 299:103535, 2021. 232

[ST04] R. S. Sutton and B. Tanner. Temporal-difference networks. In L. Saul, Y. Weiss, and L. Bottou, editors, *Advances in Neural Information Processing Systems*, volume 17. MIT Press, 2004. 348

[ST09] D. Silver and G. Tesauro. Monte-Carlo simulation balancing. In *Proceedings of the 26th Annual International Conference on Machine Learning*, pages 945–952, 2009. 347

[Sta22] StabilityAI. Stable diffusion public release, 2022. `https://stability.ai/blog/stable-diffusion-public-release`. 430

[Ste17] T. F. Sterkenburg. A Generalized Characterization of Algorithmic Probability. *Theory of Computing Systems*, 61(4):1337–1352, 2017. 423

[Str00a] M. Strens. A Bayesian framework for reinforcement learning. In *Proc. 17th International Conf. on Machine Learning*, pages 943–950. Morgan Kaufmann, San Francisco, CA, 2000. 248

[Str00b] M. Strens. A Bayesian framework for reinforcement learning. In *ICML*, volume 2000, pages 943–950, 2000. 348

[Sut88] R. S. Sutton. Learning to predict by the methods of temporal differences. *Machine Learning*, 3:9–44, 1988. 381

[SV01] G. Shafer and V. Vovk. *Probability and Finance: It's Only a Game!* Wiley-Interscience, New York, NY, 2001. 117

[SV10] D. Silver and J. Veness. Monte-Carlo planning in large POMDPs. *Advances in Neural Information Processing Systems*, 23:2164–2172, 2010. 341, 347, 348

[SvHH+16] D. Silver, H. van Hasselt, M. Hessel, T. Schaul, A. Guez, T. Harley, G. Dulac-Arnold, D. Reichert, N. Rabinowitz, A. Barreto, and T. Degris. The Predictron: End-To-End Learning and Planning. Technical report, 2016. 381

[SVK+22] R. Shah, V. Varma, R. Kumar, M. Phuong, V. Krakovna, J. Uesato, and Z. Kenton. Goal misgeneralization: Why correct specifications aren't enough for correct goals. *arXiv:2210.01790*, 2022. 411

[SVS+14] G. Story, I. Vlaev, B. Seymour, A. Darzi, and R. Dolan. Does temporal discounting explain unhealthy behavior? A systematic review and reinforcement learning perspective. 8, 2014. 233

[Sze10] C. Szepesvari. *Algorithms for Reinforcement Learning*. Number 9 in Synthesis Lectures on Artificial Intelligence and Machine Learning. Morgan & Claypool, San Rafael, Calif., 2010. 231

[SZW97]   J. Schmidhuber, J. Zhao, and M. Wiering. Shifting inductive bias with success-story algorithm, adaptive Levin search, and incremental self-improvement. *Machine Learning*, 28(1):105–130, 1997. 346

[T+95]   G. Tesauro et al. Temporal difference learning and TD-gammon. *Communications of the ACM*, 38(3):58–68, 1995. 232, 430

[Tay16]   J. Taylor. Quantilizers: A safer alternative to maximizers for limited optimization. In *Workshops at the Thirtieth AAAI Conference on Artificial Intelligence*, 2016. 406

[TBMK16]   G. Tononi, M. Boly, M. Massimini, and C. Koch. Integrated information theory: from consciousness to its physical substrate. *Nature Reviews Neuroscience*, 17(7):450–461, 2016. 417

[Tes94]   G. Tesauro. TD-Gammon, a self-teaching backgammon program, achieves master-level play. *Neural Computation*, 6(2):215–219, 1994. 232, 296, 430

[TG96]   G. Tesauro and G. Galperin. On-line policy improvement using Monte-Carlo search. *Advances in Neural Information Processing Systems*, 9:1068–1074, 1996. 318

[TH10]   M.-N. Tran and M. Hutter. Model selection by loss rank for classification and unsupervised learning. Technical Report arXiv:1011.1379, NUS and ANU, Singapore and Australia, 2010. 120

[Tho47]   R. Thornton. The age of machinery. In *The Expounder of Primitive Christianity*, volume 4, page 281. Ann Arbor, Michigan, 1847. 409

[Tho33]   W. R. Thompson. On the likelihood that one unknown probability exceeds another in view of the evidence of two samples. *Biometrika*, 25(3/4):285–294, 1933. 271, 283

[Thr02]   S. Thrun. Probabilistic robotics. *Communications of the ACM*, 45(3):52–57, 2002. 381

[Tip95]   F. J. Tipler. *The Physics of Immortality*. Macmillan, 1995. 409

[Ton12]   G. Tononi. The integrated information theory of consciousness: an updated account. *Archives italiennes de biologie*, 150(2/3):56–90, 2012. 417

[Tre76]   C. P. Tremaux. A version of depth-first search as a strategy for solving mazes, 1876. Ecole polytechnique of Paris. 349

[TSG20]   J. Tětek, M. Sklenka, and T. Gavenčiak. Performance of bounded-rational agents with the ability to self-modify. *arXiv:2011.06275*, 2020. 410

[TSS+19]   A. M. Turner, L. Smith, R. Shah, A. Critch, and P. Tadepalli. Optimal policies tend to seek power. *arXiv:1912.01683*, 2019. 411

[TSW93a]   T. J. Tjalkens, Y. M. Shtarkov, and F. M. Willems. Context tree weighting: Multi-alphabet sources. In *Symposium on Information Theory in the Benelux*, pages 128–128. Technische Universiteit Delft, 1993. 197

[TSW93b]   T. J. Tjalkens, Y. M. Shtarkov, and F. M. Willems. Sequential weighting algorithms for multi-alphabet sources. In *6th Joint Swedish-Russian International Workshop on Information Theory*, pages 230–234. Citeseer, 1993. 197

[Tur36]   A. M. Turing. On computable numbers, with an application to the Entscheidungsproblem. *Proc. London Mathematical Society*, 2(42):230–265, 1936. 81, 119

[Tur04]   A. Turing. Intelligent machinery (1948). *B. Jack Copeland*, page 395, 2004. 81, 414

[Tur09]   A. M. Turing. Computing machinery and intelligence. In *Parsing the Turing test*, pages 23–65. Springer, 2009. 432

[TYLC16]   J. Taylor, E. Yudkowsky, P. LaVictoire, and A. Critch. Alignment for advanced machine learning systems. *Ethics of Artificial Intelligence*, pages 342–382, 2016. 410

[TZXS19]   A. Trott, S. Zheng, C. Xiong, and R. Socher. Keeping your distance: Solving sparse reward tasks using self-balancing shaped rewards. *Advances in Neural Information Processing Systems*, 32, 2019. 232

[UKK+20]   J. Uesato, R. Kumar, V. Krakovna, T. Everitt, R. Ngo, and S. Legg. Avoiding tampering incentives in deep RL via decoupled approval. *arXiv:2011.08827*, 2020. 411

[Ula58]   S. Ulam. Tribute to John von Neumann. *Bulletin of the American Mathematical Society*, 64(3 II):1–49, 1958. 409

[UV98]   W. T. Uther and M. M. Veloso. Tree based discretization for continuous state space reinforcement learning. *AAI/IAAI*, 98:769–774, 1998. 343

[VBC+19]   O. Vinyals, I. Babuschkin, W. M. Czarnecki, M. Mathieu, A. Dudzik, J. Chung, D. H. Choi, R. Powell, T. Ewalds, P. Georgiev, et al. Grandmaster level in starcraft II using multi-agent reinforcement learning. *Nature*, 575(7782):350–354, 2019. 296, 421, 430

[VBH+15]   J. Veness, M. Bellemare, M. Hutter, A. Chua, and G. Desjardins. Compress and control. In *Proc. 29th AAAI Conference on Artificial Intelligence (AAAI'15)*, pages 3016–3023, Austin, USA, 2015. AAAI Press. 284, 349

[VBM+23]   M. Varadi, D. Bertoni, P. Magana, et al. AlphaFold Protein Structure Database in 2024: Providing structure coverage for over 214 million protein sequences. *Nucleic Acids Research*, page gkad1011, 2023. 428

[VCH18]   B. N. Vellambi, O. Cameron, and M. Hutter. Universal compression of piecewise i.i.d. sources. In *Proc. Data Compression Conference (DCC'18)*, Snowbird, Utah, USA, 2018. IEEE Computer Society. 197, 202

[VH12]   J. Veness and M. Hutter. Sparse sequential Dirichlet coding. Technical Report arXiv:1206.3618, UoA and ANU, 2012. 197

[VH18]   B. N. Vellambi and M. Hutter. Convergence of binarized context-tree weighting for estimating distributions of stationary sources. In *Proc. IEEE International Symposium on Information Theory (ISIT'18)*, pages 731–735, Vail, USA, 2018. IEEE. 197

[VHOB15] J. Veness, M. Hutter, L. Orseau, and M. Bellemare. Online learning of k-CNF boolean functions. In *Proc. 24th International Joint Conf. on Artificial Intelligence (IJCAI'15)*, pages 3865–3873, Buenos Aires, Argentina, 2015. AAAI Press. 154

[Vin93]   V. Vinge. The coming technological singularity. *Vision-21 Symposium, NASA Lewis Research Center and the Ohio Aerospace Institute, 30 to 31 March 1993 and Winter issue of Whole Earth Review*, 1993. 409

[VLB+21] J. Veness, T. Lattimore, D. Budden, A. Bhoopchand, C. Mattern, A. Grabska-Barwinska, E. Sezener, J. Wang, P. Toth, S. Schmitt, and M. Hutter. Gated linear networks. In *Proc. 35th AAAI Conference on Artificial Intelligence (AAAI'21)*, volume 35, Virtual, Earth, 2021. AAAI Press. 438

[VLCU07] F. Van Lishout, G. Chaslot, and J. W. Uiterwijk. Monte-Carlo tree search in backgammon. In *Computer Games Workshop*, 2007. 319

[VNB+66] J. Von Neumann, A. W. Burks, et al. Theory of self-reproducing automata. *IEEE Transactions on Neural Networks*, 5(1):3–14, 1966. 391

[VNH+11] J. Veness, K. S. Ng, M. Hutter, W. Uther, and D. Silver. A Monte-Carlo AIXI approximation. *Journal of Artificial Intelligence Research*, 40:95–142, 2011. Honorable Mention for the 2014 IJCAI-JAIR Best Paper Prize. 317, 336, 346, 347

[VNHB12] J. Veness, K. S. Ng, M. Hutter, and M. Bowling. Context tree switching. In *Proc. Data Compression Conference (DCC'12)*, pages 327–336, Snowbird, Utah, USA, 2012. IEEE Computer Society. xix, 197, 199, 201, 203, 205

[VNHS10] J. Veness, K. S. Ng, M. Hutter, and D. Silver. Reinforcement learning via AIXI approximation. In *Proc. 24th AAAI Conference on Artificial Intelligence*, pages 605–611, Atlanta, USA, 2010. AAAI Press. xviii, xix, 197, 316, 323, 328, 329, 331, 333, 334, 336, 337, 343, 346

[VNM47]   J. Von Neumann and O. Morgenstern. *Theory of Games and Economic Behavior, 2nd rev.* Princeton University Press, 1947. 286, 303

[Vov89]   V. G. Vovk. Prediction of stochastic sequences. *Problems in Information Transmission*, pages 285–296, 1989. 142

[VSBU09] J. Veness, D. Silver, A. Blair, and W. Uther. Bootstrapping from game tree search. *Advances in Neural Information Processing Systems*, 22, 2009. 347

[VSFT19] H. Van Seijen, M. Fatemi, and A. Tavakoli. Using a logarithmic mapping to enable lower discount factors in reinforcement learning. *Advances in Neural Information Processing Systems*, 32, 2019. 233

[VSH12]   J. Veness, P. Sunehag, and M. Hutter. On ensemble techniques for AIXI approximation. In *Proc. 5th Conf. on Artificial General Intelligence (AGI'12)*, volume 7716 of *LNAI*, pages 341–351. Springer, Heidelberg, 2012. 346

[VSK+22] P. Vamplew, B. J. Smith, J. Källström, G. Ramos, R. Rădulescu, D. M. Roijers, C. F. Hayes, F. Heintz, P. Mannion, P. J. Libin, et al. Scalar reward is not enough: A response to Silver, Singh, Precup and Sutton (2021). *Autonomous Agents and Multi-Agent Systems*, 36(2):1–19, 2022. 232

[VSP+17] A. Vaswani, N. Shazeer, N. Parmar, J. Uszkoreit, L. Jones, A. N. Gomez, L. Kaiser, and I. Polosukhin. Attention is all you need, 2017. 427, 428, 429

[VW98]   V. G. Vovk and C. Watkins. Universal portfolio selection. In *Proc. 11th Conf. on Computational Learning Theory (COLT'98)*, pages 12–23. ACM Press, New York, 1998. 120, 121

[VWBG13] J. Veness, M. White, M. Bowling, and A. György. Partition tree weighting. In *Proc. Data Compression Conference (DCC'13)*, pages 321–330, Snowbird, Utah, USA, 2013. IEEE Computer Society. xix, 197, 199, 206, 208, 209, 210

[Wal37]   A. Wald. Die Widerspruchsfreiheit des Kollektivbegriffs in der Wahrscheinlichkeitsrechnung. In *Ergebnisse eines Mathematischen Kolloquiums*, volume 8, pages 38–72, 1937. 117

[Wal91]   P. Walley. *Statistical Reasoning with Imprecise Probabilities*. Chapman and Hall, London, 1991. 118

[Wal05]   C. S. Wallace. *Statistical and Inductive Inference by Minimum Message Length*. Springer, Berlin, 2005. 104, 120

[Wal16]   M. M. Waldrop. The chips are down for Moore's law. *Nature News*, 530(7589):144, 2016. 426

[Wan96]   Y. Wang. *Randomness and Complexity*. PhD thesis, Universität Heidelberg, 1996. 117

[Was10]   L. Wasserman. *All of Statistics: A Concise Course in Statistical Inference*. Springer Texts in Statistics. Springer, corr. 2. print., [repr.] edition, 2010. 117

[Wat89]   C. J. C. H. Watkins. *Learning from delayed rewards*. PhD thesis, Royal Holloway, University of London, 1989. 324, 327

[WB68]   C. S. Wallace and D. M. Boulton. An information measure for classification. *Computer Journal*, 11(2):185–194, 1968. 120

[WBC⁺17] T. Wängberg, M. Böörs, E. Catt, T. Everitt, and M. Hutter. A game-theoretic analysis of the off-switch game. In *Proc. 10th Conf. on Artificial General Intelligence (AGI'17)*, volume 10414 of *LNAI*, pages 167–177, Melbourne, Australia, 2017. Springer. 411

[WD92] C. J. Watkins and P. Dayan. Q-learning. *Machine Learning*, 8(3):279–292, 1992. 347, 381

[WDA⁺23] L. K. Wenliang, G. Déletang, M. Aitchison, M. Hutter, A. Ruoss, A. Gretton, and M. Rowland. Distributional bellman operators over mean embeddings. *arXiv:2312.07358*, 2023. 349, 381

[Wei66] J. Weizenbaum. ELIZA — a computer program for the study of natural language communication between man and machine. *Communications of the ACM*, 9(1):36–45, 1966. 428

[Wel84] T. A. Welch. A technique for high-performance data compression. *Computer*, 17(06):8–19, 1984. 119

[WG07] Y. Wang and S. Gelly. Modifications of UCT and sequence-like simulations for Monte-Carlo go. In *2007 IEEE Symposium on Computational Intelligence and Games*, pages 175–182. IEEE, 2007. 318, 319

[Wik23] Wikipedia. History of artificial intelligence. *https://en.wikipedia.org/wiki/History_of_artificial_intelligence*, 2023. 428

[Wil91a] D. Williams. *Probability with Martingales*. Cambridge university press, 1991. 151

[Wil91b] R. N. Williams. An extremely fast Ziv-Lempel data compression algorithm. In *1991 Data Compression Conference*, pages 362–363. IEEE Computer Society, 1991. 119

[Wil92] R. J. Williams. Simple statistical gradient-following algorithms for connectionist reinforcement learning. *Machine Learning*, 8:229–256, 1992. 355, 381

[Wil98] F. M. Willems. The context-tree weighting method: Extensions. *IEEE Transactions on Information Theory*, 44(2):792–798, 1998. 197, 316

[Wil11] S. M. Wilson. *Stanford-Binet Intelligence Scales*, pages 1436–1439. Springer US, Boston, MA, 2011. 436

[WLBS05] T. Wang, D. Lizotte, M. Bowling, and D. Schuurmans. Bayesian sparse sampling for on-line reward optimization. In *Proceedings of the 22nd International Conference on Machine Learning*, pages 956–963, 2005. 248, 348

[WM97] D. H. Wolpert and W. G. Macready. No free lunch theorems for optimization. *IEEE Transactions on Evolutionary Computation*, 1(1):67–82, 1997. 146, 423

[Wol83] S. Wolfram. Statistical mechanics of cellular automata. *Reviews of Modern Physics*, 55(3):601, 1983. 84

[Wol23] D. H. Wolpert. The Implications of the No-Free-Lunch Theorems for Meta-induction. *Journal for General Philosophy of Science*, 54(3):421–432, 2023. 423

[WPH22] S. Wäldchen, S. Pokutta, and F. Huber. Training characteristic functions with reinforcement learning: Xai-methods play connect four. In *International Conference on Machine Learning*, pages 22457–22474. PMLR, 2022. 232

[WS96] M. A. Wiering and J. Schmidhuber. Solving POMDPs with Levin search and EIRA. In *Proc. 13th International Conf. on Machine Learning*, pages 534–542, Bari, Italy, 1996. 360

[WSB⁺20] J. Wang, E. Sezener, D. Budden, M. Hutter, and J. Veness. A combinatorial perspective on transfer learning. In *Advances in Neural Information Processing Systems (NeurIPS'20)*, volume 33, pages 918–929, Cambridge, MA, USA, 2020. Curran Associates. 210, 433

[WSH11] I. Wood, P. Sunehag, and M. Hutter. (Non-)equivalence of universal priors. In *Proc. Solomonoff 85th Memorial Conference*, volume 7070 of *LNAI*, pages 417–425, Melbourne, Australia, 2011. Springer. 149, 154

[WST93] F. M. Willems, Y. M. Shtarkov, and T. J. Tjalkens. Context tree weighting: A sequential universal source coding procedure for FSMX sources. In *1993 IEEE International Symposium on Information Theory*, page 59. Institute of Electrical and Electronics Engineers, 1993. 197

[WST95] F. M. Willems, Y. M. Shtarkov, and T. J. Tjalkens. The context-tree weighting method: Basic properties. *IEEE Transactions on Information Theory*, 41(3):653–664, 1995. 158, 159, 160, 161, 168, 175, 197

[WST96] F. M. Willems, Y. M. Shtarkov, and T. J. Tjalkens. Context weighting for general finite-context sources. *IEEE Transactions on Information Theory*, 42(5):1514–1520, 1996. 197

[WST97] F. Willems, Y. Shtarkov, and T. Tjalkens. Reflections on "the context tree weighting method: Basic properties". *Newsletter of the IEEE Information Theory Society*, 47(1), 1997. 197

[WTS00] F. M. Willems, T. Tjalkens, and Y. M. Shtarkov. Context-tree maximizing. In *Proceedings 34th Annual Conference on Information Sciences and Systems, March 15-17, 2000, Princeton, New Jersey*, pages TP6–7, 2000. 211

[Wu17] Y. Wu. Lecture notes on information-theoretic methods for high-dimensional statistics. *Lecture Notes for ECE598YW (UIUC)*, 16, 2017. 121

[WvO12] M. Wiering and M. van Otterlo. *Reinforcement Learning*. Springer, 2012. 232

[WWH⁺22] H. Wang, H. Wu, Z. He, L. Huang, and K. W. Church. Progress in machine translation. *Engineering*, 18:143–153, 2022. 428

[WWS+22] J. Wei, X. Wang, D. Schuurmans, M. Bosma, F. Xia, E. Chi, Q. V. Le, D. Zhou, et al. Chain-of-thought prompting elicits reasoning in large language models. *Advances in Neural Information Processing Systems*, 35:24824–24837, 2022. 438

[XCG+23] Z. Xi, W. Chen, X. Guo, W. He, Y. Ding, B. Hong, M. Zhang, J. Wang, S. Jin, E. Zhou, et al. The rise and potential of large language model based agents: A survey, September 2023. 438

[XRLM21] S. M. Xie, A. Raghunathan, P. Liang, and T. Ma. An explanation of in-context learning as implicit Bayesian inference. *arXiv:2111.02080*, 2021. 438

[XZC23] P. Xu, X. Zhu, and D. A. Clifton. Multimodal learning with transformers: A survey. *IEEE Transactions on Pattern Analysis and Machine Intelligence*, pages 1–20, 2023. 438

[Yam16] R. V. Yampolskiy. *Artificial Superintelligence: A Futuristic Approach*. Taylor & Francis, CRC Press, 2016. 388

[Yam24] R. V. Yampolskiy. *AI: Unexplainable, Unpredictable, Uncontrollable*. Taylor & Francis, CRC Press, 2024. 388, 410

[YJT+23] J. Yang, H. Jin, R. Tang, X. Han, Q. Feng, H. Jiang, B. Yin, and X. Hu. Harnessing the power of LLMs in practice: A survey on chatgpt and beyond. *arXiv:2304.13712*, 2023. 429

[YYZ+23] S. Yao, D. Yu, J. Zhao, I. Shafran, T. L. Griffiths, Y. Cao, and K. Narasimhan. Tree of thoughts: Deliberate problem solving with large language models. *arXiv:2305.10601*, 2023. 438

[YZNH24] S. Yang-Zhao, K. S. Ng, and M. Hutter. Dynamic knowledge injection for AIXI agents. *arXiv:2312.16184*, 2024. 346

[YZWN22] S. Yang-Zhao, T. Wang, and K. S. Ng. A direct approximation of AIXI using logical state abstractions. In *Advances in Neural Information Processing Systems*, 2022. 346, 380

[YZZC23] Z. Yang, X. Zeng, Y. Zhao, and R. Chen. AlphaFold2 and its applications in the fields of biology and medicine. *Signal Transduction and Targeted Therapy*, 8(1):115, 2023. 428

[Zad65] L. A. Zadeh. Fuzzy sets. *Information and Control*, 8:338–353, 1965. 118

[Zad78] L. A. Zadeh. Fuzzy sets as a basis for a theory of possibility. *Fuzzy Sets and Systems*, 1:3–28, 1978. 118

[Zah75] A. Zahavi. Mate selection—a selection for a handicap. *Journal of theoretical Biology*, 53(1):205–214, 1975. 303

[Zen19] H. Zenil. Compression is Comprehension, and the Unreasonable Effectiveness of Digital Computation in the Natural World. Technical report, 2019. 432

[ZGHS06] V. Zhumatiy, F. Gomez, M. Hutter, and J. Schmidhuber. Metric state space reinforcement learning for a vision-capable mobile robot. In *Proc. 9th International Conf. on Intelligent Autonomous Systems (IAS'06)*, pages 272–281. IOR Press, 2006. 381

[ZH02] M. Zaffalon and M. Hutter. Robust feature selection by mutual information distributions. In *Proc. 18th International Conf. on Uncertainty in Artificial Intelligence (UAI-2002)*, pages 577–584. Morgan Kaufmann, San Francisco, CA, 2002. 155

[ZH05] M. Zaffalon and M. Hutter. Robust inference of trees. *Annals of Mathematics and Artificial Intelligence*, 45:215–239, 2005. 118

[Zim91] H.-J. Zimmermann. *Fuzzy Set Theory–And Its Applications*. Kluwer, Dordrecht, 2nd edition, 1991. 118

[ZL70a] A. K. Zvonkin and L. A. Levin. The complexity of finite objects and the development of the concepts of information and randomness by means of the theory of algorithms. *Russian Mathematical Surveys*, 25(6):83–124, 1970. 117, 119, 120

[ZL70b] A. K. Zvonkin and L. A. Levin. The complexity of finite objects and the development of the concepts of information and randomness by means of the theory of algorithms. *Russian Mathematical Surveys*, 25(6):83, 1970. 141, 149

[ZL77] J. Ziv and A. Lempel. A universal algorithm for sequential data compression. *IEEE Transactions on Information Theory*, 23(3):337–343, 1977. 119, 347

[ZL78] J. Ziv and A. Lempel. Compression of individual sequences via variable-rate coding. *IEEE Transactions on Information Theory*, 24(5):530–536, 1978. 119, 197

[ZLW+23] J. Zhang, X. Li, Z. Wan, C. Wang, and J. Liao. Text2nerf: Text-driven 3d scene generation with neural radiance fields. *arXiv:2305.11588*, 2023. 428

[ZSW+19] D. M. Ziegler, N. Stiennon, J. Wu, T. B. Brown, A. Radford, D. Amodei, P. Christiano, and G. Irving. Fine-tuning language models from human preferences. *arXiv preprint arXiv:1909.08593*, 2019. 438

[Zus72] K. Zuse. *Der Plankalkül*. PhD thesis, 1945/1972. 349

[ZVSL18] B. Zoph, V. Vasudevan, J. Shlens, and Q. V. Le. Learning transferable architectures for scalable image recognition. In *Proceedings of the IEEE Conference on Computer Vision and Pattern Recognition*, pages 8697–8710, 2018. 428

[ZZL+23] W. X. Zhao, K. Zhou, J. Li, et al. A survey of large language models, 2023. 429

Table of Notation

The following is a list of commonly used notations. The first column is the symbol itself, the second column is its corresponding name and/or explanation. Most notations are formally defined/introduced in this book, e.g. ξ and $K(x)$, while some standard ones are used without being formally defined, e.g. $\mathbb{R}$.

We have chosen to separate the notation by chapter, and it is in roughly chronological order of when the notation is first used/defined. There are some repeated listings of a specific notation for different meanings (and sometimes the same meaning), most notably Γ is used in four different contexts for four different meanings: Cylinder set Γ_x, Gamma function $\Gamma(a)$, model cost Γ_D, and discount normalization factor Γ_t.

Symbol	Description
Global Abbreviations, Acronyms, and Initialisms	
[C35s]	classification of problems
[HQC24]	paper, book or other reference
(6.7.7)	label/reference for a formula/theorem/definition/...
AI	Artificial Intelligence
AGI	Artificial General Intelligence
ASI	Artificial Super-Intelligence
AGSI	Artificial General Super-Intelligence
UAI	Universal Artificial Intelligence
AIXI	Maximally intelligent Universal AI agent
CTW	Context Tree Weighting
MCTS	Monte Carlo Tree Search
MC-AIXI-CTW	AIXI with MCTS planning and CTW model class
AIμ	Optimal agent in known environment μ
MDP	Markov Decision Process
POMDP	Partially Observable Markov Decision Process
RL	Reinforcement Learning
TM	Turing Machine
UTM	Universal Turing Machine
MDL	Minimum Description Length
MAP	Maximum A-Posteriori
KL	Kullback–Leibler
KT	Krichevsky–Trofimov
NFL	No Free Lunch
MCTS	Monte Carlo Tree Search

KT	Krichevsky–Trofimov (estimator)
KL	Kullback–Leibler (divergence)
PST	Prediction Suffix Tree
l.h.s.	left-hand side
r.h.s.	right-hand side
w.r.t.	with respect to
w.l.g.	without loss of generality
e.g.	exempli gratia (Latin), for example
i.e.	id est (Latin), that is
etc.	et cetera (Latin), and so forth
cf.	confer (Latin, imperative of conferre), compare with
et al.	et alii (Latin), and others
i.i.d.	independent identically distributed (random variables)
iff	if and only if
w.p.1/a.s./i.p.	with probability 1 / almost surely / in probability
i.m./i.m.s.	in mean^2 / in mean^2 sum
Standard Math Operations and Spaces	
$0,1,2,...,\infty$	zero, one, two, ..., infinity
$\{a,...,z\}$	set containing elements $a,b,...,y,z$. $\{\}$ is the empty set
$[a,b)$	interval on the real line, closed at a and open at b
$\cap,\dot\cup,\cup,\Delta,\setminus,\in$	set intersection, (disjoint) union, (symmetric) difference, membership
$\wedge,\vee,\neg$	Boolean conjunction (and), disjunction (or), negation (not)
$\subseteq,\subset$	subset, proper subset
$\implies$	implies
$\iff$	equivalence, if and only if, iff
$\blacksquare,\blacklozenge,\bullet$	end of proof,example,remark
$\forall,\exists$	for all, there exists
$\ll,\gg$	much smaller/greater than
$\propto$	proportional to
$=,\neq,:=,\equiv$	equal to, not equal, definition, equal by definition
$+,-,\cdot,/$	standard arithmetic operations: sum, difference, product, ratio
$\square^2$	footnote or exponent
$\sqrt{}$	square root
$\leq,\geq,<,>$	standard inequalities
$|\mathcal{S}|,|a|$	size/cardinality of set $\mathcal{S}$, absolute value of a
$\to$	mapping, approaches, Boolean implication
$\lim_{n\to\infty}$	limiting value of argument for n tending to infinity
$\rightsquigarrow$	replace with
$\lceil x \rceil$	ceiling of x: smallest integer larger or equal than x
$\lfloor x \rfloor$	floor of x: largest integer smaller or equal than x
δ_{ab}	Kronecker symbol, $\delta_{ab}=1$ if $a=b$ and 0 otherwise
$\sum_{k=1}^{n}$	summation from $k=1$ to n
$\prod_{k=1}^{n}$	product from $k=1$ to n
$\partial f/\partial x, df/dx$	partial/total derivative of f w.r.t. x
$\int,\int_a^b dx$	Lebesgue integral, integral from a to b over x
min/max	min-/maximal element of set: $\min_{x\in\mathcal{X}} f(x)=\min\{f(x):x\in\mathcal{X}\}$

arg min	$\arg\min_x f(x)$ is the set of global minima of f
log	logarithm to some basis
$\log_b$	logarithm to basis b
ln	natural logarithm to basis e $=2.71828...$
e	base of natural logarithm e $=2.71828...$
π	Archimedes' constant $\pi=3.1415...$ (but mostly agent policy)
$*$	wildcard for some string (prefix, finite, or infinite)
$\langle o \rangle$	coding of object o
$\langle x,y \rangle$	uniquely decodable pairing of x and y
$O(),o()$	big and small oh-notation. Grow is slower (or equal)
$\mathbb{B}$	Binary set $\{0,1\}$
$\mathbb{N}^+$	Positive natural number $\{1,2,...\}$
$\mathbb{N}_0$	Natural numbers including zero $\{0,1,2,...\}$
$\mathbb{Z}$	set of integers $\mathbb{Z}=...,-2,-1,0,1,2,...$
$\mathbb{Q}$	Rationals
$\mathbb{R}$	Real numbers
General	
i,j	Generic natural number indices
$i \leq t \leq k \leq n \leq m$	natural number time index for sequences
x,y,z	finite strings
f,g	Functions
$f_n \to g_n$	converge to each other, $f_n-g_n \to 0$ for $n \to \infty$
$x_{1:n}$	$=x_{\leq n}=x_{<n+1}=x_1 x_2...x_n$
$\ell(\cdot)$	Length of a finite string or program
ϵ	The empty string. Not to be confused with ε
ε	A small positive real number. Not to be confused with ϵ
δ	Nonnegative real number
$\cup, \cap$	Union, intersection, respectively
$\mathcal{X}^*$	Set of all finite strings drawn from alphabet $\mathcal{X}$
$\mathcal{X}^\infty$	Set of all infinite sequences drawn from alphabet $\mathcal{X}$
K	Kolmogorov complexity
π	An agent or policy
μ	True/sampling probability measure $\in \mathcal{M}$
$\mathcal{M}$	The set of environments, a class of distributions, the model class; a countable set of probability semimeasures on strings
p	A program
$\mathcal{A}$	A set of actions
$\mathcal{O}$	The set of possible observations
$\mathcal{R}$	The set of possible rewards
$\mathcal{E}$	The set of all percepts $\mathcal{E}:=\mathcal{O}\times\mathcal{R}$
$\mathcal{H}$	The set of all histories $\mathcal{H}:=(\mathcal{A}\times\mathcal{E})^*$
o_t	Observation at time t
r_t	Reward at time t
e_t	A percept $=o_t r_t$ (observation and reward together)
a_t	Action at time t
$h_{<t}$	History $=a_1 e_1...a_{t-1} e_{t-1}$
ν	An environment/semimeasure $\in \mathcal{M}$

ρ	An environment/semimeasure not necessarily $\in \mathcal{M}$
ν^π	Induced probability measure on histories from π and ν
$\mathbf{E}[X]$	Expected value of X
P	Probability
P_ν^π	The probability of an event following policy π and environment ν
ξ	Bayes mixture over $\mathcal{M}$
$w_\nu \in (0,1]$	The weight of ν; the prior belief that ν is the true distribution μ
γ	A discount function $\in [0,1)$
$\mathcal{M}_{sol}$	The set of lower semicomputable (chronological) semimeasures
$\mathcal{M}_{comp}$	The set of computable (chronological) measures
Chapter 2	**Background**
MLE	Maximum Likelihood Estimation/Estimator
MSE	Mean-Square Error
$x = x_1 x_2 ... x_n$	Finite binary string of length n
0,1	Characters zero, one in a string of bits (not the numbers 0 and 1)
$x_{i:j}$	The length $(j-i+1)$ segment $x_i x_{i+1} ... x_j$
xy	Concatenation of x and y
$\mathbb{B}^n$	The set of all binary strings of length n
$\langle \cdot \rangle$	Bijection between finite binary strings and natural numbers
$b(x)$	Takes binary string x and returns the natural number it represents
$\lfloor \ \rfloor$	Floor function
x,y,z	Strings in $\mathcal{X}^*$ or $\mathbb{B}^*$
$x \sqsubseteq y,\ x \sqsubset y$	x is a (proper) prefix of y
c	Prefix free code
$\mathcal{P}$	A prefix-free set; a prefix code
$E_i(x)$	ith order prefix codeword
$\overline{x} \ / \ x'$	First-order / second-order prefix code of x
$\omega \in \mathbb{B}^\infty$	Infinite sequence
$r \in [0,1]$	A real number
$(x_{1:n})^\infty$	Infinite sequence comprised of repeating the finite string $x_{1:n}$ infinitely many times
Γ_x	Cylinder set of x
Ω	Sample space
$\mathcal{F}$	A σ-algebra on Ω
$\varnothing$	The empty set
$A \in \mathcal{F}$	An event
$\mathrm{P} : \mathcal{F} \to [0,1]$	A probability measure
$\Delta \mathcal{S}$	The set of all probability distributions on $\mathcal{S}$
$\Delta' \mathcal{S}$	The set $\{p \in [0,1]^{\mathcal{S}} : \sum_{s \in \mathcal{S}} p_s \leq 1\}$ of all semi-probabilities on $\mathcal{S}$
θ	$P(\text{heads})$ on a biased coin
H, T	Heads, tails
$\mathcal{S}$	A set of subsets of Ω
$\sigma(\mathcal{S})$	The σ-algebra generated by $\mathcal{S}$
$\omega \in \Omega$	Singleton event
μ	A semimeasure
D_n	Pairwise disjoint events
A,B,D,E	Events

$\{H\}_{i=1}^{\infty}$	A partition of Ω
$X,Y,Z:\Omega\rightarrow\mathbb{R}$	Random variables
$F_X(x)$	The cumulative distribution function
$\mathcal{E}$	A measurable space
$p_X, \mathrm{P}(X=x)$ or $p(x)$	Probability mass function
t,a,b	Real number variables
B	A set (comprised of countable unions of intervals)
r_i,p_i	Real numbers
I	An interval
$\mathcal{X}$	A sample space
$[\![...]\!]$	Indicator function
J	Subset of I
S_n	$=X_1+...+X_n$
$\mathbf{E}[f(X)]$	expectation of f
$\mathrm{P}[A]$	probability of event/outcome/predicate A
$\mathbf{Var}[f(X)]$	Variance of f
S	A predicate on events
$p\in[0,1]$	Probability
D_N	Set of events
$\xrightarrow{L^2}$	Converges in mean^2
$\xrightarrow{a.s}$	Converges almost surely
$\xrightarrow{P}$	Converges in probability
$\xrightarrow{i.m.s}$	Converges in mean^2 sum
θ	Parameter of Bernoulli distribution
Θ	A set of parameters
$x_{1:n}$	A sample of observations
$w(\theta\|...)$	A p(oste)rior probability density function
$t(x_{1:n})$	An estimator
$L(\theta)$	The likelihood function for parameter θ
$\hat{\theta}_{ML}$	The maximum likelihood estimator (MLE)
τ	Reparameterization
T_n	Estimator
$\mathrm{Bias}(T_n)$	Bias
$\mathrm{MSE}_\theta[T_n]$	Mean squared error of an estimator T_n
V	The score V of X
$\mathcal{I}(\theta)$	The Fisher information
p	Real number between 0 and 1
a,b	KT estimator counts of `1`s and `0`s respectively
$\Gamma(k)$	The gamma function
$Beta(\theta;\alpha,\beta)$	A Beta distribution
$B(\alpha,\beta)$	The Beta function
$\bar{x}$	The sample mean
$\mathcal{X},\mathcal{Y}$	Sample spaces
$h(x)$	The Shannon information content
$H(X)$	Entropy of a random variable X
$N=1,2,...,n$	Set of the first n positive natural numbers

$E:N\to\mathbb{B}^*$	A binary code
C	Binary code
$D_{\mathrm{KL}}(P\|Q)$	The Kullback–Leibler (KL) divergence
$b_1,b_2,...$	A finite or infinite sequence of binary strings
$\mathcal{C}$	A prefix code
I_x	An interval
$\emptyset$	The empty set
$\lambda([a,b))$	The length of an interval in $\mathbb{R}$
$\ell(b_n)$	Length of string b_n
$L_{D,P}$	The average codeword length
L_P^*	The minimal average codeword length
l_x	$\lceil -\log_2 P(x)\rceil$
Σ	Alphabet
Q	A finite set of states $\{q0,q1,...,qf\}$
Γ	Finite set of tape symbols
L,R	Left and right
δ	Transition function
B	blank
T	A Turing machine
$\perp$	Output when a Turing machine does not halt
L,S	Language of a Turing Machine
P	Turing Machine
$\phi(x,k)$	A recursive = finitely computable function
A	A set
$\Delta_1^0,\Sigma_1^0,\Pi_1^0,\Delta_2^0$	Set of finitely,lower,upper,limit computable relations|functions
$\Sigma_n^0,\Pi_n^0,\Delta_n^0$	Arithmetic hierarchy
η	A computable relation
U	A universal Turing machine
K_T,K	Kolmogorov complexity
F_1,F_2	Formal systems
$\overset{+}{\leq}$ / $\overset{\times}{\leq}$	Less than within an additive / multiplicative constant
$\overset{+}{=}$ / $\overset{\times}{=}$	Equal within an additive / multiplicative constant
p,q,r	Programs
Chapter 3	**Bayesian Sequence Prediction**
a,s,h,d	instantaneous absolute, square, hellinger^2, KL distances of μ and ξ
A,S,H,D	total Absolute, Square, Hellinger^2, KL Distances of μ and ξ
$\hat{\mu}$	A distribution that is approximately equal to μ
$y\in\mathcal{Y}$	An action from a set of actions $\mathcal{Y}$
$x\in\mathcal{X}$	An observation from a set of observations $\mathcal{X}$
$loss(x_t,y_t)$	Loss when predicting x_t if outcome is y_t
$\mathrm{Loss}_{1:n}(\Lambda)$	Total expected loss of predictor Λ
Λ_ρ	The predictor which minimizes ρ-expected loss
$\{y_1,...,y_N\}$	Elements of an N-dimensional probability simplex
$\{z_1,...,z_N\}$	Elements of an N-dimensional semi-probability simplex
Chapter 4	**The Context Tree Weighting Algorithm**
$\mathrm{P}_{KT}(a,b)$	KT estimator
$r_{\hat{\rho}}$	Redundancy of $\hat{\mu}$

$\mathcal{S}$	A suffix set (a set of binary strings)
$\Psi_{\mathcal{S}}$	A suffix tree (a binary tree)
$\beta_{\mathcal{S}}(x)$	The suffix function
$\Theta_{\mathcal{S}}$	A parameter vector
$\Psi_{\mathcal{S},\Theta_{\mathcal{S}}}$	A prediction suffix tree (a binary tree)
s	A string (a suffix)
a_s, b_s	Counts of 1s and 0s following s
a, b	1s and 0s counts
$\mathcal{C}_D$	The model class
$\mathcal{E}_D(\mathcal{S}), \mathcal{E}_D(\Psi)$	The encoding of a suffix set/tree Ψ
D	The length of the longest suffix in $\mathcal{S}$, depth
C	Prefix-free set
$\Gamma_D(\mathcal{S})$	The model cost (of a suffix set $\mathcal{S}$)
$\mathrm{P}_{\mathcal{S},\mathrm{KT}}$	The PST-KT probability
$\mathrm{P}_{\mathcal{S},\Theta_{\mathcal{S}}^i}$	The conditional PST probability
$\hat{\Theta}_{\mathcal{S}}^i$	A family of parameter vector estimates
P_D^{CTW}	$:= P_w(\epsilon)$
P_w	The weighted probability
θ	The KT probability $:= \mathrm{P}_{kt}(a_s, b_s)$
$\mathcal{V}, \mathcal{W}$	$\in \mathcal{C}_{D-d}$
Γ_{D-d+1}	Model cost (different depth)
$\mathcal{W}\{1\}$	$\{w1 : w \in \mathcal{W}\}$
$\Gamma_{D-d+1}(\{\epsilon\})$	Model cost of empty tree
$\gamma(k)$	$:= \frac{1}{2}\log_2(k)+1$ for $k \geq 1$ and $:= k$ for $0 \leq k < 1$
$\oplus$	Log-sum operator
Chapter 5	**Variations on CTW**
PTW	Partition Tree Weighting
CTS	Context Tree Switching
CTM	Context Tree Maximization
FMN	Forget Me Not
γ	Discount
$c, \alpha \in [0,1)$	Real valued parameters
k_s	The number of occurrences of context n up to time t
$\mathcal{L}(n)$	The set of leafs from n
$\tau_\alpha(x_{1:n})$	The switch distribution
α_k	The switch rate
$x_{1:n}^s$	All elements that follow the substring s in $x_{1:n}$
c	Context; node
$\mathrm{P}_{c,D}^{CTS}$	Context Tree Switching method
n_c	$= \ell(x_{1:n}^c)$
$t_c(k)$	$= min\{t | \ell(x_{1:t}^c) = k\}$
$\bar{\mathcal{S}}$	The set of contexts that index the internal nodes of $\mathcal{S}$
Γ	Model cost
$d(\mathcal{S})$	Maximum depth of any context in $\mathcal{S}$
$\mathcal{P}_n$	A temporal partition
$\mathcal{B}_D$	The set of all binary temporal partitions
ρ	The base model

P_D^{PTW}	The partition tree weighting
$\mathrm{MSCB}_D(t)$	The most significant changed context bit
μ	A piecewise stationary data-generating source
$\mathcal{G}$	A class of bounded memory data-generating sources
g	A non-negative, monotonically non-decreasing concave function
$\mathcal{M}$	A growing set of stored base model states $\rho(\cdot\|s)$
$\mathrm{FMN}_d(...\|...)$	Forget me not predictive distribution
Chapter 6	**Agency**
Π	A policy class
Γ_t	The discount normalization factor
$H_t(\varepsilon)$	The ε-effective horizon
m	Maximum lifetime
$V_{\nu,\gamma}^{\pi,m}$	γ-discounted value of policy π in environment ν with horizon m
$\pi_\nu^{*,m}$	Optimal policy w.r.t. $V_{\nu,\gamma}^{\pi,m}$
$Q_\nu^{\pi,m,\gamma}$	The Q-value, also called action-value
Chapter 7	**Universal Artificial Intelligence**
$\pi^{\mathrm{AIXI}},\pi_\xi^*$	AIXI,AIξ policy
$\tilde{\pi}$	A self-optimizing policy
$\pi^{\mathrm{AIXI}},\pi_\xi^*$	AIXI policy
w_ν^U	Universal prior $2^{-K(\nu)}$
ξ_U	Universal Bayes mixture over $\mathcal{M}_{sol}$
$T(p'a_{...})\to e_{...}$	Chronological Turing machine
$M(e_{1:m}\|a_{1:m})$	Chronological version of Solomonoff's distribution
Chapter 8	**Optimality of Universal Agents**
$\mathrm{Regret}_m(\pi,\mu)$	The regret of a policy π in environment μ
$\tilde{\pi}$	Policy
Chapter 9	**Other Universal Agents**
KSA	Knowledge-Seeking Agent
π^o	Optimistic policy
$\mathcal{M}_0$	Initial Set of Environments
π_{TS}	Thompson sampling policy
IG	Information gain
$V_{\mathrm{IG}}^{\pi,m}$	Information gain value function
π_{IG}^*	Optimal information gaining policy
π_{BE}	BayesExp policy
π_{Inq}	Inquisitive agent policy
π_S	Self-AIXI policy
Chapter 10	**Multi-Agent Setting**
$\prec$	Strict agent's preference relation
N	A finite set $\{1,...,N\}$ of N of players
$\mathcal{A}_i$	The set of actions for player i
$\succeq_i$	The preference relation on $\mathcal{A}$ for player i
$f:I\to\mathcal{A}_i$	An (action) profile
$u_i:\mathcal{A}\to\mathbb{R}$	A utility function
$a_{\neq j}$	Vector a without the ith element
$B_i(a_{\neq i})$	Player i's best response function

X	A compact convex subset of $\mathbb{R}^n$
U_i	The expected value of u_i
α_i	mixed strategy for player i
$\Delta \mathcal{A}_i$	set of mixed strategies for player i
θ	Real number (probability) between 0 and 1
$\pi_{1:n}$	policies/strategies of n agents/players
σ	multi-agent environment
$\sigma^{\pi_{1:n}}$	history distribution induced by policies $\pi_{1:n}$ acting in σ
σ_i	subjective environment of agent i
$\mathcal{T}$	The set of probabilistic Turing machines
T	A probabilistic Turing machine
ν_T	A conditionally lower semicomputable semimeasure
O	(Reflective) oracle
T^O	T when run with oracle O
ν_T^O	The semimeasure induced by T^O
x	An input string to a Turing machine
z	A rational number
$\overline{\xi}$	Completion of ξ into a measure
$\mathcal{M}_r^O$	the class of all reflective-oracle-computable environments
Chapter 11	**AIXI-MDP**
$\mathfrak{R}$	Known deterministic reward matrix
$\theta \in [0,1]^4$	Parameterization of MDP environment for 2×2 matrix games
ξ_{MDP}	A mixture over the class of MDPs
n_{ao}	Number of occurrences of action-obsersation pairs (a,o)
Chapter 12	**A Monte-Carlo AIXI Approximation**
UCB	Upper Confident B
UCT	Upper Confident bound for Trees
$\hat{V}(h)$	An estimate value of a history
Ω	Action space
$\hat{Q}(h_{<t},\cdot)$	An approximation of the Q-value function $Q(h_{<t},\cdot)$
Ψ	action-observation search tree
$\hat{V}(h)$	An estimate of the value function
q	A q-value
$N, N(h)$	The visit count
C	A positive exploration-exploitation parameter
s	A state
Q_{UCT}	UCT Q-value
R	# Rollout
ϵ	The root node of a tree $\mathcal{S}$
$l_{\mathcal{A}}$	Bit length of action space $\mathcal{A}$ encoding
$l_{\mathcal{E}}$	Bit length of percet space $\mathcal{E}$ encoding
$[\![a]\!]$	Bit representation of a
$\Theta_{\mathcal{S}}$	Parameters of prediction suffix trees $\mathcal{S}$
$\theta_s \in \Theta_{\mathcal{S}}$	The probability of a node $s \in \mathcal{S}$
$\mathcal{C}_D$	The set of all models of prediction suffix trees
Γ_D	A natural encoding of prediction suffix trees

$L(\mathcal{S})$	The leaf nodes of $\mathcal{S}$
d	Depth of node in tree
$D_i := D + I - 1$	The variable depth
$\mathcal{S}_j$	A Suffix tree
ϕ	An observation function, A legal action function
$\mathrm{P}_D^{\mathrm{CTW}}(e_{...}\|a_{...})$	CTW probability of $e_{...}$ given $a_{...}$
Chapter 13	**Computational Aspects**
$\succ$	A total order
$\bigwedge$	Logical and
l	Length variable
$\tilde{V}_k^{\pi_V}$	π_V's approximation of the value function of π_V
ℓ	Length function
π_ξ^{tl}	ξ-optimal policy among all length l- and time t-bounded policies
$\dot{\pi} : \mathcal{H} \to \mathcal{A} \times [0,1]$	Extended (deterministic) self-evaluating policy
$\mathrm{VA}(\dot{\pi})$	Valid approximation predicate
$\tilde{t}$	Maximal time
$\tilde{l}$	Maximal length
l_P	The maximum length of proofs
Π_{VA}	Set of valid policies
$\preceq_c$	Effective intelligence order relation
Chapter 14	**Feature Reinforcement Learning**
CTM	Context Tree Maximization
$\mathcal{S}$	State space
$\phi : \mathcal{H} \to \mathcal{S}$	Feature map
$\bar{\pi}, \bar{\mu}$	MDP policy and MDP environment
s	A state in a Markov environment
μ_ϕ	A feature environment
$B(h\|sa)$	Dispersion probability
$\langle \cdot \rangle_B$	B-average
$c \lesseqgtr a \pm b$	$c \le a+b$ and $c \ge a-b$
ΦMDP	MDP built with feature map ϕ
$N(sar's', \bar{h})$	Number of times state transition $sar's'$ occurred in $\bar{h}$
$Cost(\phi\|h)$	Cost function how well ϕ reduces h to a small MDP
$CL(...)$	Code length
$\bar{h}$	Reduced state history
ΦDBN	Feature Dynamic Bayesian Networks
$R(s)$	Reward for a next state
κ	Ordered set of action-percept pairs
n_κ	Number of occurrences of κ
$\mathrm{P}_e^{\kappa\|sa}$	Block probability estimate
s_n	nth (=last) state of $\bar{h}_{1:n}$
$\mathcal{S}$	Suffix state set
l_a, l_e	Minimum number of bits needed to encode a,e
$\mathcal{T}$	Context tree
$\mathcal{S}$	Suffix set
Chapter 15	**ASI Safety**

u	Instantaneous utility
$\tilde{u}$	Total utility
o^d, r^d	Death observation and reward
$\Pi, \mathcal{P}$	Set of (names of) policies
$\mathcal{U}$	Set of utility functions
Q^{he}, Q^{ig}, Q^{re}	hedonistic, ignorant, realistic Q-value
$d: \mathcal{E} \to \mathcal{E}$	Delusion box
A_{rl}, A_g, A_p, A_k	RL, goal-, prediction-, knowledge-seeking agent
$\mathcal{M}_{\mathrm{RS}}, \xi_{\mathrm{RS}}$	Reward Summable environment class and Bayes mixture
V_{SM}, V_{ST}, V_{STE}	Self-Modifying, Space(-Time)-Embedded value
Chapter 16	**Philosophy of AI**
PCT	Physical Church-Turing Thesis
AIQ	Artificial Intelligence Quotient
LLM	Large Language Model
ALE	Atari Learning Environment
$\overline{\Upsilon}, \Upsilon$	(Upper bound on) Legg–Hutter Intelligence measure

Index